Drug Discovery and Development

Drug Discovery and Development

Third Edition

Edited by
James J. O'Donnell III, MS, PhD
John Somberg, MD
Vincent Idemyor, PharmD
James T. O'Donnell, PharmD

CRC Press
Taylor & Francis Group
Boca Raton London New York

CRC Press is an imprint of the
Taylor & Francis Group, an **informa** business

CRC Press
Taylor & Francis Group
6000 Broken Sound Parkway NW, Suite 300
Boca Raton, FL 33487-2742

First issued in paperback 2021

ISBN 13: 978-1-03-208424-4 (pbk)
ISBN 13: 978-1-138-08026-3 (hbk)

Library of Congress Cataloging-in-Publication Data

Names: O'Donnell, James J., III, editor.
Title: Drug discovery and development / edited by James J. O'Donnell, III, MS, PhD, John Somberg, MD, Vincent Idemyor, PharmD, James T. O'Donnell, PharmD.
Description: Third edition. | Boca Raton, Florida : CRC Press, 2019. | Includes bibliographical references and index.
Identifiers: LCCN 2019031692 (print) | LCCN 2019031693 (ebook) | ISBN 9781138080263 (hardback) | ISBN 9781315113470 (ebook)
Subjects: LCSH: Drugs—Research—History. | Drugs—Design—History.
Classification: LCC RM301.25 .S55 2019 (print) | LCC RM301.25 (ebook) | DDC 615.1072/4—dc23
LC record available at https://lccn.loc.gov/2019031692
LC ebook record available at https://lccn.loc.gov/2019031693

Visit the Taylor & Francis Web site at
http://www.taylorandfrancis.com

and the CRC Press Web site at
http://www.crcpress.com

Publisher's Note
The publisher has gone to great lengths to ensure the quality of this reprint but points out that some imperfections in the original copies may be apparent.

CHARLES G. SMITH, PhD

(1928–2013)

I had the distinct pleasure and great opportunity to serve as a co-editor with Dr. Smith for the Second Edition of The Process of New Drug Discovery and Development *(2005). Dr. Smith was the sole author of the first edition of* New Drug Discovery and Development. *In 2002, when planning the second edition, he emphasized that the subject could no longer be covered by a single author. The second edition had 70 (!) contributors, as does the third edition. Charles was certainly a renaissance scientist – he made his mark in many important areas of pharmaceutical development. When we first met personally, he shared why he chose scientific research – "I want to cure cancer, not just treat one patient at a time". I've heard that motivating remark from many young scientists; the world owes scientific discovery to that motivation and dream. Dr. Smith was actively involved in drug discovery, research, development, and management of more than 20 new drug products in his career at several major Pharma companies. That work continued after his retirement through international consulting and his zeal to share the message – the first edition – which was published in 1992, 6 years after his retirement, and the second edition, almost 30 years into his retirement.*

The Editors of the Third Edition continue to feel privileged to carry on the message of Dr. Smith.

James T. O'Donnell, PharmD

Contents

SECTION I   Overview

SECTION II   Drug Discovery

SECTION IV   Regulations

Acknowledgments

The Editors would like to thank Stephanie Tedford, PhD, for her very able editorial content assistance; Gerald Walsh for his organizational assistance in communication and document control; and Hilary LaFoe, Jessica Poile, and Kirsten Barr of CRC Press/Taylor and Francis Group for their technical publishing guidance in preparing the book manuscript.

PEER REVIEWERS

James H. McKerrow, PhD, MD
Dean, Skaggs School of Pharmacy and Pharmaceutical Sciences
Associate Vice Chancellor for Health Sciences
University of California, San Diego

Dario Neri, PhD Professor
Department of Chemistry and Applied Biosciences
Swiss Federal Institute of Technology (ETH Zürich)

Rahul Deshmukh, BPharm, PhD
Assistant Professor, Department of Pharmaceutical Sciences
College of Pharmacy, Rosalind Franklin University of Medicine & Science

Introduction to the First Edition (1992)

INTRODUCTION

Charles G. Smith

Prior to the 20th century, the discovery of drug substances for the treatment of human diseases was primarily a matter of "hit-or-miss" use in humans, based on folklore and anecdotal reports. Many, if not most, of our earliest therapeutic remedies were derived from plants or plant extracts that had been administered to sick humans (e.g., quinine from the bark of the cinchona tree for the treatment of malaria in the mid-1600s and digitalis from the foxglove plant in the mid-1700s for the treatment of heart failure, to name two). Certainly, some of these early medications were truly effective (e.g., quinine and digitalis) in the sense that we speak of effective medications today. On the other hand, based on the results of careful studies of many such preparations over the years, either in animals or in man, one is forced to come to the conclusion that most likely, the majority of these plant extracts was not pharmacologically active, but rather they were perceived as effective by the patient because of the so-called placebo effect. Surprisingly, placebos (substances that are known not to be therapeutically efficacious, but that are administered so that all the psychological aspects of consuming a "medication" are presented to the patient) have been shown to exert positive effects in a wide range of disease states, attesting to the "power of suggestion" under certain circumstances. There still exist today practitioners of so-called homeopathic medicine, which is based on the administration of extremely low doses of substances with known or presumed pharmacologic activities. For example, certain poisons, such as strychnine, have been used as a "tonic" for years in various countries at doses that are not only nontoxic but that in the eyes of most scientifically trained medical and pharmacological authorities, could not possibly exert an actual therapeutic effect. Homeopathy is practiced not only in underdeveloped countries but also in certain well-developed countries, including the United States, albeit on a very small scale. Such practices will, most likely, continue since a certain number of patients who require medical treatment have lost faith, for one reason or another, in the so-called medical establishment. More will be said about proving drug efficacy in Chapters 8 to 10.

Pioneers in the field of medicinal chemistry such as Paul Ehrlich (who synthesized salvarsan, the first chemical treatment for syphilis, at the turn of the 20th century) were instrumental in initiating the transition from the study of plants or their extracts with purported therapeutic activities to the deliberate synthesis, in the laboratory, of a specific drug substance. Certainly, the discovery of the sulfa drugs in the 1930s added great momentum to this concept, since they provided one of the earliest examples of a class of pure chemical compounds that could be unequivocally shown to reproducibly bring certain infectious diseases under control when administered to patients by mouth. During World War II, the development of penicillin stimulated an enormous and highly motivated industry aimed at the random testing (screening) of a variety of microbes obtained from soil samples for the production of antibiotics. This activity was set into motion by the discovery of Alexander Fleming and others in England in 1929 that a *Penicillium* mold produced tiny amounts of a substance that was able to kill various bacteria that were exposed to it in a test tube. When activity in experimental animal test systems and in human patients was demonstrated, using extremely small amounts of purified material from the mold broth (penicillin), it was immediately recognized that antibiotics offered a totally new route to therapeutic agents for the treatment of infectious diseases in human beings. In addition to the scientific interest in these findings, a major need existed during World War II for new medications to treat members of the armed forces. This need stimulated significant activity on the part of the US Government and permitted collaborative efforts among pharmaceutical companies (which normally would be highly discouraged or prohibited by antitrust legislation from such in-depth cooperation) to pool resources so that the rate of discovery of new antibiotics would be increased. Indeed, these efforts resulted in accelerated rates of discovery and the enormous medical and commercial potential of the antibiotics, which were evident as early as 1950, assured growth and longevity to this important new industry. Major pharmaceutical companies such as Abbott Laboratories, Eli Lilly, E. R. Squibb & Sons, Pfizer Pharmaceuticals, and The Upjohn Company in the United States, to name a few, were particularly active in these endeavors and highly successful, both scientifically and commercially, as a result thereof (as were many companies in Europe and Japan). From this effort, a wide array of new antibiotics, many with totally unique and completely unpredictable chemical structures and mechanisms of action, became available and were proven to be effective in the treatment of a wide range of human infectious diseases.

In the 1960s and 1970s, chemists again came heavily into the infectious diseases' arena and began to modify the chemical structures produced by the microorganisms, giving rise to the so-called semi-synthetic antibiotics, which form a very significant part of the physicians' armamentarium in this field today. These efforts have proved highly valuable to patients requiring antibiotic therapy and to the industry alike. The truly impressive rate of discovery of the 'semi-synthetic' antibiotics was made possible by the finding that, particularly in the penicillin and cephalosporin classes of antibiotics, a portion of the entire molecule (the so-called 6-APA in the case of penicillin and 7-ACA in the case of cephalosporin) became available in large quantities from fermentation sources. These complex structures were not, in and of themselves, able to inhibit the growth of bacteria, but they provided to the chemist the central core of a very complicated molecule (via the fermentation process), which the chemist could then modify in a variety of ways to produce compounds that were fully active (hence the term "semi"-synthetic antibiotics). Certain advantages were conferred upon the new molecules by virtue of the chemical modifications such as improved oral absorption, improved pharmacokinetic characteristics, and expanded spectrum of organisms that were inhibited, to name a few. Chemical analogs of antibiotics, other than the penicillin and cephalosporins, have also been produced. The availability of truly efficacious antibiotics to treat a wide variety of severe infections undoubtedly represents one of the primary contributors to prolongation of life in modern society, as compared to the situation that existed in the early part of this century.

Coincidental with the above developments, biomedical scientists in pharmaceutical companies were actively pursuing purified extracts and pure compounds derived from plants and animal sources (e.g., digitalis, rauwolfia alkaloids, and animal hormones) as human medicaments. Analogs and derivatives of these purified substances were also investigated intensively in the hope of increasing potency, decreasing toxicity, altering absorption, securing patent protection, etc. During this period, impressive discoveries were made in the fields of cardiovascular, central nervous system, and metabolic diseases (especially diabetes); medicinal chemists and pharmacologists set up programs to discover new and, hopefully, improved tranquilizers, antidepressants, antianxiety agents, antihypertensive agents, hormones, etc. Progress in the discovery of agents to treat cardiovascular and central nervous system diseases was considerably slower than was the case with infectious diseases. The primary reason for this delay is the relative simplicity and straightforwardness of dealing with an infectious disease as compared to diseases of the cardiovascular system or of the brain. Specifically, infectious diseases are caused by organisms that, in many cases, can be grown in test tubes, which markedly facilitates the rate at which compounds that inhibit the growth of, or actually kill, such organisms can be discovered. Not only was the testing quite simple when carried out in the test tube but also the amounts of compounds needed for laboratory evaluation were extremely small as compared to those required for animal evaluation. In addition, animal models of infectious diseases were developed very early in the history of this aspect of pharmaceutical research, and activity in an intact animal as well as toxicity could be assessed in the early stages of drug discovery and development. Such was not the case in the 1950s as far as cardiovascular, mental, or certain other diseases were concerned because the basic defect or defects that lead to the disease in man were quite unknown. In addition, early studies had to be carried out in animal test systems, test systems which required considerable amounts of the compound and were much more difficult to quantitate than were the *in vitro* systems used in the infectious-disease field. The successes in the antibiotic field undoubtedly showed a carry-over or 'domino' effect in other areas of research as biochemists and biochemical pharmacologists began to search for *in vitro* test systems to provide more rapid screening for new drug candidates, at least in the cardiovascular and inflammation fields. The experimental dialog among biochemists, pharmacologists, and clinicians studying cardiovascular and mental diseases led, in the 1960s, to the development of various animal models of these diseases that increased the rate of discovery of therapeutic agents for the treatment thereof. Similar research activities in the fields of cancer research, viral infections, metabolic diseases, AIDS, inflammatory disease, and many others have, likewise, led to *in vitro* and animal models that have markedly increased the ability to discover new drugs in those important fields of research. With the increased discovery of drug activity came the need for increased regulation, and from the early 1950s on, the Food and Drug Administration (FDA) expanded its activities and enforcement of drug laws with both positive and negative results, from the standpoint of drug discovery.

In the later quarter of the 20th century, an exciting new technology emerged into the pharmaceutical scene, namely, biotechnology. Using highly sophisticated, biochemical genetic approaches, significant amounts of proteins, which, prior to the availability of so-called genetic engineering, could not be prepared in meaningful quantities, became available for study and development as drugs. Furthermore, the new technology permitted scientists to isolate, prepare in quantity, and chemically analyze receptors in and on mammalian cells, which allows one to actually design specific effectors of these receptors.

As the drug discovery process increased in intensity in the mid- to late 20th century, primarily as a result of the major screening and chemical synthetic efforts in the pharmaceutical industry in industrialized countries worldwide, but also as a result of the biotechnology revolution, the need for increased sophistication and efficacy in (a) how to discover new drugs, (b) how to reproducibly prepare bulk chemicals, (c) how to determine the activity and safety of new drug candidates in preclinical animal models prior to their administration to human beings, and finally, (d) how to establish their efficacy and safety in man became of paramount importance. Likewise, the ability to reproducibly prepare extremely pure material from natural sources or biotechnology reactors on a large scale and to deliver stable and sophisticated pharmaceutical preparations to the pharmacists and physicians also became significant.

The above brief history of early drug use and discovery is intended to be purely illustrative, and the reader is referred to an excellent treatise by Mann[1] to become well informed on the history of drug use and development from the earliest historic times to the present day.

REFERENCE

1. Mann, R.D., *Modern Drug Use: An Enquiry on Historical Principles,* MTP Press, Lancaster, England, 1984, pp. 1–769.

Introduction to the Second Edition (2006)

OVERVIEW OF THE CURRENT PROCESS OF NEW DRUG DISCOVERY AND DEVELOPMENT

Charles G. Smith and James T. O'Donnell

The first edition[1] of this book was published approximately 13 years ago. Its primary objective was to present an overview and a "roadmap" of the process of new drug discovery and development, particularly oriented to individuals or companies entering the pharmaceutical field. It was written by one of the authors (Smith), with no contributors, and drawn on Smith's experiences in the industry and field over the course of nearly 40 years. In the second edition, the scope of the first book has been expanded, and technical details in the form of hard data have been included. In addition to the editors' own commentary and contributions, the major part of the book is the result of contributions of experts in the industry. New chapters on risk assessment, international harmonization of drug development and regulation, dietary supplements, patent law, and entrepreneurial startup of a new pharmaceutical company have been added. Some of the important, basic operational aspects of drug discovery and development (e.g., organizational matters, staff requirements, and pilot plant operations) are not repeated in this book but can be found in the first edition.

In the 1990s and the new millennium, major changes have occurred in the pharmaceutical industry from the vantage points of research and development as well as commercial operations. New technologies and processes such as "high throughput screening" and "combinatorial chemistry" were widely embraced and developed to a high state of performance during this period. The very impressive rate of throughput testing the hundreds of thousands of compounds required micronization of operations, resulting in the reduction of screening reaction mixtures from milliliters to microliters. The systems are generally controlled by robots, and testing plates can accommodate a wide spectrum of biological tests. Combinatorial chemistry, a process in which a core molecule is modified with a broad spectrum of chemical reactions in single or multiple reaction vessels, can produce tens of thousands of compounds for screening. The objective of both approaches is to provide very large numbers of new chemical entities to be screened for biological activity *in vitro*. The use of computers to design new drug candidates has been developed to a significant level of sophistication. By viewing on the computer, the "active site" to which one wants the drug candidate to bind, a molecule can often be designed to accomplish that goal. The true impact of these approaches on the actual rate of discovering new drugs is yet to be established. Some have questioned the utility of these new screening methods, claiming that no new molecular entities (NME) have resulted from these new screening methodologies, despite hundreds of millions invested by the industry.

Studies in the last few years in the fields of genomics and proteomics have made available to us an unprecedented number of targets with which to search for new drug candidates. While knowledge of a particular gene sequence, for example, may not directly point to a specific disease when the sequences are first determined, investigations of their presence in normal and diseased tissues could well lead to a quantitative *in vitro* test system that is not available today. The same can be said for the field of proteomics, but final decisions on the value of these new technologies cannot be made for some years to come.

Thanks to advances in genomics, animal models can now be derived using gene manipulation and cloning methods that give us never-before available *in vivo* models to be used in new drug screening and development. A large number of mammalian cell culture systems have also been developed not only to be used in primary screening but also for secondary evaluations. For example, the *in vitro* Caco 2 system shows some very interesting correlation with drug absorption *in vivo*. A test such as this is mandatory when one is dealing with several thousands of compounds or mixtures in a given experiment. More time will be needed to be absolutely certain of the predictability of such test systems but, appropriately, Caco 2 is widely used today in screening and prioritizing new drug candidates. As is always the case, the ultimate predictability of all the *in vitro* tests must await extensive studies in humans, which will occur several years henceforth.

In addition to the discussion are metabonomics that relate to their unique position within the hierarchy of cell function and their propensity to cross-membranes and organs. Thus, many metabolites are found in bodily fluids that are accessible to measurement in humans using relatively noninvasive technologies. The study of metabolomics provides the pragmatic link from the macromolecular events of genomics and proteomics to those events recognized in histology. Applications of such strategies can potentially translate discovery and preclinical development to those metabolites measured traditionally, as first-in-human studies are performed earlier in drug discovery and development process, especially where no animal models are adequate.

During the past decade, clinical trial methodology has been expanded, improved, and, in large measure, standardized. The clinical testing phase of new drug development is the most expensive single activity performed. In addition to cost, it is very time-consuming since, with chronic diseases, one must investigate the new drug candidate in a significant number of patients over a period of months or years, in randomized, double-blind, placebo-controlled or active-drug–controlled studies. The search for surrogate endpoints continues, as it should, because a surrogate endpoint can markedly increase the rate of progression in clinical investigations with new drug candidates in certain disease states. Modern advances in molecular biology, receptor systems, cellular communication mechanisms, genomics, and proteomics will, according to our belief, provide researchers with new approaches to the treatment of a variety of chronic diseases. Significantly improved prescription medications are sorely needed in many fields. In the past decade, we have witnessed very impressive advances in the treatment of AIDS, for example. There is no question that life expectancy has been increased, albeit accompanied by significant drug toxicity and the need to use a "cocktail" of drugs in combination. The ability of the AIDS virus to mutate and become drug resistant presents a major and imminent threat to all patients afflicted with this disease. Serious efforts are under way in the pharmaceutical industry to find new drugs, across the entire infectious disease spectrum, which are not cross-resistant with existing therapies.

Cancer and AIDS vaccines are also under investigation using new technologies, and, hopefully, the day will come when we can prevent or ameliorate some of these debilitating and fatal diseases by vaccination. In the cancer field, new methodologies in science have, again, given us new targets with which to search for chemotherapeutic agents. The humanization of monoclonal antibodies has resulted in the marketing of some truly impressive drugs that are much better tolerated by the patient than are cytotoxic agents. In the case of certain drug targets in cancer, impressive results have been seen in the percentage of leukemia and lymphoma patients who can be brought into complete remission. In addition, biological medications to increase red and white blood cells have become available. Unfortunately, drug resistance once again plagues the cancer field, as are the cases with AIDS and various infectious diseases. As a result, researchers are seeking compounds that are not cross-resistant with existing therapies. Very significant advances in drug discovery are also expected to be seen in central nervous system, cardiovascular, and other chronic diseases as a result of breakthrough research in these fields.

Although the focus of this book is the research and development side of the pharmaceutical industry, certain commercial considerations are worth mentioning because of the major impact they may have on new drug research. These opinions and conclusions are based solely on decades of experience in the field by editors, working in the industry within companies and as an independent consultant (Smith), and also as a healthcare worker and academic (O'Donnell). No financial incentive for these statements has been received from the pharmaceutical industry. As the result of the very complicated nature of drug discovery and development, unbelievable costs accrue in order to bring a new therapeutic agent to market. Increasing costs are incurred, in part, from (a) shifting disease targets from more rapidly evaluable, acute diseases to those with poor endpoints and chronicity and (b) the emergence and rapid spread of serious diseases in society (e.g., AIDS, certain cancers, hepatitis C). In addition to increasing cost, the time required to gather sufficient data to be able to prove, to a statistically valid endpoint, that the drug has indeed been effective in a given disease has risen. The cost for the development of a major drug has been widely stated to be US $800 million per new therapeutic agent placed on the market.[1] This figure incorporates, of course, the cost of "lost" compounds that did not make the grade during preclinical or clinical testing. It has recently been reported that, while historically 14% of drugs that entered phase I clinical trials eventually won approval, now only 8% succeed. Furthermore, 50% of the drug candidates fail in the late stage of phase III trials compared to 20% in past years. More details on these points can be found in the literature (cf. Refs. [2–8]).

The average time from the point of identifying a clinical candidate to approval of a new drug is approximately 10 years. There is an understandable clamor in the population and in our legislative bodies to lower the price of prescription drugs. The cost of some prescription drugs is, to be sure, a serious problem that must be addressed, but some of the solutions, suggested and embraced by certain legislators, could have serious negative impact on new drug discovery and development in the future. For example, allowing the importation of prescription drugs from Canada or other non-U.S. countries (25 around the world have been mentioned) may well reduce the price of new drugs in this country to the point of significantly decreasing profits that are needed to support the tremendous cost of new drug discovery and development. The record clearly shows that countries that control drug prices, frequently under socialist governments, do not discover and develop new prescription drugs. The reason is obvious since the cost and time factors for new drug discovery can only be borne in countries in which the pharmaceutical companies are clearly profitable. Our patent system and lack of price controls are the primary reasons for the huge industrial success of new product development in this country, in and out of the pharmaceutical arena. If we undercut that system in the prescription drug field, the cost of drugs will certainly go down in the United States in the short term but, without the necessary profits to invest heavily in new drug discovery and development, the latter will also surely drop. Since it requires a decade from the time of initial investigation to marketing of a new drug, this effect would not be evident immediately after (a) allowing re-importation, (b) overriding patent protection, or (c) implementing price controls but, within a period of 5–10 years, we would certainly see pipelines of new medications beginning to dry up. Indeed, if such a system were allowed to continue for several years, new drug development as we know it would, in our opinion, be seriously impeded. When legislators look to Canada as an example of successful government subsidy of drugs, they should also consider whether a country like Canada could ever produce a steady stream of major new drugs, as does the United States. Research budgets have never been larger, we have never had as many innovative and exciting targets on which to focus, and this enormous effort cannot be afforded unless the companies selling the drugs can realize an adequate profit. If our pipelines of new prescription drugs dry up, you can be rest assured that the deficit will not be satisfied elsewhere in the world. It has been reported that 10 years ago, drug companies in Europe produced a significantly greater percentage of prescription drugs than is the case today. Society simply cannot afford to risk a marked reduction in new drug discovery in this country. Patients must join the fight to see that activities to impose price controls, which will inevitably reduce the rate of discovery of many potential drugs, are not based on political motives on the part of legislators. At this point in history, U.S. science stands in the forefront of new drug discovery and development. As noted above, never before have we had such an array of biological targets and synthetic and biotechnological methods with which to seek new medications. Hopefully, our government, in collaboration with the pharmaceutical industry, will find more suitable methods to solve the question of the cost of new pharmaceuticals than to impose price controls equal to those in countries that have socialized medicine. There can be no question as to whether the primary loser in such moves will be patients.

In addition to the question of the rate of drug discovery and development, we must be concerned about the quality of drugs available by mail or over the internet. The FDA cannot possibly afford to check all drugs flowing into America from as many as 25 foreign countries from which our citizens might be allowed to buy prescription drugs. It will be interesting to compare the regulatory requirements for FDA approval in the United States with those of the least stringent of the foreign countries from which some of our legislators want to approve importation of drugs. Would Congress be prepared to mandate a lowering of FDA standards to the same level in order to reduce the cost of drug discovery and development in this country? We certainly hope not! Indeed, there have been reports that drugs imported and sold on the internet are counterfeit and frequently contain little or no labeled active ingredients, and further may contain adulterants.

Another new topic chapter in the second edition of this book discusses the so-called dietary supplements, contributed by a recognized authority in Health Fraud. Over the past few years and, especially, since the passage of the DSHEA Act by Congress,[9] the use of such products has increased dramatically, and they are made widely available to the public with little or no FDA regulation. Although the law prevents manufacturers from making a medical treatment claim on the label of these preparations, such products generally have accompanying literature citing a variety of salutary effects in patients with various ills, the majority of which have not been proven by FDA type-randomized, double-blind, placebo-controlled clinical studies, of the kind that must be performed on prescription drugs and some "over-the-counter" drugs in this country. Published studies on quality control defects in some of these dietary supplement products (cf. ConsumerLab.com) indicate the need for tightening up of this aspect of product development. FDA is currently promulgating GMPs for dietary supplements. An enhanced enforcement of the dietary supplement regulations now exists.[9] A small segment of the dietary supplement industry has been calling for GMPs and increased FDA regulation.[10]

BASIC SCIENTIFIC DISCOVERY AND APPLICATION TO NEW DRUG DEVELOPMENT

In an apparent attempt to determine whether the American taxpayer is getting fair benefits from research sponsored by the federal government, the Joint Economic Committee of the U.S. Senate (for history, see Ref. [7]) has been considering this question. Historically, basic research has been funded by the NIH and various philanthropic foundations to discover new concepts and mechanisms of bodily function, in addition to training scientists. The role of industry has been to apply the basic research findings to specific treatments or prevention of disease. This is the appropriate manner in which to proceed. The industry cannot afford to conduct sufficient basic research on new complicated biological processes in addition to discovering new drugs or vaccines. The government does not have the money, time, or required number of experts to discover and develop new drugs.

The process that plays out in real life involves the focus of pharmaceutical industry scientists on desirable biological targets that can be identified in disease states, and to set up the program to discover specific treatments that will show efficacy in human disease. The compounds that are developed successfully become drugs on which the company holds patents. In this manner, the enormous cost of discovering and developing a new drug (estimated at $800 million plus over a period of some 10 years[1]) as noted above can be recouped by the founding company since no competitors can sell the product as long as the patent is in force. Without such a system in place, drug companies simply could not, in our opinion, afford to bring new prescription drugs to the market.

In the course of reviewing the matter, the Joint Economic Committee examined a list of 21 major drugs, which was put together apparently as an example of drug products that might justify royalty to the government. One of these agents, captopril (trade name Capoten), was discovered and developed by E.R. Squibb & Sons in the 1970s. At that time, Charles Smith (one of the authors/

editors) was vice president for R&D at The Squibb Institute for Medical Research. One of Squibb's academic consultants, Professor Sir John Vane of the Royal College of Surgeons in London brought the idea of opening a new pathway to treat the so-called essential hypertension by inhibiting an enzyme known as the angiotensin-converting enzyme (ACE). This biochemical system was certainly known at that time but, in Squibb's experience in the field of hypertension treatment, was not generally thought to play a major role in the common form of the disease, then known as "essential hypertension." The company decided to gamble on finding a treatment that was not used at the time and that would be proprietary to the company. Professor Vane (Nobel laureate in medicine in 1982) had discovered a peptide in snake venom that was a potent inhibitor of ACE. Squibb decided to pursue the approach he espoused, resulting in the development of a unique treatment for this very prevalent and serious disease.

In the first phase of their research, Squibb tested a short-chain peptide isolated from the venom of the viper *Bothrops jararaca*, with which Vane was working in the laboratory, in human volunteers and showed that it did, indeed, inhibit the conversion of angiotensin I to angiotensin II after intravenous injection. The peptide was also shown to reduce blood pressure in patients when injected. Since the vast majority of peptides cannot be absorbed from the GI tract, Squibb scientists set out to prepare a nonpeptide compound that could be used orally and manufactured at acceptable cost. The design of a true peptidomimetic that became orally active had not been accomplished at that time. Squibb then carried out a full-blown clinical program on a worldwide basis, which led to FDA approval of Squibb's drug Capoten (captopril), an ACE inhibitor. Mark also marketed an ACE inhibitor in the same time frame. This work opened a new area of research that has resulted in a bevy of new drugs that share this mechanism of action for use as antihypertensive drugs (for more detail, see Refs. [11–15]).

In the minds of pharmaceutical researchers and, hopefully, the public at large, the above example illustrates the unique role of pharmaceutical companies in making good use of basic research to discover new treatments for serious diseases. The huge costs to discover and develop a new drug could not be borne unless the companies knew that, if their gamble worked (which is not the case in the majority of situations), they would be assured of a good financial return for their shareholders. This system has served the country well in many fields of endeavor, in and out of the drug arena, and should be retained as such.

REGULATION OF NEW DRUG DEVELOPMENT

Drug development will come to a crashing halt without approval of the U.S. FDA, authorized by Congress to approve, license, and monitor the drugs sold to the American public. We are fortunate to have two contributors from the FDA, an acting associate commissioner for operations, and also CDER's (Center for Drug Evaluation and Research) associate director for International Conference on Harmonisation (ICH). These authors describe the FDA's new critical pathway initiative, pharmacists' risk management contributions, as well as the Common Technical Document (eCTD), which will enable a sponsor to file in one of the cooperating ICH partners, and receive approval for almost global marketing of the new agent. A very important chapter on pharmacogenetics and pharmacogenomics includes numerous FDA contributors.

LIABILITY AND LITIGATION

Last and the most unpopular topic in any industry, especially in the pharmaceutical industry, is the topic of liability and litigation. We have elected to include a chapter on this topic so that workers from all scientific disciplines involved in drug discovery and development can learn from history, and, hopefully, avoid being involved in the devastation of life (due to toxicity of inadequately manufactured drugs or drugs with inadequate warnings for safe use) and destruction of companies and careers that follows in the aftermath of drug product litigation.

REFERENCES

1. Smith, C.G., *The Process of New Drug Discovery and Development*, 1st ed., CRC Press, Boca Raton, FL, 2002.
2. Di Masi, J.A., Hansen, R.W., and Grabowski, H.G., The price of innovation: New estimates of drug development costs, *J. Health Econ.*, 22, 151–185, 2003.
3. Reichert, J.M. and Milne, C.-P., Public and private sector contributions to the discovery and development of 'impact' drugs, *Am. J. Therapeut.*, 9, 543–555, 2002.
4. Di Masi, J., Risks in new drug development. Approval success rates for investigational drugs, *Clin. Pharmacol. Therapeut.*, 69, 297–307, 2001.
5. Di Masi, J., New drug development in the United States from 1963 to 1999, *Clin. Pharmacol. Therapeut.*, 69, 286–296, 2001.
6. Grabowski, H., Vernon, J., and Di Masi, J.A., Returns on research and development for 1990s new drug introductions, *Pharmacol. Econ.*, 20 (suppl. 3), 11–29, 2002.
7. Reichert, J.M. and Milne, C.-P., Public and private sector contributions to the discovery and development of "impact" drugs, A Tufts Center for the Study of Drug Development White Paper, May 2002.
8. Hardin, A., More compounds failing phase I, *Scientist Daily News*, Aug. 6, 2004, p. 1.
9. Dietary Supplement Health and Education Act of 1994, Publ. no. 103-417, 108 Stat. 4325 codified 21 U.S.C. 321, et seq. (suppl. 1999).
10. FDA links: (a) www.cfsan.fda.gov/~dms/ds-warn.html (b) www.cfsan.fda.gov/~lrd/hhschomp.html (c) http://www.fda.gov/ola/2004/dssa0608.html
11. Smith, C.G. and Vane, J.R., The discovery of captopril, *FASEB J.*, 17, 788–789, 2003.
12. Gavras, H., The discovery of captopril: Reply, *FASEB J.*, 18, 225, 2004.
13. Erdas, E.G., The discovery of captopril: Reply, *FASEB J.*, 18, 226, 2004.
14. Pattac, M., From viper's venom to drug design: Treating hypertension, *FASEB J.*, 18, 421, 2004.
15. Smith, C.G. and Vane, J.R., The discovery of captopril: Reply, *FASEB J.*, 18, 935, 2004.

Editors

James J. O'Donnell III, MS, PhD, is a Pharmacologist and Chemist who serves as an Assistant Professor at Rosalind Franklin University of Medicine and Science, where he teaches pharmacology in the Chicago Medical School and the College of Pharmacy. He received his bachelor's degree in electrical engineering and computer sciences from the Massachusetts Institute of Technology (MIT), a master's degree in chemistry from the University of Wisconsin-Madison, and a doctor of philosophy in pharmacology from Rush University in Chicago. His medical research has principally focused on inflammatory diseases, but he has also conducted research in inorganic chemistry and the microbiome and consulted in the area of pharmacometrics. He is co-editor of *O'Donnell's Drug Injury* Fourth Edition, and has authored 17 peer-reviewed journal publications and 17 book chapters.

John Somberg, MD, obtained his BA from New York University and his MD from New York Medical College, where he was a Resident in Medicine and then Chief Medical Resident. He completed his cardiology fellowship at Harvard Medical School, Peter Bent Brigham Hospital, and was then an Instructor in Medicine at Harvard. He moved to Albert Einstein College of Medicine as Director of the Cardiac Arrhythmia Service and Assistant Professor of Medicine and Pharmacology. In 1988, he became a Professor of Medicine and Chief of Cardiology at the Chicago Medical School. In 1998, he moved to Rush University as a Professor of Medicine, Cardiology, and Pharmacology and Chief of the Division of Clinical Pharmacology. He founded the Masters in Clinical Research Program at Rush and directed the program. He has conducted over 50 clinical trials as Principal Investigator and has served as an FDA consultant and member of the Circulatory Device Advisory Panel. He has advised industry and been the initiator of programs leading to two NDAs and one ANDA, and obtaining two Orphan Drug Applications. He has authored four book chapters, three books, and over 300 peer-reviewed papers. He was the Editor of the *Journal of Clinical Pharmacology* for ten years and the founding Editor of the *American Journal of Therapeutics* and its Editor for 20 years. He holds 14 patents in the pharmaceutical area and currently actively consults in medical and regulatory affairs for industry.

Vincent Idemyor, PharmD, has been active in the healthcare industry and academic profession for over 25 years in the United States. He has amassed experience in teaching, clinical practice, and clinical research, specifically, in the realm of infectious diseases pharmacotherapy and the development and implementation of disease state management. His previous academic appointments include clinical faculty member, Assistant Dean, MacArthur Foundation Visiting Senior Fellow, National Universities' Commission LEADS Scholar, and Association of African Universities' Exchange Fellow, and is currently Professor at the University of Port Harcourt, Nigeria. Dr. Idemyor has held various professional management positions in affiliated community teaching hospitals in the United States where he built and managed clinical programs, as well as provided didactic and student/residency training. He has conducted research on a variety of topics and has assisted in securing extramural funding in excess of $5 million (USD) to support various educational programs and research projects nationally and internationally. He is the recipient of many awards for his work, including the 2001 Research and Publications Achievement award, 2003 Illinois Department of Public Health "Red Ribbon Award", 2004 City of Chicago Commission on Human Relations Outstanding Achievement Award, and the 2007 Medical Award of the Year from the National Technical Association. His scientific literature work was ranked in the top one percentile among the over 239,000 scientific authors dated March 16, 2006, in the world AIDS and HIV research literature database.

James T. O'Donnell, PharmD, is an Associate Professor at the Rush Medical College involved in pharmacology education, Institutional Review Board, and graduate student collaboration. After a nearly 20-year period of establishing frontier clinical pharmacy services, he switched his focus to clinical pharmacology, pharmaceutical industry, institutional, and forensic consulting. He has extensive experience as a scientific editor. He was a co-editor of *The Process of New Drug Discovery and Development* Second Edition (2006), the founding editor of the *Journal of Pharmacy Practice*, the editor of five additional books, and has more than 300 publications.

Contributors

Darrell R. Abernethy
Office of Clinical Pharmacology
US Food and Drug Administration
Silver Spring, Maryland

Loyd V. Allen, Jr.
International J Pharmaceutical Technology
Edmond, OK

Duncan Armstrong
Pre-Clinical Safety
Novartis Institutes for Biomedical Research
Cambridge, Massachusetts

Crista Brawley
Rush University Medical Center
Chicago, Illinois

Gilbert Burckart
Office of Clinical Pharmacology
US Food and Drug Administration
Silver Spring, Maryland

Janelle M. Burnham
Office of Clinical Pharmacology
US Food and Drug Administration
Silver Spring, Maryland

Annie Delaunois
UCB Pharmaceuticals
Anderlecht, Belgium

David J. Dow
Translational Research, Cell and Gene
    Therapy, Medicines Research Centre,
    GlaxoSmithKline R&D
Stevenage, United Kingdom

S. Albert Edwards
eSubmissions University
Lincolnshire, IL

Virneliz Fernandez-Vega
Scripps Research
Jupiter, Florida

Raphael M. Franzini
The University of Utah
Salt Lake City, Utah

Jennifer Garcia
Rush University Medical Center
Chicago, Illinois

Jack A. Gilbert
University of California, San Diego
San Diego, California

Vitalina Gryshkova
UCB Pharmaceuticals
Anderlecht, Belgium

Stephanie Guzik
Rush University Medical Center
Chicago, Illinois

Tim G. Hammond
Preclinical Safety Consulting Limited
Muttenz, Switzerland

Allecia Harley
Rush University Medical Center
Chicago, Illinois

Kathleen Heneghan
American College of Surgeons
Chicago, Illinois

Ismael J. Hidalgo
Absorption Systems
Exton, Pennsylvania

Elle Simone Hill
University of Chicago
Chicago, Illinois

Steven J. Howe
Process Research, Cell and Gene
    Therapy, Medicines Research Centre,
    GlaxoSmithKline R&D
Stevenage, United Kingdom

Vincent Idemyor
University of Port Harcourt
Port Harcourt, Nigeria

Thierry Jean
Levitha
Brussels, Belgium

Laura A. Johnson
Translational Medicine, Oncology Cell
    Therapy DPU, GlaxoSmithKline R&D
Collegeville, Pennsylvania

Erin Kampschmidt
Rush University Medical Center
Chicago, Illinois

Madeline Kim
University of Chicago
Chicago, Illinois

Thomas Kuntz
University of Chicago
Chicago, Illinois

S.W. Johnny Lau
Office of Clinical Pharmacology
US Food and Drug Administration
Silver Spring, Maryland

Chih-Shia Lee
Laboratory of Cancer Biology and Genetics,
    Center for Cancer Research
National Cancer Institute, National Institutes of
    Health
Bethesda, Maryland

Jibin Li
Absorption Systems
Exton, Pennsylvania

Ji Luo
Laboratory of Cancer Biology and Genetics,
    Center for Cancer Research
National Cancer Institute, National Institutes of
    Health
Bethesda, Maryland

A.M. Lynch
ToxPlus Consulting
Haymarket, Virginia

Sabyasachi Maiti
Department of Pharmacy
Indira Gandhi National Tribal University
Amarkantak, India

John McClatchy
Rush University Medical Center
Chicago, Illinois

Daniel Mufson
Apotherx
Napa, California

Lori Nesbitt
Compass Point Research
Franklin, TN

Sarfaraz K. Niazi
University of Illinois
Urbana, Illinois
and
University of Houston
Houston, Texas
and
Karyo Biologics, LLC
Hoffman Estates, Illinois

James J. O'Donnell III
Rosalind Franklin University of Medicine and
    Science
North Chicago, Illinois

James T. O'Donnell
Rush University Medical Center
Chicago, Illinois

Jeff Oswald
Rush University Medical Center
Chicago, Illinois

Damani Parran
Nouryon Chemicals
Farmsum, Netherlands

Gourang Patel
Rush University Medical Center
Chicago, Illinois

J. Pena
Department of Molecular Medicine
Scripps Research
Jupiter, Florida

Olivia Perez
Department of Molecular Medicine
Scripps Research
Jupiter, Florida

Madhu Pudipeddi
Parakam Pharma LLC
Wilmington, Delaware

Lyn Rosenbrier Ribeiro
Discovery Safety, Drug Safety and Metabolism,
    IMED Biotech Unit
AstraZeneca
Cambridge, United Kingdom

Paul F. Richardson
Pfizer-La Jolla, Medicinal Sciences
La Jolla, California

Marie-Luce Rosseels
UCB Pharmaceuticals
Anderlecht, Belgium

Martha M. Rumore
Maurice A. Deane School of Law at Hofstra
    University
Hempstead, New York

Chandrahas Sahajwalla
Office of Clinical Pharmacology
US Food and Drug Administration
Silver Spring, Maryland

Mariana C. Salas Garcia
University of Chicago
Chicago, Illinois

Louis Scampavia
Department of Molecular Medicine
Scripps Research
Jupiter, Florida

William Schmidt
Sorell, Lenna & Schmidt, LLP
Hauppauge, New York

Kalyan Kumar Sen
Department of Pharmaceutics
Gupta College of Technological Sciences
Asansol, India

Abu T.M. Serajuddin
St. John's University
Jamaica, New York

Ankita V. Shah
Freund-Vector Corporation
Tokyo, Japan

John C. Somberg
Rush University Medical Center
Chicago, Illinois

Timothy Spicer
Department of Molecular Medicine
Scripps Research
Jupiter, Florida

Mala K. Talekar
Clinical Development, Oncology,
    GlaxoSmithKline R&D
Collegeville, Pennsylvania

Stephanie E. Tedford
Rush University Medical Center
Chicago, Illinois

Willis C. Triplett
W Triplett Consulting LLC
Seattle, Washington

Jean-Pierre Valentin
Development Science
UCB-Biopharma SPRL
Braine l'Alleud, Belgium
and
UCB Pharmaceuticals
Anderlecht, Belgium

Shrijay Vijayan
Rush University Medical Center
Chicago, Illinois

Mary Jane Welch
Rush University Medical Center
Chicago, Illinois

Krishna Yeshwant
Google Ventures
Mountain View, California

Jonathan C. Young
School of Public Health
West Virginia University
Morgantown, West Virginia

1 Introduction to Drug Discovery and Development

James J. O'Donnell III
Rosalind Franklin University of Medicine and Science

John C. Somberg and James T. O'Donnell
Rush University Medical Center

CONTENTS

In the 15 years since the release of Smith and O'Donnell's *The Process of New Drug Discovery and Development, Second Edition* (2005), the fields of drug discovery and development have seen tremendous changes. Advances in understanding of human biology and disease have uncovered fresh territory for drugs to target, and this progression of knowledge has been accelerated by the invention of new investigational tools. Further, new platforms to efficiently sift through the drug candidates have made it easier to find that needle in the haystack, the drug that will treat a disease safely and effectively. What's more, the processes of drug discovery and drug development, once separated in independent silos of sorts, have become increasingly integrated. No longer is the drug candidate handed off from drug inventor to drug developer like runners in a track relay meet. These changes in the drug discovery and development processes have reverberated throughout the biotech world, significantly impacting the scientific methodology employed for drug discovery initiatives as well as the institutional platforms underlying biomedical research.

The third edition of this book builds on the information published in the earlier editions and presents a thorough up-to-date overview of the field and a review of new developments. More than 60% of the chapter topics are new, and many other chapters have new authors offering fresh perspectives. Some chapters from the second edition were removed, as their subject matter was no longer relevant to drug discovery and development today. We are fortunate to have recruited more than 70 contributors – scientists, clinicians, and a few business types – from industry, academia, and government (FDA and NIH), as well as international representatives. While highly technical, our objective is for each chapter to be useful and current to experts in that particular subfield but still accessible to someone not familiar with that aspect of drug discovery and development.

The book is divided into four sections:

- Overview
- Drug Discovery
- Drug Development
- Regulations.

As leaders in the subfield detailed in each chapter, authors are uniquely positioned not only to review the literature but also to make generalizations informed by personal experience, comment on areas under debate, and make predictions on how the subfield is likely to change in the future. In keeping with this theme, the "Overview" section contains extended editorialization from two experts, Dr. John Somberg (an author of this chapter and an editor of this book) and Dr. Krishna Yeshwant. Dr. Somberg has been a basic and clinical investigator in anti-arrhythmic pharmacology for 40 years and has served as a Principal Investigator in over 100 Phase I, Phase II, and Phase III clinical trials. Dr. Yeshwant was trained as a physician and programmer in addition to business, and he currently serves as General Partner at Google Ventures; his chapter offers the rare perspective of a venture capitalist on trends and best practices in drug discovery and development. Another unique perspective of pharmacology is shown in a chapter detailing one case study in the discovery and early development of the cancer drug, lorlatinib, which was approved by the FDA in 2018. Instead of the traditional organization of a pharmaceutical journal article with a listing of positive results only, this chapter takes a more chronological approach, showing the twists and turns of drug discovery from the point of view of one of the investigators, Dr. Paul Richardson of Pfizer. As such, this success story demonstrates how the facets of drug discovery and development fit together in real life, and how one may lead such a project and overcome the challenges that commonly emerge.

A significant portion of the book is devoted to coverage of "hot" scientific fields that are fertile ground for new drug candidates. Examples of this that are highlighted in the book include the microbiome and the potential of probiotics as drugs, nanotechnology for diagnostic and therapeutic applications, and the harnessing of the immune system to combat cancer. The emergence of the gene-editing tool CRISPR and its applications to drug discovery is also the subject of a chapter.

Another large section of the book details technical processes for screening compounds and evaluating promising drug candidates for efficacy and safety. High-throughput screening of compounds, a staple of drug discovery for decades which has undergone continual innovation, is the subject of one chapter. DNA-encoded libraries, the subject of another chapter, is a novel technology that has quickly become entrenched in the drug discovery process in the past decade. In this technology, many test compounds can be tested for affinity to target protein in the same well of a multi-well plate since amplification of the DNA tag reveals the identity of the binding compound after the fact. The evaluation of the safety of drugs is the specific focus of three chapters on preclinical toxicology, safety pharmacology, and *in vitro* pharmaceutical profiling of drug selectivity.

The viability of drugs is not limited to activity at the target site – whether the drug can reach the target site is also critical to determining "drugability." As such, one chapter is devoted to pharmacokinetics as well as pharmacodynamics. Another chapter on the Caco-2 cell model focuses on one aspect of pharmacokinetics, absorption, which is the main determinant of bioavailability.

Also presented are extended discussions of clinical research, both inside and outside the academic research institution. In addition to broad chapters on clinical trials methodology, biomedical research in academia, and ethics in clinical trials, more specialized chapters focus on the role of the pharmacist in research as well as the special challenges associated with clinical testing of HIV drugs, respectively. Another chapter is a comprehensive treatment of drug product development as a whole, discussing the progression from lead candidate selection to life-cycle management. This chapter integrates many of the other topics that are the subject of extended discussion in other chapters in the "Drug Development" section.

Since drug development does not exist in a vacuum, the regulatory hurdles which drug development aims to surmount are also discussed. Chapters in this section include a comprehensive treatment of important regulations in drug development and chapters on regulations in special populations, regulations governing compounding and pharmaceutics, and regulations guiding orphan drug development.

Patents, the protectors of the financial renumeration which drives companies to pursue drug development, are the subject of another chapter. "Repurposing" of drugs for new indications, a valuable strategy to avoid the costs of developing a brand-new compound, is covered in another chapter. This unique contribution was provided by a nurse clinician–turned researcher who led initiatives in drug repurposing, and it includes a case study of the repurposing of loperamide as a topical anesthetic for blood sample needle sticks in neonates.

The following abstracts provide a preview of the chapters offered.

SECTION I. OVERVIEW

CHAPTER 2. CURRENT OPINIONS ON THE TRAJECTORY OF THE PHARMACEUTICAL DEVELOPMENT

J. Somberg

Dr. John Somberg, a uniquely experienced clinician, academician, researcher, and president of his own pharmaceutical (drug development) company, provides a good historical review as well as a look to the future. If one goes back a couple of decades reviewing the volumes of a medical journal, one will pass through a decade or so of clinical trials from large, adequately powered studies, to small trials and then retrospective clinical data reviews often with historic controls. Before the 1970s, one would see articles dealing with therapeutic approaches relating a clinicians' experience with a therapy and recalling the results of prior therapeutic approaches. Drug development and discovery is a most exciting field. It is creative, intellectually challenging, and organizationally demanding. Those involved are to be congratulated for undertaking efforts that are usually anonymous but affect clinical therapeutics to a considerable degree. The drug discovery process has gone through the botanical phase and the synthetic chemistry phase, and is now into a most exciting era of biotechnology and gene manipulation.

CHAPTER 3. INNOVATION IN DRUG DEVELOPMENT: PERSPECTIVES OF A VENTURE CAPITALIST

K. Yeshwant

Drug discovery and development is an expensive enterprise, requiring extensive funding and support. Current reports show that development of a new molecular entity through market release costs ~$2.6 billion. As the cost to discover and develop new drugs increases, the aggregate return on investment in each new project naturally decreases. Some studies suggest that the industry is currently at a point where it no longer makes financial sense to invest in the discovery of new drugs since the internal rate of return is approaching 0% or may even be negative. Such concerns do not merely affect the profit margin. Approaches that do not bear fruit in a cost-effective way are not feasible, and thus, the types of drugs that can be discovered and diseases that can be cured are limited. This chapter is a critical commentary on the methodological approaches to deal with these challenges, discussing the merits and drawbacks of each. This includes a discussion of both technical approaches to drug discovery and development, and financial approaches to drug development. Technical approaches discussed include how best to search a large chemical space more efficiently and the merits of different biological assays to predict whether drug candidates will be successful in clinical trials. The scope of the discussion of financial approaches includes the challenges in funding tools, platforms, and disease models; company incubation models; financing through value creation milestones; and insourcing from Pharma/Megaround. This unique contribution is provided by Dr. Krishna Yeshwant, a venture capitalist trained in medicine and computer science as well as business.

SECTION II. DRUG DISCOVERY

CHAPTER 4. HIGH-THROUGHPUT SCREENING

T. Spicer

From early research to FDA approval, each drug candidate encounters many obstacles in order to ensure efficacy, potency, and safety. For every compound that becomes a drug candidate, there are tens of thousands of compounds that have failed testing. Even compounds that pass years of preclinical testing to become drug candidates sometimes never go to market. "Experience has shown that only approximately 1 out of 15–25 drug candidates survives the detailed safety and efficacy testing (in animals and humans) required for it to become a marketed product" (Lombardino and Lowe 2004). High-throughput screening (HTS) is a process in early drug discovery that uses automation, miniaturization, and robotics to increase the number of assays tested against large drug-like compound collections all the while decreasing the volume needed for each test.

CHAPTER 5. DNA-ENCODED COMPOUND LIBRARIES: AN EMERGING PARADIGM IN DRUG HIT DISCOVERY

R. Franzini

The development of new therapeutic agents critically depends on the availability of technologies that enable their effective development. High-throughput screening has enabled the identification of potent hit molecules for structurally and functionally diverse protein targets, but the high costs associated with screening and maintaining large compound collections limit its utility. Especially, high-throughput screens are generally performed only for proteins for which there already is compelling evidence of their pharmaceutical potential. DNA-encoded libraries have become a routine hit discovery method in pharmaceutical research, and most large pharmaceutical research companies, as well as an increasing number of academic research groups, routinely use DNA-encoded libraries to discover protein binders. The growing use of DNA-encoded libraries has translated into a rapidly expanding number of screening hits identified from such libraries. Many of the hits have intriguing properties including new chemotypes and unprecedented binding modes, including first-in-class compounds and allosteric regulators. Furthermore, screening protocols are becoming increasingly sophisticated. Complementing simple hit discovery efforts, parallel screening experiments with DNA-encoded libraries provide valuable information on protein binding and target selectivity. These advancements may open entirely new possibilities in drug discovery and will possibly disrupt current medicinal chemistry workflows.

CHAPTER 6. BIO-TARGETED NANOMATERIALS FOR THERANOSTIC APPLICATIONS

S. Maiti and K.K. Sen

Nanotheranostics consist of nanotechnology which can be used to simultaneously treat and diagnose a disease. This chapter reviews the current state of true nanotheranostics, as well as nanotechnology which can only be used for one of these two applications (either diagnostic or therapeutic). Different nanotechnology platforms possess different advantages and disadvantages, and interchanging the ligands attached to the nanoparticle core leads to different indications. A discussion of how different nanotechnological characteristics lead to indication is undertaken in the context of a variety of diseases, including cancer, inflammatory bowel disease, diabetes, ocular diseases, and cardiovascular diseases.

CHAPTER 7. THE DEVELOPMENT OF ADOPTIVE T CELL IMMUNOTHERAPIES FOR CANCER: CHALLENGES AND PROSPECTS

D. Dow, S.J. Howe, M.K. Talekar, and L.A. Johnson

The field of cancer immunotherapy began in the 19th century with William Coley's studies injecting live *Streptococcus bacilli* bacteria and heat-killed bacterial extracts into primary tumors resulting in an induced durable remission of inoperable sarcoma. Fast forward to today where the successful development of CTLA-4 and PD-1 checkpoint inhibitors, along with the recent FDA approval of Kymriah and Yescarta, two CAR-T cell immunotherapies, heralds a new era in cancer medicine.

CHAPTER 8. CRISPR IN DRUG DISCOVERY

C. Lee and J. Luo

The Human Genome Project has provided a blueprint of the human chromosomes to annotate all genetic elements and their involvement in biological processes and in human diseases. In parallel with the physical mapping of the human genome, advances in high-throughput genetic screens have dramatically accelerated the pace of target discovery and target validation for drug development. Recent breakthroughs in the discovery and utilization of the CRISPR/Cas9 system for genome editing have once again transformed the field of functional genomics. Owing to its specificity and programmability, CRISPR/Cas9-mediated genome editing has become a powerful tool for elucidating gene function, for high-throughput target discovery and validation, and for the development of gene therapy in various human diseases. In this chapter, we present an overview of the discovery, development, and application of CRISPR/Cas9-mediated genome editing in the context of target identification and drug discovery. We focus our discussion on genetic screens using various CRISPR/Cas9 approaches for target identification. Examples in cancer biology are used to illustrate the power and promise of CRISPR/Cas9 in drug discovery.

CHAPTER 9. PROBIOTICS IN THE WORLD: "BUGS AS DRUGS."

T. Kuntz, M. Kim, E. Hill, M.C. Salas Garcia, and J.A. Gilbert

Disease is often what comes to mind when considering the presence of microbes inside the human body. Often overlooked are the commensal and even mutualistic interactions of microbes and humans. The importance of these interactions to human health is increasingly being recognized and is underscored by the sheer number of microbes in the human body. There are approximately as many microbial cells as human cells, mainly in the gut. The microbial taxa (mainly bacterial) associated with humans are known as the microbiota, and these organisms and their gene content, much of which supplements human physiology, are collectively known as the microbiome. Like other organs of the body (albeit of a vastly unique physiology), the microbiome constitutes a druggable target in cases of disease. Antibiotics are a clear example of a class of drugs with profound effects on the microbiota. One alternative and burgeoning approach to drugging the microbiome is treatment with microbes themselves. This chapter will explore the discovery and development of probiotics in clinical practice, including new horizons of probiotics as "bugs as drugs." Emphasis is placed on establishing the foundations of this paradigm and presenting a wide breadth of ideas in the probiotic space to highlight the exciting developments and challenges this field currently faces.

P.F. Richardson

This chapter is a case study in the discovery and early development of lorlatinib, a drug for non–small cell lung cancer (NSCLC) that received FDA approval in 2018. This chapter is written from the point of view of one of the discoverers, and it details the twists and turns involved in the discovery of this drug. NSCLC had previously been linked to the activity of a receptor tyrosine kinase, ALK, and thus, this became a target for NSCLC drugs. Although there was already an FDA-approved drug (crizotinib) which demonstrates robust efficacy in ALK-positive NSCLC patients, progression during treatment eventually develops with resistant patient samples revealing a variety of point mutations in the ALK kinase domain. In addition, some patients progress due to cancer metastasis in the brain. Herein, the discovery and development of the next-generation ALK inhibitor lorlatinib is described with the goal to address these challenges. The strategies for optimizing efficacy and selectivity, and achieving central nervous system (CNS)-penetration are disclosed, while the challenges associated with the large-scale synthesis of a macrocyclic compound are also addressed.

SECTION III. DRUG DEVELOPMENT

A. Serajuddin, M. Pudipeddi, A. Shah, and D. Mufson

The dosage-form design is guided by the properties of the drug candidate. If a new chemical entity (NCE) does not have suitable physical and chemical properties or pharmacokinetic attributes, the development of a dosage form (product) may be difficult and may sometimes be even impossible. Any heroic measures to resolve issues related to physicochemical and biopharmaceutical properties of drug candidates add to the time and cost of drug development. Therefore, the interaction between discovery and development scientists increased greatly to maximize the opportunity to succeed. This chapter presents a comprehensive treatment of drug product development as a whole, discussing the progression from lead candidate selection to life-cycle management.

J.P. Valentin, D. Armstrong, L.R. Ribeiro, and T. Jean

Drug attrition can be caused by inadequate efficacy, unacceptable safety risks, poor ADME (absorption, distribution, metabolism, and excretion) characteristics, or pharmaco-economic considerations. *In vitro* secondary pharmacological profiling is the investigation of the pharmacological effects of a drug at molecular targets distinct from the intended therapeutic molecular target. It is also described as selectivity screening, pharmacological profiling, or secondary pharmacology. It is increasingly being used earlier in the drug discovery process to identify undesirable off-target activities that could hinder or halt the development of candidate drugs, or even lead to market withdrawal if discovered after drug approval. The rationale, strategies, and methodologies for *in vitro* secondary pharmacological profiling are presented and illustrated with examples of their impact on the drug discovery process. In addition, approaches for early prediction of bioavailability are also discussed; this is a key parameter of the drug discovery process, giving an indication of the drug-gability of the compound, and helps in interpreting *in vitro* secondary pharmacological profiling data by considering the potential availability of the compound at its active site(s).

Chapter 13. Pharmacokinetics–Pharmacodynamics in New Drug Development

S. Niazi

Pharmacokinetic/pharmacodynamics (PK/PD) modeling, an integral component of the drug development process, is a mathematical technique for predicting the effect and efficacy of drug dosing over time. Broadly speaking, pharmacokinetic models describe how the body manipulates a drug in terms of absorption, distribution, metabolism, and excretion. Pharmacodynamic models describe how a drug affects the body by linking the drug concentration to an efficacy (or safety) metric. A well-characterized PK/PD model is an important tool in guiding the design of future experiments and trials.

Chapter 14. The Evolving Role of the Caco-2 Cell Model to Estimate Intestinal Absorption Potential and Elucidate Transport Mechanisms

J. Li and I.J. Hidalgo

Since the introduction of Caco-2 cells as an intestinal permeability model in 1989, this model has been widely applied to many areas of pharmaceutical research. Although the initial attraction to Caco-2 cells was due to the lack of alternative experimental models to screen the increasing numbers of drug candidates; it was the utility of the model in the estimation of absorption potential that drove its subsequent adaptation to other applications such as drug-transporter interactions, permeation enhancers, mechanisms of intestinal permeation, and formulation development. As a result of its widespread use in the academic, industrial, and regulatory environments across the world, it has become, by default, the gold standard for *in vitro* permeability studies. The Caco-2 cell model in pharmaceutical research has been cemented by its application in the context of the biopharmaceutics classification system (BCS) and transporter-mediated drug–drug interactions (DDI) guidelines. More recently, Caco-2 cells applied with novel systems such as the dissolution/permeation (D/P) chambers and *in vitro* dissolution absorption systems (IDAS), and using physiologically relevant experimental conditions, should help narrow the *in vitro–in vivo* gap, by yielding *in vitro* data of greater translatability to the *in vivo* situation.

Chapter 15. Preclinical Toxicology

D. Parran

The goals of the preclinical safety evaluation generally include a characterization of toxic effects with respect to target organs, dose dependence, relationship to exposure, and, when appropriate, potential reversibility. Preclinical toxicological testing encompasses animal test model selection and selection of dose and routes of administration. Study types to assess general toxicology include acute/dose-range finding toxicity studies and acute/repeat-dose screening studies. Chronic and sub-chronic toxicity studies are also employed. US, European, and Japanese drug regulations guide preclinical toxicological studies in these respective regions.

Chapter 16. Safety Pharmacology: Past, Present, and Future

J.P. Valentin, A. Delaunois, M.L. Rosseels, and T.G. Hammond

Safety pharmacology is a discipline that has rapidly grown and evolved during the last 20 years, and is now facing new challenges on scientific, technological, regulatory, and human fronts. Improvements in safety pharmacology hopefully reduce drug attrition. The reasons for drug attrition have evolved over the years; over the last 25 years, lack of safety (both non-clinical and clinical) remains the major cause of attrition during clinical development, which accounts for approximately 20%–30% of all drug discontinuation. More worrying is the fact that there is no clear trend toward a reduction of the attrition owing to safety reasons. The fundamentals of safety pharmacology stud-ies include such general considerations and principles as species/gender selection and animal status; selection of dose/concentration, route of administration, and duration; good laboratory practice; statistical analysis; testing of isomers, metabolites, and finished products; and timing of studies in relation to clinical development. A main objective of studies is to assess impact on function of vital organ systems such as the cardiovascular, respiratory, and central nervous systems. Non-vital organ functions such as the gastrointestinal and renal systems are also commonly assessed.

CHAPTER 17. ETHICS IN CLINICAL RESEARCH

J. Young and L. Nesbitt

Drug development in the United States is both risky and expensive. The cost of developing a drug has increased from approximately $500 million in 1990 to more than $2.6 billion in 2015. At the same time, the research and development (R&D) costs of the pharmaceutical industry have doubled from approximately 12% of sales to 21%. As of 2017, only about 12% of drugs with clinical trials will be successful, and the drug development process takes at least 10 years. Given the growing number of clinical trials required for FDA approval, opportunities are numerous in the provision of clinical research services. In addition, the FDA is becoming more vigilant in enforcing the ethi-cal conduct of clinical research and the protection of research participants. Lastly, in an effort to avoid conflicts of interest or perceived improprieties, pharmaceutical and device manufacturers frequently outsource all or part of the clinical trial process to niche service providers. For these rea-sons, the clinical trial industry has become segmented. Each segment or service provider performs a necessary step in the clinical trial value chain. In addition to niche service providers, the grow-ing clinical trial industry has created a need for service organizations, publications, and websites devoted to the specialized field. To assess function and potential toxicity, functional or biochemical biomarkers may be tracked which are specific to the organ system under study.

CHAPTER 18. CLINICAL TRIALS METHODOLOGY

J. Somberg

For drugs to enter the first phase of clinical development, they have to be accepted by the FDA under an IND (investigational new drug) application. The non-clinical data submitted with that applica-tion assures the FDA, and the clinical scientists involved in the subsequent phases of development, that the product in question is safe and effective enough to be tested in humans. One must remem-ber that *in vitro* and animal model-derived information is limited in its potential extrapolation to humans. The purpose of clinical development is to explore and confirm the non-clinically acquired knowledge and expand upon it to the human condition.

CHAPTER 19. ACADEMIC RESEARCH ENTERPRISE

C. Brawley, M.J. Welch, J. Oswald, E. Kampschmidt, J. Garcia, A. Harley, S. Vijayan, J. McClatchy, S. Guzik, and S. Tedford

The complexities of taking a drug from conception to FDA approval is manifested in the complexi-ties of managing a research site in general and an academic medical center (AMC) specifically. The multiple requirements of the pharmaceutical industry, the various agencies that regulate research, the continued fiscal viability of the AMC, and most importantly, the medical needs of the patient create a tension that results in partnerships that often have competing outcomes. To manage these many competing demands, AMCs have created research systems across their enterprise. This chap-ter describes many of the activities of the Office of Research Affairs at the Rush University Medical

Center in Chicago. The research contributions of faculty generate meaningful publications, values-driven care, lifesaving therapies, and leading-edge innovations ranging from omics-based discoveries to the next breakthrough drug or device. Continuously, our research teams progress toward measured, peer-affirmed outcomes that improve quality of life, and our understanding of the human condition. The Office of Research Affairs (ORA) exists to partner with faculty and staff as they seek funding, propose clinical studies, establish collaborations, steward funds, submit grants, negotiate industry contracts, and secure patents and licensing agreements.

Chapter 20. Clinical Testing Challenges in HIV/AIDS Research

V. Idemyor

The human immunodeficiency virus (HIV) is responsible for acquired immune deficiency syndrome (AIDS), and this virus was first described in 1983. HIV continues to create formidable challenges to the biomedical research and public health communities around the world. The profile of the epidemic has shifted over the past 30 years. We have seen declining numbers of new infections and AIDS-related mortality throughout the 2000s. The expansion of antiretroviral therapy has improved survival in individuals infected with HIV. The current state of treatment for HIV/AIDs is surveyed, and the challenges associated with clinical testing of this disease are presented.

Chapter 21. The Evolving Role of the Pharmacist in Clinical, Academic, and Industry Sectors

G. Patel and S.E. Tedford

The evolution of the pharmacist has transformed the position from a limited product and task-oriented role to expert leads in research development and integrated, essential members of the healthcare team, critical for ensuring best practice outcomes for patients. The dynamic training received by the pharmacist allows for a varying range of applicability in many healthcare environments, whether in hospital, academic, or industry sectors. These evolutionary changes in the practical aspects of the pharmacist's role to a research and/or patient-oriented position have raised the caliber of the profession and continue to expand on the potential impact the pharmacist can have in the healthcare field.

Chapter 22. Intellectual Property in the Drug Discovery Process

M. Rumore and W. Schmidt

Intellectual property (IP) protections have affected drug development and are considered one of the most important assets of any corporate entity or research organization. The four main types of IP rights are patents, trademarks, copyrights, and trade secrets. IP affects everything from the choice of brand name to stock valuation. Patents are a *quid pro quo* "right to exclude" which encourage innovation by providing financial incentives for research and development (R&D) activities.

Chapter 23. Drug Repurposing: Academic Clinician Research Endeavors

K. Heneghan and S.E. Tedford

The process of developing drugs de novo is extensive, laborious, expensive, and time-consuming. Estimates for the transition of a potential therapeutic molecule to an approved drug product average 10–15 years with a success rate of only ~2% for developing a viable product. Associated costs can advance into the billions with no guarantee for approval pending outcomes in the later phases of clinical testing. In an effort to reduce the cost, risk, and time constraint associated with bringing a new drug to market and ultimately integration into beneficial patient care, increased focus has been placed on developing new paths for existing pharmaceutical agents in industry and academic settings alike. Drug repurposing, also referred to as repositioning, is an innovative strategy that capitalizes on prior investments and established research, and shows a more favorable risk–benefit ratio. These drugs may already be approved for other applications or are currently shelved having not succeeded in previous preclinical/clinical testing for other targeted indications. The considerable research these drugs have already undergone may provide important details on the drug's profile including pharmacology, safety, and efficacy information, thereby reducing the cost of necessary research activities. Thus, the recycling of these already established drugs makes repurposing an appealing avenue for drug discovery initiatives. While the end goal is to receive regulatory approval for repurposing efforts on a new indication, this does not necessarily need to be the hallmark for success. A repurposed drug with enough evidence to demonstrate efficacy of an approved drug for a new indication provides medical professionals with enough information to consider off-label use of the drug. Drug repurposing significantly reduces the cost, risk, and time compared to traditional drug development strategies. A case study is presented of the repurposing of loperamide as a topical anesthetic for blood sample needle sticks in neonates, including demonstration of efficacy for this indication and development of a patent.

SECTION IV. REGULATIONS

CHAPTER 24. THE ROLE OF THE REGULATORY AFFAIRS PROFESSIONAL IN GUIDING NEW DRUG RESEARCH, DEVELOPMENT, AND APPROVAL

S.A. Edwards

Before embarking on the description and discussion of the roles of Regulatory Affairs Professionals (RAP) in drug development, this chapter briefly reviews the US Food and Drug Administration (FDA) and other major contributors that often have major roles in influencing the regulatory path and framework of drug development. While the legislative and legal mandates, given to the FDA, require it to ensure that new drugs are safe and effective, it does not have the responsibility to develop new drugs itself. FDA's purpose, regarding new drugs, is to determine if the drug is safe enough to be tested in humans, during the early phases of drug development. During the later stages, FDA decides on whether or not there is substantial evidence from two adequate and well-controlled studies to grant marketing approval, such that the drug can be sold to the public, and what the label should say about directions for use, side effects, warnings, etc.

CHAPTER 25. ORPHAN DRUG DEVELOPMENT AND REGULATIONS

A.M. Lynch

Currently, the number of rare diseases that have no treatment option available is estimated to be between 4,000 and 5,000 worldwide. Establishment of the Orphan Drug Act (ODA) has provided companies with the incentives to develop treatment options. Over 600 orphan drugs and biologics have been approved by the FDA since the ODA was enacted in 1983. In the past 10 years, there has been a steady increase in both US FDA designations and approvals of orphan drugs, and one would expect this trend to continue into 2020 and beyond. Areas of growth resultant from the ODA include cancer and pediatric drugs, the latter of which is likely related to the fact that half of rare diseases afflict children. The volume of orphan drug prescriptions is relatively low at 0.4% in 2017, but orphan drugs are extremely costly (median annual cost $46,800 in 2017; cost of top ten indications at $1,000). However, treatment has resulted in cost savings related to reduced hospital stays and related costs. The US orphan drug application process may be filed at any point in development prior to NDA, and it includes the following: statement of orphan drug designation, information about the rare disease and the drug, clinical superiority justification, justification of designation based on a subset population of those with a common disease, regulatory status and marketing history, and disease prevalence. International regulations for orphan drugs are discussed, including policies in Europe, Japan, and Australia. Challenges for orphan drug development include general knowledge of rare diseases, diagnostic methods for such diseases, and high costs of orphan drugs. As regulations for rare diseases and orphan drugs evolve, the market potential for orphan drugs will be expected to increase worldwide.

CHAPTER 26. DEVELOPMENT OF DRUG PRODUCTS FOR OLDER ADULTS: CHALLENGES, SOLUTIONS, AND REGULATORY CONSIDERATIONS

S.W.J. Lau, C. Sahajwalla, and D.R. Abernethy

According to the United States Census Bureau, persons aged 65 years and above will be the fastest growing segment of the population in the United States for the next four decades due primarily to the migration of the Baby Boom generation into this age group with a steadily increasing life expectancy. Aging affects both pharmacokinetics and pharmacodynamics of the medications that older adults consume. Because older adults have more comorbidities, they need concurrent multiple pharmacotherapies for their multimorbidities, which can lead to polypharmacy and/or potential inappropriate medications. While deprescribing is one emerging approach to manage polypharmacy, a multiprofessional team approach is recommended. Underrepresentation of older adults in clinical trials is common, but many measures can be implemented to boost enrollment of geriatric patients in clinical trials. Currently, there are a multitude of FDA guidance documents for drug development in older adults. The ICH E7 document is another valuable resource. Future directions for drug development in older adults may include sharing of data among academia, industry, regulators, and patients; expanding role of modeling and simulation; similar strategies as in pediatric drug development; and early communications with regulators in the drug development process. Challenges, solutions, and regulatory considerations of drugs for geriatric patients are presented in this chapter.

CHAPTER 27. CLINICAL PHARMACOLOGY AND REGULATORY CONCERNS FOR DEVELOPING DRUG PRODUCTS FOR PEDIATRIC PATIENTS

J.M. Burnham and G.J. Burckart

The understanding of the use of medicines for children over the past 50 years has dramatically increased. However, it has taken many decades to truly understand the science behind pediatric diseases, determine the best drug therapy to treat the disease, plan for effective dosing in children, and design a neonatal or pediatric clinical trial that would appropriately establish safety and efficacy of the drug. The regulatory process for including children in drug development studies has taken a parallel slow and plodding process. Together, the science and the regulatory process have made real progress in the understanding of the dosing, efficacy and safety of drug use in pediatric patients since 1997.

CHAPTER 28. PHARMACY COMPOUNDING REGULATIONS

L. Allen and W.C. Triplett

Pharmaceutics-based pharmaceutical compounding is an integral part of providing pharmaceuticals and is essential to the provision of contemporary health care. Pharmaceutical compounding has an interesting relationship with the development of new drugs and is supported by the pharmaceutical sciences. Compounding can be as simple as the addition of a liquid to a manufactured drug powder, or as complex as the preparation of a multicomponent parenteral nutrition solution or a multicomponent trans-dermal gel. In general, compounding differs from manufacturing in that compounding involves a specific practitioner–patient–pharmacist relationship, the preparation of a relatively small quantity of medication, and different conditions of sale (i.e., specific prescription orders). The pharmacist is responsible for compounding preparations of acceptable strength, quality, and purity with appropriate packaging and labeling in accordance with good pharmacy practices, official standards, and current scientific principles (pharmaceutics). Laws, regulations, and standards affecting pharmaceutical compounding are presented. Adverse compounding events due to lapses in adherence to such standards are highlighted, with a special focus on the New England Compounding Center (NECC) tragedy. Components to the US Pharmacopeia related to compounding are summarized. Finally, the status of 503B outsourcing facilities is detailed. Section 503B of the Federal Food, Drug, and Cosmetic Act defines requirements for outsourcing facilities for compounding.

Section I

Overview

2 Current Opinions on the Trajectory of the Pharmaceutical Development

John C. Somberg
Rush University Medical Center

CONTENTS

2.1 INTRODUCTION

The pharmaceutical industry is undergoing revolutionary change with the advancement of new technologies, genomics, biotechnology, and proteomics, all offering the potential for significant change in therapeutics. The forces of technology are like fast-running rivers, forever changing the landscape. Change is a part of the human condition, inevitable and often disruptive. During the last 100 years, the drug discovery process has undergone considerable change, but all that which has gone before will pale in comparison with the dramatic changes the information age will cause. This will also alter the drug development process.

Even the most powerful and financially stable companies involved in drug discovery and development need to recognize the growing evolution towards change. Advancements in technology gave the International Business Machine (IBM) the opportunity to expand and alter its business from analog systems such as adding machines, to punch cards, and then to complex computer systems. As the computer age evolved, IBM became a leader in the market and laid the foundation for the personal computer. The lead was then lost by IBM when "software" became the platform of the information age, along with the chips that permit the exponential growth in machine computing performance. Another software company, Microsoft, recognized the potential in this opportunity and now is considered more dominant in today's information age than IBM. IBM still remains a leader in technology advances, in new, fundamental patents, and a formidable presence in the information technology (IT) departments of large corporations and in the business computer applications areas. It kept up in technology, but its management failed to perceive the salient change in the information age from large computers to small, and then, the paramount importance of the "software" that controls information processing, analysis, and communication. In fact, history may be repeating itself with the rise of Apple and Google with their markets and stock evaluations far outpacing Microsoft's.

Similar to the dynamic observed in the software industry, drug development is in an analogous situation. We have observed the evolution from the age of botanicals to the age of chemicals and synthetic discovery and, now, to the age of continuing advances in biotechnology and gene manipulation. Each area has much opportunity for continued growth, but the shift in direction is fundamental to scientific development. Taking these major changes into consideration, drug discovery and development will be fundamentally influenced by the information age. The construct of the information age had been coined by Alvin Toffler, and he correctly perceived it to be revolutionary in its effects on society. The agricultural and industrial revolutions brought about fundamental changes in society, as will the information age. In his book entitled *Future Shock*, Toffler describes an era when the pace of change in modern life is so great as to disenfranchise individuals from society.[1] This is a significant problem for society – a problem with profound political dimensions. Failure of our institutions and corporate structures to adjust may result in considerable societal and economic disruption. For these reasons, an understanding of the evolution of drug discovery and development and how this evolution will be affected by the information age is essential for those working in this area. This is especially true with the considerable percentage of the economy being involved in health care and the discovery of new pharmaceutical therapies; a driving force in the health care industry.

2.2 DRUGS FROM PLANTS

The utility of herbals was recognized from the earliest times, dating back at least to the hunter and gatherer societies. While the true intent of these herbal uses cannot be deciphered between uses as foods, items of religious significance, or medication, as civilization progressed, remedies from plants further developed. Advancements in plant knowledge led to an increasing number of new therapies. Earlier folklore relates stories of plant medicinals. The Bible contains passages alluding to medicinal herbs and plants. In fact, all the major religions discuss plant remedies as part of their sacred heritage.

The use of medicinal plants and the work of herbalists laid the foundation in the early discovery of drugs. For example, William Withering, a physician from Birmingham, England, was on his charity rounds in Shropshire and saw that an herbal potion was used to treat a woman with dropsy (congestive heart failure), who then showed improvement. Withering's botanical training in Edinburgh permitted him to identify the probable active ingredient, the leaf of the foxglove plant. After 10 years of clinical evaluation, employing a series of dose-ranging case studies, he categorized the adverse side-effect profile of the digitalis leaf, its potentially life-threatening toxicities, as well as its efficacy. Withering noted the adverse outcomes and carefully chronicled the conditions in which the drug was most useful. Although he thought the agent increased urine volume and, therefore, had diuretic properties, he commented in his thesis that the drug had a "powerful action on the motion of the heart" and, thus, recognized its "cardiotonic" action years before this was scientifically proven. Withering was a masterful botanist who chronicled the plants of Great Britain later in his life. He was an exemplary clinical pharmacologist and demonstrated a mastery in botanical drug discovery and testing given the time period he existed. Of note is that Withering's observations may not be unique. The effect of the foxglove plant on disease was known to be part of European plant folklore. The use of these glycoside-yielding plants and the skin of the toad for medicinal purposes goes back to ancient Egypt and is mentioned in Chinese herbal writings. Confucius talks of glycoside plants for edematous states, and cardiac glycosides are a significant component of Chinese herbal medicine. Although Withering's observations were a defining moment for modern medicine, botanicals of similar action have been used for more than 5,000 years.

Clearly, botanicals have been an important component to therapeutics. Whether we are discussing digitalis, atropine, quinidine, aspirin, or any number of other drugs, plants have contributed much to our initial explorations into therapeutics. Using quinidine in atrial fibrillation or quinine to treat malaria are other examples of the importance botanicals have played in therapeutics. In the 1700s and 1800s, botanicals were the only source of drugs for treatment of disease. The sourcing of anti-infective agents has depended on extracts from molds and fungus for a very long time. More recent therapies are still derived from natural botanical sources with some chemical modifications to improve effectiveness or to decrease toxicity.

We often think of the age of botanicals as one that has gone by. Indeed, it was the first step in the field of drug discovery and development, but one that continues to this day to play a major role. Although reserpine was used for 1,000 years in India and parts of China, it was only in the 1950s that it was purified and used as an effective antihypertensive agent. The recent use of Taxol in oncology is an example of a botanical product developed for clinical use. Because the drug was in very short supply, a synthetic pathway for commercial production was developed, sparing the bark of the yew tree from which Taxol is obtained. The excessive harvesting of the yew tree led to it being endangered. The antirejection drug sirolimus (Rapamune, Wyeth Laboratories, Collegeville, PA, USA) is an even more recent example of a soil sample taken years earlier from an atoll in the Pacific Ocean that has led to a major new therapy. The therapy is used both as an antirejection drug in transplantation surgery as well as a coating for coronary artery stents to prevent restenosis.

There was a trend for large companies to form alliances with botanical gardens and/or countries and to explore remote areas to find new therapies from botanical sources. A company called Shaman Pharmaceuticals (South San Francisco, CA, USA) was founded with a corporate purpose to discover and develop pharmaceuticals from botanical sources based on the folklore of "medicine men." However, these ventures and alliances were not quick to deliver new therapies, and the impetus to continue this aspect of discovery has faded. This is unfortunate because this type of discovery has proven fruitful in the past. The one discouraging factor imperiling botanical sources for drugs is the rapid "drop off" in the planet's biodiversity that has been caused by the relentless expansion of human development with the emphasis on corporate agriculture, the clearing of the Amazon jungle, international travel contaminating the biomass, and the worldwide, rampant growth of "nonnative species."

To successfully deal with the challenge of drug development from botanicals, modern day techniques must be applied, the most important of which may well be those related to handling the vast amount of information that can be collected. Computer-based applications for exploration of the plant world include automating processes for analysis, plant and chemical categorization with innovative storage, organization, and retrieval, which will all be required to make the drug discovery process cost-effective. The systematic computerization of knowledge in ethnobotany and pharmacognosy, with emphasis on the use of plants categorized across primitive societies, will be helpful in sustaining the discovery process. Using sophisticated computer techniques to look for similarities in medicinal plants and their use among primitive peoples to ascertain potentially useful therapeutics can greatly aid the ethnobotanist. Hopefully, these computerized techniques will replace the hundreds, if not thousands, of years that are needed for serendipitous observations such as those made by William Withering 200 years ago, which led to the introduction of the digitalis into clinical medicine.

2.3 SYNTHETIC BIOLOGY

The majority of new compounds that have been discovered over the last 75 years are products that come from synthetic chemistry. In cardiology, the β-blockers and calcium channel blockers have revolutionized cardiovascular therapeutics. β-Agonists in respiratory therapy and H_2 antagonists in gastrointestinal ulcer disease therapy are a few examples of the work of the synthetic chemist that has greatly changed our treatment strategies in the clinic. The proven model of finding a useful transmitter in a physiological system, then finding a receptor to which the transmitter interacts, and then modifying the agonist structure to find a specific antagonist has worked well for the drug discovery process.

Advances in technology will continue to reveal new receptors and new physiological system targets. Therefore, the synthetic chemist will surely continue to make considerable contributions to the field of drug discovery. The process is ongoing. For example, as the precise role of the endothelium is further understood, its impact on pharmaceutical research greatly expands. What was once called endothelial-derived relaxing factor (EDRF) has been characterized as a locally released gas, known as nitric oxide. Studies on endothelium function have found endogenous substances involved in the modulation of vasodilation and vasoconstriction at the most basic level of the vasculature. There are endogenous substances opposing the vasodilating properties of nitric oxide. Endothelin is one of these transmitters, and the development of specific endothelin antagonists was thought to have great potential. Unfortunately, studies thus far with these endothelin antagonists have not shown them to be clinically effective therapies in angina, hypertension, or heart failure. The discovery of a new signaling pathway or a new transmitter may not be enough for a breakthrough therapy. Other physiological systems or pathways may compensate, undoing the therapeutic effects of the antagonists. Still, the process of synthetic discovery of drugs, combined with physiological transmitter research, is one with great potential.

Even here, with well-established approaches, we see the influence of the information age. Software is now available to aid in determining receptor structure, and thus, possible receptor blockers have become essential tools in the discovery process. Computer-assisted drug synthesis also has great potential. The revolution in this aspect of synthetic chemistry is analogous to the revolution that computers caused in the animation industry. Where once dozens of artists were necessary to create thousands of picture "cells" for animation, computers have now replaced them, creating "life-like" animations that were not previously feasible. The same type of revolution is occurring in the chemical drug synthetic industry.

In addition to design, computerization can readily be applied to categorizations of synthetic pathways. I believe the application of computer sciences to chemistry will lead to considerable advances in this field. The application of computers to the steps beyond modeling systems to identify chemical structures and then to develop synthetic approaches will have considerable implications. Synthetic antagonists with optimum potency can be developed from a host of chemical possibilities. The development of these agents with superior receptor selectivity and potency will increase the yield of these approaches. The industry should target both improved therapies as well as new and novel therapies. A more potent, less-toxic agent can be as useful and financially rewarding in the right indication, potentially providing clinical benefits to large populations in need of novel therapy.

However, the information age applications to synthetic modeling will be inherently limited unless we can improve our screening techniques. For many years, I have given considerable thought to the link between drug synthesis, discovery, and development. Almost 30 years ago, I had the good fortune to visit Janssen Pharmaceutica (Beerse, Belgium) and discuss the drug discovery process with the late Paul Janssen, a genius in the field of pharmaceutical chemistry and drug discovery. I was most impressed with his knowledge of pharmaceutical chemistry, his diverse therapeutic interests, and his unparalleled success in the discovery of novel therapies. Jansen was a chemist looking for novel compounds that could then be assessed to find biological activity. A new, promising compound would be processed through hundreds of models, looking for possible pharmacologic activity. The question arose about the ability to screen for biological activity. This is a critical linkage point in the discovery and development process, one to which the great potential of the information age can be effectively applied to good purpose.

Jansen's approach was most impressive and fascinating, but to this day that visit caused some abiding concerns. I have been interested in the field of antiarrhythmic drug development and have participated in all stages of discovery and development in this area, from chemical synthesis to clinical electrophysiology, to the acute and chronic treatment of patients. I was particularly interested in

Jansen's approach as it applied to antiarrhythmic pharmacology. I had been working on lorcainide, a drug Jansen developed at Bersa complex in Belgium and wondered how this compound was discovered and how it compared to other agents screened. Jansen employed a costly and time-consuming dog model of premature ventricular contraction (PVC) suppression post coronary artery ligation. Lorcainide, a Vaughan Williams Ic antiarrhythmic agent, was a sodium channel blocker. With this profile, lorcainide was predicted to be effective in the PVC-suppression model, and indeed, it was effective. But PVC suppression and the Ic agents have not shown prolongation of life in post myocardial infarction (MI) in clinical studies.[2] The type III agents appear to be the most effective clinically, although not shown in some studies to prolong life in randomized controlled trials (including the European Myocardial Infarction Amiodarone Trial and the Canadian Amiodarone Myocardial Infarction Trial).[3,4] However, a meta-analysis of drugs of the type III variety, specifically amiodarone, has suggested it to be far more effective with much less proarrhythmia than the sodium channel blockers and, in fact, for amiodarone to show a reduction in sudden death.[5,6] This leads one to consider what the effect would be of the clinically valuable agent amiodarone in the screening model that Jansen was using to pick out his antiarrhythmic drugs for testing and then into clinical development. In fact, the records at Bersa were so accurate that the scientists in the cardiology department could look up the results in a few minutes and describe the actions of other known antiarrhythmic agents in the dog model. The answer they reported was that amiodarone was much less effective and, in fact, hardly effective at all in the model in which lorcainide was extremely effective. It is no wonder that the pharmaceutical industry in the 1980s found a host of Ic agents (flecainide, encainide, lorcainide, propafenone, indecainide, and ethmozine), because that is what their assays were best at picking up as an active, effective agent.

The model, therefore, is critically important in the drug discovery process, and it will often determine development. We could synthesize thousands of compounds and then select a few for development that may not be optimum for clinical therapy. These agents, however, would fit the characteristics being sought by the model employed in the screening process. This is a major caveat that has not been given enough consideration. We can only think of the possibility that there may be hundreds, if not thousands, of compounds buried in "analytical hoppers" such as Jansen's Bersa research establishment that have the potential to be extremely useful but were discarded because they were not identified as biologically active in a flawed screening model.

In addition to the models used in drug screening, there is a fundamental difference in discovery between mass screening and receptor-targeted research. The latter has proven more successful in the past decade, but some major advances have come out of pure chemistry and follow-up screening to determine biological activity. Can the revolution of the information age and computer sciences be applied to synthesis and screening? These questions will challenge us in the coming years. I believe a revolution will occur in this area. Synthesis on a grand scale will be tied to automated, focused biological-activity screening that will permit the evaluation of tens of thousands of molecules on a daily basis. Clearly, how we screen will determine the validity of this approach. Additionally, our ability to combine screening with gene expression targeting will further increase the potential for success.

Although we are in transition from the age of synthetic chemistry to the age of biotechnology and gene manipulation, synthetic discovery will still play a major role in advancing the therapeutic armamentarium.

2.4 BIOTECHNOLOGY, GENOMICS, AND PROTEOMICS

As a society, we are currently in the age of advanced exploration into therapeutics from biotechnology and gene manipulation. As technology advances in these areas, large companies are taking note and are busy positioning themselves for acquisition of the biotechnology companies, which are often small, start-up enterprises. The pharmaceutical companies are undertaking these acquisitions so that they will be prepared to benefit from the anticipated outpouring of therapies from biotechnology and gene manipulation. Biotechnology has not advanced as rapidly as predicted, but a considerable number of therapies have been introduced, especially in the field of cancer. The science has made tremendous strides, but a number of factors have limited the advances and commercialization.

The scaleup and commercialization of biotechnology processes are limited by technologic difficulties and the expense that technology imposes. The first generation of compounds were effective, as demonstrated by growth hormone and recombinant tissue plasminogen activator (rTPA). However, there have been major failures, such as the antibodies to counteract the effects of septic shock. Combinatorial robotic chemistry permits the creation of many compounds that are not effective in a patient. This dichotomy stems from our imperfect knowledge of the pathophysiology of disease states, such as is the case with Gram-negative sepsis and ensuing shock. We may target a signaling pathway that is redundant in man or one that may be "bypassed" if blocked, thus failing to affect the disease process.

Genentech (South San Francisco, CA, USA) and Amgen (Thousand Oaks, CA, USA) have been successful in bringing drugs to the marketplace, but even the most successful of companies have struggled to continue to innovate, undertaking the research and development for the next generation of products. The hundreds, if not thousands, of smaller companies may not fare as well, and it is safe to predict that only a small fraction will indeed find a successful product. Besides the discrepancy between the ability to make a compound and its clinical efficacy lie the problems of corporate capitalization and effective drug development. The mergers of biotech concerns and the established pharmaceutical industry improved capitalization and will bring with it more expertise in drug development and the regulatory approval process. But there are other impediments to success. Many of the products of biotechnology synthesis are proteins that are not orally active. A major area of research is going to be to convert the protein-based therapies requiring injection into ones with alternative routes of administration. Innovative drug delivery systems to overcome the problems of lack of oral bioavailability will be important. Carrier molecules, topical transport enhancers, nasal absorption methodologies, and needleless syringes are but a few of the possible solutions to the drug delivery problems that considerably hamper the biotechnology field. Another approach has been the development of chemical molecules that have similar key structural elements that may permit the chemical compound to act like the protein molecules. We may find ourselves using the tools of biotechnology to enhance the drug discovery process through chemical synthesis and the treatment of disease with small molecules.

Despite these caveats, the biotechnology field will greatly increase the possible therapies available for development and, in fact, promote development in many clinical areas that have lacked effective therapies. The initial cost and the pressures for successful development are so great that the critical elements of the development process will need to be more effectively used if we are not to repeat the mistakes of the past. For example, demonstrating the blood clot lysing capabilities of rTPA and the reversal of an acute MI *in evolution* were not enough for commercial viability of the product. Genentech persisted and initiated a comparative study of rTPA with streptokinase, demonstrating superiority of the rTPA product. The superiority of the rTPA combined with an aggressive marketing strategy permitted Genentech to dominate the thrombolytic market for many years. In the development process of biotechnology products, their value and place in the therapeutic armamentarium may be as important as the demonstration of efficacy in a pivotal trial. Discussion of the pharmacoeconomic impact of new therapies must take into consideration the great expense of a biotechnology-derived drugs, the benefits of the drug, its place in therapy, and especially its cost-benefit ratio as critical factors in the product's success. An example is the PCSk9 drugs to lower cholesterol. They are very effective, lowering cholesterol to a far greater extent than the statin agents, but their cost and a lack of a clear improved survival benefit with those drugs have made their commercial introduction very difficult and a financial disappointment.

Recently, the area of oncology has been most promising for success in translating biotechnology into successful products. A number of specific products targeted to cancer have proven successful in clinical trials. Specifically, using antibodies to target specific receptors in order to turn off cell growth appear to be a fruitful approach. The potential for proteins to affect genes and modify disease is tremendous, and we are just touching the surface of the myriad of potential opportunities this field has to offer. Another very interesting field is that of angiogenesis, which involves inhibiting angiogenesis factor and, thus, tumor growth. The alternative has potential as well, increasing angiogenesis and, thus, decreasing organ ischemia, such as that seen with coronary artery disease. Modifying cell growth through inhibition of angiogenesis has proven to be highly successful with some cancers. Maintaining a balance between promoting angiogenesis, decreasing ischemia, while avoiding the potential of tumor growth has proven a challenge.

2.5   GENE THERAPY

Genetic engineering is most often associated with agricultural products such as the foods that we eat; however, genetic engineering also plays a critical role in the development of novel therapeutics. Gene modification, substitution, and inhibition are a promising array of new strategies for the effective treatment of disease. That a single gene may be responsible for a metabolic disease, like gout or homocystinuria, seems reasonable. That a single gene mutation could cause a metabolic condition like Ehlers–Danlos syndrome also seems reasonable. However, the notion that breast cancer, lupus erythematosus, or coronary artery disease might be caused by a single, abnormal gene is less likely to hold true. Yet, a mounting body of evidence supports the critical role of genetics in human disease. The developments in gene therapy are exciting and may represent a new age of effective therapies for some of the most difficult conditions to treat in humans. The identification of the gene itself, although an important first step, is only the initial step in a long process for the effective treatment and possible cure of diseases in man. The recent introduction of the CRISPR/Cas9 (clustered regulatory interspaced short palindromic repeats) techniques for gene editing offers considerable potential.[7] The technique can be precisely targeted and is relatively simple and cost-effective, which greatly increases the potential for gene editing as a useful therapy. The techniques for gene modification are in an early stage but will have a major impact on disease therapy in the future.

An area of cardiology in which gene therapy would be promising is restenosis following acute angioplasty. Angioplasty entails placing a catheter in the coronary vessel, inflating a balloon at the tip of the catheter and pushing aside the atherosclerotic lesion. This is a successful interventional technique; however, a major problem limiting the success of angioplasty has been restenosis. Restenosis is the narrowing of an artery or valve following corrective surgery. At the time of the initial angioplasty, there are stimuli that initiate cell proliferation of the media, which leads to restenosis. The medial cells that proliferate are very homogeneous, and this process seems to occur quite rapidly in about 20%–60% of individuals having a single-vessel angioplasty. Currently, a therapy utilizing a stent and a drug coating to prevent medial hyperplasia is employed. The possibility of avoiding a stent is still a viable area for therapeutic development. But even this simple model for gene therapy has proven a difficult target. Antisense therapy was once seen as a potential therapy, but the need for a viral vector to insert the material in the medial cells to turn off protein synthesis is limited by concern about the use of viral vectors and the toxic side effect the viral vector may cause. Major questions arise. Can the virus replicate? Will the gene be correctly inserted, or will additional genetic material of the virus be inserted? The safety aspects are formidable and have markedly slowed the development process. Adverse hepatic effects of the viral vector have been seen, slowing the process of this type of research. After several years, using viral vector in clinical research has just restarted after the deaths of some study subjects. As experience increases in conducting clinical trials, the overall time for developing gene manipulation strategies will decrease.

Besides gene editing, another approach can be modifying gene activity with micro RNAs. This approach appears primarily in early studies and can either block the synthesis of protein by blocking mRNA or using the mRNA to increase protein synthesis at a specific site by "turning on a specific gene."

One area of difficulty surrounding this advancement in biotechnology is the ability to patent a gene. Supporters of gene patenting believe that the proprietary nature of a patent will spur commercial development. However, a gene is not a drug but part of the disease process. We do not patent diseases. One therapy directed at a given gene may be different from another. An antibody could block gene expression, a chemical could block a protein from being made by a gene, or a gene's action could be promoted instead of inhibited. Genes could be genetically modified using the CRISPR/Cas9 technique. But if the gene is patented, then many facets of research can be blocked by the patent owners, blocked not just by the cost of working with the gene from royalty payments, but deliberately, by the gene's owner, who wants to ensure that their therapy directed at the gene is the only one available. Patenting genes will, I believe, be a major inhibitor of research in this area and a grave error on the part of the US Patent and Trademark Office and the courts. So far, the US Supreme Court has blocked the concept of patenting genes. Let's hope that this approach continues.

2.6 THE DEVELOPERS

Along with the evolution of the discovery process and the tremendous influence the revolution in information handling will have on drug development, changes in the participants in the development process will also have a considerable effect. There is an evolution occurring in those involved to the development process. Observing the trends makes me think of theories about the origins of the universe with oscillations in mass accumulating, exploding, and re-accumulating, forming large aggregates and small break-off components. The drug discovery process started, perhaps, with the entrepreneurs who led the field successfully and developed the large corporate giants of today. Merck & Company (Whitehouse Station, NJ, USA), Hoffmann-La Roche (Basel, Switzerland), and Pfizer (New York, NY, USA) are examples of one-man entrepreneur-driven companies expanding into major, international concerns. In fact, the major companies dominate the pharmaceutical industry to an unparalleled extent.

The largest ten companies represent more than 90% of pharmaceutical sales. During the last 20 years, mergers have continued, and the last few years have seen even further consolidation of the pharmaceutical industry. It is said that for a company to survive, it must be able to compete with the major players in the pharmaceutical field. This is not just because of funding requirements for drug development programs but because of the development impediments established by these very large competitors. Impediments can be something simple, like the number of patients exposed to a new entity, or more complex, such as a survival study or the use of experimental ancillary technologies that are prohibitively expensive and would not be automatically required for the development of a new therapy. These impediments can create an impression, both to the FDA and within other companies, that they are requirements for developing a second or third agent in the field, making such development much more difficult, time consuming, and expensive. The time factor is especially important because the longer it takes to develop a compound, the more dominance in the market the first drug has gained. Time is equivalent to money, and the loss of a product's position in the approval process can all but destroy the market potential for an agent.

Another result of industry consolidation is the tremendous pressure to develop "blockbuster" products. To sustain the pharmaceutical behemoths, one billion dollar compounds are mandatory. Compounds that only gross 50–100 million are no longer of interest to the large companies, and the "niche" diseases they treat are no longer attractive targets for therapy. A company that loses one or two compounds in its pipeline can rapidly fall from favor and become a possible takeover target. However, the focus on "blockbuster" products has created an opportunity for small companies to develop drugs in the vacuum created by large multinational companies, an opportunity not just to develop "niche" compounds but drugs that can approach "blockbuster" revenues. "Big science" is not the substitute for new ideas, new approaches, and "thinking out of the box," which often is the requisite for new discoveries and blockbuster drugs. Often it is small- to medium-sized companies that can take on the risks of a novel product resulting in breakthrough therapy.

Pharmaceutical pricing is a very controversial area with companies selecting astronomical prices for their products. Some companies purchase old products and markedly increase their prices, turning an old money-losing drug into a profit leader. Instead of recouping corporate investment over a number of years, companies expect to maximize their returns. This practice of "over pricing" can lead to price controls, or the lowering of development requirements, permitting generic competition to force price reductions. Either approach to the cost of drugs may be harmful to the drug development process and ultimately the pharmaceutical consumers.

2.7 SMALL DEVELOPMENT COMPANIES

Considering these obstacles to development and the considerable regulatory maze, the trend to combine with bigger and bigger companies is not surprising. What is indeed surprising is the simultaneous opposing trend of the development of the very small niche start-up companies proliferating along with the ever-increasing size of the major players in pharmaceutical industry. In fact, it is not just how small these companies are that is important; it is that they only encompass an aspect of the drug discovery and development process. Some companies focus on discovery; others specialize in clinical development, and others in sales and marketing.

Some companies plan to license their product to a larger firm for marketing once the new drug application (NDA) is granted. Then, there are other companies that neither discover nor develop drugs but rather market developed pharmaceuticals. One may argue that the niche enterprises are doomed to failure. But a number of factors combine to make this approach viable. The total overhead of the very large companies limits development to compounds expected to have sales of less than $50–$100 million per year. At times, a drug will have a smaller market, and a large company will develop the product for public relations, because of interest in the field, to bolster sales of existing products in their product line or out of sheer miscalculation of market potential. However, the small companies look at a potential market of $5–$50 million a year as a bonanza. Their costs are far lower, permitting the recoup of adequate profit margins after developing, marketing, and discovery costs are accounted for, and expenses for sales are covered. Small companies cannot carry out clinical trials at the same level expected of large companies such as a Pfizer or a Merck. Thus, the studies are fewer, smaller, and aimed at proving the efficacy and safety as directly as possible. However, the niche company can play a significant role in bringing forth new therapies. They service areas that are not considered appropriate in terms of market size by the larger companies. They represent the dynamic growth of academic entrepreneurs who look to the commercialization of their ideas, especially in biotechnology and gene manipulation fields. Small companies can also develop product in the anti-microbial field, or for rare pediatric disease, obtaining a voucher from FDA that can be bought by a larger company to accelerate drug development (6 months review vouchers). These vouchers sell for a hundred million dollars or more.

2.8 ACADEMICIANS AND ENTREPRENEURS

Many companies are investor-driven and, thus, have an intense dedication to development and success. But can a company with one or two products compete? It appears it can. However, often the process requires merger or acquisition by larger firms to succeed. Some companies start out as "spin offs" of academic research, others license a product being developed by "big pharma," and others start as marketing or generic firms that grow from their success.

King Pharmaceuticals (Bristol, TN, USA) started small, made some brilliant acquisitions, and through mergers, moved into research and development and then being managed with larger concerns. Biovail (Mississauga, Canada) is an example of a generic company developing its own products and becoming a small, viable concern and then running into stock issues before being bought out by Valeant Pharmaceuticals. In the biotechnology field, many companies are very small. The biotechnology industry seems most appropriate for the small-company start-up approach.

Whether companies besides Genentech and Amgen can climb into the "big" pharma group remains to be seen. Centocor (Malvern, PA, USA), Genzyme, and US Bioscience (Philadelphia, PA, USA) have all made attempts to move into the second tier of "pharma," some more successful than others; it is difficult to develop a viable product and then to sustain research and development and continue to grow. But the trend is clearly established.

Academicians with a novel idea no longer publish their results and go on to the next project. Rather, patents are obtained, and a company may be started. I marvel at the recent reports of a new technique in cardiothoracic surgery being performed using the laparoscopic approach. Instead of reading about the advances in a medical/surgical journal, the information is presented in a business forum, in this case, a recent issue of the *Wall Street Journal*, focusing on the possible initial public offering (IPO) that will be forthcoming for a company making the instruments essential for the procedure.[8]

Entrepreneurs and academicians are forming alliances that may speed a procedure or chemical entity into a viable product for development. Although the free exchange of ideas may be limited, and scientific discourse suffers a bit, the possibility of widespread clinical use facilitated by commercial development is enhanced. The pros and cons of this approach are not for us to debate but rather to accept as a trend that is ongoing and growing. I do think that the fast-moving "nimbleness" of these small, dynamic companies, coupled with their lower overhead costs, offers considerable benefit to pharmaceutical discovery and development. Drugs are being developed that the larger concerns would not have considered. The advancement in niche areas, such as orphan drugs, is for the most part being pursued by smaller companies. I believe this is a healthy trend and one that will provide opportunities for other small-market targeted therapies to be pursued and commercialized.

Combined with the trend of small niche companies in drug development has come the parallel corporate trend of downsizing and the hesitancy to expand divisions to take on short-term projects. A great number of large companies are outsourcing for critical aspects of drug discovery and development. Compounds can be manufactured under contract. Consultants can put together manufacturing specifications, and preclinical testing and stability work can be done under contract. Clinical studies are performed by clinical research organizations (CROs) with the data handled by contract statistical analysts. A consulting team can put together an NDA under the supervision of a small, core group at corporate headquarters. This can be done for the small company or the very largest of the pharmaceutical giants. Parts of a project can be subcontracted. Indeed, it is not uncommon for intermediate-to-small projects at the largest companies to be entirely subcontracted. For these reasons, the CROs and other contract service organizations (CSOs) have become successful. A bonanza of new business has created exponential growth for these companies. The companies are competitive, and the work is relentless, but results are what make the industry thrive, and drug development has, in some instances, sped up using this "piecework" approach. Another model I have recently come across is a company taking on development for a larger company, financing the development, and then either returning the product to the investor company for an agreed-upon cost, absorbing the loss if the project fails, or out licensing to another company, pocketing the profit. Clearly, the "virtual" pharmaceutical company has come into existence.

There are dangers with this fragmented approach. Outside companies can be less dependable, projects can fall apart when the capitalization of the company is inadequate, and the company "goes under." Less than favorable schedules can sometimes develop because the project is not necessarily the highest priority of the contracting company. The fragmented approach can create situations in which the contracting company may be less alert to important clinical findings that should alter the development team or to serious toxicities that need to be taken into account.

If studies are performed outside the United States, as they often are, the quality of the data may suffer and the important clinical observations made by US investigators may be unavailable or missed. Important information about the drug may not be passed along to the developing company, and this can seriously impede the development process. In addition, corporate rapport with the site investigators may be lost, and the important "seeding" of the market with experienced investigators for a product may not occur when contract organizations are involved and non–US study sites are employed. However, there may be significant cost savings and increased patient accession with the CRO and foreign study approach that may make using them advisable. An example of outside the United States (OUS) studies that did not work out is the evaluation of a drug for heart failure with preserved ejection fraction where the work of the US investigators showed success, but the overall findings, which included Eastern European and Russian sites, found no effect of the drug. An overall negative result was disastrous for the field and a major loss for a new therapeutic option.

Clearly, a balanced program that gives careful consideration to the limitations of using CROs to run the studies, provides the statistical analyses necessary, monitors the study sites, and tracks accession of subjects in the study can be most helpful. The utilization of study sites outside the United States needs to be evaluated against the more traditional approaches to drug development, the latter of which cost more but are perhaps more reliable.

2.9 GOVERNMENT, REGULATION, AND DRUG DEVELOPMENT

The influences of the federal government are pervasive in our society, from our tax structure to the actions of regulatory agencies. All aspects of industry, and especially the pharmaceutical industry, are greatly influenced by government. In the 1990s, some manipulative politicians targeted the pharmaceutical industry in their rhetoric to pander to voters. This populist approach has continued, especially from leading Democrats (but from Republicans, the current administration, as well) – all attracted to this approach to garner votes. But, for the most part, there is a balanced tension between the Democrats, representing more government, and the Republicans, representing less government and increased deregulation. This is, of course, a simplification, but one with historic justification. In the last few years, we saw a trend towards regulation with the federal government as the provider of solutions to most economic problems. This trend has changed drastically with deregulation in many areas and a de-emphasis on government as a solution for all societal problems. How the pendulum swings is anyone's guess. The left of the Democratic Party is moving ever closer to socialism, especially that of a single payer health care system with price controls on pharmaceuticals.

Even with the current movement away from government and regulation, the effect of overregulation of the pharmaceutical industry remains substantial. The loosening of Occupational Safety and Health Administration (OSHA) regulations and environmental impact statements are helpful to pharma. The election of President Trump and the nomination and confirmation of Dr. Gottlieb as FDA Commissioner may bring a new era of reduced regulation and facilitation of drug development by FDA. However, the push to reduce prices of drugs through government action must temper this optimism.

Over the past few years, the FDA and especially Center for Drug Evaluation and Research (CDER) at FDA has striven to be an active facilitator of research. The Orphan Drug Program started in the 1980s has been one of the first proactive incentives to facilitate pharmaceutical development. The recent Pediatric and Infectious Disease Voucher Program is a "follow on" program, to incentivize development of drugs for pediatric patients and in the infectious disease area, areas seeing little development for years. Recently, the FDA has identified drugs off patent protection that could be the target of generic drug development to facilitate market competition and possible more competitive pricing. We will see more and more of these efforts under the Gottlieb/Trump era. However, the FDA still imposes considerable obstacles. For small companies, the eCTD electronic filing and the Electronic Gateway for submissions has created obstacles and significant costs for small companies to communicate with FDA. While there is an effort to streamline generic drug development, the FDA's generic division can take years to approve a relatively straightforward application. Drug shortages are frequently created by FDA citations of manufacturing when the concerns are theoretical, but the FDA interventions cause real disruptions of drug supplies and patient care. Shortages in bicarbonate, IV epinephrine, atropine, and sublingual nitroglycerine are a few of the many shortages that can be cited.

However, "Big Pharma" often supports FDA vigorously. Big "pharma" operates successfully within its framework, and in a way, the FDA has become part of the process of limiting competition and diminishing the effectiveness of smaller companies unable to compete against the more-formidable pharmaceutical giants in navigating the FDA requirements. FDA also serves the industry well, particularly in the scientific arena, ensuring efficacy and safety and instilling high, public confidence in pharmaceutical products. In addition, FDA stands at the "gate," the guardian of "big pharma's" market – both have a stake in preventing a more relaxed, expedited drug and device review and approval process.

The cost of pharmaceutical research is staggering and the success rate is poor, which makes pharmaceutical research a very risky business. Today the US "free market" is what drives the engine of drug discovery and development worldwide. US, European, and Japanese companies all look to "make it" in the United States because of the price controls in Europe and Japan. European and Japanese systems are going to have to pay more for pharmaceuticals if the United States is going to pay less for new drugs. In a way, this is a trade issue. Currently, US consumers are subsidizing drug development. With the new, massive intrusion of the federal government into the pharmaceutical purchasing arena, there will be tremendous pressure to reduce the cost of drugs in the United States. This is a laudable goal as long as research continues to be funded through an incentive-based system. The US government is going to face a major issue with Europe and Japan; we either find ways to balance the incentive to industry with OUS sources contributing or lose the worldwide economic engine supporting drug discovery and development that the United States serves as its primary sustaining force.

2.10   GOVERNMENT RESEARCH

The importance of defense and space-related technology on drug development has been minimal. The National Institutes of Health (NIH), the National Science Foundation (NSF), and the National Cancer Institute (NCI) make a very substantial contribution to basic research and its translation into drug discovery and development. Private funding still provides the majority of research funds for pharmaceutical discovery. It is estimated that over a third of funding is indeed federal at the early stages. Although the NSF, NIH, and NCI granting mechanisms are imperfect, they are far better at supporting the advancement of knowledge than the military or NASA. Whether the government can improve its effectiveness is unpredictable. However, the trend towards very big scientific projects has slowed, and a more reasonable, decentralized approach is taking shape under the current Congressional leadership and the administration. This is encouraging because by supporting new programs, small programs, and diverse projects, we are more likely to see important advances, as opposed to the results seen when the established, industrialized, scientific complex, and its bureaucracy are the only recipients of support. Programs that are especially helpful to the device and drug discovery and development are the small business administration (SBIR – Small Business Innovation Research and STTR – Small Business Technology Transfer) programs. Their funding should be increased, especially to support development projects in niche areas.

Governmental support for pharmaceutical-related research, clinical pharmacology research, or research related to drug development is minimal. This is unfortunate because tremendous public health benefit could be obtained. This is not to suggest that government should compete with industry, but in areas that industry is ignoring or in fundamental research that leads to discovery and development or that is seminal to drug discovery and development, government could and should play an important role. However, a major component of the nation's public health remains solely funded by for-profit pharmaceutical enterprises. The federal government's genome project is an example of great promise for the biotechnology and gene therapy sector. That information is fundamental and will form the information base of many discoveries in the future. However, the program's successful outcome was driven by the commercial sector competition. We need to see more public–private joint efforts in both discovery of new agents and their translation into clinical therapeutics. The recent NIH support for translational research is to be commended. Studies supporting comparative effectiveness research are invaluable and greatly help in the selection of optimum therapies. However, the push for community-based studies, with data mining over controlled clinical trials, is ill-advised. Big data has great potential, but it cannot replace the information obtained from controlled, randomized clinical trials.

2.11 THE FDA REGULATORS

The drug development process is highly regulated by the FDA. From initial clinical testing in phase 1 to later phase 2 and 3 clinical trials, the FDA has considerable influence. Unlike European agencies, the scientific personnel at FDA are accessible from before work starts on investigational new drug (IND) applications, phase 2 trials, or NDA meetings. The FDA can provide meaningful guidance to a drug development program. The FDA is the judge of the data presented and, for the most part, the creator of the regulations governing approval. FDA's assistance comes from experience in the drug development process. The medical division chiefs and other senior individuals at FDA see a tremendous number of clinical trials, have encountered a variety of clinical development problems, and can without disclosing confidential information, provide considerable assistance to those involved in drug development. Whereas an individual within a company may be involved in four or five compounds during a career, the FDA senior people may see that many compounds in a week and see issues from many different perspectives. Clearly, the FDA is the nexus of pharmaceutical development information and training that, unfortunately, has not been used as effectively as it should be.

Those involved in drug development must work in concert with FDA. The FDA and industry, working together on a product, will often bring about a development program that is more effective and more efficient in time and resources. However, too many companies take FDA's advice as dictum. There are regulatory advisors, what could be termed "the shadow FDA," who tell industry what FDA requires and wants, and whose recommendations are all too often distortions and impediments to effective drug development. FDA should be relied on as an important resource. Those who are pivotally involved in drug development in industry should communicate directly with FDA and not use intermediaries. Regulatory advisors, consultants, past regulators, facilitators, and legal advisors all have their place, but they should not be interposed between those at the companies who are running a development program and those at FDA overseeing the clinical trials. No advice should be binding, everything needs to be discussed, and reasonable approaches need to be taken.

The individuals at FDA are not omniscient. A development plan may not work and may need substantial modifications. Failure to realize this and blindly going forth after an FDA meeting can lead to failure. Coming back to FDA and saying, "But this is what we were instructed to do" is foolish and, in a sense, undercuts the free and open exchange of ideas between the regulators and the developers. Advice is what is given, and later reproach because of changing circumstances, developments in the field, or just the lack of efficacy of a compound is counterproductive. In fact, it may deter the critical assistance from FDA that can be so very helpful to a drug development team.

These impressions, of course, need to be modified in the context of the divisions and the individuals involved. There are differences among and between divisions and individuals at FDA, and this needs to be factored into the equation. But, clearly, the companies most successful in development have created a working relationship with the FDA and have made use of the extensive scientific experience individuals at FDA have with the drug development process. Having been the organizer of a course on Cardiovascular Drug Development, Protocol Design, and Methodology for 15 years, I can attest to the unselfish assistance of many senior individuals at FDA. Their knowledge of the drug development field, their interest in successful drug development, and their desire to find scientific truth are clear. The course involved many leading academics and industry physicians who had considerable knowledge in drug development, and each year the symposium demonstrated that senior FDA participants consistently offered a broad and indispensable knowledge of the field of drug development.

FDA can facilitate drug development further than it does currently. There are times that the delays are needless, that the debate is unhelpful, but the era of the "drug lag" behind Europe that so severely crippled therapeutics in the 1950s, 1960s, and 1970s no longer exists. Excessive drug regulation should not be the goal. Rather, more expeditious, less-costly drug development should be the goal of the FDA review. Separately chronicling each data point and a meticulous review for efficacy and toxicity by a junior officer is an immense undertaking and one with little need in our information age. Having the primary reviewer recreate the NDA piece by piece and then producing his or her own summary is a laborious process that takes months. The NDA is put together by many individuals, highly trained in the pharmaceutical industry, and having one or two people go through this on a line-by-line basis, checking every data point, reanalyzing the presentations, is going to be a most arduous and time-consuming process that may not be necessary.

Quality assurance techniques are in place to the accuracy and integrity of an NDA database. The FDA could make use of these techniques to make analyses of the NDA as expeditious as possible. To keep up with the information age, the FDA will be one of the links in the drug development process that is most stressed by forthcoming change. User fees and more FDA revenues are not the answer. Placing the cost of submission beyond the capacity of small start-up companies is ill-advised. Using "user fees" for more and more reviewers, thus expanding the laborious approach of data review is a mistake. Many of these programs are not objected to by the giants of the pharmaceutical industry and, in fact, are encouraged, in order to the place impediments on the more formative, small companies trying to compete. This is another of the anticompetitive practices in which the FDA has become an unwitting ally.

The statement that the FDA needs to "take its time," to "plow through each data point," to protect the public is one that is heard often. By never approving a drug, FDA would be the most protective, because no adversity would ensue from approved drugs. However, the adverse effects that would result from having no therapeutic advances would be intolerable. A "common ground" – between the regulators charged to protect the public and the public's need for new, effective therapies – must to be struck. The use of the information revolution to facilitate drug development is needed. We are at the beginning of this exciting period, and the government will evolve more slowly, perhaps, than other centers in the development process, but it will evolve.

A number of approaches are possible. The use of quality assurance techniques for data verification on a random basis certainly needs to be validated and then applied. Perhaps data analysis performed by certified groups that are paid for by the company but, at the same time, are licensed by the FDA, would facilitate the review process and not overburden FDA. Focusing on critical, pivotal studies and ascertaining their data's veracity need to be the priority. With the acceptance of efficacy, rapid computerized analysis of the data for the product's toxicity and comparison of the results to those obtained with other agents could permit an estimation of the agent's potential benefits and toxicity. This could facilitate early presentation of the NDA material to an advisory committee that would be able to understand the drugs' place in the therapeutic armamentarium and decide whether a more prolonged and thorough evaluation was needed or an early release could be considered. Of course, an early release might be combined with a more prolonged preliminary period, where information is collected on adverse experiences and efficacy, and this information could then be used for continued drug evaluation.

The process of approving a drug, getting little additional information after the initial approval and allowing the drug to remain on the market forever is as wrong as a very slow and time-consuming initial regulatory review process. Because it is so difficult to get a drug off the market and we have so little post-marketing information, these problems reinforce the regulators' need to make the initial approval very stringent. It would be far better to look to a system with an initial provisional approval. A detailed program of post-marketing surveillance that encourages participation by practitioners would then be requisite. We must understand that drugs can be marketed, and then knowledge and information may develop that changes our initial impression. A drug could be severely limited in its labeling, have warnings issued to physicians who will be prescribing the drug, or a drug could be taken off the market if our knowledge about the product changes. A drug withdrawal should not be looked at as a criticism of the FDA, or the developing company, but as a realization that our knowledge continues to grow. When a drug is first approved, perhaps 1,000–3,000 patients are exposed. It could take thousands more for much longer period of time to see an unexpected adverse drug effect. A greater exposure is needed to increase the probability of seeing an unexpected toxicity.

Unfortunately, transcripts exist of Congressional committees which severely criticize regulators when adversity is later discovered from an agent that was approved on a small exposure but with a meritorious application. If we can't get the "ground rules" straight and be able to understand that our knowledge base expands constantly with greater clinical exposure, we will never move beyond the counterproductive "got ya" culture. With this understanding of the need for prolonged exposure and increased number of patients exposed to evaluate pharmaceuticals, we can accept the early approval and later withdrawal of a drug, without faulting our regulatory colleagues. This approach is necessary if we are to fundamentally change the drug development process for the better. This is a difficult change, because so much of the process is developed by lawyers who view drug development with an eye to litigation, looking for guilt or innocence, rather than a pursuit of science, scientific knowledge that is continuously developing as more information is collected. There is no guilt or innocence, no liability to determine. We need new treatments, and we need to update our use and cautions associated with new drugs. The use of electronic medical records for drug surveillance post-approval will prove a very useful approach, helping with the exposure number and the duration of exposure.

2.12 THE ACADEMIC AND CLINICAL INVESTIGATOR

If one goes back a couple of decades reviewing the volumes of a medical journal, one will pass through a decade or so of clinical trials from large, adequately powered studies, to small trials and then retrospective clinical data reviews often with historic controls. Before the 1970s, one would see articles dealing with therapeutic approaches relating a clinician's experience with a therapy and recalling the results of prior therapeutic approaches. The clinical trial is a recent introduction and those involved have changed over the years. From the clinicians often in a private office, to the academic investigator at a university hospital involved in clinical trials, the investments have dramatically changed. The clinical trial investigator needs training and knowledge in multiple fields: medicine, statistics, trial design, ethics, epidemiology, and, more recently, bioinformatics (meta-analysis techniques and large data set utilization and analysis). The clinical researcher does not always reside in academia but can work in a CRO, the pharmaceutical industry, or FDA. In fact, some individuals migrate from one entity to another over time. The demands of clinical practice and the skills needed to be a clinical investigator are evolving. Often clinicians are deferring clinical research to physicians spending more time devoted to clinical trials. The age of the trialists has commenced. It is important that academic medical centers committed to the practice of evidence-based medicine facilitate the training of the trialists needed for the translational research and drug and device development. While many of us have learned the process of clinical research by experience, more and more future investigators are looking for a more codified process of education preparing for a clinical investigative career.

The search for new knowledge in the basic sciences has always had a training process for the Ph.D. of the future. Clinical, translational research is moving in that direction with mentorship progress and the Masters in Clinical Research. The future, I believe, will require even more advanced training in clinical research. There are a few Ph.D. programs in clinical research, but more are needed, and these programs must be compatible with a simultaneous clinical career. As a basic scientist attends classes early on and then develops a project in a programmatic way, we will see a similar process evolve in the clinical sciences. It will start with the clinical trainings and then classes in statistics, trial design, and epidemiology; programs seen in the Masters in Clinical

Research Programs; and then progress to a systemic clinical research program in a specific area, addressing in a step-by-step way a research question and culminating in a doctoral thesis. The future will demand a more highly trained clinical researcher, able to undertake the clinical investigative programs of the future.

2.13   CONCLUSION

Drug development and discovery is a most exciting field. It is creative, intellectually challenging, and organizationally demanding. Those involved are to be congratulated for undertaking efforts that are usually anonymous but that affect clinical therapeutics to a considerable degree. The drug discovery process has gone through the botanical phase and the synthetic chemistry phase and has now emerged into a most exciting era of biotechnology and gene manipulation.

Our understanding of medicine, disease processes, as well as the design and statistical evaluation of clinical trials has evolved tremendously. These are evolving areas that are important and that dynamically affect the drug discovery process. These areas are being markedly affected by the third wave, the information age. We can only think back to when a chapter like this would be handwritten, typed on a typewriter with carbon paper, corrections made, and a copy sent off to a publisher. Word processing has revolutionized this approach and will continue to revolutionize it in the coming years. This information processing revolution will totally change the drug development process, facilitating it, offering economies of time, scale, and money. This will change how we approach drug and device development. All phases of drug development will undergo significant change, and we will be better for it. This is a most exciting era and one in which participation will be both rewarding as well as demanding.

REFERENCES

1. Toffler A, *Future Shock*, Random House, New York, 1970.
2. Echt DS, Liebson PR, Mitchell LB, et al. Mortality and morbidity in patients receiving encainide, flecainide or placebo: the Cardiac Arrhythmia Suppression Trial. *The New England Journal of Medicine.* 1991;324:781–8.
3. Julian DG, Camm AJ, Frangin G, Janse MJ, Munoz A, Schartz PJ, Simon P. Randomized trial of effect of amiodarone on mortality in patients with left-ventricular dysfunction after recent myocardial infarction. EMIAT. European Myocardial Infarct Amiodarone Trial Investigators. *Lancet.* 1997;349(9053):667–74.
4. Cairns JA, Connolly SJ, Roberts R, Gent M. Randomized trial of outcome after myocardial infarction in patients with frequent or repetitive ventricular premature depolarizations: CAMIAT. Canadian Amiodarone Myocardial Infarction Arrhythmia Trial Investigators. *Lancet.* 1997;349(9053):675–82.
5. Sim I, McDonald KM, Lavori PW, Norbutas CM, Hlatky MA. Quantitative overview of randomized trials of amiodarone to prevent sudden cardiac death. *Circulation.* 1997;96(9):2823–9.
6. Amiodarone Trials Meta-Analysis Investigators. Effect of prophylactic amiodarone on mortality after acute myocardial infarction and in congestive heart failure: Meta-analysis of individual data from 6500 patients in randomized trials. *Lancet.* 1997;350(9089):1417–24.
7. Doudna JA, Charpentier E. The new frontier of genome engineering with CRISPR-Cas9. *Science.* 2014;346:1258096.
8. Tanouye E, Does corporate funding influence research? *Wall Street Journal*, B1, B, January 8, 1998.

3 Innovation in Drug Development

Perspectives of a Venture Capitalist

Krishna Yeshwant
Google Ventures

CONTENTS

3.1 INTRODUCTION

Drug development is among the most expensive, time-consuming, and technically challenging endeavors in science. Yet, despite the difficulty involved, collaboration between academia, startups, and the pharmaceutical industry has consistently produced miraculous therapeutics that treat and, in some cases, cure profound human diseases.

We find ourselves in a remarkable moment when discovering and developing drugs are on the one hand accelerated by a constant flow of chemical and biological discoveries, while on the other hand hobbled by the rapidly increasing cost incurred for each new therapy. Most academics, startup executives, and pharmaceutical leaders naturally emphasize the technical progress in the field. Yet, according to one analysis,[1] despite all our technical achievements, the cost of each new drug (including the costs of failed programs) has been on a trajectory of roughly doubling every 9 years over the last several decades (Figure 3.1).[2]

These studies suggest that it currently costs ~$2.6 billion to bring each new drug to market. As the cost to discover and develop new drugs increases, the aggregate return on investment in each new project naturally decreases. Some studies suggest that the industry is currently at a point where it no longer makes financial sense to invest in the discovery of new drugs since the internal rate of return is now approaching 0% or may even be negative.[3,4]

But The Number of New Drugs Approved Per $bn of R&D Spending Has Fallen by Around 100-fold Since The 1950s

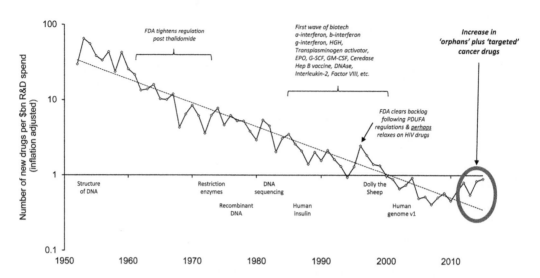

FIGURE 3.1 From Jack Scannell's "Damn the Compass" talk.[40] Note: R&D costs are based on the PhRAM Annual Survey (2011) and Munos (2009). PhRAM is a trade association that does not include all drug and biotechnology companies so the PhRAM figure understates R&D spending at an industry level. Total industry expenditure since 2004 has been 30%–40% higher than the PhRAM member's total spend, which formed the basis of the figure. The NME count, on the other hand, is the total number of small molecule and biologic approvals by the FDA from all sources, not just PhRAM members. The overall picture seems fairly robust to the precise details of the cost and inflation calculation. (Source: Monos 2009, PhRAM Annual Surveys, FDA, and Bernstein analysis.)

The pharmaceutical industry has had an array of responses ranging from:

- Refocusing on diseases with higher likelihood of discovering a drug[5]
- Developing or acquiring phenotypic and genetic data sets to increase confidence in new therapeutic efforts[6–8]
- Investing in improving internal capabilities,[9] access to new modalities[10–12]
- Acquiring companies in the startup ecosystem that have sufficiently de-risked their programs (though several of these programs fail in clinical trials as well)
- Exploring tactics that extend patent life to increase the overall payback period for a therapeutic[13]
- Increasing drug prices to make up for revenue shortfalls caused by competition that arises when patents on lucrative drugs expire.

These approaches have successfully protected pharmaceutical company profit margins to date, though in many cases at the expense of investment in the core resources, people, and processes that historically enabled the industry's most celebrated clinical successes.

At its core, drug discovery is the process of searching a large chemical space for a rare molecule that will successfully make its way through clinical trials and ultimately have positive clinical impact for patients. One broad approach to optimizing this process is to develop techniques to more rapidly search chemical space. Another approach is to develop representations of biology (assays) that are highly predictive of how likely a given drug candidate is to be successful in making its way through clinical trials, regulatory approval, and clinical use.

In an important pair of papers, Scannell et al.[5,14] note that there have been dramatic improvements in brute force screening methods and basic science methods that have increased investigator's efficiency in searching chemical space. Yet, the fact that drug candidates seem to fail in clinical trials at approximately the same rate over the last 50 years[15] suggests that the industry has optimized the wrong set of research and development process. They note several potential causes for the lack of improvement in the rate of success of molecules entering clinical trials. One particularly salient conclusion they emphasize is based on a decision theory framework: there is a need for better predictive models in drug discovery since greater chemical screening efficiency does not generate a higher probability of success either theoretically or empirically based on the last 50 years of experience.

On an initial review it would seem like an obvious area for companies and investors to focus on. After all, if we don't understand the pathophysiology of a disease then how can we cure it? Why has this not happened to date? The first reason is that it is very hard to judge whether a predictive model really is predictive until *after* a drug has already moved through clinical trials. An assay may have subtle features that are not representative of human biology (or perhaps not representative of a subpopulation's biology) that may be challenging to identify until a drug has actually been tried in humans. It may only be when unexpected clinical results arise that an investigator would have the data needed to begin to identify the discrepancy between a model system and a human biology.

Secondly, there is no repository or even ongoing maintenance of disease models that were used for either successful or failed drugs. After a drug has been submitted for clinical trials and ultimately FDA approval, many of the learnings attached to the assays that helped identify and develop the molecular entity are discarded or archived but not mined for further learning. Animal models that represent human diseases do tend to be maintained, often by commercial entities that provide these animals to other investigators.

Notably the most recent period of increased drug discovery productivity has been marked by two trends – an increased focus on cancer and orphan diseases. Each of these areas is differentiated in the continuum of disease by the availability of high-quality disease models or a readier path to development of such models. Both arenas have also increasingly been characterized by genetic markers that allow for subcategorization of patients carrying a clinical diagnosis that may actually be the composition of several closely related but genetically (and potentially pathophysiologically) distinct entities. In many cases, a successful therapeutic comes along with a companion diagnostic that screens a patient in or out as a candidate for the drug.

This progress in cancer and orphan diseases could offer a path towards a brighter future for the trajectory of drug development productivity as well as a brighter path for patients. Identifying disease areas where there are high-quality models available, or a path towards developing them, would seem like an initially logical step. Some authors have suggested utility in a disease model repository,[5] so that learnings from one development process could be shared across investigators in multiple institutions. At the very least it would seem to be valuable to have such a compendium available inside even a single institution.

The last decade has demonstrated a variety of new approaches to developing disease models. The jury is still out as to whether these disease models are *actually* more predictive (as noted earlier, we may need for new drugs to be developed using these models before reaching this conclusion), though they do seem to intuitively point in exciting directions. In this chapter, we will review a selection of the techniques enabling new disease models and we will touch on some of the recent innovations in funding many of these scientific approaches. While the science is central, there are still large budgets that need to be funded to properly ask and answer the questions posed.

There is palpable excitement in the drug development world around each of these technologies – enough that the drug development industry has sustained a multiyear bull run despite several of the fundamental productivity questions noted earlier. There is no doubt greater scientific progress to come. We will have to wait and see how the markets respond!

3.2   INNOVATIONS IN DISEASE MODELS

As we have discussed in the prior section, there is a developing understanding of the tremendous leverage that a highly predictive disease model can have at the earliest stages of drug development. However, it has been challenging for companies working purely on disease models to gain funding. One reason may be that it is often unclear whether the disease models are actually representative of human pathophysiology until after a drug has been run through expensive clinical trials. Also, while predictive disease models may be critical to moving a program towards the right targets, there is still significant development work that needs to be done before a molecule can be considered a drug. Given this complexity, it is often challenging to retrospectively assign how much value a disease model ultimately contributed.

With that said, it is clearly valuable to set a drug development program moving in the right direction, and highly predictive disease models are the only guidance we have in the early discovery period. In this section, we will discuss some of the innovations occurring in the disease model space.

3.2.1   PHENOTYPIC DISEASE MODELS

Phenotypic models involve using cellular assays to recapitulate a disease process in a dish ("disease-in-a-dish") that can be studied and used as a screening system to identify new drugs. These models provide an *in vitro* system that attempts to recapitulate *in vivo* biology.

Phenotypic models generally involve using biopsies, cadaveric tissue, human embryonic pluripotent stem cells (hESC), or induced pluripotent stem cells (iPSC) to grow a two- or three-dimensional construct. There are several opportunities and issues with each of these approaches which we will review.

3.2.2   TWO-DIMENSIONAL PHENOTYPIC MODELS (hESC AND iPSC)

Two-dimensional phenotypic models have been used for more than 100 years.[16] A two-dimensional phenotypic model is essentially a plate with a particular type of cell, or potentially multiple types of cells, that have been grown and exposed to a variety of factors to coax them to grow in a desired fashion and then maintained with nutrients bathed over them. These cells can exhibit complex characteristics that appear to be similar in several cases to the human biology they attempt to mimic.

Where do these cells come from? Historically investigators have used cadaveric tissue or biopsies from various tissues to better understand pathophysiology. However, these tissue sources are logistically challenging to obtain and generally provide a low yield of cells. Human embryonic stem cells (hESCs) offered a path towards a more accessible and adaptable source of cells.[17] Using directed differentiation hESCs can be differentiated into different lineages, and since hESCs are capable of self-renewal, they can provide a consistent source of cells. Despite their advantages, hESCs faced ethical issues since they are obtained from embryos. In 2006, Shinya Yamanaka discovered a set of factors that could be applied to reprogram somatic cells derived from a patient to induce pluripotency similar to the capacity of hESCs.[18] This approach was transformative to the field since it enabled a readily available source of material using a technique that was accessible to most laboratories. Yamanaka proceeded to win the Nobel Prize for this work in 2012. Despite the amazing progress in developing iPSC science since this publication, several studies also demonstrated differences in genetic and epigenetic characterization of iPSCs when compared to hESCs, which is an ongoing area of investigation.[19] While iPSC-derived cell lines have been shown to recapitulate several important diseases,[20] it is still not fully appreciated how much these genetic and epigenetic differences hamper iPSCs as high-fidelity representations of human pathophysiology, though investigators continue to make progress.[21]

The iPSC field has made tremendous progress since Yamanka's initial discovery including the development of a variety of protocols to guide pluripotent stem cells to a diversity of cell types. However, several fundamental challenges remain.[22] First, iPSC's protocols are sensitive, even the slightest deviation from protocol can lead to a cohort of cells developing into subtly or more profoundly different types of cells. Second is that, while iPSCs may be valuable in modeling diseases present early in human development, they may not be able to recapitulate diseases that appear later in human development like Alzheimer's disease or nonalcoholic steatotic hepatitis (NASH). Investigators have attempted to address this by exposing cells to environmental toxins (e.g. oxidative stress) to artificially age them, but these approaches have not been particularly successful to date. The third limitation is that most diseases don't involve just one cell, but rather a complex interaction between cells in a milieu comprised of a variety of agents that proximate and distant cells help produce. Most of the disease models to date have likely underrepresented the complexity of actual human biology. This is partially due to our inability to differentiate cells to all the different types that are present in an actual human and partially due to the fact that we do not know all the cells that may be involved in a particular disease. Finally, the fourth limitation in iPSCs is that the reprogramming process used to develop iPSC-derived cell lines currently involves removing epigenetic alterations. Although this is a less understood area, there are suggestions epigenetic markers may be significant in a variety of diseases, so their loss may be significant.

There are of course a variety of techniques investigators in the field are developing to counter each of these limitations, though we will not review all of them here.[17,20–28] However, one promising technique leads us to the conversation of organoids.

3.2.3 Three-Dimensional Phenotypic Models (Organoids)

The term "organoid" has been broadly used to describe three-dimensional structures made up of multiple types of cells (usually multiple cell lineages) along with a tissue architecture that represents human biology.[29] In organoid systems, cells spontaneously self-organize into structures that are similar to human *in vivo* tissues in both cellular architecture and function as well as other ways. Investigators have developed organoid models of retinal, cerebral, kidney, intestinal, and stomach (among others) to date. These tissues can be derived from hESCs, iPSCs, or adult-derived stem/progenitor cells (AdSCs). Unlike iPSCs, AdSC-derived organoids may be better at representing diseases associated with aging since they begin with cells from an adult.

Recent progress in 3d culture systems along with a better understanding of cell development protocols have enabled an expansion in organoid work, which has resulted in a number of potentially useful disease modeling systems. These models recapitulate the tissue-specific cells, cell–cell interactions, and cell–matrix interactions, which appear to be relevant to pathophysiology in cancer, immune disorders, and other pathologies.[30–34]

Similar to the progress in iPSCs noted earlier, the organoid field has been rapidly developing but still faces significant challenges. For example, *in vivo* tissues are fed by a network of vasculature involving arteries, veins, and capillaries. Currently there is not a robust way to generate this sort of network in an organoid, though there are several groups who have made progress. Similarly, *in vivo* tissues have an integrated peripheral nervous system that is not present in current organoid models. Furthermore, organoid models have yet to fully incorporate stromal cells like immune cells, limiting their use in modeling inflammatory or certain cancer systems. Also, while organoids are unique in their capacity to model cell–cell and cell–matrix interactions, it is still challenging for investigators to control the dynamics of these interactions in a reproducible fashion. This makes it difficult to model a variety of environmental qualities of a tissue like pH, oxygenation, nutrient transport, metabolic state, and other important characteristics.[35]

3.2.4 Genetically Engineered Phenotypic Models

In addition to the amazing progress that has occurred in the iPSC and organoid arenas over the last decade, there has also been stunning progress in the gene-editing space. In particular, the development of an easy-to-use and reproducible editing system known as CRISPR/Cas9 has been transformative to many aspects of medical research.

One particularly intriguing application of gene-editing approaches involves editing iPSC and organoid models. This approach is particularly useful when an investigator would like to produce a disease model that is representative of a patient's pathophysiology but does not have access to a patient from whom to obtain a sample. In these cases, an iPSC or organoid model can be developed that closely resembles that patient and then a CRISPR/Cas9, or other editing system, can be used to make specific genetic or epigenetic alterations. Using a deactivated Cas9 attached to functional effectors can enable investigators to artificially upregulate, downregulate, image, or otherwise manipulate the biology of the model system.

3.2.5   Computational Models of Cells/Organisms/Chemistry

In addition to the biological disease modeling systems described earlier, there has been an ongoing effort to model biological and chemical systems *in silico* (i.e. in a computer system) over the last several decades. These efforts have had mixed success to date.[36] We should quickly note that the successful efforts in this space have been focused on chemical simulation rather than attempting to represent the full complexity of a biological system *in silico*. Thus, most of the successful efforts have focused on developing therapeutic candidates for targets that have already been well validated in other model systems but may be proving challenging to drug, rather than attempting to use *in silico* methods to discover new mechanisms of action.

The field of computational chemistry has gone through prior eras of intense growth. In the 1980s and 1990s, there was excitement about computer-aided drug design and the concept that computational resources that were coming online could accelerate the discovery process through virtual screens which would rapidly improve investigators' capacity to find molecules with high potency and selectivity. While the core thesis was and continues to be sound, there are several significant unanswered questions including: Do computational methods enable a large-enough chemical search space to offer new insights? Can computational methods add new approaches to a chemist's toolbox or do they recapitulate existing methods? If there are virtual hits, can we actually synthesize them?

These along with several other questions have hampered the field's progress. However, in the intervening years, we have seen increases in computational processing power, software efficiency, and approaches to machine learning that have significantly accelerated progress that will hopefully lead to greater efficiencies in drug discovery over the coming years.

3.3   INNOVATION IN FINANCING DRUG DEVELOPMENT

3.3.1   Challenges in Funding Tools, Platforms, and Disease Models

There are myriad approaches that investigators and entrepreneurs can use to fund a concept from discovery through development into commercialization. Most academics fund early-stage work through traditional government grants. There have been some remarkable successes in which translational research was funded by patient advocacy groups and other forms of nonprofits.[37] We are occasionally also seeing large corporate entities fund early-stage research work when it is particularly transformative.[38] More commonly, early-stage discovery work performed in an academic lab is translated into a formalized drug development process through angel, venture capital, and/or other forms of private equity.

This approach has overall worked well, though it has significantly evolved over the last decade. While venture-backed therapeutics startups have been a relatively profitable asset class over the last several decades,[39] the activity in this space slowed down significantly during the 2008 credit crisis. This was problematic for several large pharmaceutical companies who had been looking to the startup ecosystem as part of their pipeline for new drugs. In response, several pharmaceutical companies bolstered their existing venture funds or started new funds. The role of corporate venture funds in therapeutics has grown significantly over the ensuing decade and has also led to growth in pharmaceutical companies partnering earlier with venture-stage companies to help bring needed skill sets and capital into areas of developing science. For instance, a pharmaceutical company can provide insights into regulatory process, clinical trial design, or manufacturing requirements that would traditionally be challenging to access in an early-stage startup, and in some cases, pharmaceutical companies can bring sufficient dollars to help a startup actually implement those insights at scale earlier than the company might otherwise be able to do.

Following the credit crisis of 2008, there were several years in which the early-stage biotech industry relied heavily on pharmaceutical companies to help fund science through both corporate venture investing and partnerships. However, this investment was richly rewarded in the years following when a robust initial public offering (IPO) market brought forth a slew of well-funded companies that were able to raise large amounts of capital and retain independence on the public markets. While the stakes pharmaceutical companies held in these companies grew in value and they continued to grow a robust crop of companies that could continue feeding their pipelines, on the other hand, the active IPO market made it challenging for pharmaceutical companies to acquire companies since those startups have had financing alternatives that allow them to maintain their independence.

It is currently unknown whether therapeutic assets developed in small companies or by a select group of people are produced in a systematically more efficient way than drugs developed by large pharmaceutical companies. Per our earlier discussion, we know that the industry as a whole has faced productivity challenges, but are there approaches and pockets of the industry that may perform better?

One model for therapeutics that has worked well in recent years is the "platform" company. In this approach, a novel area of biology is developed that entrepreneurs and scientists agree could be a rich source of multiple drugs. That area of biology could be a novel approach to screening, a new source of chemical matter, a new therapeutic modality (e.g. gene-editing mechanisms, cell therapy), a new mechanism of action that may apply in multiple indications, and so on. While these new areas of biology are inherently exciting, they are usually not inherently valuable in their own right. It is hard for the public markets or for a therapeutics company to fully value a biological platform without knowing that there is a drug that works that has been produced by this new platform. In many cases, the drug itself ends up driving the value of the whole company rather than the platform. But, in some ways, the platonic ideal of a platform company in therapeutics would be a well-developed biological platform that has produced multiple drugs that have worked in humans based on insights derived from that platform that would have been hard to come by in other ways. In this state, a company would theoretically be able to raise money to drive more programs through its biological platform that would turn into successful drugs that would be worth a premium to the dollars invested. There are few, if any, real examples of this sort of company, but it is an ideal that many groups continue to try to achieve, since a failure mode may be a drug that works via a novel mechanism of action, which is valuable in its own right.

Platform companies are intriguing because they help solve the problem noted earlier around how to invest in better screening platforms and potentially other technologies that may be hard to fund by themselves. For example, a screening platform may not be particularly valuable by itself until it is validated as having been able to identify a novel drug. However, if the platform company working on this new technology were to raise sufficient capital to both develop their underlying platform technology and use that platform to discover a drug, then they may be able to get through the "valley of death" where many companies find themselves in which they have a mostly functioning platform but can't truly validate it since they haven't produced a drug.

3.3.2 COMPANY INCUBATION MODELS

Several venture funds have taken this approach over the years, but Third Rock Ventures, based in Boston, MA, has scaled this approach further than most. Unlike most traditional biotech venture funds who may start a company with a $10–15 million investment, Third Rock has historically founded companies with $60 million or more. This money comes in a segmented fashion, with each segment (known as a "tranche") unlocked when the company achieves particular milestones. This tranched approach allows the venture fund to limit overall capital exposure in case there is a major scientific or leadership failure, while also allowing the company to have sufficient capital committed at the beginning to be able to complete both a large platform build as well as a full (often multiprogram) drug discovery effort. This approach has been incredibly successful having produced multiple companies that have brought impressive therapeutics to market.

This model was particularly effective during the 2008–2012 timeframe during which many venture capital funds stopped investing in biotech due to the capital constraints caused by the broader credit crisis occurring during that time. Biotechnology investments are often challenging to sustain during extended financial downturns because they usually require a very large amount of capital for many years before bringing in their first dollar of revenue. Investors often do not want to take on this extended and large-scale risk when they are not sure what the broader market will look like over that timeframe. With that said, humans will always have diseases that need to be treated and cured, so there will always be value to producing effective drugs. One of the significant advantages to funding a company with sufficient capital to get to an unequivocal value creation event is that the company has a better chance of withstanding unpredictable market conditions during those times.

3.3.3   Financing through Value Creation Milestone

Many other venture funds also incubate companies using different financing models. In a more traditional model, a venture fund will put forward a relatively small amount of funding (e.g. anywhere from $100,000 to <$5,000,000) for an entrepreneur or group of scientists to develop a specific concept to the point where they may be able to start a company. The art of determining whether the science is sufficiently "derisked" is an art in its own right and is usually achieved through multiple rounds of debate between the core group of founders, scientists, and investors around a clear series of milestones that, if achieved, would define a successful potential program. If the milestones are all achieved, then the venture fund will be more likely to provide larger dollar amounts to move the overall company and science forward. This usually involves hiring an executive team, finding office space, and launching forth on a broader exploration of the science.

One of the benefits to this approach is that a company raises money more frequently and can ideally command a higher price for each share of the company since it will have incremental progress to show along the way. If the science allows itself to be successively derisked in a way that a new investor might be able to efficiently assess the state of the company, then this sort of progressive funding based on milestone achievements may be a preferable approach. On the other hand, if it may not be clear whether the company's science is really progressing until a much more significant amount of risk is retired, then the larger rounds of funding as described in the previous section may be more appropriate.

3.3.4   Insourcing from Pharma/Megarounds

One intriguing model that a few investment groups have been finding success with involves licensing assets out of large pharmaceutical companies to develop in the context of a startup. This is an unusual model since we usually think of large pharmaceutical companies licensing assets in from startups to help fill their drug development pipelines. However, the door can indeed swing both ways! In these scenarios, a large pharmaceutical company may not have sufficient internal resources to fund a program that does not quite meet the internal bar needed to progress forward. There may not be any specific scientific reason other than that another set of programs seems more likely to be successful or more likely to provide a commercial return given the expenditures required. Or it may be that the company has other financial pressures that are causing it to be careful around increasing internal expenses. Or of course, the company may just not understand what an exciting internal program it may actually have (though this is less likely).

In these cases, purpose-built venture funds can work with the pharmaceutical company to spin those assets, sometimes along with the teams working on them, into a startup. These startups are often funded with hundreds of millions of dollars since they are often bringing in later-stage drugs that need to undergo Phase II or Phase III development, which is usually a very expensive endeavor. In many cases, the pharmaceutical company will have a right to buy the drug back at a premium if the trial is successful.

This approach has offered several benefits to the market including bringing drugs to market that might not otherwise have been moved forward due to resource constraints. It enables a team to focus all of their efforts on a drug in the context of a startup rather than trying to manage the large array of projects a larger company has to prosecute. Also, on occasion, this approach allows assets from multiple large companies to be aggregated into one company where development and potential commercialization efficiencies can be realized.

3.4 CONCLUSION

As has been true for decades now, there are ever more tools and techniques with which we can understand and manipulate biology. And while our industry has produced some truly miraculous therapeutics that in some cases can cure patients of profound diseases, we still have not figured out how to control for the complexity of human biology. This leads us to some fundamentally uncomfortable questions – like is the global biopharma industry financially viable? Yet, we also know that there are few things in life that are as important as our health and few things that are more terrifying than confronting an illness. There is an urgency to gain better control of our biology and to continue creating better tools to enable us to achieve that control. To do this, we will need to continue to find innovative ways to understand and to fund risky science and to ensure that we will continue to make progress in addressing diseases for which we have no approaches today. This will require a deeper partnership between all parts of the healthcare system. Certainly, it will require deeper connectivity between academia and industry. But it will also require partnership between pharmaceutical companies, insurance companies, provider systems, and of course patients.

There are many challenges ahead, but as all good scientists and entrepreneurs know, it is in the challenges that the opportunities lie as well.

REFERENCES

1. DiMasi JA, Grabowski HG, Hansen RW. Innovation in the pharmaceutical industry: New estimates of R&D costs. *Journal of Health Economics*. 2016;47:20–33. www.sciencedirect.com/science/article/pii/S0167629616000291. doi: 10.1016/j.jhealeco.2016.01.012.
2. DiMasi JA, Hansen RW, Grabowski HG. The price of innovation: New estimates of drug development costs. *Journal of Health Economics*. 2003;22(2):151–185. www.sciencedirect.com/science/article/pii/S0167629602001261. doi: 10.1016/S0167-6296(02)00126-1.
3. Stott K. Pharma's broken business model - part 1: An industry on the brink of terminal decline. www.linkedin.com/pulse/pharmas-broken-business-model-industry-brink-terminal-kelvin-stott/. Updated 2017. Accessed May 10, 2018.
4. Stott K. Pharma's broken business model - part 2: Scraping the barrel in drug discovery. www.linkedin.com/pulse/pharmas-broken-business-model-part-2-scraping-barrel-kelvin-stott/. Updated 2018.
5. Scannell JW, Bosley J. When quality beats quantity: Decision theory, drug discovery, and the reproducibility crisis. *PLoS One*. 2016;11(2):e0147215. www.openaire.eu/search/publication?articleId=od_____267::2a46ae703137cff6ed0288b6fece097e. doi: 10.1371/journal.pone.0147215.
6. deCODE genetics. www.decode.com/. Accessed April 10, 2018.
7. Regeneron genetics center. www.regeneron.com/genetics-center. Accessed April 10, 2018.
8. 23andMe. www.23andme.com/. Accessed April 10, 2018.
9. Mayr LM, Bojanic D. Novel trends in high-throughput screening. *Current Opinion in Pharmacology*. 2009;9(5):580–588. www.clinicalkey.es/playcontent/1-s2.0-S1471489209001283. doi: 10.1016/j.coph.2009.08.004.

10. Herper M. Merck teams with moderna to take on one of the toughest targets in cancer, forbes.com Web site. www.forbes.com/sites/matthewherper/2018/05/03/merck-teams-with-moderna-to-take-on-one-of-the-toughest-targets-in-cancer/#5c9c97669939. Updated 2018. Accessed May 10, 2018.

11. de la Merced MJ. Gilead to buy kite, maker of cancer treatments, for $11.9 billion, nytimes.com Web site. www.nytimes.com/2017/08/28/business/dealbook/gilead-kite-gene-therapy.html. Updated 2017. Accessed April 10, 2018.

12. Bennett J. Celgene buys juno therapeutics: A risky $9B bet, forbes.com Web site. www.forbes.com/sites/johannabennett/2018/01/22/celgene-buys-juno-therapeutics-a-risky-9b-bet/#6afc1ff317c7. Updated 2018. Accessed April 10, 2018.

13. Thomas K. How to protect a drug patent? Give it to a native american tribe, nytimes.com Web site. www.nytimes.com/2017/09/08/health/allergan-patent-tribe.html. Updated 2017. Accessed April 10, 2018.

14. Scannell JW, Blanckley A, Boldon H, Warrington B. Diagnosing the decline in pharmaceutical R&D efficiency. *Nature Reviews Drug Discovery*. 2012;11(3):191–200. www.ncbi.nlm.nih.gov/pubmed/22378269. doi: 10.1038/nrd3681.

15. Dunwiddie CT, Munos BH, Lindborg SR, et al. How to improve R&D productivity: The pharmaceutical industry's grand challenge. *Nature Reviews Drug Discovery*. 2010;9(3):203–214. doi: 10.1038/nrd3078.

16. Kelava I, Lancaster MA. Dishing out mini-brains: Current progress and future prospects in brain organoid research. *Developmental Biology*. 2016;420(2):199–209.

17. Thomson JA, Itskovitz-Eldor J, Shapiro SS, et al. Embryonic stem cell lines derived from human blastocysts. *Science*. 1998;282(5391):1145–1147. www.sciencemag.org/cgi/content/abstract/282/5391/1145. doi: 10.1126/science.282.5391.1145.

18. Takahashi K, Tanabe K, Ohnuki M, Narita M, Ichisaka T, Tomoda K, Yamanaka S. Induction of pluripotent stem cells from adult human fibroblasts by defined factors. *Cell*. 2007;131(5):861–867.

19. Hussein SM, Batada NN, Vuoristo S, et al. Copy number variation and selection during reprogramming to pluripotency. *Nature*. 2011;471(7336):58–62.

20. Ebert A, Liang P, Wu J. Induced pluripotent stem cells as a disease modeling and drug screening platform. *Journal of Cardiovascular Pharmacology*. 2012;60(4):408–416. www.ncbi.nlm.nih.gov/pubmed/22240913. doi: 10.1097/FJC.0b013e318247f642.

21. Rashid ST, Corbineau S, Hannan N, et al. Modeling inherited metabolic disorders of the liver using human induced pluripotent stem cells. *The Journal of Clinical Investigation*. 2010;120(9):3127–3136. www.ncbi.nlm.nih.gov/pubmed/20739751. doi: 10.1172/JCI43122.

22. Saha K, Jaenisch R. Technical challenges in using human induced pluripotent stem cells to model disease. *Cell Stem Cell*. 2009;5(6):584–595. www.sciencedirect.com/science/article/pii/S1934590909005827. doi: 10.1016/j.stem.2009.11.009.

23. Lister R, Pelizzola M, Kida YS, et al. Hotspots of aberrant epigenomic reprogramming in human induced pluripotent stem cells. *Nature*. 2011;471(7336):68–73. www.ncbi.nlm.nih.gov/pubmed/21289626. doi: 10.1038/nature09798.

24. Bock C, Tomazou EM, Brinkman A, et al. Genome-wide mapping of DNA methylation: A quantitative technology comparison. *Nature Biotechnology*. 2010;28(10):1106–1114. www.openaire.eu/search/publication?articleId=dedup_wf_001::160cedb0a97bebfd4c384baf0dc793ec. doi: 10.1038/nbt.1681.

25. Ghosh Z, Wilson KD, Wu Y, Hu S, Quertermous T, Wu JC. Persistent donor cell gene expression among human induced pluripotent stem cells contributes to differences with human embryonic stem cells. *PLoS One*. 2010;5(2):e8975. www.openaire.eu/search/publication?articleId=od_____267::c600ef04 2944bc536a888dd91a58aca7. doi: 10.1371/journal.pone.0008975.

26. Gore A, Li Z, Fung H, et al. Somatic coding mutations in human induced pluripotent stem cells. *Nature*. 2011;471(7336):63–67. www.ncbi.nlm.nih.gov/pubmed/21368825. doi: 10.1038/nature09805.

27. Ortmann D, Vallier L. Variability of human pluripotent stem cell lines. *Current Opinion in Genetics and Development*. 2017;46:179. www.ncbi.nlm.nih.gov/pubmed/28843810.

28. Yiangou L, Ross ADB, Goh KJ, Vallier L. Human pluripotent stem cell-derived endoderm for modeling development and clinical applications. *Cell Stem Cell*. 2018;22(4):485–499. www.sciencedirect.com/science/article/pii/S193459091830122X. doi: 10.1016/j.stem.2018.03.016.

29. Ho BX, Pek NMQ, Soh B. Disease modeling using 3D organoids derived from human induced pluripotent stem cells. *International Journal of Molecular Sciences*. 2018;19(4):936. www.ncbi.nlm.nih.gov/pubmed/29561796. doi: 10.3390/ijms19040936.

30. Huch M, Koo B. Modeling mouse and human development using organoid cultures. *Development (Cambridge, England)*. 2015;142(18):3113–3125. www.ncbi.nlm.nih.gov/pubmed/26395140. doi: 10.1242/dev.118570.

31. Boj S, Hwang C, Baker L, et al. Organoid models of human and mouse ductal pancreatic cancer. *Cell*. 2015;160(1–2):324–338. www.sciencedirect.com/science/article/pii/S009286741401592X. doi: 10.1016/j.cell.2014.12.021.

32. Hindley CJ, Cordero-Espinoza L, Huch M. Organoids from adult liver and pancreas: Stem cell biology and biomedical utility. *Developmental Biology*. 2016;420(2):251–261. www.sciencedirect.com/science/article/pii/S0012160616302779. doi: 10.1016/j.ydbio.2016.06.039.

33. Crespo M, Vilar E, Tsai S, et al. Colonic organoids derived from human pluripotent stem cells for modeling colorectal cancer and drug testing. *Nature Medicine*. 2017;24(4):526. www.openaire.eu/search/publication?articleId=od_____267::fda7e748a6b99e2ece196554e62a86d7. doi: 10.1038/nm.4355.

34. Li X, Nadauld L, Ootani A, et al. Oncogenic transformation of diverse gastrointestinal tissues in primary organoid culture. *Nature Medicine*. 2014;20(7):769–777. www.ncbi.nlm.nih.gov/pubmed/24859528. doi: 10.1038/nm.3585.

35. Gao D, Vela I, Sboner A, et al. Organoid cultures derived from patients with advanced prostate cancer. *Cell*. 2014;159(1):176–187. www.sciencedirect.com/science/article/pii/S0092867414010472. doi: 10.1016/j.cell.2014.08.016.

36. Jordan AM. Artificial intelligence in drug design: The storm before the calm? *ACS Medicinal Chemistry Letters*. 2018. doi: 10.1021/acsmedchemlett.8b00500.

37. CYSTIC FIBROSIS FOUNDATION. CF Foundation Venture Philanthropy Model. www.cff.org/About-Us/About-the-Cystic-Fibrosis-Foundation/CF-Foundation-Venture-Philanthropy-Model/ Web site. Accessed August 23, 2018.

38. Penn Medicine News. University of pennsylvania and novartis form alliance to expand use of personalized T cell therapy for cancer patients. www.pennmedicine.org/news/news-releases/2012/august/university-of-pennsylvania-and. Updated 2012. Accessed August 28, 2018.

39. Das S, Rousseau RF, Adamson PC, Lo AW. The challenge of pediatric oncology: New business models to accelerate innovation. *Journal of Clinical Oncology*. 2018;36(15_suppl):10528. doi: 10.1200/JCO.2018.36.15_suppl.10528.

40. Scannell J. "Damn the compass, full speed ahead": Why quality beats quantity in drug R&D. *Standing Together: Health Care for Our Common Good*." goldlabfoundation.org/symposia/7th-annual-goldlab-symposium/. 2016.

Section II

Drug Discovery

4 High-Throughput Screening

Olivia Perez, J. Pena, Virneliz Fernandez-Vega,
Louis Scampavia, and Timothy Spicer
Scripps Research

CONTENTS

4.1 INTRODUCTION TO HIGH-THROUGHPUT SCREENING AND DRUG DISCOVERY

Many patients can attest to the outrageous price of prescription drugs. How can it cost so much to make each pill and why are prices so high? From early research to FDA approval, each drug candidate encounters many obstacles in order to ensure efficacy, potency, and safety. For every compound that becomes a drug candidate, there are tens of thousands of compounds that have failed testing. Even compounds that pass years of preclinical testing to become drug candidates sometimes never go to market. "Experience has shown that only approximately 1 out of 15–25 drug candidates survives the detailed safety and efficacy testing (in animals and humans) required for it to become a marketed product" (Lombardino and Lowe 2004). High-throughput screening (HTS) is a process in early drug discovery that uses automation, miniaturization, and robotics to increase the number of assays tested against large drug-like compound collections all the while decreasing the volume needed for each test. Thus, HTS decreases both waste and cost per test while increasing the number

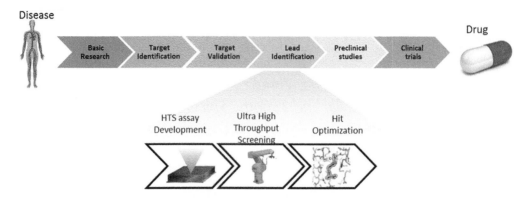

FIGURE 4.1   The drug discovery process and where HTS is involved.

of data points generated. Some of the most efficient HTS labs can run up to 200,000 samples in a single day. Ultimately, it is the high level of drug candidate attrition that adds cost to consumers because pharmaceutical companies not only pay for the compound that advances to market but also the failed attempts. In addition, the cost of development and clinical trials is very expensive for each drug product. The drug discovery process for one drug product takes anywhere from 7 to 15 years and can cost more than $1–$2 billion (Hughes et al. 2011). This chapter analyzes the process of early drug discovery in the United States, and the positive impact that HTS can have on this process (Figure 4.1).

4.2   TARGET IDENTIFICATION

Research into a potential drug begins when there is an unmet need for clinical treatment of a disease. In the drug discovery process, priority is given to research that impacts a disease affecting large populations without an effective treatment. The first step is to determine the cause of a disease and, thus, a potential biological target. A target can be any biological entity from RNA to a protein to a gene that is "druggable" or accessible to binding with a drug-like compound. Drug interactions with a biological target change the shape or confirmation of some facet of the target when bound to a small molecule and alter the target's ability to function. This conformational change ideally triggers a desired biological response involved in the particular disease process being studied (Hughes et al. 2011). A target is usually identified before testing to discover drug-like compounds that alter a disease process. Zeroing in on a particular aspect of a target protein function and substrate interactions is known as a target-based approach and is typically applied when scientists have a thorough understanding of a disease process. A target may also be identified after a cell-based disease model is tested with drug-like compounds. This case is called a phenotypic cell-based screen since drug selection is based on a desired biological response without knowing the exact interaction or target that is responsible. Once the desired outcome against a disease is discovered, scientists can reverse engineer and test the individual cellular processes versus the hits to deduce the biological component most responsible for the disease. Often target-based biochemical approaches are implemented when there is a greater understanding of the disease, but phenotypic screening is important in diseases that are not yet completely understood. In both cases, target identification is critically important to understanding which drug-like compounds to use when testing for reactivity and to understanding the disease process. However, for the most high-priority diseases, both target-based screening and phenotypic-based HTS are often used in parallel with each other to help drive the drug discovery process faster allowing for a rapid analysis of both outcomes providing a synergistic drug discovery portfolio.

4.3   TARGET VALIDATION

In the best scenario, target validation often occurs *in vivo*. Using transgenic animals, scientists can ensure that their target affects the desired biological change. By knocking out a gene, for example, a biological target can be confirmed. Not only does the gene knockout method validate a target, but it also gives insight into how a drug should affect a disease. If a protein is over expressed, knocking out the genes associated with this protein may stop the disease process. This means that scientists will look for compounds that can stop/modulate the activity of a particular protein. This can also be done through the use of antibodies, dominant negative controls, antisense oligonucle-otides, ribozymes, and small-interfering RNAs (Sioud 2007). Antibodies are desirable as they can be specific to an antigen or a protein being studied. To validate an antibody's response, a scientist can profile protein expression patterns using Western blotting to determine whether or not the anti-body is bound to the desired antigen (Bordeaux et al. 2010). Dominant negative controls work by finding a mutant that will block function of a specific protein. By blocking function, the effect of inhibiting this protein can be examined in regard to the disease process (Sheppard 1994). Antisense oligonucleotides are pieces of nucleic acid that stop protein formation by either degrading mRNA, affecting the translational machinery, or rerouting a bound protein for degradation. The effect of removing that protein can then be studied in order to validate the target (Dias and Stein 2002). Ribozymes are RNA molecules that can act as enzymes and can be used to target the mRNA that synthesizes the protein being studied. Small-interfering RNA or si-RNA is a cost-effective method that uses RNA to specifically target the gene involved in the disease process. Targeting this gene allows for protein knockout or knockdown and, thus, target validation (Sioud 2007). All of these methods allow for study of specific proteins in the disease process by removing them or inhibiting their function. Conversely, many disease processes function by downregulation of a particular gene or protein expression. Hence, upregulation or turning on gene expression increases protein function, which may be the desired outcome and targets can be validated accordingly, that is, by looking at disease phenotype upon modulation and upregulation of the concordant transcription and transla-tion processes. Regardless of the method, if the protein is found to be important for the disease process, the target can be said to be validated. Thus, once the target is validated, an assay can be developed and be considered for HTS.

4.4   AUTOMATED versus NONAUTOMATED DRUG DISCOVERY

Before the automation of drug discovery utilizing HTS was invented, nonautomated techniques were used. These traditional methods, however, involved a much larger sample size, individual test tubes, and assays for each compound (Lee and Kerns 1999). However, what used to take a week to screen ~100–1,000 compounds by hand now can easily be done in 1 day with automated HTS methods. Since automation has been implemented, scientists have the opportunity to screen up to 1 million samples in a day. Automation of HTS would not be as opportunistic and fast without the assisting equipment and machinery (Figure 4.2).

There are three key pieces of automation required to do HTS: (a) bulk reagent dispensers, (b) compound transfer devices, and (c) plate readers. Of course, most assays require incubation steps, but the fundamental hardware pieces needed to perform automated HTS are these. The table below provides the examples and descriptions of the dispenser, compound transfer machine, and reader machines that are used at Scripps Florida and in other large-scale screening centers around the world, but there are many others as well. A few other companies that produce and vend their own similar machines are Genomic Novartis Foundation, HighRes Biosolutions, Labcyte, Inc., Beckman Coulter Diagnostics, and PerkinElmer, Inc. While it is rare to find fully automated large-scale screening plat-forms in an academic setting, several examples exist including those at the Scripps Florida, CALIBR, and NCATs. It is less rare to find examples in large pharmaceutical companies with Bristol–Myers Squibb, Amgen, Merck, and Forma Therapeutics all with substantial systems in place (Table 4.1).

FIGURE 4.2   Stäubli Robotic Arm in Scripps Florida.

TABLE 4.1
Examples of the Tools that Scripps Florida Uses to Assist in the Screening and Creation of Samples Every Day

Instrument	Image	Purpose
Dispenser: FRD		Dispense small volumes quickly and accurately into a plate
Compound Transfer: Pintool		Fastest and least expensive way to transfer small quantities
Reader: Tecan		Measures and analyzes plates by reading monochromator-based absorbance and fluorescence intensity

(Continued)

TABLE 4.1 (*Continued*)
Examples of the Tools that Scripps Florida Uses to Assist in the Screening and Creation of Samples Every Day

Instrument	Image	Purpose
Reader: Viewlux/envision		Measures and analyzes plates by reading luminescence, BRET, absorbance, fluorescence intensity, FP, time-resolved fluorescence (TRF), FRET, TR-FRET, AlphaScreen, and AlphaLISA
Reader: Cell Insight		Measures and analyzes plates using high-content analysis
Reader: FLIPR		Measures and analyzes plates by reading kinetic assays, fluorescence intensity, and FRET

4.5 ASSAY OPTIMIZATION FOR HTS

The process of HTS begins with standard 96-, 384-, or 1,536-well microtiter plates (Figure 4.3). Each well holds a particular compound, and every assay is tested against a protein (biochemical assay) or a whole cell (cell-based assay). Before the assay ever reaches the robotics system, the assay and all of its components must be prepared and optimized by the HTS scientists. The scientist or compound manager prepares the drug-like compounds in the chosen microtiter plate. Before the biological targets and drug-like compounds are tested, the biologists in the lab prepare solutions to be added to this assay plate. These solutions consist of many variables that may be an enzyme, cofactors, a substrate, and a mix of nutrients and buffers needed for cell survival and so on. Assay optimization is critically important to ensure that the developed assay works at a miniaturized level. Lower density test plates are used as a point of head-to-head comparison so that changes that affect the outcome can be understood. The appropriate positive and negative controls are always included on each individual test plate at an N of 24 per plate. An example of a positive control is using the IC100 concentration of a known pharmacologic inhibitor while the negative control might be vehicle

(a)   (b)   (c)

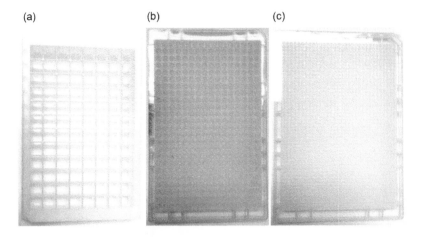

FIGURE 4.3   (a) 96-, (b) 384-, and (c) 1,536-well (from left to right) microtiter white plates.

only. There are literally hundreds of methods to use as a readout to determine the biological effect on an assay. Listing and explaining all of them would require an additional chapter, but you can think about all of them in a sense that they typically exploit the physical properties of light energy. An alternative to that is to use radioactive isotopes to determine a rise or fall in assay signal. Many times tests are done using fluorescence and luminescence tags to determine which assay is the most conducive to survival of the biological target and its activity. Here, HTS scientists want to maximize the Z′-factor which assesses the statistical reliability of the assays performance for HTS. A Z′-factor close to 1 is ideal, and any value above a 0.5 is considered a high-quality HTS assay. Once the assay is optimized, it can be prepared to enter into the robotics system so that reactivity between the drug-like compound and biological target can be measured. By using absorbance, fluorescence, luminescence, time-resolved fluorescence resonance energy transfer (TR-FRET), Flou-Pol, Ca^{++} detection, bioluminescence (BRET), high-content analysis, etc., HTS scientists can determine if there is activity between the biological target and drug-like compound. If activity is found, then the drug-like compound and biological target are said to be active.

4.6   HTS PROCESS OVERVIEW

The process begins with compound management. Here, compound plates with different chemicals are made that will ultimately be tested against a protein or cell and run in the robotics system. Compound managers not only make the compound plates but also ensure the purity, quality, and maintenance of the compounds used in the process. This includes not only the physical transfers of compounds but also the database which keeps track of each compound. Microtiter plates are used in this process. Each plate is standardized to be compatible with the robotics system and other automation in the lab. Within each plate, there are 96, 384, or 1,536 wells. Compound libraries in HTS labs range from hundreds of thousands in academia to millions in pharmaceutical companies. Every compound or compound plate is traced with a barcode to ensure that no mistakes are made in the plate making process. The dilution and transfer of compounds are aided by machines which evenly distribute compounds, verify these volumes, and speed up the process all the while limiting the chance for human error.

Compound source plates are delivered to the HTS scientists who test their proteins or cells against the diverse range of chemicals. Here, biological assays are also prepared in either 384, 1,536, or sometimes even 3,456-well microtiter plates depending on the level of miniaturization needed and the complimentary format of the compound plates. The approximate working volume of a standard 384 plate is 40–80 μL while the approximate working volume of a low-volume 384 plate is ~20 μL. The recommended working volume for 1,536-well plates is 10 μL, but assays can be run with as little as 3 μL. The 3,456-well plates, only available in commercial labs, use approximately 2 μL. Bulk dispensers allow quick dispensing of liquid. For 384 plates specifically, most often a peristaltic pump is used for continuous dispensing and can recycle reagents that are left in the tubing after the dispensing is complete. Alternatively, Inkjet dispensers, also known as flying reagent dispensers (FRDs), allow for small volumes to be dispensed quickly and accurately.

These interactions can be measured using spectrophotometric detection (fluorescence, absorbance, and luminescence) or other methods. Fluorescence is the most commonly used method. The biologists want to maximize the statistical resolution of hits from non-hits (Z-factor) and the signal-to-noise ratio. In other words, the scientists want an interaction that elicits a strong response which can be easily detected over the normal activity of a cell. Since HTS often deals with living cells, there will always be some activity in the process. If the hit is statistically found to be three standard deviations or more from the normal (i.e. non-hits), it is considered a statistically relevant hit. The hits from the primary HTS are cherry-picked and reformatted for retesting in triplicate. These source plates are then used in confirmation screens which are run in order to ensure that the hit is reproducible. Then, counter screens and secondary screens are tested against the hit to confirm that the hit is selective (non-promiscuous) and or is not cytotoxic. These secondary screens involve additional tests of the same chemical with the homologs to the protein or different cells to ensure that the results selectively hit its target. This step also helps rule out any readout artifacts. Once confirmed and selective molecules are identified, cherry-picking commences and a dose–response titration effort is also conducted in triplicate to determine the concentration–response curve and the IC_{50} or EC_{50} of each molecule, hence measuring the potency.

Technology and robotics are used throughout the process of HTS. In compound management, automation is used to store the compounds, make the plates, and verify that the plates are set up properly. In the following steps, the use of robotics aids in the distribution of cells or proteins as well as the incubation, tracking, and measuring of results such as fluorescence. This technology is the reason that HTS labs can output more data than most standard laboratory operations (Figure 4.4).

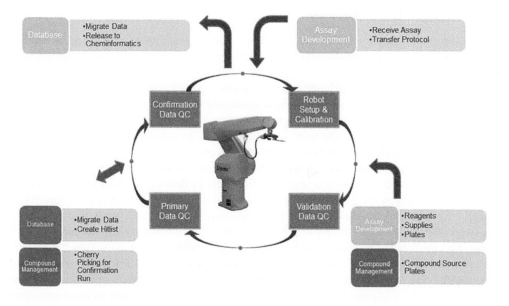

FIGURE 4.4 Example of HTS process in the Scripps Florida Lead Identification Lab.

4.7 COMPOUND MANAGEMENT

As mentioned, compound management is the first step in screening and provides core functionality for the solubilization, cherry-picking, storage, reformatting, transfer, and replication of compounds. These processes are supported by laboratory automation which allows work to be performed in microtiter plate densities up to 1,536-well format with transfer volumes as low as 500 nL. Typically, these processes are utilized to support medicinal chemistry, HTS, and assay development efforts for both in-house and external collaborator's projects.

At Scripps Florida, the compound management automation allows samples to be solvated and transferred between various labware types, densities, and formats. Common liquid-handling processes include performing compound titrations, reformatting compounds from lower to higher density labware, microtiter plate replication, compound transfer, and cherry-picking of compound libraries. All physical liquid-handling procedures are matched with appropriate data management processes to ensure the fidelity of data associated with all samples and labware which pass through the compound management facilities.

All processes utilize a barcode-based inventory system to ensure compound integrity throughout the life of the sample. This system tracks sample genealogies, liquid chromatography mass spectrometry quality control (LC-MS QC) reports, and sample availability. In addition, the inventory system has been integrated with sample storage automation which provides seamless sample access to end users and ensures samples are available on demand.

A proprietary setup based on Agilent's Rapid Resolution HT technology allows for fast high-performance liquid chromatography (HPLC) separations using small sample injections yet generating high-resolution chromatograms. For sampling compounds directly from microtiter plates, ≤1 μL of sample is needed for a 5-min separation and analysis time. Data collected includes UV and total ion chromatograms (TIC), full UV peak spectrum analysis, Atmospheric pressure chemical ionization (APCI), and ± electrospray (ES) mass spectral fragmentation analysis. Estimated compound purity as well as mass ion identification is automatically calculated using custom software, and key QC attributes (compound ID, structure, compound mass, and purity) are exported and stored in an HTS database.

4.8 ASSAY IMPLEMENTATION

While compound management is ongoing, HTS scientist diligently develop an assay or biological experiment that best corresponds with the particular target. This depends on the properties of the target.

A biochemical assay, or target-based approach, is one type of assay in which a biologist is targeting a specific interaction. This could include compound/drug interactions between a protein and a ligand such as receptors or enzymes or protein–protein interactions. Cell-based screening, on the other hand, targets living cells. The focus of a cell-based assay can differ. For example, a scientist might want to see whether there is growth or no growth on the cell as a whole or may want to test influxes of ions which would focus on secondary messenger signaling in the cell. A biochemical assay can be done after cell-based screening is complete to determine which protein in the cascade is most affected by the ligand (Martis 2010). One way to determine activity in either assay is through fluorescence detection (Baker 2010). Once activity can be consistently reproduced, then drug-like compounds can be added to see whether there is inhibition, increased activity, or no change. This process has been aided by the introduction of HTS as it has allowed biologists to run many tests at one time. Using standardized plates (e.g. 384 wells and 1,536 wells), assays can be tested for different reagent concentrations or cell quantities to determine the optimal environment as measured by the Z'-factor. Many variables can affect the outcome, such as incubation times; substrate fluorophore sensitivity; and pH of the buffers, proteins, and detergents in the buffers or medias, all of which are individually optimized to provide the best Z' and consistency well to well, plate to plate, and experiment to experiment. Finally, dimethyl sulfoxide (DMSO) is used as the universal solvent to prepare compounds which once in solution are at 75% DMSO and 25% water. Assays aren't always tolerant of high levels of DMSO which come as part of the test compound addition. So this is monitored via % DMSO tolerance determination which with appropriate HTS qualified assays and transfer devices is found to be ~2% DMSO tolerance or less, a condition very amenable to HTS. Assay development and assay optimization are critical for successful HTS and meaningful informatics.

4.9 IDENTIFYING A HIT

Drug discovery scientists must confirm activity in an assay in order to validate a "hit" as a statistically significant reaction. Many reactions will give false positives for reasons such as organic impurities, reader interference, and nonspecific binding (Hermann et al. 2013). Thus, secondary screens are conducted by HTS scientists to ensure that all activity is indeed due to an interaction between the drug-like compound and its biological target. HTS scientists also have to determine the statistical relevance of the activity that constitutes a hit. Hits are identified post screening in the HTS campaign. The hit can be identified using a fluorescence output or transient output, such as kinetic-based observations using image readers such as in the Molecular Devices FLIPR Tetra. Criteria must be laid out as to what value constitutes a hit. In some cases, a hit may be classified when the signal is at least three standard deviations from the average field values and/or negative controls (Schoenen et al. 2013). When a hit is identified, a confirmation screen is done. This confirmation screen establishes the reproducibility of the hit by repeating the test with the same compound.

As mentioned, the Z-factor is commonly used to determine the quality of high-throughput assay screens. A Z coefficient of 1 is ideal, and any value above 0.5 shows a high-quality assay. As shown in Equation 4.1, the Z-factor uses the means (μ) and standard deviations (σ) of both the positive and negative values. The Z coefficient reflects the variation of data as well as the signal strength. It is dimensionless and allows a uniform means for comparing the quality of HTS assays. The Z' prime factor, on the other hand, is used to find the quality of the assay development and its optimization. Using the same equation as Equation 4.1, the Z' equation replaces the mean and standard deviation with those measured from the control (Zhang 1999). Thus, Z' helps HTS scientists determine the quality of a particular assay for an individual experiment, and the Z-factor is an indicator of the overall quality of an HTS assay.

Equation for Z-factor used in HTS

$$\text{Z-factor} = 1 - \frac{3\left(\text{Standard deviation of sample} + \text{Standard deviation of control}\right)}{\left|\text{Mean of sample} - \text{mean of control}\right|} \tag{4.1}$$

4.10 HIT TO LEAD

Once a hit has been established, HTS scientists must advance the hits into leads. A drug-like compound is classified as a lead when it has been deemed selective for a particular pathway or protein and does not have adverse cytotoxic properties. Selectivity is important as many compounds will work on a particular protein but will also affect many other proteins in a cell. This means that the drug is not selective and may bring about side effects if progressed to *in vivo* testing. In addition, this counter screen often determines whether or not the compound is toxic to cells. The cytotoxicity of a drug-like compound is imperative to understanding whether or not a drug will be toxic to the cell and potentially the patient. The next step is to determine how different concentrations of the compound affect the biological activity. This is known as dose–response titration or concentration–response curve and is recorded on a semilog graph which gives a sigmoidal curve. A broad hillslope means that a slight increase of dosage has little effect on the activity of the drug.

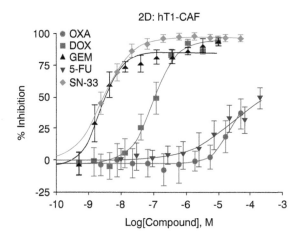

FIGURE 4.5   Five drugs with different EC50 and potencies within the same assay.

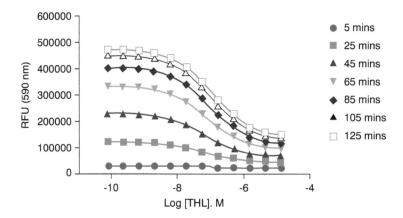

FIGURE 4.6   One drug over seven different time periods.

A steep hillslope means that a slight increase of dosage has a significant effect on the activity of the drug. The potency of the drug is determined by the EC50 which is the concentration at which the drug elicits 50% of its maximum response. A potent drug is one in which a small dosage is needed to elicit 50% of its maximal response. The maximum is the point at which increasing the dose of the drug no longer elicits any additional response. A higher maximum response is important in determining a compound's effectiveness or efficacy. A more effective compound will elicit a higher percent response at a particular dosage. The potency, efficacy, and safety of a drug can be observed using multiple assays and their respective dose–response curves, and this information is necessary to advance from a hit to a lead (Figures 4.5 and 4.6).

4.11   INFORMATICS

Bioinformatics is the science of organizing, managing, and analyzing large sets of data. In HTS labs, bioinformatics is critical in properly storing and analyzing all aspects of compound heredity through results from screening. Incorporating the fields of computer science and biology, bioinformatics allows scientists to store and access large amounts of data. Bioinformatics allows data mining, a computational process of discovering patterns in large data sets. Pharmaceutical companies have implemented programs to allow for rational and systematic screening of drug-like compounds against biological targets. Using data, scientists can determine which biological targets should be studied and what drug-like compounds would most likely affect this target. This process requires efficient technology and an organized data source but will become increasingly important as new screening techniques gather more and more data (Babu, 2000).

Genomics is the study of an organism's genome. Thanks to the Human Genome Project, scientists now have a comprehensive list of genes. This is important in identifying new targets for drug discovery (Emilien et al. 2000). Using the known genome, scientists can target a specific gene or genes that are most likely involved in a disease process. This process again incorporates additional technology into the field of biology as databases are used to categorize and analyze the genes. A genomic approach to identify potential drug targets is currently used against neglected/orphan diseases and inherited disorders with the hope that drugs can be quickly discovered and developed in a cost-effective manner against diseases that affect small populations.

Proteomics is the study of the protein products that come from certain genes as well as their function and interactions. Proteomics is important in drug discovery as it is most often an abnormality associated with a protein that contributes to the disease process. Thus, if a scientist knows the gene that controls the suspect protein, then drug-like compounds can be used to probe its biology in order to understand the function and interactions of the particular protein (Burbaum et al. 2002). All three of these fields combine science and technology to better organize data and allow scientists to make educated predictions of reactivity between drug-like compounds and biological targets.

4.12 MODULATING BIOLOGICAL RESPONSES: AGONISTS, ANTAGONISTS, INVERSE AGONISTS, PAMS, AND NAMS

An agonist, or activator, is a compound that stimulates activity higher than the basal or normal level by activating receptors. An antagonist, or inhibitor, on the other hand, takes this stimulated activity and lowers it. An antagonist does not lower the activity below the basal level since its role is to compete with the agonist. An antagonist by itself has no effect on the activity of a reaction. An antagonist can be competitive, noncompetitive, or uncompetitive. A competitive antagonist binds to the same site as the ligand; thus, it competes for the same binding site. A noncompetitive antagonist binds to a different site on the receptor. This changes the function or properties of that receptor and does not allow the natural ligand to bind. An uncompetitive antagonist cannot act on the receptor until a ligand has bound. Thus, a natural ligand or some agonist must already be activating the receptor for an uncompetitive ligand to inhibit activity. Binding of an agonist or antagonist can be either reversible or irreversible. When an agonist and reversible antagonist compete, the dose of the agonist must be increased in order to achieve the same activity. Adding additional agonist to an irreversible antagonist will have no effect as the antagonist will bind to the receptor permanently. If an irreversible agonist binds, no amount of antagonist can reverse this effect. An inverse agonist which is also an inhibitor is a compound that lowers activity below the basal level. This is done by binding to the same receptor sites as the agonist but eliciting an effect opposite to the agonist (Guzman 2014). In research, agonists, antagonists, and inverse agonists can be used to study specific interactions in assays; for example, compound/drug interactions between a protein and a ligand. However, in certain assays, development of and work with a certain ligand can be difficult. In these situations, additional compounds that react and bind allosterically to the protein are developed. Such compounds are referred to as allosteric modulators and can either inhibit or enhance ligand binding to the protein (Gregory et al. 2013).

Negative allosteric modulators (NAMs) and positive allosteric modulators (PAMs) are the two most common allosteric modulators. They are currently being used in assays with metabotropic glutamate receptors and muscarinic acetylcholine receptors and have potential utility for treatment of cognitive disorders, schizophrenia, Alzheimer's disease, and addiction (Gentry et al. 2010) (Figure 4.7).

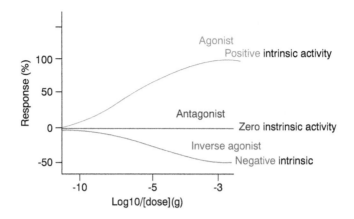

FIGURE 4.7 Comparison of agonist, antagonist, and inverse agonist.

4.13   TARGET-BASED VERSUS PHENOTYPIC-BASED HTS

We've already touched on this a bit, but target-based HTS involves understanding the biological process involved in the disease you are targeting. This information usually comes from animal disease models and clinical patient observation. With this understanding, a specific enzyme or receptor can be targeted to control the process. Phenotypic screening is valuable when little is known about the disease process. Phenotypic HTS involves introducing compounds into a cell-based system and measuring the response. Unlike the targeted approach, there is no identified biological component for the drug interaction but rather selection is based on a desired elicited response (e.g. cancer cell apoptosis, protein trafficking, receptor translocation). Thus, a process or pathway is tested instead of a specific targeted enzyme or receptor. Hence, any component along the process pathway that produces the sought results can result in potential therapeutic leads. After a compound is found to have a desired result, the process can be isolated to determine which step the compound effects and, thus, which biological target is the major contributor to the disease (Lowe 2012). In HTS, roughly 50%–60% of the assays are cell-based assays (Figures 4.8 and 4.9).

4.14   HOW ARE RESULTS DETERMINED?

Drug efficacy can be measured in HTS using absorbance, luminescence, and fluorescence. These three techniques are used to determine the optimal conditions for the assay and to measure the extent to which an interaction occurred between the compound and its biological target. Absorbance is the extent to which a sample absorbs light at a given wavelength. Luminescence occurs when excited electrons in a molecule fall to the ground state and energy is emitted in the form of light. The excitation of the electrons can occur after the molecule absorbs energy in the form of light or during a chemical reaction. Among the different types of luminescence, we find bioluminescence. It is found in some living organisms (e.g. fireflies) and occurs when light is emitted after a biochemical reaction within the organism. Fluorescence is another form of luminescence that allows some atoms and molecules to absorb light at a shorter wavelength which excites an electron to a higher state. When this electron returns to its ground state, it releases energy, in the form of light, at a longer wavelength. Absorbance, luminescence, and fluorescence measurements can be used to determine whether or not an interaction occurred between a drug-like compound and its biological target (Figure 4.10).

The Fluorescent Imaging Plate Reader (FLIPR) measures kinetic changes such as ion influxes simultaneously for each well. This allows for a quick reading of all wells, up to 1,536 at the same time, and the reading occurs in real time which limits the variable of time. This is also important

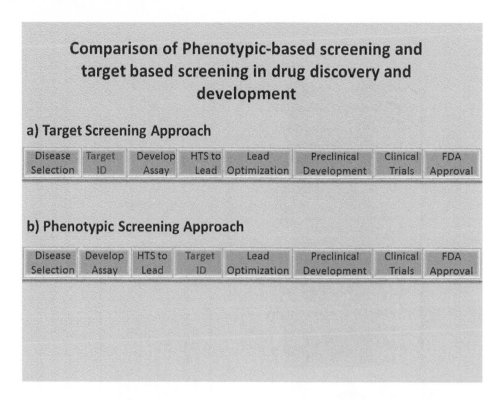

FIGURE 4.8 Difference in target identification between target-based screening and phenotypic-based screening in HTS.

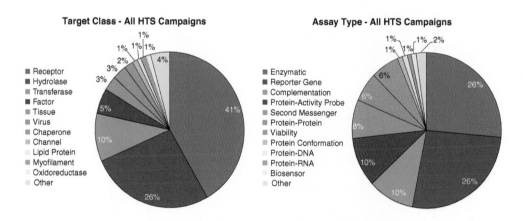

FIGURE 4.9 Pie charts illustrating division of HTS campaigns at the Scripps Research Institute in Florida.

as it can accurately measure processes such as calcium flux which occur quickly (FLIPR, 2011). Ion fluxes are measured via fluorescent tagging. These dyes are introduced to the assay so that intracellular ion changes can be measured such as calcium fluxes, an important signaling method within cells. These dyes are fluorescent ligands which attach to calcium. The signal emits light at a higher wavelength when calcium is bound. Thus, the wavelength at which light is absorbed can tell the HTS scientists the location of calcium in the assay giving information about the activity of the cell (Shuck 1997) (Figure 4.11).

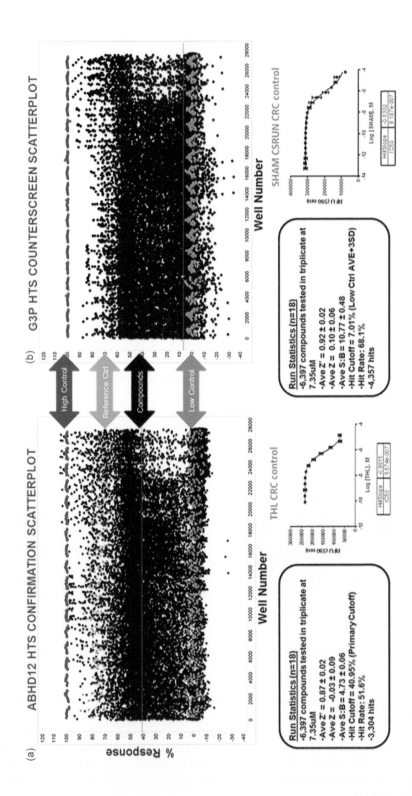

FIGURE 4.10 An example scatterplot of results from an assay. Each panel (a and b) of the figure is a scatterplot of the compounds tested. Each dot graphed represents the activity result of a well containing test compound (black dots) or controls (red, or green). The solid red line represents the cutoff for determining "hit" compounds. Overall screening statistics are described in the box below each scatterplot, with control concentration response curve (CRC) shown and discussed in the text.

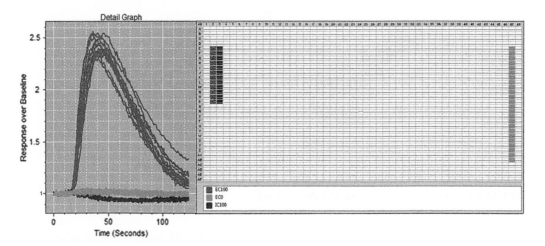

FIGURE 4.11 FLIPR readout for 1,536-well plate.

TR-FRET is used as a tag in many assays to determine activity. TR-FRET is based on the transfer of energy between two entities. Each part is labeled with a fluorescent marker. When the donor is excited, it releases energy which is picked up by the acceptor. This means that the donor and acceptor have to be close in proximity. This acceptor then emits an alternate yet specific fluorescence at a given wavelength. Thus, the interaction of proximal proteins can be measured. In practice, this assay occurs in the presence of buffers, proteins, lysate, or other components. Each of these also emits some fluorescence, but since the decay is in the nanosecond domain and the signal fluorophore is in the microsecond time domain, the signals can be separated by time, in other words, time-resolved (TR) as shown below. Thus, TR-FRET can be used to eliminate this quick measurement of fluorescence so that the target molecules alone can be measured (Figure 4.12).

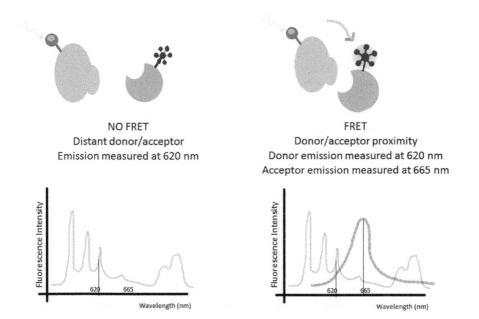

FIGURE 4.12 These two graphs illustrate that FRET measures the proximity of donor and acceptor and the output is fluorescence (Rathi 2014).

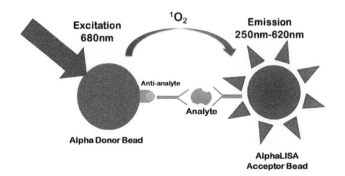

FIGURE 4.13   Demonstration of how AlphaScreen works.

Amplified Luminescent Proximity Homogeneous Assay (AlphaScreen) is a bead-based assay which is used to study biomolecular interactions in a microtiter plate format. There is both a donor bead and an acceptor bead in each assay. When a molecule is bound to the bead, an energy transfer occurs from one bead to another which produces a luminescent/fluorescent signal. When the donor bead is hit with a light at 680 nm, a photosensitizer in the donor bead converts oxygen to an excited state of O_2. In this excited state, the oxygen can diffuse 200 nm in solution. If an acceptor bead lies within this distance, it will accept that energy. The acceptor bead will then produce light at 520–620 nm. If there is no acceptor in that distance, the oxygen will simply lose that energy and it will fall back down to ground state without eliciting a response. Similar to TR-FRET, this is a method for measuring whether or not an interaction took place. The advantage of this technique is the removal of autofluorescent biologics and compounds that could create the discovery of false hits (Figure 4.13).

Fluorescence polarization (FP) is a testing method that allows analysis of activity with drug-like compounds and biological targets. This method stems from the theory that the degree of a fluorophore's polarization is inversely related to molecular rotation. Thus, it measures the difference between light intensity that is perpendicular and parallel to the light that was used to excite the fluorophore. FP is not used for measuring the concentration of fluorophore or the intensity of light but rather the difference between the planes of light. It can be used in a variety of interactions such as protein–protein, protein–DNA, and protein–ligand interactions as well as G protein-coupled receptors, nuclear receptors, and enzymes. FP is a favorable method as few, inexpensive reagents are necessary. This is in keeping with the low volume desired in HTS. It also does not destroy samples, so more than one reading can be conducted on one plate. With its wide range of application and accurate results, FP is popular in HTS as a method of analyzing activity between drug-like compounds and biological targets (Lea and Simeonov 2011).

Analytical techniques using MS, LC-MS, and nuclear magnetic resonance (NMR) are also useful in determining assay results. Each method has analytical figures of sensitivity, speed of analysis, selectivity, and cost-effectiveness that are desirable (Pereira and Williams 2007). LC-MS in particular is a method used in the Lead Identification Lab at Scripps Florida. In the quality control process, LC is utilized for purity detection, then MS spectra analysis is used for compound identification which is entered then into a searchable database as results (Figure 4.14).

4.15   MEDICINAL CHEMISTRY

Medicinal chemistry is the field that deals with designing and synthesizing new drug candidates. Medicinal chemists often work in a team with biologists, toxicologists, pharmacologists, and many other disciplines. This is necessary to ensure that the drug has a reasonable structure, interacts well

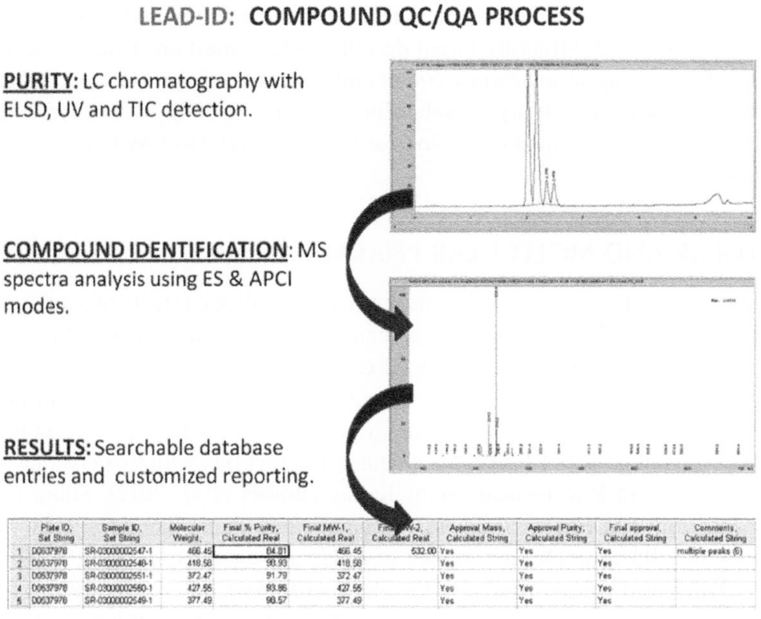

LEAD-ID: COMPOUND QC/QA PROCESS

PURITY: LC chromatography with ELSD, UV and TIC detection.

COMPOUND IDENTIFICATION: MS spectra analysis using ES & APCI modes.

RESULTS: Searchable database entries and customized reporting.

	Plate ID, Set String	Sample ID, Set String	Molecular Weight	Final % Purity, Calculated Real	Final MW-1, Calculated Real	Final MW-2, Calculated Real	Approval Mass, Calculated String	Approval Purity, Calculated String	Final approval, Calculated String	Comments, Calculated String
1	D0637978	SR-03000002547-1	466.45	64.81	466.45	532.00	Yes	Yes	Yes	multiple peaks (6)
2	D0637978	SR-03000002548-1	418.58	98.99	418.58		Yes	Yes	Yes	
3	D0637978	SR-03000002551-1	372.47	91.79	372.47		Yes	Yes	Yes	
4	D0637978	SR-03000002560-1	427.55	93.86	427.55		Yes	Yes	Yes	
5	D0637978	SR-03000002549-1	377.49	98.57	377.49		Yes	Yes	Yes	

FIGURE 4.14 A complete flowchart of the compound management process.

with the target protein, and is not toxic to cells. Medicinal chemists will synthesize novel compounds to be tested in HTS campaigns to discover any biological efficacy. Medicinal chemists also take a lead discovered from an HTS campaign and develop it to improve its chances of becoming a drug candidate (Lombardino and Lowe 2004). This field strives to find compounds that are easily synthesized and have adequate selectivity, potency, and efficacy to be a potential new drug candidate.

In medicinal chemistry, drug databases have become important. These libraries allow scientists to discover patterns between drug-like compounds. In the Comprehensive Medicinal Chemistry database, the molecular weight of drug-like compounds is typically found to be between 160 and 480 mw with 357 mw being the average, nonaromatic rings have been found to be more common than aromatic, and benzene rings have been found to be the most common structure in the database (Ghose et al. 1999). This information can be helpful in the synthesis and design of new drug candidates. Empirical rules based on common chemical properties currently found in approved drugs were also developed. These rules specify molecular mass, lipophilicity, hydrogen-bond donors, polar surface area, and hydrogen-bond acceptors (Lombardino and Lowe 2004). Although not every chemist and biology agrees with these rules, it is an important point of consideration when developing a drug-like compound library. This information can also help determine how a drug will react *in vivo*. Medicinal chemistry will continue to expand these databases to improve their understanding about drug-like compounds.

The goal of medicinal chemists is to make a compound that can be synthesized readily and have biological activity. Their role is important in the beginning of HTS as they develop many of the drug-like compounds that are initially tested against a biological assay. The difficulty with designing new molecules is that the structure must allow for selective interaction with the biological target while also having adequate efficacy and limited toxicity. Medicinal chemistry is also important after HTS in converting the leads discovered to drug candidates to be used in animal and human clinical testing. They can either synthesize a new molecule or use the existing compound and manipulate it. Changing one substituent on a compound can completely change its potency or interaction. Medicinal chemists must also make a compound that can be quickly and cost-effectively synthesized. Structure-based design is often based on structure–activity relationships or SARs. By looking at analogous compounds, a chemist can more accurately determine the compounds' efficacy and activity (Lombardino and Lowe 2004). SARs give insight into how a compound interacts with its intended biological target and how to resynthesize derivative compounds to improve it.

4.16   CELLULAR AND MOLECULAR PHARMACOLOGY

The purpose of *in vitro* drug activity is to determine potential side effects of a compound before it reaches living organisms. The compound is screened against a wide range of targets from receptors to ion channels to enzymes in an artificial environment outside of a living organism. This identifies any additional reactions that may occur when the drug candidate is introduced into an animal then possibly humans. One required assay screens for possible interaction with the hERG channel which is a voltage-gated potassium channel. Interactions with this channel are unwanted as these interactions can lead to heart arrhythmias (Bowes et al. 2012). Some side effects or adverse drug reactions (ADRs) are mild such as a rash, but others like heart arrhythmias can be life threatening and illustrate the importance of *in vitro* drug safety testing. This screening can also be done earlier in the drug discovery testing to eliminate potentially harmful drugs before additional time and money is spent. An *in vitro* study is also important before *in vivo* study can take place as it is a faster way to test for safety and can catch problems before animal testing begins. It also uses fewer compounds than *in vivo* studies, again saving time and money. *In vitro* studies strive to emulate *in vivo* processes which are important when understanding side effects during clinical trials.

4.17   PRECLINICAL DEVELOPMENT

With the high rate of failure in clinical trials due to limitations in efficacy or high toxicity, pharmacokinetics in preclinical development is important to ensure that only the strongest lead candidates proceed to the next stage of development. Pharmacokinetics essentially studies what the body does to the drug as it processes it. It studies how a drug is absorbed, distributed, metabolized, and excreted (ADME). Pharmacokinetics also studies the toxicity of a drug and its metabolism (ADMET) (Singh 2006). Some drugs fail not because of their function but because of the toxic nature of their byproducts. The study of pharmacokinetics usually begins with *in vitro* testing to ensure that the drug-like compound is safe enough to advance to animal testing. Next, ADMET can be studied in animals to ensure that it will not be toxic in humans and to determine a predicted dosage. Pharmacokinetics is also studied in clinical trials as absorption, distribution, metabolism, excretion, and toxicity may be slightly different in humans. Cytotoxicity screens are being used earlier in the process during lead optimization to encounter and hopefully solve problems well before animal testing or clinical trials (Zhang et al. 2012).

After preliminary *in vitro* testing, a compound can advance to *in vivo* testing. Because of their cost-effectiveness, rats are often the first line of *in vivo* testing. These animals also require low dosages of the compound. These experiments illustrate how quickly the drug is cleared from the system, safe dosages, and the breakdown of the compound. All of this is important to understanding the pharmacokinetics of the potential drug candidate (Zhang et al. 2012). Even with animal testing, some side effects may not be found due to the complexity of the human body. *In vivo* testing allows for additional understanding of how a drug functions with the wide and diverse range of targets in an organism, but animal testing will never give a complete understanding of interactions in the human body. Thus, many drugs that advance to clinical testing may not succeed in human trials.

4.18 SAFETY PHARMACOLOGY AND TOXICOLOGY

Safety pharmacology is a field that allows the use of basic pharmacology principles to determine a risk/benefit ratio. Toxicology studies the toxic effect of a drug while safety pharmacology also takes into account the side effects. Both fields strive to determine whether a drug is safe before it proceeds to clinical trials. Safety pharmacology uses both *in vitro* and *in vivo* testing to determine safety, but even then, some adverse effects may not be found until Phase III when the drug candidate is introduced to a large group of diverse patients. *In vitro* testing is used to target specific organs such as the liver which may be affected by the compounds as well as contribute to the breakdown of compounds in the body. Animal testing is used to determine how the compound will interact with a complex system (Pugsley et al. 2008). Toxicology helps to determine the therapeutic and toxic dosages of the particular compound while safety pharmacology weighs both the therapeutic benefits and the side effects to determine whether the drug is worth the risk. If the compound shows no substantial risk in preclinical trials, an application to the FDA can be submitted making the compound a drug candidate.

4.19 NEW FRONTIERS IN HTS

Along with the advancements in drug discovery, HTS techniques and research have become more advanced. Chemical library development, specifically, is one contribution towards the success and advancements in HTS. By providing more compounds to screen against targets, there is a larger potential for lead identifications and drug discoveries. In efforts to expand libraries, the Molecular Libraries Program (MLP) launched the Molecular Libraries Probe Production Centers Network (MLPCN) in 2004. This focused on advancing small molecular screening and bringing it into academic settings (Schreiber et al. 2015). Over the course of the program and with the help of multiple screening centers, the library grew to its final size of over 390,000 compounds. The molecular screening center at the Scripps Research Institute in Florida helped develop 14 out of the 18 licensed MLPCN probes as well as completed over 300 ultra-high-throughput screening (uHTS) campaigns. UHTS technologies are practiced now in efforts to more rapidly screen the large compound libraries, screening up to 100,000 samples per day (Hodder 2005/2006).

Additionally, high-content screening (HCS) has gained significant attention as another development in complex screening technologies. HTS provides a single read out of activity for individual assays, whereas HCS can provide data and measurements for multiple features of individual cells in assays all at once. HCS has the potential to both identify lead compounds and predict compound toxicity which has specific prominence in drug discovery (Buchser et al. 2014). Predicting toxicity and efficiency is extremely powerful and part of what makes HCS so complex.

More recently, research with HCS has even been enhanced with the introduction of 3D culture of cell lines as opposed to working with the traditional 2D monolayer cell cultures. Three dimensionsal (3D) spheroid models are being used in assays and in screening because scientists believe that they better simulate the cells *in vivo* physiological context (Kota et al. 2018). At Scripps Florida, a project was completed by looking at selective inhibitors and factors in the KRAS oncogene which would not have been possible without the technique of 3D cell culture. The 3D culture format was implemented with an HTS approach and found specifically to be more resistant to cytotoxicity and have higher IC50 (inhibitory concentration at 50%) values (Hou et al. 2018). Another project was completed by working with PANC-1 and hT_1 cell lines to compare the toxicity of 2D versus 3D cells. Using homogenous spheroids across an entire screening plate, the 3D and 2D assays presented significantly different responses to drug controls. The 2D tests generally demonstrated less efficacy supporting the idea of 3D culturing as a method of better representing cells in vivo (Kota et al. 2018) (Figures 4.15 and 4.16).

(a)   (b)

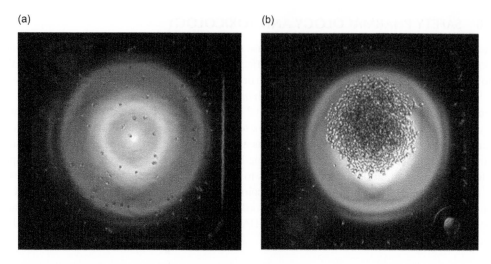

FIGURE 4.15 The difference between traditionally cultured 2D monolayer cells (a) and 3D cultured cell spheroids (b) under 4X microscope in individual microtiter plate wells.

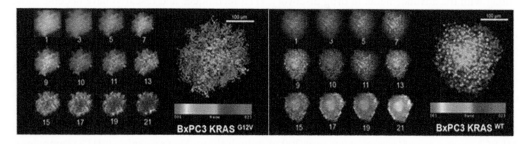

FIGURE 4.16 3D cultured cell spheroids of BxPC-3-KRASWT and BxPC-3-KRASG12V determined using light microscopy and confirmed by confocal microscopy of Hoechst-stained spheroids.

Additionally, HTS is being widely incorporated across different fields of scientific research. CRISPR is a recently discovered genome-editing technique allowing direct alterations of the eukaryotic genome. In efforts to advance CRISPR research, an HTS strategy has been proposed that allows parallel screening of multiple samples or clones (Bell et al. 2014). HTS technology when compared to previously used detection assays provides much more accurate data and estimates of activity due to its high sensitivity in detection (Sentmanat et al. 2018).

Increasing in its complexity, HTS can also be applied with label-free detection methods such as MS. These techniques are becoming more common as they are highly sensitive as well as extremely useful in ligand detection. Without using MS, false positives and negatives can be created because of molecular labels that can change or alter the binding between targeted proteins and ligands (Imaduwage et al. 2017). To avoid this, MS differentiates ligands based on masses and can be screened. Additionally, the complexity of MS-based HTS methods allows for a separation step that gives the potential for assessing more than one compound at a time.

A final frontier that is being investigated with the help of HTC is to implement primary human cells to produce induced neurons (iNs) for research (Rana et al. 2017). This alleviates the use of mouse primary neurons and takes us one step further towards humanized models for neurologic disorder's HTS. We are practicing this at Scripps, producing iNs from induced pluripotent stem cells and using them in functional screening assays to measure neurite outgrowth. Using both HTS and HCS, these iNs could be screened for toxicity and against multiple compounds for cell activity in a miniaturized, cost, and time-efficient throughput method (Figure 4.17).

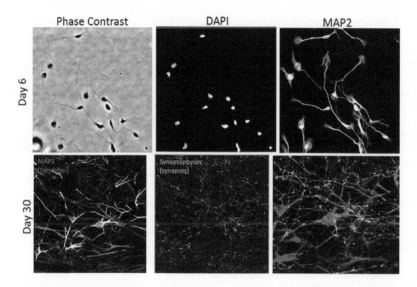

FIGURE 4.17 Representative images of immunocytochemical staining of iNs showing positive for dendritic marker MAP2, positive for pre-synaptic marker Synaptophysin at day *xx* of neuronal differentiation. In all images, nuclei were labeled by Hoechst and scale bar = *xx* um.

4.20 CONCLUSION

The drug discovery process is long, complicated, and expensive. Pharmaceutical companies can spend over $1 billion and 12 years in developing a new drug. In addition, very few drug candidates actually make it to market. The drug discovery process has changed dramatically with the implementation of new technologies such as genomics, proteomics, and HTS, as well as advancements in bioinformatics and *in silico* modeling. New information in the form of data will continue to expand as these *state-of-the-art* fields advance. This will be beneficial in every stage from bioinformatics in target identification to compound libraries and SARs in medicinal chemistry. Collaboration will be important in ensuring that these databases are kept current. Finding trends in targets and drug-like compounds is important for future research and for developing computer models to test potential interactions. One change that has already begun and will continue to advance is repurposing FDA-approved drugs. This is important for rare and neglected diseases as it would allow a low-cost solution to treat or lessen the symptoms of some diseases. Information is already available on the mechanism of action, safety, potency, and toxicity. Only tests concerning efficacy on a different target would need to be done. The drug discovery process will continue to adapt with new information and technology, but the basic principles will remain the same and the end goal of curing disease will remain at the heart of drug discovery.

REFERENCES

Babu, M. 2000. Integrating bioinformatics, medicinal sciences, and drug discovery. In *National Conference on Medical Informatics*. https://pdfs.semanticscholar.org/0d71/2c0590ff3a15cd68d002d80601f4c5750 49d.pdf

Baker, M. 2010. Academic screening goes high-throughput. *Nature Methods* 7 (10):787–92. doi: 10.1038/ nmeth1010-787.

Bell, C. C., G. W. Magor, K. R. Gillinder, and A. C. Perkins. 2014. A high-throughput screening strategy for detecting CRISPR-Cas9 induced mutations using next-generation sequencing. *BMC Genomics* 15:1002.

Bordeaux, J., A. Welsh, S. Agarwal, E. Killiam, M. Baquero, J. Hanna, V. Anagnostou, and D. Rimm. 2010. Antibody validation. *Biotechniques* 48 (3):197–209. doi: 10.2144/000113382.

Bowes, J., A. J. Brown, J. Hamon, W. Jarolimek, A. Sridhar, G. Waldron, and S. Whitebread. 2012. Reducing safety-related drug attrition: The use of in vitro pharmacological profiling. *Nature Reviews Drug Discovery* 11 (12):909–22. doi: 10.1038/nrd3845.

Buchser, W., M. Collins, T. Garyantes, R. Guha, S. Haney, V. Lemmon, Z. Li, and O. Joseph Trask. 2014. Assay development guidlines for image-based high content screening, high content analysis and high content imaging. In: *Assay Guidance Manual*, edited by G. Sitta Sittampalam, N. P. Coussens, K. Brimacombe, A. Grossman, M. Arkin, D. Auld, C. Austin, J. Baell, B. Bejcek, T. D. Y. Chung, J. L. Dahlin, V. Devanaryan, T. L. Foley, M. Glicksman, M. D. Hall, J. V. Hass, J. Inglese, P. W. Iversen, S. D. Kahl, S. C. Kales, M. Lal-Nag, Z. Li, J. McGee, O. McManus, T. Riss, O. Joseph Trask, J. R. Weidner, M. Xia, and X. Xu, 88. Bethesda, MD: Eli Lilly and Company and the National Center for Advanging Translational Sciences.

Burbaum, J. and G. M. Tobal. 2002. Proteomics in drug discovery. *Current Opinion in Chemical Biology* 6 (4):427–33. doi: 10.1016/S1367-5931(02)00337-X.

Dias, N. and C. A. Stein. 2002. Antisense oligonucleotides: Basic concepts and mechanisms. *Molecular Cancer Therapeutics* 1:347–55.

Emilien G., M. Ponchon, C. Caldas, O. Isacson, and J. M. Maloteaux. 2000. Impact of genomics on drug discovery and clinical medicine. *QJM* 93:391–423.

FLIPR. 2011. FLIPR screening. Aurelia Bioscience, accessed June 19th, 2014. www.aureliabio.com/dev2/specialised-technologies/flipr-screening/.

Gentry, P. R., M. Kokubo, T. M. Bridges, J. S. Daniels, C. M. Niswender, E. Smith, P. Chase, P. S. Hodder, H. Rosen, P. J. Conn, J. Engers, K. A. Brewer, M. R. Wood, and C. W. Lindsley. 2010. Development of the first potent, selective and CNS penetrant M5 Negative Allosteric Modulator (NAM). In: *Probe Reports from the NIH Molecular Libraries Program*. Bethesda, MD: National Center for Biotechnology Information (US).

Ghose, A. K., V. N. Viswanadhan, and J. J. Wedoloski. 1999. A knowledge based approach in designing combinatorial or medicinal chemistry libraries for drug discover. *Journal of Combinatorial Chemistry* 1 (1):55–68.

Gregory, K. J., E. D. Nguyen, S. D. Reiff, E. F. Squire, S. R. Stauffer, C. W. Lindsley, J. Meiler, and P. J. Conn. 2013. Probing the metabotropic glutamate receptor 5 (mGlu(5)) positive allosteric modulator (PAM) binding pocket: Discovery of point mutations that engender a "molecular switch" in PAM pharmacology. *Molecular Pharmacology* 83 (5):991–1006. doi: 10.1124/mol.112.083949.

Guzman, F. M. D. 2014. Pharmacodynamics animation: Full agonists, partial agonists, inverse agonists, competitive antagonishts, and irreversible antagonists. *Pharmacology Corner*, accessed June 17th. http://pharmacologycorner.com/pharmacodynamics-animation-full-agonists-partial-agonists-inverse-agonists-competitive-antagonists-and-irreversible-antagonists/?full-site=true.

Hermann, J. C., Y. Chen, C. Wartchow, J. Menke, L. Gao, S. K. Gleason, N. E. Haynes, N. Scott, A. Petersen, S. Gabriel, B. Vu, K. M. George, A. Narayanan, S. H. Li, H. Qian, N. Beatini, L. Niu, and Q. F. Gan. 2013. Metal impurities cause false positives in high-throughput screening campaigns. *ACS Medicinal Chemistry Letters* 4 (2):197–200. doi: 10.1021/ml3003296.

Hodder, P. 2005/2006. Building a uHTS laboratory. *Drug Discovery World* 7 (Winter Issue):31–42.

Hou, S., H. Tiriac, B. P. Sridharan, L. Scampavia, F. Madoux, J. Seldin, G. R. Souza, D. Watson, D. Tuveson, and T. P. Spicer. 2018. Advanced development of primary pancreatic organoid tumor models for high-throughput phenotypic drug screening. *SLAS Discovoery* 23 (6):574–84. doi: 10.1177/2472555218766842.

Hughes, J. P., S. Rees, S. B. Kalindjian, and K. L. Philpott. 2011. Principles of early drug discovery. *British Journal of Pharmacology* 162 (6):1239–49. doi: 10.1111/j.1476-5381.2010.01127.x.

Imaduwage, K. P., J. Lakbub, E. P. Go, and H. Desaire. 2017. Rapid LC-MS based high-throughput screening method, affording no false positives or false negatives, identifies a new inhibitor for carbonic anhydrase. *Scientific Reports* 7 (1):10324. doi: 10.1038/s41598-017-08602-w.

Kota, S., S. Hou, W. Guerrant, F. Madoux, S. Troutman, V. Fernandez-Vega, N. Alekseeva, N. Madala, L. Scampavia, J. Kissil, and T. P. Spicer. 2018. A novel three-dimensional high-throughput screening approach identifies inducers of a mutant KRAS selective lethal phenotype. *Oncogene*. doi: 10.1038/s41388-018-0257-5.

Lea, W. A. and A. Simeonov. 2011. Fluorescence polarization assays in small molecule screening. *Expert Opinion on Drug Discover* 6 (1):17–32. doi: 10.1517/17460441.2011.537322.

Lee, M. S. and E. H. Kerns. 1999. *LC/MS Applications in Drug Development*. Pennington, NJ: John Wiley & Sons, Inc.

Lombardino, J. and J. Lowe. 2004. The role of the medicinal chemist in drug discovery then and now. *Nature* 3 (10):853–62.

Lowe, D. 2012. In the Pipeline. In Corante. http://pipeline.corante.com/archives/2012/07/18/ the_best_rings_ to_put_in_your_molecules.php

Martis, E. 2010. High throughput screening: The hits and leads of drug discovery: An overview. *Journal of Applied Pharmaceutical Science* 1 (1):2–10.

Pereira, D. A. and J. A. Williams. 2007. Origin and evolution of high throughput screening. *British Journal of Pharmacology* 152 (1):53–61. doi: 10.1038/sj.bjp.0707373.

Pugsley, M. K., S. Authier, and M. J. Curtis. 2008. Principles of safety pharmacology. *British Journal of Pharmacology* 154 (7):1382–99. doi: 10.1038/bjp.2008.280.

Rana, P., G. Luerman, D. Hess, E. Rubitski, K. Adkins, and C. Somps. 2017. Utilization of iPSC-derived human neurons for high-throughput drug-induced peripheral neuropathy screening. *Toxicol in Vitro* 45 (Pt 1):111–8. doi: 10.1016/j.tiv.2017.08.014.

Rathi, A. 2014. Why there may be fewer truly new drugs hitting the market. *Epoch Times*, February 7th, 2014. www.theepochtimes.com/n3/494172-why-there-may-be-fewer-truly-new-drugs-hitting-the-market/.

Schoenen, F. J., W. S. Weiner, P. Baillargeon, C. L. Brown, P. Chase, J. Ferguson, V. Fernandez-Vega, P. Ghosh, P. Hodder, J. P. Krise, D. S. Matharu, B. Neuenswander, P. Porubsky, S. Rogers, T. Skinner-Adams, M. Sosa, T. Spicer, J. To, N. A. Tower, K. R. Trenholme, J. Wang, D. Whipple, J. Aubé, H. Rosen, E. Lucile White, J. P. Dalton, and D. L. Gardine. 2013. Inhibitors of the Plasmodium falciparum M18 aspartyl aminopeptidase. Probe Reports from the NIH Molecular Libraries Program. Bethesda, MD: National Center for Biotechnology Information (US).

Schreiber, S. L., J. D. Kotz, M. Li, J. Aube, C. P. Austin, J. C. Reed, H. Rosen, E. L. White, L. A. Sklar, C. W. Lindsley, B. R. Alexander, J. A. Bittker, P. A. Clemons, A. de Souza, M. A. Foley, M. Palmer, A. F. Shamji, M. J. Wawer, O. McManus, M. Wu, B. Zou, H. Yu, J. E. Golden, F. J. Schoenen, A. Simeonov, A. Jadhav, M. R. Jackson, A. B. Pinkerton, T. D. Chung, P. R. Griffin, B. F. Cravatt, P. S. Hodder, W. R. Roush, E. Roberts, D. H. Chung, C. B. Jonsson, J. W. Noah, W. E. Severson, S. Ananthan, B. Edwards, T. I. Oprea, P. J. Conn, C. R. Hopkins, M. R. Wood, S. R. Stauffer, K. A. Emmitte, and NIH Molecular Libraries Project Team. 2015. Advancing biological understanding and therapeutics discovery with small-molecule probes. *Cell* 161 (6):1252–65. doi: 10.1016/j.cell.2015.05.023.

Sentmanat, M. F., S. T. Peters, C. P. Florian, J. P. Connelly, and S. M. Pruett-Miller. 2018. A survey of validation strategies for CRISPR-Cas9 editing. *Scientific Reports* 8 (1):888. doi: 10.1038/s41598-018-19441-8.

Sheppard, D. 1994. Dominant negative mutants: Tools for the study of protein function *in vitro* and *in vivo*. *American Journal of Respiratory Cell and Molecular Biology* 11:1–6.

Shuck, P. 1997. Use of surface plasmon resonance to probe the equilibrium and dynamic aspects of interactions between biological macromolecules. *Annual Review of Biophysics and Biomolecular Structure* 26:541–66.

Singh, S. S. 2006. Preclinical pharmacokinetics: An approach towards safer and efficacious drugs. *Current Drug Metabolism* 7 (2):165–82.

Sioud, M. 2007. *Target Discovery and Validation Reviews and Protocols*, Vol. 1. Totowa, NJ: Humana Press Inc.

Zhang, D., G. Luo, X. Ding, and C. Lu. 2012. Preclinical experimental models of drug metabolism and disposition in drug discovery and development. *Acta Pharmaceutica Sinica B* 2 (6):549–61. doi: 10.1016/j.apsb.2012.10.004.

Zhang, J. H. 1999. A simple statistical parameter for use in evaluation and validation of high throughput screening assays. *Journal of Biomolecular Screening* 4 (2):67–73. doi: 10.1177/108705719900400206.

5 DNA-Encoded Compound Libraries

An Emerging Paradigm in Drug Hit Discovery

Raphael M. Franzini
The University of Utah

CONTENTS

5.1 INTRODUCTION

The development of new therapeutic agents critically depends on the availability of technologies that enable their effective development. A typical drug discovery workflow starts from studying the biology of a disease to identify prospective drug targets followed by the validation of these proteins. Subsequently, medicinal chemists search for molecules that bind to the target and modulate its activity and perform hit-to-lead optimization.[1] The discovery of chemical target binders is a key challenge in such endeavors, and technologies that are routinely used for hit discovery include high-throughput screening, fragment-based screening, and structure-guided drug discovery.[2] High-throughput screening has enabled the identification of potent hit molecules for structurally and functionally diverse protein targets,[3] but the high costs associated with screening and maintaining large compound collections limit its utility. Especially, high-throughput screens are generally performed only for proteins for which there already is compelling evidence of their pharmaceutical potential. As a result, medicinal chemistry efforts are strongly biased to a fraction of the proteome

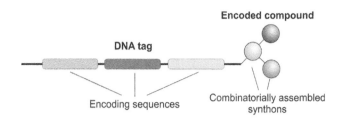

FIGURE 5.1   General structure of a DNA-encoded chemical library.

containing proteins known prior to the sequencing of the human genome, and a wide range of potentially medically relevant targets remain unexplored.[4,5] In contrast, discovery of affinity binders based on biomolecules (e.g., proteins, peptides, antibodies, RNA and DNA aptamers) is uncomplicated.[6] Streamlined affinity-selection protocols enable the discovery of such binders for most proteins in few weeks, at low costs, and using simple infrastructure. Undoubtedly, the availability of a screening technology that allowed sampling large libraries of synthetic compounds with the speed, cost-efficiency, and throughput of biological affinity screening methods would be of considerable interest to biomedical sciences. Encoding compounds with appended DNA sequences that define the structural information of the library compounds is a promising approach to achieve this goal.[7–9] Such DNA-encoded chemical libraries are collections of small molecules each modified with oligonucleotide tags enabling their identification (Figure 5.1).[7] The concept of using nucleic acids to encode chemical libraries was first proposed by Lerner and Brenner in a seminal theoretical paper in 1992.[10] The first examples of such libraries were reported soon thereafter in the form of DNA-encoded one-bead-one-compound libraries.[11,12] In the early 2000s, several approaches for generating DNA-encoded libraries were developed, and the possibility to identify affinity binders was demonstrated in proof-of-concept studies.[13–16] The development of high-throughput DNA sequencing technologies[17] tremendously boosted the potential of DNA-encoded libraries because it was suddenly possible to decipher millions of sequence reads simultaneously.[18,19]

With the main technical challenges overcome, DNA-encoded libraries have become a routine hit discovery method in pharmaceutical research, and most large pharmaceutical research companies, as well as an increasing number of academic research groups, routinely use DNA-encoded libraries to discover protein binders.[8,20] The growing use of DNA-encoded libraries has translated into a rapidly expanding number of screening hits identified from such libraries.[21,22] Many of the hits have intriguing properties including new chemotypes and unprecedented binding modes, including first-in-class compounds and allosteric regulators. Furthermore, screening protocols are becoming increasingly sophisticated. Complementing simple hit discovery efforts, parallel screening experiments with DNA-encoded libraries provide valuable information on protein binding and target selectivity. These advancements may open entirely new possibilities in drug discovery and will possibly disrupt current medicinal chemistry workflows.[23]

This book chapter summarizes major developments in the area of DNA-encoded libraries. A special focus is put on technical aspects such as methods for library encoding, on-DNA compound synthesis, selection protocols, and data analysis. Highlights in terms of discovered hits and prospective applications are outlined.

5.2   PREPARATION OF DNA-ENCODED CHEMICAL LIBRARIES

5.2.1   ASSEMBLY STRATEGIES

DNA offers unique properties as readable barcode for chemical libraries. Nucleic acids store information at high density and allow to conveniently amplify and read this information by PCR and DNA sequencing. The extensive toolbox of enzymes for processing nucleic acids greatly facilitates library assembly, and the polyanionic backbone imparts excellent solubility to conjugates of small molecules. The utility of nucleic acids to encode combinatorial compound libraries is well precedented by display technologies used for the discovery of biomolecules.[6] These methods use simple affinity-selection protocols combined with sequencing of the linked genome (or other coding nucleic acid) to discover target binders. Phage display is a prototypical example of such a technology. In phage display, bacteriophages are engineered to express polypeptides on the filamentous protein coat based on sequences engineered into the phage's genome.[24–26] Panning phages over immobilized targets are followed by amplification of the retained viruses in host organisms to provide binder-enriched libraries. Iterative screening cycles are repeated until the high-affinity phages can be identified by sequencing a representative number of genomes. Related technologies include ribosome display,[27] mRNA display,[28] and SELEX.[29,30] However, the dependence of these technologies on processes for transferring the genetic information into phenotypic proteins (i.e., transcription and translation) limits the value of such methods for medicinal chemistry.

The prospect of discovering low-molecular-weight protein binders with the efficacy, throughput, and economics of phage display motivated the development of several methods to assemble compound libraries with conjugated DNA barcodes.[31] A key challenge in the conception of DNA-encoded libraries (DECL) as a drug discovery technology was the invention of methods to link structures of library molecules to readable barcodes (Figure 5.2).[10] In contrast to biological display technologies, the synthetic nature of molecules in DNA-encoded libraries prevents the use of the biological transcription and translation machinery for library synthesis. Instead, it was necessary to develop means to (bio)chemically assemble such libraries.[31,32] DECL synthesis methods can be divided into three categories. One approach referred to as DNA-recorded synthesis achieves library assembly by iterative cycles of compound synthesis and encoding steps.[18,33] A second approach termed DNA-directed synthesis relies on mechanisms that convert DNA sequences into molecular structures.[14,16,34] The third approach harnesses hybridization to self-assemble sub-libraries of DNA-fragment conjugates into DNA-encoded libraries.[35,36]

A fundamental advantage of DNA-directed relative to DNA-recorded library synthesis is the conceptual possibility to iterate screening-amplification cycles because of the option to convert amplified sequences into molecular structures.[37] Such a translation step is incompatible with DNA-recorded library synthesis and, in case of self-assembled libraries, depends on the library design.[38] Initially, it was assumed that series of affinity-selection cycles would be required for effective hit discovery in large libraries. Tremendous advances in DNA sequencing technologies[17] made it, however, possible to achieve sufficient sequencing depth that a single affinity screening cycle (which may include several panning steps) is sufficient for hit identification obviating the need for information transfer from DNA to molecules.[18,19]

5.2.2 ENCODING MOLECULES WITH DNA

The most widely used approach for the preparation of DNA-encoded libraries is a split-and-pool synthesis scheme involving cycles of DNA elongation and chemical reaction steps (Figure 5.2a–d).[18,19] In these approaches, individual sequences with a chemical handle are modified in separate reaction vessels with a panel of chemical building blocks. Additional cycles of pooling, chemical modification, and encoding of the incoming molecular entities will provide the desired combinatorial library with appended DNA barcodes. Several methods are available for extending the coding oligonucleotides during library preparation, including (a) parallel solid-phase synthesis of oligonucleotides and compounds (Figure 5.2a), (b) enzymatic or chemical DNA ligation (Figure 5.2b,d), and (c) polymerase extension (Figure 5.2c).

Protocols for encoding molecules by DNA solid-phase synthesis were used in the earliest examples of combinatorial one-bead-one-compound libraries with readable DNA tags.[11,12] The approach was first reduced to practice in the synthesis of a library of peptides on 10-µm beads, in which the introduced amino acids were recorded by parallel synthesis of DNA strands using

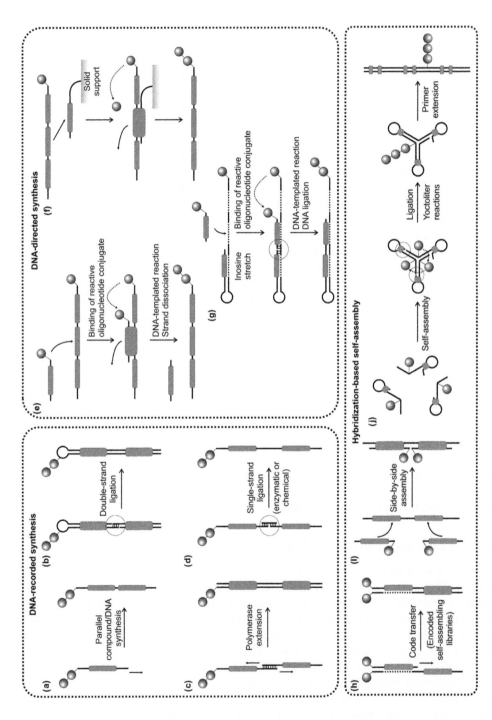

FIGURE 5.2 Strategies for encoding of DNA-encoded chemical libraries. (a)–(j) described in Section 5.2.2.

phosphoramidite chemistry.[11] The imperfect orthogonality of DNA and peptide chemistry[39] and the incompatibility of DNA with many organic reactions[40] constitute major limitations for encoding libraries by DNA synthesis. The same strategy was also implemented to synthesize compound libraries encoded with peptide nucleic acid[41] (PNA) tags.[42] Cleavage of PNA-compound conjugates prepared by parallel synthesis of the PNA-barcodes and the synthetic compounds can afford libraries for screening in solution or for assembly on DNA microarrays.[13] Relative to DNA, PNA chemistry offers a larger set of protecting group strategies that are orthogonal to peptide synthesis[43–45] and generally offers higher chemical inertness.[43] For example, depurination of DNA in acidic conditions is a limitation for DNA[40] but unproblematic for PNA.[43] The use of 7-deazapurines[11] and oligo-T strands[46] can provide acid resistance for at least the first synthesis cycle.

Enzymatic ligation is the prevalent method to introduce DNA codes during the preparation of DNA-recorded libraries. One approach is the enzymatic ligation of single-stranded DNA codes (Figure 5.2d). In single-strand ligation, a splint oligonucleotide juxtaposes the incoming coding sequence and the DNA-compound conjugate, and serves as a template for enzymatic strand joining by DNA ligase. Ligation requires pairs of adjoining 5′-phosphate and 3′-hydroxy termini. Removal of the template is possible by enzymatic or chemical digestion, or by chromatography.[47] Similarly, one can encode libraries by ligation of double-stranded DNA (Figure 5.2b).[19,48] Short overhang sequences ("sticky-ends") allow sequences to align for enzymatic ligation.[19,49] An advantage of this approach, in addition to obviating the need for a template sequence, is that double-stranded DNA provides protection for nucleobases from potentially damaging reagents during compound synthesis.[19] Ligation-based encoding has also been applied to one-bead-one-compound libraries.[50]

Chemical ligation has been introduced as an alternative for the synthesis of DNA-encoded chemical libraries.[51] This approach harnesses the well-established rate-enhancing effect of hybridizing two nucleic acids with adjoining reactive termini on a common strand to ligate them without the need of an enzyme.[52] For example, DNAs containing a highly nucleophilic phosphorothioate group at the 3′-end ligate spontaneously with oligonucleotides modified with a 5′-terminal leaving group (e.g., tosyl,[53] iodide[54]) when juxtaposed on a template sequence.[52] The similarity of the formed thiophosphate connection to the natural phosphodiester allows polymerases to read through such links.[55] Similarly, Cu(I)-catalyzed [3 + 2]-Huisgen-type cycloaddition can connect oligonucleotides with terminal azide and alkyne groups forming triazole linkers,[56] and certain polymerases can read through the resulting triazoles.[55,57,58]

Another method to encode chemical structures is to extend two partially hybridized DNA sequences by enzymatic replication (Figure 5.2c).[18] The incoming coding strand and the oligonucleotide linked to the synthetic compound have conserved non-coding sequences that form a short duplex, whereas the coding sequences remain single-stranded; a polymerase (e.g., Klenow fragment) bidirectionally extends the duplex establishing a fully double-stranded DNA.[18] Encoding additional fragments requires separation of the strands or invasion of the duplex. Encoding by polymerase extension can be combined with DNA ligation as exemplified in a recent study of a three-cycle library in which the first fragment was encoded in the original synthetic strand, the second code was introduced by single-strand DNA ligation, and the third code was introduced by polymerase extension.[47]

The programmable hybridization of DNAs allows for the self-assembly of DNA-encoded libraries, and several such strategies have been reported.[35] Melkko et al. first outlined DNA-encoded libraries consisting of non-covalently associated DNA-fragment conjugates.[15] This technology, termed "encoded self-assembling chemical libraries", relies on subsets of conjugates of single-stranded DNA with diverse fragments and which contain conserved sequence elements that guide library assembly by base-pairing.[15] Terminal overhangs on the DNA sequences allow for quantification of the compound-containing duplexes on DNA microarrays.[15,59] To be compatible with hit identification by DNA sequencing instead of microarrays, it is necessary that a single DNA sequence encodes the information for all combined fragments. The use of abasic spacer sequences is one possibility for the required inter-strand code transfer (Figure 5.2h).[60] Instead of direct pairwise hybridization, it is feasible to arrange DNA-appended fragments by adjacent binding to a carrier sequence (Figure 5.2i).[61] In this method, post-selection sequencing of the carrier is the readout for compound enrichment. Furthermore, amplification of the carrier sequence makes it possible to reassemble a binder-enriched library for repeat affinity screening experiments.[38] This possibility is especially appealing for PNA-encoded libraries because the transfers of molecular information from PNA-tags to DNA sequences open the possibility of using PCR amplification and DNA sequencing for analysis and overcome limitations in library size of PNA-encoded librairies.[38,62] In some approaches, PNA sequences protect DNAs from nuclease degradation followed by PCR amplification and readout.[62] The assembly of oligonucleotide-fragment conjugates directly on DNA microarrays enables the straightforward monitoring of protein binding.[42,63–65] Sideways binding of oligonucleotide conjugates further allows sampling the optimal distance between encoded fragments for target binding simply by varying the numbers of interjacent nucleobases.[66–68] Ligating self-assembled library members generates a single readable sequence containing all information for affinity selection and hit discovery.[69] An illustrative example is the Yoctoliter technology developed by Vipergen, in which hybridization-based self-assembly of reactant-oligonucleotide conjugates into DNA nanostructures occurs (Figure 5.2j).[70] The confinement of the reactants in this complex drives compound synthesis, whereas ligation of the strands and enzymatic replication leads to a readable double-stranded barcode.[70–71]

Two principal strategies have been disclosed for the preparation of DNA-encoded libraries by DNA-directed synthesis, relying either on DNA-templated chemical reactions[72] or on hybridization-based routing of intermediates.[37] In both cases, the original DNAs contain coding sequences to instruct the synthesis of the compounds. In the DNA-templated synthesis approach pioneered by the group of David Liu, the coding strands recruit sequence-defined DNA-conjugates of reactants and act as templates for chemical reactions (Figure 5.2e). In this way, a single DNA strand with several coding sequences can coordinate the multi-step synthesis of complex organic compounds.[14,34,72]

Diverse chemical reactions compatible with DNA-templated synthesis have been reported.[73–77] In principle, it is possible to add the different reactive conjugates simultaneously and to use hybridization to guide parallel reactions in a single batch[75]; however, as of today library synthesis generally involved performing the individual reactions separately to allow for better synthesis control and facile purification of the products.[34] One challenge with DNA-directed synthesis is the necessity of generating libraries of coding nucleic acid sequences. An elegant solution to this problem is to use a template carrier strand containing stretches of inosine nucleobases.[78] Inosine forms stable base pairs with all nucleobases.[79] During library synthesis, coding sequences bind to specific positions on the template strand pairing with the inosine stretches flanked by conserved sequences; linkage of the incoming DNA sequences to the coding sequence to the growing strand provides the coding stretches for DNA sequencing (Figure 5.2f).[79]

The second DNA-directed library synthesis strategy uses DNA-dependent routing invented by the group of Pehr Harbury.[16,37] Libraries of coding DNA strands are divided into separate compartments by hybridization to defined capture tags[80] for modification with diverse synthons (Figure 5.2g).[81] Elution of the conjugates,[81] pooling, and hybridization-based redistribution into reaction compartments will be repeated for translating the coding sequences into chemical structures.[16] Considerable refinement of technical protocols allows highly parallelized synthesis with large fragment sets using mesofluidic devices.[82,83]

In conclusion, diverse methods are available for linking chemical structures to encoding nucleic acids.[84] Most of these approaches have been successfully implemented for library synthesis and researchers interested in practicing this technology have several options for library assembly at their disposition. Each technology has advantages and disadvantages for specific applications. For example, DNA-recorded split-and-pool synthesis is arguably technically the simplest method and allows accessing large fragment sets. On the other hand, DNA-directed approaches may require larger upfront investments but ultimately allow synthesizing more complex molecules (e.g., macrocycles[14]) in higher quality. Understanding the intellectual property protection landscape for different technologies is important when establishing DNA-encoded libraries for commercial applications.

5.2.3 On-DNA Organic Synthesis

Another key challenge when linking chemical structures to genetic information is performing organic synthesis on DNA.[85] For DNA modification, reactions should be tolerant to at least traces of water. Furthermore, nucleic acids are multifunctional molecules susceptible to chemical damage and degradation.[40,86] Lability of the glycosidic bond to acidic conditions[87] limits the reaction scope, and nucleobases are sensitive to acylating, alkylating, oxidative agents, and transition metals.[86] Consequently, most organic reactions are incompatible with synthesis of such libraries.[40] Preparing DNA-encoded libraries further requires that reactions are high-yielding for a broad array of structurally diverse substrates and typically requires extensive optimization of reaction conditions.[46,88–91]

Although only a fraction of chemical transformations are applicable to DNA-encoded library synthesis, the available reactions provide chemists with sufficient versatility to access a vast chemical space of pharmaceutically relevant molecules (Figure 5.3).[92] Analysis of reported on-DNA reactions and their use in library preparation reveals that some classes of reactions and functional groups are privileged for DECL synthesis.[22,85] Most reported libraries made use of reactions between amines and electrophilic groups. Transformations include formation of amides, sulfonamides, and ureas by reactions of amines with activated carboxylic acids, sulfonyl chlorides and isocyanates, respectively.[18,19] Zr(IV)-catalyzed additions of amines to DNA-appended epoxides are also possible.[93] Reductive amination[89] and alkylation[94] as well as nucleophilic aromatic substitutions[94] are frequently used. Attachment of fragments via amide-bond formation is arguably the most prevalent chemical reaction in DNA-encoded library synthesis.[22] It can be implemented by either reacting activated carboxylic acids with amine-modified DNAs or by the reaction of amines with activated carboxylic acids on DNA. Amide-bond formation is pivotal for the synthesis of peptide-like libraries but is also the most widely used bond-forming reaction in medicinal chemistry.[95] A distinct benefit of amide-bond formation is that carboxylic acids and amines are prevalent classes of building blocks, which provides access to structurally diverse fragments at modest costs.[96] Typically, carboxylic acids are activated with DNA-compatible coupling reagents such as EDC or DMT-MM.[89,90] If available, N-hydroxysuccinimide esters are very effective in forming amide bonds on DNA.[97] The downside of these relatively mild reaction conditions is that for certain fragments, the reaction yields may be unsatisfactory.[89,90] Performing amide-bond formation on protected DNAs attached to the solid support used for library synthesis is one approach to increase yields at least for the first reaction cycle.[98] Such a reaction strategy, however, necessitates that the chemical structures are stable to DNA-deprotection conditions.[99] Despite the wide use of amide-bond formation in DECL synthesis, it has limitations. Reactions with secondary amines, especially sterically encumbered ones, and aromatic amines tend to be low-yielding, which precludes access to potentially valuable structures.[89,90] Reactions with other electrophiles (e.g., sulfonyl chlorides, isocyanates) are often used in combination with amide-bond–forming reaction to generate maximal diversity on amine functionalities.[19] Reductive amination is another privileged reaction for the synthesis of DNA-encoded libraries.[89,100] Reductive amination offers the combined benefits of high reaction yields and excellent DNA compatibility.[89] Aldehyde-modified DNAs are amenable to incorporation of diverse sets of amines including secondary and aromatic ones.[89,100] Inversely, amine-modified DNAs can be mono- or bis-alkylated depending on the amines and the experimental parameters.[94] An appealing feature of reductive amination is that formed secondary amines can serve as handles for attachment of electrophilic fragments (e.g., activated carboxylic acids) generating compact structures.[101] Nucleophilic aromatic substitution can also provide compact structures relevant to medicinal chemistry. This reaction was most extensively elaborated in the development of triazine-based libraries by sequential reactions of cyanuric chloride with amines.[19,102,103]

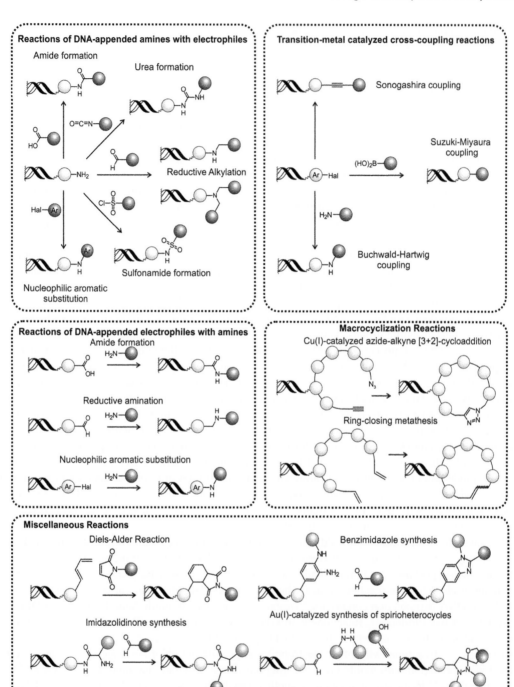

FIGURE 5.3   Summary of key reactions used for the synthesis of DNA-encoded chemical libraries.

Transition-metal–catalyzed bond-forming reactions compatible with DNA are of great interest for the synthesis of DNA-encoded chemical libraries.[104] Considering the extensive use of Cu(I)-catalyzed azide-alkyne [3 + 2] cycloaddition reactions in chemical biology,[105] it is not surprising that this reaction is useful for the synthesis of DNA-encoded libraries.[47,106] Yields for Cu(I)-catalyzed azide-alkyne [3 + 2] cycloaddition are generally high,[106] and the formed 1,2,3-triazoles are of interest to medicinal chemists.[107] The relatively small set of commercially available azides and alkynes is a limitation, and such building blocks tend to be expensive.[96] In situ preparation of these reactants from readily available haloalkanes[106] (azides) and aldehydes[94] (alkynes) is one approach to overcome this hurdle and is applicable to other hard-to-access building blocks.[33,94] Although the Cu(I)-catalyzed azide-alkyne cycloaddition is a prototypical example of a bioorthogonal reaction, careful control of reaction conditions is necessary to prevent damage to the DNA like for other DNA-conjugation reactions.[40,108] The impact of cross-coupling reactions on DNA-encoded library synthesis is rapidly growing. Especially, the on-DNA Suzuki–Miyaura reaction has received considerable attention because of the possibility to generate compact, drug-like structures from commercially available boronic acid building blocks.[58,94,100,109–112] Recent advances include, for example, catalyst systems suitable for modifying aryl chloride.[110] Sonogashira[94] and Stille[113,114] couplings on DNA have been reported on DNA. Brunschweiger and his group reported gold-catalyzed and diversity-generating reactions for synthesis of DNA-encoded chemical libraries.[46,88,115] Particularly intriguing is the possibility to generate DNA-encoded libraries based on palladium-catalyzed C–N coupling of amines because of the availability of these building blocks and the structural appeal of the formed molecules.[91] This on-DNA reaction has been reported and improved C–N cross-coupling protocols would boast tremendous potential for future library synthesis efforts.[91] A recent report described Ru-catalyzed alkene metathesis on DNA for the synthesis of DNA-encoded libraries.[116] This reaction could be especially useful for the synthesis of macrocycles by ring-closing metathesis.[116] Macrocycles are privileged scaffolds for targeting challenging proteins and are typically underrepresented in conventional high-throughput libraries.[117] Other macrocyclic DNA-encoded chemical libraries relied on Cu(I)-catalyzed azide-alkyne cycloadditons[118] or DNA-templated synthesis[14,34,119,120] for their preparation.

The programmable assembly of sets of chemical building blocks requires scaffolds and connector molecules.[22] Bifunctional molecules are often used to introduce both diversity and handles for subsequent modification. Amino acids are widely used in DNA-encoded library synthesis because they can be effectively coupled to DNA-appended amines, diverse structures are commercially available, and the deprotected amine can be capped with varied molecules. Effective DNA-compatible deprotection conditions for Fmoc,[19] Boc,[94] and alloc groups[94] among others on DNA are established. Protocols for the conversion of azides[47,121] and nitro groups[94,122] to amines have also been reported. Other examples of bifunctional building blocks are haloaryl-containing carboxylic acids[100,109,110] and formyl-carboxylic acids[123] for subsequent modification with boronic acids and amines. Trifunctional building blocks are more challenging to access and generally need custom synthesis.[112,124,125]

In addition to the mainstay reactions of amines with electrophiles and transition-metal–catalyzed reactions, there is a rapidly growing number of transformations that can be used for preparing DNA-encoded libraries. The use of Diels–Alder reactions was an early example of using unconventional building blocks for library preparation.[33,49] Reactions forming heterocycles (e.g., benzimidazole, imidazolinone) have received increasing attention because the formed structures often have drug-like structures.[94,122] Additional DNA-compatible reactions include the Knoevenagel, Mannich, and Wadsworth–Emmons reactions.[126–128]

In summary, the rapidly expanding scope of on-DNA organic synthesis allows accessing a vast chemical space of structurally complex molecules. A broad set of reactions for DNA-templated multi-step organic synthesis have further been reported.[73–75,77] In this approach, the proximity of two reagents on a DNA-template accelerates the rate of reactions and enables to perform transformations that would be challenging to perform in solution.[75] A disadvantage of DNA-templated multi-step synthesis is the need for preparing reactive oligonucleotide conjugates, which becomes prohibitive for large sets of synthons.

Although the DNA tag reduces the toolbox of synthetic chemists, the available transformations cover the key reactions used in medicinal chemistry. In fact, of the top ten reactions identified as the most widely used transformations in medicinal chemistry, nine can be implemented in a DNA-encoded chemical library context albeit at a reduced substrate scope.[95] Continued efforts to expand the range of DNA-compatible chemistry and to improve experimental protocols will further broaden the molecular horizon of drug discovery. One area of interest that remains underdeveloped is the use of enzymatic reactions in DNA-encoded chemical library synthesis. The feasibility of this approach was exemplified by the enzymatic synthesis of carbohydrates on DNA.[129]

5.3 DNA-ENCODED CHEMICAL LIBRARIES: FISHING IN A SEA OF CHEMICAL DIVERSITY

5.3.1 Screening Protocols

Encoding compounds with DNA sequences enables the deconvolution of library screening results with unparalleled ease.[10] In most embodiments, screening DNA-encoded libraries involves an affinity-selection protocol, in which conjugates with high target affinity are enriched for identification by DNA sequencing.[18,19] Typically, the DNA-encoded library is incubated with surface-immobilized target proteins, followed by washing steps to elute low-affinity binders and accumulation of affinity binders (Figure 5.4). Hits are recognized by PCR amplification of the codes and deciphered by sequencing. Quantitative PCR can be used to assess post-selection sequence content.[130] The technical details of this general protocol (e.g., choice of solid support, mode of protein immobilization) differ from research group to research group. Reported immobilization matrices include sepharose resins,[18] magnetic beads,[131] and coated pipette tips.[132] Methods for protein immobilization compatible with DECL screening protocols encompass chemical attachment to reactive supports (e.g., cyanogen bromide), binding of proteins with hexa-histidine tags to matrices displaying nitrile triacetic acid (NTA), and association of proteins modified with biotin or streptavidin-binding tags[133] to streptavidin-coated surfaces. Screening procedures can be automated,[20,134] and ready-to-use protocols have been reported.[131] Similar protocols are also effective for the discovery of covalent binders.[65,135] In fact, the possibility to stringently wash solid supports combined with the exceptional amplification power of PCR makes screening DNA-encoded libraries particularly effective for the discovery of irreversible inhibitors.[65,135,136]

Certain proteins are incompatible with conventional methods of target immobilization. Transmembrane proteins, many of which are important drug targets, are especially problematic for screening on surfaces. One possibility is to interrogate such proteins directly on the outer membrane of cells. Cell-based screening was first described by Svensen et al. for the discovery of peptides binding to cell-surface markers,[62,138] and researchers at GSK have used this approach to discover potent tachykinin neurokinin 3 receptors antagonists.[139] Achieving the high protein densities necessary for hit discovery, however, makes it necessary to carefully optimize the expression system and may be intolerable for many target proteins. Equally attractive is the possibility to embed membrane proteins in nanodiscs[140] for surface immobilization. Successful implementation of this method was reported for the discovery of negative allosteric modulators of the β_2-adrenergic receptor.[87]

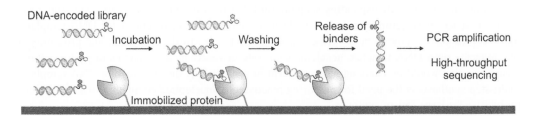

FIGURE 5.4 General protocol of DNA-encoded library screening with surface-immobilized proteins. (Reproduced with permission from Gentile et al.[137])

If these approaches prove generally applicable, screening of surface proteins will undoubtedly boast tremendous potential for drug discovery.

Another challenge with screening on solid matrices is that immobilization may alter the structure of target proteins or even unfold them, which in turn may result in false negatives and screening artifacts. It is therefore advisable to test the functionality of targets upon surface immobilization, for example, by enzyme assays. Screening DNA-encoded libraries in homogeneous solution would overcome the aforementioned problems, and several solution-phase screening methods have been proposed. Examples of methods that obviate immobilization of proteins make use of immuno-precipitation[141] and capillary electrophoresis[142] to separate conjugates from the library. Several methods rely on linking the encoded sequence to the target of interest.[31] Liu and co-workers developed interaction-dependent PCR to target proteins with DNA-encoded libraries directly in cell lysates.[143] Alternatively, the binding of DNA-conjugates to a protein can be used to link the two by photo-cross-linking[144,145] and reactive handles.[146] Rapid emulsification of mixtures of DNA-encoded libraries and target proteins has further been used to trap pairs of molecules with mutual affinity to identify binders.[71,147] Recently, dynamic DNA-encoded libraries have been developed for solution-phase screening.[148,149]

One of the earliest methods for screening DNA-encoded libraries relied on DNA microarrays either for post-selection screening analysis[15,150,151] or to assemble molecules on surfaces for direct interrogation with fluorophore-labeled proteins.[13,141,152] Although the use of microarray has been largely superseded by DNA sequencing,[18,19] they remain the method of choice for PNA-encoded libraries.[42] Microarrays have the disadvantage of limiting library sizes to <10^6 molecules, but they offer the benefit of rapid and inexpensive screening readout. Libraries of 10^5–10^6 encoded compounds are sufficient for many applications especially for focused libraries that target specific proteins of interest.[63] Similarly, protein binding can be detected by fluorescence staining on compound-displaying beads encoded with DNA.[11,153] A key limitation of bead-based technologies is the limited number of beads that can be assessed simultaneously using microscopy methods. Relative to other one-bead-one-compound library designs, tagging with DNA permits using significantly smaller beads (i.e., $10\,\mu m$), which translates into larger sampling sizes.[11,153] Fluorescence-activated bead sorting has been used for such libraries,[11] and sophisticated microfluidics systems are in development to further enhance sample throughput.[50,153–155]

In summary, several methods have been described and successfully reduced to practice for effectively identifying hit structures from DNA-encoded libraries.

5.3.2 Data Analysis

With increasing screening throughput, library sizes, and hit discovery rates, the streamlined analysis of the wealth of sequencing data becomes a main challenge. The first step involves the processing of crude DNA sequencing files into formats suitable for statistical interpretation (Figure 5.5). Erroneous sequences (e.g., containing misincorporated nucleotides) need to be filtered off, and typically, the data is converted into n-dimensional histograms in which sequence reads are binned into specific combination of fragments. Additionally, PCR codes used for amplification of the sequencing codes may contain identifier sequences that enable to unambiguously assign each sequence read to a particular screening experiment. In cases where libraries are pooled for screening, the coding sequences need to contain identification tags that allow assigning reads to particular libraries.[134] The resulting histograms can then be conveniently handled by standard data-processing software. Typically, screening data is visualized in scatter plots. In case of two-cycle libraries, visualization can be achieved in three-dimensional plots in which the x/y-axes define compound composition and the z-axis indicates the number of sequences.[131] Alternatively, two-dimensional plots may be used in which the sequence counts are visualized using the size and/or color of markers. Similar three-dimensional plots are possible for three-cycle libraries, whereas for libraries with four or more cycles, visualization becomes challenging and only parts of the libraries can be shown. An open-access

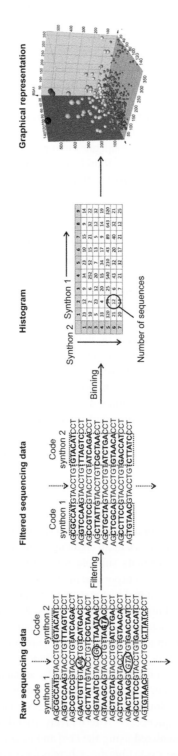

FIGURE 5.5 Workflow used for processing sequencing data obtained from DNA-encoded chemical library screens.

software useful for this purpose is DataWarrior[156] (www.openmolecules.org/datawarrior/index.html), whereas Spotfire appears to be most widely used in industry.[157,158] From the obtained data, hits will need to be defined. The simplest approach is to triage library members based on sequence counts. Alternatively, p-values derived from sequence counts relative to expected count distribution can be used. For large undersampled libraries, Poisson statistics was suggested,[19] and for smaller oversampled libraries, binominal statistic may be used.[159] Instead of looking for the most highly enriched compound, one may cluster the most prominent hits into clusters of structurally related compounds. Such clusters are frequently observed in DNA-encoded library screens. Typically, the top compounds of each cluster are synthesized and validated experimentally. A first step of such a process may be the elimination of compounds that show no sequence enrichment to simplify the data pool followed by hierarchical clustering of the compounds.[157] A challenge with clustering is to simultaneously take into account sequence enrichment patterns as well as structural similarity. The choice of descriptors of structural resemblance is another key aspect because popular Tanimoto distances may be suboptimal for combinatorial library molecules.

An exciting prospect of screening DNA-encoded libraries is the possibility of deriving a global perspective of the chemical ligand space of targets of interest. The value of such libraries extends beyond the simple identification of a set of hit compounds. Instead, the semi-quantitative readout for each library member (i.e., sequence counts) lends itself to chemoinformatic analysis to identify structural patterns and trends. The derived information may aid prioritizing hits and provide guidance for hit-to-lead development.

Molecular clusters provide facile access to structure-activity information such as preferred substitution patterns on aromatic rings or distances between target-binding fragments. Ideally, sequence counts correlate with target affinity. Technical problems that impede with this goal include heterogeneous distribution of conjugates, unequal synthesis yields, formation of truncated compounds and side products, PCR bias of sequence tags, different binding of DNA-appended and off-DNA molecules, and structural changes of proteins upon surface immobilization.[160] Correlations between sequence enrichment and binding affinity were observed for related molecules in screens for human serum albumin and tankyrase 1[98,101,161] but were absent in other studies.[103,162]

Sequence counts can also serve as quantitative parameters for comparing screening results for panels of proteins. Parallel screening of proteins should enable the discovery of hits that are specific to a defined target or alternatively pan-family inhibitors. This concept was first demonstrated using serum albumins of three different species showing that side-by-side comparison of screening results enabled the discovery of molecules that bound to a specific albumin or to all three tested ones.[161] By pharmaceutically blocking defined protein pockets during screening, one can discover molecules that target specific binding sites (e.g., allosteric pocket). Such an approach was used to identify Bruton's tyrosine kinase inhibitors which, depending on screening conditions, were either ATP-competitive or not.[163] Also comparing different isoforms, mutants, or post-translationally modified proteins may be possible with parallel library screening. Isoform-selectivity for some compounds was for instance achieved for inhibitors of PI3Kα from screening wild-type and mutant proteins in parallel.[164]

Quantitative analysis requires homogeneous libraries and sufficient sampling depths,[160] which is easier to achieve with two-cycle libraries than with multi-cycle libraries. The homogeneity of libraries decreases with number of cycles,[160] and very large libraries result in undersampling of libraries considering the limited number of reads available with current DNA sequencing technologies.[158,165]

Two philosophies are emerging with regard to library design and data analysis. One approach focuses on generating the maximal possible diversity by making large libraries with the goal of increasing the probability of discovering a hit while compromising on the ability to extract structural information from screening data. Screening such libraries is especially effective if the resources to synthesize and evaluate numerous hits are available. Hits based on truncated structures and even side products are acceptable as long as it is possible to elucidate their structures. In the second philosophy, the goal is to generate libraries and screening protocols to achieve the highest technically possible hit reliability. Such libraries tend to be smaller but will be more suitable for chemoinformatic data mining. Which strategy to pursue depends on the environment and available resources. Large research entities aiming to identify a hit for a limited number of high-value target will likely favor the first approach because it is associated with a higher chance of finding an ideal hit for lead development and because the resources for synthesizing and testing many molecules of hits are available. A small research group with limited resources will require a certain level of accuracy to economically map out binders for multiple targets and may favor the second approach.[166] Independent of these considerations, the development of new data analysis algorithms and innovative application of comparative screening protocols will undoubtedly be a pivotal area of future research in the area of DNA-encoded libraries. It will also be interesting to see how DNA-encoded library screens can be integrated with other approaches such as fragment-based screening, computational binding predictions, and structural biology approaches.

5.4 CHEMICAL SPACE OF ENCODED LIBRARIES

Ultimately, the value of a DNA-encoded chemical library is determined by its ability to provide hit compounds for a defined panel of targets of interests. Understanding the parameters that govern screening productivity is pivotal for the design of productive libraries.

5.4.1 COMBINATORIAL LIBRARIES DONE RIGHT

Intrinsically, every DNA-encoded chemical library has a combinatorial design in which sets of building blocks are assembled in combinatorial ways.[167] The library structure will depend on the number of cycles, the chemical reactions used for assembly, the size of the fragment sets, and ultimately the selected fragments.[22] Understanding the influence of these design parameters is essential for designing libraries with high screening productivity and the properties of the discovered hits.

5.4.2 THE EFFECT OF LIBRARY SIZE

It is intuitive to assume that the screening productivity of libraries increases with the number of encoded compounds in a library. Assuming that hit discovery is a stochastic process in which the probability of discovering a binder correlates with the number of testable hypothesis (i.e., potential compound–target interactions), one would indeed expect the screening success rate to correlate with library size. However, the combinatorial nature of most DNA-encoded libraries has important ramifications on the rules of library design.[22] First, the quantity of encoded compounds is primarily a function of the number of reaction cycles used for library assembly because of the exponential relationship of reaction cycles and combinatorial structures. Consequently, it is relatively straightforward to generate four-cycle libraries with >10,000,000 molecules, whereas two-cycle libraries of the same size would require sets of >10,000 synthons per cycle. Although both hypothetical libraries contain the same number of structures, it is obvious that the two-cycle library covers a larger chemical space and is structurally more diverse. From a technical perspective, the compounds in the two-cycle library will (assuming similar building blocks were used for synthesis) have preferable physicochemical properties with regard to peroral bioavailability because the molecular mass of library compounds increases with number of assembled building blocks.[22] Truncated compounds as hits are often observed in DECL screens, which can overcome size problems because such compounds have lower molecular weight relative to the encoded structure.[168]

Empirical data supports the argument that library size does not necessarily result in improved screening success. An analysis of reported screening studies revealed that three-cycle libraries tend to generate hits with the highest ligand efficiency,[22] although the conclusion is limited by the small dataset and because many hits from two-cycle libraries originated from academic libraries with relatively small synthon sets. An informative study analyzed the number of identified clusters of hits from a panel of structurally diverse DNA-encoded libraries with regard to two undisclosed targets.[168] No correlation between library size and number of discovered clusters was observed.[168] Indeed, the largest library consisting of four fragment sets and containing 8.1×10^{10} compounds afforded no binders, whereas two hit clusters were obtained with the smallest two-cycle library even though it was 67,500-fold smaller than the largest library.[168] Although two targets are insufficient to deduce general trends, the study provides strong evidence that the number of compounds in a library should not be regarded as the main parameter of library quality.

5.5 DNA-ENCODED LIBRARIES IN DRUG DISCOVERY

5.5.1 Overview of Reported DNA-Encoded Library Screening Hits

The use of DNA-encoded libraries has provided numerous hit compounds for protein targets that are structurally and functionally diverse (Figure 5.6).[8,22] Protein classes include kinases,[19,65,112,135,137,147,164,169,170] phosphate-ester hydrolyzing enzymes,[171,172] poly (ADP-polymerases),[47,98,101] sirtuin deacetylases,[173] proteases,[103,174–176] and G-protein–coupled receptors,[139,177,178] among others. A key study that for the first time revealed the potential of DNA-encoded chemical libraries to discover potent inhibitors to actual drug targets reporting the identification of nanomolar inhibitors of p38 and Aurora kinase.[19] In the past years, the number of disclosed hit compounds has grown rapidly,[8] and a comprehensive review of these molecules and their properties is beyond the scope of this book chapter. Importantly, with increasing refinement of the technology, the hits have improved in terms of ligand efficiency and physicochemical properties. While it was a major concern early in the initial phase of developing the technology that it would be challenging to discover drug-like molecules,[22,167] many reported hits have features suitable for lead development. Careful library design[101] and emergence of truncated hit molecules afforded compounds with desired properties (e.g., compliance with rule-of-five criteria[179]).

A benefit of DNA-encoded libraries is that the vast, structurally diverse, and chemically unbiased compound collections often provide hits with novel chemotypes and in some cases unprecedented binding modes. Illustrative examples are inhibitors of glycogen synthase kinase 3[137] and Bruton's tyrosine kinase that have unusual geometries of binding.[163] Additionally, DNA-encoded libraries can effectively provide binders to allosteric pockets as illustrated by the discovery of a negative allosteric regulator for the β_2-adrenergic receptor.[177] Screening in the presence of a competitor with known binding site can direct the discovery of molecules that bind to specific pocket as demonstrated for inhibitors of Bruton's tyrosine kinase.[163] The prospect of DNA-encoded libraries to afford bona fide lead compounds is shown by the advancement of at least two hits into clinical trials including inhibitors for soluble epoxide hydrolase[102,180] and receptor-interacting kinase 1.[181]

These examples clearly establish the potential of this technology for drug discovery. It is safe to assume that the number of compounds that have been identified with this method largely exceeds the published cases[20] and that the use of DNA-encoded libraries in pharmaceutical sciences will continue to grow rapidly.

5.5.2 Targeting Challenging Proteins

One exciting potential use of DNA-encoded chemical libraries is to target proteins with low chemical tractability. This prospect is based on the idea that DNA-encoded chemical libraries cover such a large chemical space that it is possible to find molecules to even proteins lacking well-defined small-molecule binding pockets. This hypothesis remains, however, to be convincingly proven, and there is evidence to the contrary. Study results indicated that screening productivity of DNA-encoded chemical libraries correlates with chemical tractability.[20,132] One study tested 29 antibiotic targets from *Staphylococcus aureus* comparing high-throughput screening and DNA-encoded library screening. Of the 22 targets for which high-throughput screening was unsuccessful, 17 also did not

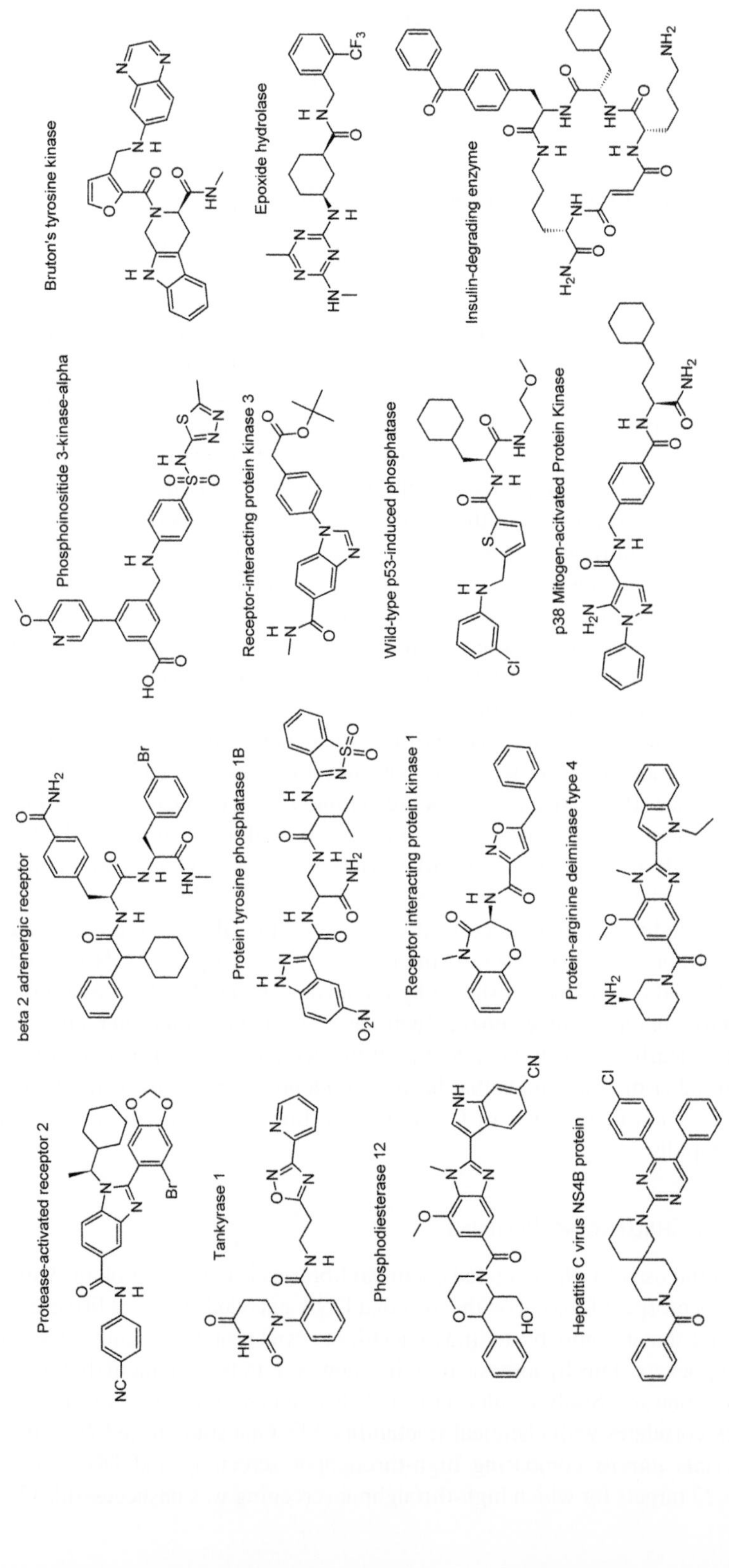

FIGURE 5.6 Representative examples of screening hits identified from DNA-encoded chemical libraries.

lead to hits with the used DNA-encoded library platform. In contrast, four of the seven proteins that provided hits with high-throughput screening were also active in DNA-encoded library screens. This result shows that (a) DNA-encoded libraries are of similar efficacy for the discovery of hits as high-throughput screens and (b) screening success correlated for the two methods. Inspired by the outcome, the authors used global sequence enrichment to predict chemical tractability and to prioritize targets for medicinal chemistry efforts.[132]

On the other hand, it may be possible to design DNA-encoded chemical libraries specifically to target proteins that lack good pockets for small-molecule binding. DNA-encoded libraries of 10^8–10^{11} compounds should in principle enable to cover significant portions of the relevant chemical space of molecules with >500 Da, which is impossible with conventional libraries. The large interaction surface of such molecules may be favorable for the discovery of binders of, for example, protein–protein interaction surfaces. Examples of successful hit discovery efforts with DNA-encoded chemical libraries include molecules that bind to tumor-necrosis factor α,[159] B-cell lymphoma extra large,[162] and interleukin 2,[182] and the Src homology domain 3 of proto-oncogen c-Crk.[16] Macrocycles have received special attention for this purpose because they combine multiple contact points with structural pre-organzation.[183,184] Several macrocyclic libraries have been disclosed[14,34,118–120,185] and afforded hits for challenging targets.[119,120,169,175,186] An illustrative example is the discovery of a potent and in vivo active macrocyclic inhibitor of insulin-degrading enzyme.[175] The evaluation of DNA-encoded chemical libraries for the discovery of binders to featureless surfaces and understanding of design parameters associated with screening success for such challenging targets will remain a focus area.

5.5.3 PORTABLE PROTEIN BINDERS

Modification with oligonucleotides has the disadvantage that the DNA tags may affect the target binding of encoded compounds. Conjugation to DNA restricts the possible orientations with which a molecule can bind to a protein. On the other hand, hits from DNA-encoded library screens provide the benefit of a readily identifiable site for chemical modification. It is therefore straightforward to convert DNA-encoded library hits for bioconjugation, surface immobilization, and modification with payloads.[187] An informative example is the discovery of portable binders of human serum albumin from a DNA-encoded library.[151] A 4-(4-iodophenyl)butyric acid fragment linked to lysine was discovered, and its affinity to human serum albumin was largely unaffected by conjugation to molecules including proteins.[151,188] This portable albumin binder (Albu-tag) has since found widespread use for the half-life extension of molecules with rapid renal clearance. For example, conjugation of Albu-tag to small tumor-homing molecules based on folic acid[189] or ligands of prostate-specific membrane antigen[190–192] resulted in enhanced pharmacokinetic properties. More recent DNA-encoded library screens provided molecules that bind human serum albumin with orders of magnitude higher affinities than Albu-tag,[47,98,161] but their utility for serum half-life extension remains to be evaluated in vivo.

Similarly, ligands of tumor-associated antigens can be modified with radionuclides for tumor imaging and radiotherapy or cytotoxic agents.[187] It was shown that DNA-encoded libraries can provide peptide-like molecules that bind to surfaces of specific cells.[62,138] Screening DNA-encoded libraries provided improved ligands for carbonic anhydrase IX,[60] a surface antigen ubiquitously expressed in clear cell renal cell carcinoma and absent from most healthy tissues.[60] The discovered ligand could be modified with fluorophores, cytotoxic agents, and radionuclides without significantly loosing affinity to the target protein. Intriguingly, these reagents showed excellent tumor uptake and high treatment efficacy as small-molecule cytotoxics and radiopharmaceuticals in murine cancer models.[60,193–195]

An area of emerging interest is the use of DNA-encoded chemical libraries to develop proteolysis-targeting chimeras (PROTACs). PROTACs are bivalent molecules that bind to a target protein and as well as effector ubiquitin ligase.[196] In this way, the ubiquitin ligase is brought into proximity of the target and inducing its degradation by polyubiquitination. A challenge with PROTAC is the discovery of molecules that bind to target proteins with high affinity and contain an inconspicuous attachment site. DNA-encoded chemical libraries are privileged for the discovery of such ligands, and it is utility for PROTACs will likely expand.

5.5.4 Encoded Libraries for Substrate Profiling

In addition to their utility for the identification of drug lead compounds, DNA-encoded chemical libraries will certainly find applications in other areas of chemical biology.[197] As one example, the use of such libraries for the discovery of enzyme substrates has been proposed. Libraries of protease[198] and kinase substrates[141] on DNA microarrays enabled the rapid discovery of substrates of members of these enzyme classes.[141,199] Recently, it was further shown that DNA encoding can be used to identify substrates with DNA-conjugates in solution[200,201] enabling larger library sizes. These applications hint at the tremendous untapped opportunities of such libraries in biomedical sciences.

5.6 CONCLUSIONS

Tremendous progress has been achieved in recent years in the aspiration to develop DNA-encoded chemical libraries as an enabling technology for drug discovery. Many of the key challenges have been successfully addressed, and the technology is now routinely used in drug discovery efforts as illustrated by the ever-growing number of hits discovered by screening DNA-encoded chemical libraries. These libraries have the potential to disrupt the standard medicinal chemistry workflow, and it will be exciting to see what new applications will be developed in the years to come.

REFERENCES

1. Lombardino, J. G. and Lowe, J. A., The role of the medicinal chemist in drug discovery--then and now, *Nat. Rev. Drug Discov.*, *3*, 853–862, 2004.
2. Janzen, W. P., Screening technologies for small molecule discovery: The state of the art, *Chem. Biol.*, *21*, 1162–1170, 2014.
3. Macarron, R., Banks, M. N., Bojanic, D., Burns, D. J., Cirovic, D. A., Garyantes, T., Green, D. V., Hertzberg, R. P., Janzen, W. P., Paslay, J. W., et al., Impact of high-throughput screening in biomedical research, *Nat. Rev. Drug Discov.*, *10*, 188–195, 2011.
4. Fedorov, O., Muller, S., and Knapp, S., The (un)targeted cancer kinome, *Nat. Chem. Biol.*, *6*, 166–169, 2010.
5. Edwards, A. M., Isserlin, R., Bader, G. D., Frye, S. V., Willson, T. M., and Yu, F. H., Too many roads not taken, *Nature*, *470*, 163–165, 2011.
6. Bradbury, A. R., Sidhu, S., Dubel, S., and McCafferty, J., Beyond natural antibodies: The power of in vitro display technologies, *Nat. Biotechnol.*, *29*, 245–254, 2011.
7. Franzini, R. M., Neri, D., and Scheuermann, J., DNA-encoded chemical libraries: Advancing beyond conventional small-molecule libraries, *Acc. Chem. Res.*, *47*, 1247–1255, 2014.
8. Goodnow, R. A., Jr., Dumelin, C. E., and Keefe, A. D., DNA-encoded chemistry: Enabling the deeper sampling of chemical space, *Nat. Rev. Drug Discov.*, *16*, 131–147, 2017.
9. Zimmermann, G. and Neri, D., DNA-encoded chemical libraries: Foundations and applications in lead discovery, *Drug Discov. Today*, *21*, 1828–1834, 2016.
10. Brenner, S. and Lerner, R. A., Encoded combinatorial chemistry, *Proc. Natl. Acad. Sci. U. S. A.*, *89*, 5381–5383, 1992.
11. Needels, M. C., Jones, D. G., Tate, E. H., Heinkel, G. L., Kochersperger, L. M., Dower, W. J., Barrett, R. W., and Gallop, M. A., Generation and screening of an oligonucleotide-encoded synthetic peptide library, *Proc. Natl. Acad. Sci. U. S. A.*, *90*, 10700–10704, 1993.
12. Nielsen, J., Brenner, S., and Janda, K. D., Synthetic methods for the implementation of encoded combinatorial chemistry, *J. Am. Chem. Soc.*, *115*, 9812–9813, 1993.

13. Winssinger, N., Harris, J. L., Backes, B. J., and Schultz, P. G., From split-pool libraries to spatially addressable microarrays and its application to functional proteomic profiling, *Angew. Chem. Int. Ed. Engl.*, *40*, 3152–3155, 2001.

14. Gartner, Z. J., Tse, B. N., Grubina, R., Doyon, J. B., Snyder, T. M., and Liu, D. R., DNA-templated organic synthesis and selection of a library of macrocycles, *Science*, *305*, 1601–1605, 2004.

15. Melkko, S., Scheuermann, J., Dumelin, C. E., and Neri, D., Encoded self-assembling chemical libraries, *Nat. Biotechnol.*, *22*, 568–574, 2004.

16. Wrenn, S. J., Weisinger, R. M., Halpin, D. R., and Harbury, P. B., Synthetic ligands discovered by in vitro selection, *J. Am. Chem. Soc.*, *129*, 13137–13143, 2007.

17. Heather, J. M. and Chain, B., The sequence of sequencers: The history of sequencing DNA, *Genomics*, *107*, 1–8, 2016.

18. Mannocci, L., Zhang, Y., Scheuermann, J., Leimbacher, M., De Bellis, G., Rizzi, E., Dumelin, C., Melkko, S., and Neri, D., High-throughput sequencing allows the identification of binding molecules isolated from DNA-encoded chemical libraries, *Proc. Natl. Acad. Sci. U. S. A.*, *105*, 17670–17675, 2008.

19. Clark, M. A., Acharya, R. A., Arico-Muendel, C. C., Belyanskaya, S. L., Benjamin, D. R., Carlson, N. R., Centrella, P. A., Chiu, C. H., Creaser, S. P., Cuozzo, J. W., et al., Design, synthesis and selection of DNA-encoded small-molecule libraries, *Nat. Chem. Biol.*, *5*, 647–654, 2009.

20. Arico-Muendel, C. C., From haystack to needle: Finding value with DNA encoded library technology at GSK, *Med. Chem. Commun.*, *7*, 1898–1909, 2016.

21. Lewis, H. D., Liddle, J., Coote, J. E., Atkinson, S. J., Barker, M. D., Bax, B. D., Bicker, K. L., Bingham, R. P., Campbell, M., Chen, Y. H., et al., Inhibition of PAD4 activity is sufficient to disrupt mouse and human NET formation, *Nat. Chem. Biol.*, *11*, 189–191, 2015.

22. Franzini, R. M. and Randolph, C., Chemical space of DNA-encoded libraries, *J. Med. Chem.*, *59*, 6629–6644, 2016.

23. Lerner, R. A. and Brenner, S., DNA-encoded compound libraries as open source: A powerful pathway to new drugs, *Angew. Chem. Int. Ed. Engl.*, *56*, 1164–1165, 2017.

24. Hammers, C. M. and Stanley, J. R., Antibody phage display: Technique and applications, *J. Invest. Dermatol.*, *134*, 1–5, 2014.

25. Kang, A. S., Barbas, C. F., Janda, K. D., Benkovic, S. J., and Lerner, R. A., Linkage of recognition and replication functions by assembling combinatorial antibody Fab libraries along phage surfaces, *Proc. Natl. Acad. Sci. U. S. A.*, *88*, 4363–4366, 1991.

26. McCafferty, J., Griffiths, A. D., Winter, G., and Chiswell, D. J., Phage antibodies: Filamentous phage displaying antibody variable domains, *Nature*, *348*, 552–554, 1990.

27. Hanes, J. and Pluckthun, A., In vitro selection and evolution of functional proteins by using ribosome display, *Proc. Natl. Acad. Sci. U. S. A.*, *94*, 4937–4942, 1997.

28. Wilson, D. S., Keefe, A. D., and Szostak, J. W., The use of mRNA display to select high-affinity protein-binding peptides, *Proc. Natl. Acad. Sci. U. S. A.*, *98*, 3750–3755, 2001.

29. Tuerk, C. and Gold, L., Systematic evolution of ligands by exponential enrichment: RNA ligands to bacteriophage T4 DNA polymerase, *Science*, *249*, 505–510, 1990.

30. Ellington, A. D. and Szostak, J. W., In vitro selection of RNA molecules that bind specific ligands, *Nature*, *346*, 818–822, 1990.

31. Shi, B., Zhou, Y., Huang, Y., Zhang, J., and Li, X., Recent advances on the encoding and selection methods of DNA-encoded chemical library, *Bioorg. Med. Chem. Lett.*, *27*, 361–369, 2017.

32. Mannocci, L., Leimbacher, M., Wichert, M., Scheuermann, J., and Neri, D., 20 years of DNA-encoded chemical libraries, *Chem. Commun.*, *47*, 12747–12753, 2011.

33. Buller, F., Mannocci, L., Zhang, Y., Dumelin, C. E., Scheuermann, J., and Neri, D., Design and synthesis of a novel DNA-encoded chemical library using Diels-Alder cycloadditions, *Bioorg. Med. Chem. Lett.*, *18*, 5926–5931, 2008.

34. Tse, B. N., Snyder, T. M., Shen, Y., and Liu, D. R., Translation of DNA into a library of 13,000 synthetic small-molecule macrocycles suitable for in vitro selection, *J. Am. Chem. Soc.*, *130*, 15611–15626, 2008.

35. Scheuermann, J. and Neri, D., Dual-pharmacophore DNA-encoded chemical libraries, *Curr. Opin. Chem. Biol.*, *26*, 99–103, 2015.

36. Pianowski, Z. and Winssinger, N., Nucleic acid encoding to program self-assembly in chemical biology, *Chem. Soc. Rev.*, *37*, 1330–1336, 2008.

37. Wrenn, S. J. and Harbury, P. B., Chemical evolution as a tool for molecular discovery, *Annu. Rev. Biochem.*, *76*, 331–349, 2007.

38. Daguer, J. P., Ciobanu, M., Alvarez, S., Barluenga, S., and Winssinger, N., DNA-templated combinatorial assembly of small molecule fragments amenable to selection/amplification cycles, *Chem. Sci.*, 2, 625–632, 2011.

39. Lu, K., Duan, Q. P., Ma, L., and Zhao, D. X., Chemical strategies for the synthesis of peptide-oligonucleotide conjugates, *Bioconjugate Chem.*, 21, 187–202, 2010.

40. Malone, M. L. and Paegel, B. M., What is a "DNA-compatible" reaction? *ACS Comb. Sci.*, 18, 182–187, 2016.

41. Nielsen, P. E. and Egholm, M., An introduction to peptide nucleic acid, *Curr. Issues Mol. Biol.*, 1, 89–104, 1999.

42. Daguer, J. P., Zambaldo, C., Ciobanu, M., Morieux, P., Barluenga, S., and Winssinger, N., DNA display of fragment pairs as a tool for the discovery of novel biologically active small molecules, *Chem. Sci.*, 6, 739–744, 2015.

43. Chouikhi, D., Ciobanu, M., Zambaldo, C., Duplan, V., Barluenga, S., and Winssinger, N., Expanding the scope of PNA-encoded synthesis (PES): Mtt-protected PNA fully orthogonal to fmoc chemistry and a broad array of robust diversity-generating reactions, *Chem. Eur. J.*, 18, 12698–12704, 2012.

44. Pothukanuri, S., Pianowski, Z., and Winssinger, N., Expanding the scope and orthogonality of PNA synthesis, *Eur. J. Org. Chem.*, 2008, 3141–3148, 2008.

45. Awasthi, S. K. and Nielsen, P. E., Parallel synthesis of PNA-peptide conjugate libraries, *Comb. Chem. High Throughput Screening*, 5, 253–259, 2002.

46. Skopic, M. K., Salamon, H., Bugain, O., Jung, K., Gohla, A., Doetsch, L. J., Dos Santos, D., Bhat, A., Wagner, B., and Brunschweiger, A., Acid- and Au(I)-mediated synthesis of hexathymidine-DNA-heterocycle chimeras, an efficient entry to DNA-encoded libraries inspired by drug structures, *Chem. Sci.*, 8, 3356–3361, 2017.

47. Favalli, N., Biendl, S., Hartmann, S., Piazzi, J., Sladojevich, F., Gräslund, S., Brown, P. J., Näreoja, K., Schüler, H., Scheuermann, J., et al., A DNA-encoded library of chemical compounds based on common scaffolding structures reveals the impact of ligand geometry on protein recognition, *ChemMedChem*, 13(13), 1303–1307, 2018.

48. Kinoshita, Y. and Nishigaki, K., Enzymatic synthesis of code regions for encoded combinatorial chemistry (ECC), *Nucleic Acids Symp. Ser.*, 34, 201–202, 1995.

49. Buller, F., Steiner, M., Frey, K., Mircsof, D., Scheuermann, J., Kalisch, M., Buhlmann, P., Supuran, C. T., and Neri, D., Selection of carbonic anhydrase IX inhibitors from one million DNA-encoded compounds, *ACS Chem. Biol.*, 6, 336–344, 2011.

50. MacConnell, A. B., McEnaney, P. J., Cavett, V. J., and Paegel, B. M., DNA-encoded solid-phase synthesis: Encoding language design and complex oligomer library synthesis, *ACS Comb. Sci.*, 17, 518–534, 2015.

51. Keefe, A. D., Clark, M. A., Hupp, C. D., Litovchick, A., and Zhang, Y., Chemical ligation methods for the tagging of DNA-encoded chemical libraries, *Curr. Opin. Chem. Biol.*, 26, 80–88, 2015.

52. Di Pisa, M. and Seitz, O., Nucleic acid templated reactions for chemical biology, *ChemMedChem*, 12, 872–882, 2017.

53. Herrlein, M. K., Nelson, J. S., and Letsinger, R., A covalent lock for self-assembled oligonucleotide conjugates, *J. Am. Chem. Soc.*, 117, 10151–10152, 1995.

54. Xu, Y. and Kool, E. T., A novel 5′-iodonucleoside allows efficient nonenzymatic ligation of single-stranded and duplex DNAs, *Tetrahedron Lett.*, 38, 5595–5598, 1997.

55. Litovchick, A., Clark, M. A., and Keefe, A. D., Universal strategies for the DNA-encoding of libraries of small molecules using the chemical ligation of oligonucleotide tags, *Artif. DNA PNA XNA*, 5, e27896, 2014.

56. Kukwikila, M., Gale, N., El-Sagheer, A. H., Brown, T., and Tavassoli, A., Assembly of a biocompatible triazole-linked gene by one-pot click-DNA ligation, *Nat. Chem.*, 9, 1089–1098, 2017.

57. El-Sagheer, A. H. and Brown, T., Synthesis and polymerase chain reaction amplification of DNA strands containing an unnatural triazole linkage, *J. Am. Chem. Soc.*, 131, 3958–3964, 2009.

58. Litovchick, A., Dumelin, C. E., Habeshian, S., Gikunju, D., Guie, M. A., Centrella, P., Zhang, Y., Sigel, E. A., Cuozzo, J. W., Keefe, A. D., and Clark, M. A., Encoded library synthesis using chemical ligation and the discovery of sEH inhibitors from a 334-million member library, *Sci. Rep.*, 5, 10916, 2015.

59. Melkko, S., Dumelin, C. E., Scheuermann, J., and Neri, D., Lead discovery by DNA-encoded chemical libraries, *Drug Discov. Today*, 12, 465–471, 2007.

60. Wichert, M., Krall, N., Decurtins, W., Franzini, R. M., Pretto, F., Schneider, P., Neri, D., and Scheuermann, J., Dual-display of small molecules enables the discovery of ligand pairs and facilitates affinity maturation, *Nat. Chem.*, 7, 241–249, 2015.

61. Winssinger, N., DNA display of PNA-tagged ligands: A versatile strategy to screen libraries and control geometry of multidentate ligands, *Artif. DNA PNA XNA*, *3*, 105–108, 2012.
62. Svensen, N., Diaz-Mochon, J. J., and Bradley, M., Decoding a PNA encoded peptide library by PCR: The discovery of new cell surface receptor ligands, *Chem. Biol.*, *18*, 1284–1289, 2011.
63. Barluenga, S., Zambaldo, C., Ioannidou, H. A., Ciobanu, M., Morieux, P., Daguer, J. P., and Winssinger, N., Novel PTP1B inhibitors identified by DNA display of fragment pairs, *Bioorg. Med. Chem. Lett.*, *26*, 1080–1085, 2016.
64. Daguer, J. P., Ciobanu, M., Barluenga, S., and Winssinger, N., Discovery of an entropically-driven small molecule streptavidin binder from nucleic acid-encoded libraries, *Org. Biomol. Chem.*, *10*, 1502–1505, 2012.
65. Zambaldo, C., Daguer, J. P., Saarbach, J., Barluenga, S., and Winssinger, N., Screening for covalent inhibitors using DNA-display of small molecule libraries functionalized with cysteine reactive moieties, *Med. Chem. Commu.*, *7*, 1340–1351, 2016.
66. Eberhard, H., Diezmann, F., and Seitz, O., DNA as a molecular ruler: Interrogation of a tandem SH_2 domain with self-assembled, bivalent DNA-peptide complexes, *Angew. Chem. Int. Ed. Engl.*, *50*, 4146–4150, 2011.
67. Gorska, K., Huang, K. T., Chaloin, O., and Winssinger, N., DNA-templated homo- and heterodimerization of peptide nucleic acid encoded oligosaccharides that mimick the carbohydrate epitope of HIV, *Angew. Chem. Int. Ed. Engl.*, *48*, 7695–7700, 2009.
68. Scheibe, C., Wedepohl, S., Riese, S. B., Dernedde, J., and Seitz, O., Carbohydrate-PNA and aptamer-PNA conjugates for the spatial screening of lectins and lectin assemblies, *ChemBioChem*, *14*, 236–250, 2013.
69. Cao, C., Zhao, P., Li, Z., Chen, Z., Huang, Y., Bai, Y., and Li, X., A DNA-templated synthesis of encoded small molecules by DNA self-assembly, *Chem. Commun.*, *50*, 10997–10999, 2014.
70. Hansen, M. H., Blakskjaer, P., Petersen, L. K., Hansen, T. H., Hojfeldt, J. W., Gothelf, K. V., and Hansen, N. J., A yoctoliter-scale DNA reactor for small-molecule evolution, *J. Am. Chem. Soc.*, *131*, 1322–1327, 2009.
71. Blakskjaer, P., Heitner, T., and Hansen, N. J., Fidelity by design: Yoctoreactor and binder trap enrichment for small-molecule DNA-encoded libraries and drug discovery, *Curr. Opin. Chem. Biol.*, *26*, 62–71, 2015.
72. Gartner, Z. J., Kanan, M. W., and Liu, D. R., Multistep small-molecule synthesis programmed by DNA templates, *J. Am. Chem. Soc.*, *124*, 10304–10306, 2002.
73. Gartner, Z. J., Kanan, M. W., and Liu, D. R., Expanding the reaction scope of DNA-templated synthesis, *Angew. Chem. Int. Ed. Engl.*, *41*, 1796–1800, 2002.
74. Sakurai, K., Snyder, T. M., and Liu, D. R., DNA-templated functional group transformations enable sequence-programmed synthesis using small-molecule reagents, *J. Am. Chem. Soc.*, *127*, 1660–1661, 2005.
75. Calderone, C. T., Puckett, J. W., Gartner, Z. J., and Liu, D. R., Directing otherwise incompatible reactions in a single solution by using DNA-templated organic synthesis, *Angew. Chem. Int. Ed. Engl.*, *41*, 4104–4108, 2002.
76. Oberhuber, M. and Joyce, G. F., A DNA-templated aldol reaction as a model for the formation of pentose sugars in the RNA world, *Angew. Chem. Int. Ed. Engl.*, *44*, 7580–7583, 2005.
77. Li, X., Gartner, Z. J., Tse, B. N., and Liu, D. R., Translation of DNA into synthetic N-acyloxazolidines, *J. Am. Chem. Soc.*, *126*, 5090–5092, 2004.
78. Li, Y., Zhao, P., Zhang, M., Zhao, X., and Li, X., Multistep DNA-templated synthesis using a universal template, *J. Am. Chem. Soc.*, *135*, 17727–17730, 2013.
79. Watkins, N. E. and SantaLucia, J., Nearest-neighbor thermodynamics of deoxyinosine pairs in DNA duplexes, *Nucleic Acids Res.*, *33*, 6258–6267, 2005.
80. Halpin, D. R. and Harbury, P. B., DNA display I. Sequence-encoded routing of DNA populations, *PLoS Biol.*, *2*, E173, 2004.
81. Halpin, D. R., Lee, J. A., Wrenn, S. J., and Harbury, P. B., DNA display III. Solid-phase organic synthesis on unprotected DNA, *PLoS Biol.*, *2*, E175, 2004.
82. Weisinger, R. M., Marinelli, R. J., Wrenn, S. J., and Harbury, P. B., Mesofluidic devices for DNA-programmed combinatorial chemistry, *PLoS One*, *7*, e32299, 2012.
83. Weisinger, R. M., Wrenn, S. J., and Harbury, P. B., Highly parallel translation of DNA sequences into small molecules, *PLoS One*, *7*, e28056, 2012.
84. Li, G., Zheng, W., Liu, Y., and Li, X., Novel encoding methods for DNA-templated chemical libraries, *Curr. Opin. Chem. Biol.*, *26*, 25–33, 2015.

85. Luk, K. C. and Satz, A. L., DNA-compatible chemistries. In: *A Handbook for DNA-Encoded Chemistry: Theory and Applications for Exploring Chemical Space and Drug Discovery*, Goodnow, R. A., Jr (ed). Wiley: Hoboken, NJ, 67–98, 2014.

86. Gates, K. S., An overview of chemical processes that damage cellular DNA: Spontaneous hydrolysis, alkylation, and reactions with radicals, *Chem. Res. Toxicol.*, 22, 1747–1760, 2009.

87. An, R., Jia, Y., Wan, B., Zhang, Y., Dong, P., Li, J., and Liang, X., Non-enzymatic depurination of nucleic acids: Factors and mechanisms, *PLoS One*, 9, e115950, 2014.

88. Skopic, M. K., Willems, S., Wagner, B., Schieven, J., Krause, A., and Brunschweiger, A., Exploration of a Au(I)-mediated three-component reaction for the synthesis of DNA-tagged highly substituted spiro-heterocycles, *Org. Biomol. Chem.*, 15, 8648–8654, 2017.

89. Franzini, R. M., Samain, F., Abd Elrahman, M., Mikutis, G., Nauer, A., Zimmermann, M., Scheuermann, J., Hall, J., and Neri, D., Systematic evaluation and optimization of modification reactions of oligo-nucleotides with amines and carboxylic acids for the synthesis of DNA-encoded chemical libraries, *Bioconjugate Chem.*, 25, 1453–1461, 2014.

90. Li, Y., Gabriele, E., Samain, F., Favalli, N., Sladojevich, F., Scheuermann, J., and Neri, D., Optimized reaction conditions for amide bond formation in DNA-encoded combinatorial libraries, *ACS Comb. Sci.*, 18, 438–443, 2016.

91. Lu, X., Roberts, S. E., Franklin, G. J., and Davie, C. P., On-DNA Pd and Cu promoted C-N cross-coupling reactions, *Med. Chem. Commu.*, 8, 1614–1617, 2017.

92. Kontijevskis, A., Mapping of drug-like chemical universe with reduced complexity molecular frame-works, *J. Chem. Inf. Model*, 57, 680–699, 2017.

93. Fan, L. and Davie, C. P., Zirconium(IV)-catalyzed ring opening of on-DNA epoxides in water, *ChemBioChem*, 18, 843–847, 2017.

94. Satz, A. L., Cai, J., Chen, Y., Goodnow, R., Gruber, F., Kowalczyk, A., Petersen, A., Naderi-Oboodi, G., Orzechowski, L., and Strebel, Q., DNA compatible multistep synthesis and applications to DNA encoded libraries, *Bioconjugate Chem.*, 26(8), 1623–1632, 2015.

95. Roughley, S. D. and Jordan, A. M., The medicinal chemist's toolbox: An analysis of reactions used in the pursuit of drug candidates, *J. Med. Chem.*, 54, 3451–3479, 2011.

96. Kalliokoski, T., Price-focused analysis of commercially available building blocks for combinatorial library synthesis, *ACS Comb. Sci.*, 17, 600–607, 2015.

97. Franzini, R. M., Biendl, S., Mikutis, G., Samain, F., Scheuermann, J., and Neri, D., "Cap-and-Catch" purification for enhancing the quality of libraries of DNA conjugates, *ACS Comb. Sci.*, 17, 393–398, 2015.

98. Franzini, R. M., Ekblad, T., Zhong, N., Wichert, M., Decurtins, W., Nauer, A., Zimmermann, M., Samain, F., Scheuermann, J., Brown, P. J., et al., Identification of structure-activity relationships from screening a structurally compact DNA-encoded chemical library, *Angew. Chem. Int. Ed. Engl.*, 54, 3927–3931, 2015.

99. Yuen, L. H. and Franzini, R. M., Stability of oligonucleotide-small molecule conjugates to DNA-deprotection conditions, *Bioconjugate Chem.*, 28, 1076–1083, 2017.

100. Ding, Y., Franklin, G. J., DeLorey, J. L., Centrella, P. A., Mataruse, S., Clark, M. A., Skinner, S. R., and Belyanskaya, S., Design and synthesis of biaryl DNA-encoded libraries, *ACS Comb. Sci.*, 18, 625–629, 2016.

101. Samain, F., Ekblad, T., Mikutis, G., Zhong, N., Zimmermann, M., Nauer, A., Bajic, D., Decurtins, W., Scheuermann, J., Brown, P. J., et al., Tankyrase 1 inhibitors with drug-like properties identified by screening a DNA-encoded chemical library, *J. Med. Chem.*, 58, 5143–5149, 2015.

102. Podolin, P. L., Bolognese, B. J., Foley, J. F., Long, E., Peck, B., Umbrecht, S., Zhang, X., Zhu, P., Schwartz, B., Xie, W., et al., In vitro and in vivo characterization of a novel soluble epoxide hydrolase inhibitor, *Prostaglandins Other Lipid Mediat.*, 104–105, 25–31, 2013.

103. Ding, Y., O'Keefe, H., DeLorey, J. L., Israel, D. I., Messer, J. A., Chiu, C. H., Skinner, S. R., Matico, R. E., Murray-Thompson, M. F., Li, F., et al., Discovery of potent and selective inhibitors for ADAMTS-4 through DNA-encoded library technology (ELT), *ACS Med. Chem. Lett.*, 6, 888–893, 2015.

104. Defrancq, E. and Messaoudi, S., Palladium-mediated labeling of nucleic acids, *ChemBioChem*, 18, 426–431, 2017.

105. Amblard, F., Cho, J. H., and Schinazi, R. F., Cu(I)-catalyzed Huisgen azide-alkyne 1,3-dipolar cycloaddition reaction in nucleoside, nucleotide, and oligonucleotide chemistry, *Chem. Rev.*, 109, 4207–4220, 2009.

106. Skopic, M. K., Bugain, O., Jung, K., Onstein, S., Brandherm, S., Kalliokoski, T., and Brunschweiger, A., Design and synthesis of DNA-encoded libraries based on a benzodiazepine and a pyrazolopyrimidine scaffold, *Med. Chem. Commu.*, 7, 1957–1965, 2016.

107. Agalave, S. G., Maujan, S. R., and Pore, V. S., Click chemistry: 1,2,3-triazoles as pharmacophores, *Chem. Asian J.*, 6, 2696–2718, 2011.
108. Abel, G. R., Jr., Calabrese, Z. A., Ayco, J., Hein, J. E., and Ye, T., Measuring and suppressing the oxidative damage to DNA during Cu(I)-catalyzed azide-alkyne cycloaddition, *Bioconjugate Chem.*, 27, 698–704, 2016.
109. Ding, Y. and Clark, M. A., Robust Suzuki-Miyaura cross-coupling on DNA-linked substrates, *ACS Comb. Sci.*, 17, 1–4, 2015.
110. Ding, Y., DeLorey, J. L., and Clark, M. A., Novel catalyst system for Suzuki-Miyaura coupling of challenging DNA-linked aryl chlorides, *Bioconjugate Chem.*, 27, 2597–2600, 2016.
111. Lercher, L., McGouran, J. F., Kessler, B. M., Schofield, C. J., and Davis, B. G., DNA modification under mild conditions by Suzuki-Miyaura cross-coupling for the generation of functional probes, *Angew. Chem. Int. Ed. Engl.*, 52, 10553–10558, 2013.
112. Deng, H., Zhou, J., Sundersingh, F. S., Summerfield, J., Somers, D., Messer, J. A., Satz, A. L., Ancellin, N., Arico-Muendel, C. C., Sargent Bedard, K. L., et al., Discovery, SAR, and X-ray binding mode study of BCATm inhibitors from a novel DNA-encoded library, *ACS Med. Chem. Lett.*, 6, 919–924, 2015.
113. Krause, A., Hertl, A., Muttach, F., and Jaschke, A., Phosphine-free Stille-Migita chemistry for the mild and orthogonal modification of DNA and RNA, *Chem. Eur. J.*, 20, 16613–16619, 2014.
114. Wicke, L. and Engels, J. W., Postsynthetic on column RNA labeling via Stille coupling, *Bioconjugate Chem.*, 23, 627–642, 2012.
115. Klika Skopic, M., Willems, S., Wagner, B., Schieven, J., Krause, N., and Brunschweiger, A., Exploration of a Au(I)-mediated three-component reaction for the synthesis of DNA-tagged highly substituted spiroheterocycles, *Org. Biomol. Chem.*, 15, 8648–8654, 2017.
116. Lu, X., Fan, L., Phelps, C. B., Davie, C. P., and Donahue, C. P., Ruthenium promoted on-DNA ring-closing metathesis and cross-metathesis, *Bioconjugate Chem.*, 28, 1625–1629, 2017.
117. Terrett, N. K., Drugs in middle space, *Med. Chem. Comm.*, 4, 474–475, 2013.
118. Zhu, Z., Shaginian, A., Grady, L. C., O'Keeffe, T., Shi, X. E., Davie, C. P., Simpson, G. L., Messer, J. A., Evindar, G., Bream, R. N., et al., Design and application of a DNA-encoded macrocyclic peptide library, *ACS Chem. Biol.*, 13, 53–59, 2018.
119. Seigal, B. A., Connors, W. H., Fraley, A., Borzilleri, R. M., Carter, P. H., Emanuel, S. L., Fargnoli, J., Kim, K., Lei, M., Naglich, J. G., et al., The discovery of macrocyclic XIAP antagonists from a DNA-programmed chemistry library, and their optimization to give lead compounds with in vivo antitumor activity, *J. Med. Chem.*, 58, 2855–2861, 2015.
120. Zhang, Y., Seigal, B. A., Terrett, N. K., Talbott, R. L., Fargnoli, J., Naglich, J. G., Chaudhry, C., Posy, S. L., Vuppugalla, R., Cornelius, G., et al., Dimeric macrocyclic antagonists of inhibitor of apoptosis proteins for the treatment of cancer, *ACS Med. Chem. Lett.*, 6, 770–775, 2015.
121. Chen, Y., Kamlet, A. S., Steinman, J. B., and Liu, D. R., A biomolecule-compatible visible-light-induced azide reduction from a DNA-encoded reaction-discovery system, *Nat. Chem.*, 3, 146–153, 2011.
122. Du, H. C. and Huang, H., DNA-compatible nitro reduction and synthesis of benzimidazoles, *Bioconjugate Chem.*, 28, 2575–2580, 2017.
123. Fernandez-Montalvan, A. E., Berger, M., Kuropka, B., Koo, S. J., Badock, V., Weiske, J., Puetter, V., Holton, S. J., Stockigt, D., Ter Laak, A., et al., Isoform-selective ATAD2 chemical probe with novel chemical structure and unusual mode of action, *ACS Chem. Biol.*, 12, 2730–2736, 2017.
124. Encinas, L., O'Keefe, H., Neu, M., Remuinan, M. J., Patel, A. M., Guardia, A., Davie, C. P., Perez-Macias, N., Yang, H., Convery, M. A., et al., Encoded library technology as a source of hits for the discovery and lead optimization of a potent and selective class of bactericidal direct inhibitors of Mycobacterium tuberculosis InhA, *J. Med. Chem.*, 57, 1276–1288, 2014.
125. Estevez, A. M., Gruber, F., Satz, A. L., Martin, R. E., and Wessel, H. P., A carbohydrate-derived trifunctional scaffold for DNA-encoded libraries, *Tetrahedron Assym.*, 28, 837–842, 2017.
126. Tran-Hoang, N. and Kodadek, T., Solid-phase synthesis of β-amino ketones via DNA-compatible organocatalytic mannich reactions, *ACS Comb. Sci.*, 20, 55–60, 2018.
127. Tian, X., Basarab, G. S., Selmi, N., Kogej, T., Zhang, Y., Clark, M. A., and Goodnow, R. A., Development and design of the tertiary amino effect reaction for DNA-encoded library synthesis, *Med. Chem. Commu.*, 7, 1316–1322, 2016.
128. Pels, K., Dickson, P., An, H., and Kodadek, T., DNA-compatible solid-phase combinatorial synthesis of β-cyanoacrylamides and related electrophiles, *ACS Comb. Sci.*, 20, 61–69, 2018.
129. Thomas, B., Lu, X., Birmingham, W. R., Huang, K., Both, P., Reyes Martinez, J. E., Young, R. J., Davie, C. P., and Flitsch, S. L., Application of biocatalysis to on-DNA carbohydrate library synthesis, *ChemBioChem*, 18, 858–863, 2017.

130. Li, Y., Zimmermann, G., Scheuermann, J., and Neri, D., Quantitative PCR is a valuable tool to monitor the performance of DNA-encoded chemical library selections, *ChemBioChem*, *18*, 848–852, 2017.

131. Decurtins, W., Wichert, M., Franzini, R. M., Buller, F., Stravs, M. A., Zhang, Y., Neri, D., and Scheuermann, J., Automated screening for small organic ligands using DNA-encoded chemical libraries, *Nat. Protoc.*, *11*, 764–780, 2016.

132. Machutta, C. A., Kollmann, C. S., Lind, K. E., Bai, X., Chan, P. F., Huang, J., Ballell, L., Belyanskaya, S., Besra, G. S., Barros-Aguirre, D., et al., Prioritizing multiple therapeutic targets in parallel using automated DNA-encoded library screening, *Nat. Commun.*, *8*, 16081, 2017.

133. Keefe, A. D., Wilson, D. S., Seelig, B., and Szostak, J. W., One-step purification of recombinant proteins using a nanomolar-affinity streptavidin-binding peptide, the SBP-Tag, *Protein Expr. Purif.*, *23*, 440–446, 2001.

134. Roman, J. P., Haro, R., de Blas, J., Jessop, T. C., and Castanon, J., Design and development of a technology platform for DNA-encoded library production and affinity selection, *SLAS Discov.*, *23* (5), 387–396, 2018.

135. Zimmermann, G., Rieder, U., Bajic, D., Vanetti, S., Chaikuad, A., Knapp, S., Scheuermann, J., Mattarella, M., and Neri, D., A specific and covalent JNK-1 ligand selected from an encoded self-assembling chemical library, *Chem. Eur. J.*, *23*, 8152–8155, 2017.

136. Daguer, J. P., Zambaldo, C., Abegg, D., Barluenga, S., Tallant, C., Muller, S., Adibekian, A., and Winssinger, N., Identification of covalent bromodomain binders through DNA display of small molecules, *Angew. Chem. Int. Ed. Engl.*, *54*, 6057–6061, 2015.

137. Gentile, G., Merlo, G., Pozzan, A., Bernasconi, G., Bax, B., Bamborough, P., Bridges, A., Carter, P., Neu, M., Yao, G., et al., 5-aryl-4-carboxamide-1,3-oxazoles: Potent and selective GSK-3 inhibitors, *Bioorg. Med. Chem. Lett.*, *22*, 1989–1994, 2012.

138. Svensen, N., Diaz-Mochon, J. J., and Bradley, M., Encoded peptide libraries and the discovery of new cell binding ligands, *Chem. Commun.*, *47*, 7638–7640, 2011.

139. Wu, Z., Graybill, T. L., Zeng, X., Platchek, M., Zhang, J., Bodmer, V. Q., Wisnoski, D. D., Deng, J., Coppo, F. T., Yao, G., et al., Cell-based selection expands the utility of DNA-encoded small-molecule library technology to cell surface drug targets: Identification of novel antagonists of the NK3 tachykinin receptor, *ACS Comb. Sci.*, *17*, 722–731, 2015.

140. Borch, J. and Hamann, T., The nanodisc: A novel tool for membrane protein studies, *Biol. Chem.*, *390*, 805–814, 2009.

141. Pouchain, D., Diaz-Mochon, J. J., Bialy, L., and Bradley, M., A 10,000 member PNA-encoded peptide library for profiling tyrosine kinases, *ACS Chem. Biol.*, *2*, 810–818, 2007.

142. Bao, J., Krylova, S. M., Cherney, L. T., Hale, R. L., Belyanskaya, S. L., Chiu, C. H., Arico-Muendel, C. C., and Krylov, S. N., Prediction of protein-DNA complex mobility in gel-free capillary electrophoresis, *Anal. Chem.*, *87*, 2474–2479, 2015.

143. McGregor, L. M., Gorin, D. J., Dumelin, C. E., and Liu, D. R., Interaction-dependent PCR: Identification of ligand-target pairs from libraries of ligands and libraries of targets in a single solution-phase experiment, *J. Am. Chem. Soc.*, *132*, 15522–15524, 2010.

144. Shi, B., Deng, Y., Zhao, P., and Li, X., Selecting a DNA-Encoded Chemical Library against non-immobilized proteins using a "Ligate-Cross-Link-Purify" strategy, *Bioconjugate Chem.*, *28*, 2293–2301, 2017.

145. Zhao, P., Chen, Z., Li, Y., Sun, D., Gao, Y., Huang, Y., and Li, X., Selection of DNA-encoded small molecule libraries against unmodified and non-immobilized protein targets, *Angew. Chem. Int. Ed. Engl.*, *53*, 10056–10059, 2014.

146. Denton, K. E. and Krusemark, C. J., Crosslinking of DNA-linked ligands to target proteins for enrichment from DNA-encoded libraries, *Med. Chem. Comm.*, *7*, 2020–2027, 2016.

147. Peterson, L. K., Blakskjaer, P., Chaikuad, A., Christensen, A. B., Dietvorst, J., Holmkvist, J., Knapp, S., Korinek, M., Larsen, L. K., Pedersen, A. E., et al., Novel p38α MAP kinase inhibitors identified from yoctoReactor DNA-encoded small molecule library, *Med. Chem. Commu.*, *7*, 1332–1339, 2016.

148. Li, G., Zheng, W., Chen, Z., Zhou, Y., Liu, Y., Yang, J., Huang, Y., and Li, X., Design, preparation, and selection of DNA-encoded dynamic libraries, *Chem. Sci.*, *6*, 7097–7104, 2015.

149. Reddavide, F. V., Lin, W., Lehnert, S., and Zhang, Y., DNA-encoded dynamic combinatorial chemical libraries, *Angew. Chem. Int. Ed. Engl.*, *54*, 7924–7928, 2015.

150. Melkko, S., Zhang, Y., Dumelin, C. E., Scheuermann, J., and Neri, D., Isolation of high-affinity trypsin inhibitors from a DNA-encoded chemical library, *Angew. Chem. Int. Ed. Engl.*, *46*, 4671–4674, 2007.

151. Dumelin, C. E., Trussel, S., Buller, F., Trachsel, E., Bootz, F., Zhang, Y., Mannocci, L., Beck, S. C., Drumea-Mirancea, M., Seeliger, M. W., et al., A portable albumin binder from a DNA-encoded chemical library, *Angew. Chem. Int. Ed. Engl.*, *47*, 3196–3201, 2008.

152. Urbina, H. D., Debaene, F., Jost, B., Bole-Feysot, C., Mason, D. E., Kuzmic, P., Harris, J. L., and Winssinger, N., Self-assembled small-molecule microarrays for protease screening and profiling, *ChemBioChem*, *7*, 1790–1797, 2006.

153. MacConnell, A. B., Price, A. K., and Paegel, B. M., An integrated microfluidic processor for DNA-encoded combinatorial library functional screening, *ACS Comb. Sci.*, *19*, 181–192, 2017.

154. MacConnell, A. B. and Paegel, B. M., Poisson statistics of combinatorial library sampling predict false discovery rates of screening, *ACS Comb. Sci.*, *19*, 524–532, 2017.

155. Erharuyi, O., Simanski, S., McEnaney, P. J., and Kodadek, T., Screening one bead one compound libraries against serum using a flow cytometer: Determination of the minimum antibody concentration required for ligand discovery, *Bioorg. Med. Chem. Lett.*, *28*(16), 2773–2778.

156. Sander, T., Freyss, J., von Korff, M., and Rufener, C., DataWarrior: An open-source program for chemistry aware data visualization and analysis, *J. Chem. Inf. Model*, *55*, 460–473, 2015.

157. Castanon, J., Roman, J. P., Jessop, T. C., De Blas, J., and Haro, R., Design and development of a technology platform for DNA-encoded library production and affinity selection, *SLAS Discov.*, *23*, 387–396, 2018.

158. Kuai, L., O'Keefe, H., and Arico-Muendel, C., Randomness in DNA encoded library selection data can be modeled for more reliable enrichment calculation, *SLAS Discov.*, 2018.

159. Buller, F., Zhang, Y., Scheuermann, J., Schafer, J., Buhlmann, P., and Neri, D., Discovery of TNF inhibitors from a DNA-encoded chemical library based on Diels-Alder cycloaddition, *Chem. Biol.*, *16*, 1075–1086, 2009.

160. Satz, A. L., DNA encoded library selections and insights provided by computational simulations, *ACS Chem. Biol.*, *10*, 2237–2245, 2015.

161. Franzini, R. M., Nauer, A., Scheuermann, J., and Neri, D., Interrogating target-specificity by parallel screening of a DNA-encoded chemical library against closely related proteins, *Chem. Commun.*, *51*, 8014–8016, 2015.

162. Melkko, S., Mannocci, L., Dumelin, C. E., Villa, A., Sommavilla, R., Zhang, Y., Grutter, M. G., Keller, N., Jermutus, L., Jackson, R. H., et al., Isolation of a small-molecule inhibitor of the antiapoptotic protein Bcl-xL from a DNA-encoded chemical library, *ChemMedChem*, *5*, 584–590, 2010.

163. Cuozzo, J. W., Centrella, P. A., Gikunju, D., Habeshian, S., Hupp, C. D., Keefe, A. D., Sigel, E. A., Soutter, H. H., Thomson, H. A., Zhang, Y., and Clark, M. A., Discovery of a potent BTK inhibitor with a novel binding mode by using parallel selections with a DNA-encoded chemical library, *ChemBioChem*, *18*, 864–871, 2017.

164. Yang, H., Medeiros, P. F., Raha, K., Elkins, P., Lind, K. E., Lehr, R., Adams, N. D., Burgess, J. L., Schmidt, S. J., Knight, S. D., et al., Discovery of a potent class of PI3Kα inhibitors with unique binding mode via encoded library technology (ELT), *ACS Med. Chem. Lett.*, *6*, 531–536, 2015.

165. Satz, A. L., Hochstrasser, R., and Petersen, A. C., Analysis of current DNA encoded library screening data indicates higher false negative rates for numerically larger libraries, *ACS Comb. Sci.*, *19*, 234–238, 2017.

166. Yuen, L. H. and Franzini, R. M., Achievements, challenges, and opportunities in DNA-encoded library research: An academic point of view, *ChemBioChem*, *18*, 829–836, 2017.

167. Kodadek, T., The rise, fall and reinvention of combinatorial chemistry, *Chem. Commun.*, *47*, 9757–9763, 2011.

168. Eidam, O. and Satz, A. L., Analysis of the productivity of DNA encoded libraries, *Med. Chem. Commu.*, *7*, 1323–1331, 2016.

169. Georghiou, G., Kleiner, R. E., Pulkoski-Gross, M., Liu, D. R., and Seeliger, M. A., Highly specific, bisubstrate-competitive Src inhibitors from DNA-templated macrocycles, *Nat. Chem. Biol.*, *8*, 366–374, 2012.

170. Mandal, P., Berger, S. B., Pillay, S., Moriwaki, K., Huang, C., Guo, H., Lich, J. D., Finger, J., Kasparcova, V., Votta, B., et al., RIP3 induces apoptosis independent of pronecrotic kinase activity, *Mol. Cell.*, *56*, 481–495, 2014.

171. Gilmartin, A. G., Faitg, T. H., Richter, M., Groy, A., Seefeld, M. A., Darcy, M. G., Peng, X., Federowicz, K., Yang, J., Zhang, S. Y., et al., Allosteric Wip1 phosphatase inhibition through flap-subdomain interaction, *Nat. Chem. Biol.*, *10*, 181–187, 2014.

172. Wood, E. R., Bledsoe, R., Chai, J., Daka, P., Deng, H., Ding, Y., Harris-Gurley, S., Kryn, L. H., Nartey, E., Nichols, J., et al., The role of phosphodiesterase 12 (PDE12) as a negative regulator of the innate immune response and the discovery of antiviral inhibitors, *J. Biol. Chem.*, *290*, 19681–19696, 2015.

173. Disch, J. S., Evindar, G., Chiu, C. H., Blum, C. A., Dai, H., Jin, L., Schuman, E., Lind, K. E., Belyanskaya, S. L., Deng, J., et al., Discovery of thieno[3,2-d]pyrimidine-6-carboxamides as potent inhibitors of SIRT1, SIRT2, and SIRT3, *J. Med. Chem.*, *56*, 3666–3679, 2013.

174. Deng, H., O'Keefe, H., Davie, C. P., Lind, K. E., Acharya, R. A., Franklin, G. J., Larkin, J., Matico, R., Neeb, M., Thompson, M. M., et al., Discovery of highly potent and selective small molecule ADAMTS-5 inhibitors that inhibit human cartilage degradation via encoded library technology (ELT), *J. Med. Chem.*, *55*, 7061–7079, 2012.

175. Maianti, J. P., McFedries, A., Foda, Z. H., Kleiner, R. E., Du, X. Q., Leissring, M. A., Tang, W. J., Charron, M. J., Seeliger, M. A., Saghatelian, A., and Liu, D. R., Anti-diabetic activity of insulin-degrading enzyme inhibitors mediated by multiple hormones, *Nature*, *511*, 94–98, 2014.

176. Mannocci, L., Melkko, S., Buller, F., Molnar, I., Bianke, J. P., Dumelin, C. E., Scheuermann, J., and Neri, D., Isolation of potent and specific trypsin inhibitors from a DNA-encoded chemical library, *Bioconjugate Chem*, *21*, 1836–1841, 2010.

177. Ahn, S., Kahsai, A. W., Pani, B., Wang, Q. T., Zhao, S., Wall, A. L., Strachan, R. T., Staus, D. P., Wingler, L. M., Sun, L. D., et al., Allosteric "beta-blocker" isolated from a DNA-encoded small molecule library, *Proc. Natl. Acad. Sci. U. S. A.*, *114*, 1708–1713, 2017.

178. Cheng, R. K. Y., Fiez-Vandal, C., Schlenker, O., Edman, K., Aggeler, B., Brown, D. G., Brown, G. A., Cooke, R. M., Dumelin, C. E., Dore, A. S., et al., Structural insight into allosteric modulation of protease-activated receptor 2, *Nature*, *545*, 112–115, 2017.

179. Lipinski, C. A., Rule of five in 2015 and beyond: Target and ligand structural limitations, ligand chemistry structure and drug discovery project decisions, *Adv. Drug Deliv. Rev.*, *101*, 34–41, 2016.

180. Belyanskaya, S. L., Ding, Y., Callahan, J. F., Lazaar, A. L., and Israel, D. I., Discovering drugs with DNA-encoded library technology: From concept to clinic with an inhibitor of soluble epoxide hydrolase, *ChemBioChem*, *18*, 837–842, 2017.

181. Harris, P. A., King, B. W., Bandyopadhyay, D., Berger, S. B., Campobasso, N., Capriotti, C. A., Cox, J. A., Dare, L., Dong, X., Finger, J. N., et al., DNA-encoded library screening identifies Benzo[b][1,4] oxazepin-4-ones as highly potent and monoselective receptor interacting protein 1 kinase inhibitors, *J. Med. Chem.*, *59*, 2163–2178, 2016.

182. Leimbacher, M., Zhang, Y., Mannocci, L., Stravs, M., Geppert, T., Scheuermann, J., Schneider, G., and Neri, D., Discovery of small-molecule interleukin-2 inhibitors from a DNA-encoded chemical library, *Chem. Eur. J.*, *18*, 7729–7737, 2012.

183. Yudin, A. K., Macrocycles: Lessons from the distant past, recent developments, and future directions, *Chem. Sci.*, *6*, 30–49, 2015.

184. Driggers, E. M., Hale, S. P., Lee, J., and Terrett, N. K., The exploration of macrocycles for drug discovery—an underexploited structural class, *Nat. Rev. Drug Discov.*, *7*, 608–624, 2008.

185. Connors, W. H., Hale, S. P., and Terrett, N. K., DNA-encoded chemical libraries of macrocycles, *Curr. Opin. Chem. Biol.*, *26*, 42–47, 2015.

186. Kleiner, R. E., Dumelin, C. E., Tiu, G. C., Sakurai, K., and Liu, D. R., In vitro selection of a DNA-templated small-molecule library reveals a class of macrocyclic kinase inhibitors, *J. Am. Chem. Soc.*, *132*, 11779–11791, 2010.

187. Krall, N., Scheuermann, J., and Neri, D., Small targeted cytotoxics: Current state and promises from DNA-encoded chemical libraries, *Angew. Chem. Int. Ed. Engl.*, *52*, 1384–1402, 2013.

188. Trussel, S., Dumelin, C., Frey, K., Villa, A., Buller, F., and Neri, D., New strategy for the extension of the serum half-life of antibody fragments, *Bioconjugate Chem.*, *20*, 2286–2292, 2009.

189. Muller, C., Struthers, H., Winiger, C., Zhernosekov, K., and Schibli, R., DOTA conjugate with an albumin-binding entity enables the first folic acid-targeted[177]Lu-radionuclide tumor therapy in mice, *J. Nucl. Med.*, *54*, 124–131, 2013.

190. Benesova, M., Umbricht, C. A., Schibli, R., and Muller, C., Albumin-binding PSMA ligands: Optimization of the tissue distribution profile, *Mol. Pharm.*, *15*(6), 2297–2306, 2018.

191. Choy, C. J., Ling, X., Geruntho, J. J., Beyer, S. K., Latoche, J. D., Langton-Webster, B., Anderson, C. J., and Berkman, C. E., ^{177}Lu-labeled phosphoramidate-based PSMA inhibitors: The effect of an albumin binder on biodistribution and therapeutic efficacy in prostate tumor-bearing mice, *Theranostics*, *7*, 1928–1939, 2017.

192. Kelly, J. M., Amor-Coarasa, A., Nikolopoulou, A., Wustemann, T., Barelli, P., Kim, D., Williams, C., Jr., Zheng, X., Bi, C., Hu, B., et al., Dual-target binding ligands with modulated pharmacokinetics for endoradiotherapy of prostate cancer, *J. Nucl. Med.*, *58*, 1442–1449, 2017.

193. Cazzamalli, S., Dal Corso, A., Widmayer, F., and Neri, D., Chemically defined antibody- and small molecule-drug conjugates for in vivo tumor targeting applications: A comparative analysis, *J. Am. Chem. Soc.*, *140*, 1617–1621, 2018.

194. Minn, I., Koo, S. M., Lee, H. S., Brummet, M., Rowe, S. P., Gorin, M. A., Sysa-Shah, P., Lewis, W. D., Ahn, H. H., Wang, Y., et al., [64Cu]XYIMSR-06: A dual-motif CAIX ligand for PET imaging of clear cell renal cell carcinoma, *Oncotarget*, *7*, 56471–56479, 2016.

195. Yang, X., Minn, I., Rowe, S. P., Banerjee, S. R., Gorin, M. A., Brummet, M., Lee, H. S., Koo, S. M., Sysa-Shah, P., Mease, R. C., et al., Imaging of carbonic anhydrase IX with an ^{111}In-labeled dual-motif inhibitor, *Oncotarget*, *6*, 33733–33742, 2015.

196. Neklesa, T. K., Winkler, J. D., and Crews, C. M., Targeted protein degradation by PROTACs, *Pharmacol. Ther.*, *174*, 138–144, 2017.

197. Salamon, H., Klika Skopic, M., Jung, K., Bugain, O., and Brunschweiger, A., Chemical biology probes from advanced DNA-encoded libraries, *ACS Chem. Biol.*, *11*, 296–307, 2016.

198. Winssinger, N., Damoiseaux, R., Tully, D. C., Geierstanger, B. H., Burdick, K., and Harris, J. L., PNA-encoded protease substrate microarrays, *Chem. Biol.*, *11*, 1351–1360, 2004.

199. Harris, J., Mason, D. E., Li, J., Burdick, K. W., Backes, B. J., Chen, T., Shipway, A., Van Heeke, G., Gough, L., Ghaemmaghami, A., et al., Activity profile of dust mite allergen extract using substrate libraries and functional proteomic microarrays, *Chem. Biol.*, *11*, 1361–1372, 2004.

200. Jetson, R. R. and Krusemark, C. J., Sensing enzymatic activity by exposure and selection of DNA-encoded probes, *Angew. Chem. Int. Ed. Engl.*, *55*, 9562–9566, 2016.

201. Krusemark, C. J., Tilmans, N. P., Brown, P. O., and Harbury, P. B., Directed chemical evolution with an outsized genetic code, *PLoS One*, *11*, e0154765, 2016.

6 Bio-Targeted Nanomaterials for Theranostic Applications

Sabyasachi Maiti
Indira Gandhi National Tribal University

Kalyan Kumar Sen
Gupta College of Technological Sciences

CONTENTS

6.1 INTRODUCTION

Nanotheranostics integrates nanomaterials with theranostics. The word "theranostics" refers to the simultaneous integration of diagnosis and therapy (Sumner and Gao 2008). By combining both therapeutic and diagnostic functions in one carrier, theranostic agents "enable disease diagnosis, therapy, and real-time monitoring of treatment progress and efficacy" (Jo et al. 2016). In theranostic nanocarriers of 10–1,000 nm size, the diagnostic and therapeutic agents are adsorbed, conjugated, entrapped, and encapsulated in nanomaterials for diagnosis and treatment simultaneously at cellular and molecular level (Muthu et al. 2014a).

The therapeutic agents in theranostic nanomedicine include hydrophobic drugs, proteins, peptides, and genetic materials. In addition to therapeutic agents, diagnostic agents commonly used in theranostic nanomedicine include fluorescent dyes or quantum dots, superparamagnetic iron oxides (SPIOs), radionuclides, and iodine. This strategy can provide better theranostic effects and fewer side effects by co-delivering diagnostic and therapeutic agents in sustained, controlled manner in targeted tissues. Nanotheranostics can be promising even for the fatal diseases such as cancer, inflammatory bowel disease (IBD), cardiovascular diseases (CVDs), diabetes, ocular diseases, and AIDS, improving the prognosis while making the treatment more tolerable for patients and saving resources (Muthu et al. 2014b).

Advanced theranostic nanomedicine is multifunctional in nature, capable of diagnosis and delivery of therapy to the diseased cells with the help of a targeting ligand and biomarkers (Xie et al. 2013, Yu et al. 2010).

Polymeric nanoparticles responsive to external pH stimuli have attracted much attention for a wide range of applications, especially in biology, such as imaging-guided drug delivery, molecular sensing, and tissue engineering. Moreover, with the help of nanotechnology, theranostics

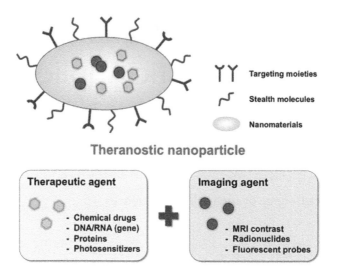

Theranostic nanoparticle

FIGURE 6.1 Schematic illustration of theranostic nanoparticles (e.g., micelles, liposomes, nanospheres, dendrimers, quantum dots, iron oxide particles) embedded with therapeutic and imaging agents. (Adapted from Nanotheranostics, 1, Gupta, A.S. et al., Recent strategies to design vascular theranostic nanoparticles, 166–77, 2017.)

may promote the diversification of therapeutic approaches such as photothermal therapy (PTT)), photodynamic therapy (PDT), and immunotherapy (Ascierto et al. 2015).

To develop theranostic agents, imaging materials were simply conjugated to delivery carriers loaded with drugs, and a variety of imaging methods were used to visualize the agent. The most widely employed noninvasive imaging methods include magnetic resonance imaging (MRI), computed tomography (CT), positron emission tomography (PET), and optical imaging to monitor the biodistribution, drug release kinetics, and therapeutic efficacy of theranostic nanomaterials. The design of nanotheranostic materials is shown in Figure 6.1.

Progress in the development of nanomaterials has spurred the discovery of numerous theranostic agents possessing both imaging and therapeutic functions to diagnose and treat a variety of diseases, respectively. This chapter provides insight into the potential role of nanotheranostics in various diseases.

6.2 THERANOSTIC POTENTIALS OF NANOMATERIALS

6.2.1 Cancer

As the second leading cause of death worldwide, cancer has a major impact on society (World Health Organization 2018). The expected number of new cancer cases per year is to rise from 14.1 million in 2012 to 23.6 million by 2030 (Stewart and Wild 2014). These statistics underscore the alarming situation regarding the increase in prevalence of cancer. Under this situation, immense efforts have been exercised to expand new strategies such as nanomedicine for efficient and personalized cancer therapy. Ideally, theranostics would enable the ability to visualize and monitor diseased tissues, delivery kinetics, and anticancer efficacy, while simultaneously providing the possibility to tune the therapeutic strategy in a controlled and rational manner. Most anticancer drugs, such as doxorubicin (DOX) and paclitaxel (PTX), have toxic effects on normal cells. Cancer cell–targeted delivery can lessen this problem; nanomedicine offers hope in this regard. Incorporation of near-infrared (NIR) fluorescent probes in nano drug carriers enables real-time *in vivo* tracking of the nanoparticles.

The mechanism of drug targeting to cancer cells is illustrated in Figure 6.2. Once the nanocarriers enter the cancer cells, the diagnostic agent emits the signal, and that signal is processed into images to detect the various stages of diseases by MRI, CT scan, ultrasound, etc.

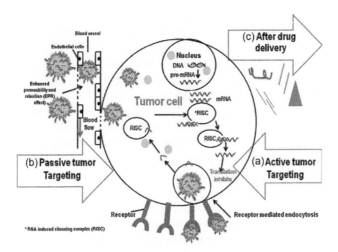

FIGURE 6.2 Fate of polymeric nanoparticle-based targeted drug delivery system inside the tumor cell [a drug-loaded nanoparticle enters the tumor cell through receptor-mediated endocytosis (a) or through an enhanced permeability and retention effect (b), and the degradation of the residual carrier components after drug delivery to the tumor site (c)]. (Reprinted from *Materials Science and Engineering C*, 60, Masood, F., Polymeric nanoparticles for targeted drug delivery system for cancer therapy, 569–578, Copyright (2016), with permission from Elsevier.)

Theranostic applications mean various combinations of therapeutic and imaging modalities used together. Several noninvasive imaging methods (like optical, magnetic resonance (MR), CT, PET, and ultrasound (US) imaging techniques) are utilized to monitor the biodistribution, pharmacokinetics, and therapeutic efficacy of theranostic nanoparticles.

Presently, squamous cell carcinoma (SCC) covers more than 90% of all head and neck cancers (HNCs). A plasmonic-based nanotheranostics platform was developed for combined, ultrasensitive *in vivo* spectroscopic detection and targeted therapy of HNC. In this study, NIR spectroscopy of gold nano-rods (GNRs) was used that selectively target and attach to SCC–HNC cells through an immune complex. These were used in a spectral shift analysis method for detection; the spectral shift is generated by interparticle-plasmon-resonance patterns of the specifically targeted GNRs used as diagnostic tools. The study showed that both *in vitro* and *in vivo*, those immune-targeted GNRs can target SCC–HNC cancer cells with high specificity and facilitate the differentiation between cancerous and noncancerous tissues. This targeted, noninvasive, and nonionizing spectroscopic detection method can provide a highly sensitive, simple, and inexpensive diagnostic tool for micrometastasis (Betzer et al. 2015).

In another study, X-rays were used as the irradiation source to initiate a PDT process in deep tissues. A novel $LiGa_5O_8$:Cr (LGO:Cr)-based nanoscintillator was used, which emits persistent, NIR X-ray luminescence encapsulated with a photosensitizer (2,3-naphthalocyanine) into mesoporous silica nanoparticles. The nanoparticles were conjugated with cetuximab and systemically injected into H1299 orthotopic non–small-cell lung cancer tumor models. This permits deep-tissue optical imaging that can be employed to guide irradiation. Specifically, the nanoconjugates can efficiently accumulate in tumors in the lungs, confirmed by monitoring the X-ray luminescence from LGO:Cr. Guided by the imaging, external irradiation was applied, leading to efficient tumor suppression while minimally affecting normal tissues.

MRI is a powerful noninvasive imaging method using magnetic fields and radio waves to acquire cross-sectional tomographic images of tissues. Compared to other imaging techniques, the advantages of MRI include high spatial resolution, deep penetration, and the use of nonionizing radiation. Chen et al. (2013) reported the synthesis and characteristics of a novel dual-modality MRI and NIR fluorescence (NIRF) probe. This MRI-NIRF probe was comprised of a SPIO nanoparticle coated with a liposome to which a tumor-targeted agent and an NIRF dye were conjugated. In this case, the tumor-targeting agent was an Arg-Gly-Asp peptide (RGD), and the NIRF dye was indocyanine green (ICG). The feasibility of this probe was verified in nude mouse models with liver cancer. This study demonstrated that both MRI and fluorescent images show clear tumor delineation after probe injection (SPIO@Liposome-ICG-RGD). These novel MRI-NIRF dual-modality probes are promising for the achievement of more accurate liver tumor detection and resection.

Wu et al. (2018) reported another nanotheranostic employing MRI and NIRF technology, in this case a defect-rich, multifunctional, Cu-doped, layered double hydroxide (Cu-LDH) nanoparticle. The imaging by T_1-MRI and the amount of heat generated by NIR are both amplified by low pH, which is ideal for targeting cancer since the tumor environment is more acidic. This feature allows one to identify the region of the body with the tumor via MRI. Subsequently, an NIR laser focused on that region will more selectively kill cancer cells since the tumor microenvironment will enhance photothermal conversion, generating more heat. The defects of the Cu microstructure yield pH sensitivity of the T_1-MRI, while the pH-sensitive nature of the photothermal conversion via NIR derived from the exposure of hydroxyl groups in the CuOH octahedra on the LDH surface. Combined with chemotherapy, this PTT leads to "nearly complete" eradication of tumor tissues *in vivo* at low doses of injection.

MRI-guided PTT has also been performed using polypyrrole nanoparticle (PPy)-based theranostic agents. In one study, nanocomposites PEGylated PPy@Fe^{3+}-chelated PDA (PPDEs) based on polydopamine (PDA), specifically PEGylated PPy@Fe^{3+}-chelated PDA, were designed and developed. PPDE with a uniform core–shell structure was obtained by adjusting the ratio of dopamine and PPys. In this nanocomplex, the shells confer good biocompatibility and MRI signal-enhancing ability to the nanoparticles. Moreover, the PPy cores play a role in photothermal ablation of tumors. After intravenous injection, the PPDEs exhibited tumor accumulation, as shown by MRI and verified by biodistribution analysis. Under NIR irradiation, the PPDEs showed highly effective photothermal ablation of $4T_1$ cells (Yang 2017).

Computed tomography (CT) is one of the most widely used noninvasive diagnostic imaging modalities that utilizes computerized X-ray images to produce cross-sectional tomographic, three-dimensional images of tissues of interest (Lusic and Grinstaff 2013). Use of contrast agents in functional imaging enables one to highlight structures despite a lack of natural contrast.

Iodine-containing echogenic glycol chitosan (GC) nanoparticles are recently being used for X-ray CT and US as dual-mode imaging of tumor diagnosis (Choi et al. 2018). The preparation includes two steps. Initially, iodine-containing diatrizoic acid (DTA) was conjugated with GC, and stable nanoparticles were formed for the CT imaging. In the next step, a US imaging agent, perfluoropentane (PFP), was physically encapsulated into GC-DTA NPs to form GC-DTA-PFP nanoparticles (GC-DTA-PFP NPs). Effective accumulation of the GC-DTA-PFP NPs in the tumor site was achieved due to enhanced permeation and retention effects. These GC-DTA-PFP NPs showed X-ray CT and US signals in tumor tissues after intratumoral and intravenous injection, respectively, enabling a comprehensive and accurate diagnosis of tumors.

A recent study reported a peptide (LyP-1)-labeled ultra-small semimetal nanoparticles of bismuth (Bi-LyP-1 NPs) as multifunctional theranostic agents (Yu et al. 2017). These agents exhibited a high level of accumulation at the tumor site, and they are able to absorb both ionizing radiation and the second near-infrared (NIR-II) window laser radiation. These properties helped facilitate dual-modal CT/photoacoustic imaging and efficient synergistic NIR-II photothermal/radiotherapy of tumors. This study also reveals that these multifunctional nanoparticles can be used as competent theranostic agents for cancer treatment with low toxicity. These agents also exhibited fast clearance, a pharmacokinetic property which lowers toxicity through reduction of drug exposure.

Gold nanoparticles (AuNPs) are also emerging. Ashton et al. focused on the role of AuNPs to influence tumor vascular permeability as potential tools for cancer therapy and CT imaging. "AuNPs absorb x-rays and subsequently release low-energy, short-range photoelectrons" during external beam radiation therapy (RT), increasing the local radiation dose (Ashton et al. 2018). "When AuNPs are near tumor vasculature, the additional radiation dose can lead to increased vascular permeability" (Ashton et al. 2018). Accumulation of both liposomal iodine and AuNPs in sarcoma tumors following AuNP-augmented RT was quantified using dual-energy CT (iodine). A noteworthy increase in vascular permeability was seen for all groups. Combination therapy of liposomal DOX and targeted AuNPs with RT led to a significant tumor growth delay (Ashton et al. 2018).

Although several studies highlighted that CT has several advantages related to fast diagnosis, high spatial resolution, and unlimited depth penetration, it has potential side effects caused by radiation and contrast agents (Brenner and Hall 2007).

PET is a technique that produces three-dimensional functional images of the body. Advantages such as high sensitivity, noninvasiveness, and quantitative real-time imaging capability make PET an ideal method for tracking and quantifying physiologic processes *in vivo* (Jo et al. 2016, Phelps 2000, Law and Wong 2014). PET uses biologically active positron-emitting radiotracers (such as fluorine-18) that decay and cause the total destruction of a positron and an electron producing two γ-rays. PET can be used to track radiolabeled nanomolecules potentially used for diagnosis of cancer or other indications. In one study, ^{18}F-labeled PEGylated liposomes were successfully prepared and their biodistribution was analyzed under the PET imaging format (Marik et al. 2007). Andreozzi et al. (2011) used chelator 6-[p-(bromoacetami- do) benzyl]-1, 4, 8, 11-tetraazacyclotetra-decane-N, N′, N″, N‴-tetraacetic acid (BAT) to label ^{64}Cu onto solid lipid nanoparticles (SLNs) for PET imaging.

6.2.2 INFLAMMATORY BOWEL DISEASE (IBD)

IBD represents a promising frontier in the field of drug targeting since it is characterized by segmental inflammation of the bowel. IBD consists of ulcerative colitis and Crohn's disease. In ulcerative colitis, mucosal lesions are seen in rectum and extend some distance up the colon (and rarely to the terminal ileum), while in Crohn's disease, inflammation is usually present only in short segments throughout the entire bowel. Thus, in IBD, most of the bowel is healthy and should not be exposed to any drug to reduce or avoid systemic side effects. In fact, the major problem in reducing IBD recurrences is to maintain an effective concentration of the drug in the inflamed mucosa (Frieri et al. 2005).

The systemic toxicity, poor targeting of conventional therapeutic agents, and low adherence to prescription drugs represent frequent therapeutic challenges. Recent observations suggest that nanotechnology could be advantageous in modifying the pharmacokinetics and efficacy of existing drugs within the intestinal inflammatory cells (Viscido et al. 2014). The Brownian motion of the nanoparticles in the luminal content may increase their mucoadhesive characteristics (Tamura et al. 2006). Because the nanoparticles remain unaffected by luminal streaming, this could further enhance adhesion and penetration abilities to the intestinal mucosa. Under inflammatory conditions, the integrity of intestinal epithelial cell monolayer is disrupted, thus losing its barrier function. The gaps or holes created at the epithelial cell membrane greatly enhance the entry of nanoparticles into mucosa (Collnot et al. 2012). Moreover, the small size could facilitate endocytosis and transcytosis, responsible for uptake of particles of <100 and 500nm in diameter, respectively (Yun et al. 2013). Taken together, these characteristics allow a nanoparticle absorption rate up to 15–250 folds higher compared to larger size particles. Once entered into tissue, the small size enhances the retention time at the target site via bioadhesion, and the nanoparticles can also directly enter into phagocytic cells that populate the inflamed tissue.

Lamprecht et al. (2001) reported about fivefold higher deposition of 1-μm particles in ulcerative colitis compared to controls, with the highest deposition in inflamed tissue for 100-nm particles. A recent study indicated that both nanoparticles (250nm) and microparticles (3μm) concentrated to a large extent in the areas of epithelial lesions, though the transport of nanoparticles was more rapid than microparticles from the surface to a deeper layer of the intestinal wall of IBD patients with mild-to-moderate inflammation (Schmidt et al. 2013). Until now, colon-specific drug delivery systems are mostly based on action of bacteria, luminal pH, and sustained release (Caprilli et al. 2009). However, the drug delivery systems cannot prevent the exposure of a certain amount of drug onto normal mucosa nor assure its distribution over the whole area of inflamed tissues. The inflammatory features such as disruption of mucosal barrier, increased permeability, and increased production of mucus represent a target that could be selectively reached by nanocarriers (Laroui et al. 2011).

An increased adherence of particles to the inflamed tissue is observed at the thicker mucus layer and in ulcerated regions, influencing in turn increase targeting and retention of drug delivery systems in ulcerative colitis (Serra et al. 2009). In view of their physicochemical characteristics, oral pharmacokinetics, and their ability to discriminate between diseased and non-diseased sites, the nanoparticles would appear an ideal carrier for effective delivery of therapeutic agents in IBD.

Because the carbohydrates of colonic mucins are substituted with numerous sulfate and sialic acid residues, they carry negative charge (Carasi et al. 2014). Cationic nanocarriers may adhere to the mucosal surface within inflamed tissue due to the ionic interaction with negatively charged intestinal mucosa. The mucoadhesive NPs could promote better contact with the mucosal surface for cellular uptake and drug release and reduce their clearance in case of increased intestinal motility, common in IBD (Han et al. 2012). Coco et al. (2013) conducted *ex vivo* studies on inflamed mouse colon tissue using mucoadhesive trimethylchitosan (TMC)-coated, pH-sensitive Eudragit® S100 nanoparticles, and sustained release formulations using poly-(d,l-lactic-co-glycolic acid) (PLGA) polymer-based nanoparticles. Ovalbumin was used as a model drug. TMC nanoparticles demonstrated the highest apparent permeability in healthy tissue. However, in the inflamed model, there were no difference in permeability between TMC, PLGA-based, and Eudragit® nanoparticles. PLGA nanoparticles grafted with mannose, a substance that targets immune cells, showed the highest accumulation of OVA in inflamed colon. These results suggest the viability of targeting immune cells in the colon using nanomolecules in the treatment of IBD.

Chitosan-coated (Takeuchi et al. 2003, Manconi et al. 2013) and pectin-coated (Thirawong et al. 2008) liposomes have demonstrated enhanced drug uptake *in vivo* using animal models of colitis, compared to uncoated liposomes. Thirawong et al. (2008) evaluated pectin-liposome nanocomplexes (PLNs) in fasted healthy Wistar rats for their mucoadhesion and oral uptake. Despite their increased residence in GI tract mucosa, the PLNs mostly accumulated in the small intestine, with a very small uptake in colon. Conversely, anionic nanocarriers have been developed for preferential adhesion to inflamed tissue. A high amount of eosinophil cationic protein and transferrin has been found in inflamed colon of IBD patients (Tirosh et al. 2009). Therefore, electrostatic interaction of the nanoparticle with inflamed tissue is possible. Another advantage is that anionic nanoparticles can interdiffuse through mucus due to less electrostatic attraction.

Colonic inflammation may change the expression of receptors on the cellular surface of tissues, and therefore, the selective drug accumulation at inflamed sites may be improved by attaching ligands to the surface of NPs (Chen et al. 2013). The interaction between ligands and specific receptors expressed predominantly at inflamed sites could improve mucoadhesion of the NPs to specific cells and increase the extent for endocytosis. Mannose receptors and macrophage galactose-type lectin (MGL) are highly expressed by activated macrophages under inflammatory conditions (Wileman et al. 1986). Xiao et al. (2013) observed a significant macrophage-targeting ability of mannosylated bioreducible cationic polymer NPs. The principle of electrostatic interaction between cationic polymer, sodium triphosphate (TPP), and TNF-α siRNA was used for the preparation of NPs. The uptake of NPs by colon macrophages was 29.5%; however, uptake was insignificant by epithelial cells. In another study, the highest accumulation was noted in DSS-induced colitis model for mannose-grafted PLGA NPs compared to TMC, PLGA, and Eudragit® S-100 NPs (Coco et al. 2013). However, the colon-targeting ability of NPs in an *in vivo* colitis model was not tested by the investigators. The transferrin receptor (TfR) represents another target which is overexpressed in inflamed colon tissue (Tirosh et al. 2009). Harel et al. (2011) conjugated anti-TfR antibodies onto the surface of immunoliposomes and evaluated their *ex vivo* mucoadhesion capacity to inflamed tissue. They reported a fourfold higher uptake in inflamed colon tissue in 2,4,6-trinitrobenzene sulfonic acid (TNBS) induced colitis model than non-inflamed colon tissue. This result suggested that mucosal TfR can be targeted in IBD; however, the prime concern is the premature degradation of lipid-based immune liposomes in GI tract.

Researchers have begun to use nanomolecules to deliver existing or experimental IBD drugs to inflamed intestinal tissue. Examples of IBD agents targeted using nanomolecules include prednisolone, budesonide, curcumin (CC), and TNFα siRNA. Prednisolone is one of the two most commonly used oral glucocorticoids for IBD, working through broad anti-inflammatory actions. It can be used for mild, moderate, or severe IBD depending on the route of administration. Budenoside is a synthetic analog of prednisolone with a high affinity for the glucocorticoid than prednisolone but with lower oral bioavailability, so it is only used topically on the intestinal lumen. Cucurmin is a plant-derived experimental anti-inflammatory agent that may have efficacy for IBD and other inflammatory diseases. TNFα is a cytokine which mediates much of the inflammatory damage induced by IBD; siRNA specific to TNFα degrades the mRNA encoding TNFα, thereby inhibiting it.

Prednisolone-loaded Eudragit® S100 (567 nm) nanocapsules with pH-dependent drug release properties under normal physiological pH condition were studied by Gupta et al. (2013). The plasma drug concentration peaked at 3 h after oral administration in healthy rats. This indicated that the nanocapsules underwent dissolution and released the drug in the colon after this lag period.

Beloqui et al. (2013) observed that anionic budesonide-loaded lipid carriers of 200 nm size reduced inflammation in dextran-sulfate-induced colitis model significantly after oral administration. In another *in vivo* study, Beloqui et al. (2014) showed that the lipid nanocarriers localized in small intestine and retained in the underlying epithelium, thus allowing further uptake by epithelial cells. It has been noted that coating of NPs with small molecular weight, hydrophilic poly (ethylene glycol) (PEG) could provide an accelerated translocation into the leaky inflamed intestinal epithelium, an ideal delivery platform for colitis-targeted drug delivery (Lautenschläger et al. 2013). It was thought that the hydrophilic, uncharged PEG molecules minimized strong interaction with mucus and increased particle translocation (Cu and Saltzman 2009). Lautenschläger et al. (2013) reported that PEG-coated PLGA nanoparticles (300 nm) could significantly enhance particle translocation and deposition in inflamed human intestinal mucosa compared to chitosan-functionalized or nonfunctionalized PLGA NPs. *In vivo* studies have also been reported the possible use of pH-dependent polymer-based nanocarriers for colon targeting in IBD. The budesonide-loaded pH-sensitive PLGA/Eudragit® S100 nanospheres (260–290 nm) demonstrated superior therapeutic efficacy in alleviating colitis than conventional enteric-coated about 2 μm size microcarriers in TNBS colitis model (Makhlof et al. 2009). The NPs accumulated in colon to a large extent, exhibited low systemic bioavailability, and further, adhered to the ulcerated and inflamed mucosal tissue of the rat colon.

Recently, Beloqui et al. (2014) attempted to deliver CC using pH-sensitive PLGA/Eudragit® S100 NPs (166 nm). Drug permeation across Caco-2 cell monolayers, a model for intestinal permeability, was greatly enhanced compared to drug suspension. The local delivery of CC-loaded nanocarriers to the inflamed colon was evaluated *in vivo* by visualizing CC accumulation in both healthy and inflamed tissues. Higher accumulation of CC was observed in colon sections of both healthy and murine dextran sulfate (DSS)-treated mice after encapsulation into NPs compared to drug suspension (Figure 6.3). (DSS treatment is a standard IBD model.)

TNF-α siRNA is another experimental anti-IBD agent which was targeted to macrophages through nanoparticles. This targeting was achieved through cationic nanoparticles loaded with mannose, whose receptor is upregulated on macrophages when they are activated as during inflammation. (This is described more fully in discussion of receptor targeting above.) The NPs significantly inhibited TNF-α synthesis and secretion in tissue samples in a DSS-induced colitis model.

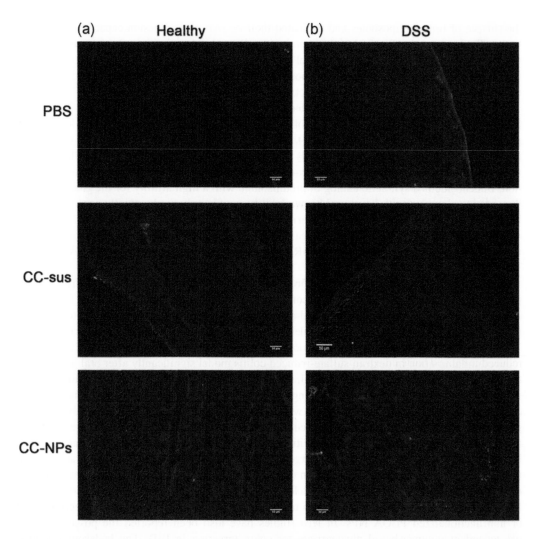

FIGURE 6.3 *In vivo* localization of CC 12 h post-administered as suspension or as pH-sensitive nanoparticles in healthy (a) and inflamed (b) colon sections. Green corresponds to CC. Cell nuclei are stained in blue (magnification 10×). Scale bars = 50 μm. (Reprinted from *International Journal of Pharmaceutics*, 473, Beloqui, A., Coco, R., Memvanga, P.B., Ucakar, B., des Rieux, A., Préat V., pH-sensitive nanoparticles for colonic delivery of CC in inflammatory bowel disease, 203–212, Copyright (2014), with permission from Elsevier.)

6.2.3 Diabetes

Diabetes mellitus or diabetes is caused by low insulin secretion by pancreatic islet cells, leading to hyperglycemia. In Type 1 (insulin-dependent) diabetes mellitus, there is a loss of insulin-producing beta cells of islets of Langerhans in pancreas, thereby leading to deficiency of insulin. In contrast, Type 2 (non-insulin-dependent) diabetes is caused by insulin resistance or reduced insulin secretion. Glycemic control can be achieved through diet, physical activity, insulin therapy, and oral medications. Nanomedicine research is directed towards the applications of nanoparticles for the treatment of Type 1 diabetes through effective insulin delivery (Subramani et al. 2012).

In Type 1 and many Type 2 diabetes patients, three or more subcutaneous injections or insulin are required to maintain an acceptable serum glucose level (American Diabetes Association 2004). This can cause psychological stress leading to poor patient compliance. As an alternative, the most popular oral route of administration has been tested. The major obstacles in developing oral insulin formulations are either enzymatic barriers or physical barriers. Insulin is degraded by gastric pH and intestinal enzymes, and the intestinal epithelial cell membranes serve as a formidable absorption barrier itself. Together, this results in <1% bioavailability of total oral insulin intake (Lowman et al. 1999, Binder et al 1984). Although significant advancement has been made for effective oral insulin delivery, sufficient commercial development is not achieved yet. To overcome these challenges, nano-carriers have been considered as the best-suited vehicle for oral delivery of insulin (Jin et al. 2012).

In addition to the complications of acidity and digestive enzymatic action in the stomach, the hydrophilic nature of insulin accounts for its poor absorption through the intestinal epithelium (Binder et al. 1984). Tight junctions between intestinal epithelial cells also prohibit the paracellular transport of insulin and other hydrophilic molecules (Ballard et al. 1995). Therefore, improving the delivery of hydrophilic molecules through intestinal epithelium using nanotechnology has been the focus of diabetes research. Absorption of drugs and molecules in the intestine depends on particle size. Nanoparticles <200 nm in diameter are absorbed by enterocytes via endocytosis. Larger particles may be absorbed via phagocytosis (1 μm in diameter) or transported through lymphatic islands in the intestinal tract called Peyer's patches (<10 μm in diameter) (O'Hagan 1996). Though nanoparticles generally are <200 nm, studies show that nanoparticles ranging from 2 to 1,000 nm (1 μm) may be effective therapeutic or imaging agents (Singh and Lillard 2009).

Lowman et al. (1999) protected the insulin molecule from the acidity and digestive enzymes of the stomach by encasing it in a hydrogel (Lowman et al. 1999, Sharma et al. 2015). This poly(methacrylic-g-ethylene glycol) hydrogel was pH-sensitive, so it released the insulin molecule in the neutral and basic environment of the intestine. Oral delivery of insulin utilizing this hydrogel showed sustained efficacy, decreasing blood glucose level for 8 h with 25 IU/kg insulin dose. The onset of action. Hypoglycemic actions of insulin with this delivery strategy were dose-dependent and could achieve strong hypoglycemic effects within 2 h of administration in both healthy and diabetic rats.

Chitosan-based nanomolecules can enhance paracellular transport across intestinal epithelium and can prolong small intestinal residence time, both of which augment absorption. Intestinal residence time is increased because chitosan can adhere to the mucosal surface. Paracellular transport is increased because it can open the tight junctions tethering epithelial cells together (Pan et al. 2002). In this way, chitosan acts as a permeation agent, enabling chitosan-based nanomolecules to deliver hydrophilic molecules across the barrier and thus increase their absorption. Exploiting these properties, Lin et al. (2007) examined a chitosan-based nanoparticle as an insulin delivery system, showing increases in both insulin residence time in the small intestine and increases in paracellular transport across intestinal epithelium. However, polyelectrolyte chitosan-insulin complexes spontaneously dissociate in the acidic environment of the stomach, exposing it to proteolytic enzymes and thus limiting the bioavailability (Ma et al. 2015). It was reported that insulin-loaded chitosan/alginate polyelectrolyte complex NPs could undergo internalization through intestinal mucosa (Lin et al. 2007, Nam et al. 2010).

Sarmento et al. (2007a,b) designed an efficient oral insulin delivery system using DSS and chitosan nanoparticles to encapsulate insulin. Mukhopadhyay et al. (2015) also tested core-shell CS-alginate nanoparticles (100–200 nm) for as an oral insulin carrier. According to their report, the particles retained almost entire amount of insulin in simulated gastric fluid and gradually discharged the encapsulated protein in simulated intestinal condition. Only ~8.11% relative bioavailability was attained along with a significant hypoglycemic effect.

Cui et al. (2007) investigated nanoparticles based on PLGA, an FDA-approved synthetic polymer frequently used for drug delivery (Cui et al. 2007, Sharma et al. 2015). *In vivo* studies on reduction of serum glucose levels in diabetic rats were conducted using insulin-loaded PLGA (PNP) and a pH-sensitive derivative PLGA-Hp55 (PHNP). Hp55 is a pH-sensitive cellulose coating which protects against gastric acid but which dissolves in the higher pH environment of the small intestine, thus releasing its insulin load. Hp55 derivative PLGA nanoparticles had nearly twice the bioavailability of PLGA without Hp55. In another approach, the bioavailability of insulin-PLGA nanoparticles was augmented using a mucoadhesive chitosan derivative (Zhang et al. 2012, Sharma et al. 2015). In another approach, Zion et al. (2003) synthesized a nanoparticle which releases insulin on contact with glucose. The nanoparticle itself was a polymeric glucose cross-linked with the glucose-binding protein Concanavalin A (Con A). Interaction of Con A with free glucose triggers the release of polymeric glucose and further binds to free glucose, inducing the disintegration of the hydrogel nanoparticle (Sharma et al. 2015).

SLNs can be taken up by the lymphoid tissues in the Peyer's patches. Oral administration of lectin-modified SLNs with loaded insulin demonstrated reduced enzymatic degradation and enhanced oral absorption (Zhang et al. 2006). Lectins are proteins possessing affinity for specific carbohydrates. The prevalence of glycosylation of proteins and lipids in the gastrointestinal tract suggests that nanoparticles with lectin may target the intestinal tract and have increased uptake due to greater residence time.

In addition to drug delivery, the progression of Type 1 or 2 diabetes may be tracked through molecular imaging and biomedical imaging tools bolstered by nanotechnology (Naesens and Sarwal 2010). Though β-cell mass decreases as diabetes progresses (Matveyenko and Butler 2008), direct quantification of β-cell mass is not practical. To overcome this obstacle, nanoprobes are being developed with β-cell specificity and high contrast (Mailaisse and Maedler 2012). This technology could allow the quantification of endogenous β-cell mass *in vivo* (Lamprianou et al. 2011).

Supermagnetic iron oxide nanoparticles (SPIONs) represent a nanotechnology platform whose metal nature can be exploited for targeting, controlled release, and imaging. Magnetism can target SPIONS to the tissue of interest or trigger drug release, and the SPIONs may be imaged through MRI. Infiltration of immune cells has been tracked using SPION technology as an early detection tool for diabetes (Fu et al. 2012). SPIONs are readily taken up by immune cells such as macrophages. Further, a pilot clinical study on pancreatitis associated with diabetes demonstrated that a SPION-based MRI contrast-imaging agent enabled the visualization of the pancreas, and moreover, a twofold difference in T_2 relaxation time was found between nondiabetic healthy volunteers and recent-onset diabetes patients. This difference was due to islet inflammation in the diabetes patients. This suggests that diabetes may be diagnosed using this technology (Gaglia et al. 2011).

6.2.4 OCULAR DISEASES

Significant achievements have been made in the management of ocular diseases. However, due to the special physiological barriers and anatomical structures of the human eye, diagnoses and treatments of these disorders can suffer from low efficiency and lack of specificity. The current therapeutic methods seldom can completely restore vision loss or detect severe ocular diseases at an early stage (Schoenfeld et al. 2001, Weng et al. 2017). Therefore, the development of improved diagnostics and therapeutics for ocular disease is receiving intense attention.

The human corneal epithelium has multilayers of epithelial cells interconnected by tight junctions and, thus, severely limits ocular penetration of drugs, especially many types of hydrophilic molecules. Further, topical drug administration to the anterior segment of the eye is often limited by clearance mechanisms including eye blinking, tear film, tear turnover, solution drainage, and lacrimation (Gipson and Argüeso 2003). Most topically administered drugs are washed away within a few seconds after instillation because human tear film has a restoration time of only 2–3 min. If the volume of instillation of eye drops for treating common ocular anterior diseases such as corneal injury, dry-eye, keratitis, conjunctivitis, and cataract becomes >30 μL, most of the drug undergoes nasolacrimal drainage or gravity-induced drainage (Gaudana et al. 2009). Overall, the efficacy of the total administered drugs is <5%, suggesting the poor bioavailability of ocular drugs (Barar et al. 2008). Nanosystems represent an interesting strategy to circumvent this problem. Based on bioadhesive enhancement, sustainable release, stealth function, specifically targeted delivery, and stimuli-responsive release, several multi-functional nanosystems have been developed for the management of ocular diseases with the aim to improve drug delivery to both the anterior and posterior segments of the eye (Zarbin et al. 2010).

Vega et al. (2006) reported that PLGA nanoparticles (200 nm) could release flurbiprofen in a sustained manner and improve anti-inflammatory effect as compared to commercial flurbiprofen eye drops on the rabbit ocular inflammation model. Pignatello et al. (2002) reported that flurbiprofen-loaded acrylate polymer nanosuspensions (100 nm size) exhibited an equivalent inhibitory effect on the miotic response in a rabbit surgical trauma model even at a lower dosage than commercial eye drops. Natarajan et al. (2014) developed nano-unilamellar vesicles for glaucoma therapy in a sustainable fashion following single subconjunctival injection. Zhang et al. (2009) formulated dexamethasone (Dex)-loaded PLGA NPs (Dex-NPs) and evaluated their pharmacokinetics and tolerability in rabbits after intravitreal injection. They claimed that the injection of Dex-NPs induce no abnormalities even after 50 days post-injection in rabbits, while the Dex-NPs were able to maintain a sustained liberation of drug up to 50 days in vitreous with a fairly constant level of drug for up to 30 days. The rabbits treated with Dex-NPs showed significantly higher bioavailability of the drug as compared the control group injected with Dex alone. Jo (2012) studied the anti-angiogenic impacts of silicate nanoparticles (SiNPs) on the retinal neovascularization and showed that the SiNPs did not impose any direct toxicity in the retinal tissues. The intravitreal injection of NPs was able to substantially reduce the anomalous retinal angiogenesis in oxygen-induced retinopathy mice.

To treat the macular degeneration and diabetic retinopathy, intravitreal injections every 4–8 weeks are inevitable – a treatment modality that is undoubtedly considers as an invasive uncomfortable retinal damaging intervention. Hence, Huu et al. (2015) developed a novel stimuli-responsive NP-based reservoir platform for the delivery of nintedanib (BIBF 1120) which is a small molecule angiogenesis inhibitor. To this end, the researchers capitalized on a far ultraviolet (UV) light-degradable polymer to be able to trigger the liberation of the drug molecules on demand. Once injected, the NSs were found to be able to keep the encapsulated drug molecules in the vitreous for up to 30 weeks without inducing significant inadvertent side effects, while the liberation of cargo drug molecules was plausible through emission of far UV. They showed that the choroidal neovascularization (CNV) in rats can be suppressed 10 weeks after injection of nintedanib carrying NPs. In addition to biological barriers in the posterior segment, the specific microenvironment of intraocular cancers (retino blastoma (Rb) and uveal melanoma) is another therapeutic obstacle. Based on reports that folate receptors are overexpressed in Rb cells, folate-linked PLGA and chitosan nanoparticles have been proposed with sustainable, controllable, and targeted delivery of anticancer drug – DOX to Rb cells (Gupta et al. 2009, Parveen and Sahoo 2010). Despite great efforts devoted to the intraocular cancer therapy, the current studies are mainly limited in the stage of *in vitro* assessment, due to the lack of mature intraocular cancer animal models.

PDT is an emerging therapeutic strategy which consists of three functional modules: a light-activated photosensitizer, an energy laser beam to induce activation, and a surrounding oxygen environment with the ability to produce a toxic compound. One commercial drug Visudynes used for age-related macular degeneration (AMD) treatment is a typical PDT product. The active ingredient of Visudynes is a photo-activated drug – verteporfin. Upon a 689 nm laser depositing with a proper intensity, the drug can generate reactive oxygen species (ROS) and induce neovascular endothelial cell death, resulting in vessel occlusion and ending the growth of choroidal neovascular cells (Laville et al. 2006, Wilson, and Patterson 2008).

Recently, researchers have designed carbohydrate-targeted mesoporous silica nanoparticles (MSNP) encapsulated with both anticancer drug camptothecin (CPT) and one- or two-photon photosensitizers. Encouraging results were achieved showing that the MSNP nanoparticles presented an interesting therapeutic property by killing Rb cells efficiently *in vitro* (Gallud et al. 2013). Similar results were found in Wang et al.'s work, in which dendrimeric nanocarriers were developed with excellent cellular uptake, significant photo-efficiency, and superior phototoxicity in Rb cells (Wang et al. 2012). Although PDT showed great promising potential in some cancer treatments, more efforts are required on the development of nanosystems to implement PDT in ocular applications. There are several approaches employed for clinical ocular disease diagnoses, such as optical coherence tomography (OCT), fundus photography, fluorescein angiography, PET, MRI, ultrasonography, and confocal microscopy. They have played a significant role in monitoring disease recovery. For example, MRI is useful for monitoring progress of ocular diseases such as diabetic retinopathy, AMD, and ocular tumor angiogenesis by *in vivo* imaging of neovascularization (Yang et al. 2002, Townsend et al. 2008, De Potter et al. 1992).

However, due to poor imaging sensitivity or imaging resolution, each of these approaches has limited advantages for disease diagnosis. For example, PET has high sensitivity but limited spatial resolution, while MRI has good spatial resolution but low sensitivity (Kiyosawa et al. 1996, Finger et al. 2005). In order to overcome these drawbacks, nanotechnology seems to provide multiple options. Anderson et al. (2000) developed a Gd-perfluorocarbon nano-particulate emulsion linked with a biotinylated anti-αvβ3 monoclonal integrin antibody DM101.The system showed a site-directed contrast enhancement of angiogenic vessels in a rabbit corneal neovasculature model. After administrating the targeted agent for 90 min, the average MRI signal intensity was enhanced by 25% *in vivo*.

Although nanotechnologies in tumor diagnosis and therapy have been developed and evaluated in recent years, there are only limited studies focusing on ocular disease application. Yet strategies used in other diseases can also guide the treatment and diagnosis in ocular disease. Recently, Hosoya et al. (2016) developed a hydrogel nanosystem that combined tumor targeting, triggered drug delivery, and photon-to-heat conversion together to enable multimodal imaging and also controlled release of therapeutic cargo in human tumor xenografts. In this study, peptide-targeted phage particles, heat sensitive–based liposome (HSL), mesoporous silica nanoparticles (MSNPs), and photon-to-heat conversion were integrated into a hydrogel system. The HSL and MSNPs could generate heat after NIR laser illumination. The heat induced release of hydrogel contents, and meanwhile, the loaded drugs were controlled to release at tumor site 95. The nanoplatforms exhibited great potential for clinical application or diagnostic therapeutic monitoring and targeted delivery to malignant tumors and ocular diseases.

The systemic delivery of ocular drugs into posterior segment of the eye often fails because of the excellent barrier function of blood-retinal barrier (BRB). Since the current strategies upon efficiently delivery of the ocular drugs to the site of action within the eye and treat the ocular diseases provide limited successes, intraocular drug delivery and targeted therapy of the ophthalmic diseases appear to be very challenging.

6.2.5 CARDIOVASCULAR DISEASES

CVD encompasses all pathologies of the heart or circulatory system, including coronary heart disease, peripheral vascular disease, and stroke. The primary conditions underlying the great majority of CVDs are dyslipidemia, atherosclerosis, and hypertension. Angiotensin-converting enzyme inhibitors, angiotensin receptor blockers, anticlotting drugs, and beta blockers are the examples of drug classes which are commonly used in CVD. Up to date, these drugs are mostly introduced into the market in conventional formulations such as tablets or capsules. Recently, nanomedicine is gaining importance for the treatment of CVD. Since CVD has been the top killers for human beings, rapid and accurate diagnosis of CVD is critically important to save lives. The imaging techniques routinely used for CVD diagnosis are electrocardiography (ECG), chest X-ray, echocardiography, cardiac catheterization, and blood tests (Kakoti and Goswami 2013). These imaging techniques could only detect changes in the appearance of tissues when the symptoms were relatively advanced. Today, nanotechnology-based imaging applications are being refined with the goal of detecting disease as early as possible. Nanoparticles have the potential for imaging from its current anatomy-based level to the molecular level. Hence, nanoparticle imaging techniques cover advanced optical and luminescence imaging and spectroscopy, ultrasound, and X-ray imaging, MRI, and nuclear imaging with radioactive tracers most of which depend on targeting agents or contrast agents that have been introduced into the body to mark the disease site (Godin et al. 2010).

Nowadays, magnetically sensitive nanoparticles have become important tool in diagnostic application of detecting disease as early as possible (Gupta 2011). FDA has been approved Gadolinium (Gd) as a T_1 relaxation paramagnetic MRI contrast agent. In a study by Winter et al. (2006), Gd-decorated $\alpha v \beta 3$-integrin (overexpressed in angiogenesis)–targeted perflurocarbon nanoparticles were loaded with the anti-angiogenesis drug fumagillin. MRI was used to evaluate the response to treatment by measuring changes in signal enhancement in atherosclerotic plaques. Decreased signal enhancement was seen after treatment, which indicated reduced angiogenesis. Inhibiting angiogenesis has been shown to halt the progression of atherosclerosis (Moulton 2006). Winter et al. (2008) developed a prolonged antiangiogenesis therapy regimen based on theranostic $\alpha v \beta 3$–targeted nanoparticles. For this purpose, fumagillin was incorporated into perfluorooctylbromide nanoparticles to elicit acute antiangiogenic effects. The impetus for this study is the observation in animal models that chronically high systemic doses of a water-soluble version of fumagillin resulted in a decrease in neovascularization and plaque development; thus, a targeted nanoagent may allow for localized delivery requiring decreased dosing.

Smooth muscle cell (SMC) proliferation in the vessel wall is believed to promote restenosis, and therefore, inhibition of this process has been investigated as a strategy to prevent restenosis. In a study by Lanza et al. (2002), perfluorocarbon nanoparticles were used to deliver DOX and PTX, two potent anti-proliferation drugs, to SMCs in the arterial wall of restenotic vessels. The nanoparticles, with a size of 250 nm, were targeted to tissue factor expressed by SMCs, and their inhibited proliferation was observed *in vitro*. T_1-weighted MRI detected the uptake of particles while 19F MRI spectroscopy allowed distinguishing the nanoparticles from other tissue. Although no *in vivo* data were shown, this study exemplifies an interesting theranostic nanoparticle system for CVD. Due to the inflammatory nature of atherosclerotic plaques, anti-inflammatory drugs such as glucocorticoids have been proposed as a treatment, though the systemic side effects impede its application in clinical medicine. Lobatto et al. (2010) used paramagnetically labeled liposomes, with a size of 103 nm, as drug carriers to deliver glucocorticoids to atherosclerotic lesions in order to improve therapeutic efficacy. T_1-weighted MRI showed that liposomes could be detected in atherosclerotic plaques. To evaluate therapeutic efficacy, ^{18}F-fluoro-deoxy-glucose PET (18F-FDG-PET), a noninvasive imaging modality that can detect plaque inflammation, showed reduced uptake of FDG in plaques. McCarthy et al. (2010) developed a novel theranostic platform to target macrophages. Dextran-coated iron oxide nanoparticles were loaded with NIR fluorophores and phototoxic agents and resulted in a particle size of 49.9 nm. The nanoparticles specifically targeted macrophages and were activated by light to induce apoptosis in the targeted macrophages. In an ApoE KO mouse model, the nanoparticles were shown to localize in atherosclerotic plaques by intravital fluorescence microscopy. *Ex vivo* histology staining showed that the nanoparticles induced massive death of macrophages, while they showed less skin toxicity than free phototoxic agents.

The most important clinical applications of nanotechnology are in the area of pharmaceutical development, and pharmaceutical nanoparticles have gained great importance for the treatment of CVD. Soujanya et al. (2012) developed Glycoprotein Ib (GPIb)-conjugated dexamethasone-loaded biodegradable PLGA nanoparticles. Glycoprotein Ib (GPIb) was chosen as the targeting ligand and conjugated to nanoparticles because its role in platelet adhesion to the vascular wall. The results demonstrate that conjugation of GPIb to PLGA nanoparticles increased particle adhesion onto targeted surfaces and increased cellular uptake of these nanoparticles by activated endothelial cells under shear flow. Further, these nanoparticles provided a controlled release of the model drug. Therefore, these drug-loaded, GPIb-conjugated PLGA nanoparticles could be used as a targeted and controlled drug delivery system to the site of vascular injury for treatment of CVDs. Song et al. (1998) investigated the potential usefulness of biodegradable nanoparticles for the local intraluminal therapy of restenosis. NPs containing a water-insoluble anti-proliferative agent U-86983 were formulated using PLGA and used heparin, didodecylmethylammonium bromide, fibrinogen, or combinations as specific additives after particle formation, to enhance arterial retention. The *in vivo* studies on a new dog model for arterial angioplasty supported that the modified nanoparticles along with optimized infusion conditions could enhance arterial wall drug concentrations of agents to treat restenosis. Klugherz et al. (1999) have demonstrated transcatheter local delivery of the anti-restenotic agent probucol-loaded PLGA particles, in rabbit iliac arteries, for enhanced retention, sustained release, and increased therapeutic effects. Liposomes are made up of natural lipids and/or synthetic polymers and cholesterol as building blocks to form the bilayer structure surrounding a hydrophilic core. Some liposomal technologies have been reported for the treatment of CVD. Joner et al. (2008) developed cationic liposomal nanoparticles of prednisolone that specifically bind to chondroitin sulfate proteoglycans that are expressed within the subendothelial matrix but not vascular endothelial cells. *In vivo* studies were conducted with atherosclerotic New Zealand white Rabbits which were implanted with bare metal stents. Results showed that site-specific targeting by this nanoparticles steroid in injured atherosclerotic areas might be a valuable and cost-effective approach for the prevention of in-stent restenosis. The main nanocarrier classes researched as therapeutic and "theranostic" agents for atherosclerosis are liposomes with different surface characteristics. Hua et al. (2010) have developed the targeting perfluoropropane-containing liposomes to activated platelets via a peptide (RGDS) derived from the α-chain of fibrinogen. Initially, the peptide and rtPA were each modified with distinct fluorescent labels to obtain tracking of the microbubbles and thrombolytic drug. When injected into healthy rabbits, the thrombolytic drug was visualized within the liver by using ultrasound imaging. This may demonstrate the additive effect of microbubble rupture on the ultimate lysis of clots. Abraham et al. (2017) reported a non-toxic, cell-compatible cationic DC-cholesterol/(1,2-dioleoyl-sn-glycero-3-phosphoethanolamine) nanoliposomes as nanotheranostic agents to successfully deliver therapeutic mRNA-eGFP mRNA or the therapeutic anti-inflammatory, CD39 mRNA. No toxicity was found using these nano-plexes, but high cell viability was noted after transfection into CHO, HEK293, and A549 cell lines. Nanoplexes for the transfection of eGFP mRNA showed an increase in fluorescence signals on microscopy as compared to the mRNA control after 24 h in Chinese hamster ovary (CHO) cells. Nanoplexes for the transfection of CD39 mRNA showed increased CD39 expression as compared to the mRNA control after 24 h of using CHO cells. They stated that nanoliposome preparations for the delivery of therapeutic mRNA hold promise for future nanotheranostics in diseases such as inflammatory and CVDs, as well as cancer. The clinical utility of this approach, however, remains to be proven. CVDs are not as much benefited from nanotechnology in terms of drug delivery as the field of cancer and others. There are still unmet challenges in cardiovascular drug delivery that need to be overcome.

6.3 CONCLUSION

In this chapter, we have elaborated on the evolution and state of the art of disease-specific nanotheranostics, with an emphasis on clinical impact and translation. Nanoparticles are attractive candidates for better therapeutic applications for various diseases due to their controlled release properties, cell/tissue-targeting ability, and selectiveness. The surface of nanoparticles can easily be modified with ligands to make them useful for active targeting and consequently improve their targeting efficiencies. Imaging agents can be introduced with the drug carriers to make treatment much less troublesome and prognosis bright. A nanotheranostic can be a general approach in which diagnosis and therapy are interwoven to solve clinical issues and improve treatment outcomes. In most cases, the interesting results of theranostic nanomedicines in the literature are available only for *in vitro* studies. There are more challenges to be faced for their *in vivo* applications to pre-clinical and clinical levels. It appears that the advanced nanotheranostic becomes an approach of future generation. Much research efforts are needed towards the development of versatile and smart theranostic materials.

REFERENCES

Abraham, M.-K., Peter, K., Michel, T., Wendel, H.P., Krajewski, S., Wang, X. 2017. Nanoliposomes for safe and efficient therapeutic mRNA delivery: A step toward nanotheranostics in inflammatory and cardiovascular diseases as well as cancer. *Nanotheranostics* 1:154–65.

American Diabetes Association. 2004. Insulin administration. *Diabetes Care*, 27 (suppl 1): S106–S107.

Anderson, S.A., Rader, R.K., Westlin, W.F., et al. 2000. Magnetic resonance contrast enhancement of neovasculature with αvβ3-targeted nanoparticles. *Magn Reson Med* 44:433–39.

Andreozzi, E., Seo, J. W., Ferrara, K., Louie, A. 2011. Novel method to label solid lipid nanoparticles with 64cu for positron emission tomography imaging. *Bioconjug Chem*, 22(4):808–18.

Ascierto, M.L., Melero, I., Ascierto, P.A. 2015. Melanoma: from incurable beast to a curable bet. The success of immunotherapy. *Front Oncol.* doi: 10.3389/fonc.2015.00152.

Ashton, J.R., Castle, K.D., Qi, Y., Kirsch, D.G., West, J.L., Badea, C.T. 2018. Dual-energy CT imaging of tumor liposome delivery after gold nanoparticle-augmented radiation therapy. *Theranostics*, 8(7):1782–97.

Ballard, S.T., Hunter, J.H., Taylor, A.E. 1995. Regulation of tight-junction permeability during nutrient absorption across the intestinal epithelium. *Annu Rev Nutr* 15:35–55.

Barar, J., Javadzadeh, A.R., Omidi, Y. 2008. Ocular novel drug delivery: Impacts of membranes and barriers. *Expert Opin Drug Del* 5:567–81.

Beloqui, A., Coco, R., Memvanga, P.B., Ucakar, B., des Rieux, A., Préat, V. 2014. pH-sensitive nanoparticles for colonic delivery of curcumin in inflammatory bowel disease. *Int J Pharm*, 473(1–2):203–12. doi: 10.1016/j.ijpharm.2014.07.009.

Beloqui, A., Solinís, M.Á., Gascón, A.R., del Pozo-Rodríguez, A., des Rieux, A., Préat, V. 2013. Mechanism of transport of saquinavir-loaded nanostructured lipid carriers across the intestinal barrier. *J Control Release*, 166(2):115–23. doi: 10.1016/j.jconrel.2012.12.021.

Betzer, O., Ankri, R., Motiei, M., Popovtzer, R. 2015. Theranostic approach for cancer treatment: Multifunctional gold nanorods for optical imaging and photothermal therapy. *J Nanomat.* https://dl.acm.org/citation.cfm?id=2975086

Binder, C., Lauritzen, T., Faber, O., Pramming, S. 1984. Insulin pharmacokinetics. *Diabetes Care* 7:188–99.

Brenner, D.J., Hall, E.J. 2007. Computed tomography--an increasing source of radiation exposure. *N Engl J Med* 357:2277–84.

Caprilli, R., Frieri, G. 2009. The dyspeptic macrophage 30 years later: An update in the pathogenesis of Crohn's disease. *Dig Liver Dis*, 41(2):166–8. doi: 10.1016/j.dld.2008.09.012.

Carasi, P., Ambrosis, N.M., De Antoni, G.L., Bressollier, P., Urdaci, M.C., Serradell Mde, L. 2014. Adhesion properties of potentially probiotic Lactobacillus kefiri to gastrointestinal mucus. *J Dairy Res*, 81(1):16–23. doi: 10.1017/S0022029913000526.

Chen, K., Wu, H., Hua, Q., Chang, S., Huang, W. 2013. Enhancing catalytic selectivity of supported metal nanoparticles with capping ligands. *Phys Chem Chem Phys*, 15(7):2273–7. doi: 10.1039/c2cp44571a.

Choi, D., Jeon, S., You, D.G., et al. 2018. Iodinated echogenic glycol chitosan nanoparticles for x-ray CT/US dual imaging of tumor. *Nanotheranostics* 2:117–27.

Coco, R., Plapied, L., Pourcelle, V., Jérôme, C., Brayden, D.J., Schneider, Y.J., Préat, V. 2013. Drug delivery to inflamed colon by nanoparticles: Comparison of different strategies. *Int J Pharm*, 440(1):3–12. doi: 10.1016/j.ijpharm.2012.07.017.

Collnot, E.M., Ali, H., Lehr, C.M. 2012. Nano- and microparticulate drug carriers for targeting of the inflamed intestinal mucosa. *J Control Release*, 161(2):235–46.

Cu, Y., Saltzman, W.M. 2009. Mathematical modeling of molecular diffusion through mucus. *Adv Drug Deliv Rev*, 61(2):101–14. doi: 10.1016/j.addr.2008.09.006.

Cui, F.D., Tao, A.J., Cun, D.M., Zhang, L.Q., Shi, K. 2007. Preparation of insulin loaded PLGA-Hp55 nanoparticles for oral delivery. *J Pharm Sci* 96:421–27.

De Potter, P., Shields, C.L., Shields, J.A., Flanders, A.E., Rao, V.M. 1992. Role of magnetic resonance imaging in the evaluation of the hydroxyapatite orbital implant. *Ophthalmology* 99:824–30.

Finger, P.T., Kurli, M., Reddy, S., Tena, L.B., Pavlick, A.C. 2005. Whole body PET/CT for initial staging of choroidal melanoma. *Brit J Ophthalmol* 89:1270–74.

Frieri, G., Pimpo, M., Galletti, B., et al. 2005. Long-term oral plus topical mesalazine in frequently relapsing ulcerative colitis. *Dig Liver Dis*, 37:92–6.

Fu, W., Wojtkiewicz, G., Weissleder, R., Benoist, C., Mathis, D. 2012. Early window of diabetes determinism in NOD mice, dependent on the complement receptor CRIg, identified by noninvasive imaging. *Nature Immunol* 13:361–68.

Gaglia, J.L., Guimaraes, A.R., Harisinghani, M., et al. 2011. Noninvasive imaging of pancreatic islet inflammation in type 1A diabetes patients. *J Clin Invest* 121:442–45.

Gallud, A., DaSilva, A., Maynadier, M., et al. 2013. Functionalized nanoparticles for drug delivery, one-and two- photon photodynamic therapy as a promising treatment of retino- blastoma. *J Clin Exp Ophthalmol* 4:288.

Gaudana, R., Jwala, J., Boddu, S.H., Mitra, A.K. 2009. Recent perspectives in ocular drug delivery. *Pharm Res* 26:1197–216.

Gipson, I.K., Argüeso, P. 2003. Role of mucins in the function of the corneal and conjunctival epithelia. *Int Rev Cytol* 231:1–49.

Godin, B., Sakamoto, J.H., Serda, R.E., Grattoni, A., Bouamrani, A., Ferrari, M. 2010. Emerging applications of nanomedicine for the diagnosis and treatment of cardiovascular diseases. *Trends Pharmcol Sci* 31:199–5.

Gupta, A.S. 2011. Nanomedicine approaches in vascular disease: A review. *Nanomed NanotechBiolMed* 7:763–79.

Gupta, A.S., Kshirsagar, S.J., Bhalekar, M.R., Saldanha, T. 2013. Design and development of liposomes for colon targeted drug delivery. *J Drug Target*, 21(2):146–60. doi: 10.3109/1061186X.2012.734311.

Gupta, J., Boddu, S.H., Pal, D., Mitra, A.K. 2009. Targeted delivery of doxorubicin for the treatment of retinoblastoma. *Invest Ophthalmol Vis Sci* 50:5976.

Gupta, M.K., Lee, Y., Boire, T.C., Lee, J.-B., Kim, W.S., Sung, H.-J. 2017. Recent strategies to design vascular theranostic nanoparticles. *Nanotheranostics* 1:166–77.

Han, N.Y., Kim, E.H., Choi, J., Lee, H., Hahm, K.B. 2012. Quantitative proteomic approaches in biomarker discovery of inflammatory bowel disease. *J Dig Dis*, 13(10):497–503. doi: 10.1111/j.1751-2980.2012.00625.x.

Harel, E., Rubinstein, A., Nissan, A., Khazanov, E., Nadler Milbauer, M., Barenholz, Y., Tirosh, B. 2011. Enhanced transferrin receptor expression by proinflammatory cytokines in enterocytes as a means for local delivery of drugs to inflamed gut mucosa. *PLoS One*, 6(9):e24202. doi: 10.1371/journal.pone.0024202.

Hosoya, H., Dobroff, A.S., Driessen, W.H., et al. 2016. Integrated nanotechnology platform for tumor-targeted multimodal imaging and therapeutic cargo release. *Proc Natl Acad Sci USA* 113:1877–82.

Hua, X., Liu, P., Gao, Y.H., et al. 2010. Construction of thrombus-targeted microbubbles carrying tissue plasminogen activator and their in vitro thrombolysis efficacy: A primary research. *J Thromb Thrombolysis* 30:29–35.

Huu, V.A., Luo, J., Zhu, J., et al. 2015. Light-responsive nanoparticle depot to control release of a small molecule angiogenesis inhibitor in the posterior segment of the eye. *J Control Release* 200:71–7.

Jin, Y., Song, Y., Zhu, X., et al. 2012. Goblet cell-targeting nanoparticles for oral insulin delivery and the influence of mucus on insulin transport. *Biomaterials* 33:1573–82.

Jo, D.H., Kim, J.H., Yu, Y.S., Lee, T.G., Kim, J.H. 2012. Antiangiogenic effect of silicate nanoparticle on retinal neovascularization induced by vascular endothelial growth factor. *Nanomedicine* 8:784–91.

Jo, S.D., Ku, S.H., Won, Y.Y., Kim, S.H., Kwon, I.C. 2016. Targeted nanotheranostics for future personalized medicine: Recent progress in cancer therapy. *Theranostics* 6(9):1362–1377.

Joner, M., Morimoto, K., Kasukawa, H., et al. 2008. Site-specific targeting of nanoparticle prednisolone reduces in-stent restenosis in a rabbit model of established atheroma. *Arterioscler Thromb Vasc Biol* 28:1960–66.

Kakoti, A., Goswami, P. 2013. Heart type fatty acid binding protein: Structure, function and biosensing applications for early detection of myocardial infarction. *Biosens Bioelectron* 43:400–11.

Kiyosawa, M., Inoue, C., Kawasaki, T., et al. 1996. Functional neuroanatomy of visual object naming: A PET study. *Graefes Arch Clin Exp Ophthalmol* 234:110–15.

Klugherz, B.D., Meneveau, N., Chen, W., Wade, W.F., Papandreou, G., Levy, R. 1999. Sustained intramural retention and regional redistribution following local vascular delivery of polylactic-coglycolic acid and liposomal nanoparticulate formulations containing probucol. *J Cardiovasc Pharmacol Ther* 4:167–74.

Lamprecht, A., Ubrich, N., Yamamoto, H., Schäfer, U., Takeuchi, H., Maincent, P., Kawashima, Y., Lehr, C.M. 2001. Biodegradable nanoparticles for targeted drug delivery in treatment of inflammatory bowel disease. *J Pharmacol Exp Ther*, 299(2):775–81.

Lamprianou, S., Immonen, R., Nabuurs, C., et al. 2011. High-resolution magnetic resonance imaging quantitatively detects individual pancreatic islets. *Diabetes* 60:2853–60.

Lanza, G.M., Yu, X., Winter, P.M., et al. 2002. Targeted antiproliferative drug delivery to vascular smooth muscle cells with a magnetic resonance imaging nanoparticles contrast agent implications for rational therapy of restenosis. *Circulation* 106:2842–47.

Laroui, H., Wilson, D.S., Dalmasso, G., Salaita, K., Murthy, N., Sitaraman, S.V., Merlin, D. 2011. Nanomedicine in GI. *Am J Physiol Gastrointest Liver Physiol*, 300(3):G371–83. doi: 10.1152/ajpgi.00466.2010.

Lautenschläger, C., Schmidt, C., Lehr, C.M., Fischer, D., Stallmach, A. 2013. PEG-functionalized microparticles selectively target inflamed mucosa in inflammatory bowel disease. *Eur J Pharm Biopharm*, 85(3 Pt A):578–86. doi: 10.1016/j.ejpb.2013.09.016.

Laville, I., Pigaglio, S., Blais, J.C., et al. 2006. Photodynamic efficiency of diethyleneglycol–linked glycoconjugated porphyrins in human retino blastoma cells. *J Med Chem* 49:2558–67.

Law, G.-L., Wong, W.-T. 2014. An introduction to molecular imaging. In: *The Chemistry of Molecular Imaging*, eds. N. Long, W.-T. Wong, 1–24. Hoboken, NJ: John Wiley & Sons, Inc.

Lin, Y.H., Mi, F.L., Chen, C.T., et al. 2007. Preparation and characterization of nanoparticles shelled with chitosan for oral insulin delivery. *Biomacromolecules* 8:146–52.

Lobatto, M.E., Fayad, Z.A., Silvera, S., et al. 2010. Multimodal clinical imaging to longitudinally assess a nanomedical anti-inflammatory treatment in experimental atherosclerosis. *Mol Pharm* 7:2020–29.

Lowman, A.M., Morishita, M., Kajita, M., Nagai, T., Peppas, N.A. 1999. Oral delivery of insulin using pH-responsive complexation gels. *J Pharm Sci* 88:933–37.

Lusic, H., Grinstaff, M.W. 2013. X-ray computed tomography contrast agents. *Chem Rev.* doi: 10.1021/cr200358s.

Ma, Z., Lim, T.M., Lim, L.Y. 2005. Pharmacological activity of peroral chitosan insulin nanoparticles in diabetic rats. *Int J Pharm* 293:271–80.

Makhlof, A., Tozuka, Y., Takeuchi, H. 2009. pH-Sensitive nanospheres for colon-specific drug delivery in experimentally induced colitis rat model. *Eur J Pharm Biopharm*, 72(1):1–8.

Malaisse, W.J., Maedler, K. 2012. Imaging of the β-cells of the islets of Langerhans. *Diabetes Res Clin Pract* 98:11–18.

Manconi, M., Nacher, A., Merino, V., et al. 2013. Improving oral bioavailability and pharmacokinetics of liposomal metformin by glycerolphosphate–chitosan microcomplexation. *AAPS PharmSciTech* 14:485–96.

Marik, J., Tartis, M.S., Zhang, H., et al. 2007. Long-circulating liposomes radiolabeled with [18F]fluorodipalmitin ([18F]FDP). *Nucl Med Biol* 34:165–71.

Matveyenko, A.V., Butler, P.C. 2008. Relationship between β-cell mass and diabetes onset. *Diabetes Obes Metab* 10:23–31.

McCarthy, J.R., Korngold, E., Weissleder, R., Jaffer, F.A. 2010. A light activated theranostic nanoagent for targeted macrophage ablation in inflammatory atherosclerosis. *Small* 6:2041–49.

Moulton, K.S. 2006. Angiogenesis in atherosclerosis: Gathering evidence beyond speculation. *Curr Opin Lipidol* 17:548–55.

Mukhopadhyay, P., Chakraborty, S., Bhattacharya, S., Mishra, R., Kundu, P.P. 2015. pH-sensitive chitosan/alginate core-shell nanoparticles for efficient and safe oral insulin delivery. *Int J Biol Macromol* 72:640–48.

Muthu, M.S., Leong, D.T., Mei, L., Feng, S-S. 2014b. Nanotheranostics - Application and further development of nanomedicine strategies for advanced theranostics. *Theranostics*, 4(6):660–677.

Muthu, M.S., Mei, L., Feng, S-S. 2014a. Nanotheranostics: Advanced nanomedicine for the integration of diagnosis and therapy. *Nanomedicine*, 9(9):1277–1279.

Naesens, M., Sarwal, M.M. 2010. Molecular diagnostics in transplantation. *Nat Rev Nephrol* 6:614–28.

Nam, J.-P., Choi, C., Jang, M.-K., et al. 2010. Insulin-incorporated chitosan nanoparticles based on polyelectrolyte complex formation. *Macromol Res* 18:630–35.

Natarajan, J.V., Darwitan, A., Barathi, V.A., et al. 2014. Sustained drug release in nanomedicine: A long-acting nanocarrier- based formulation for glaucoma. *ACS Nano* 8:419–29.

O'Hagan, D.T. 1996. The intestinal uptake of particles and the implications for drug and antigen delivery. *J Anat* 189:477–82.

Pan, Y., Li, Y.J., Zhao, H.Y., et al. 2002. Bioadhesive polysaccharide in protein delivery system: Chitosan nanoparticles improve the intestinal absorption of insulin in vivo. *Int J Pharm* 249:139–47.

Parveen, S., Sahoo, S.K. 2010. Evaluation of cytotoxicity and mechanism of apoptosis of doxorubicin using folate-decorated chitosan nanoparticles for targeted delivery or etinoblastoma. *Cancer Nanotechnol* 1:47–62.

Phelps, M.E. 2000. PET: The merging of biology and imaging into molecular imaging. *J Nucl Med* 41:661–81.

Pignatello, R., Bucolo, C., Spedalieri, G., Maltese, A., Puglisi, G. 2002. Flurbiprofen-loaded acrylate polymer nanosuspensions for ophthalmic application. *Biomaterials* 23:3247–55.

Sarmento, B., Ribeiro, A., Veiga, F., Ferreira, D., Neufeld, R. 2007a. Oral bioavailability of insulin contained in polysaccharide nanoparticles. *Biomacromolecules* 8:3054–60.

Sarmento, B., Ribeiro, A., Veiga, F., Sampaio, P., Neufeld, R., Ferreira, D. 2007b. Alginate/chitosan nanoparticles are effective for oral insulin delivery. *Pharm Res* 24:2198–206.

Schmidt, C., Lautenschlaeger, C., Collnot, E.M., et al. 2013. Nano- and microscaled particles for drug targeting to inflamed intestinal mucosa: A first in vivo study in human patients. *J Control Release* 165:139–45.

Schoenfeld, E.R., Greene, J.M., Wu, S.Y., Leske, M.C. 2001. Patterns of adherence to diabetes vision care guidelines: Baseline findings from the diabetic retinopathy awareness program. *Ophthalmology* 108:563–71.

Serra, L., Doménech, J., Peppas, N.A. 2009. Engineering design and molecular dynamics of mucoadhesive drug delivery systems as targeting agents. *Eur J Pharm Biopharm* 71:519–28.

Singh, R., Lillard, J.W. Jr. 2009. Nanoparticle-based targeted drug delivery. *Exp Mol Pathol* 86:215–23.

Sharma, G., Sharma, A.R., Nam, J., Doss, G.P.C, Lee S.S., Chakraborty, C. 2015. Next generation efficient therapy for type 1 diabetes. *J Nanobiotechnol* 13:74.

Song, C., Labhasetwara, V., Cui, X., Underwood, T., Levy, R.J. 1998. Arterial uptake of biodegradable nanoparticles for intravascular local drug delivery: Results with an acute dog model. *J Control Release* 54:201–11.

Soujanya, K., Jing-Fei, D., Yaling, L., Jifu, T., Kytai, TN. 2012. Biodegradable nanoparticles mimicking platelet binding as a targeted and controlled drug delivery system. *Int J Pharm* 423:516–24.

Stewart, B., Wild, C.P. (eds) World Cancer Report 2014. WHO. www.iarc.fr/en/publications/books/wcr/wcr-order.php

Subramani, K., Pathak, S., Hosseinkhani, H. 2012. Recent trends in diabetes treatment using nanotechnology. *Dig J Nanomater Biostruct* 7:85–95.

Sumner, B., Gao, J. 2008. Theranostic nanomedicine for cancer. *Nanomedicine (Lond)*, 3(2):137–40. doi: 10.2217/17435889.3.2.137.

Takeuchi, H., Matsui, Y., Yamamoto, H., Kawashima, Y. 2003. Mucoadhesive properties of carbopol or chitosan-coated liposomes and their effectiveness in the oral administration of calcitonin to rats. *J Control Release* 86:235–42.

Tamura, A., Ozawa, K., Ohya, T., Tsuyama, N., Eyring, E.M., Masujima, T. 2006. Nanokinetics of drug molecule transport into a single cell. *Nanomedicine* 1:345–50.

Thirawong, N., Thongborisute, J., Takeuchi, H., Sriamornsak, P. 2008. Improved intestinal absorption of calcitonin by mucoadhesive delivery of novel pectin–liposome nanocomplexes. *J Control Release* 125:236–45.

Tirosh, B., Khatib, N., Barenholz, Y., Nissan, A., Rubinstein, A. 2009. Transferrin as a luminal target for negatively charged liposomes in the inflamed colonic mucosa. *Mol Pharm* 6:1083–91.

Townsend, K.A., Wollstein, G., Schuman, J.S. 2008. Clinical application of MRI in ophthalmology. *NMR Biomed* 21:997–2.

Vega, E., Egea, M.A., Valls, O., Espina, M., García, M.L. 2006. Flurbiprofen loaded biodegradable nanoparticles for ophthalmic administration. *J Pharm Sci* 95:2393–405.

Viscido, A., Capannolo, A., Latella, G., Caprilli, R., Frieri, G. 2014. Nanotechnology in the treatment of inflammatory bowel diseases. *J Crohn's Colitis* 8:903–18.

Wang, Z.J., Chauvin, B., Maillard, P., et al. 2012. Glycodendrimeric phenylporphyrins as new candidates for retinoblastoma PDT: Blood carriers and photodynamic activity in cells. *J Photochem Photobiol B* 115:16–24.

Weng, Y., Liu, J., Jin, S., Guo, W., Liang, X., Hu, Z. 2017. Nanotechnology-based strategies for treatment of ocular disease. *Acta Pharm Sinica B* 7:281–91.

Wileman, T.E., Lennartz, M.R., Stahl, P.D. 1986. Identification of the macrophage mannose receptor as a 175-kDa membrane protein. *Proc Natl Acad Sci USA* 83:2501–5.

Wilson, B.C., Patterson, M.S. 2008. The physics, biophysics and technology of photodynamic therapy. *Phys Med Biol* 53:R61.

Winter, P.M., Caruthers, S.D., Zhang, H., Williams, T.A., Wickline, S.A., Lanza, G.M. 2008. Antiangiogenic synergism of integrin-targeted fumagillin nanoparticles and atorvastatin in atherosclerosis. *JACC Cardiovasc Imaging* 1:624–34.

Winter, P.M., Neubauer, A.M., Caruthers, S.D., et al. 2006. Endothelial alpha(v)beta3 integrin-targeted fumagillin nanoparticles inhibit angiogenesis in atherosclerosis. *Arterioscler Thromb Vasc Biol* 26:2103–9.

World Health Organization. Cancer. www.who.int/news-room/fact-sheets/detail/cancer. Updated September 12, 2018. (accessed April 22, 2019).

Wu, B., Lu, S-T., Yu, H., et al. 2018. Gadolinium-chelated functionalized bismuth nanotheranostic agent for in vivo MRI/CT/PAI imaging-guided photothermal cancer theapy. *Biomaterials*, 159:37–47.

Xiao, B., Laroui, H., Ayyadurai, S., et al. 2013. Mannosylated bioreducible nanoparticle-mediated macrophage-specific TNF-alpha RNA interference for IBD therapy. *Biomaterials* 34:7471–82.

Xie, Z., Chen, G., Zhang, X., et al. 2013. Salivary microRNAs as promising biomarkers for detection of esophageal cancer. *PLoS One*, 8(4):e57502. doi: 10.1371/journal.pone.0057502.

Yang, M.S., Hu, Y.J., Lin, K.C., Lin, C.C. 2002. Segmentation techniques for tissue differentiation in MRI of ophthalmology using fuzzy clustering algorithms. *Magn Reson Imaging* 20:173–79.

Yang, X. 2017. Science to practice: Enhancing photothermal ablation of colorectal liver metastases with targeted hybrid nanoparticles. *Radiology*, 285(3):699–701.

Yu, X., Li, A., Zhao, C., Yang, K., Chen, X., Li, W. 2017. Ultrasmall semimetal nanoparticles of bismuth for dual-modal computed tomography/photoacoustic imaging and synergistic thermoradiotherapy. *ACS Nano* 11:3990–4001.

Yu, Y., Jin, H., Holder, D., et al. 2010. Urinary biomarkers trefoil factor 3 and albumin enable early detection of kidney tubular injury. *Nat Biotechnol*, 28(5):470–7. doi: 10.1038/nbt.1624.

Yun, Y., Cho, Y.W., Park, K. 2013. Nanoparticles for oral delivery: Targeted nanoparticles with peptidic ligands for oral protein delivery. *Adv Drug Deliv Rev* 65:822–32.

Zarbin, M.A., Montemagno, C., Leary, J.F., Ritch, R. 2010. Nanotechnology in ophthalmology. *Can J Ophthalmol* 45:457–76.

Zhang, L., Li, Y., Zhang, C., Wang, Y., Song, C. 2009. Pharmacokinetics and tolerance study of intravitreal injection of dexamethasone-loaded nanoparticles in rabbits. *Int J Nanomed* 4:175–83.

Zhang, N., Ping, Q., Huang, G., Xu, W., Cheng, Y., Han, X. 2006. Lectin-modified solid lipid nanoparticles as carriers for oral administration of insulin. *Int J Pharm* 327:153–59.

Zhang, X., Sun, M., Zheng, A., Cao, D., Bi, Y., Sun, J. 2012. Preparation and characterization of insulin-loaded bioadhesive PLGA nanoparticles for oral administration. *Eur J Pharm Sci* 45:632–38.

Zion, T.C., Tsang, H.H., Ying, J.Y. 2003. Glucose-sensitive nanoparticles for controlled insulin delivery. http://hdl.handle.net/1721.1/3783 (accessed July 06, 2018).

7 The Development of Adoptive T-Cell Immunotherapies for Cancer
Challenges and Prospects

David J. Dow, Steven J. Howe,
Mala K. Talekar, and Laura A. Johnson
GlaxoSmithKline R&D

CONTENTS

7.1 INTRODUCTION

7.1.1 Cancer Immunotherapy

The field of cancer immunotherapy began in the 19th century with William Coley's studies injecting live *Streptococcus bacilli* bacteria and heat-killed bacterial extracts into primary tumors resulting in an induced durable remission of inoperable sarcoma (Burdick, 1937; Decker and Safdar, 2009). Fast forward to today where the successful development of CTLA-4 and PD-1 checkpoint inhibitors, along with the recent FDA approval of Kymriah and Yescarta, two CAR-T cell immunotherapies, heralds a new era in cancer medicine.

Along this journey, there were many notable discoveries. In 1909, Nobel Prize winner Paul Ehrlich proposed that the immune system can suppress tumor formation, and this subsequently became known as the "cancer immune surveillance" hypothesis. In the 1980s, Boon, De Plaen, and Lurquin demonstrated that cytotoxic T lymphocytes (CTL) could recognize a mutated version of a normal protein and that the CTL recognized a peptide bearing the mutated residue (De Plaen et al., 1988). This was direct evidence that the immune system can survey the genome and that genomic mutations, a hallmark of cancer, can cause cancer cells to be immunogenic. In 2001, Schreiber, Old, and others provided experimental evidence of Ehrlich's hypothesis showing that mice lacking certain immune cell types had increased susceptibility to B-cell lymphomas (Shankaran et al., 2001).

Today there are three main modalities being developed in the cancer immunotherapy field. These are checkpoint inhibitors, adoptive T-cell therapies, and cancer vaccines (Lohmueller and Finn, 2017). While briefly mentioning the important advance of checkpoint inhibitors, this chapter will focus exclusively on the development of adoptive T-cell therapies.

7.1.2 Checkpoint Inhibitors

The recent discovery and development of immune checkpoint inhibitors represents a major advance in cancer immunotherapy. Two groups independently reported evidence in mice that a receptor on T lymphocytes called CTLA-4 had a negative regulatory function on T lymphocytes (Waterhouse et al., 1995; Tivol et al., 1995). In the following year, Allison and Krummel reported that a monoclonal antibody (mAb) directed against CTLA-4 resulted in rejection of tumors in a mouse melanoma model and importantly conferred immunity to a second exposure of the same tumor cells (Leach et al., 1996). This work demonstrated that blocking the inhibitory effects of CTLA-4 can result in an effective immune response against tumor cells. Naturally this important finding led to significant interest in identifying further inhibitory regulators of T-cell function. Nishimura and colleagues demonstrated that a gene knockout of the PD-1 receptor in mice resulted in autoimmune syndromes, suggesting that the PD-1 receptor is also an inhibitory regulator of immune responses (Nishimura et al., 1999). These basic discoveries in T-cell biology were subsequently developed into marketed medicines. The CTLA-4-specific mAb, ipilimumab (Yervoy®), demonstrated a survival advantage in patients with advanced melanoma (Hodi et al., 2010). Following this, pembrolizumab (Keytruda®) was granted accelerated approval for advanced melanoma, making it the first PD-1 inhibitor approved by the FDA (Khoja et al., 2015). Today, many other trials are ongoing in a multitude of different tumor types, combining checkpoint inhibitors together and with other traditional cancer therapies such as chemotherapy and radiation therapy.

7.1.3 Adoptive T-Cell Therapy

In another approach using T cells, pioneering studies of tumor infiltrating lymphocytes (TILs) led to the development of cellular immunotherapies and a process called adoptive T-cell therapy. In adoptive TIL therapy, TILs are extracted from patient tumors and cultured outside the body *in vitro*

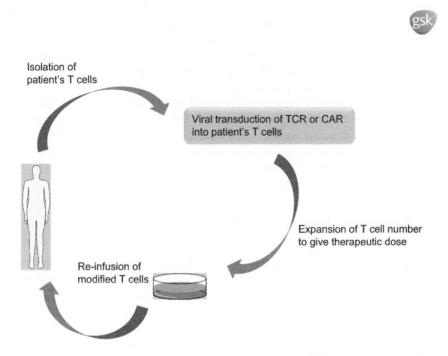

FIGURE 7.1 Adoptive T-cell therapies. The medicine is made from the patient's own cells. T cells are extracted from the patient's blood, often an engineered TCR is introduced to the cells *in vitro* using a viral vector. Alternatively, TILs are isolated from tumor. The patient undergoes a "pre-conditioning regime" where existing cells are removed to encourage engraftment of sufficient numbers of therapeutic cells. The cells are activated and expanded in number *in vitro* before being re-infused into the patient. As a result, autologous therapies are very personalized medicines and a manufacturing run is required for each patient.

(also called *ex vivo*) before returning the cultured cells to the patient (see Figure 7.1). Employing this method led to tumor regressions in patients with advanced melanoma (Dudley et al., 2002). This approach is highly personalized, since the patient's own cells form the foundation of the medicine. As we will describe later, the T cells can also undergo molecular engineering to give them new properties. By using the patient's own cells instead of those from a different individual, immune rejection in the form of graft versus host disease (GvHD) can be avoided.

7.2 TCRS AND CARS

7.2.1 T Cell Biology

T lymphocytes are important components of the adaptive immune system. Through a process of genetic rearrangement, each T cell carries a unique receptor, similar in structure to an antibody, whose function is to recognize and respond to antigens. Collectively the T cell population has a highly diverse set of T-cell receptors (TCRs), capable of recognizing foreign and self-proteins, including viral proteins and proteins from cancer cells (Sewell, 2012). The main type of TCR is a heterodimer composed of an α and a β polypeptide and typically recognizes 8–10 amino acid peptide bound to the surface of an major histocompatibility complex (MHC) class I molecule or up to 30 amino acids when bound to MHC class II molecules (Attaf et al., 2015). TCRs bind to peptide-MHC through a complex with endogenous CD3 subunits and in CTL, an additional complex with the CD8 co-receptor. This leads to recruitment of further proteins resulting in signal transduction through intracellular signaling pathways, ending in lysis of the target cancer cell.

The human leukocyte antigen (HLA) complex is polymorphic in the human population. The most frequent form of HLA is HLA-A*0201 (HLA-A2) which is present in 40%–50% of the Caucasian population. For this reason, many TCRs in clinical development are restricted to patients with this HLA-A2 type, as this is the most common sub-type giving the largest treatable patient population. Treating patients with other less common HLA types requires the development of additional TCRs, as the TCR is specific for the peptide-HLA complex.

There are two main classes of T lymphocyte defined by their cell surface markers and function. CD4 helper T cells respond to peptide antigens presented by MHC class II and have diverse functions, including activating macrophages and helping antibody-producing B cells respond to antigens. CD8 T cells recognize foreign peptides displayed by MHC class I on antigen-presenting cells. Following activation and division, these cells become cytotoxic CD8 T cells that recognize the same peptide-MHC class I complex on tumor cells, resulting in tumor cell killing.

7.2.2   Tumor Infiltrating Lymphocytes (TILs)

Several studies have reported on the association between the presence of TILs in tumors and clinical prognosis in several tumor types including colon and breast (Clemente et al., 1996; Galon et al., 2006). The development of TIL therapy began some 30 years ago when Rosenberg et al. showed that T cells isolated from tumors recognized tumor-derived antigens, and these T cells, when subsequently expanded with the addition of interleukin (IL)2, could be effective at reducing pulmonary and hepatic metastatic tumors in mice (Shiloni et al., 1986). One of the challenges of this approach is that to dose a patient, a large number of cells are typically required, in the order of up to 10^{11} T cells (Rosenberg and Restifo, 2015). Extracting small numbers of TILs from tumor and expanding these to give such a number can be a complex and challenging task and not all tumors have sufficient numbers of TILs.

Over the years, this method has evolved to incorporate sophisticated molecular techniques. The advent of high-throughput deep sequencing technologies, in particular, exome sequencing, has enabled the identification of a patient's individual tumor mutations on a genome-wide scale. Coupled with *in vitro* systems that transcribe mRNA or pulse mutation containing peptides for presentation by autologous antigen-presenting cells, this results in the identification of which mutations are immunogenic (Tran et al., 2017). This approach has revealed that most patients with melanoma and epithelial cancers mount immune responses to these so-called neoantigens (Tran et al., 2017).

It is also now possible to isolate T cells that are specific for those mutations, culture to sufficient quantities, and return to the patient as a personalized TIL therapy. Subsequent treatment of patients with this TIL method has yielded some promising clinical results. In a recent study, a patient with metastatic colorectal cancer was treated with TILs recognizing a KRAS G12D neoantigen and showed regression of all seven metastatic lung lesions. A single lesion progressed 9 months after the therapy and was shown by molecular analysis to contain a deletion of the HLA region, providing evidence for resistance through antigen escape in response to TIL therapy (Tran et al., 2016).

7.2.3   Engineered T Cells and TCRs

As an alternative to the extraction and propagation of TILs, a TCR, often engineered to be affinity enhanced, can be introduced into the patient's T cells. In this approach, a viral vector is used to deliver α and β TCR chains into the T cells (Kuball et al., 2005). Early work by Linda Sherman's group in mice, then work on a p53 TCR (Cohen et al., 2005), and a MART-1 TCR (Hughes et al., 2005) translated into first successful solid tumor TCR clinical trial in metastatic melanoma (Morgan et al., 2006). Subsequently, studies showing that increased affinity TCRs work better *in vitro*

(Johnson et al., 2006) translated into increased efficacy in patients (Johnson et al., 2009). However, the method also carried with it increased toxicity to shared antigen on normal tissues.

Engineering TCRs, using specific nucleotide replacement in binding regions to increase peptide-MHC affinity, may help to overcome the challenge of the low frequency of endogenous high-affinity T cells to tumor antigens. This could perhaps provide clinical benefit to patients who do not respond to treatment with checkpoint inhibitors. This approach has been shown to increase the affinity of the wild-type TCRs for their natural ligand peptide/MHC class I complex by 10–1,000 fold *in vitro* for several antigens including gp100, MART-1, and NY-ESO-1 (Johnson et al., 2009; Robbins et al., 2008). Higher affinity TCRs allow T cells to respond to lower levels of antigen. This could prove important where the tumor has adapted by reducing both antigen expression and expression of MHC class I molecules (Barrett and Blazar, 2009; Marincola et al., 2000).

7.2.4 CHIMERIC ANTIGEN RECEPTORS (CARs)

Another major advance in the T-cell cancer immunotherapy field was the discovery and development of chimeric antigen receptors (CARs). CARs were first described by Zelig Eshhar who introduced the concept of "T-bodies" in 1989 (Gross et al., 1989). While TCRs recognize intracellular-processed antigens as cleaved peptides bound to MHC molecules, CARs recognize only cell surface antigens and are not limited by MHC restriction. Initially referred to as "immunoglobulin-TCR chimeric molecules," the first CAR-Ts were made by incorporating the heavy and light chain variable regions of an mAb and a TCR constant region into a T lymphocyte cell line. Later, the same group modified their approach to generate a single-chain variable fragment (scFv) encoding both heavy and light regions physically joined together by an amino acid linker sequence, negating the need for multiple gene transfers to achieve the combined variable heavy and light chain antibody receptor specificity (Eshhar et al., 1996). So-called first-generation CARs included only the scFv joined to an intracellular CD3ζ TCR signaling motif. Upon scFv binding to cognate antigen, the signal transduction via CD3ζ would activate the T cell to release cytokines and lyse the target cell. For use as a cancer therapeutic, the first clinical trials were run using first-generation constructs targeting folate receptor in patients with ovarian cancer (Kershaw et al., 2006), carbonic anhydrase IX (CAIX) in patients with renal cancer (Lamers et al., 2006), and CD171/L1-CAM in pediatric patients with neuroblastoma (Park et al., 2007). While these early trials did not demonstrate antitumor efficacy or durable cell engraftment, the CAIX trial did lead to reports of unexpected biliary tract toxicity, with follow-up studies finding CAIX expression in this normal tissue; this was to be the first of several on-target, off-tumor toxicities.

The first successful clinical trial demonstrating direct antitumor CAR effects was published by Martin Pule and Malcolm Brenner's group (Pule et al., 2008). Using a first-generation CAR targeting disialoganglioside (GD2) in pediatric patients with neuroblastoma, they were able to show radiographic antitumor responses and detection of engineered T cells in the periphery 24 h later, albeit at low levels (0.1%). Preclinical studies from multiple groups followed evaluating the benefit of adding in a "second signal" to support the CAR T-cell survival and engraftment, and the benefits of using CD28 or 4-1BB were demonstrated to provide a more powerful T cell that survived longer. A further follow-up included adding two different costimulatory stimuli, resulting in "third-generation" CARs; however, these do not appear to have added benefits over second-generation CARs, and the combination may have a net negative effect on T-cell metabolic cost and survival. Even in light of the small early successes of CAR- and TCR-engineered T cells to treat patients with cancers, the high costs and challenging conditions required to make such a therapy kept it from becoming a mainstream treatment. Then came CD19 CAR-T therapy for B-cell malignancies.

Fairly rapidly after the identification of CD19 as an "ideal" target antigen for B-cell leukemias and lymphomas, at least three main groups embarked on use of second-/third-generation

CD19 CAR-T therapy for blood cancers. Steven Rosenberg's group at the US NCI was the first to publish their interim clinical results, in 2010 (Kochenderfer et al., 2010), followed closely in 2011 by Carl June's group at the University of Pennsylvania (Porter et al., 2011), and Michel Sadelain's team at MSKCC (Brentjens et al., 2013). These ignited a series of follow-up clinical studies published in different B-cell–based indications and treated both pediatric and adult patients. The results were pivotal; objective response rates of 40%, 50%, even 90% were regularly occurring in patients with extremely advanced and pretreated refractory malignancies that had no curative regimen.

With these unprecedented response rates, some even ventured to talk about the "C" word – Cure. While only time will tell, there are certainly patients from those early trials that remain cancer-free, living full and healthy lives, almost a decade later. With minimal on-target toxicity limited to loss of healthy circulating B cells, patients receive prophylactic injections of gamma-globulins, although some research suggests this may not even be necessary as the patient's underlying memory plasma cells remain intact. The more serious toxicity is not an on-target one but rather a systemic reaction to the high levels of T-cell activation throughout the body, termed cytokine release syndrome (CRS). While initially considered a potentially fatal toxicity, much work has centered around identifying the mechanism behind the CRS toxicities and how to treat it while maintaining patient antitumor response. One treatment that has now become standard is to block IL-6 functioning by giving patients tocilizumab, an IL-6R blocking mAb that is clinically available, as IL-6 has been demonstrated clinically to act as a mediator of CRS.

Since this time, now in 2017/2018, these CD19 CAR-T therapies have been adopted by Pharma companies and have resulted in global pivotal clinical trials and registration as a true "medicine" led by Kymriah™ by Novartis, who backed Carl June's work at the University of Pennsylvania. With support from Gilead, Kite Pharma has launched Yescarta™ based on the NCI's initial work. Follow-up from a third major group, Juno Therapeutics, launched by Michel Sadelain (MSKCC) and their team of collaborators is likely very close behind.

With these historical launches of the first regulated gene-engineered CAR-T medicines, the scientific and medical communities are asking, "What's next?" While CD19 was the ideal "low hanging fruit" with ubiquitous tumor expression on B cells and a safe normal tissue profile, the challenge now is to translate into other tumor types, in particular, solid cancers. Both the Rosenberg group and Carl June's group evaluated CARs targeting the tumor-specific epidermal growth factor variant III (EGFRvIII) mutation in glioblastoma (GBM) (Johnson et al., 2015; Morgan et al., 2012; Sampson et al., 2014). However, the NCI group has closed the study to enrolment, and while University of Pennsylvania group published encouraging biomarker-driven data from the trial, ultimately there were no patient objective responses observed (O'Rourke et al., 2017).

City of Hope (Duarte, CA) recently published the results of their clinical trial for patients with GBM, using a "zetakine" CAR comprised of a recombinant IL-13 molecule tethered to the intracellular TCRζ signal. Of note, though it was a single-patient response reported. This patient initially received CAR-T via an intravenous route (usual for CAR-T) and with no response observed, received additional infusions via intrathecal injection into the central nervous system (CNS). Notably, following the intrathecal injections, the patient underwent an object radiographic tumor response. Although this response was ultimately short-lived, it demonstrated proof of concept that CAR-T cells can work in patients to eliminate solid tumors.

Over the past decade, much more has become known about the suppressive factors involved in the solid tumor microenvironment. There is not one but many different axes upon which tumors have evolved to bypass or actively suppress endogenous tumor immunity. One of the most well-described lately is the PD-1/PD-L1/2 axis for T-cell suppression, others include soluble cytokines or chemokines including transforming growth factor (TGF)β, IL10, and additional suppressive chemicals like IDO and PGE2 produced by tumor or other infiltrating immune cells. A key challenge in the development of CAR-T technology will be to overcome this immunosuppressive environment.

Table 7.1

Comparison of the Pros and Cons for a TCR versus a CAR Approach

	Pros	Cons
TCR	• Demonstrated efficacy in solid tumors. • Intracellular targets can be targeted	• HLA restriction limits treatable patient population. • Antigen presentation can be downregulated via tumor mutations in antigen processing and presentation machinery genes
CAR	• Demonstrated efficacy in hematological tumors • Not restricted by HLA haplotype	• Unproven approach in treating solid tumors • Targets are restricted to those expressed on the cell surface

7.2.5 COMPARING TCRs AND CARs

So far we have discussed both TCRs and CARs, but which is the better approach? In reality, both approaches have advantages and disadvantages. As mentioned previously, in contrast to CARs, the TCR binds to a complex of a short peptide bound to HLA class I. Most nucleated cells present peptides on their cell surface through this HLA system. Through a process of protein turnover, intracellular peptides are generated by the proteosome to be presented on to the cell surface. TCRs can recognize intracellular cancer-specific antigens, including mutated versions of intracellular proteins, an advantage of TCRs over CARs. In fact, it has been demonstrated that T cells can recognize cancer-specific antigens that differ by only one amino acid from the normal cellular protein (Tran et al., 2017). However, tumor cells have been shown to evade the T cells by downregulation of HLA class I, which can be associated with poor prognosis (Speetjens et al., 2008).

For TCR therapies, the treatable patient population must express both the TCR target and the correct HLA type and so the treatable patient population is limited. In addition, two different diagnostic tests for these are required to establish eligibility for treatment. As CARs are HLA-independent, only one test for antigen expression is required. However, these points must be balanced by the fact that TCRs, in contrast to CARs, have demonstrated more reproducible clinical efficacy in solid tumors (Morgan et al., 2006). Table 7.1 summarizes further the pros and cons of TCRs versus CARs as an approach.

7.2.6 TARGETS FOR T-CELL IMMUNOTHERAPIES

Target discovery and validation represent a challenge for both TCR and CAR therapeutic approaches. As TCR or CAR-T cells are directed against the expression of an antigen, it is important to show clear expression or splice differences between normal and tumor tissue that can be targeted safely. The challenge is to identify those antigens expressed solely on tumors and not normal tissue. The heterogeneity of tumors is also a challenge, as in any given tumor, the target may be expressed on certain tumor cells but not others and expression levels may vary between cells. Treatment-elicited immunological responses that result in antigen spreading, where following cell death a secondary immune response is raised against tumor mutations other than the target of the therapeutic T cell, may be required for complete tumor clearance (Gulley et al., 2017).

While CAR-T cells are restricted to targets expressed on the cell surface, there are three classes of target that TCRs can recognize. These are: (a) proteins over-expressed in cancer cells compared with normal cells; (b) proteins not expressed by normal cells, but their expression is dysregulated in cancer cells – an example of this are the cancer testes antigens, including NY-ESO-1, expressed by several tumor types, whose normal expression is restricted to the immune privileged testes tissue; and (c) proteins that have been mutated in cancer cells giving rise to different amino acid sequences versus normal cells. These are called neoantigens, somatic mutations arising in tumor and not present in normal cells.

7.3   TRANSLATIONAL CHALLENGES OF T-CELL THERAPIES: CONCEPT TO CLINIC

Translational medicine encompasses the making of a medicine, from selection of a target, through *in vitro* development via bioinformatics and molecular biology, to testing in T cells *in vitro*, evaluation *in vivo*, and ultimately encompasses practical questions including actual scale-up and production of the cell product and delivery to patients in a clinical setting. While many pharmaceutical companies have been translating concepts into medicines for years and even decades, most of these have traditionally been small molecule inhibitors, antibodies, or other synthesized drugs that have readily measurable qualities and controlled parameters for delivery. Using gene-engineered T cells as a medicine faces numerous challenges, many of which need novel evaluations and solutions to answer questions that previously had straightforward answers, such as mechanism of action (MoA), known as dose–response curves, and well-established models for evaluating safety.

To consider these in order, the first, MoA, may initially seem obvious: the patient's T cell is made to transgenically express a tumor-specific receptor, conferring the ability to recognize and kill tumor cells bearing this marker. While this may be the proximal form of the medicine in the laboratory, and the effects on tumors can be directly evaluated *in vitro* by observing tumor destruction or indirectly via T-cell activation in the form of immunologic cytokine secretion including IFNγ and TNFα, the ultimate cause of tumor eradication in patients remains unknown. While laboratory testing may measure CAR- and TCR-engineered T-cell antitumor function over hours or days, in the patient, disappearance of tumor occurs over weeks and months. In the context of immunomodulators such as checkpoint blockade therapies, cytokine induction of changes in the tumor microenvironment can extend to months and years. The original cell product infused into patients is no longer the same product 1 year, 1 month, 1 week, and even 1 day later. The initial delivery of the engineered T cells into a lymphopenic environment provides excesses of the prolymphocyte homeostatic expansion factors IL-7 and IL-15. These cytokines, as well as the encounter with cognate tumor antigen act to stimulate the T cell, causing DNA replication, immune cytokine production, Fas/FasL upregulation, and molecules necessary for target cell perforation and degranulation. The activated T cell will then start to divide, continuing to do so multiple times over each 24-h period, resulting in a very different cell product and dosage than what was originally administered. Because each product is derived from a different individual's cells, they will have inherent differences in activation levels, expansion capacity, and internal programming, as well as effector, memory, and even suppressor-type T cells. These differences have required a full rewrite of the way to launch these as a "medicine" in the eyes of industry, manufacturing, and the regulatory agencies.

As pertains to pharmacokinetics and pharmacodynamics, there is currently no way to effectively track the engineered T cells in the patient after infusion in real time, beyond evaluation of persistence in peripheral blood, which may or may not reflect what is taking place at the tumor site. While still in its relative infancy, clinical observations of biomarkers in resected or biopsied CAR- and TCR-treated solid tumors from patients suggest that this type of directed cell therapy is sufficient to change the overall tumor microenvironment, including direct impacts on tumor, as well as the composition and phenotype of infiltrating immune cells.

In a recently published study by O'Rourke et al. (2017) from the University of Pennsylvania, patients with recurrent GBM, a grade IV brain tumor, were infused with autologous CAR-T engineered to express an scFv-based receptor recognizing EGFRvIII with intracellular 4-1BB and CD3ζ signaling. A subset of these patients was scheduled to undergo tumor resection at a predetermined time after CAR-T administration, thereby allowing for direct evaluation of cell trafficking and entry to tumors, as well as functional phenotype via immunohistochemistry (IHC) and RNA-ISH (*in situ* hybridization). On comparing tumors sampled from patients prior to treatment, and at various time points following CAR infusion, there were numerous changes to the tumor itself and the surrounding tumor microenvironment. Notably, in seven out of seven cases, the pretreatment tumors were immunologically "cold," with low T-cell infiltration and, in particular, low CD8+ effector T cells. After CAR treatment, tumor samples from four of seven patients evaluated demonstrated a marked increase in T-cell infiltration (both CAR-engineered and not), of both CD8 and CD4 T cells, and these showed an activated Type 1 immune signature, with granzyme B and interferon gamma production, correlated with tumor necrosis and reduction of EGFRvIII antigen expression. At later time points, however, the tumor microenvironment changed yet again. The presence of T cells was markedly reduced, in particular with loss of CD8 T cells. And while CD4 T cells remained, they were no longer effector-like, rather they expressed FoxP3, characteristic of regulatory T-suppressor cells.

Perhaps surprising to some is that the single biggest barrier to establishing this as an effective large-scale treatment for cancer is actually not the science and biology of the T cells working on the tumors, as Steve Rosenberg's group at the US NCI Surgery Branch has been showing that since 1998 (Dudley et al., 2002) and 2006 (Morgan et al., 2006) for endogenous T cells and gene-engineered T cells, respectively. Rather, the ability to manufacture the "drug" at the scale required for patient treatment, outside of boutique leading academic institutions and industry, represents a significant challenge as we will describe in the next section.

An additional challenge to translating gene-engineered T-cell therapy as a widely available medicine includes the multiple different regulatory requirements to register in each country, including aspects as diverse as the demographic inclusion requirements of the associated clinical trials through varied requirements for vector productions, shipping and handling of cell products and reimbursement for payment of the therapy.

7.4 MANUFACTURING MODIFIED T CELLS

7.4.1 OVERVIEW OF MANUFACTURING STEPS

There are many different approaches to modifying T cells for immunotherapy. The majority, especially those in commercial use, currently use a virus to modify a patient's own (autologous) cells *ex vivo*. The complexity of this procedure and number of unit operations required represent a significant challenge for manufacturing teams.

A generalized autologous *ex vivo* process involves the following major steps:

1. Removal of cells from patient.
2. Isolation and/or enrichment of the target cell population.
3. Culturing and activation of the cells.
4. Modification of cells, most frequently using a viral vector, to deliver a genetic construct (CAR or TCR, for example).
5. Expansion of the engineered cells to produce the drug substance.
6. Formulation of the drug product.
7. Release testing.
8. Clinical delivery of modified T cells to the patient.

This process is summarized in Figure 7.2; challenges and progress will be discussed in more detail.

7.4.2 PRODUCTION OF VIRAL VECTORS

Engineered viruses, or vectors, most frequently gamma-retroviruses or lentiviruses from the retrovirus family, are used to deliver genes for T-cell modification and application in immunotherapy (Vormittag et al., 2018). Application of these vectors in early gene therapy clinical trials for rare disease provided the foundation from which *ex vivo* T-cell processing for oncology assets has been further developed over the last three decades (Blaese et al., 1995). Vectors are disabled,

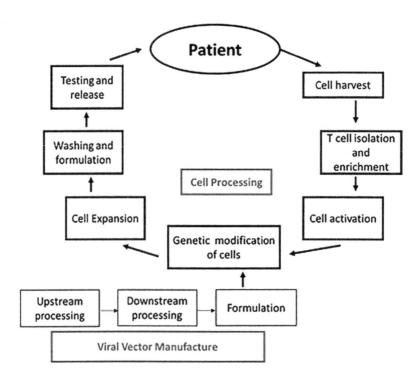

FIGURE 7.2 An outline of the manufacturing process to generate modified T cells. Each stage comprises multiple unit operations. Viral vectors for gene delivery are produced in a separate process. Analytics and quality control systems are embedded throughout the process.

non-pathogenic delivery vehicles which cannot replicate but effectively transduce hematopoietic cells (including T cells) with a genetic construct of choice. The capacity of retroviruses is generally sufficient to deliver CAR or TCR cassettes expressed from a heterologous promoter (Kumar et al., 2001) and the "payload" is integrated into the target cell chromosome following cell entry enabling long-term gene expression (Lesbats et al., 2016).

Production of sufficient quantities of vector is a current limitation in late-phase clinical trials or commercial application of modified T cells for common oncology applications; a fact that reached public attention through an article in the "*New York Times*" recently (see the "References" section). The most widespread method to generate vector, at research or small industrial scale, is from producer cells grown in adherent cultures in tissue culture flasks. These cells are normally a variation of Human Embryonic Kidney 293 cell (Graham et al., 1977), engineered to produce vector in a transient process where plasmid DNA is transfected into cells to deliver genes encoding all the necessary components; the genetic backbone, or transfer vector, along with the structural and enzymatic proteins to package functional viral particles (Howe and Chandrashekran, 2012). Viral components are assembled at the cell membrane before budding into the growth media for collection and downstream processing. This method is effective but has limitations, including lack of scalability, reproducibility, and cost. Clinical-grade DNA plasmids are required to be generated by specialist commercial vendors, which can then be transferred to one of the limited number of cGMP-approved suppliers able to generate and produce clinical-grade vector.

Teams of highly trained personnel are involved in the process and much of the upstream work is "open" and requires significant manual input. In recent years, production has been scaled up through the development and implementation of multilayered cell factories, but as producer cells are adherent, increase in scale is limited by the surface area of the vessel. For this reason, many manufacturers are developing upstream processes where HEK 293 cells have been adapted for growth in hollow fiber reactors or suspension vessels such as stirred tank bioreactors. Here, vector production is in three dimensions and so limited only by volume. Due to the nature of transient transfection, delivery of packaging DNA needs to be performed close to the point of harvest so the plasmids are not diluted out from the producer cells as they grow. Reproducible transfection of plasmid DNA in large volumes is challenging and the main driver behind efforts to generate stable producer cell lines.

Due to the potential cytotoxicity of different vector components, the production of high titer, self-inactivating, stable producer cell lines has been elusive, especially for lentiviruses. However, cell line development efforts are now focusing on high-throughput automated approaches to generate clonal producer lines with inducible expression of vector components. Suspension culture of stable producer cell lines simplifies scalability of vector production using Good Manufacturing Practice (GMP) beyond the tens of liters per batch. Additional benefits may include perfusion feed, continual processing approaches, or other methods to generate long-term lines to increase the number of harvests per campaign if there is sufficient demand.

Downstream processing remains a major challenge, balancing the requirement for removing unwanted contaminants such as plasmid DNA and cellular protein whist maximizing vector purity, viability, and yield. Production of larger batches will, depending on patient numbers, reduce the frequency of vector manufacture, so the formulation of the vector at the end of the downstream process is important. The formulation should be designed and tested to improve stability during any freeze thaw steps (for instance, if fill–finish from bulk into final containers must be carried out at a remote site) and to extend the shelf life of the vector in the selected storage conditions to ensure it is stable for a sufficient period. The biology and production of retroviral vectors were extensively reviewed recently (Merten and Wright, 2016). Analytical testing is vital to determine the quality and quantity of vector produced, enabling accurate dosing to obtain sufficient numbers of transduced cells expressing the amount of TCR or CAR required.

7.4.3 CELL PROCESSING

Peripheral blood mononuclear cells are collected from the patient by apheresis to generate an autologous product (Figure 7.2). T cells are isolated and particular lineages, such as CD4 or CD8, can be enriched if necessary. The cells are then activated with a cocktail of cytokines or growth factors to initiate cell division, which will increase cell number but can also facilitate improved gene delivery by viral vectors. This transduction step genetically modifies the T cells to enhance their ability to recognize, and kill, tumor cells. Following further expansion and washing of the cells, they are formulated and released for infusion into the patient. These steps are described in greater detail by Levine et al. (2017). Depending on how and where the cells are manufactured, shipping may be required to and from the clinical site.

Compared to the manufacture of small molecules or biopharmaceuticals where one batch of drug product is sufficient to treat many patients, the concept of a production run to treat a single patient with an autologous product requires considerable innovation to treat large numbers of oncology patients with a commercially viable product. The reproducibility and robustness of the process remain vital but made complex by the inherent variability of input material (cells) from different patients with wide-ranging medical histories, collected from a number of different medical centers, often using varying procedures. Once the optimum cell product has been defined, manufacturing processes must ensure that the heterogeneous population of transduced and non-transduced cells meets the same specification every time.

Early-phase clinical trials to date have generally employed manual cell handling processes, that require considerable expertise to harvest cells from the patient, isolate or enrich the T-cell population required, activate the cells to improve vector transduction, transduce with vector, and expand cell numbers to ensure the necessary amounts of material are produced for treatment (see Figure 7.1). Generation of sufficient cell numbers can take ~2 weeks, highlighting the resource required to complete the complex and varied standard operating procedures. Each step comprises multiple unit operations, and there are various established and developing technologies for carrying out cell isolation, activation, and expansion, as summarized by Vormittag et al. (2018) currently performed in clean rooms using open processes. While it is possible to scale out this approach to treat a greater number of patients, it is economically challenging, and producing vector or processing cells under GMP conditions requires specialist knowledge, equipment, facilities, and experience (Abou-El-Enein et al., 2013). GMPs must be applied throughout to ensure consistent production of autologous T cells within regulatory frameworks, adhering to strict quality standards to minimize risks inherent in pharmaceutical manufacture.

Building, maintaining, and validating large clean room facilities are very expensive, and recruiting a large workforce of highly skilled operatives is not trivial. There are also risks to product quality when using open systems with multiple, complex manual steps. The length of the process coupled with the potential to be handling multiple patient samples concurrently, in the same facility, represents a challenge for ensuring product identity, which is of paramount importance for autologous therapies.

Increasing patient numbers in later phase trials and moving towards the more demanding manufacturing regulation required for commercialization have prompted many to pursue automation and methods for closing the cell process. Equipment suppliers are generating platforms to perform multiple cell handling processes and reduce the requirement for operator interaction by producing modular or all-in-one systems that carry out the tens or hundreds of the necessary unit operations. In T-cell therapy manufacture, multiple approaches have been summarized by Wang and Riviere (2016). Closed systems and devices greatly reduce the risk to product sterility and reduce the potential for human error which should reduce batch failure rates; assuming the variation in starting material can be accommodated. Concurrently, automation should reduce the complexity of technology transfer, making manufacture at multiple sites a possibility. They could also potentially reduce the level of cleanroom classification required and unlock the potential to manufacture multiple products simultaneously in the same facility, reducing cost and the number of operators required. The risks associated with such an approach need to be fully assessed and the optimal operational capacity modeled.

7.4.4 Analytics and Quality Control

Analytical testing is an essential component of developing and improving manufacturing processes in drug manufacture. The strategy and assays applied during development usually change as the product moves through clinical testing phases and into commercial production; the wide array of tests used to initially understand biological parameters and develop processes normally become refined to use faster, more accurate, validated assays tailored to measure increasingly defined critical quality attributes of the product. Measuring and controlling various parameters such as cell division, viability, and cell lineages within the population are important to ensure a consistent product. Further testing is required at the end of the process to ensure the drug product is safe and effective before returning cells to the patient.

Improvement of many assays used in cell and gene therapy is required. For example, sterility testing can take up to 2 weeks currently, which creates a problem when releasing "fresh" cell products that have not been cryopreserved. Such cells have a very short shelf life and release of the product may be necessary before full test results are obtained. Similarly, assays to demonstrate vector safety, such as tests for replication competent lentiviruses (RCLs), can take months and require significant amounts of material (Cornetta et al., 2018). The price of many tests, especially when outsourced to specialist laboratories, can contribute to a considerable proportion of the cost of goods. As processes are improved and subsequently altered throughout the drug development pathway, extensive analytical testing is required to prove comparability of the medicine generated by the modified process.

The potential for deviations in the production of T cell therapies is high, due in part to the nature of the patient-derived input material and the way in which it is collected at multiple sites. As well as characterizing the starting cells and other raw materials, quality control testing of vector and drug product stability over time, the purity, potency, and identity of the product are regulatory requirements to ensure patient safety. Validation of assays and equipment, rigorous staff training, regulated documentation, and assurance of the manufacturing facilities are required to safeguard product quality and safety.

7.4.5 MANUFACTURING STRATEGIES AND SUPPLY CHAIN

Different models exist for the manufacture of T-cell therapies, and the benefits and risks must be carefully considered when selecting the strategy, taking market location and size, clinical phase, and location of contract manufacturing organizations into account. The production of vectors is generally centralized, as these can be frozen and shipped along with other starting/raw materials to the cell processing site. Cell processing can be carried out at a single central facility, although a decentralized model is also possible; multiple smaller manufacturing sites may be preferred if the target population is large or widely distributed geographically or the process requires the use of fresh cells. Manufacturing at two or more sites also de-risks the possibility of a catastrophic event closing single sites that could block drug supply (Bethencourt, 2009). Alternatively, the complexity in shipping patient material could be reduced by cell processing close to, or at, clinical sites if the necessary facilities, manufacturing processes (preferably closed and highly automated), and staff are available. All require a robust and highly regulated supply chain, both of raw materials entering the process to ensure product availability and of the released drug product back to the patient. Depending on the model used, cryopreservation of the drug product may be advantageous, and so, correct formulation of the transduced cells is required to ensure sufficient product stability for a defined length of time in controlled storage/shipping conditions.

As a mix-up of autologous products could have potentially fatal consequences for the patient, accurate labeling and tracking of cells between the manufacturing site and the patient are of paramount importance. If more than one site is required to produce the T-cell products, manufacture between the sites must be harmonized to ensure medicines of the same quality are produced. Multicenter production will require effective quality control and quality assurance to ensure medicinal products meet regulatory requirements, and a robust technology transfer process will be needed when using new contract manufacturing organizations (CMOs) or opening new sites.

7.4.6 FUTURE DIRECTIONS

Large-scale manufacture of autologous gene therapy products is a major challenge that is generating rapid improvements in techniques, equipment, and strategies to treat large populations of patients. Alternatives to viral gene delivery in T cells are being tested for improved efficacy and safety (Monjezi et al., 2017). While autologous therapies are currently the state of the art with outstanding efficacy, the difficulty and cost of manufacturing one batch per patient will further drive innovation, and significant effort is being applied to developing allogeneic or "off-the-shelf" products (Qasim et al., 2017) or direct *in vivo* treatments of cells within the body (Smith et al., 2017), where eventually hundreds or thousands of patients may be treated from a single production run.

7.5 CLINICAL DEVELOPMENT AND SAFETY

The bench to bedside translation of adoptive immunotherapy with T cells essentially involves the isolation of T cells from either the tumor itself (TILs), or from peripheral blood, *ex vivo* genetic engineering of these T cells to equip them with tumor-specific antigens (TCRs or CARs), *in vitro* amplification of these modified T cells with cytokine support and stimulatory antibodies and then re-introduction of these engineered T cells back into the patient. Further modification of these T cells by proinflammatory cytokines and or tethered ligands have generated "armored" T cells capable of secreting cytokines or expressing soluble or tethered ligands that intend to further improve the clinical efficacy (Brentjens and Curran, 2012). Although the majority of adoptive T-cell therapies have involved the use of autologous T cells, proof-of-concept clinical activity has been demonstrated with allogeneic or off-the-shelf CAR-Ts that come with the promise of a single donor providing competent-modified CAR-T's for several patients, albeit at the added risk of host responses to allogeneic products.

The foray of CAR-T cells and TCR immunotherapies has revolutionized the treatment of difficult-to-treat and refractory malignancies. CAR-T therapy, with its unprecedented success, has opened a new frontier in the treatment of particularly aggressive, relapsed/ refractory hematologic malignancies such as pediatric and young adult acute B-lymphoblastic leukemias (B-ALL) and diffuse large B-cell lymphomas (DLBCL) in adults. TCRs have similarly demonstrated clinical activity in solid tumors as has been seen with use of a TCR targeting the cancer testis antigen NY-ESO-1 in patients with metastatic melanoma and metastatic synovial sarcoma and some hematologic malignancies such as multiple myeloma.

Improved understanding of the genetic engineering mechanisms, evolving laboratory technologies, and emerging clinical data from the ongoing clinical trials in the CD-19-targeted CAR-T cell therapy space has paved the way for advancement in the basic structure of the CAR-T cells (Figure 7.3), raising the hope of improved efficacy and limited/manageable toxicities.

Most of the clinical data that currently exists in the CAR-T space arises from clinical trial experience with CD19-directed CAR-T therapies and to some extent with CD22-targeted CAR-T therapies in hematologic malignancies (leukemia – ALL and chronic lymphocytic leukemia (CLL); and non-Hodgkin lymphoma – mostly DLBCL). The clinical safety discussion regarding CAR-Ts herein, therefore, mostly derives from the CD19-directed CAR-T clinical experience.

While CD-19-targeted CAR-T cell therapies have provided long-term remission to patients with relapsed or refractory B-cell malignancies, the treatment comes with the risk of immuno-inflammatory cascade of host responses that can be potentially life-threatening. The two principal

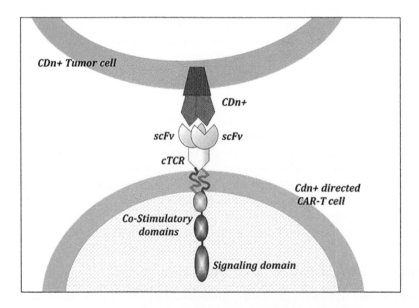

FIGURE 7.3 CDn-directed CAR-T cell representative construct where CD, Cluster of Differentiation; "n" represents a tumor cell surface antigen such as CD19; cTCR, chimeric TCR; and scFv, single-chain variable fragment.

toxicities widely associated with CAR-Ts in the clinic have been the CRS – which is the most common toxicity, and, encephalopathy/neurotoxicity, also referred to as CRES (CAR-T cell-related encephalopathy syndrome).

7.5.1 CYTOKINE RELEASE SYNDROME

Although *in vivo* proliferation of CAR-T cells post-infusion is desirable, the exponential multiplication (100–100,000x) of activated T cells is capable of inducing a vigorous inflammatory response in the host that comes with marked elevations in cytokine levels. The clinical constellation of signs and symptoms resulting from this phenomenon has been termed "cytokine release syndrome" (CRS). In addition to being seen with CAR-T cells, CRS has also been seen with the clinical use of other T-cell engaging therapies such as bispecific T-cell engaging (BiTE) antibodies such as blinatumomab (Teachey et al., 2016).

Although data from several groups suggests a correlation between development of CRS and efficacy, the degree of manifest CRS has not been found to be predictive of response to therapy (Davila et al., 2014; Lee et al., 2015; Maude et al., 2014b). However, the disease burden at the time of CAR-T cell infusion has been found to correlate with the severity of CRS (Maude et al., 2014b, 2015; Teachey et al., 2016).

Clinical symptoms of CRS can range from mild and flu-like (headache, myalgias, fevers) to a severe inflammatory syndrome including vascular leak, hypotension, shock, pulmonary edema, and coagulopathy leading to multi-organ failure and even death (Maude et al., 2014a). CRS is felt to be related to elevation of cytokines resulting from recognition and engagement of the targeted tumor antigen by the T cells and their consequent activation (Porter et al., 2018). The grading of CRS (Table 7.2) has been attempted by several groups (Brudno and Kochenderfer, 2016; Lee et al., 2014; Porter et al., 2015). However, these systems have not been adapted symmetrically across CAR-T trials yet. As more information is appreciated about neurologic toxicities, such as mental status changes, aphasia, and seizures, these events may need to be graded separately or merged into

Table 7.2
CRS Grading Scale

Grade of CRS	Type of Reaction	Description
Grade 1	Mild	Symptoms/signs (S/S): Flu-like symptoms – fever, headache, nausea, vomiting Treatment (T/t): Supportive care such as anti-pyretics, anti-emetics
Grade 2	Moderate	S/S: Some signs of organ dysfunction (elevation in creatinine and/or transaminases) related to CRS and not attributable to any other condition, fevers with associated neutropenia T/t: Requiring IV therapies or parenteral nutrition; hospitalization for management of CRS-related symptoms
Grade 3	Severe	S/S: Organ dysfunction including significant elevations in creatinine/transaminases related to CRS and not attributable to any other conditions; excludes management of fever or myalgias. T/t: Hospitalization required for management of symptoms – intravenous fluids (multiple fluid boluses for blood pressure support) for hypotension or low-dose pressors, coagulopathy requiring FFP or cryoprecipitate, and hypoxia requiring supplemental O_2 (nasal cannula oxygen, high-flow O_2, CPAP, or BiPAP)
Grade 4	Life-threatening	S/S and T/t: Hypotension requiring multiple, escalating or high-dose vasopressors, hypoxia requiring mechanical ventilation, requirement for dialysis
Grade 5	Fatal	Death

Source: Adapted from the Penn Grading System for CAR-T cell-associated CRS (Porter et al., 2015).

the current CRS grading. The current scales also do not utilize serum cytokine measurements for grading as rapid turnaround for these tests are not yet available at most centers.

Most patients have evidence of measurable elevations in their cytokine levels, although they might not always be in proportion to their clinical symptoms (Klinger et al., 2012). Levels of acute phase reactants such as C-reactive protein (CRP), ferritin, lactate dehydrogenase (LDH), and transaminases and additionally those of biomarkers such as interleukins (IL-2R, IL-6, IL-10, IL-13), IFNγ, sgp130, MIP1α, and MIP1β have been shown to be consistently elevated during CRS (Chen et al., 2016; Teachey et al., 2016). Although generally correlating well with the degree of CRS, these cytokine elevations do not predict CRS. Predictive models are being developed, and there is some initial evidence that shows elevations in cytokines such as IFNγ and sgp130 could predict severe CRS when measured within the first 72h of CAR-T cell infusion (Teachey et al., 2016).

Management of CRS is challenging as one needs to weigh the benefit of accelerated T-cell proliferation against the risk of loss of efficacy/persistence of activated T cells. Cytokine-targeted approaches to treat CRS need to ensure that the elevations are involved in causing CRS and are not required for T-cell effector function so as not to inhibit T-cell activity and, thereby, clinical efficacy. IL-10, IL-6, and IFNγ have been found to be the most significantly elevated cytokines during CRS. IL-10 is a negative regulator while IFNγ is an effector cytokine released by activated T cells. IL-6, in contrast, which is strikingly elevated in CAR-T-associated CRS response, has not been found to be required for T-cell efficacy making it an amenable target. IL-6 blockade with tocilizumab, a humanized mAb against IL-6 receptor (anti-IL6R), has demonstrated efficacy in reversing life-threatening CRS without compromising clinical activity of T-cell engaging therapies (Grupp et al., 2013; Teachey et al., 2013). Although steroids could be beneficial in minimizing the inflammatory response in CRS, the potential negative impact of steroids on T-cell proliferation restricts their use to treat CRS post-CAR-T cell infusion, and further use of steroids, in the months following the infusion, has been limited to life-threatening emergencies. However, to note, a recent model-based cellular kinetic analysis of CAR-T cells did not show a significant impact on CAR-T cell expansion with administration of either tocilizumab or corticosteroids.

The underlying biology of the interplay between the inflammatory response and the homeostatic milieu in CRS is not yet completely understood, and improved understanding of cytokine activation will advance the clinical management of this potentially life-threatening toxicity.

7.5.2 NEUROTOXICITY

Neurotoxicity/CRES reported with CAR-T cells has been more challenging, both in terms of etiology and management. Reported neurologic symptoms have ranged from headaches to delirium to global encephalopathy and seizures. Although CAR-T cells have been identified in the CSF of the affected patients, imaging and lumbar punctures have not provided insight into the etiology (Maude et al., 2015). The mechanism of neurotoxicity is yet to be understood, and steroids remain the mainstay in current treatment paradigms (Neelapu et al., 2018). In addition, although CRS related to differing CAR-T constructs has been reported to behave and be managed similarly across protocols, neurotoxicity has not. CD19 CAR-T cell clinical trials for adult patients with relapsed or refractory ALL from Juno Therapeutics (JCAR015) were put on a clinical hold due to excessive neurotoxicity, specifically, fatal cerebral edema. This has led to some speculation that CRES might be higher with CAR-Ts utilizing the CD28 co-stimulatory domain (Grupp, 2018). Recent detailed JCAR015 analysis attributed fatal cerebral edema to several factors including a rapid rate of CAR-T cell proliferation and higher peak of CAR-T cell numbers. Clinical trials with Novartis' Kymriah (tisagenlecleucel), utilizing the 4-1BB costimulatory domain, have however reported neurotoxicity to be brief and self-limited, resolving over several days without apparent sequelae (Maude et al., 2018). Larger studies with longer follow ups are needed to better characterize the pathophysiology and spectrum of CRES.

7.5.3 B-Cell Aplasia

In the realm of CD19-directed cell therapies, an expected on-target/off-tumor toxicity, due to expression of CD19 on normal B cells, is the resultant B-cell aplasia, with the essential success of such therapies hinging on elimination of normal B cells as a part of the process of "purging out" the precursor malignant clones. As a result, hypogammaglobulinemia related to chronic B-cell aplasia is an anticipated outcome and has been manageable with regular immunoglobulin replacement. It, however, remains to be determined whether long-term B-cell aplasia will have any additional late effects requiring intervention.

7.5.4 Survival of the Fittest

Unfortunately, relapses have been seen with CAR-T therapies. Post–CAR-T relapses have been seen in two forms – reappearance of disease that remains CD19 positive and emergence of a CD19-negative clone. Relapse with CD19-positive disease most likely reflects poor persistence of CAR-T cells or suboptimal CAR-T function. These can be potentially prevented with further optimization of CAR constructs and manufacturing technologies. Current strategies being explored to prevent antigenic escape entail the use of serial CAR-T cell infusions directed against a second antigen (such as CAR-T against CD22 following a CD-19 directed therapy) or using a combination or dual/tandem CARs joining two different antigen moieties to a single T cell. However, these approaches need further clinical investigation and validation. Table 7.3 lists selected references that describe the toxicity profile of CD19-directed CAR-T cell therapies in hematologic malignancies.

Table 7.3

Selected References Describing the Toxicity Profile of CD19-Targeted CAR-T Cell Therapies

	Quoted References	Patient Characteristics	Key Safety Attributes
1	Porter et al. (2011)	Three ALL subjects treated	Tumor lysis syndrome (TLS), cytopenias, constitutional symptoms including fevers, chills, myalgia, headache
2	Kochenderfer et al. (2012)	Eight patients with indolent lymphoma (3 FL, 4 CLL, and 1 splenic marginal zone lymphoma) received treatment	Six subjects had signs/symptoms consistent with CRS. Four patients had associated elevated markers of inflammation. Toxicities included fever, hypotension, renal failure, obtundation, and elevated liver enzymes. There was one death, attributed to cytokine release and influenza. All patients had pancytopenia with chemotherapy
3	Davila et al. (2014)	Sixteen subjects with relapsed, refractory B-cell ALL	Seven subjects developed CRS managed with high-dose steroid and/or tocilizumab
4	Maude et al. (2014b)	Thirty children and adults with refractory B-cell ALL	All the patients developed CRS, 8 of the 30 patients had severe CRS. Tociliuzmab was given to nine patients with rapid resolution of symptoms of CRS. No deaths reported
5	Schuster et al. (2015)	Thirty-eight subjects (21 DLBCL, 14 FL, and 3 MCL) with relapsed refractory NHL	CRS occurred in 16 subjects, two patients had grade ≥3 toxicity related to CRS. Neurologic toxicity occurred in three patients, one grade 5 encephalitis
6	Lee et al. (2015)	Twenty-one subjects with relapsed CD19-positive ALL	Three subjects had grade 4 CRS, with no study-related deaths. There were correlations between CRP and IL-6 levels with severity of CRS

(Continued)

Table 7.3 (*Continued*)
Selected References Describing the Toxicity Profile of CD19-Targeted CAR-T Cell Therapies

	Quoted References	Patient Characteristics	Key Safety Attributes
7	Brudno et al. (2016)	Twenty adult subjects with measurable CD19-positive disease (ALL, NHL, or CLL)	No patients developed new onset acute graft versus host disease. Grade 3 or 4 toxicities occurred in 12 of 20 patients, there were no deaths attributed to CAR-T cells
8	Neelapu et al. (2016)	Hundred and one patients received CAR-T cells	Grade ≥3 toxicity was seen in up to 30% of subjects, with one death as a result of CRS
9	Wang et al. (2014)	Seven subjects with relapsed, refractory DLBCL	One patient died before response assessment due to massive hemorrhage. There were delayed toxicities appreciated in several subjects in this study that occurred after 4 weeks, specifically in patients with heavy tumor burden in intrapulmonary tissue and submucosa of the alimentary tract
10	Kochenderfer et al. (2015)	Subjects with chemorefractory NHL. There were nine subjects with DLBCL and six subjects with indolent B-cell malignancies	The trial was amended to decrease the cell dose from $5 \times 10e6$ cells/kg to $2.5 \times 10e6$ cells/kg and then further down to $1 \times 10e6$ cells/kg due to high levels of toxicity. Grade ≥3 toxicities were seen in the majority patients and mostly occurred within 2 weeks of infusion. CRS was generally manageable, and there was one death in the study which was not clearly related to cell infusion. CRS correlated with IL-6 levels in the blood. Neurologic toxicities ranged from confusion and obtundation to aphasia and stroke-like symptoms. Symptoms improved in all subjects
11	Turtle et al. (2016)	Eighteen subjects with CLL	CRS was seen in eight subjects; four subjects had grade >2 neurotoxicity. There were no deaths related to CAR-T toxicity
12	Frey et al. (2016)	Relapsed, refractory ALL	Subjects who received cells in a fractionated fashion over split doses experienced less severe CRS than patients who received the full dose in one infusion
13	Neelapu et al. (2017)	One hundred and eleven patients with diffuse large B-cell lymphoma, primary mediastinal B-cell lymphoma, or transformed follicular lymphoma	Grade ≥3 CRS occurred in 13% and neurologic events occurred in 28% of patients. Three patients died during treatment
14	Abramson et al. (2018)	Sixty-nine patients with R/R DLBCL NOS, PMBCL, FL grade 3B, or MCL	Twenty-one patients (30%) had CRS, with 1 serious CRS event (1%; grade 4). Neurotoxicity at any grade developed in 20%, including 14% with grade 3–4 events. There were no grade 5 CRS or NT events. No acute infusional toxicity occurred, and the majority of patients, 64% had no CRS or NT
15	Maude et al. (2018)	Single-cohort, 25-center, global study of tisagenlecleucel in pediatric and young adult patients with CD19+ relapsed or refractory B-cell ALL	Grade 3 or 4 adverse events occurred in 73% of patients. CRS occurred in 77% of patients, 48% of whom received tocilizumab. Neurologic events occurred in 40% of patients and were managed with supportive care, and no cerebral edema was reported

7.5.5 ACUTE MYELOID LEUKEMIA (AML)

Although it seems natural that the precedence of success of CD19-directed CAR-T therapies in the space of ALL, CLL, and NHL be translated and exploited for AML, the hindrance has been identification of an ideal target antigen – one that is ubiquitously expressed on "tumor" but minimally on normal essential cells. It is probably easy to manage anti-CD-19 therapy resultant B-cell aplasia by immunoglobulin replacement; however, most AML leukemic blasts share a host of cell surface molecules with normal hematopoietic progenitors and stem cells, and losing the reserve of progenitor stem cells would most likely be fatal unless rescued by an allogeneic transplant. Targets currently being explored include CD33, CD34, CD38, CD116, and CD123 (Gill et al., 2014; Kenderian et al., 2015; Nakazawa et al., 2016). Another strategy is to use CAR designs with in-built mechanisms for limiting expression/suicide genes or using combinations of CARs that could potentially modulate TCR signaling (Wang et al, 2015).

7.5.6 SOLID TUMORS

Just as the CAR-T cell therapies continue to flourish in their unprecedented success in the treatment of hematologic malignancies, TCRs are leading the way for immunotherapeutic treatment of solid tumors. Table 7.4 lists some of the solid tumor antigenic targets being explored in currently ongoing phase I/II trials of TCRs in the clinic, along with their intended indications. In contrast to hematologic malignancies, solid tumors pose an additional barrier of a complex tumor microenvironment that modulates the dynamics of antitumor responses (Thomas et al., 2018). Many solid tumor antigens are expressed on normal somatic tissue with the potential for on-target, off-tumor toxicities. From the selected list in Table 7.4, NY-ESO-1 is the most advanced in terms of clinical trials (early phase I/II). At present, it is too early to make any robust conclusions regarding the safety profile of TCRs.

Table 7.4

Selected Solid Tumor TCR Gene Therapy Clinical Trials

	Target	Agent	Indication
1	NY-ESO-1	Anti-NY-ESO-1 mTCR PBL	Several solid tumors
		NY-ESO-1c259T	Synovial sarcoma, NSCLC, Ovarian cancer
		Anti-NY-ESO-1 TCR, CRISPR to delete PD-1 and autologous TCRs	Multiple myeloma, sarcoma, melanoma
2	MAGE-A3	Anti-MAGE-A3-DP4-TCR	Several solid tumors
3	WT1	WT1: JTCR016	Mesothelioma, NSCLC
4	AFP	AFPc233T	Hepatocellular cancer
5	MAGE-A4	MAGE-A4c1032T	Several solid tumors
6	MAGE-A10	MAGE-A10c796T	NSCLC, bladder, head and neck, melanoma
7	MAGE-A3/A6	KITE-718	MAGE-A3/A6-positive tumors
8	PRAME	BPX-701	AML, MDS, uveal melanoma
9	HPV-16 E7	KITE-439	HPV-associated cancers
10	Mesothelin	Anti-mesothelin TCR	Pancreatic cancer
11	IL-13Rα2	Second-generation IL-13BBζ T cells	GBM

Source: Adapted from Garber (2018).

TCR, T-cell receptor; NY-ESO, New York esophageal squamous cell carcinoma – cancer testis antigen; CRISPR, clustered regularly interspaced short palindromic repeats; MAGE, melanoma-associated antigen; WT1, Wilms Tumor Protein; AFP, alpha-fetoprotein; PRAME, preferentially expressed antigen in melanoma; HPV, human papilloma virus; NSCLC, non–small-cell lung cancer; AML, acute myeloid leukemia; MDS, myelodysplastic syndrome; IL-13Rα2, interleukin 13 receptor alpha 2.

7.5.7 Challenges

It is needless to say that TCRs and CAR-T cell therapies are changing the current landscape of refractory/relapsed malignancies and paving the new frontier of immunotherapies, but they also bring forth several new clinical challenges. With cell therapy also comes the challenge of managing CRS, neurotoxicity, and emergence of "survival of the fittest" target-antigen-negative clones. State-of-the-art technology and new efficient delivery systems are evolving this terrain further; however, they come with the caveat of requiring comprehensive clinical care expertise in managing the toxicities. As much as it is tempting to say that "May the force be with this one," it remains to be seen whether, by harnessing these therapies, we can continue to outsmart cancer and triumph over the "dark side."

7.6 CELL SWITCHES

7.6.1 *In Vivo* Safety Control

As previously discussed, adoptive T-cell therapy has led to some early clinical successes. It has also become clear that the treatment can have significant safety risks that require mitigation. Safety events can broadly be divided into transient and permanent toxicities. Transient toxicities include CRS, neurotoxicity, and tumor lysis syndrome. For CRS and neurotoxicity, fatalities in several cell therapy trials for hematological malignancies have been observed.

Permanent toxicities can be broadly divided into two categories: on-target off-tumor and off-target toxicities. On-target off-tumor toxicity occurs when the target is expressed on normal tissues other than tumor. The first fatal adverse event due to off-tumor recognition by a CAR occurred in a patient with colorectal cancer treated with high numbers of T cells expressing a third-generation CAR targeting ERBB2/HER2. The patient developed respiratory distress and cardiac arrest shortly after T-cell transfer and died of multisystem organ failure 5 days later. It was postulated that the CAR T cells recognized ERBB2 expressed at low levels in the lung epithelium, leading to pulmonary toxicity and a cascading cytokine storm with a fatal outcome (Morgan et al., 2010). B-cell aplasia is an expected side effect of CD19 CAR-T treatments as both tumor and normal B cells express the CD19 target. B-cell aplasia in treated patients lasts from months to years. To mitigate this toxicity, patients receive monthly immunoglobulin replacement; however, long-term follow-up is needed to assess the late effects of B-cell aplasia.

Off-target toxicity arises where the T cell recognizes a protein structure similar to, but distinct from, the intended target. In one clinical trial, development of an affinity-enhanced TCR to a peptide derived from the cancer antigen MAGEA3 resulted in a fatal cardiac event. Subsequent investigation revealed a likely cross recognition of a peptide derived from the cardiac protein titin that was only detectable in a complex stem cell cardiomyocyte system and not in a panel of cardiac-derived primary cells (Cameron et al., 2013).

These illustrate that the preclinical development of CAR-Ts and TCRs requires very careful evaluation of target expression in normal human tissues, with further molecular investigations and potentially the use of more complex cell culture models to mitigate the risk of off-target binding.

7.6.2 Cell Switch Systems

In response, several approaches are being developed. These so-called "cell switch" systems attempt to give a level of control over the activity (on/off switches) or removal of therapeutic cells altogether (suicide switches). There are three main types of system in development that could potentially be used to control the activity of therapeutic cells *in vivo* – namely metabolic, dimerization inducing, and antibody mediated.

The metabolic approach converts a nontoxic prodrug to a drug with cell toxicity *in vivo*. An example of this approach uses herpes simplex virus thymidine kinase (HSV-TK), converting ganciclovir to ganciclovir phosphate, which is incorporated into replicating DNA leading to replication failure and cell death. Clinical data using this approach show that the kinetics of cell death take place over several days, so this approach is unlikely to be effective in acute toxicity scenarios (Marin et al., 2012).

An alternative approach is to engineer chimeric proteins comprising a drug-binding domain to link protein domains that activate the apoptotic cell death pathway. One such example is the inducible Caspase 9 (iCasp9) system. Addition of a small molecule dimerizes casp9 domains inducing apoptosis and subsequent cell death. In the clinic, this approach was shown to be very effective in killing therapeutic T cells with over 90% of cells removed in under an hour upon administration of the small molecule (Di Stasi et al., 2011). Antibody-mediated depletion of cells is another mechanism by which therapeutic cells could be eliminated. For example, targeting of CD20 by the clinically approved monoclonal antibody Rituximab has been demonstrated to be a successful approach (Griffioen et al., 2009). Other methods reported include using a truncated EGFR protein with defective signaling (Wang et al., 2011) and the SynNotch system for antigen-directed expression of CAR molecules (Roybal et al., 2016).

There are several challenges to overcome when employing such cell switch technologies. The addition of a switch necessitates the introduction of an additional construct into the T cells. This can complicate the manufacturing process and reduce the number of cells positive for both the CAR or TCR construct and the switch molecule, introducing extra steps and complexity into the process. In addition, any switch technology must be able to act within the desired clinical timeframe. In a scenario where a patient is experiencing an acute toxicity event, a switch system must be able to reduce activity or eliminate therapeutic cells in a short timeframe. The safety of adoptive T-cell therapies is of paramount importance. It is likely that future iterations of the technology will contain a switch system enabling the control of T cells *in vivo*.

7.7 SUMMARY AND FUTURE DIRECTIONS

The unprecedented rates of efficacy for CD19 CAR-T cell therapies in refractory hematological tumors, along with the approval of two CAR-T therapies by the FDA in 2017, will likely be viewed in years to come as a landmark in cancer medicine. Adoptive cell therapies show great promise in becoming a significant new treatment option for cancer. However, for this to be realized, significant challenges must be overcome. These include antigen loss by tumor cells, overcoming the immunosuppressive tumor microenvironment in solid tumors, identifying predictive biomarkers of efficacy, mitigating potentially serious safety events, streamlining the patient experience, and reducing the cost of making these highly personalized medicines.

REFERENCES

Abramson, J.S., Gordon, L.I., Palomba, M.L., Lunning, M.A., Arnason, J.E., Forero-Torres, A., Wang, M., Maloney, D.G., Sehgal, A., Andreadis, C., Purev, E., Solomon, S.R., Ghosh, N., Albertson, T.M., Xie, B., Garcia, J., and Siddiqi, T. (2018) Updated safety and long term clinical outcomes in TRANSCEND NHL 001, pivotal trial of lisocabtagene maraleucel (JCAR017) in R/R aggressive NHL. *Journal of Clinical Oncology* 36(15 suppl), 7505.

Abou-El-Enein, M., Romhild, A., Kaiser, D., Beier, C., Bauer, G., Volk, H.D., and Reinke, P. (2013). Good Manufacturing Practices (GMP) manufacturing of advanced therapy medicinal products: A novel tailored model for optimizing performance and estimating costs. *Cytotherapy* 15, 362–383.

Attaf, M., Legut, M., Cole, D.K., and Sewell, A.K. (2015). The T cell antigen receptor: The Swiss army knife of the immune system. *Clinical and Experimental Immunology* 181, 1–18.

Barrett, J. and Blazar, B.R. (2009). Genetic trickery--escape of leukemia from immune attack. *The New England Journal of Medicine 361*, 524–525.

Bethencourt, V. (2009). Virus stalls Genzyme plant. *Nature Biotechnology 27*, 681.

Blaese, R.M., Culver, K.W., Miller, A.D., Carter, C.S., Fleisher, T., Clerici, M., Shearer, G., Chang, L., Chiang, Y., Tolstoshev, P., et al. (1995). T lymphocyte-directed gene therapy for ADA- SCID: Initial trial results after 4 years. *Science 270*, 475–480.

Brentjens, R.J. and Curran, K.J. (2012). Novel cellular therapies for leukemia: CAR-modified T cells targeted to the CD19 antigen. *Hematology American Society of Hematology Education Program 2012*, 143–151.

Brentjens, R.J., Davila, M.L., Riviere, I., Park, J., Wang, X., Cowell, L.G., Bartido, S., Stefanski, J., Taylor, C., Olszewska, M., et al. (2013). CD19-targeted T cells rapidly induce molecular remissions in adults with chemotherapy-refractory acute lymphoblastic leukemia. *Science Translational Medicine 5*, 177ra138.

Brudno, J.N. and Kochenderfer, J.N. (2016). Toxicities of chimeric antigen receptor T cells: Recognition and management. *Blood 127*, 3321–3330.

Brudno, J.N., Somerville, R.P., Shi, V., Rose, J.J., Halverson, D.C., Fowler, D.H., Gea-Banacloche, J.C., Pavletic, S.Z., Hickstein, D.D., Lu, T.L., et al. (2016). Allogeneic T cells that express an anti-CD19 chimeric antigen receptor induce remissions of B-cell malignancies that progress after allogeneic hematopoietic stem-cell transplantation without causing graft-versus-host disease. *Journal of Clinical Oncology 34*, 1112–1121.

Burdick, C.G. (1937). William Bradley Coley 1862-1936. *Annals of Surgery 105*, 152–155.

Cameron, B.J., Gerry, A.B., Dukes, J., Harper, J.V., Kannan, V., Bianchi, F.C., Grand, F., Brewer, J.E., Gupta, M., Plesa, G., et al. (2013). Identification of a Titin-derived HLA-A1-presented peptide as a cross-reactive target for engineered MAGE A3-directed T cells. *Science Translational Medicine 5*, 197ra103.

Chen, F., Teachey, D.T., Pequignot, E., Frey, N., Porter, D., Maude, S.L., Grupp, S.A., June, C.H., Melenhorst, J.J., and Lacey, S.F. (2016). Measuring IL-6 and sIL-6R in serum from patients treated with tocilizumab and/or siltuximab following CAR T cell therapy. *Journal of Immunological Methods 434*, 1–8.

Clemente, C.G., Mihm, M.C., Jr., Bufalino, R., Zurrida, S., Collini, P., and Cascinelli, N. (1996). Prognostic value of tumor infiltrating lymphocytes in the vertical growth phase of primary cutaneous melanoma. *Cancer 77*, 1303–1310.

Cohen, C.J., Zheng, Z., Bray, R., Zhao, Y., Sherman, L.A., Rosenberg, S.A., and Morgan, R.A. (2005). Recognition of fresh human tumor by human peripheral blood lymphocytes transduced with a bicistronic retroviral vector encoding a murine anti-p53 TCR. *Journal of Immunology (Baltimore, MD: 1950) 175*, 5799–5808.

Cornetta, K., Duffy, L., Turtle, C.J., Jensen, M., Forman, S., Binder-Scholl, G., Fry, T., Chew, A., Maloney, D.G., and June, C.H. (2018). Absence of replication-competent lentivirus in the clinic: Analysis of infused T cell products. *Molecular Therapy 26*, 280–288.

Davila, M.L., Riviere, I., Wang, X., Bartido, S., Park, J., Curran, K., Chung, S.S., Stefanski, J., Borquez-Ojeda, O., Olszewska, M., et al. (2014). Efficacy and toxicity management of 19-28z CAR T cell therapy in B cell acute lymphoblastic leukemia. *Science Translational Medicine 6*, 224ra225.

De Plaen, E., Lurquin, C., Van Pel, A., Mariame, B., Szikora, J.P., Wolfel, T., Sibille, C., Chomez, P., and Boon, T. (1988). Immunogenic (tum-) variants of mouse tumor P815: Cloning of the gene of tum- antigen P91A and identification of the tum- mutation. *Proceedings of the National Academy of Sciences of the United States of America 85*, 2274–2278.

Decker, W.K. and Safdar, A. (2009). Bioimmunoadjuvants for the treatment of neoplastic and infectious disease: Coley's legacy revisited. *Cytokine and Growth Factor Reviews 20*, 271–281.

Di Stasi, A., Tey, S.K., Dotti, G., Fujita, Y., Kennedy-Nasser, A., Martinez, C., Straathof, K., Liu, E., Durett, A.G., Grilley, B., et al. (2011). Inducible apoptosis as a safety switch for adoptive cell therapy. *The New England Journal of Medicine 365*, 1673–1683.

Dudley, M.E., Wunderlich, J.R., Robbins, P.F., Yang, J.C., Hwu, P., Schwartzentruber, D.J., Topalian, S.L., Sherry, R., Restifo, N.P., Hubicki, A.M., et al. (2002). Cancer regression and autoimmunity in patients after clonal repopulation with antitumor lymphocytes. *Science 298*, 850–854.

Eshhar, Z., Bach, N., Fitzer-Attas, C.J., Gross, G., Lustgarten, J., Waks, T., and Schindler, D.G. (1996). The T-body approach: Potential for cancer immunotherapy. *Springer Seminars in Immunopathology 18*, 199–209.

Frey, N.V., Shaw, P.A., Hexner, E.O., Gill, S., Marcucci, K., Luger, S.M., Mangan, J.K., Grupp, S.A., Maude, S.L., Ericson, S., et al. (2016). Optimizing chimeric antigen receptor (CAR) T cell therapy for adult patients with relapsed or refractory (r/r) acute lymphoblastic leukemia (ALL). *ASCO Meeting Abstracts 34*(15_suppl), 7002.

Galon, J., Costes, A., Sanchez-Cabo, F., Kirilovsky, A., Mlecnik, B., Lagorce-Pages, C., Tosolini, M., Camus, M., Berger, A., Wind, P., et al. (2006). Type, density, and location of immune cells within human colorectal tumors predict clinical outcome. *Science 313*, 1960–1964.

Garber, K. (2018). Driving T-cell immunotherapy to solid tumors. *Nature Biotechnology 36*, 215–219.

Gill, S., Tasian, S.K., Ruella, M., Shestova, O., Li, Y., Porter, D.L., Carroll, M., Danet-Desnoyers, G., Scholler, J., Grupp, S.A., et al. (2014). Preclinical targeting of human acute myeloid leukemia and myeloablation using chimeric antigen receptor-modified T cells. *Blood 123*, 2343–2354.

Graham, F.L., Smiley, J., Russell, W.C., and Nairn, R. (1977). Characteristics of a human cell line transformed by DNA from human adenovirus type 5. *The Journal of General Virology 36*, 59–74.

Griffioen, M., van Egmond, E.H., Kester, M.G., Willemze, R., Falkenburg, J.H., and Heemskerk, M.H. (2009). Retroviral transfer of human CD20 as a suicide gene for adoptive T-cell therapy. *Haematologica 94*, 1316–1320.

Gross, G., Waks, T., and Eshhar, Z. (1989). Expression of immunoglobulin-T-cell receptor chimeric molecules as functional receptors with antibody-type specificity. *Proceedings of the National Academy of Sciences of the United States of America 86*, 10024–10028.

Grupp, S. (2018). Beginning the CAR T cell therapy revolution in the US and EU. *Current Research in Translational Medicine 66*, 62–64.

Grupp, S.A., Kalos, M., Barrett, D., Aplenc, R., Porter, D.L., Rheingold, S.R., Teachey, D.T., Chew, A., Hauck, B., Wright, J.F., et al. (2013). Chimeric antigen receptor-modified T cells for acute lymphoid leukemia. *The New England Journal of Medicine 368*, 1509–1518.

Gulley, J.L., Madan, R.A., Pachynski, R., Mulders, P., Sheikh, N.A., Trager, J., and Drake, C.G. (2017). Role of antigen spread and distinctive characteristics of immunotherapy in cancer treatment. *Journal of the National Cancer Institute 109*. doi: 10.1093/jnci/djw261.

Hodi, F.S., O'Day, S.J., McDermott, D.F., Weber, R.W., Sosman, J.A., Haanen, J.B., Gonzalez, R., Robert, C., Schadendorf, D., Hassel, J.C., et al. (2010). Improved survival with ipilimumab in patients with metastatic melanoma. *The New England Journal of Medicine 363*, 711–723.

Howe, S.J. and Chandrashekran, A. (2012). Vector systems for prenatal gene therapy: Principles of retrovirus vector design and production. *Methods in Molecular Biology (Clifton, NJ) 891*, 85–107.

Hughes, M.S., Yu, Y.Y., Dudley, M.E., Zheng, Z., Robbins, P.F., Li, Y., Wunderlich, J., Hawley, R.G., Moayeri, M., Rosenberg, S.A., et al. (2005). Transfer of a TCR gene derived from a patient with a marked antitumor response conveys highly active T-cell effector functions. *Human Gene Therapy 16*, 457–472.

Johnson, L.A., Heemskerk, B., Powell, D.J., Jr., Cohen, C.J., Morgan, R.A., Dudley, M.E., Robbins, P.F., and Rosenberg, S.A. (2006). Gene transfer of tumor-reactive TCR confers both high avidity and tumor reactivity to nonreactive peripheral blood mononuclear cells and tumor-infiltrating lymphocytes. *Journal of Immunology (Baltimore, MD: 1950) 177*, 6548–6559.

Johnson, L.A., Morgan, R.A., Dudley, M.E., Cassard, L., Yang, J.C., Hughes, M.S., Kammula, U.S., Royal, R.E., Sherry, R.M., Wunderlich, J.R., et al. (2009). Gene therapy with human and mouse T-cell receptors mediates cancer regression and targets normal tissues expressing cognate antigen. *Blood 114*, 535–546.

Johnson, L.A., Scholler, J., Ohkuri, T., Kosaka, A., Patel, P.R., McGettigan, S.E., Nace, A.K., Dentchev, T., Thekkat, P., Loew, A., et al. (2015). Rational development and characterization of humanized anti-EGFR variant III chimeric antigen receptor T cells for glioblastoma. *Science Translational Medicine 7*, 275ra222.

Kenderian, S.S., Ruella, M., Shestova, O., Klichinsky, M., Aikawa, V., Morrissette, J.J., Scholler, J., Song, D., Porter, D.L., Carroll, M., et al. (2015). CD33-specific chimeric antigen receptor T cells exhibit potent preclinical activity against human acute myeloid leukemia. *Leukemia 29*, 1637–1647.

Kershaw, M.H., Westwood, J.A., Parker, L.L., Wang, G., Eshhar, Z., Mavroukakis, S.A., White, D.E., Wunderlich, J.R., Canevari, S., Rogers-Freezer, L., et al. (2006). A phase I study on adoptive immunotherapy using gene-modified T cells for ovarian cancer. *Clinical Cancer Research 12*, 6106–6115.

Khoja, L., Butler, M.O., Kang, S.P., Ebbinghaus, S., and Joshua, A.M. (2015). Pembrolizumab. *Journal for Immunotherapy of Cancer 3*, 36.

Klinger, M., Brandl, C., Zugmaier, G., Hijazi, Y., Bargou, R.C., Topp, M.S., Gokbuget, N., Neumann, S., Goebeler, M., Viardot, A., et al. (2012). Immunopharmacologic response of patients with B-lineage acute lymphoblastic leukemia to continuous infusion of T cell-engaging CD19/CD3-bispecific BiTE antibody blinatumomab. *Blood 119*, 6226–6233.

Kochenderfer, J.N., Dudley, M.E., Feldman, S.A., Wilson, W.H., Spaner, D.E., Maric, I., Stetler-Stevenson, M., Phan, G.Q., Hughes, M.S., Sherry, R.M., et al. (2012). B-cell depletion and remissions of malignancy along with cytokine-associated toxicity in a clinical trial of anti-CD19 chimeric-antigen-receptor-transduced T cells. *Blood 119*, 2709–2720.

Kochenderfer, J.N., Dudley, M.E., Kassim, S.H., Somerville, R.P., Carpenter, R.O., Stetler-Stevenson, M., Yang, J.C., Phan, G.Q., Hughes, M.S., Sherry, R.M., et al. (2015). Chemotherapy-refractory diffuse large B-cell lymphoma and indolent B-cell malignancies can be effectively treated with autologous T cells expressing an anti-CD19 chimeric antigen receptor. *Journal of Clinical Oncology 33*, 540–549.

Kochenderfer, J.N., Wilson, W.H., Janik, J.E., Dudley, M.E., Stetler-Stevenson, M., Feldman, S.A., Maric, I., Raffeld, M., Nathan, D.A., Lanier, B.J., et al. (2010). Eradication of B-lineage cells and regression of lymphoma in a patient treated with autologous T cells genetically engineered to recognize CD19. *Blood 116*, 4099–4102.

Kolata, G. Gene therapy hits a peculiar roadblock: A virus shortage. *The New York Times.* www.nytimes.com/2017/11/27/health/gene-therapy-virus-shortage.html. Updated November 27, 2017. Accessed June 1, 2019.

Kuball, J., Schmitz, F.W., Voss, R.H., Ferreira, E.A., Engel, R., Guillaume, P., Strand, S., Romero, P., Huber, C., Sherman, L.A., et al. (2005). Cooperation of human tumor-reactive CD4+ and CD8+ T cells after redirection of their specificity by a high-affinity p53A2.1-specific TCR. *Immunity 22*, 117–129.

Kumar, M., Keller, B., Makalou, N., and Sutton, R.E. (2001). Systematic determination of the packaging limit of lentiviral vectors. *Human Gene Therapy 12*, 1893–1905.

Lamers, C.H., Sleijfer, S., Vulto, A.G., Kruit, W.H., Kliffen, M., Debets, R., Gratama, J.W., Stoter, G., and Oosterwijk, E. (2006). Treatment of metastatic renal cell carcinoma with autologous T-lymphocytes genetically retargeted against carbonic anhydrase IX: First clinical experience. *Journal of Clinical Oncology 24*, e20–e22.

Leach, D.R., Krummel, M.F., and Allison, J.P. (1996). Enhancement of antitumor immunity by CTLA-4 blockade. *Science 271*, 1734–1736.

Lee, D.W., Gardner, R., Porter, D.L., Louis, C.U., Ahmed, N., Jensen, M., Grupp, S.A., and Mackall, C.L. (2014). Current concepts in the diagnosis and management of cytokine release syndrome. *Blood 124*, 188–195.

Lee, D.W., Kochenderfer, J.N., Stetler-Stevenson, M., Cui, Y.K., Delbrook, C., Feldman, S.A., Fry, T.J., Orentas, R., Sabatino, M., Shah, N.N., et al. (2015). T cells expressing CD19 chimeric antigen receptors for acute lymphoblastic leukaemia in children and young adults: A phase 1 dose-escalation trial. *Lancet (London, England) 385*, 517–528.

Lesbats, P., Engelman, A.N., and Cherepanov, P. (2016). Retroviral DNA integration. *Chemical Reviews 116*, 12730–12757.

Levine, B.L., Miskin, J., Wonnacott, K., and Keir, C. (2017). Global manufacturing of CAR T cell therapy. *Molecular Therapy Methods and Clinical Development 4*, 92–101.

Lohmueller, J. and Finn, O.J. (2017). Current modalities in cancer immunotherapy: Immunomodulatory antibodies, CARs and vaccines. *Pharmacology and Therapeutics 178*, 31–47.

Marin, V., Cribioli, E., Philip, B., Tettamanti, S., Pizzitola, I., Biondi, A., Biagi, E., and Pule, M. (2012). Comparison of different suicide-gene strategies for the safety improvement of genetically manipulated T cells. *Human Gene Therapy Methods 23*, 376–386.

Marincola, F.M., Jaffee, E.M., Hicklin, D.J., and Ferrone, S. (2000). Escape of human solid tumors from T-cell recognition: Molecular mechanisms and functional significance. *Advances in Immunology 74*, 181–273.

Maude, S.L., Barrett, D., Teachey, D.T., and Grupp, S.A. (2014a). Managing cytokine release syndrome associated with novel T cell-engaging therapies. *Cancer Journal (Sudbury, MA) 20*, 119–122.

Maude, S.L., Frey, N., Shaw, P.A., Aplenc, R., Barrett, D.M., Bunin, N.J., Chew, A., Gonzalez, V.E., Zheng, Z., Lacey, S.F., et al. (2014b). Chimeric antigen receptor T cells for sustained remissions in leukemia. *The New England Journal of Medicine 371*, 1507–1517.

Maude, S.L., Laetsch, T.W., Buechner, J., Rives, S., Boyer, M., Bittencourt, H., Bader, P., Verneris, M.R., Stefanski, H.E., Myers, G.D., et al. (2018). Tisagenlecleucel in children and young adults with B-cell lymphoblastic leukemia. *The New England Journal of Medicine 378*, 439–448.

Maude, S.L., Teachey, D.T., Porter, D.L., and Grupp, S.A. (2015). CD19-targeted chimeric antigen receptor T-cell therapy for acute lymphoblastic leukemia. *Blood 125*, 4017–4023.

Merten, O.W. and Wright, J.F. (2016). Towards routine manufacturing of gene therapy drugs. *Molecular Therapy Methods and Clinical Development 3*, 16021.

Monjezi, R., Miskey, C., Gogishvili, T., Schleef, M., Schmeer, M., Einsele, H., Ivics, Z., and Hudecek, M. (2017). Enhanced CAR T-cell engineering using non-viral Sleeping Beauty transposition from minicircle vectors. *Leukemia 31*, 186–194.

Morgan, R.A., Dudley, M.E., Wunderlich, J.R., Hughes, M.S., Yang, J.C., Sherry, R.M., Royal, R.E., Topalian, S.L., Kammula, U.S., Restifo, N.P., et al. (2006). Cancer regression in patients after transfer of genetically engineered lymphocytes. *Science 314*, 126–129.

Morgan, R.A., Johnson, L.A., Davis, J.L., Zheng, Z., Woolard, K.D., Reap, E.A., Feldman, S.A., Chinnasamy, N., Kuan, C.T., Song, H., et al. (2012). Recognition of glioma stem cells by genetically modified T cells targeting EGFRvIII and development of adoptive cell therapy for glioma. *Human Gene Therapy 23*, 1043–1053.

Morgan, R.A., Yang, J.C., Kitano, M., Dudley, M.E., Laurencot, C.M., and Rosenberg, S.A. (2010). Case report of a serious adverse event following the administration of T cells transduced with a chimeric antigen receptor recognizing ERBB2. *Molecular Therapy 18*, 843–851.

Nakazawa, Y., Matsuda, K., Kurata, T., Sueki, A., Tanaka, M., Sakashita, K., Imai, C., Wilson, M.H., and Koike, K. (2016). Anti-proliferative effects of T cells expressing a ligand-based chimeric antigen receptor against CD116 on CD34(+) cells of juvenile myelomonocytic leukemia. *Journal of Hematology and Oncology 9*, 27.

Neelapu, S.S., Locke, F.L., Bartlett, N.L., Lekakis, L., Miklos, D., Jacobson, C.A., Braunschweig, I., Oluwole, O., Siddiqi, T., Lin, Y., et al. (2016) Kte-C19 (anti-CD19 CAR T Cells) induces complete remissions in patients with refractory diffuse large B-cell lymphoma (DLBCL): Results from the pivotal phase 2 Zuma-1. *Blood 128*, LBA-6.

Neelapu, S.S., Locke, F.L., Bartlett, N.L., Lekakis, L.J., Miklos, D.B., Jacobson, C.A., Braunschweig, I., Oluwole, O.O., Siddiqi, T., Lin, Y., et al. (2017). Axicabtagene ciloleucel CAR T-cell therapy in refractory large B-cell lymphoma. *The New England Journal of Medicine 377*, 2531–2544.

Neelapu, S.S., Tummala, S., Kebriaei, P., Wierda, W., Locke, F.L., Lin, Y., Jain, N., Daver, N., Gulbis, A.M., Adkins, S., et al. (2018). Toxicity management after chimeric antigen receptor T cell therapy: One size does not fit 'ALL'. *Nature Reviews Clinical Oncology 15*, 218.

Nishimura, H., Nose, M., Hiai, H., Minato, N., and Honjo, T. (1999). Development of lupus-like autoimmune diseases by disruption of the PD-1 gene encoding an ITIM motif-carrying immunoreceptor. *Immunity 11*, 141–151.

O'Rourke, D.M., Nasrallah, M.P., Desai, A., Melenhorst, J.J., Mansfield, K., Morrissette, J.J.D., Martinez-Lage, M., Brem, S., Maloney, E., Shen, A., et al. (2017). A single dose of peripherally infused EGFRvIII-directed CAR T cells mediates antigen loss and induces adaptive resistance in patients with recurrent glioblastoma. *Science Translational Medicine 9*. doi: 10.1126/scitranslmed.aaa0984.

Park, J.R., Digiusto, D.L., Slovak, M., Wright, C., Naranjo, A., Wagner, J., Meechoovet, H.B., Bautista, C., Chang, W.C., Ostberg, J.R., et al. (2007). Adoptive transfer of chimeric antigen receptor re-directed cytolytic T lymphocyte clones in patients with neuroblastoma. *Molecular Therapy 15*, 825–833.

Porter, D., Frey, N., Wood, P.A., Weng, Y., and Grupp, S.A. (2018). Grading of cytokine release syndrome associated with the CAR T cell therapy tisagenlecleucel. *Journal of Hematology and Oncology 11*, 35.

Porter, D.L., Hwang, W.T., Frey, N.V., Lacey, S.F., Shaw, P.A., Loren, A.W., Bagg, A., Marcucci, K.T., Shen, A., Gonzalez, V., et al. (2015). Chimeric antigen receptor T cells persist and induce sustained remissions in relapsed refractory chronic lymphocytic leukemia. *Science Translational Medicine 7*, 303ra139.

Porter, D.L., Levine, B.L., Kalos, M., Bagg, A., and June, C.H. (2011). Chimeric antigen receptor-modified T cells in chronic lymphoid leukemia. *The New England Journal of Medicine 365*, 725–733.

Pule, M.A., Savoldo, B., Myers, G.D., Rossig, C., Russell, H.V., Dotti, G., Huls, M.H., Liu, E., Gee, A.P., Mei, Z., et al. (2008). Virus-specific T cells engineered to coexpress tumor-specific receptors: Persistence and antitumor activity in individuals with neuroblastoma. *Nature Medicine 14*, 1264–1270.

Qasim, W., Zhan, H., Samarasinghe, S., Adams, S., Amrolia, P., Stafford, S., Butler, K., Rivat, C., Wright, G., Somana, K., et al. (2017). Molecular remission of infant B-ALL after infusion of universal TALEN gene-edited CAR T cells. *Science Translational Medicine 9*. doi: 10.1126/scitranslmed.aaj2013.

Robbins, P.F., Li, Y.F., El-Gamil, M., Zhao, Y., Wargo, J.A., Zheng, Z., Xu, H., Morgan, R.A., Feldman, S.A., Johnson, L.A., et al. (2008). Single and dual amino acid substitutions in TCR CDRs can enhance antigen-specific T cell functions. *Journal of Immunology (Baltimore, MD: 1950) 180*, 6116–6131.

Rosenberg, S.A. and Restifo, N.P. (2015). Adoptive cell transfer as personalized immunotherapy for human cancer. *Science 348*, 62–68.

Roybal, K.T., Williams, J.Z., Morsut, L., Rupp, L.J., Kolinko, I., Choe, J.H., Walker, W.J., McNally, K.A., and Lim, W.A. (2016). Engineering T cells with customized therapeutic response programs using synthetic notch receptors. *Cell 167*, 419–432.e416.

Sampson, J.H., Choi, B.D., Sanchez-Perez, L., Suryadevara, C.M., Snyder, D.J., Flores, C.T., Schmittling, R.J., Nair, S.K., Reap, E.A., Norberg, P.K., et al. (2014). EGFRvIII mCAR-modified T-cell therapy cures mice with established intracerebral glioma and generates host immunity against tumor-antigen loss. *Clinical Cancer Research 20*, 972–984.

Schuster, S.J., Svoboda, J., Nasta, S.D., Porter, D.L., Chong, E.A., Landsburg, D.J., Mato, A.R., Lacey, S.F., Melenhorst, J.J., and Chew, A., et al. (2015) Sustained remissions following chimeric antigen receptor modified T cells directed against CD19 (CTL019) in patients with relapsed or refractory CD19+ lymphomas. *Blood, 126*(23), 183.

Sewell, A.K. (2012). Why must T cells be cross-reactive? *Nature Reviews Immunology 12*, 669–677.

Shankaran, V., Ikeda, H., Bruce, A.T., White, J.M., Swanson, P.E., Old, L.J., and Schreiber, R.D. (2001). IFNgamma and lymphocytes prevent primary tumour development and shape tumour immunogenicity. *Nature 410*, 1107–1111.

Shiloni, E., Lafreniere, R., Mule, J.J., Schwarz, S.L., and Rosenberg, S.A. (1986). Effect of immunotherapy with allogeneic lymphokine-activated killer cells and recombinant interleukin 2 on established pulmonary and hepatic metastases in mice. *Cancer Research 46*, 5633–5640.

Smith, T.T., Stephan, S.B., Moffett, H.F., McKnight, L.E., Ji, W., Reiman, D., Bonagofski, E., Wohlfahrt, M.E., Pillai, S.P.S., and Stephan, M.T. (2017). In situ programming of leukaemia-specific T cells using synthetic DNA nanocarriers. *Nature Nanotechnology 12*, 813–820.

Speetjens, F.M., de Bruin, E.C., Morreau, H., Zeestraten, E.C., Putter, H., van Krieken, J.H., van Buren, M.M., van Velzen, M., Dekker-Ensink, N.G., van de Velde, C.J., et al. (2008). Clinical impact of HLA class I expression in rectal cancer. *Cancer Immunology, Immunotherapy: CII 57*, 601–609.

Teachey, D.T., Lacey, S.F., Shaw, P.A., Melenhorst, J.J., Maude, S.L., Frey, N., Pequignot, E., Gonzalez, V.E., Chen, F., Finklestein, J., et al. (2016). Identification of predictive biomarkers for cytokine release syndrome after chimeric antigen receptor T-cell therapy for acute lymphoblastic leukemia. *Cancer Discovery 6*, 664–679.

Teachey, D.T., Rheingold, S.R., Maude, S.L., Zugmaier, G., Barrett, D.M., Seif, A.E., Nichols, K.E., Suppa, E.K., Kalos, M., Berg, R.A., et al. (2013). Cytokine release syndrome after blinatumomab treatment related to abnormal macrophage activation and ameliorated with cytokine-directed therapy. *Blood 121*, 5154–5157.

Thomas, R., Al-Khadairi, G., Roelands, J., Hendrickx, W., Dermime, S., Bedognetti, D., and Decock, J. (2018). NY-ESO-1 based immunotherapy of cancer: Current perspectives. *Frontiers in Immunology 9*, 947.

Tivol, E.A., Borriello, F., Schweitzer, A.N., Lynch, W.P., Bluestone, J.A., and Sharpe, A.H. (1995). Loss of CTLA-4 leads to massive lymphoproliferation and fatal multiorgan tissue destruction, revealing a critical negative regulatory role of CTLA-4. *Immunity 3*, 541–547.

Tran, E., Robbins, P.F., Lu, Y.C., Prickett, T.D., Gartner, J.J., Jia, L., Pasetto, A., Zheng, Z., Ray, S., Groh, E.M., et al. (2016). T-cell transfer therapy targeting mutant KRAS in cancer. *The New England Journal of Medicine 375*, 2255–2262.

Tran, E., Robbins, P.F., and Rosenberg, S.A. (2017). 'Final common pathway' of human cancer immunotherapy: Targeting random somatic mutations. *Nature Immunology 18*, 255–262.

Turtle, C.J., Hanafi, L.-A., Li, D., Conrad Liles, W., Wurfel, M.M., López, J.A., Chen, J., Chung, D., Harju-Baker, S., Cherian, S., et al. (2016). CD19 CAR-T cells are highly effective in ibrutinib refractory chronic lymphocytic leukemia. *Blood 128*(22), 56.

Vormittag, P., Gunn, R., Ghorashian, S., and Veraitch, F.S. (2018). A guide to manufacturing CAR T cell therapies. *Current Opinion in Biotechnology 53*, 164–181.

Wang, E., Wang, L.C., Tsai, C.Y., Bhoj, V., Gershenson, Z., Moon, E., Newick, K., Sun, J., Lo, A., Baradet, T., et al. (2015). Generation of potent T-cell immunotherapy for cancer using DAP12-based, multichain, chimeric immunoreceptors. *Cancer and Immunology Research 3*, 815–826.

Wang, X., Chang, W.C., Wong, C.W., Colcher, D., Sherman, M., Ostberg, J.R., Forman, S.J., Riddell, S.R., and Jensen, M.C. (2011). A transgene-encoded cell surface polypeptide for selection, in vivo tracking, and ablation of engineered cells. *Blood 118*, 1255–1263.

Wang, X. and Riviere, I. (2016). Clinical manufacturing of CAR T cells: Foundation of a promising therapy. *Molecular Therapy Oncolytics 3*, 16015.

Wang, Y., Zhang, W.Y., Han, Q.W., Liu, Y., Dai, H.R., Guo, Y.L., Bo, J., Fan, H., Zhang, Y., Zhang, Y.J., et al. (2014). Effective response and delayed toxicities of refractory advanced diffuse large B-cell lymphoma treated by CD20-directed chimeric antigen receptor-modified T cells. *Clinical immunology (Orlando, FL) 155*, 160–175.

Waterhouse, P., Penninger, J.M., Timms, E., Wakeham, A., Shahinian, A., Lee, K.P., Thompson, C.B., Griesser, H., and Mak, T.W. (1995). Lymphoproliferative disorders with early lethality in mice deficient in Ctla-4. *Science 270*, 985–988.

8 CRISPR in Drug Discovery

Chih-Shia Lee and Ji Luo
National Cancer Institute

CONTENTS

8.1 INTRODUCTION

The Human Genome Project [1] has provided a blueprint of the human chromosomes to annotate all genetic elements and their involvement in biological processes and in human diseases [2,3]. In parallel with the physical mapping of the human genome, advances in high-throughput genetic screens have dramatically accelerated the pace of target discovery and target validation for drug development. Recent breakthroughs in the discovery and utilization of the CRISPR/Cas9 system for genome editing have once again transformed the field of functional genomics. Owing to its specificity and programmability, CRISPR/Cas9-mediated genome editing has become a powerful tool for elucidating gene function, for high-throughput target discovery and validation, and for the development of gene therapy in various human diseases. In this chapter, we will present an overview of the discovery, development, and application of CRISPR/Cas9-mediated genome editing in the context of target identification and drug discovery. We will focus our discussion on genetic screens using various CRISPR/Cas9 approaches for target identification. We will use examples in cancer biology to illustrate the power and promise of CRISPR/Cas9 in drug discovery.

8.2 DISCOVERY OF THE CRISPR/CAS9 SYSTEM AND ITS ADAPTATION FOR MAMMALIAN GENE EDITING

8.2.1 FROM MYSTERY SEQUENCES IN THE BACTERIA GENOME TO A POWERFUL TOOL IN BIOTECHNOLOGY

While attempting to sequence the *iap* gene in *Escherichia coli* to identify the primary structure of its putative gene product, Ishino and colleagues at Osaka University discovered an unusual structure near the 3′-end of the gene that consists of five highly homologous repeats of 29 nucleotides (nt), with each repeat flanked by non-repetitive sequences of 32 nt in length [4]. At the time, the biological function of these repetitive sequences was not known. As more and more microbial genomes were sequenced, additional repeat elements analogous to the *E. coli* ones were found in other bacterial and archaeal strains [5]. They began to draw the attention of microbiologists, and they were later named Clustered Regularly Interspaced Short Palindromic Repeats (CRISPR) [6].

In 2002, CRISPR was found to be transcribed as a small noncoding RNA in archaea [7], and the putative proteins encoded by the CRISPR-associated (*cas*) gene clusters adjacent to the CRISPR array were recognized to harbor helicase and exonuclease motifs [6]. However, the significance of these genetic elements was not appreciated until 2005 when the non-repetitive spacer sequences separating the repetitive sequences were bioinformatically mapped to similar sequences in bacteriophages and in conjugative plasmids [8–10]. Strikingly, viruses that carry the spacer sequences in their genomes fail to infect archaeal cells [8]. These findings suggested that the spacers are DNA fragments derived from phage genomes and that CRISPR might function as a form of adaptive immunity in bacteria. Direct evidence supporting this hypothesis came in 2007 when it was demonstrated that *cas* genes and spacers are critical for the phage-resistant phenotype [11]. The mechanism of several CRISPR systems as a bacterial adaptive immunity was further elucidated in the next few years and was reviewed recently [12]. Interest was soon focused on the type II CRISPR system because it was discovered that only three critical components are required to reconstitute the active nuclease system: the CRISPR RNA (crRNA), the trans-activating crRNA (tracrRNA), and the Cas9 nuclease [13,14]. Together, these three components form a ribonucleoprotein complex that can recognize and cleave invading phage DNA in a sequence-specific manner (Figure 8.1a). The realization that the type II CRISPR/Cas9 system can be artificially programmed to edit any host genome when expressed in a heterologous fashion led to an explosion in CRISPR technology in the next few years.

8.2.2 ADAPTING A DEFENSE SYSTEM IN BACTERIA TO A GENOME-EDITING TOOL IN EUKARYOTIC CELLS

Unlike other types of CRISPR system that use multicomponent protein complexes for target DNA recognition and cleavage, the type II CRISPR system requires only one protein, the Cas9 endonuclease, to cleave target DNA. This makes the CRISPR/Cas9 system an attractive tool for gene editing. Several research teams simplified the CRISPR/Cas9 system by combining the crRNA and tracrRNA elements into a single, chimeric RNA called the single guide RNA (sgRNA) and demonstrating that synthetic sgRNA could direct purified Cas9 to cleave its target DNA *in vitro* [15,16]. It was further shown that when Cas9 and sgRNA are co-expressed in the cell, they can be programmed to introduce target-specific deletion, insertion, and other types of mutation in the human genome [17,18]. These seminal works converted the bacterial type II CRISPR/Cas9 system to a highly versatile genome-editing tool.

The mechanism of CRISPR/Cas9-mediated DNA cleavage has been characterized. The Cas9/sgRNA complex recognizes its DNA target site with two specific sequences: a target-specific 20-nt sequence that is complementary to the spacer sequence in the crRNA, and a protospacer-adjacent

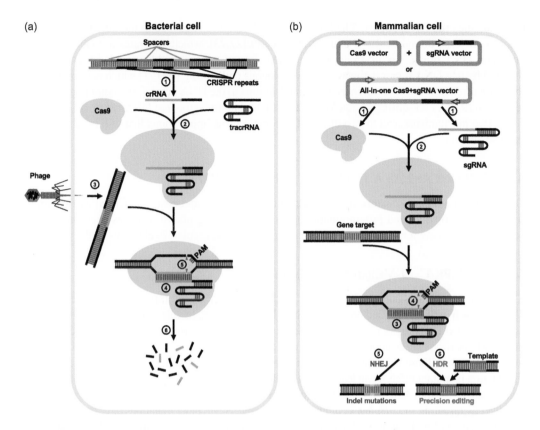

FIGURE 8.1 Mechanism of bacterial and engineered Type II CRISPR/Cas9 systems. (a) Simplified mechanism of the bacterial Type II CRISPR/Cas9 system. The spacer sequences in the bacteria genome are integrated DNA fragments of bacterial phages. These foreign DNA fragments are retained in the genome between flanking CRISPR repeats, and they serve as "immunological memory" of previous infections. Spacers and CRISPR repeats are transcribed into crRNAs (Step 1), which hybridizes with tracrRNA through the homologous CRISPR repeat region to form a crRNA:tracrRNA duplex. Loading of these RNAs into the Cas9 protein forms the functional CRISPR/Cas9 ribonucleoprotein complex (Step 2). Bacteriophages infect bacteria cells by injecting their genetic material into the host (Step 3). The CRISPR/Cas9 complex binds to the phage DNA by recognizing the PAM sequence and unwinding the phage DNA upstream of the PAM. This allows strand invasion by the crRNA and homology pairing with the target site (Step 4). Stable binding activates the endonuclease domains of Cas9 which generate double-stranded, blunt-end DNA break 3-nt upstream of the PAM (Step 5). Ultimately, this results in the degradation of the phage DNA (Step 6). (b) The genome-editing mechanism of Type II CRISPR/Cas9 system in mammalian cells. Cas9 protein and target site-specific sgRNA, which represents a hybrid of crRNA and tracrRNA, can be expressed from either separate expression vectors or an all-in-one vector in (Step 1). Loading of sgRNA into Cas9 forms the functional CRISPR/Cas9 ribonucleoprotein complex (Step 2). The CRISPR/Cas9 complex binds to its DNA target site in the genome by recognizing the PAM sequence and unwinding the DNA double-helix upstream of the PAM. This allows strand invasion by the sgRNA and homology pairing with the target site (Step 3). Table binding activates the endonuclease domains of Cas9 which generate double-stranded, blunt-end DNA break 3-nt upstream of the PAM (Step 4). These double-stranded breaks are repaired by two endogenous DNA repair pathways. The imprecise NHEJ pathway which introduces small insertions and deletions (indel) mutations at the target sites (Step 5). The precise HDR pathway uses a homologous DNA template to generate error-free DNA repair (Step 6).

motif (PAM) that flanks the 3′ DNA target site. The PAM is a critical element of the CRISPR/Cas system for dictating target recognition and it is Cas ortholog-specific. The most commonly used Cas9 variant, the *Streptococcus pyogenes* Cas9 (SpCas9) [19], primarily recognizes an "NGG" PAM although it can also use an "NAG" PAM at a lower efficiency [20,21]. Upon binding to the PAM,

Cas9 unwinds the DNA duplex. The binding of crRNA space sequence to the target site activates the two nuclease domains, RuvC and HNH, in Cas9 [22,23], which generates a double-stranded, blunt-end DNA break 3-nt upstream of the PAM. In cells, these double-stranded breaks can be repaired by two DNA repair pathways: nonhomologous end joining (NHEJ) which is imprecise and therefore often introduces small insertions and deletions to the DNA breaks, and homology-directed repair (HDR) which requires a homologous DNA template for error-free DNA repair (Figure 8.1b). These two mechanisms lead to two different genome-editing applications that will be discussed in the next section.

8.3   CRISPR/CAS9 AS A TOOL FOR GENOME EDITING

Owing to its modularity feature in DNA recognition, binding, and cleavage, CRISPR/Cas9 represents a highly flexible and programmable genome-engineering platform. It has been modified to several distinct tools for exploring the functions of genes and regulatory elements in the genome.

8.3.1   CRISPR/CAS9-MEDIATED GENE KNOCKOUT

The most common application of CRISPR/Cas9 is to turn off a gene through gene knockout (KO). This approach will be referred to as "CRISPR KO" hereafter. In this application, Cas9-mediated double-stranded DNA break is repaired by the error-prone NHEJ mechanism, which introduces various small insertion and deletion (indel) mutations to the cleavage site. When targeted to exons of protein coding genes, these indel mutations often frame-shift the open reading frame. Early translational termination in mutant mRNA can lead to its degradation through the nonsense-mediated mRNA decay mechanism or the translation of a non-functional, truncated gene product. Better KO efficiency is often achieved by targeting the first few coding exons of a gene. Since indel mutations alter the sequence of the target site and make it no longer recognizable by the sgRNA/Cas9 complex, CRISPR KO could achieve near-complete elimination of the target gene over time. In the past, knocking out a gene often required months of work for targeting vector design and optimization and was often met with low efficiency. Using CRISPR/Cas9, this can be easily achieved in a week because only the 20-nt target-specific sequence within the sgRNA needs to be customized. In our recent study, we demonstrated that CRISPR/Cas9-mdeiated gene KO happens in the first 5 days in 80%–90% of the transduced cells [24]. The potency of the sgRNA sequence is the dominant factor in determining KO efficiency, whereas the copy number of the target and the expression level of Cas9 plays a less prominent role [24].

8.3.2   CRISPR/CAS9-DIRECTED PRECISION GENOME EDITING

When coupled to HDR, CRISPR/Cas9 can be used to introduce precise modifications in DNA sequence at a specific target site. In the presence of a homologous repair template, the double-stranded DNA break caused by Cas9 can be repaired through the error-free HDR pathway. These modifications include virtually all possible sequence alterations from point mutations to large deletions and insertions. However, since the endogenous HDR machinery usually works at a lower efficiency and is dependent on the expression of HDR proteins during the G2 phase, it has been challenging to achieve a high editing rate, and clonal selection is often required to identify the correctly edited clones. It has been shown that small molecules may enhance HDR efficiency by inhibiting the NHEJ pathway [25,26].

8.3.3   GENE SILENCING VIA CRISPRi

The CRISPR/Cas9 system can be further repurposed to modulate gene expression. One such application, called CRISPR interference (CRISPRi), is to achieve gene silencing by interfering with RNA transcription. This approach takes advantage of endonuclease-deficient mutants of Cas9 called dCas9 that retain the RNA-guided homing feature of Cas9 but do not cut the target DNA. When directed to the transcription start site or the 5′ gene body, dCas9 can suppress transcriptional initiation or elongation, likely through a hindering mechanism against the RNA polymerase machinery [27]. The repressive effect of dCas9 can be dramatically enhanced by coupling dCas9 to various transcriptional repressor domains such as the Krüppel-associated box (KRAB) domain and the mSin3 interaction domain SID4X [28] (Figure 8.2a). These dCas9 fusion proteins can recruit repressive chromatin-modifying complexes to the target site and, thus, stably silence gene expression. CRISPRi can be made inducible, and the silencing effect has been shown to be both reversible and highly specific [27,29].

8.3.4 GENE ACTIVATION VIA CRISPRA

Similar to the concept of CRISPRi, dCas9 can also be modified to be gene activators by fusing it with transcriptional activator proteins such as VP64 and the p65 activation domain [29]. This gene activation approach is named CRISPR activator (CRISPRa). Gene activation efficiency can be further enhanced by coupling dCas9-VP64 fusion protein with a modified sgRNA that contains MS2 protein-interacting RNA aptamers in its stem loops to recruit MS2-p65 fusion protein as transactivator [30,31] (Figure 8.2b). This approach can achieve robust overexpression of most genes in the genome which was not possible in the past using cDNA library due to the size limitation for cDNA cloning into expression vectors.

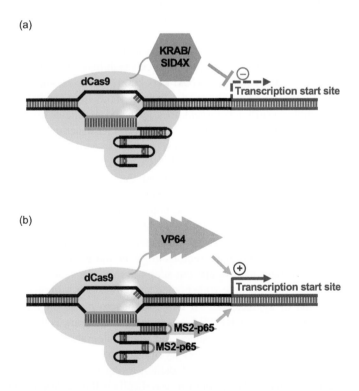

FIGURE 8.2 CRISPRi and CRISPRa as gene expression modulators. (a) In the CRISPRi configuration, an endonuclease-deficient mutant form of Cas9 called dCas9 was used in the CRISPR/Cas9 complex. dCas9 retains the RNA-guided targeting feature but cannot cut DNA. When fused with a transcription repressor, such as the KRAB domain or the SID4X domain, this dCas9-repressor fusion protein can be directed to the regulatory element of a gene to repress its expression. (b) In the CRISPRa configuration, dCas9 is fused with transcriptional activators, such as the VP64 or the p65 activator domains, and directed to the regulatory element of a gene to activate its expression. Gene activation efficiency can be enhanced by using a modified sgRNA that contains the MS2 protein-binding RNA aptamers in its stem loops to recruit MS2-p65 fusion proteins.

The above CRISPR/Cas9 tools have been widely adopted for exploring gene functions and for functional genomics. Since the PAM sequence requirement for SpCas9, "NGG," is relatively common throughout the genome, CRISPR/Cas9 can target virtually any region in the genome. A single 20-nt guide sequence can be easily synthesized and cloned into an all-in-one vector containing all the essential CRISPR/Cas9 components. In addition, it has been shown that CRISPR/Cas9-based gene targeting can be multiplexed to achieve either multiple-gene KO [17,32–35] or to enhance CRISPRi/CRISPRa effects [27,29,30,36,37]. Large libraries of guide sequences can be created using array-based oligonucleotide synthesis to construct either customized or genome-wide lentiviral CRISPR libraries. By combining CRISPR KO, CRISPRi, and CRISPRa approaches with sgRNA library screening, it is possible to activate or inactivate almost any genomic locus of interest. This enables large-scale functional annotation of the genome.

8.4 CRISPR/CAS9 SCREENS FOR DRUG TARGET DISCOVERY

High-throughput genetic screens on a genome-wide scale have been extensively employed for identifying drug targets. These genetic screens are helpful to identify direct drug targets or genes that play critical roles in key mechanisms responsible for a disease phenotype. Depending on the gene manipulation method, genetic screens fall into two categories, loss of function (LOF) screens and gain of function (GOF) screens, each aims to identify genes whose downregulation and upregulation, respectively, result in desired phenotypic changes. These two approaches are complementary to each other. Using cancer cell survival as an example, in an LOF screen, the silencing of tumor-promoting genes and tumor-suppressor genes will reduce or enhance cancer cell viability, respectively. On the other hand, in a GOF screen, the activation of tumor-promoting genes and tumor-suppressor genes will enhance or reduce cancer cell viability, respectively.

In the past, LOF screens often employed RNAi knockdown libraries, and GOF genetic screens mostly employed cDNA overexpression libraries. These methods have several limitations. RNAi-based screens often encounter insufficient target knockdown and significant off-target effects, which lead to false-negative and false-positive results, respectively. Constructing a cDNA library is both time consuming, and cDNA libraries are often biased towards smaller transcripts due to difficulties with cloning and expressing large cDNAs. CRISPR/Cas-based approaches can overcome these issues. CRISPR often leads to homozygous gene KO and, thus, created true nulls rather than hypomorphs. It has been shown that CRISPR KO and CRISPRi give lower noise and less off-target effect than RNAi in genetic screens [38,39], likely due to both the additional PAM sequence requirement and a more stringent requirement for target sequence match for triggering Cas9 endonuclease cleavage [40]. CRISPRa can theoretically upregulate any open reading frame regardless of the size of the gene, although genes in heterochromatic regions might be more difficult to activate. Therefore, CRISPR/Cas9-based screens have gain significant adoption in recent years.

Most CRISPR screens employ genome-wide pooled lentiviral sgRNA libraries. A pooled sgRNA library can be constructed by synthesizing all guide sequences on an oligonucleotide array and cloning them as a pool into lentiviral CRISPR/Cas9 vectors. The library is then packaged as a pool of viruses, which serve as a medium to transduce and stably integrate the CRISPR/Cas9 elements into the host genome for expression. The multiplicity of infection (MOI) is typically kept below one to ensure most of the transduced cells are each infected by a single viral particle only and, therefore, receiving only one guide sequence. The pool of infected cells is then cultured and selected for the phenotype of interest, such as proliferation, death, or changes in the level of a sortable reporter/biomarker. After the phenotypic selection, cells are subjected to deep sequencing to identify changes in library composition as a result of the selection process. Depending on the configuration of the CRISPR/Cas9 library and their target genes' functional relationship to the phenotypic readout, guide RNAs can experience either depletion or enrichment in response to selection pressure imposed by the assay (Figure 8.3a). By combining different CRISPR/Cas9 tools (i.e. CRISPR KO/CRISPRi for LOF screens and CRISPRa for GOF screens) with either positive or negative selection pressure,

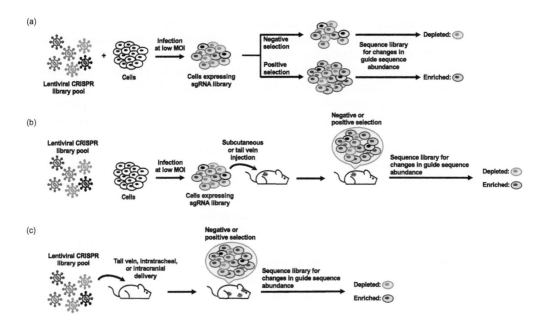

FIGURE 8.3 *In vitro* and *in vivo* CRISPR screens. (a) For *in vitro* CRISPR screens in cell culture, pooled lentiviral sgRNA library is first transduced into cells at low MOI. Cells are then selected for the phenotype of interest under either a negative or positive selection pressure. After the selection, cells are subjected to deep sequencing to identify changes in sgRNA library composition. Guide RNAs can experience either depletion or enrichment in response to the selection pressure imposed by the assay. (b) For *in vivo* CRISPR screens using transplant models, cells are transduced with sgRNA library *in vitro*, then transplanted into mice, usually via subcutaneous or tail vein injection, to allow engraftment in mice. The transplanted cells are selected for a phenotype of interest (for example, tumor growth). Guide RNAs that are either depleted or enriched by the selection process can be identified by deep sequencing of the engrafted cells. (c) For *in vivo* CRISPR screens using autochthonous mouse models, the pool of lentivirus carrying the sgRNA library are directly injected into the organs. Using mouse models of cancer as an example, tail vein, intratracheal and intracranial deliveries of the library has been used in models of liver, lung, and brain tumors, respectively. Tumors initiated from these organs are harvested and analyzed for changes in guide sequence abundance in the tumor.

CRISPR/Cas9-based genetic screens are a powerful approach for identifying drug targets whose functions are responsible for a disease phenotype. Below we will use selected examples LOF and GOF CRISPR screens to illustrate the value of this approach in drug discovery.

8.4.1 LOF Screens Using CRISPR KO and CRISPRi

CRISPR KO and CRISPRi screens have been carried out in human cancer cells to identify cancer dependency genes that might serve as potential drug targets. An important consideration is that genes that are critical for essential cellular processes may not serve as good targets as the inhibition of their protein products could be too toxic to normal cells to offer a sufficient therapeutic window. LOF CRISPR screens have been employed to identify essential genes that regulate the same core cellular functions across various cell types such as DNA replication, gene transcription, protein translation, cell-cycle progression, and protein degradation [41–45]. This provides a better functional annotation of the druggable genome and a catalog of essential genes that require more care when being considered for drug development. By screening multiple cancer cell lines with different tissue origins and distinct patterns of driver mutations, cancer type- and mutation-specific dependency genes can be identified. Owing to their ability to confer context-specific vulnerabilities, these genes could be valuable cancer drug targets with a reasonable therapeutic window [44,45].

In addition to identifying cell-autonomous dependencies, CRISPR KO screens are powerful tools for dissecting heterotypic interaction between cancer cells and its microenvironment for target discovery. A CRISPR KO screen aiming to identify genes that regulate tumor-specific T-cell activity uncovered the transmembrane protein CMTM6 as a novel positive regulator for PD-L1 expression on the surface of cancer cells. Silencing CMTM6 enhances tumor-specific T-cell activity by downregulating PD-L1 expression [46]. In this case, the CRISPR screen not only furthered our understanding of immune checkpoint but identified a potential therapeutic target to overcome resistance to cancer immune therapy.

Combined with a positive selection pressure, CRISPR KO and CRISPRi screens have been used to elucidate the mechanism of acquired drug resistance. Oncogenic mutants in BRAF is a major driver in more than 60% of human melanomas [47]. Mutant BRAF^{V600E} strongly activates its downstream mitogen-activated protein kinase (MAPK) pathway, a major signaling pathway that drives cell-cycle entry. Targeting the MAPK pathway components BRAF, MEK, and ERK has been a rational and attractive strategy to treat BRAF^{V600E} melanomas [48]. The US Food and Drug Administration (FDA) has approved the use of RAF inhibitor and MEK inhibitor for metastatic BRAF^{V600E} melanoma [49]. However, despite great initial responses to RAF and MEK inhibition, BRAF^{V600E} melanoma patients almost invariably develop resistance quickly, and this limits the durability of the therapy. To better understand the mechanism for acquired resistance to RAF inhibitors, a CRISPR KO screen was performed in human BRAF^{V600E} melanoma cells treated with a BRAF inhibitor, vemurafenib. This study identified both known and novel negative regulators of the MAPK pathway whose LOF render cells resistant to RAF inhibition, thus providing valuable mechanistic insights to drug resistance [42]. Interestingly, BRAF^{V600E} melanoma cells that have developed resistance to BRAF and MEK inhibitors became "addicted" to drug treatments and massive cell death ensued following drug withdrawal. To elucidate the mechanism for this intriguing phenotypic switch, a CRISPR KO screen was carried out in BRAF inhibitor-addicted melanoma cells upon drug withdrawal to select for surviving cells. In this case, the drug withdraw is the positive selection pressure. This study discovered that drug withdrawal causes an ERK-dependent transcription reprogramming event that downregulates the microphthalmia-associated transcription factor (MITF), which in turn, shuts down the expression of the pro-survival gene BCL2. This study demonstrated that RAF/MEK inhibitor withdrawal synergizes with MITF inhibitor to kill resistant cells, thus suggesting a potential approach to overcome drug resistance in melanoma patients [50]. Analogous CRISPR screens have been employed to identify genes that are responsible for resistance mechanism to other cancer therapies including 6-thioguanine, etoposide [43], ATR inhibitor [51], receptor tyrosine inhibitors [52], and T-cell response [53]. These studies have demonstrated the power of using CRISPR screens to elucidate the mechanism of drug resistance and to identify strategies to overcome drug resistance.

CRISPR screens have been employed to study the action of microbial toxins. CRISPR KO and CRISPRi screens have been used to identify genes whose LOF results in resistance to bacterial toxins. This approach not only led to new understanding about the underlying mechanism of toxin action, they should also facilitate the development of preventive therapeutics [41,54,55]. In infectious disease research, CRISPR KO screen has been deployed to identify host cell components that are required for Zika virus and Dengue virus infection and replication. These screens identified proteins involved in endocytosis, transmembrane protein processing, and endoplasmic reticulum-associated multiprotein complex to play a role in the viral life cycle [56]. In addition to host factors, CRISPR screens can be performed in microorganisms to identify pathogen factors that are critical for pathogenesis. A CRISPR/Cas9 library targeting all annotated genes in the *Toxoplasma gondii* genome was screened in the parasite for genes that control its infection and replication in human cells and for genes that regulate resistance to antiparasitic drugs. These genes could potentially serve as new drug targets for treating *Toxoplasma* infection [57].

8.4.2 GOF Screens Using CRISPRa

Complementary to CRISPR KO and CRISPRi-based LOF screens, CRISPRa-based GOF screens can reveal genes whose activation alter a desired phenotype or confers drug resistance. For example, a genome-wide CRISPRa screen was performed in human BRAF^{V600E} mutant melanoma cells in the presence of BRAF inhibitor to identify genes whose overexpression confer drug resistance. This study identifies both known and novel drug-resistant genes that could serve as potential drug targets to overcome BRAF inhibitor resistance [31]. A CRISPRa screen was carried out to identify genes that render cells resistant to *diphtheria* toxin, and this work revealed host factors in various biogenesis pathways as positive mediator for toxin entry [41]. Worth noting in this study is that both CRISPRi and CRISPRa were employed as complementary approaches to interrogate host pathways that both negatively and positively regulate toxin entry, thus demonstrating the power of using different CRISPR screening tools to dissect a cellular process and identify potential drug targets.

8.4.3 *In Vivo* CRISPR Screens

Until recently, genetic screens using various CRISPR approaches have been primarily performed in cultured cells *in vitro*. Genetic screens performed in animal models should more closely recapitulate aspects of the human disease including tissue environment, immune response, and metabolic status. Several *in vivo* CRISPR screens have been developed in mice, nearly all of these were carried out in the context of cancer biology. These screens involve either tumor transplant models or autochthonous cancer models. Similar to *in vitro* screens, *in vivo* screens can be LOF or GOF using various CRISPR platforms and both negative and positive selections can be implemented.

An *in vivo* CRISPR screen with a transplant model consists of two steps. The first step is to introduce Cas9 and the CRISPR library into cells cultured *in vitro*, and the second step is to transplant the library-carrying cells, as a pool, into mice. These cells are grown and selected for a desired phenotype *in vivo*, and changes in library composition in response to the selection pressure are analyzed from recovered tumor cells (Figure 8.3b). The first *in vivo* CRISPR screen used a subcutaneous transplant model to study tumor metastasis. In this study, a genome-wide CRISPR KO library was first introduced into a non-metastatic human lung cancer cell line in culture. The pool of cells was subcutaneously injected into nude mice and monitored for metastasis to the lung. By comparing the difference in sgRNA abundance between the primary tumor and lung metastasis, tumor suppressor genes *NF2*, *PTEN*, and *CDKN2A* were identified as negative regulators of metastasis [58]. Other studies using *in vivo* assays have identified tumor suppressor genes for Burkitt's lymphoma [59] and for liver cancer [60].

Potential drug targets can be directly identified using *in vivo* CRISPR screens. A genome-wide CRISPR KO screen performed in transplant models identified potential drug targets for ovarian cancer including known oncogenes *ERBB2* and *RAF1*, as well as a novel target KPNB1 which is a transporter involved in the nuclear localization of client proteins [61]. An *in vivo* CRISPR KO screen was performed in isogenic *KRAS* mutant and wild-type HCT116 colorectal cancer cell lines [62]. This isogenic system has been used in genetic screens *in vitro* for identifying KRAS synthetic lethal partners as drug targets to selectively kill KRAS mutant cells [63]. KRAS mutant and wild-type HCT116 cells carrying a genome-wide CRISPR library were subcutaneously injected into nude mice to allow tumor growth. By comparing changes in library composition between KRAS mutant and wild-type tumors, new KRAS synthetic lethal partners including metabolic enzymes were identified [64].

In addition to target identification, *in vivo* CRISPR screens are valuable tools to elucidate mechanism of resistance to therapies. Using a transplant mouse model, CRISPRi and CRISPRa screens were performed to identify genes that alter the sensitivity of lymphoma cells to chemotherapy [65].

Similarly, an *in vivo* CRISPR KO screen was recently carried out to identify genes that modulate the sensitivity of melanoma cells to immunotherapy and identified potential targets that may increase the efficacy of checkpoint blockade [66].

Although mouse xenograft models involving subcutaneous tumor cell injection tend to be the easiest *in vivo* model for screening, there are two limitations associated with this approach that do not fully recapitulate the process of tumor development. First, tumors are growing from millions of cancer cells at an irrelevant site rather than growing from a single transformed cell in its native tissue of origin. Second, the interaction between tumor and host immunity cannot be addressed in these immunocompromised models. In order to recapitulate the process of tumorigenesis more faithfully, autochthonous mouse models, in which the tumor cells are derived from its natural tissue of origin in an immunocompetent animal, can be used for *in vivo* CRISPR screens. In this setting, genetically engineered mouse models of cancer are employed, and the CRISPR/Cas9 system is directly delivered to the organ site to introduce mutagenesis in tumor cells (Figure 8.3c). For example, a pioneer study utilized hydrodynamic tail vein injection of plasmids to deliver Cas9 gene and sgRNAs into the mouse liver cells. In this study, CRISPR KO was carried out to model hepatocellular carcinoma (HCC) and intrahepatic cholangiocarcinoma (ICC) by mutating ten known HCC/ICC-associated tumor suppressors and oncogenes [67]. The creation of germline Cas9 transgenic mice has greatly facilitated CRISPR screens in autochthonous models [68–70]. Using the Cas9 mouse, a study compared the effect of inactivating 11 different tumor-suppressor genes, using intratracheal lentiviral delivery of pooled sgRNAs, in a KRAS-driven lung adenocarcinoma model [71]. Autochthonous CRISPR KO screens have also been performed in astrocytes and in liver cells to interrogate the functional of 49 common tumor-suppressor genes in glioblastoma and HCC, respectively [72,73]. These studies have demonstrated the feasibility of *in vivo* CRISPR/Cas9 screens in sophisticated mouse cancer models for target discovery.

8.4.4 Target Validation

Although high-throughput CRISPR screens are a powerful approach to identify potential targets for drug development, extensive target validation is necessary to evaluate a gene's potential for entering the drug development pipeline. A comprehensive analysis of the small-molecule drug development projects from 2005 to 2010 at AstraZeneca found that 40% of the projects that were terminated owing to efficacy-related issues were associated with a lack of strong evidence linking the drug targets to disease models [74]. Therefore, hit validation is a crucial step following genetic screens. CRISPR screen hits can be validated using various approaches. Using additional guide sequences and performing rescue experiments are always necessary to rule out off-target effects. Different screens can serve as cross validation for each other. For example, targeting the same gene with different gene inactivation mechanisms, such as RNAi, CRISPR KO, and CRISPRi, should give the same phenotype [42,46,56,61,75]. If available, small molecule inhibitors targeting the same hits would be expected to validate CRISPR screen results [76,77]. By employing both CRISPRi and CRISPRa approaches, antagonistic pathways and mechanisms should be revealed [41]. In addition, larger panel of cell lines and autochthonous disease models are needed to further evaluate the potential therapeutic window of a candidate drug targets.

8.5 LIMITATIONS OF CRISPR/CAS9 AND STRATEGIES FOR IMPROVEMENT

To maximize the utility of a screening technology, it is important to understand its limitations. Like RNAi, the CRISPR/Cas9 system relies on a targeting mechanism that is dependent on a short guide sequence. As a result, sgRNA off-target effect is always a concern despite it being less pronounced than that seen with RNAi [38,39]. Several strategies have been implemented to overcome sgRNA off-target effects. One strategy is to include more guide sequences per gene in a CRISPR library. This could lead to large library size that may not be amenable to certain assay types, particularly *in vivo* screens. By generating CRISPR libraries that consist of multiple small sub-pools, libraries of any desirable complexity can be quickly customized to match assay needs [78]. Several algorithms have been developed to identify guide RNA sequences with reduced off-target effects [21,38,78–82]. Cas9 mutants that are rationally designed to enhance sgRNA targeting specificity have been shown to reduce sequence-based off-target effects [83,84]. For HDR-based gene editing, using two guide sequences, each targeting a different DNA strand flanking the targeting site, together with a Cas9 nickase in which one of the two endonuclease domains is inactivated can also decrease off-target effects [85,86]. In addition to sequence-based off-target effects, Cas9 can efficiently generate multiple on-target cuts when directed to a highly amplified gene locus [24]. This may lead to target-independent lethality and the nomination of false-positive candidate genes in a cell viability screen [87,88]. This specific concern can be mitigated by using CRISPRi and RNAi as complementary approaches for hit validation.

In addition to false-positive and false-negative results, some factors may result in low efficiency of CRISPR/Cas9-mediated genome editing. To generate a KO allele, CRISPR/Cas9 relies on frame-shift mutations created by NHEJ-mediated repair of the target site. At certain frequencies, NHEJ may lead to in-frame indels and the translation of a functional, albeit mutant, gene products. To enhance the rate at which non-functional mutant is generated, sgRNAs can be targeted to conserved functional domains in proteins [89]. For HDR-based gene editing, low HDR efficiency is often a problem, especially in cells with low mitotic index since homologous recombination preferentially occurs in the G2/M phase of the cell cycle. Inhibiting the NHEJ pathway has been shown to enhance HDR efficiency and, thus, facilitate the recovery of edited clones [25,26]. Chromatin accessibility has been shown to affect Cas9 binding efficiency, with genes located in heterochromatin regions being more difficult to manipulate by CRISPR/Cas9 [90]. In this situation, strategies that can improve DNA access by Cas9 are needed to improve CRISPR/Cas9-mediated targeting of heterochromatin sequence elements. In mammalian cells, gene paralog redundancy, as well as pathway feedback and cross compensation could confound single-gene phenotype in screens. To overcome this limitation, CRISPR/Cas9 has been adapted to simultaneously target multiple gene paralogs [17,32–35] and combinatorial CRISPR screens have been conducted to assess functional interactions between two or more genes [91,92].

Because high-content screens are generally not compatible with pooled CRISPR libraries, efforts have been directed towards developing arrayed CRISPR libraries. It was demonstrated that transfection of synthetic guide RNAs into Cas9-expressing cells, an approach that is analogous to siRNA transfection, can be used to carry out high-content screens in an arrayed format and is scalable [93].

8.6 SUMMARY

CRISPR/Cas has emerged as a revolutionary technology for gene editing and functional genomics studies. Its wide adoption as a screening tool in a variety of assay contexts has accelerated the pace of target discovery in drug development. In this chapter, we have selected representative examples of target discovery effort driven by various CRISPR/Cas9 screening platforms to illustrate the power of this system. Further advances in this technology, especially in animal models, will lead to new advances in biomedical research and drug development.

DISCLAIMER

This work was supported by the Intramural Research Program of the National Institutes of Health and the National Cancer Institute. The content is solely the responsibility of the authors and does not necessarily represent the official views or policies of the National Institutes of Health and the Department of Health and Human Services. The mention of trade names, commercial products or organizations does not imply endorsement from the US Government.

REFERENCES

1. International Human Genome Sequencing Consortium, Finishing the euchromatic sequence of the human genome. *Nature* **431**, 931–945 (2004).
2. E. S. Lander, Initial impact of the sequencing of the human genome. *Nature* **470**, 187–197 (2011).
3. J. Shendure et al., DNA sequencing at 40: Past, present and future. *Nature* **550**, 345–353 (2017).
4. Y. Ishino, H. Shinagawa, K. Makino, M. Amemura, A. Nakata, Nucleotide sequence of the iap gene, responsible for alkaline phosphatase isozyme conversion in Escherichia coli, and identification of the gene product. *J Bacteriol* **169**, 5429–5433 (1987).
5. F. J. Mojica, C. Diez-Villasenor, E. Soria, G. Juez, Biological significance of a family of regularly spaced repeats in the genomes of Archaea, Bacteria and mitochondria. *Mol Microbiol* **36**, 244–246 (2000).
6. R. Jansen, J. D. Embden, W. Gaastra, L. M. Schouls, Identification of genes that are associated with DNA repeats in prokaryotes. *Mol Microbiol* **43**, 1565–1575 (2002).
7. T. H. Tang et al., Identification of 86 candidates for small non-messenger RNAs from the archaeon Archaeoglobus fulgidus. *Proc Natl Acad Sci U S A* **99**, 7536–7541 (2002).
8. F. J. Mojica, C. Diez-Villasenor, J. Garcia-Martinez, E. Soria, Intervening sequences of regularly spaced prokaryotic repeats derive from foreign genetic elements. *J Mol Evol* **60**, 174–182 (2005).
9. C. Pourcel, G. Salvignol, G. Vergnaud, CRISPR elements in Yersinia pestis acquire new repeats by preferential uptake of bacteriophage DNA, and provide additional tools for evolutionary studies. *Microbiology* **151**, 653–663 (2005).
10. A. Bolotin, B. Quinquis, A. Sorokin, S. D. Ehrlich, Clustered regularly interspaced short palindrome repeats (CRISPRs) have spacers of extrachromosomal origin. *Microbiology* **151**, 2551–2561 (2005).
11. R. Barrangou et al., CRISPR provides acquired resistance against viruses in prokaryotes. *Science* **315**, 1709–1712 (2007).
12. P. D. Hsu, E. S. Lander, F. Zhang, Development and applications of CRISPR-Cas9 for genome engineering. *Cell* **157**, 1262–1278 (2014).
13. J. E. Garneau et al., The CRISPR/Cas bacterial immune system cleaves bacteriophage and plasmid DNA. *Nature* **468**, 67–71 (2010).
14. E. Deltcheva et al., CRISPR RNA maturation by trans-encoded small RNA and host factor RNase III. *Nature* **471**, 602–607 (2011).
15. M. Jinek et al., A programmable dual-RNA-guided DNA endonuclease in adaptive bacterial immunity. *Science* **337**, 816–821 (2012).
16. G. Gasiunas, R. Barrangou, P. Horvath, V. Siksnys, Cas9-crRNA ribonucleoprotein complex mediates specific DNA cleavage for adaptive immunity in bacteria. *Proc Natl Acad Sci U S A* **109**, E2579–E2586 (2012).
17. L. Cong et al., Multiplex genome engineering using CRISPR/Cas systems. *Science* **339**, 819–823 (2013).
18. P. Mali et al., RNA-guided human genome engineering via Cas9. *Science* **339**, 823–826 (2013).
19. F. J. Mojica, C. Diez-Villasenor, J. Garcia-Martinez, C. Almendros, Short motif sequences determine the targets of the prokaryotic CRISPR defence system. *Microbiology* **155**, 733–740 (2009).
20. W. Jiang, D. Bikard, D. Cox, F. Zhang, L. A. Marraffini, RNA-guided editing of bacterial genomes using CRISPR-Cas systems. *Nat Biotechnol* **31**, 233–239 (2013).
21. P. D. Hsu et al., DNA targeting specificity of RNA-guided Cas9 nucleases. *Nat Biotechnol* **31**, 827–832 (2013).
22. K. S. Makarova, N. V. Grishin, S. A. Shabalina, Y. I. Wolf, E. V. Koonin, A putative RNA-interference-based immune system in prokaryotes: Computational analysis of the predicted enzymatic machinery, functional analogies with eukaryotic RNAi, and hypothetical mechanisms of action. *Biol Direct* **1**, 7 (2006).
23. R. Sapranauskas et al., The Streptococcus thermophilus CRISPR/Cas system provides immunity in Escherichia coli. *Nucleic Acids Res* **39**, 9275–9282 (2011).
24. G. Yuen et al., CRISPR/Cas9-mediated gene knockout is insensitive to target copy number but is dependent on guide RNA potency and Cas9/sgRNA threshold expression level. *Nucleic Acids Res* **45**, 12039–12053 (2017).
25. C. Yu et al., Small molecules enhance CRISPR genome editing in pluripotent stem cells. *Cell Stem Cell* **16**, 142–147 (2015).
26. T. Maruyama et al., Increasing the efficiency of precise genome editing with CRISPR-Cas9 by inhibition of nonhomologous end joining. *Nat Biotechnol* **33**, 538–542 (2015).
27. L. S. Qi et al., Repurposing CRISPR as an RNA-guided platform for sequence-specific control of gene expression. *Cell* **152**, 1173–1183 (2013).

28. S. Konermann et al., Optical control of mammalian endogenous transcription and epigenetic states. *Nature* **500**, 472–476 (2013).
29. L. A. Gilbert et al., CRISPR-mediated modular RNA-guided regulation of transcription in eukaryotes. *Cell* **154**, 442–451 (2013).
30. P. Mali et al., CAS9 transcriptional activators for target specificity screening and paired nickases for cooperative genome engineering. *Nat Biotechnol* **31**, 833–838 (2013).
31. S. Konermann et al., Genome-scale transcriptional activation by an engineered CRISPR-Cas9 complex. *Nature* **517**, 583–588 (2015).
32. K. Xie, B. Minkenberg, Y. Yang, Boosting CRISPR/Cas9 multiplex editing capability with the endogenous tRNA-processing system. *Proc Natl Acad Sci U S A* **112**, 3570–3575 (2015).
33. T. Sakuma, A. Nishikawa, S. Kume, K. Chayama, T. Yamamoto, Multiplex genome engineering in human cells using all-in-one CRISPR/Cas9 vector system. *Sci Rep* **4**, 5400 (2014).
34. A. M. Kabadi, D. G. Ousterout, I. B. Hilton, C. A. Gersbach, Multiplex CRISPR/Cas9-based genome engineering from a single lentiviral vector. *Nucleic Acids Res* **42**, e147 (2014).
35. J. F. Li et al., Multiplex and homologous recombination-mediated genome editing in Arabidopsis and Nicotiana benthamiana using guide RNA and Cas9. *Nat Biotechnol* **31**, 688–691 (2013).
36. P. Perez-Pinera et al., RNA-guided gene activation by CRISPR-Cas9-based transcription factors. *Nat Methods* **10**, 973–976 (2013).
37. M. L. Maeder et al., CRISPR RNA-guided activation of endogenous human genes. *Nat Methods* **10**, 977–979 (2013).
38. J. G. Doench et al., Optimized sgRNA design to maximize activity and minimize off-target effects of CRISPR-Cas9. *Nat Biotechnol* **34**, 184–191 (2016).
39. B. Evers et al., CRISPR knockout screening outperforms shRNA and CRISPRi in identifying essential genes. *Nat Biotechnol* **34**, 631–633 (2016).
40. X. Wu et al., Genome-wide binding of the CRISPR endonuclease Cas9 in mammalian cells. *Nat Biotechnol* **32**, 670–676 (2014).
41. L. A. Gilbert et al., Genome-scale CRISPR-mediated control of gene repression and activation. *Cell* **159**, 647–661 (2014).
42. O. Shalem et al., Genome-scale CRISPR-Cas9 knockout screening in human cells. *Science* **343**, 84–87 (2014).
43. T. Wang, J. J. Wei, D. M. Sabatini, E. S. Lander, Genetic screens in human cells using the CRISPR-Cas9 system. *Science* **343**, 80–84 (2014).
44. T. Hart et al., High-resolution CRISPR screens reveal fitness genes and genotype-specific cancer liabilities. *Cell* **163**, 1515–1526 (2015).
45. T. Wang et al., Identification and characterization of essential genes in the human genome. *Science* **350**, 1096–1101 (2015).
46. M. L. Burr et al., CMTM6 maintains the expression of PD-L1 and regulates anti-tumour immunity. *Nature* **549**, 101–105 (2017).
47. H. Davies et al., Mutations of the BRAF gene in human cancer. *Nature* **417**, 949–954 (2002).
48. D. B. Solit et al., BRAF mutation predicts sensitivity to MEK inhibition. *Nature* **439**, 358 (2006).
49. A. M. Menzies, G. V. Long, Dabrafenib and trametinib, alone and in combination for BRAF-mutant metastatic melanoma. *Clin Cancer Res* **20**, 2035–2043 (2014).
50. X. Kong et al., Cancer drug addiction is relayed by an ERK2-dependent phenotype switch. *Nature* **550**, 270–274 (2017).
51. S. Ruiz et al., A genome-wide CRISPR screen identifies CDC25A as a determinant of sensitivity to ATR inhibitors. *Mol Cell* **62**, 307–313 (2016).
52. P. Hou et al., A genome-wide CRISPR screen identifies genes critical for resistance to FLT3 inhibitor AC220. *Cancer Res* **77**, 4402–4413 (2017).
53. S. J. Patel et al., Identification of essential genes for cancer immunotherapy. *Nature* **548**, 537–542 (2017).
54. H. Koike-Yusa, Y. Li, E. P. Tan, C. Velasco-Herrera Mdel, K. Yusa, Genome-wide recessive genetic screening in mammalian cells with a lentiviral CRISPR-guide RNA library. *Nat Biotechnol* **32**, 267–273 (2014).
55. Y. Zhou et al., High-throughput screening of a CRISPR/Cas9 library for functional genomics in human cells. *Nature* **509**, 487–491 (2014).
56. G. Savidis et al., Identification of Zika virus and dengue virus dependency factors using functional genomics. *Cell Rep* **16**, 232–246 (2016).
57. S. M. Sidik et al., A genome-wide CRISPR screen in toxoplasma identifies essential apicomplexan genes. *Cell* **166**, 1423–1435. e1412 (2016).

58. S. Chen et al., Genome-wide CRISPR screen in a mouse model of tumor growth and metastasis. *Cell* **160**, 1246–1260 (2015).

59. A. Katigbak et al., A CRISPR/Cas9 functional screen identifies rare tumor suppressors. *Sci Rep* **6**, 38968 (2016).

60. C. Q. Song et al., Genome-wide CRISPR screen identifies regulators of mitogen-activated protein kinase as suppressors of liver tumors in mice. *Gastroenterology* **152**, 1161–1173. e1161 (2017).

61. M. Kodama et al., In vivo loss-of-function screens identify KPNB1 as a new druggable oncogene in epithelial ovarian cancer. *Proc Natl Acad Sci U S A* **114**, E7301–E7310 (2017).

62. S. Shirasawa, M. Furuse, N. Yokoyama, T. Sasazuki, Altered growth of human colon cancer cell lines disrupted at activated Ki-ras. *Science* **260**, 85–88 (1993).

63. J. Luo et al., A genome-wide RNAi screen identifies multiple synthetic lethal interactions with the Ras oncogene. *Cell* **137**, 835–848 (2009).

64. E. H. Yau et al., Genome-wide CRISPR screen for essential cell growth mediators in mutant KRAS colorectal cancers. *Cancer Res* **77**, 6330–6339 (2017).

65. C. J. Braun et al., Versatile in vivo regulation of tumor phenotypes by dCas9-mediated transcriptional perturbation. *Proc Natl Acad Sci U S A* **113**, E3892–E3900 (2016).

66. R. T. Manguso et al., In vivo CRISPR screening identifies Ptpn2 as a cancer immunotherapy target. *Nature* **547**, 413–418 (2017).

67. J. Weber et al., CRISPR/Cas9 somatic multiplex-mutagenesis for high-throughput functional cancer genomics in mice. *Proc Natl Acad Sci U S A* **112**, 13982–13987 (2015).

68. R. J. Platt et al., CRISPR-Cas9 knockin mice for genome editing and cancer modeling. *Cell* **159**, 440–455 (2014).

69. L. E. Dow et al., Inducible in vivo genome editing with CRISPR-Cas9. *Nat Biotechnol* **33**, 390–394 (2015).

70. S. H. Chiou et al., Pancreatic cancer modeling using retrograde viral vector delivery and in vivo CRISPR/Cas9-mediated somatic genome editing. *Genes Dev* **29**, 1576–1585 (2015).

71. Z. N. Rogers et al., A quantitative and multiplexed approach to uncover the fitness landscape of tumor suppression in vivo. *Nat Methods* **14**, 737–742 (2017).

72. R. D. Chow et al., AAV-mediated direct in vivo CRISPR screen identifies functional suppressors in glioblastoma. *Nat Neurosci* **20**, 1329–1341 (2017).

73. G. Wang et al., Mapping a functional cancer genome atlas of tumor suppressors in mouse liver using AAV-CRISPR-mediated direct in vivo screening. *Sci Adv* **4**, eaao5508 (2018).

74. D. Cook et al., Lessons learned from the fate of AstraZeneca's drug pipeline: A five-dimensional framework. *Nat Rev Drug Discov* **13**, 419–431 (2014).

75. R. M. Deans et al., Parallel shRNA and CRISPR-Cas9 screens enable antiviral drug target identification. *Nat Chem Biol* **12**, 361–366 (2016).

76. J. Barretina et al., The Cancer Cell Line Encyclopedia enables predictive modelling of anticancer drug sensitivity. *Nature* **483**, 603–607 (2012).

77. M. J. Garnett et al., Systematic identification of genomic markers of drug sensitivity in cancer cells. *Nature* **483**, 570–575 (2012).

78. A. Read, S. Gao, E. Batchelor, J. Luo, Flexible CRISPR library construction using parallel oligonucleotide retrieval. *Nucleic Acids Res* **45**, e101 (2017).

79. J. G. Doench et al., Rational design of highly active sgRNAs for CRISPR-Cas9-mediated gene inactivation. *Nat Biotechnol* **32**, 1262–1267 (2014).

80. F. Heigwer, G. Kerr, M. Boutros, E-CRISP: Fast CRISPR target site identification. *Nat Methods* **11**, 122–123 (2014).

81. H. Liu et al., CRISPR-ERA: A comprehensive design tool for CRISPR-mediated gene editing, repression and activation. *Bioinformatics* **31**, 3676–3678 (2015).

82. M. A. Moreno-Mateos et al., CRISPRscan: Designing highly efficient sgRNAs for CRISPR-Cas9 targeting in vivo. *Nat Methods* **12**, 982–988 (2015).

83. I. M. Slaymaker et al., Rationally engineered Cas9 nucleases with improved specificity. *Science* **351**, 84–88 (2016).

84. B. P. Kleinstiver et al., High-fidelity CRISPR-Cas9 nucleases with no detectable genome-wide off-target effects. *Nature* **529**, 490–495 (2016).

85. F. A. Ran et al., Genome engineering using the CRISPR-Cas9 system. *Nat Protoc* **8**, 2281–2308 (2013).

86. Y. Fu, J. D. Sander, D. Reyon, V. M. Cascio, J. K. Joung, Improving CRISPR-Cas nuclease specificity using truncated guide RNAs. *Nat Biotechnol* **32**, 279–284 (2014).

87. A. J. Aguirre et al., Genomic copy number dictates a gene-independent cell response to CRISPR/Cas9 targeting. *Cancer Discov* **6**, 914–929 (2016).
88. D. M. Munoz et al., CRISPR screens provide a comprehensive assessment of cancer vulnerabilities but generate false-positive hits for highly amplified genomic regions. *Cancer Discov* **6**, 900–913 (2016).
89. J. Shi et al., Discovery of cancer drug targets by CRISPR-Cas9 screening of protein domains. *Nat Biotechnol* **33**, 661–667 (2015).
90. C. Kuscu, S. Arslan, R. Singh, J. Thorpe, M. Adli, Genome-wide analysis reveals characteristics of off-target sites bound by the Cas9 endonuclease. *Nat Biotechnol* **32**, 677–683 (2014).
91. K. Han et al., Synergistic drug combinations for cancer identified in a CRISPR screen for pairwise genetic interactions. *Nat Biotechnol* **35**, 463–474 (2017).
92. J. P. Shen et al., Combinatorial CRISPR-Cas9 screens for de novo mapping of genetic interactions. *Nat Methods* **14**, 573–576 (2017).
93. J. Tan, S. E. Martin, Validation of synthetic CRISPR reagents as a tool for arrayed functional genomic screening. *PLoS One* **11**, e0168968 (2016).

9 Probiotics in the World
"Bugs-as-Drugs"

Thomas Kuntz, Madeline Kim, Elle Simone Hill, and Mariana C. Salas Garcia
University of Chicago

Jack A. Gilbert
University of California

CONTENTS

9.1 INTRODUCTION

Disease is often what comes to mind when considering the presence of microbes inside the human body. This is not surprising, given the tight coupling of the discovery of microbes with germ theory by Robert Koch and Louis Pasteur and the ensuing focus on the pathogenicity of microbes in both the scientific and popular literature. Infectious diseases remain a huge global health burden (Dye 2014) and a salient aspect of day-to-day life. Often overlooked are the commensal and even mutualistic interactions of microbes and humans. However, these interactions are not just important to human health; they are, in fact, fundamental to what it means to be human. This is because our bodies harbor approximately as many microbial cells as human cells, mainly in the gut (Sender, Fuchs, and Milo 2016).

The microbial taxa (mainly bacterial) associated with humans are known as the microbiota, and these organisms and their gene content, much of which supplements human physiology, are collectively known as the microbiome. Note that while microbiome sometimes strictly refers to gene content, in analogy to genome, the previous looser definition is more common in the literature (Ursell et al. 2013). Speculations on the existence of a microbiome first began when Antonie van Leeuwenhoek observed oral and fecal microbes in the 1700s (Dobell 1920). However, advances in culturing and sequencing technology in the last 20 years have begun to reveal the true scope of the microbiome (Morgan and Huttenhower 2012). The gut microbiome is ultimately a presence so important that many have taken to considering it an organ of the body (Baquero and Nombela 2012).

Like other organs of the body (albeit of a vastly unique physiology), the microbiome constitutes a druggable target in cases of disease. Antibiotics are a clear example of a class of drugs with profound effects on the microbiota. The on-target effects of antibiotics on pathogens are wildly successful and constituted a revolution in healthcare; however, off-target effects on commensal microbiota are becoming better understood and, in some cases, may be dire, such as in *Clostridium difficile* (*C. diff*) infection (Surawicz et al. 2013). Furthermore, antibiotic resistance is an increasingly serious public health concern, fueled by overuse, which should be combated not solely through discovery of novel antibiotics. This approach has proven difficult and does little to address the underlying problem of resistance (Aminov, Otto, and Sommer 2010). Lastly, with respect to the microbiome, killing microbes is certainly not the only desired outcome and even antibiotics that could perfectly target pathogens would be unable to treat disease states caused by imbalances in the microbiome, often called dysbiosis, that influence the microbial metabolism and host-interaction equilibrium (Carding et al. 2015).

One alternative and burgeoning approach to drugging the microbiome is treatment with microbes themselves. Etymologically (and in many ways functionally) contrasting antibiotics, these are referred to as probiotics (Lilly and Stillwell 1965). However, as they are live organisms, probiotics pose unique challenges and novel possibilities compared to traditional drugs. Nonetheless, probiotics are considered by many to be close enough to traditional pharmaceuticals to warrant the moniker "bugs as drugs" (Shanahan 2010).

Notably, there has been knowledge of microbes as health improving agents for some time. While there may have been some appreciation of beneficial microbes in antiquity, mainly in the form of naturally occurring fermented milk, Eli Metchnikoff, who is widely considered the father of probiotics, kick-started the field with his rational for the health benefit of certain food-borne microbes in 1907. Specifically, he suggested, through food, "to adopt measures to modify the flora in our bodies and to replace the harmful microbes by useful microbes" (Metchnikoff 1907). Currently, the most widely accepted definition for probiotics is "live microorganisms which when administered in adequate amounts confer a health benefit to the host," which has been adopted by the World Health Organization and Food and Agriculture Organization (Araya et al. 2002).

This chapter will explore the discovery and development of probiotics in clinical practice, including new horizons of probiotics as "bugs as drugs." Emphasis is placed on establishing the foundations of this paradigm and presenting a wide breadth of ideas in the probiotic space to highlight the exciting developments and challenges this field currently faces.

9.2 THE HUMAN MICROBIOME

The microbiome comprises of a highly interconnected network of microorganisms, primarily located in the environmentally suitable descending colon, though with complex biogeography (Donaldson, Lee, and Mazmanian 2015). In addition to bacteria (the primary focus of this chapter), archaea, yeast, and vast numbers of viruses and bacteriophages inhabit the gut and contribute to microbiome function. The bacteria are dominated by the phyla Firmicutes and Bacteroides, with major genera in many healthy individuals being *Bacteroides*, *Bifidobacterium*, *Clostridium*, *Eubacterium*, *Fusobacterium*, *Peptococcus*, *Peptostreptococcus*, and *Ruminococcus* (Shen, Obin, and Zhao 2013). At the species level, however, the notion of a core microbiome is weak, with very high variability between individuals, despite its significance to human physiology. Despite the variance in taxonomic composition, there remains a relatively stable functional potential between people, suggesting that most microbial metabolisms are conserved (Lozupone et al. 2012). The lack of a compositional core microbiome may be related to a lack of genetic control over gut microbial content. A recent study placed the average genetic heritability of gut microbial taxa at 1.9% (Rothschild et al. 2017), while environmental variability described 20% of the variance in microbiome composition. Furthermore, the microbiome is malleable, rapidly changing in response to dietary changes (David et al. 2013), and differing across age and geography (Yatsunenko et al. 2012). This is not to say the microbiome is completely unstable; one study found an average of 60% of strains were conserved across the first 5 years of life (Faith et al. 2013). Certain early gut colonizers and certain taxa, especially those belonging to the phyla Bacteroidetes and Actinobacteria, may be particularly persistent across time and within a human population. However, the variability of the stability of each microbe is poorly understood, despite being extremely important to assessing clinical relevance of each strain.

The mutable characteristics of the microbiome are a positive benefit for probiotic approaches, as they allow for the possibility of targeted and lasting alterations to the microbiome, but they have also led to health problems as our lifestyle behaviors and environment have diverged from our ancestors, which may carry potential negative consequences for host–microbe relationships. The microbiome co-evolved with humans who had a vastly different diet than humans living in the developed world, with far more nondigestible carbohydrates and fibers (Ley et al. 2008). In particular, the "western diet," characterized by high-fat and high-simple sugar consumption, is known to have deleterious effects on the microbiome and, thus, overall health (Martinez, Leone, and Chang 2017). Other environmental risk factors include pollution (Salim, Kaplan, and Madsen 2013) and C-section delivery (Dominguez-Bello et al. 2016). Sterility in the built environment, often allowing for establishment of pathogens (Kelley and Gilbert 2013), can prevent the natural acquisition of and exposure to microbes (Rook, Raison, and Lowry 2014), which may also contribute to poor outcomes in host health, especially if experienced during early phases of development (i.e. in utero and early life). The concept of an interruption in the interaction between our developing immune system and potentially beneficial microbes is often known as the "hygiene hypothesis" (Okada et al. 2010). Thus, the challenges and conditions of modern life may well require microbial supplementation and alteration for ideal health. It is important to appreciate that selection pressures on the microbiome imposed by environment, behavior, disease, and pharmaceutical use, are so crucial to human health because of how closely coupled the microbiome is with its host, to the point of the combination of human and microbiome being termed a "superorganism" (Sleator 2018). This tight relationship causes astonishingly profound physiological effects both within and far beyond the gut. We will now turn the discussion to some of the primary functions of the microbiome.

First, the microbiome supplements and improves human metabolism and nutrient absorption by degrading of indigestible compounds such as complex polysaccharides, especially those from plants (Devaraj, Hemarajata, and Versalovic 2014). The result is a mutually beneficial partnership: energy that would otherwise be wasted is harvested by the microbiota, and the host is able to absorb some of the metabolic byproducts. Chief among these metabolites are short-chain fatty acids (SCFAs), which have profound systemic effects, including roles in appetite, gut integrity, glucose homeostasis, lipid metabolism, and immune function (Morrison and Preston 2016). Positive effects have even been reported on neural functioning and plasticity (Bourassa et al. 2016). In addition, gut microbiota synthesize a number of vitamins that can be utilized by the host, transform bile acids, and ultimately produce metabolites in ways still being discovered and explored (Vernocchi, Del Chierico, and Putignani 2016). Lastly, the microbiome may also mediate gaseous compounds within the host, affecting host physiology (Pimentel, Mathur, and Chang 2013).

Another major function of the microbiome is its network of interactions with the intestinal epithelium, the barrier between the gut luminal contents and the underlying immune cells. The complex homeostatic processes between host and microbiome crucially involve the mucosal layer and innate immune system to regulate the commensal microbiota while preventing invasion or overgrowth of pathogens (Okumura and Takeda 2017), indicating a delicate interplay between the host and the microbiota. Importantly, gut microbes can modulate intestinal permeability and junctions between epithelial cells, which, in a disease state, can break down, leading to unwanted translocation of substances across the gut barrier, termed "leaky gut" (Hollander 1999). In addition, gut microbiota can moderate the immune system even as it acts upon them (Thaiss et al. 2016). This is the interface where pathogens are often recognized and rejected, or conversely, take hold, so it is key that these systems work correctly. A healthy microbiome moderates immune responses and gut physiology to exclude pathogens without overstimulating or under-stimulating the immune system or damaging the gut. This is especially important developmentally, as the effects of early-life dysregulation may be persistent and likely increase the risk of disease later in life (Wang, Monaco, and Donovan 2016).

Perhaps one of the most surprising major functions of the microbiome is its interactions with the nervous and neuroendocrine systems, a set of pathways known as the gut-brain axis. Gut microbiota are known to produce and transform numerous neurologically active metabolites, including gamma-aminobutyric acid, histamine, serotonin, and tryptophan (Clarke et al. 2014). The gut-brain axis encompasses these signaling molecules and other communications acting locally via the gut walls (and connected blood vessels) and the enteric nervous system (including the vagus nerve) to ultimately affect even central nervous system functioning (Carabotti et al. 2015). Furthermore, this signaling is strongly suggested to be bi-directional; for example, stress-associated gut alterations can increase the proliferation and virulence of pathogens (Cogan et al. 2007). This link is critical developmentally, with anxiety and motor control neural circuits being especially implicated (Heijtz et al. 2011).

The microbiome is crucial in pathogen resistance. This is most obvious in infections contracted following antibiotic use; the microbiome is significantly disrupted allowing for pathogens to take hold. These pathogens are either externally acquired or are members of the normal microbiome that become pathogenic (pathobionts). There are several mechanisms by which pathogens flourish. Immune dysregulation can occur; for example, communication of T lymphocytes with the microbiota is impaired. Direct microbial colonization resistance is hampered as well with lower cell densities, less competition for resources, and less production of pathogen-inhibiting chemicals allowing for growth (Blaser 2016). While they are more difficult to study and determine, disease states, especially inflammatory bowel disease (IBD) (Chow, Tang, and Mazmanian 2011), and general dysbiosis (Stecher and Hardt 2011), may allow for pathobiont expansion through similar mechanisms. Interestingly, IBD symptoms are reduced if bacterial load in the intestine is decreased through antibiotics, supporting the aforementioned hypothesis on the role of pathobionts (Videla et al. 1994). Clearly, given their profoundly important roles in human health, microbiome states have been shown to cause or correlate with a number of diseases, which include autism, cancer, chronic heart disease, diabetes, IBD, liver disease, and obesity (Zhang et al. 2015). Thus, there is a great health need to develop treatments for these diseases, and microbiome-based interventions present a promising opportunity. Possibly the "lowest hanging fruit" in this search is the use of probiotics.

9.3 PROBIOTICS FOUNDATIONS

Probiotics are live, typically orally ingested microbes with the primary aim to treat disease as well as prevent it and improve overall health. Since by current law, a drug is anything used for "diagnosis, cure, mitigation, treatment, or prevention" of disease or "intended to affect the structure or any function of the body of humans or other animals" (Food, Drug and Cosmetic Act of 1938), probiotics can typically be considered drugs, though unique pharmacological and regulatory principles apply. While probiotics (especially those discussed here) are mainly still investigational, interest in probiotics is steadily growing; a bibliometric study of pediatric probiotics showed about a 90-fold increase from 1994 to 2014 (Sweileh et al. 2016).

Additionally, there are already some products on the market that have had significant success. For example, Yakult, a Japanese fermented skimmed milk product with a *Lactobacillus casei* strain is quite popular and well studied (Spanhaak, Havenaar, and Schaafsma 1998). One of the reasons that most probiotics remain experimental is that probiotic action varies greatly between members of the same genus, or even species. Only microbes that have a long history of association with humans are typically employed, as they have been thoroughly confirmed as safe. Nonetheless, there are a number of common principles and mechanisms by which many probiotics act. These may better elucidate the actions of existing probiotics and fuel discovery and development of new ones, ultimately ushering in a new age of probiotic utilization.

Probiotic properties can be somewhat separated into three meaningful categories: widespread, frequent (species level), and rare (strain level) (Hill et al. 2014). Widespread effects are ones that are in nearly all heavily studied probiotics, though likely less so in novel genera or families. These include pathogen resistance and exclusion, bile acid metabolism, SCFA production, and regulation of intestinal transit. These are often general traits of positive gut bacteria as opposed to neutral or pathogenic, which are then increased by probiotic supplementation. Frequent effects include a wide number of possibilities, such as vitamin synthesis and bile salt metabolism, and often reflect the ecological niches of the specific microbes. Rare effects are very specific outcomes, such as the production of a single bioactive molecule, and are quite particular genetically or even bioengineered. These so-called frequent and rare effects should be the target of further research in order to further our scientific understanding of the mechanisms behind probiotic influence on health. What falls into each category and how much they overlap will continue to evolve as more probiotics are discovered and studied.

The mechanisms by which probiotics achieve their therapeutic properties are varied, but there are certain pathways and effectors that are common enough to warrant a brief summary. These can be divided into two factors: "adaptation" and "probiotic."

The microbes survive transit through the gastrointestinal tract (GIT), adhere in the intestinal mucosal layer, are adapted metabolically to thrive on available nutrients, and survive among other microbiota. Clearly, adaptation is not desired in pathogenic bacteria, however. While classic probiotic genera tend to succeed in adaptation, this is not always true especially for nontraditional or next-generation probiotics. In particular, survival while passing through the stomach is a significant hurdle as it constitutes a major defense of the body against foreign substances. Humans secrete approximately 2.5 L of acidic gastric juice each day, generating a fasting pH of 1.5, though with an increase to pH 3–5 during food intake. The intestinal pH however is closer to neutral, pH 7. Thus, microbiota that typically function best in this neutral environment must have defenses against the brief acidic assault from the stomach (Lund, Tramonti, and De Biase 2014). Additionally, an important digestive mixture, bile, which is an aqueous solution of bile acids, cholesterol, phospholipids, and biliverdin (a pigment), has strong antimicrobial effects. Bile has detergent properties: bacterial membrane dissolution, for example, as well as subtler cell-damaging mechanisms. Defense against environmental attacks or biotransformation of bile to more harmless forms is therefore essential for survival (Begley, Gahan, and Hill 2005). In probiotics with desirable qualities but low stress resistance, synthetic means of surviving the stomach are often required. This is a widely explored field with high commercial interest as it allows for both novel probiotics and for delivery of higher concentrations of bacteria that can but do not necessarily survive transit (Govender et al. 2014).

Once at least some of the ingested microbes have survived their trip to their destination, they must adhere or at least stay in transit for some time to have a significant effect on the FIhost.

Adherence is not always a desired quality, especially in experimental probiotics where one might desire or require an effect or dosing schedule typical to a traditional drug, but the physiological impact on the host is generally considered to be a core feature of a probiotic organism, rendering longevity in the patient problematic. Where along the lower GIT a microbe settles is determined by physiological variation, including pH, oxygen, specific nutrients, and host immunity. Typically, the small intestine has lower diversity and density due to harsher conditions, especially the presence of oxygen, which does not suit the anaerobic majority of microbiome members. Short transit time in the small intestine further limits microbes that cannot tightly adhere and rapidly uptake and metabolize available nutrients (Zoetendal et al. 2012). Nonetheless, the families *Lactobacillaceae* and *Enterobacteriaceae* are enriched here, at least in the tractable mouse model (Gu et al. 2013).

The colon houses the majority of the microbiome, having far more hospitable conditions, though not without barriers to colonization. Particularly important, to paradoxically both the adherence of microbes and to the host's defenses against them is a mucus layer secreted by goblet cells, specialized epithelial cells. In the small intestine, this is a single thick layer, but in the colon, an outer loose layer that is often sloughed off tops an inner layer protective of the epithelial cells. The loose layer is furthermore metabolized by a number of taxa, most notably *Akkermansia muciniphilia*, providing nutrients both for those bacteria and others associated with them, while the thick layer provides protection for microbes that can tolerate its conditions, particularly immune regulation and antimicrobial molecules. This ultimately creates a host habitat aimed at harboring beneficial microbes in favorable numbers and conditions, while dysregulation or loss of the mucosal layer can lead to pathogen invasion or translocation across the epithelium (Johansson et al. 2011). The microbial adherence necessary to sustain a beneficial relationship is provided by a number of effectors.

These primarily involve the cell envelope, as it is the point of contact with the host. Bacterial cell envelopes are Gram-positive or Gram-negative, which due to differing cell wall characteristics present different surface molecules. The Gram-positive bacteria have a thick peptidoglycan cell wall containing teichoic acids (WTAs) and their lipid-linked counterparts, lipoteichoic acids (LTA) with varying structures. Gram-negative bacteria have an outer lipid membrane which greatly alters the ability of substrates to enter and present mainly lipopolysaccharides (LPS) on their surfaces. While characterized as a prototypical bacterial toxin, LPS molecules likely have more complex immunomodulatory effects, and not all Gram-negative bacteria are detrimental to human health (d'Hennezel et al. 2017). Bacteria of both cell envelope types can additionally present a wide array of largely unexplored exopolysaccharides (EPS) (Schmid, Sieber, and Rehm 2015), S-layer proteins (Sara and Sleytr 2000), and a great number of general wall-anchored proteins, including specific mucus-binding proteins (MacKenzie et al. 2010), as well as pili and other adhesive factors. All of these cell envelope features can differ greatly among microbes and importantly can be involved in direct adherence or immune system tolerance (without which adherence is impossible).

Metabolic compatibility is another key adaptation principle. Availability of energy sources is fundamental to where microbiota settle and how they thrive and coexist. This obviously depends strongly on the host diet and environment, which could potentially be improved to alter probiotic effectiveness. However, the study of adaptation of probiotics to such complex conditions and dynamics of the human gut across populations remains difficult, theoretically and experimentally. Instead, the focus is on general principles, especially foods that best ensure the survival of target microbes. While basic nutrients and a balanced diet are necessary for a healthy microbiome, a specific and tenable scrutiny has been placed on what are known as prebiotics. Unfortunately, a precise characterization of prebiotics is currently heavily debated, for good reason (Hutkins et al. 2016). One commonly accepted, though somewhat broad definition is "a nondigestible compound that, through its metabolization by microorganisms in the gut, modulates composition and/or activity of the gut microbiota, thus conferring a beneficial physiological effect on the host" (Bindels et al. 2015). Dissenting characterizations require prebiotics to have better defined effects on specific microbes or limit them to smaller classes of compounds. The tradeoff is generality for making research and regulation simpler. Either way, shifting from whole diet and environmental effects to only those compounds which directly impact the microbiome may aid probiotic research, and these factors may ultimately be more responsible for the success or failure of a specific probiotic (Rastall and Gibson 2015). Prebiotics not only affect the thriving of intestinal microbes but are also utilized to aid survival through the GIT and adherence (Pandey, Naik, and Vakil 2015).

Lastly, even when a probiotic organism has made it to the appropriate location, adhered properly, and is receiving the proper nutrients, it needs to survive in the complex gut ecosystem. This is likely far easier for "autochthonous species," those with a long-term association with a particular host, forming a stable population of characteristic size in a particular region of the gut, and has a demonstrable ecological function (Tannock 2004). Much of ecological success overall involves the previous three factors, though competition with other microbes is another major element. Utilization of metabolites by microbes can both inhibit and foster the growth of others as can secondary metabolites produced by the microbiota (Cardona et al. 2016). Furthermore, bacteria can communicate directly in a process called quorum sensing (QS), where small diffusible molecules are released and interpreted by both intraspecies and interspecies targets. The elicited bacterial dynamics can allow group colonization, enhanced adaptation to host environment changes, and defense against competition (Miller and Bassler 2001). However, virulence can also be triggered in this fashion. One of the primary competitive mechanisms that bacteria possess is the production of bacteriocins, a nearly universal trait. These antimicrobial compounds are also incredibly diverse, indicating their importance in microbe population evolution and mediation (Riley and Wertz 2002). Interestingly, it's predicted that ecologically too much pure cooperation would destabilize the microbiome; varied interactions including many competitive ones are key to stability and, ultimately, to host health (Coyte, Schluter, and Foster 2015).

In addition to SCFA metabolism, many direct probiotic factors that make a microbe suitable for beneficial consumption overlap significantly with the microbe's ability to adapt to survival in the host. A crucial difference is that probiotics must have a positive health effect on the host, as opposed to more neutral commensals or pathogens. While surviving transit through the GIT is simply a necessary condition, the adherence, metabolism, and microbial ecological effects of a microbe can largely determine its probiotic function. For example, the cell surface proteins presented when adhering to the host often have probiotic effects, especially in immune system modulation and epithelial barrier protection (Lebeer, Vanderleyden, and De Keersmaecker 2008). Intestinal epithelial cells actively participate in the immune response, in concert with dendritic cells and macrophages, they constantly "sense" the microbiome to ensure protection of the host. Immediate innate responses can include the production of antimicrobial peptides and the secretion of immune activation molecules such as interleukins. The adaptive immune response to probiotic organisms is mediated in part through the mucosal lymphoid tissue, part of the gut-associated lymphoid tissue (GALT).

Direct microbial metabolism may alter gut homeostasis (Kumar et al. 2012), though secondary metabolites and products are mostly responsible for probiotic effects, in some cases, strain specific and in others, wide ranging (for example, SCFAs). Ecological perturbations induced by probiotics typically involve pathogen exclusion but may also restructure the microbiome in more significant ways by cooperating or competing for resources or niches. This latter effect remains more difficult to study and quantify but has been clearly shown to be a significant probiotic effect in a simpler organismal system (Yang et al. 2017). Other important probiotic factors include the inability to become pathogenic or transfer pathogenic potential to other microbes, regulation or alteration of the gut–brain axis and endocrine systems, as well as more specific effects as bioengineered and next-generation probiotics continue to be developed. Thus, the science of probiotics consists of finding or designing microbes that best fit these requirements or have unique desirable aspects and then tuning parameters such as delivery method to best optimize their actions as probiotics. Unfortunately, there are currently no specific probiotic supplements that encompass all of the potential effector mechanisms, which reveals the necessity for rational design of new probiotic formulations that can influence a precise clinical end point (van Baarlen, Wells, and Kleerebezem 2013). Indeed, in order to do so, probiotics and the human microbiota at large have to be characterizable and explorable.

9.4 MICROBIOME METHODS

The technologies for growing and querying bacteria in both the natural environment and the human body are necessarily strongly tied to the discovery and development of probiotics. The earliest techniques for culturing the human microbiota were crude, especially with respect to survival of oxygen intolerant anaerobes, which make up the vast majority of the microbiome.

It wasn't until the 1970s when what are considered to be the first proper explorations of the human gut microbiome took place, following significant advancements in anaerobic techniques as well as identification methods (Lagier et al. 2015). Not only were studies done detailing the composition of the microbiome (Moore and Holdeman 1974) but intriguing work on diet (Finegold, Attebery, and Sutter 1974) and even a tentative gut–brain connection (Holdeman, Good, and Moore 1976) proceeded shortly after. However, a significant problem remained: the vast majority of gut microorganisms were still unculturable, despite many more of them being observable under a microscope. This has been termed the "great plate anomaly" (Staley and Konopka 1985) and remains a significant problem even today. However, in the 1990s, a revolution began in our understanding of the human microbiome: the development of culture-independent techniques able to describe microbiome compositions as they are *in vivo* (Shendure and Ji 2008). These techniques mainly focus on the 16S rRNA gene present in all bacteria (and archaea, chloroplasts, and mitochondria). This gene contains both highly conserved and highly variable regions, allowing for extremely general probes that can nonetheless differentiate microbes to a species or even strain level. Due to the slow evolution of this gene and advanced computational analysis techniques, it is therefore possible to construct phylogenies and characterize bacterial sequences from uncultured bacteria (Snel, Bork, and Huynen 1999).

In one of the first large-scale 16S rDNA studies, a huge amount of uncultured bacterioplankton were found in the Sargasso Sea, and a novel group was observed, expanding ecological understanding of that habitat (Giovannoni et al. 1990). Since then, innovations such as next-generation sequencing (NGS) techniques have lowered the price of sequencing to the point that a veritable explosion in microbiome sequencing studies took place (Shendure and Ji 2008). These include numerous detailed studies of the gut microbiome, which form the basis of our understanding of that ecology. However, simply knowing what microbes are present hardly explains the microbiome, especially with relation to probiotic action and discovery. A better understanding is often framed as three questions that can be answered to varying degrees by existing technologies: who's there, what can they do, and what are they doing (Young 2017). 16S sequencing does much to answer who's there, though other technologies such as qPCR and fluorescent in situ hybridization (FISH) can supplement it with more specific microbe targeting (Matsuda et al. 2009).

The capability of members of the microbiome influencing their host, also called functional potential, is explored through genome sequencing to construct metabolic and other pathways. This can be done on cultured bacteria relatively easily, though these bacteria are a small fraction of the human microbiome. Furthermore, culture-independent single-cell isolation techniques where the cells aren't required to propagate have been developed, such as microfluidics and flow cytometry (Tolonen and Xavier 2017). Incredibly, genomes can also be characterized by sequencing complex communities. This is done through metagenomic analysis, whereby DNA from a sample is amplified in an unbiased fashion and genomes are reconstructed bioinformatically from the resulting data (Thomas, Gilbert, and Meyer 2012). Metagenomics is currently a promising though expensive and computationally intensive process. As such, results are often limited to high-abundance microbes and small datasets, though a number of large-scale experiments have taken place by leveraging sizeable research consortia (Huttenhower et al. 2012). As a partial remedy, alternative bioinformatic methods have been developed to utilize specific marker sequences in the same metagenomic data to find both species compositions and overall gene contents for the sample (Abubucker et al. 2012). As metagenomic sequencing and computing costs lower, it is likely these techniques will become more dominant; however, it is important to remember in order to determine the functional potential of the microbiome, understanding what these genes do is paramount, which culturing and other *in vitro* experiments are an integral part of. Importantly, metagenomics are now able to inform and refine culture methods and vice versa, accelerating the progress of both (Tramontano et al. 2018).

Finally, what the microbiome is doing, or *in situ* function, is answered by a trio of techniques: metabolomics, metatranscriptomics, and metaproteomics. These omic techniques are named after what they interrogate: metabolites, RNA transcripts, and proteins. While metatranscriptomics (Bashiardes, Zilberman-Schapira, and Elinav 2016) and especially metaproteomics (Petriz and Franco 2017) are theoretically powerful and likely necessary techniques for a full understanding of *in situ* function, they are in their infancies with regard to the microbiome, requiring further methodological and technological improvements to be feasible and economical. However, metabolomics is quickly becoming a widely adopted technique with significant successes. Two flavors of metabolomics exist: targeted and untargeted. Targeted metabolomics assay for specific metabolites under a hypothesis-driven framework, such as that higher intake of fruits and whole grains, is correlated with lower metabolic syndrome risk and higher hippurate levels, likely via gut microbial metabolism of polyphenols (Pallister et al. 2017). This technique can importantly confirm whether desired *in vitro* effects of a drug or probiotic are being recapitulated in the human host. Untargeted metabolomics are more utilized for discovery of novel effects, sacrificing stronger signals and better differentiation of specific metabolites (and increasing costs) for casting a wider net, including metabolites, one might simply not expect to be significant. For example, it is known that red meat is a risk factor for the development of colorectal cancer, but determining which metabolites might contribute to the disease is a daunting task, though an untargeted approach recently implicated several kynurenine pathway metabolites (Rombouts et al. 2017). One limitation of metabolomics is that human and microbial-derived compounds, when both sources are available, are conflated; however, combination with more direct microbiome compositional and functional profiling, as well as careful study design, can greatly mitigate this effect. Ultimately, metabolomics is a powerful technique for measuring physiological effects of altering the microbiome in a way 16S and metagenomics cannot, making it an important tool for confirming effects of microbial interventions.

These major technologies are currently used to explore the microbiome, including the probiotics that are used and are under development. They can also be used to interrogate how a probiotic will interact with the microbiome and, importantly, with the host. Specific probiotics should be probed by a huge library of existing technologies for *in vitro* study as well, especially to determine the roles or mechanisms of specific proteins and metabolites (Hover et al. 2018), though these crucially might not accurately reflect what is happening in the gut environment. Ultimately, when combined with animal models and human studies, these techniques can advance the discovery and development of both microbiome science and of probiotics as drugs, ultimately with the goal of translating these discoveries to the commercial or (personalized) therapeutic spaces. The following sections highlight the discovery and development of the two dominant bacterial groups used as probiotics: lactobacilli and bifidobacteria.

9.5   LACTOBACILLI

The *Lactobacillus* genus is part of a group of so-called lactic acid bacteria (LAB), which produce lactic acid as a major fermentation product. For this reason, they are ubiquitous in food preservation and flavoring, metabolizing carbohydrates in products such as yogurt and sauerkraut. From a safety as well as a historical perspective, it follows that microbes involved in food preservation and protection from pathogenic bacteria might be considered ideal candidate for probiotics. And indeed, they are some of the most studied and certified; lactobacilli make up 84 of the 195 species with certified beneficial use (Bourdichon et al. 2012). However, some have questioned the rational of focusing on lactobacilli for probiotic usage, as they only make up a few percent of the composition of the human gut (Heeney, Gareau, and Marco 2018). It is nonetheless important to remember that functional importance is not entirely dictated by percent abundance; rare organisms can have a disproportionate impact on the function of a system (Shade et al. 2014). Furthermore, there are a number of general characteristics of lactobacilli that are believed to be important to their functionality as probiotics as well as evidence of their efficacy. It is worth noting that *Lactobacillus* is an extremely genetically diverse genus, equaling that of a typical family of bacteria (Claesson, van Sinderen, and O'Toole 2007). While these characteristics are mostly observed in those already analyzed for probiotic usage, they are likely to be largely conserved, depending on how the genus is refined in the future. Furthermore, additional species and strain-level effects and mechanisms are common in lactobacilli because of this large diversity. The most studied and discussed species include *L. acidophilus*, *L. casei*, *L. reuteri*, and *L. rhamnosus* (Reid 1999), but far more are being explored and cataloged down to the strain level (Salvetti, Torriani, and Felis 2015).

The first important trait of *Lactobacillus* species is their ability to survive ingestion and transit in the GIT, either having evolved for this niche or because of their overall stress tolerance (Hussain et al. 2013). Nonetheless, there is still ongoing research into improving survival, especially given food processing and microbe storage conditions, such as microentrapment in whey protein (Reid et al. 2007) and direct compression into tablets (Villena et al. 2015). Unfortunately, lactobacilli do not seem to fair as well with regards to adherence; 16S studies find their abundance drops dramatically after discontinuation of probiotic treatment (Walter 2008). Exactly why this is the case is of great interest, as better-performing strains or techniques for greater adhesion could greatly improve the efficacy of this promising class of probiotics. The majority of current studies are *in vitro* for feasibility reasons, but a few *in vivo* studies exist (Valeur et al. 2004). Despite these limitations, a number of common and strain-specific adhesion mechanisms (often somewhat overlapping) have been elucidated. These can be classified as either specific molecular interactions with the host or nonspecific bindings. The latter appears more conserved both among lactobacilli and Gram-positive bacteria generally. LTA contribute to both the negative charge and hydrophobicity of the cell wall, promoting adhesion to many surfaces generally (Delcour et al. 1999). Extracellular polysaccharide (EPS) likely has a similar effect albeit with greater differentiation due to the variety of EPS structures, which will take a significant effort to untangle (Ruas-Madiedo et al. 2006). Specific host tissue binding mechanisms mainly consist of mucus-binding domains, and occasionally S-layer proteins (Avall-Jääskeläinen and Palva 2005), though with enormous strain to strain variability makes characterization for *in vivo* activity difficult (van Tassell and Miller 2011). Nonetheless, as techniques, especially *in silico*, continue to advance, so too does knowledge of strain-specific adhesins in a way that hopefully also elucidates more general principles (Chatterjee et al. 2018). One avenue for interrogating or selecting for better adhesion properties may be to study autochthonous *Lactobacillus* species, as these established members of the gut flora likely possess better adhesion and niche occupying abilities, though with less of the desired probiotic properties (Grover et al. 2011). Similarly, in infants, lactobacilli are highly overrepresented, in part, because of diet and exposure and also due to the selective pressure of infant gut environment, making this environment an opportunity to identify new probiotics based on selection and subsequent development of the infant gut (Xanthopoulos, Litopoulou-Tzanetaki, and Tzanetakis 2000). Despite current hurdles in adaptation characteristics of *Lactobacillus* probiotics, there is still clear evidence that they work and a growing understanding of the mechanisms by which they achieve probiotic efficacy.

Once established, lactobacilli carry out their probiotic role in a number of ways. Primary to both the definition and probiotic effects of lactobacilli is the production of lactic acid as a major catabolic end product of glucose fermentation. This makes them a subset of the LAB which additionally predominantly includes *Leuconostoc*, *Pediococcus*, *Lactococcus*, and *Streptococcus*; however, the majority of probiotic strains currently belong to *Lactobacillus*. Lactic acid production is a well-known mode of providing spoilage protection in fermented food products by lowering the pH to inhibit the growth of other bacteria (Rawat 2015). In the gut, lactic acid can likewise cause local pH reduction which can provide resistance against pathogen colonization and proliferation, potentially by favoring commensal bacteria adapted to this environment over invading microbes (Engevik et al. 2013). Lactic acid is also toxic to Gram-negative bacteria and to some extent, Gram-positive bacteria via penetration of the bacterial cytoplasmic membrane and subsequent buildup and free-radical formation (Alakomi et al. 2005). Other antimicrobial substances, such as hydrogen peroxide and bacteriocins, have been shown to strengthen this effect and those of lowered pH (Fayol-Messaoudi et al. 2005). Many lactobacilli possess these capabilities, with hydrogen peroxide being considered particularly important in vaginal health and bacteriocins in regulating niche population. Lactobacilli bacteriocins are mostly of the class II type, which induce membrane permeabilization and have a narrow spectrum of action towards closely related bacteria. How much this, combined with QS, constitutes a system of self-regulation to avoid overgrowth in low nutrient conditions, compared to pathogen protection, is unknown. However, one landmark study demonstrated the ability of *Lactobacillus salivarius* to protect against *Listeria monocytogenes*, a significant human pathogen, in a bacteriocin-mediated manner (Corr et al. 2007). Another factor by which lactobacilli may have probiotic action is competitive exclusion; however, it is difficult to determine if they effectively employ this as a result of their low prevalence in the gut. A few intriguing case studies do exist though, including mannose-specific adhesin-mediated exclusion of *Escherichia coli* (Mack et al. 2003) and pathogen coaggregation (which hinders adherence and facilitates clearance) inducing surface proteins (Schachtsiek, Hammes, and Hertel 2004).

Likely more promising are the direct probiotic effects lactobacilli may have on the host. These are again diverse and often strain specific, but some significant insights have been made from both focused and systems level studies. Regarding beneficial metabolic potential, one exciting study utilized a mouse model with a defined seven-strain consortium that replaced the dominant metabolic features of the entire microbiome (Martin et al. 2008). This revealed many altered pathways both in the microbiome and the host, importantly including butyrate production and altered hepatic lipid metabolism. These findings have been corroborated and expanded by various full-human microbiome studies, such as investigation of the age-related effect of *L. plantarum* on SCFAs and bile acids (Wang et al. 2014), but it remains a unique approach with high potential for hypothesis generation. Similar to metabolic effects, lactobacilli appear to generally have a protective effect on epithelial cells and barrier function, but the pathways by which this is achieved are quite varied and involve both direct cell contact and secreted products, as evidenced by a complex genetic response to *L. rhamnosus* GG administration in the duodenal mucosa (Dicario et al. 2005). This response included significant modulation of the immune system. This result and others (Galdeano and Perdigon 2006) may indicate that immune regulation is a paramount probiotic effect of lactobacilli. Yet again, owing partially to the extreme genetic diversity among the *Lactobacillus* genus, the immunological pathways on which these bacteria act are quite varied, though intensely researched (van Baarlen, Wells, and Kleerebezem 2013). It is suggested, however, that they are particularly connected to the "hygiene hypothesis" as microbes with which humans have increasingly lost sufficient contact, and these types of microbes are thought to be particularly important in priming regulatory responses (hence a developmental importance and prevalence of lactobacilli in early life) and in induction of regulatory dendritic cells (Christensen, Frokiaer, and Pestka 2002) and T cells (Rook 2005).

An intriguing idea that has followed from immunological research of lactobacilli is that some strains are good probiotics because of and not in spite of their low adherence. These allochthonous bacteria, recognized in some part as novel antigens, activate the immune system but in a controlled and beneficial way. Clearly the mechanisms and uses of the existing probiotics require further elucidation and novel lactobacilli studied (down to the strain level), mining what is clearly a rich source of probiotics. It is clear that there are a myriad of effects, and many of these depend not only on the probiotic but the host as well. Ultimately, despite difficulties in unraveling all the probiotic effectors and properties of lactobacilli, real clinical and commercial results are beginning to manifest. A recent example is in the use of *Lactobacillus plantarum* plus fructooligosaccharide in the prevention of childhood sepsis in at-risk populations in India (Panigrahi et al. 2017). As more strains are identified, and their mechanisms of action determined, we will likely see a greater array of clinically relevant interventional trials to determine efficacy.

9.6 BIFIDOBACTERIA

In 2002, the International Dairy Federation (IDF) and the European Food and Feed Cultures Association compiled a comprehensive list of microorganisms with documented use in food. This list, known as the "2002 IDF Inventory," has become an industry-wide authoritative reference on microbial food cultures. Compared to lactobacilli, of which there are 84 reported species, bifidobacteria make up only 8 of the 195 bacterial species in the 2011 update of the inventory (Bourdichon et al. 2012). However, bifidobacteria, along with lactobacilli, are among the most commonly used genera of bacteria in probiotics (Fijan 2014). Bifidobacteria were first isolated in 1899 from fecal samples taken from breastfed infants by Henri Tissier, who named his discovery *Bacillus bifidus*, after the "bifid" or V-shaped morphology of the bacteria. The name was later edited as more species were discovered and reclassified until they were filed under the overarching genus *Bifidobacterium* (Lee and O'Sullivan 2010).

Bifidobacteria are non-motile, Gram-positive commensal anaerobes that, while generally noninvasive, have been implicated in cases of bacteremia (Esaiassen et al. 2017). Bifidobacteria are more genomically coherent as a unit than lactobacilli and have been mostly isolated from the colon of humans and animals (Lukjancenko, Ussery, and Wassenaar 2012). The bifidobacteria strains currently used as probiotics have been classified as "generally recognized as safe." Additionally, strains such as *Bifidobacterium animalis* strain DN-173 010 have demonstrated high survivability in the GIT and probiotic benefits in the colon and are widely used in dairy products (Picard et al. 2005).

The benefits that particular bifidobacteria strains convey to their human hosts are well documented and include their ability to prevent infectious diarrhea through their resistance to pathogen colonization, their influence on the immune system, as well as their potential to counteract carcinogenic activity from other gut microbes (O'Callaghan and van Sinderen 2016). Given the ability of certain strains to survive passage through the GI tract, as well as their potential benefits to their hosts, bifidobacteria are strong probiotic candidates for combating disease and maintaining a healthy human body.

Bifidobacteria within the greater ecosystem of the surrounding gut play important roles in the health of the host through direct interactions with the intestinal barrier and growth control of other microorganisms. Bifidobacteria contribute to the structure of the gut microbiome through their ability to bind to extracellular matrix proteins with their pili, indicating their role in bacterial aggregation (Vlasova et al. 2016). By competing with other microbiota, particularly pathogens, for space in the mucus layer, bifidobacteria control access to mucosa and influence the composition of microorganisms that interact with the intestinal mucus. Evidently, the viable colonization of beneficial strains of bifidobacteria within the GI tract is an important component of normal, healthy gut function.

In total, 80% of the genera that colonize human beings early in life are bifidobacterial. Subsequent microbial succession and competition with other species ultimately reduce the proportion of bifidobacteria in the gut. Likely as a result of this early life dominance, *Bifidobacterium* spp. have become not only the focus of many microbiome studies but also one of the most important probiotic agents in commercial use today. Bifidobacteria have been investigated as a potential therapeutic in the prevention of animal and human gastrointestinal diseases during early development (Picard et al. 2005). Establishment of the infant gut microbiome begins at the moment of passage through the mother's birth canal and continues to develop with changes in the infant's diet. Contact with the maternal vaginal microbiome and the components in breast milk serves to shape the composition of microbiota that establish stably in the gut. Vaginally delivered infants are colonized with microbes belonging to genera that include *Bifidobacterium*, *Lactobacillus*, *Prevotella*, *Bacteroides*, and *Escherichia/Shigella*, all of which have been identified in vaginal and fecal samples from adult mothers. Conversely, the microbial dysbiosis experienced by cesarean section in infants is dominated by human skin and oral bacteria, including *Corynebacterium*, *Staphylococcus*, *Streptococcus*, *Veillonella*, and *Propionibacterium* (Dominguez-Bello et al. 2010). Therefore, the introduction of bacteria that are predominant in potential healthy infants has been identified as strategy for treating childhood diseases. Bifidobacteria have a wide distribution across human organs, including the oral cavity and outer mucus layer of the colon, and are well adapted to the metabolism of the long chain sugars in breast milk (Donaldson, Lee, and Mazmanian 2015). For instance, *Bifidobacterium infantis* present in breastfed babies delivered by cesarean section promotes a healthy fecal microbiota that can be maintained with breastfeeding until weaning at 6 months (Mikami, Kimura, and Takahashi 2012). *Bifidobacterium longum* can exploit fucosylated oligosaccharides in breast milk as carbon sources, outcompeting other gut microbiota such as *E. coli* and *Clostridium perfringens* (Donaldson, Lee, and Mazmanian 2015). In another double-blind, placebo-controlled study, *Bifidobacterium animalis* subsp. lactis BB-12, was introduced as a probiotic agent and resulted in infants having fewer respiratory tract infections when compared to controls (Taipale et al. 2016).

Additionally, evidence from another experiment reported that probiotic supplementation with bifidobacteria early in life may reduce the risk of some neuropsychiatric disorder, deficit hyperactivity disorder, and autism spectrum disorder (ASD), development later in childhood, possibly by mechanisms not limited to the impact of the probiotic on endogenous gut microbiota composition (Pärtty et al. 2015).

Bifidobacteria may play a role in modulating intestinal barrier function by secreting bioactive factors that increase epithelial resistance and regulate expression of tight-junction proteins ZO-1, occludin, and claudin-2 (Ewaschuk et al. 2008). The intestinal barrier refers to the single-cell epithelial layer that lines the digestive tract and serves as a key physical defense against the external environment. While the intestinal epithelium regulates the absorption of water and dissolved nutrients and electrolytes, it also prevents the entry of toxins, antigens, and enteric microbiota into the bloodstream. The integrity of the intestinal barrier depends on the network of adhesive molecules – primarily desmosomes, adherens junctions, and tight junctions – linking the cytoskeletons of these epithelial cells to each other and to the extracellular matrix. Disruption of this barrier has been implicated in several autoimmune and inflammatory diseases (Groschwitz and Hogan 2009). In animal intestinal epithelium injury models, pretreatment with bifidobacteria slowed intestinal permeability, preserved assembly and localization of tight-junction proteins, and suppressed production of pro-inflammatory cytokines IL-6 and TNF-α (Ling et al. 2016). In another study, supplementation of bifidobacteria after intestinal thermal injury in mice decreased bacterial translocation, increased expression of secretory immunoglobulin A (sIgA), and mitigated mucosal damage compared to controls (Wang et al. 2006). The exact mechanisms and pathways by which bifidobacteria influence the barrier function of the intestine remain unclear, but growing evidence suggests that probiotic treatment with bifidobacteria not only reduces damage prior to intestinal injury but also after injury has taken place, indicating that bifidobacteria may be a potent supplement to gut barrier health.

There is also substantial evidence that orally administered LAB increase the systemic immune response (Tojo 2014). In mice, bifidobacteria have been reported to have protective roles against intestinal pathogens such as *E. coli*, *S. typhimurium*, and rotavirus (Kawahara et al. 2017). Additionally, the administration of bifidobacteria has been shown to have protective effects against rotavirus infection in children (Pugliese 1995). Bifidobacteria are unique in that they can stimulate Ig secreting cells without inciting an overactive immune systemic immune response, reducing the risk of developing autoimmune diseases. High levels of IgA in the feces of infants in early life has been associated with fewer cases of atopic disease (Kukkonen et al. 2010). Therefore, if children at risk for immune system disorders due to dysbiosis of the microbiome receive supplemental prebiotics and probiotics during development, many of which are found in breast milk, they reduce their risk of developing allergic disease (Lee and O'Sullivan 2010).

Certain gut microbiota also form an important partnership with the host intestinal defense system to control the growth and spread of pathogenic microbes (Servin 2004). The commensal microflora can produce their own antimicrobial agents to kill pathogenic organisms directly or to render their immediate environment unsuitable for the growth of other species (for example, lowering the pH). They have also been implicated in regulating various components of the host immune system in response to enteric pathogens (Ohland and MacNaughton 2010).

Bifidobacteria have been implicated in pathogen exclusion through their ability to inhibit growth and survival of pathogens and to outcompete other species for space in the mucus layer and on the intestinal epithelium. One study found that certain strains of bifidobacteria release unknown factors that kill *Salmonella typhimurium in vitro* and reduce viability of *Staphylococcus aureus*, *Klebsiella pneumoniae*, *E. coli*, and *Yersinia pseudotuberculosis* (Lievin 2000). High iron-sequestering bifidobacteria have been shown to inhibit growth of *E. coli* and *S. typhimurium* by preventing these enteropathogens from adhering to the intestinal mucus-producing cell lines (Vazquez-Gutierrez et al. 2016). Other studies have confirmed bifidobacteria's role in inhibiting bacterial translocation and potential infection by occupying and dominating the intestinal mucus layer and epithelial surface (Matsumoto et al. 2002). Given their significance in protecting their host from enteropathogen invasions (Gopal et al. 2001), bifidobacteria show promise as probiotics to supplement the intestinal defense system and a prophylactic treatment for certain infections.

A symbiotic microbial community is crucial for the proper metabolic, immune, and physiological function of the human body. Each human has an individualized microbiome, and each body site has a unique composition of bacteria. Disruption of the microbiome away from the average for a given population is directly correlated with numerous diseases; for example, *C. difficile* infections. Rather than treating *C. difficile* infections with antibiotics, which kills both beneficial and pathogenic bacteria alike, a less complex treatment option is fecal transplantation. Bifidobacteria are crucial in a healthy, functional GIT. However, one major complication in harnessing the benefits of bifidobacteria is finding a successful delivery method to allow the bacteria to travel to and colonize the intestine. Unlike lactobacilli which evolved to endure harsh environments, particularly acidic conditions, bifidobacteria lack many adaptations towards this purpose (Ruiz et al. 2011). Fecal transplants may be a suitable delivery method for this fragile bacterium, as it bypasses stomach acid. The population of bifidobacteria in human intestinal microflora decreases with age (Kato et al. 2017). This newfound bacterial delivery method may be able to compensate the decrease in the bifidobacteria composition over time.

Many commercially available probiotics contain different bacterial families in combination. These formulations, rather than competing against each other, work together to regulate the gut microbiome. For example, bifidobacteria can work in conjunction with other probiotics, such as lactobacilli, to improve adhesion to the intestinal mucosa, as shown in a study in which *B. lactis Bb12* and *L. delbrueckii subsp. Bulgaricus* enhances the binding of bifidobacteria *in vitro* (Ouwehand et al. 2000). Adhesion of the probiotic microorganisms to the intestinal mucosa is crucial for successful colonization, allowing for stimulation of the immune system and enhancing protection against enteropathogen colonization. These bacteria likely undergo coaggregation, which allows for the survival and replication of both genera within the same niche space, likely mediated by the secretion of EPS. EPS are exocellular polymers that are present on the surface of many LAB, such as *lactobacillus* and *bifidobacterium*. EPS are also utilized against bacteriophage attack, antimicrobial compounds, and phagocytosis (Salazar et al. 2008). This provides evidence for the symbiotic ability of bacterial genera.

9.7   WHAT'S NEXT?

While approved probiotics mainly consist of those classes of microorganisms discovered in food – primarily because these are generally recognized as safe, both the scope of microorganisms considered as probiotic and the methods in which they are administered are changing. In a similar pattern to the discovery of chemical drugs, the field of probiotics is progressing from employing readily available probiotics in food to mining for the microbiome for novel probiotics and engineering entirely new probiotics. However, microbiome research continues into the mechanisms of action and physiology of probiotics. In tandem with improved characterization of existing probiotics and identification of novel microorganisms, our expanded understanding of the microbiome itself continues apace. This is essential, as it is absolutely necessary to understand how the endogenous microbiome "works" if we are to use probiotics to "fix" it.

Upon examination of the two most commonly used classes of probiotics, bifidobacteria and lactobacilli, it is clear there is room to expand the selection of available probiotics based on our current understanding of the human microbiome. *Lactobacillus* species tend to come from food sources and are thus known to be safe and orally available but are not always successful in colonizing the gut. Conversely, *Bifidobacterium* species are generally human-isolated and colonize well, especially early in life, but are difficult to administer orally. However, there are many bacteria types with poorly understood roles but likely equal importance to human health. While there are other probiotics currently available in the marketplace such as bacilli species, for example, *E. coli* (particularly strain 1917 Nissle), and some probiotic yeasts (O'Toole, Marchesi, and Hill 2017), it is clear from culturing and especially sequencing experiments that we have just scratched the surface of what can be administered for probiotic effects. *E. coli* presents an interesting case that may prove instructive for future probiotic organisms, often referred to as "next-gen"; its efficacy and even safety is dependent on both the genomic potential of the particular strain and the environment in which it resides. This presents challenges for future development but also points towards the promise and ultimate necessity for better control and understanding of the numerous bacteria, archaea, fungi, protists, and even viruses that make up our microbiomes and play distinct roles, many of which may be beneficial, or pathogenic if the environmental selective pressures support such activity. It should be noted that even traditional probiotics, such as *Lactobacillus rhamnosus,* are not without cases of virulence, particularly in immunocompromised subjects (Mackay et al. 1999).

The complexity of the relationship between microbiota and host is well exemplified by a particular genus poised on the frontier of microbiome research, *Bacteroides.* Certain members of this genus may be beneficial in cancer and intestinal inflammation models and generally have a beneficial but complex relationship with humans. However, they have been also shown to be pathogenic in cases where they have multiple antibiotic resistances and manage to escape the gut (Wexler 2007). Likewise, the genus *Akkermansia* may be key in regulating mucus production and homeostasis with the host but additionally has both positive correlations with increasing pathology in some disease states, such as Parkinson's (Heintz-Buschart et al. 2018), and decreasing pathology in others, such as obesity (Dao et al. 2016). Like *Bifidobacterium,* this genus has been mined from the human microbiome, but with increasingly sophisticated culturing and analysis techniques, it is increasingly likely that a new tranche of probiotic organisms designed to treat specific ailments and so approved by the Federal Drug Administration (FDA) are right around the corner.

REFERENCES

Abubucker, Sahar, Nicola Segata, Johannes Goll, Alyxandria M Schubert, Jacques Izard, Brandi L Cantarel, Beltran Rodriguez-Mueller, et al. 2012. Metabolic reconstruction for metagenomic data and its application to the human microbiome. Edited by Jonathan A. Eisen. *PLoS Computational Biology* 8 (6): e1002358. doi:10.1371/journal.pcbi.1002358.

Alakomi, Hanna-Leena, E Skyttä, Maria Saarela, and Iikka M Helander. 2005. Lactic acid permeabilizes gram-negative bacteria by disrupting the outer membrane lactic acid permeabilizes gram-negative bacteria by disrupting the outer membrane. *Applied and Environmental Microbiology* 66 (5): 2000–2005. doi:10.1128/AEM.66.5.2001-2005.2000.

Aminov, Rustam I, Morten Otto, and Alexander Sommer. 2010. A brief history of the antibiotic era : Lessons learned and challenges for the future. *Frontiers in Microbiology* 1 (December): 1–7. doi:10.3389/fmicb.2010.00134.

Araya, Magdalena, Lorenzo Morelli, Gregor Reid, Mary Ellen Sanders, Catherine Stanton, Maya Pineiro, and Peter Ben Embarek. 2002. Guidelines for the evaluation of probiotics in food. Joint FAO/WHO Working Group Report on Drafting Guidelines for the Evaluation of Probiotics in Food, London, Ontario, Canada, 1–11. doi:10.1111/j.1469-0691.2012.03873.

Avall-Jääskeläinen, Silja and Airi Palva. 2005. *Lactobacillus* surface layers and their applications. *FEMS Microbiology Reviews* 29 (3): 511–29. doi:10.1016/j.femsre.2005.04.003.

Baquero, Fernando and Cesar Nombela. 2012. The microbiome as a human organ. *Clinical Microbiology and Infection* 18 (SUPPL. 4). European Society of Clinical Infectious Diseases: 2–4. doi:10.1111/j.1469-0691.2012.03916.x.

Bashiardes, Stavros, Gili Zilberman-Schapira, and Eran Elinav. 2016. Use of metatranscriptomics in microbiome research. *Bioinformatics and Biology Insights* 10 (April). Libertas Academica: 19–25. doi:10.4137/BBI.S34610.

Begley, Máire, Cormac G M Gahan, and Colin Hill. 2005. The interaction between bacteria and bile. *FEMS Microbiology Reviews* 29 (4): 625–51. doi:10.1016/j.femsre.2004.09.003.

Bindels, Laure B, Nathalie M Delzenne, Patrice D Cani, and Jens Walter. 2015. Opinion: Towards a more comprehensive concept for prebiotics. *Nature Reviews Gastroenterology and Hepatology* 12 (5). Nature Publishing Group: 303–10. doi:10.1038/nrgastro.2015.47.

Blaser, Martin J. 2016. Antibiotic use and its consequences for the normal microbiome. *Science* 352 (6285): 544 LP–545. http://science.sciencemag.org/content/352/6285/544.abstract.

Bourassa, Megan W, Ishraq Alim, Scott J Bultman, and Rajiv R Ratan. 2016. Butyrate, neuroepigenetics and the gut microbiome: Can a high fiber diet improve brain health? *Neuroscience Letters* 625. Elsevier Ireland Ltd: 56–63. doi:10.1016/j.neulet.2016.02.009.

Bourdichon, François, Serge Casaregola, Choreh Farrokh, Jens C Frisvad, Monica L Gerds, Walter P Hammes, James Harnett, et al. 2012. Food fermentations: Microorganisms with technological beneficial use. *International Journal of Food Microbiology* 154 (3). Elsevier B.V.: 87–97. doi:10.1016/j. ijfoodmicro.2011.12.030.

Carabotti, Marilia, Annunziata Scirocco, Maria Antonietta Maselli, and Carola Severi. 2015. The gut-brain axis: Interactions between enteric microbiota, central and enteric nervous systems. *Annals of Gastroenterology* 28 (2): 203–9. doi:10.1038/ajgsup.2012.3.

Carding, Simon, Kristin Verbeke, Daniel T Vipond, Bernard M Corfe, and Lauren J Owen. 2015. Dysbiosis of the gut microbiota in disease. *Microbial Ecology in Health and Disease* 26: 1–9. doi:10.3402/mehd.v26.26191.

Cardona, Cesar, Pamela Weisenhorn, Chris Henry, and Jack A Gilbert. 2016. Network-based metabolic analysis and microbial community modeling. *Current Opinion in Microbiology* 31 (June): 124–31. doi:10.1016/j.mib.2016.03.008.

Chatterjee, Maitrayee, Anju Choorakottayil Pushkaran, Anil Kumar Vasudevan, Krishna Kumar N Menon, Raja Biswas, and Chethampadi Gopi Mohan. 2018. Understanding the adhesion mechanism of a mucin binding domain from *Lactobacillus* fermentum and its role in enteropathogen exclusion. *International Journal of Biological Macromolecules* 110 (April): 598–607. doi:10.1016/j.ijbiomac.2017.10.107.

Chow, Janet, Haiqing Tang, and Sarkis K Mazmanian. 2011. Pathobionts of the gastrointestinal microbiota and inflammatory disease. *Current Opinion in Immunology* 23 (4): 473–80. doi:10.1016/j.coi.2011.07.010. Pathobionts.

Christensen, Hanne R, Hanne Frokiaer, and James J Pestka. 2002. Lactobacilli differentially modulate expression of cytokines and maturation surface markers in murine dendritic cells. *The Journal of Immunology* 168 (1): 171–78. doi:10.4049/jimmunol.168.1.171.

Claesson, Marcus J, Douwe van Sinderen, and Paul W O'Toole. 2007. The genus *Lactobacillus*--a genomic basis for understanding its diversity. *FEMS Microbiology Letters* 269 (1): 22–28. doi:10.1111/j.1574-6968.2006.00596.x.

Clarke, Gerard, Roman M Stilling, Paul J Kennedy, Catherine Stanton, John F Cryan, and Timothy G Dinan. 2014. Minireview: Gut microbiota: The neglected endocrine organ. *Molecular Endocrinology* 28 (8): 1221–38. doi:10.1210/me.2014-1108.

Cogan, T A, A O Thomas, L E N Rees, A H Taylor, M A Jepson, P H Williams, J Ketley, and T J Humphrey. 2007. Norepinephrine increases the pathogenic potential of campylobacter jejuni. *Gut* 56 (8): 1060–65. doi:10.1136/gut.2006.114926.

Corr, Sinead C, Yongtao Li, Christian U Riedel, Paul W O'Toole, Colin Hill, and Cormac G M Gahan. 2007. Bacteriocin production as a mechanism for the antiinfective activity of *Lactobacillus salivarius* UCC118. *Proceedings of the National Academy of Sciences* 104 (18): 7617–21. doi:10.1073/pnas.0700440104.

Coyte, Katharine Z, Jonas Schluter, and Kevin R Foster. 2015. The ecology of the microbiome: Networks, competition, and stability. *Science* 350 (6261): 663–66. doi:10.1126/science.aad2602.

d'Hennezel, Eva, Sahar Abubucker, Leon O Murphy, and Thomas W Cullen. 2017. Total lipopolysaccharide from the human gut microbiome silences toll-like receptor signaling. Edited by Catherine Lozupone. *MSystems* 2 (6): e00046-17. doi:10.1128/mSystems.00046-17.

Dao, Maria Carlota, Amandine Everard, Judith Aron-Wisnewsky, Nataliya Sokolovska, Edi Prifti, Eric O Verger, Brandon D Kayser, et al. 2016. Akkermansia muciniphila and improved metabolic health during a dietary intervention in obesity: Relationship with gut microbiome richness and ecology. *Gut* 65 (3): 426–36. doi:10.1136/gutjnl-2014-308778.

David, Lawrence A, Corinne F Maurice, Rachel N Carmody, David B Gootenberg, Julie E Button, Benjamin E Wolfe, Alisha V Ling, et al. 2013. Diet rapidly and reproducibly alters the human gut microbiome. *Nature* 505 (7484): 559–63. doi:10.1038/nature12820.

Delcour, Jean, Thierry Ferain, Marie Deghorain, Emmanuelle Palumbo, and Pascal Hols. 1999. The biosynthesis and functionality of the cell-wall of lactic acid bacteria. *Antonie van Leeuwenhoek* 76 (1–4): 159–84. www.ncbi.nlm.nih.gov/pubmed/10532377.

Devaraj, Sridevi, Peera Hemarajata, and James Versalovic. 2014. The human gut microbiome and body metabolism: Implications for obesity and diabetes. *Clinical Chemistry* 59 (4): 617–28. doi:10.1373/clinchem.2012.187617.

Dicario, S, H Tao, A Grillo, C Elia, G Gasbarrini, A Sepulveda, and A Gasbarrini. 2005. Effects of *Lactobacillus* GG on genes expression pattern in small bowel mucosa. *Digestive and Liver Disease* 37 (5): 320–29. doi:10.1016/j.dld.2004.12.008.

Dobell, Clifford. 1920. The discovery of the intestinal protozoa of man. *Proceedings of the Royal Society of Medicine* 13 (Sect Hist Med): 1–15. www.ncbi.nlm.nih.gov/pmc/articles/PMC2151982/.

Dominguez-Bello, Maria G, Elizabeth K Costello, Monica Contreras, Magda Magris, Glida Hidalgo, Noah Fierer, and Rob Knight. 2010. Delivery mode shapes the acquisition and structure of the initial microbiota across multiple body habitats in newborns. *Proceedings of the National Academy of Sciences* 107 (26): 11971–75. doi:10.1073/pnas.1002601107.

Dominguez-Bello, Maria G, Kassandra M De Jesus-Laboy, Nan Shen, Laura M Cox, Amnon Amir, Antonio Gonzalez, Nicholas A Bokulich, et al. 2016. Partial restoration of the microbiota of cesarean-born infants via vaginal microbial transfer. *Nature Medicine* 22 (3): 250–53. doi:10.1038/nm.4039.

Donaldson, Gregory P, S Melanie Lee, and Sarkis K Mazmanian. 2015. Gut biogeography of the bacterial microbiota. *Nature Reviews Microbiology* 14 (1). Nature Publishing Group: 20–32. doi:10.1038/nrmicro3552.

Dye, Christopher. 2014. After 2015: Infectious diseases in a new era of health and development. *Philosophical Transactions of the Royal Society of London. Series B, Biological Sciences* 369 (1645): 20130426. doi:10.1098/rstb.2013.0426.

Engevik, Melinda A, Annelies Hickerson, Gary E Shull, and Roger T Worrell. 2013. Acidic conditions in the NHE2-/- mouse intestine result in an altered mucosa-associated bacterial population with changes in mucus oligosaccharides. *Cellular Physiology and Biochemistry* 32 (7): 111–28. doi:10.1159/000356632.

Esaiassen, Eirin, Erik Hjerde, Jorunn Pauline Cavanagh, Gunnar Skov Simonsen, and Claus Klingenberg. 2017. *Bifidobacterium* bacteremia: Clinical characteristics and a genomic approach to assess pathogenicity. Edited by Nathan A Ledeboer. *Journal of Clinical Microbiology* 55 (7): 2234–48. doi:10.1128/JCM.00150-17.

Ewaschuk, Julia B, Hugo Diaz, Liisa Meddings, Brendan Diederichs, Andrea Dmytrash, Jody Backer, Mirjam Looijer-van Langen, and Karen L Madsen. 2008. Secreted bioactive factors from *Bifidobacterium infantis* enhance epithelial cell barrier function. *American Journal of Physiology-Gastrointestinal and Liver Physiology* 295 (5): G1025–34. doi:10.1152/ajpgi.90227.2008.

Faith, Jeremiah J, Janaki L Guruge, Mark Charbonneau, Sathish Subramanian, Henning Seedorf, Andrew L Goodman, Jose C Clemente, et al. 2013. The long-term stability of the human gut microbiota. *Science* 341 (6141). http://science.sciencemag.org/content/341/6141/1237439.abstract.

Fayol-Messaoudi, Domitille, Cédric N Berger, Vanessa Liévin-le Moal, and Alain L Servin. 2005. pH-, lactic acid-, and non-lactic acid-dependent activities of probiotic lactobacilli against *Salmonella enterica* serovar Typhimurium. *Applied and Environmental Microbiology* 71 (10): 6008–13. doi:10.1128/AEM.71.10.6008.

Fijan, Sabina. 2014. Microorganisms with claimed probiotic properties: An overview of recent literature. *International Journal of Environmental Research and Public Health* 11 (5): 4745–67. doi:10.3390/ijerph110504745.

Finegold, Sydney M, H R Attebery, and Vera L Sutter. 1974. Effect of diet on human fecal flora: Comparison of Japanese and American Diets. *The American Journal of Clinical Nutrition* 27 (12): 1456–69. www.ncbi.nlm.nih.gov/pubmed/4432829.

Galdeano, Carolina M and Gabriela Perdigon. 2006. The probiotic bacterium *Lactobacillus casei* induces activation of the gut mucosal immune system through innate immunity. *Clinical and Vaccine Immunology* 13 (2): 219–26. doi:10.1128/CVI.13.2.219-226.2006.

Giovannoni, Stephen J, Theresa B Britschgi, Craig L Moyer, and Katharine G Field. 1990. Genetic diversity in Sargasso Sea bacterioplankton. *Nature* 345 (6270): 60–3. doi:10.1038/345060a0.

Gopal, Pramod K, Jaya Prasad, John Smart, and Harsharanjit S Gill. 2001. In vitro adherence properties of *Lactobacillus rhamnosus* DR20 and *Bifidobacterium lactis* DR10 strains and their antagonistic activity against an enterotoxigenic Escherichia coli. *International Journal of Food Microbiology* 67 (3): 207–16. doi:10.1016/S0168-1605(01)00440-8.

Govender, Mershen, Yahya E Choonara, Pradeep Kumar, Lisa C du Toit, Sandy van Vuuren, and Viness Pillay. 2014. A review of the advancements in probiotic delivery: Conventional vs. non-conventional formulations for intestinal flora supplementation. *AAPS PharmSciTech* 15 (1): 29–43. doi:10.1208/s12249-013-0027-1.

Groschwitz, Katherine R and Simon P Hogan. 2009. Intestinal barrier function: Molecular regulation and disease pathogenesis. *Journal of Allergy and Clinical Immunology* 124 (1): 3–20. doi:10.1016/j.jaci.2009.05.038.

Grover, Sunita, Yudhishthir Singh Rajput, Raj Kumar Duary, and Virender Kumar Batish. 2011. Assessing the adhesion of putative indigenous probiotic lactobacilli to human colonic epithelial cells. *The Indian Journal of Medical Research* 134 (5): 664. doi:10.4103/0971-5916.90992.

Gu, Shenghua, Dandan Chen, Jin-Na Zhang, Xiaoman Lv, Kun Wang, Li-Ping Duan, Yong Nie, and Xiao-Lei Wu. 2013. Bacterial community mapping of the mouse gastrointestinal tract. Edited by Colin Dale. *PLoS One* 8 (10): e74957. doi:10.1371/journal.pone.0074957.

Heeney, Dustin D, Mélanie G Gareau, and Maria L Marco. 2018. Intestinal *Lactobacillus* in health and disease, a driver or just along for the ride? *Current Opinion in Biotechnology* 49: 140–47. doi:10.1016/j.copbio.2017.08.004.

Heijtz, Diaz, Rochellys, Shugui Wang, Farhana Anuar, Yu Qian, Britta Björkholm, Annika Samuelsson, Martin L Hibberd, Hans Forssberg, and Sven Pettersson. 2011. Normal gut microbiota modulates brain development and behavior. *Proceedings of the National Academy of Sciences of the United States of America* 108 (7): 3047–52. doi:10.1073/pnas.1010529108.

Heintz-Buschart, Anna, Urvashi Pandey, Tamara Wicke, Friederike Sixel-Döring, Annette Janzen, Elisabeth Sittig-Wiegand, Claudia Trenkwalder, Wolfgang H Oertel, Brit Mollenhauer, and Paul Wilmes. 2018. The nasal and gut microbiome in Parkinson's disease and idiopathic rapid eye movement sleep behavior disorder. *Movement Disorders* 33 (1): 88–98. doi:10.1002/mds.27105.

Hill, Colin, Francisco Guarner, Gregor Reid, Glenn R Gibson, Daniel J Merenstein, Bruno Pot, Lorenzo Morelli, et al. 2014. Expert consensus document: The international scientific association for probiotics and prebiotics consensus statement on the scope and appropriate use of the term probiotic. *Nature Reviews Gastroenterology and Hepatology* 11 (8): 506–14. doi:10.1038/nrgastro.2014.66.

Holdeman, L V, I J Good, and W E Moore. 1976. Human fecal flora: Variation in bacterial composition within individuals and a possible effect of emotional stress. *Applied and Environmental Microbiology* 31 (3): 359–75. www.ncbi.nlm.nih.gov/pubmed/938032.

Hollander, Daniel. 1999. Intestinal permeability, leaky gut, and intestinal disorders. *Current Gastroenterology Reports* 1 (5): 410–16. doi:10.1007/s11894-999-0023-5.

Hover, Bradley M, Seong-Hwan Kim, Micah Katz, Zachary Charlop-Powers, Jeremy G Owen, Melinda A Ternei, Jeffrey Maniko, et al. 2018. Culture-independent discovery of the malacidins as calcium-dependent antibiotics with activity against multidrug-resistant gram-positive pathogens. *Nature Microbiology* 3 (4): 415–22. doi:10.1038/s41564-018-0110-1.

Hussain, Malik A, Marzieh Hosseini Nezhad, Yu Sheng, and Omega Amoafo. 2013. Proteomics and the stressful life of lactobacilli. *FEMS Microbiology Letters* 349 (1): 1–8. doi:10.1111/1574-6968.12274.

Hutkins, Robert W, Janina A Krumbeck, Laure B Bindels, Patrice D Cani, George Fahey, Yong Jun Goh, Bruce Hamaker, et al. 2016. Prebiotics: Why definitions matter. *Current Opinion in Biotechnology* 37 (February): 1–7. doi:10.1016/j.copbio.2015.09.001.

Huttenhower, Curtis, Dirk Gevers, Rob Knight, Sahar Abubucker, Jonathan H Badger, Asif T Chinwalla, Heather H Creasy, et al. 2012. Structure, function and diversity of the healthy human microbiome. *Nature* 486 (7402): 207–14. doi:10.1038/nature11234.

Johansson, Malin E V, Daniel Ambort, Thaher Pelaseyed, André Schütte, Jenny K Gustafsson, Anna Ermund, Durai B Subramani, et al. 2011. Composition and functional role of the mucus layers in the intestine. *Cellular and Molecular Life Sciences* 68 (22): 3635–41. doi:10.1007/s00018-011-0822-3.

Kato, Kumiko, Toshitaka Odamaki, Eri Mitsuyama, Hirosuke Sugahara, Jin-Zhong Xiao, and Ro Osawa. 2017. Age-related changes in the composition of gut *Bifidobacterium* species. *Current Microbiology* 74 (8): 987–95. doi:10.1007/s00284-017-1272-4.

Kawahara, Tomohiro, Yutaka Makizaki, Yosuke Oikawa, Yoshiki Tanaka, Ayako Maeda, Masaki Shimakawa, Satoshi Komoto, Kyoko Moriguchi, Hiroshi Ohno, and Koki Taniguchi. 2017. Oral administration of *Bifidobacterium* bifidum G9-1 alleviates rotavirus gastroenteritis through regulation of intestinal homeostasis by inducing mucosal protective factors. Edited by Francesco Cappello. *PLoS One* 12 (3): e0173979. doi:10.1371/journal.pone.0173979.

Kelley, Scott T and Jack A Gilbert. 2013. Studying the microbiology of the indoor environment. *Genome Biology* 14 (2): 1–9. doi:10.1186/gb-2013-14-2-202.

Kukkonen, Kaarina, Mikael Kuitunen, Tari Haahtela, Riitta Korpela, Tuija Poussa, and Erkki Savilahti. 2010. High intestinal IgA associates with reduced risk of IgE-associated allergic diseases. *Pediatric Allergy and Immunology* 21 (1–Part–I): 67–73. doi:10.1111/j.1399-3038.2009.00907.x.

Kumar, Manoj, Ravinder Nagpal, Rajesh Kumar, R Hemalatha, Vinod Verma, Ashok Kumar, Chaitali Chakraborty, et al. 2012. Cholesterol-lowering probiotics as potential biotherapeutics for metabolic diseases. *Experimental Diabetes Research* 2012: 1–14. doi:10.1155/2012/902917.

Lagier, Jean-Christophe, Sophie Edouard, Isabelle Pagnier, Oleg Mediannikov, Michel Drancourt, and Didier Raoult. 2015. Current and past strategies for bacterial culture in clinical microbiology. *Clinical Microbiology Reviews* 28 (1): 208–36. doi:10.1128/CMR.00110-14.

Lebeer, Sarah, Jos Vanderleyden, and Sigrid C J De Keersmaecker. 2008. Genes and molecules of lactobacilli supporting probiotic action. *Microbiology and Molecular Biology Reviews* 72 (4): 728–64. doi:10.1128/MMBR.00017-08.

Lee, Ju-Hoon and Daniel J O'Sullivan. 2010. Genomic insights into bifidobacteria. *Microbiology and Molecular Biology Reviews* 74 (3): 378–416. doi:10.1128/MMBR.00004-10.

Ley, Ruth E, Micah Hamady, Catherine Lozupone, Peter J Turnbaugh, Rob Roy Ramey, J Stephen Bircher, Michael L Schlegel, et al. 2008. Evolution of mammals and their gut microbes. *Science* 777 (June): 1647–52. doi:10.1126/science.1155725.

Lievin, V. 2000. *Bifidobacterium* strains from resident infant human gastrointestinal microflora exert antimicrobial activity. *Gut* 47 (5): 646–52. doi:10.1136/gut.47.5.646.

Lilly, D M and R H Stillwell. 1965. Probiotics: Growth-promoting factors produced by microorganisms. *Science (New York, N.Y.)* 147 (3659): 747–48. www.ncbi.nlm.nih.gov/pubmed/14242024.

Ling, Xiang, Peng Linglong, Du Weixia, and Wei Hong. 2016. Protective effects of *Bifidobacterium* on intestinal barrier function in LPS-induced enterocyte barrier injury of Caco-2 monolayers and in a rat NEC model. Edited by Tony T. Wang. *PLoS One* 11 (8): e0161635. doi:10.1371/journal.pone.0161635.

Lozupone, Catherine, Jesse Stomabaugh, Jeffrey Gordon, Janet Jansson, and Rob Knight. 2012. Diversity, stability and resilience of the human gut microbiota. *Nature* 489 (7415): 220–30. doi:10.1038/nature11550. Diversity.

Lukjancenko, Oksana, David W Ussery, and Trudy M Wassenaar. 2012. Comparative genomics of *Bifidobacterium*, *Lactobacillus* and related probiotic genera. *Microbial Ecology* 63 (3): 651–73. doi:10.1007/s00248-011-9948-y.

Lund, Peter, Angela Tramonti, and Daniela De Biase. 2014. Coping with low PH: Molecular strategies in neutralophilic bacteria. *FEMS Microbiology Reviews* 38 (6): 1091–1125. doi:10.1111/1574-6976.12076.

Mack, David R, Siv Ahrne, Lucie Hyde, S Wei, and Michael A Hollingsworth. 2003. Extracellular MUC3 mucin secretion follows adherence of *Lactobacillus* strains to intestinal epithelial cells in vitro. *Gut* 52 (6): 827–33. www.ncbi.nlm.nih.gov/pubmed/1773687.

Mackay, Andrew D, Mark B Taylor, Christopher C Kibbler, and Jeremy M T Hamilton-Miller. 1999. *Lactobacillus* endocarditis caused by a probiotic organism. *Clinical Microbiology and Infection : The Official Publication of the European Society of Clinical Microbiology and Infectious Diseases* 5 (5): 290–92. www.ncbi.nlm.nih.gov/pubmed/11856270.

MacKenzie, Donald Alexander, Faye Jeffers, Mary L Parker, Amandine Vibert-Vallet, R J Bongaerts, Stefan Roos, Jens Walter, and Nathalie Juge. 2010. Strain-specific diversity of mucus-binding proteins in the adhesion and aggregation properties of *Lactobacillus reuteri*. *Microbiology* 156 (11): 3368–78. doi:10.1099/mic.0.043265-0.

Martin, Francois-Pierre J, Yulan Wang, Norbert Sprenger, Ivan K S Yap, Torbjörn Lundstedt, Per Lek, Serge Rezzi, et al. 2008. Probiotic modulation of symbiotic gut microbial–host metabolic interactions in a humanized microbiome mouse model. *Molecular Systems Biology* 4 (January). doi:10.1038/msb4100190.

Martinez, Kristina B, Vanessa Leone, and Eugene B Chang. 2017. Western diets, gut dysbiosis, and metabolic diseases: Are they linked? *Gut Microbes* 8 (2). Taylor & Francis: 130–42. doi:10.1080/19490976.2016.1270811.

Matsuda, Kazunori, Hirokazu Tsuji, Takashi Asahara, Kazumasa Matsumoto, Toshihiko Takada, and Koji Nomoto. 2009. Establishment of an analytical system for the human fecal microbiota, based on reverse transcription-quantitative PCR targeting of multicopy RRNA molecules. *Applied and Environmental Microbiology* 75 (7): 1961–69. doi:10.1128/AEM.01843-08.

Matsumoto, Mitsuharu, Hisanori Tani, Hiroyuki Ono, Hifumi Ohishi, and Yoshimi Benno. 2002. Adhesive property of *Bifidobacterium lactis* LKM512 and predominant bacteria of intestinal microflora to human intestinal mucin. *Current Microbiology* 44 (3): 212–15. doi:10.1007/s00284-001-0087-4.

Metchnikoff, Elie. 1907. *The Prolongation of Life : Optimistic Studies*. London: William Heinemann.

Mikami, Katsunaka, Moto Kimura, and Hidenori Takahashi. 2012. Influence of maternal bifidobacteria on the development of gut bifidobacteria in infants. *Pharmaceuticals* 5 (6): 629–42. doi:10.3390/ph5060629.

Miller, Melissa B and Bonnie L Bassler. 2001. Quorum sensing in bacteria. *Annual Review of Microbiology* 55: 165–99. doi:10.1146/annurev.micro.55.1.165.

Moore, William E and Lillian V Holdeman. 1974. Human fecal flora: The normal flora of 20 Japanese-Hawaiians. *Applied Microbiology* 27 (5): 961–79. www.ncbi.nlm.nih.gov/pubmed/4598229.

Morgan, Xochitl C and Curtis Huttenhower. 2012. Chapter 12: Human microbiome analysis. *PLoS Computational Biology* 8 (12). doi:10.1371/journal.pcbi.1002808.

Morrison, Douglas J and Tom Preston. 2016. Formation of short chain fatty acids by the gut microbiota and their impact on human metabolism. *Gut Microbes* 7 (3). Taylor & Francis: 189–200. doi:10.1080/19490976.2015.1134082.

O'Callaghan, Amy and Douwe van Sinderen. 2016. Bifidobacteria and their role as members of the human gut microbiota. *Frontiers in Microbiology* 7 (June). doi:10.3389/fmicb.2016.00925.

O'Toole, Paul W, Julian R Marchesi, and Colin Hill. 2017. Next-generation probiotics: The spectrum from probiotics to live biotherapeutics. *Nature Microbiology* 2 (5): 17057. doi:10.1038/nmicrobiol.2017.57.

Ohland, Christina L and Wallace K MacNaughton. 2010. Probiotic bacteria and intestinal epithelial barrier function. *American Journal of Physiology-Gastrointestinal and Liver Physiology* 298 (6): G807–19. doi:10.1152/ajpgi.00243.2009.

Okada, Haruhiko, Christina T Kuhn, Helene Feillet, and J. F. Bach. 2010. The 'hygiene hypothesis' for autoimmune and allergic diseases: An update. *Clinical and Experimental Immunology* 160 (1): 1–9. doi:10.1111/j.1365-2249.2010.04139.x.

Okumura, Ryu and Kiyoshi Takeda. 2017. Roles of intestinal epithelial cells in the maintenance of gut homeostasis. *Experimental and Molecular Medicine* 49 (5). Nature Publishing Group: e338. doi:10.1038/emm.2017.20.

Ouwehand, Arthur C, Erika Isolauri, Pirkka V Kirjavainen, Satu T Olkko, and Seppo J Salminen. 2000. The mucus binding of *Bifidobacterium lactis* Bb12 is enhanced in the presence of *Lactobacillus* GG and Lact. delbrueckii subsp. bulgaricus. *Letters in Applied Microbiology* 30 (1): 10–13. doi:10.1046/j.1472-765x.2000.00590.x.

Pallister, Tess, Matthew A Jackson, Tiphaine C Martin, Jonas Zierer, Amy Jennings, Robert P Mohney, Alexander MacGregor, et al. 2017. Hippurate as a metabolomic marker of gut microbiome diversity: Modulation by diet and relationship to metabolic syndrome. *Scientific Reports* 7 (1). Springer US: 1–9. doi:10.1038/s41598-017-13722-4.

Pandey, Kavita R, Suresh R Naik, and Babu V Vakil. 2015. Probiotics, prebiotics and synbiotics- a review. *Journal of Food Science and Technology* 52 (12): 7577–87. doi:10.1007/s13197-015-1921-1.

Panigrahi, Pinaki, Sailajanandan Parida, Nimai C Nanda, Radhanath Satpathy, Lingaraj Pradhan, Dinesh S Chandel, Lorena Baccaglini, et al. 2017. A randomized synbiotic trial to prevent sepsis among infants in rural India. *Nature* 548 (7668): 407–12. doi:10.1038/nature23480.

Pärtty, Anna, Marko Kalliomäki, Pirjo Wacklin, Seppo Salminen, and Erika Isolauri. 2015. A possible link between early probiotic intervention and the risk of neuropsychiatric disorders later in childhood: A randomized trial. *Pediatric Research* 77 (6): 823–28. doi:10.1038/pr.2015.51.

Petriz, Bernardo A and Octávio L Franco. 2017. Metaproteomics as a complementary approach to gut microbiota in health and disease. *Frontiers in Chemistry* 5 (January). Frontiers Media S.A.: 4. doi:10.3389/fchem.2017.00004.

Picard, C, Jean Fioramonti, Arnaud Francois, Tobin Robinson, Francoise Neant, and C Matuchansky. 2005. Bifidobacteria as probiotic agents: Physiological effects and clinical benefits. *Alimentary Pharmacology and Therapeutics* 22 (6): 495–512. doi:10.1111/j.1365-2036.2005.02615.x.

Pimentel, Mark, Ruchi Mathur, and Christopher Chang. 2013. Gas and the microbiome topical collection on neuromuscular disorders of the gastrointestinal tract. *Current Gastroenterology Reports* 15 (12). doi:10.1007/s11894-013-0356-y.

Pugliese, Gina. 1995. Formula supplemented with *Bifidobacterium* bifidum and Streptococcus thermophilus prevents diarrhea and shedding of rotavirus in infants. *Infection Control and Hospital Epidemiology* 16 (03): 185–86. doi:10.1017/S0195941700007438.

Rastall, Robert A and Glenn R Gibson. 2015. Recent developments in prebiotics to selectively impact beneficial microbes and promote intestinal health. *Current Opinion in Biotechnology* 32. Elsevier Ltd: 42–46. doi:10.1016/j.copbio.2014.11.002.

Rawat, Seema. 2015. Food spoilage: Microorganisms and their prevention. *Asian Journal of Plant Ascience and Research* 5 (4): 47–56.

Reid, Alexander A, Claude P Champagne, Nancy Gardner, Patrick Fustier, and J C Vuillemard. 2007. Survival in food systems of *Lactobacillus rhamnosus* R011 microentrapped in whey protein gel particles. *Journal of Food Science* 72 (1): M031–7. doi:10.1111/j.1750-3841.2006.00222.x.

Reid, Gregor. 1999. The scientific basis for probiotic strains of *Lactobacillus*. *Applied and Environmental Microbiology* 65(9): 3763–66.

Riley, Margaret A and John E Wertz. 2002. Bacteriocin diversity: Ecological and evolutionary perspectives. *Biochimie* 84 (5–6): 357–64. doi:10.1016/S0300-9084(02)01421-9.

Rombouts, Caroline, Lieselot Y Hemeryck, Thomas Van Hecke, Stefaan De Smet, Winnok H De Vos, and Lynn Vanhaecke. 2017. Untargeted metabolomics of colonic digests reveals kynurenine pathway metabolites, dityrosine and 3-dehydroxycarnitine as red versus white meat discriminating metabolites. *Scientific Reports* 7 (February). Nature Publishing Group: 1–13. doi:10.1038/srep42514.

Rook, G A W. 2005. Microbes, immunoregulation, and the gut. *Gut* 54 (3): 317–20. doi:10.1136/gut.2004.053785.

Rook, G A W, C L Raison, and C A Lowry. 2014. Microbial 'old friends', immunoregulation and socioeconomic status. *Clinical and Experimental Immunology* 177 (1): 1–12. doi:10.1111/cei.12269.

Rothschild, Daphna, Omer Weissbrod, Elad Barkan, Tal Korem, David Zeevi, Paul Igor Costea, Anastasia Godneva, et al. 2017. Environmental factors dominate over host genetics in shaping human gut microbiota composition. Nature 555: 210–215. doi:10.1101/150540.

Ruas-Madiedo, Patricia, Miguel Gueimonde, Abelardo Margolles, Clara G de los Reyes-Gavilán, and Seppo Salminen. 2006. Exopolysaccharides produced by probiotic strains modify the adhesion of probiotics and enteropathogens to human intestinal mucus. *Journal of Food Protection* 69 (8): 2011–15. www.ncbi. nlm.nih.gov/pubmed/16924934.

Ruiz, Lorena, Patricia Ruas-Madiedo, Miguel Gueimonde, Clara G de los Reyes-Gavilán, Abelardo Margolles, and Borja Sánchez. 2011. How do bifidobacteria counteract environmental challenges? Mechanisms involved and physiological consequences. *Genes and Nutrition* 6 (3): 307–18. doi:10.1007/s12263-010-0207-5.

Salazar, Nuria, Miguel Gueimonde, Ana M Hernandez-Barranco, Patricia Ruas-Madiedo, and Clara G de los Reyes-Gavilan. 2008. Exopolysaccharides produced by intestinal *Bifidobacterium* strains act as fermentable substrates for human intestinal bacteria. *Applied and Environmental Microbiology* 74 (15): 4737–45. doi:10.1128/AEM.00325-08.

Salim, Saad Y, Gilaad G Kaplan, and Karen L Madsen. 2013. Air pollution effects on the gut microbiota. *Gut Microbes* 5 (2): 215–19. doi:10.4161/gmic.27251.

Salvetti, Elisa, Sandra Torriani, and Giovanna Felis. 2015. A survey on established and novel strains for probiotic applications. In: Petra Ger and Chalat Santivarangkna (eds) *Advances in Probiotic Technology*, 26–44. Boca Raton, FL: CRC Press. doi:10.1201/b18807-4.

Sara, Margit and Uwe B Sleytr. 2000. S-layer proteins. *Journal of Bacteriology* 182 (4): 859–68. doi:10.1128/JB.182.4.859-868.2000.

Schachtsiek, Martina, Walter P Hammes, and Christian Hertel. 2004. Characterization of *Lactobacillus* coryniformis DSM 20001T surface protein Cpf mediating coaggregation with and aggregation among pathogens. *Applied and Environmental Microbiology* 70 (12): 7078–85. doi:10.1128/AEM.70.12.7078-7085.2004.

Schmid, Jochen, Volker Sieber, and Bernd Rehm. 2015. Bacterial exopolysaccharides: Biosynthesis pathways and engineering strategies. *Frontiers in Microbiology* 6 (May). doi:10.3389/fmicb.2015.00496.

Sender, Ron, Shai Fuchs, and Ron Milo. 2016. Are we really vastly outnumbered? Revisiting the ratio of bacterial to host cells in humans. *Cell* 164 (3): 337–40. doi:10.1016/j.cell.2016.01.013.

Servin, Alain L 2004. Antagonistic activities of lactobacilli and bifidobacteria against microbial pathogens. *FEMS Microbiology Reviews* 28 (4): 405–40. doi:10.1016/j.femsre.2004.01.003.

Shade, Aahley, Stuart E Jones, J. G. Caporaso, Jo Handelsman, Rob Knight, Noah Fierer, and Jack A Gilbert. 2014. Conditionally rare taxa disproportionately contribute to temporal changes in microbial diversity. *MBio* 5 (4). doi:10.1128/mBio.01371-14.

Shanahan, Fergus. 2010. Gut microbes: From bugs to drugs. *American Journal of Gastroenterology* 105 (2). Nature Publishing Group: 275–79. doi:10.1038/ajg.2009.729.

Shen, Jian, Martin S Obin, and Liping Zhao. 2013. The gut microbiota, obesity and insulin resistance. *Molecular Aspects of Medicine* 34 (1). Elsevier Ltd: 39–58. doi:10.1016/j.mam.2012.11.001.

Shendure, Jay and Hanlee Ji. 2008. Next-generation DNA sequencing. *Nature Biotechnology* 26 (10): 1135–45. doi:10.1038/nbt1486.

Sleator, Roy D. 2018. The human superorganism; of microbes and men. *Medical Hypotheses* 74 (2). Elsevier: 214–15. doi:10.1016/j.mehy.2009.08.047.

Snel, Berend, Peer Bork, and Martijn A Huynen. 1999. Genome phylogeny based on gene content. *Nature Genetics* 21 (1): 108–10. doi:10.1038/5052.

Spanhaak, Steven, Robert Havenaar, and Gertjan Schaafsma. 1998. The effect of consumption of milk fermented by *Lactobacillus casei* strain Shirota on the intestinal microflora and immune parameters in humans. *European Journal of Clinical Nutrition* 52 (12): 899–907. doi:10.1038/sj.ejcn.1600663.

Staley, James T and Allan Konopka. 1985. Measurement of in situ activities of nonphotosynthetic microorganisms in aquatic and terrestrial habitats. *Annual Review of Microbiology* 39: 321–46. doi:10.1146/annurev.mi.39.100185.001541.

Stecher, Bärbel and Wolf Dietrich Hardt. 2011. Mechanisms controlling pathogen colonization of the gut. *Current Opinion in Microbiology* 14 (1). Elsevier Ltd: 82–91. doi:10.1016/j.mib.2010.10.003.

Surawicz, Christina M, Lawrence J Brandt, David G Binion, Ashwin N Ananthakrishnan, Scott R Curry, Peter H Gilligan, Lynne V McFarland, Mark Mellow, and Brian S Zuckerbraun. 2013. Guidelines for diagnosis, treatment, and prevention of clostridium difficile infections. *American Journal of Gastroenterology* 108 (4). Nature Publishing Group: 478–98. doi:10.1038/ajg.2013.4.

Sweileh, Waleed M, Naser Y Shraim, Samah W Al-Jabi, Ansam F Sawalha, Belal Rahhal, Rasha A Khayyat, and Sa'ed H Zyoud. 2016. Assessing worldwide research activity on probiotics in pediatrics using scopus database: 1994–2014. *World Allergy Organization Journal* 9 (1): 25. doi:10.1186/s40413-016-0116-1.

Taipale, Teemu J, Kaisu Pienihäkkinen, Erika Isolauri, Jorma T Jokela, and Eva M Söderling. 2016. *Bifidobacterium animalis* subsp. lactis BB-12 in reducing the risk of infections in early childhood. *Pediatric Research* 79 (1): 65–69. doi:10.1038/pr.2015.174.

Tannock, Gerald W. 2004. A special fondness for lactobacilli. *Applied and Environmental Microbiology* 70 (6): 3189–94. doi:10.1128/AEM.70.6.3189-3194.2004.

Thaiss, Christoph A, Niv Zmora, Maayan Levy, and Eran Elinav. 2016. The microbiome and innate immunity. *Nature* 535 (7610): 65–74. doi:10.1038/nature18847.

Thomas, Torsten, Jack Gilbert, and Folker Meyer. 2012. Metagenomics: A guide from sampling to data analysis. *Microbial Informatics and Experimentation* 2 (1). BioMed Central Ltd: 3. doi:10.1186/2042-5783-2-3.

Tojo, Rafael. 2014. Intestinal microbiota in health and disease: Role of bifidobacteria in gut homeostasis. *World Journal of Gastroenterology* 20 (41): 15163. doi:10.3748/wjg.v20.i41.15163.

Tolonen, Andrew C and Ramnik J Xavier. 2017. Dissecting the human microbiome with single-cell genomics. *Genome Medicine* 9 (1). Genome Medicine: 7–9. doi:10.1186/s13073-017-0448-7.

Tramontano, Melanie, Sergej Andrejev, Mihaela Pruteanu, and Martina Klünemann. 2018. Nutritional preferences of human gut bacteria reveal their metabolic idiosyncrasies. *Nature Microbiology* 3 (4): 514–522. doi:10.1038/s41564-018-0123-9.

Ursell, Luke K, Jessica L Metcalf, Laura Wegener Parfrey, and Rob Knight. 2013. Defining the human microbiome. *NIH Manuscripts* 70 (Suppl 1): 1–12. doi:10.1111/j.1753-4887.

Valeur, Nana, Peter Engel, Noris Carbajal, Eamonn Connolly, and Karin Ladefoged. 2004. Colonization and immunomodulation by *Lactobacillus reuteri* ATCC 55730 in the human gastrointestinal tract. *Applied and Environmental Microbiology* 70 (2): 1176–81. doi:10.1128/AEM.70.2.1176-1181.2004.

van Baarlen, Peter, Jerry M Wells, and Michiel Kleerebezem. 2013. Regulation of intestinal homeostasis and immunity with probiotic lactobacilli. *Trends in Immunology* 34 (5): 208–15. doi:10.1016/j.it.2013.01.005.

van Tassell, Maxwell L and Michael J Miller. 2011. *Lactobacillus* adhesion to mucus. *Nutrients* 3 (5): 613–36. doi:10.3390/nu3050613.

Vazquez-Gutierrez, Pamela, Tomas de Wouters, Julia Werder, Christophe Chassard, and Christophe Lacroix. 2016. High iron-sequestrating bifidobacteria inhibit enteropathogen growth and adhesion to intestinal epithelial cells in vitro. *Frontiers in Microbiology* 7 (September). doi:10.3389/fmicb.2016.01480.

Vernocchi, Pamela, Federica Del Chierico, and Lorenza Putignani. 2016. Gut microbiota profiling: Metabolomics based approach to unravel compounds affecting human health. *Frontiers in Microbiology* 7 (July). doi:10.3389/fmicb.2016.01144.

Videla, Sebastia, Jaume Vilaseca, F Guarner, Antonia Salas, F Treserra, E Crespo, Maria Antolín, and J R Malagelada. 1994. Role of intestinal microflora in chronic inflammation and ulceration of the rat colon. *Gut* 35 (8): 1090–97. www.ncbi.nlm.nih.gov/pmc/articles/PMC1375061/.

Villena, María José Martín, Ferderico Lara-Villoslada, María Adolfina Ruiz Martínez, and María Encarnación Morales Hernández. 2015. Development of gastro-resistant tablets for the protection and intestinal delivery of *Lactobacillus* fermentum CECT 5716. *International Journal of Pharmaceutics* 487 (1–2): 314–19. doi:10.1016/j.ijpharm.2015.03.078.

Vlasova, Anastasia N, Sukumar Kandasamy, Kuldeep S Chattha, Gireesh Rajashekara, and Linda J Saif. 2016. Comparison of probiotic lactobacilli and bifidobacteria effects, immune responses and rotavirus vaccines and infection in different host species. *Veterinary Immunology and Immunopathology* 172 (April): 72–84. doi:10.1016/j.vetimm.2016.01.003.

Walter, Jens. 2008. Ecological role of lactobacilli in the gastrointestinal tract: Implications for fundamental and biomedical research. *Applied and Environmental Microbiology* 74 (16): 4985–96. doi:10.1128/AEM.00753-08.

Wang, Lifeng, Jiachao Zhang, Zhuang Guo, Laiyu Kwok, Chen Ma, Wenyi Zhang, Qiang Lv, Weiqiang Huang, and Heping Zhang. 2014. Effect of oral consumption of probiotic *Lactobacillus* planatarum P-8 on fecal microbiota, SIgA, SCFAs, and TBAs of adults of different ages. *Nutrition* 30 (7–8): 776–783.e1. doi:10.1016/j.nut.2013.11.018.

Wang, Mei, Marcia H Monaco, and Sharon M Donovan. 2016. Impact of early gut microbiota on immune and metabolic development and function. *Seminars in Fetal and Neonatal Medicine* 21 (6). Elsevier Ltd: 380–87. doi:10.1016/j.siny.2016.04.004.

Wang, Zhongtang, Guangxia Xiao, Yongming Yao, Shuzhong Guo, Kaihua Lu, and Zhiyong Sheng. 2006. The role of bifidobacteria in gut barrier function after thermal injury in rats. *The Journal of Trauma: Injury, Infection, and Critical Care* 61 (3): 650–57. doi:10.1097/01.ta.0000196574.70614.27.

Wexler, Hannah M. 2007. *Bacteroides*: The good, the bad, and the nitty-gritty. *Clinical Microbiology Reviews* 20 (4): 593–621. doi:10.1128/CMR.00008-07.

Xanthopoulos, Vassillis, Evanthia Litopoulou-Tzanetaki, and N Tzanetakis. 2000. Characterization of *Lactobacillus* isolates from infant faeces as dietary adjuncts. *Food Microbiology* 17 (2): 205–15. doi:10.1006/fmic.1999.0300.

Yang, Gang, Mo Peng, Xiangli Tian, and Shuanglin Dong. 2017. Molecular ecological network analysis reveals the effects of probiotics and florfenicol on intestinal microbiota homeostasis: An example of sea cucumber. *Scientific Reports* 7 (1): 4778. doi:10.1038/s41598-017-05312-1.

Yatsunenko, Tanya, Federico E Rey, Mark J Manary, Indi Trehan, Maria Gloria Dominguez-Bello, Monica Contreras, Magda Magris, et al. 2012. Human gut microbiome viewed across age and geography. *Nature* 486 (7402): 222–27. doi:10.1038/nature11053.

Young, Vincent B 2017. The role of the microbiome in human health and disease: An introduction for clinicians. *BMJ (Online)* 356. doi:10.1136/bmj.j831.

Zhang, Yu-Jie, Sha Li, Ren-You Gan, Tong Zhou, Dong-Ping Xu, and Hua-Bin Li. 2015. Impacts of gut bacteria on human health and diseases. *International Journal of Molecular Sciences* 16 (4): 7493–7519. doi:10.3390/ijms16047493.

Zoetendal, Erwin G, Jeroen Raes, Bartholomeus van den Bogert, Manimozhiyan Arumugam, Carien C G M Booijink, Freddy J Troost, Peer Bork, Michiel Wels, Willem M de Vos, and Michiel Kleerebezem. 2012. The human small intestinal microbiota is driven by rapid uptake and conversion of simple carbohydrates. *The ISME Journal* 6 (7): 1415–26. doi:10.1038/ismej.2011.212.

10 Discovery and Early Development of the Next-Generation ALK Inhibitor, Lorlatinib (18)

Agent for Non–Small-Cell Lung Cancer

Paul F. Richardson
Pfizer-La Jolla

CONTENTS

10.1   INTRODUCTION

Non–small-cell lung cancer (NSCLC) remains the most common cause of cancer-related deaths worldwide with over 2.1 and 1.8 million new cases and deaths, respectively, representing 12% and 18% of new cancers and cancer mortality, respectively, as estimated by the International Agency for Research on Cancer of the World Health Organization in 2018.[1] Most patients who have NSCLC present with advanced or incurable disease, and chemotherapy generally results in modest improvement of survival with typically the median survival being <1 year after diagnosis. NSCLC represents the majority (~90%) of lung cancers and consists of a number of subtypes, which are driven by various activated oncogenes.[2,3] Although recently, immunotherapy has represented a breakthrough in the treatment of NSCLC with several immune checkpoint inhibitors already on the market,[4–7] the majority of NSCLC patients do not respond to PD-1/PD-L1 inhibition, and in addition, those presenting with driver mutations appear to be even less sensitive to this approach.[8]

Regarding driver mutations, it has been recognized for over a decade that clinical outcomes can be improved in NSCLC treatment can be achieved by identifying the molecular events that underlie its pathogenesis, and it has been estimated that currently 25% of Caucasians with lung adenocarcinoma (ADC) have a targetable driver mutation.[9] With this in mind, a subset of NSCLCs have been identified harboring mutations in the epidermal growth factor receptor gene (EGFR), and these cancers have been shown to be responsive to EGFR-tyrosine kinase inhibitors (TKIs) such as gefitinib. It should be noted that EGFR mutations have been shown to be primarily associated with Asians (40%–50% of the cases) and nonsmokers. In contrast, there have been few reports of oncogenes associated for NSCLC in individuals who smoke, which constitutes the majority of cases of NSCLC.[10] However, further validation of the precision-medicine–based approach in targeting key growth drivers can be found in the case of chronic myeloid leukemia, in which treatment with the inhibitor STI571 has been shown to effectively reduce the number of cancer cells.[11]

In 2007, Mano, Soda, and co-workers demonstrated that in NSCLC cells, a small inversion within chromosome 2p resulted in the formation of a fusion gene (EML4-ALK); this gene consists of portions of the echinoderm microtubule-associated protein-like-4 (EML4) gene and the anaplastic lymphoma kinase (ALK) gene.[12,13] ALK is a receptor tyrosine kinase belonging to the insulin receptor superfamily. ALK's function in normal human tissues (primarily expressed in adult brain tissue) is unclear, but it appears to play a role in physiological development and function of the nervous system.[14,15] Previously, ALK had been identified as a part of the fusion oncogene nucleophosmin (NPM)-ALK, which is detected in approximately 75% of all ALK-positive ALCL (anaplastic large-cell lymphoma) and is implicated in the pathogenesis of ALCL.[16] Furthermore, Mano et al. report that mouse 3T3 fibroblasts forced to express this human fusion EML4-ALK kinase generated transformed foci in culture and subcutaneous tumors in nude mice. Evaluation of a cohort of NSCLC patients indicated that 6.7% (5 out of 75) harbored this EML4-ALK fusion transcript with there being no overlap with those presenting either EGFR or KRAS mutations. This data suggested that inhibition of the tyrosine kinase activity of EML4–ALK may induce cell death in tumors expressing this fusion protein. Crucially, it had been demonstrated that ALK knockout mice display no overt developmental, anatomical, or locomotor deficiencies (presenting with an "antidepressive phenotype"),[17] and as such, suppression of EML4-ALK function would be expected to be free of severe side effects in NSCLC patients. It should also be noted that other fusion genes of ALK have been identified in inflammatory myofibroblastic tumors[18] and diffuse large B-cell lymphomas,[19] while in childhood neuroblastoma, the ALK gene was amplified.[20–22]

The EML4-ALK fusion gene is expressed in 3%–5% of NSCLC patients (median age 52 years), with most of the cases being ADCs.[23] Given that ALK gene rearrangement involves a chromosomal inversion and translocation, fluorescence *in situ* hybridization (FISH) has become the method of choice for detecting all forms of ALK gene rearrangement as this method can detect particular DNA sequences in chromosomes. A Food and Drug Administration (FDA)-approved kit is available for FISH in the United States, and its use in study of ALK gene rearrangement extends to even the rare non-EML4 fusions. In fact, it was this assay that was used to detect this genetic aberration in the initial clinical trials with ALK inhibitor therapy (*vide infra*).[24] As many guidelines already recommend, Break-Apart ALK FISH is a reliable standard diagnostic method in surgical pathology because it is easily applicable to formalin-fixed, paraffin embedded (FFPE) samples, even when the exact fusion partners are not known. However, FISH is time-consuming and requires the use of expensive probes and a special fluorescence microscopy facility, and more recently, Roche Ventana has obtained approval to test for the mutation by immunohistochemistry (IHC).[25] Other techniques like reverse-transcriptase PCR (RT-PCR) can also be used to detect lung cancers with an ALK gene fusion but are not recommended, though Fujishima and co-workers have reported on a novel rapid FISH (RaFISH) method developed to facilitate hybridization. This method takes advantage of the noncontact mixing effect of an alternating current (AC) electric field. With RaFISH, the ALK test was completed within 4.5 h, as compared to 20 h needed for the standard FISH.[26] Although RaFISH produced results more promptly, the staining and accuracy of the ALK evaluation with RaFISH were equal to the standard, with 97.6% agreement found between FISH and RaFISH based on the status of the ALK signals. This suggests that RaFISH could be implemented as a clinical tool to promptly determine ALK status in the future.

ROS1 is a receptor tyrosine kinase with high homology to ALK that was first identified in 1986.[27] Although the precise role of the ROS1 protein in human development is not known, gene rearrangements involving the ROS1 gene have been detected in various cancer lines with the first being observed in glioblastoma tumors in 1987.[28] Since the observation of a ROS1 rearrangement in a lung ADC patient in 2007,[29] it has been demonstrated that in patients with NSCLC, approximately 2% are positive for a *ROS1* gene rearrangement, and these rearrangements are mutually exclusive of *ALK* rearrangements.[30] In a similar manner to ALK fusions, ROS1 fusions are more commonly found in light smokers (<10 pack years) and/or never smokers, and are typically associated with younger age (median age 49.8 years) and ADCs. The presence of ROS1 fusions can be determined by a number of techniques, and there are reports comparing FISH, IHC, and quantitative real-time reverse transcription-PCR (qRT-PCR) assays in a large number of ROS1-positive lung ADC patients. These suggest that IHC is a reliable and rapid screening tool in routine pathologic laboratories for the identification of suitable candidates for ROS1-targeted therapy and even demonstrated several cases in which IHC detected the translocation event when FISH did not.

As time progresses, an increasing number of molecular subsets of NSCLC are being discovered including genetic alterations of c-MET, NTRK translocations, and RET translocations. These discoveries suggest a potential increase in the development of selective inhibitors as viable treatment options for NSCLC in the coming years.

10.2   DISCOVERY AND DEVELOPMENT OF CRIZOTINIB (1)

Targeting mesenchymal epithelial transition (MET) signaling has long been promoted as an attractive objective for cancer therapy, though it has up to now met with limited success in the numerous clinical trials that have been conducted.[31,32] There are purported to be numerous reasons for this with the most prevalent being the fact that the biology and the clinical significance of the MET pathway activation are still not completely understood. The c-MET gene is located on chromosome 7q21-31, and it encodes a protein tyrosine kinase belonging to the hepatocyte growth factor (HGF) family, which is involved in a variety of important cellular processes including proliferation, cell cycle, differentiation, and apoptosis. With regard to NSCLC, irregular activation of the MET pathway may occur through a variety of processes with overexpression of the MET protein appearing to be the most frequent. Gene amplification is also reported to occur in both untreated NSCLC patients and in a significant percentage of patients with EGFR-mutated tumors, a group which characteristically have acquired resistance to EGFR-TKIs. A further aberrant activation of the MET pathway, which appears to currently present the most promise for clinical intervention, is through a splice mutant of MET that leads to skipping of exon 14 (METex14); this mutation induces prolonged signaling and

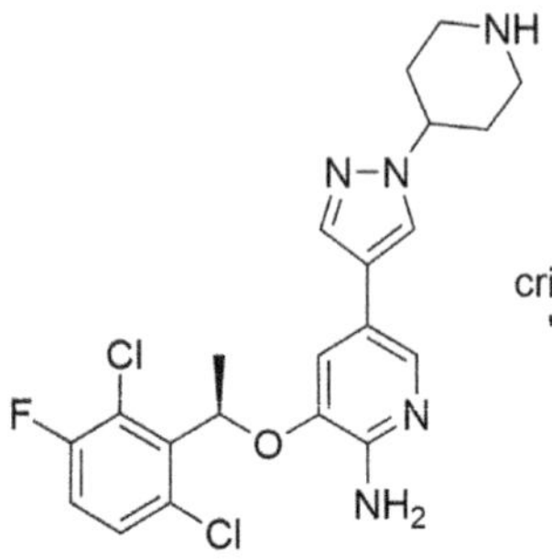

FIGURE 10.1 Crizotinib (**1**).

oncogenic capacity. Alterations in the MET protein have been reported in approximately 3%–4% of NSCLC cases, though they can present challenges from a diagnostic testing standpoint.

Regarding preclinical and clinical studies, numerous MET inhibitors have been tested including small-molecule TKIs of the receptor (either multi-kinase inhibitors, or those selective for MET), monoclonal antibodies against c-MET and against the HGF ligand. It was as a potential MET inhibitor that the human clinical Phase I trials of crizotinib (**1**) (PF-02341066 – developed by Pfizer) were initiated in 2006 (Figure 10.1).[33–35] In preclinical tumor xenograft studies, crizotinib (**1**) had been demonstrated to inhibit HGF-stimulated growth and survival of cell lines, as well as a decrease in c-MET phosphorylation. Furthermore, in MET-amplified lung cancer cell lines, crizotinib was proven to induce apoptosis and inhibit both AKT and extracellular signal-related kinase phosphorylation.

Crizotinib has been evaluated in several clinical trials with the targeting of MET-signaling being the primary objective, and these studies suggest this strategy may be efficacious for NSCLC.[36,37] For example, a multicenter retrospective analysis on patients with MET exon 14 mutant NSCLC demonstrated an improvement in overall survival for those treated with an MET inhibitor (including crizotinib).[38] In addition, analysis of the efficacy and safety data for a cohort of patients with advanced c-MET-amplified NSCLC demonstrated partial responses (PRs, 33%; 95% CI: 10%– 65%) in both the intermediate- and high-level amplification groups.[39] Finally, crizotinib has been evaluated in combination with both first- and next-generation EGFR inhibitors (erlotinib and dacomitinib), though enthusiasm for this approach was limited by observations of both minimal activity and substantial toxicity (grade 3% or 4% in 43% of patients).[40,41] Despite the challenges of not only fully elucidating the key underlying processes inherent in the MET pathway in NSCLC but also establishing either a robust diagnostic test or molecular biomarkers for patient identification, in 2018, crizotinib was granted breakthrough-therapy designation by the FDA in metastatic NSCLC with Met exon 14 alterations.

Crizotinib was approved in 2011 by the FDA as the first agent for the treatment of patients harboring the EML4-ALK translocation gene. The initial oral dose was 50 mg/day, which was escalated to 300 mg BID. From these investigations, the maximum tolerated dose was established at 250 mg BID with plasma concentrations reaching steady state within 15 days.[42,43] The approval of crizotinib (Xalkori) was based on demonstrated superiority over conventional chemotherapy in metastatic ALK-positive NSCLC in two randomized Phase III trials, achieving both higher response rates and a significantly longer progression-free survival (PFS).[44,45] The majority of adverse reactions were grades 1 and 2 with vision disorders, elevated transaminases, diarrhea, and nausea reported among other side effects. Crizotinib was approved for this indication in 2011 and is currently offered as a treatment in 70 countries. In addition, based on the single arm of a Phase II trial, crizotinib was approved in 2016 as a first-line treatment for advanced NSCLC patients with ROS1 fusions.[46] A summary of the data for the key clinical trials is provided in Table 10.1.

TABLE 10.1
Key Clinical Trials Involving Crizotinib (1)

Clinical Trial NCT00585195	# Patients	ORR[a] (95% CI[b])	Estimated PFS[c] (Range)	Median DOR (weeks)
Phase 1 – ALK	143	60.8 (52.3–68.9)	9.7 months (7.7–12.8)	49.1
Phase 2 – ALK	255	53 (47–60)	8.5 months (6.2–9.9)	43
Phase 1 – ROS1	50	72 (58–84)	19.2 months (14.4–not reached)	70.4

[a] Overall response rate.
[b] Confidence interval.
[c] Progression-free survival.

10.3   PROJECT GOALS

Despite the initial robust efficacy demonstrated by crizotinib (**1**) in the treatment of ALK-positive NSCLC, there emerged several drivers indicating the potential utility of a next-generation EML4-ALK inhibitor. These are briefly summarized in the following sections and provided the key goals for the product profile for lorlatinib (**18**).

10.3.1   Efficacy Against ALK-Resistance Mutants

It was noted that typically over 10–12 months that resistance to crizotinib (**1**) occurred, leading to the recurrence of disease. In-depth analysis of the resistance mechanisms at play indicated that although both upregulation of alternative signaling pathways (EGFR, KRAS, c-KIT, IGF1R) and amplification of the ALK-fusion gene has been observed, a prevalent factor causing resistance was mutations found in or near the long narrow groove that comprises the binding pocket of both ATP and crizotinib.[47–49] The most common mutation (observed in approximately 7% of cases) was L1196M (in the kinase gatekeeper region, and this is homologous to gatekeeper mutations observed in BCR-ABL-T315I and EGFR-T790M). The L1196M mutation has been demonstrated to alter and subsequently hinder the binding of crizotinib through steric interference. The next most prevalent mutation observed is G1269A (~4%). This residue is found next to the DFG (Asp-Phe-Gly) motif within the kinase activation loop, which is a portion of the kinase domain necessary for kinase activity. Substitution herein of the smallest possible amino-acid residue glycine with alanine again leads to an increase in steric hindrance, thus decreasing the affinity for crizotinib. Though less frequent, further mutations have also been reported, including 1151T insertion and the point mutations C1156Y, F1174L, G1202R, I1171T, S1206Y, and E1210K. Crizotinib preferentially binds to the inactive conformation of ALK, and computational modeling studies indicate that the T1151 insertion and the point mutations, G1202R and S1206Y, are at the solvent-exposed region of the kinase domain. Whereas G1202R is analogous to an imatinib-resistant BCR-ABL mutation (G340W), both S1206Y and 1151T appear to be unique to ALK and crizotinib resistance. G1202R is hypothesized to interfere with the binding of crizotinib through steric hindrance with the introduction of a large basic residue in place of guanine. S1206Y is hypothesized to destabilize the interaction of the side-chain hydroxyl of Ser1206 with the carboxylate group of D1203. In addition, the larger tyrosine side chain of S1206Y may lead to a number of conformational changes around the solvent, with the bulkier tyrosine leading to not only a clash with the ligand but also the potential destabilization of the complementary electrostatic interaction between the basic piperidine of crizotinib and the acidic E1210 residue. In contrast, 1151T insertion is not located near the binding site for crizotinib and is instead farther away at the loop of the N-terminus of α–helix carbon. This mutation is predicted to affect the affinity of ALK for ATP through the disruption of a critical hydrogen bond between T1151 and the carbonyl backbone of E1129.

Both the heterogeneity and complexity of ALK-inhibitor resistance present a major challenge in the successful treatment of patients with ALK-rearranged tumors. Though the operation of multiple-resistance mechanisms necessitates judicious decisions with respect to an individualized treatment strategy, both the development of next-generation ALK inhibitors (specifically targeting mutations of the parent protein) and rational combination strategies (to target the upregulation of compensatory activation pathways) represent a potentially promising development pathway.

10.3.2 TARGETING BRAIN METASTASES

The effective treatment of primary brain tumors and brain metastases currently represents an area of unmet medical need, with approximately 200,000 brain metastases (BM) diagnosed annually in the United States, which account for 20% of cancer mortality.[50] From peripheral tumors, metastasis to the central nervous system (CNS) occurs in as many as 40% of cases per year, with lung (46%) and breast (20%) representing the most prevalent tumors of origin.[51] In the case of NSCLC, approximately 10%–25% of patients present with brain metastases at the time of diagnosis, an additional 30%–40% during the course of diseases, and more still, at autopsy. For many ALK-positive NSCLC patients, the CNS represents not only the first site of progression for patients receiving an ALK-directed therapy but also enables disease advancement despite emerging superior treatments for systemic disease.[52]

The development of kinase inhibitors to treat CNS metastases has been recognized as an opportunity for therapeutic intervention.[53,54] Given that there has been a significant degree of scholarship establishing the optimal ranges for a molecule's physicochemical properties (MW, log D, H-bond donors, *etc.*) to achieve brain penetration, theoretically this should be achievable to some degree.[55,56] However, there are still significant obstacles to achieving this goal. From a practical standpoint, early detection of CNS metastases is often challenging due to resolution limits of current conventional imaging techniques.[57] At the early stage of disease, small aggregates of metastatic tumor cells (commonly referred to as micrometastases) can cross the blood–brain barrier in a relatively facile manner, and then they continue to grow until they reach a clinically significant size at which point they can be detected.[58] In addition, the physicochemical properties inherently required for kinase inhibition is often diametrically opposed to those required for CNS penetration. For example, it is recognized that to avoid efflux and thus achieve brain penetration, minimizing the number of H-bond donor interactions is an important consideration.[59] However, typically kinase inhibitors not only require such interactions to enable binding with the hinge of the protein but also utilize multiple H-bond donors to optimize both potency and selectivity. Given this quandary, the current cohort of kinase inhibitors are typically not freely CNS penetrant as exemplified by crizotinib, which displays poor free-brain exposure likely due to the high p-glycoprotein (P-gp)-mediated efflux measured *in vitro*. Furthermore, the small CSF-to-free plasma ratio of 0.03 measured in a patient at 250 mg BID dose of crizotinib at steady state demonstrates the poor blood–brain penetration of the drug.[60]

10.3.3 OVERCOMING ROS1-RESISTANCE MUTATIONS

As we have seen, a significant amount of work has been done on understanding and overcoming resistance mechanisms in the case of ALK-positive NSCLC, though significantly less studied is the area of ROS1-positive NSCLC. Shaw et al. have reported a study on the mechanisms of resistance in ROS1-positive NSCLC, noting initially that crizotinib-treated patients with ALK-positive NSCLC generally exhibited a significantly longer PFS compared to ROS1-positive NSCLC. This disparity is hypothesized to be due to crizotinib's greater potency against ALK compared to ROS1.[61] A second observation made within this study was that ALK rearrangements were associated with a higher cumulative incidence of BM than ROS1 rearrangements. Although the biological basis for this difference is unclear, it does again highlight the need for effective inhibitors capable of penetrating the CNS. Although several mutations of ROS1 have been identified, the most predominant by far in contributing to resistance was G2032R. This mutation occurs on the solvent-front region of the ATP-binding site (analogous to ALK G1202R), and it confers resistance to crizotinib through steric hindrance. Other less prevalent mutations identified include D2033N and S1986F. D2033N leads to reorientation of neighboring residues and subsequent loss of an electrostatic interaction with crizotinib. S1986F is proposed to lead to resistance through a shift in position of the glycine-rich loop. Despite these studies, none of the second-generation ALK TKIs with ROS1 activity are effective against G2032R. Consequently, treatment options for crizotinib-resistant *ROS1*-positive patients are limited.

10.4 SECOND-GENERATION ALK INHIBITORS

Given the issue with resistance to crizotinib (**1**), a number of second-generation ALK inhibitors have been developed, several of which have been evaluated in the clinic and have subsequently achieved approval for the treatment of crizotinib-naive and relapsed ALK-positive NSCLC patients (Figure 10.2).[62] Ceritinib (**2**) marketed as Zykadia® (LDK378, Novartis) was granted accelerated approval by the FDA in April 2014 for the treatment of ALK-positive NSCLC patients who have either progressed while on crizotinib or are insensitive to this agent. Approval was based on results of a multicenter trial, which demonstrated an objective response rate (ORR) of 44% and a median duration of response (DOR) of 7.1 months for patients receiving a daily, 750 mg oral dose of ceritinib (**2**). Ceritinib (**2**) has since undergone priority review as a first-line therapy for ALK-positive NSCLC (2017) and was granted breakthrough-therapy designation by the FDA for ALK-positive NSCLC that has metastasized to the brain. The initial lead compound TAE684 (a 5-chloro-2,4-diaminophenylpyrimidine) was identified from a cellular screen of a kinase-directed small-molecule library assembled from several different medicinal chemistry programs to identify compounds that were selectively cytotoxic to Ba/F3 NPM-ALK cells.[63] Despite showing excellent potency, TAE684 was found to form a number of reactive adducts upon metabolic oxidation, leading to concerns regarding potential toxicity issues. These issues were primarily linked to the potential for the electron-rich aromatic ring to undergo oxidation to form a highly reactive 1,4-diiminoquinone moiety. Reversal flipping the orientation of the piperidine was the primary change made to alleviate this concern, while increasing the steric bulk of the aryl ether substituent was utilized to increase kinase selectivity leading to the identification of Zykadia (**2**).[64]

Marketed as Alecensa® (CH5424802), alectinib (**3**) was developed by Chugai (part of the Hoffmann-La Roche group) and approved in Japan in July 2014 based on clinical trials in which 19.6% of patients achieved a complete response with a 2-year PFS rate of 76%. Alecensa (**3**) was granted accelerated approval by the FDA in December 2015 and full approval in November 2017, at which point the compound was also approved as a first-line treatment of patients with ALK-positive metastatic NSCLC. The unique tetracyclic scaffold was initially identified through an HTS campaign[65] and subsequently chemically optimized initially for potency and metabolic stability. Finally, it was optimized for selectivity by taking advantage of a unique amino-acid sequence in the ATP-binding site of ALK to minimize adverse effects through inhibition of off-target kinases.[66]

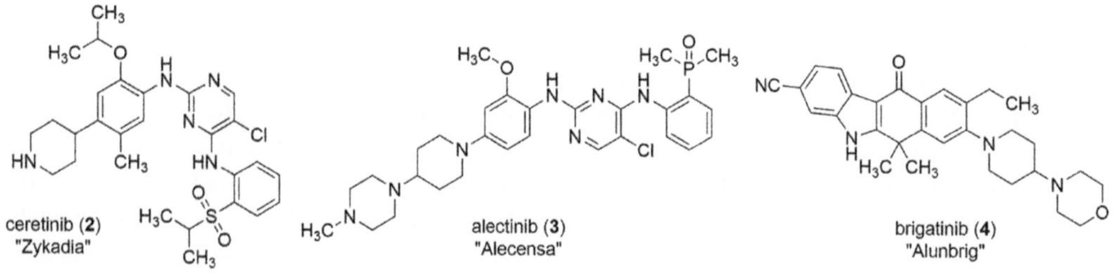

ceretinib (2)
"Zykadia"

alectinib (3)
"Alecensa"

brigatinib (4)
"Alunbrig"

FIGURE 10.2 Second-generation ALK inhibitors.

TABLE 10.2

Key Physicochemical Properties of ALK Inhibitors

Compound	HBD	TPSA (Å²)	cLogP	MW
CNS drugs ($n = 119$)	1	45	2.8	305
Crizotinib (**1**)	3	77	4.2	450
Ceritinib (**2**)	3	105	6.5	558
Alectinib (**3**)	1	72	5.2	483
Brigatinib (**4**)	2	82	5.9	529
Kinase inhibitors ($n = 34$)	2	91	4.2	483

Brigatinib (**4**), marketed as Alunbrig® (AP26113), was identified by Ariad through initial screening of a focused chemical library through a series of *in vitro* assays followed by optimization through a structure-based drug design-based approach. (Note: Ariad was acquired by Takeda in February 2017.) The pyrimidine-based compound features an unusual *ortho*-dimethylphosphine oxide on a pendant aniline substituent.[67] Alunbrig (**4**) is a dual ALK/EGFR inhibitor and shows high levels of selectivity, only inhibiting 11 of 289 kinases screened. Further, Alunbrig is potent, achieving this selective inhibition with an IC$_{50}$ of <10 nM. Clinical trials of Alunbrig (**4**) were initiated in September 2011 with breakthrough-therapy designation being granted in 2014. Orphan drug status was granted in 2016 with results from the clinic showing a 1-year overall survival rate of 100% in crizotinib-naive patients, and 81% in patients with prior crizotinib treatment. In April 2017, Alunbrig (**4**) was granted accelerated approval by the FDA as a second-line therapy for the treatment of ALK-positive NSCLC.

As has been previously noted, the optimization of several physicochemical factors has been highlighted as being important for maximal CNS penetration – specifically MW, TPSA, cLogP, and number of H-bond donors. Table 10.2 provides the values for these properties for crizotinib (**1**) and each of the approved second-generation ALK inhibitors, comparing them to the average values for not only approved CNS drugs ($n = 119$) but also for all approved kinase inhibitors ($n = 34$). Although literature reports suggest that neither efflux nor free brain penetration was consideration at the design stage for these molecules, there have been reports of positive responses in the remediation of CNS disease using these compounds. In this respect, ceritinib (**2**) has been reported to achieve responses in patients with BM,[68–70] while both alectinib (**3**) and brigatinib (**4**) have shown efficacy in preclinical CNS-based tumor models.[71,72] Positive responses have also been observed with CNS disease in the clinic for both alectinib (**3**)[73–75] and brigatinib (**4**)[76,77] in patients for which crizotinib (**1**) was ineffective. Despite these promising results, one factor which is not well established within these reports is to what degree the blood brain barrier (BBB) has been compromised because of disease progression, thus enabling CNS penetration for the compound under evaluation.

10.5 LABORATORY OBJECTIVES

Upon initiation of the project to identify our next-generation ALK inhibitor, the laboratory objectives listed in Table 10.3 were set. Crizotinib (**1**) was established as the initial benchmark for the program.

The specific challenge herein for the team would be achieving a balance between two core physiochemical parameter sets. Firstly, achieving good potency in a molecule not only for the native protein target but more importantly against as many of the clinically reported mutations as possible against which crizotinib is ineffective. Within our testing cascade, the gatekeeper mutation L1196M (both the most prevalent and resistant crizotinib mutation) was chosen in order to initially evaluate compounds. Note that in our biochemical assay, crizotinib shows approximately a tenfold loss in efficacy against this target. However, it has long been recognized that the key to successful drug design is to ensure that optimizing for potency does not take place in a vacuum, and that one seeks to progress a compound

TABLE 10.3
Laboratory Objectives

	Crizotinib (1)	Desired CNS Profile
Dosing	500 mg/day BID	<500 mg/day QD or BID
EML4-ALK-WT cell IC_{50} (nM)	80	<50
EML4-ALK-L1196M cell IC_{50} (nM)	841	<50
HLM clearance (μL/min/mg)	53	<30
RRCK P_{app} (10^{-6} cm/s)	0.79	>2.5
Efflux (MDR BA/AB)	45	<2.5

with reasonable properties of absorption, distribution, metabolism, and excretion (ADME). Utilizing a numerical index is an invaluable tool to gauge improvements in terms of overall compound quality, and for this purpose, we typically use lipophilic efficiency (LipE = pKi (or pIC_{50}) – log D) as a means to evaluate binding effectiveness per unit lipophilicity.[78–80] Provided metabolic liabilities are not introduced into a molecule during the design process, optimization through improving LipE will lead to both improved ADME properties and safety profiles of a lead series during the discovery phase.

Critical to the success for obtaining brain penetration is to avoid transporter-mediated efflux of the molecule at the blood–brain barrier. In addition, avoiding such efflux transporters at the cell surface potentially can lead to a higher tumor intracellular concentration and thus better target modulation. However, efflux presents a double-edged sword as these transporters are expressed throughout healthy cells and serve as a protective mechanism from xenotoxics. Given this, it is imperative within this program to pursue rigorous levels of selectivity particularly around closely related kinases with known or potential risks (particularly within the CNS) to both mitigate cell toxicity and improve therapeutic indices.

Given that no satisfactory *in vitro* model of the BBB is available to date, several high-throughput assays have been developed in order to assay permeability using alternative cell lines such as the industry favorite MDCK (Madin–Darby Canine Kidney), which was used here. Although this cell line is neither endothelial in nature nor does it originate from the brain, the tightness of the monolayer results in permeability values (Papp) that correlate well with *in vivo* CNS permeation. Its popularity is largely due to low background expression of endogenous transporters and stable transfection of human multi-drug resistant (MDR1) or rodent mdr1a transporter (MDR1-MDCK or Mdr1a-MDCK, respectively), enabling an accurate *in vitro* assessment of P-gp (P-glycoprotein or MDR) efflux. P-gp is expressed at high levels on the luminal face (blood side) of the brain vascular endothelium and is by far the most studied active efflux transporter at the BBB. Naturally, CNS drugs have a significantly lower incidence of P-gp-mediated efflux than non-CNS drugs. In the assay, the apparent permeability is determined in both directions (apical-to-basolateral A–B, and basolateral-to-apical, B–A) with compounds expressing efflux (or MDR) ratios of greater than 3 generally considered to be P-gp substrates, and thus not brain penetrant. Studies of cohorts of CNS drugs have demonstrated that the MW (size) of a molecule is inversely correlated with the permeability, and thus designing the smallest, most potent inhibitors will enhance our chances of identifying a molecule possessing the desired drug-like properties while accessing the CNS. As will be seen, this proved to be significantly more challenging than initially anticipated.

10.6 ACYCLIC INHIBITORS: DESIGN OF PF-06439015 (5)

Given the fact that crizotinib (1) was originally designed as a c-MET inhibitor, our initial design strategy focused on utilizing this as a starting point. Subsequently, through structure-based drug design, we evaluated potential opportunities to optimize interactions with the ALK protein while maintaining good ADME (Absorption-Distribution-Metabolism-Elimination) properties.

Crizotinib is a type 1 ATP-competitive inhibitor of both c-MET and ALK. Co-crystal structures show similar binding conformations for the two proteins. However, a tyrosine $\pi-\pi$ stacking interaction is lacking in the case of ALK, which is hypothesized to account for the slight loss of potency (Figure 10.3). Examination of co-crystal structures of crizotinib bound to wild-type (wt) and 1196M ALK also shows similar binding conformations, with the gatekeeper residue contacting the small molecule at both the exocyclic amino and the (*R*)-methyl groups. Closer inspection of the dichlorofluorophenyl "head" group demonstrates that the fluorine at the C_3 position enhances potency by filling a small hydrophobic pocket between G1269 and N1254 as well as polarizing the neighboring C_4-H bond to interact with the R1253 carbonyl oxygen.

Modifications of the aforementioned "head" group and the pyrazolopiperidine "tail" moiety were pursued as two separate strategies towards the discovery of PF-06439015 (**5**).[81] Further examination of the co-crystal of crizotinib with ALK suggested that the 2-chloro-substituent of the head group is in close contact with the backbone carbonyl of G1269, thus creating a potentially unfavorable interaction. This hypothesized unfavorable interaction is supported by the observed ~30° rotation of this group away from the inhibitor. In the case of c-MET, this negative interaction is not an issue herein, there is an A1221 residue, which presents an inversion of the amide NH and C=O orientation. In both the wt and L1196M co-crystal structures, a degree of disorder is also observed within the G-loop, which implies that the affinity is not optimal specifically in areas close to the 6-Cl substituent of the head group. Based on these observations, the 2-des-chloro analogue of crizotinib (i.e., lacking chlorine at 2-position) should lead to a removal of the steric clash with G1269 while also allowing the 2-H (polarized by the adjacent fluorine) to interact in a positive manner with the carbonyl group of G1269. This initial change also leads to a decrease in both MW and log *D*, thus increasing the likelihood for improving pharmaceutical properties.

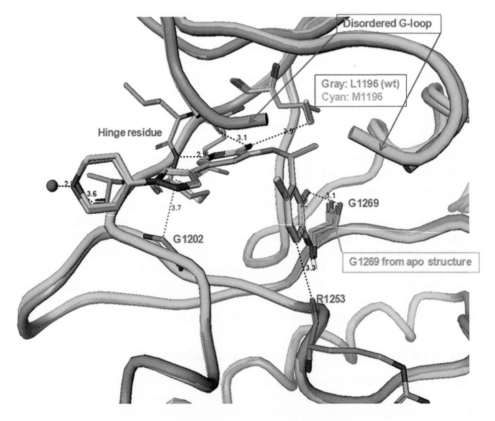

FIGURE 10.3   Binding of crizotinib (**1**) to ALK.

Within the headpiece of the molecule, attention was now switched to identifying higher LipE replacements for the 6-chloro substituent. With the des-chloro modification held constant, and a potent pyrazole tail moiety, a series of fluorophenyl groups were evaluated. Whereas enzymatic LipE values increased for larger groups such as CN (Compound (**6**)), OMe (Compound (**7**)), and a triazole (Compound (**8**)), the latter compound was shown to present the best balance of cell potency and metabolic stability. Hence, this led to the 2-triazolo-5-fluorophenyl functional group being established as the optimal head group in our quest for a next-generation ALK inhibitor (Figure 10.4). Co-crystal structures indicate that the triazole group forms a 60° torsion angle with the phenyl group providing contacts with the backbone of residues L1122-G1123, leading to a more ordered ALK G-loop.

For the tailpiece, the primary initial objective was to replace the basic pyrazolopiperidine with a neutral moiety. Various neutral cyclic moieties were designed and synthesized, from which a thiazole emerged as the optimal. Examination of a co-crystal structure indicated that the thiazole ring was rotated about 30° from the plane of the 2-aminopyridine, allowing the heteroaryl ring to form a CH donor-π interaction with G1202. It is important to note that angles over 45° create an unfavorable bump with the protein. Protein–ligand interactions suggested the addition of a pendant hydroxyl group to the ring system in order to form a hydrogen-bond interaction with the carbonyl group of D1203 (Compound (**9**)). Addition of a further hydroxyl group to the side chain enabled the NH group of the proximal D1203 to be satisfied as an acceptor while also donating either to the carbonyl side chain of D1203 or a water molecule stabilized by the carbonyls of G1201 and D1203. The robust nature of the modeling exercise undertaken is highlighted through the ability to predict that the (*R*)-enantiomer of the diol moiety will lead to a higher LipE, thus enabling the potent and pharmacokinetically favorable compound PF-06439015 (**5**) to be identified. PF-06439015 (**9**) displays potency across a broad panel of engineered ALK mutant cell lines and demonstrates both suitable preclinical pharmacokinetics and robust tumor growth inhibition (Figure 10.5).

10.7 THE CHALLENGE OF BALANCING PROPERTIES: ACYCLICS

PF-06439015 (**5**) achieved many of our initial laboratory objectives, specifically achieving significant improvements in both wild-type and gatekeeper (L1196M) ALK enzymatic assays while also demonstrating *in vivo* efficacy in a gatekeeper-mutant crizotinib-resistant cell line. However, the overall higher lipophilicity of the PF-06439015 (**5**) molecule was too high, requiring the introduction of polar groups to lower lipophilicity in order to achieve acceptable LipE and good metabolic stability. The compound also showed a high efflux ratio and very low rat CSF drug levels, the latter of which were likely due to the increased MW and added polar functionality. The degree of polarity reduces passive diffusion through the BBB, and it also increases the number of H-bond donors and with it P-gp efflux; both of these factors lower CNS penetrance.

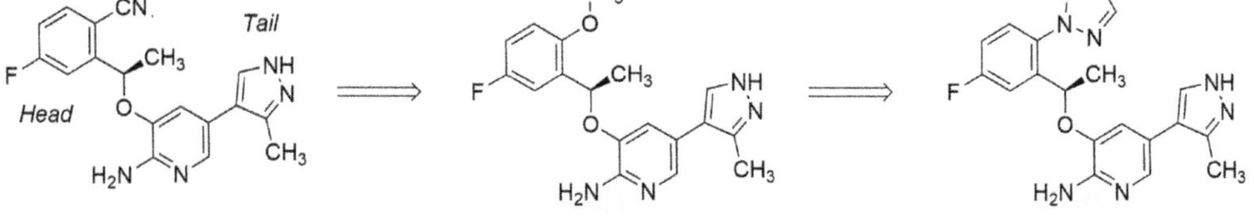

Compound (**6**)
L1196M K_i = 11 nM
ALK-L1196M cell IC_{50} = 1805 nM
HLM = 37 ml/min/kg
LipE (K_i) = 5.2

Compound (**7**)
L1196M K_i = 2 nM
ALK-L1196M cell IC_{50} = 69 nM
HLM = 65 ml/min/kg
LipE (K_i) = 5.2

Compound (**8**)
L1196M K_i = 2.6 nM
ALK-L1196M cell IC_{50} = 60 nM
HLM = 46 ml/min/kg
LipE (K_i) = 5.3

FIGURE 10.4 Headpiece modifications in acyclic series.

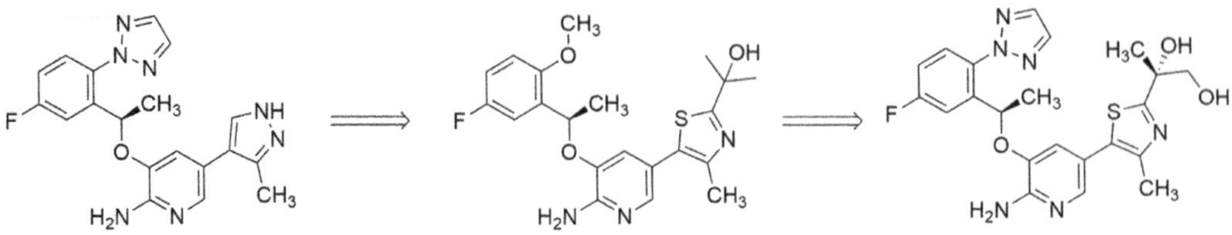

Compound (**8**)
L1196M K_i = 2.6 nM
ALK-L1196M cell IC_{50} = 60 nM
HLM = 46 ml/min/kg
LipE (K_i) = 5.3

Compound (**9**)
L1196M K_i = 0.4 nM
ALK-L1196M cell IC_{50} = 27 nM
HLM = 23 ml/min/kg
LipE (K_i) = 5.8

PF-06439015 (**5**)
L1196M K_i = 0.2 nM
ALK-L1196M cell IC_{50} = 6.6 nM
HLM = 15 ml/min/kg
MDR (BA/AB) = 10.9
LipE (K_i) = 6.8

FIGURE 10.5 Tail group modifications – PF-06439015 (**5**).

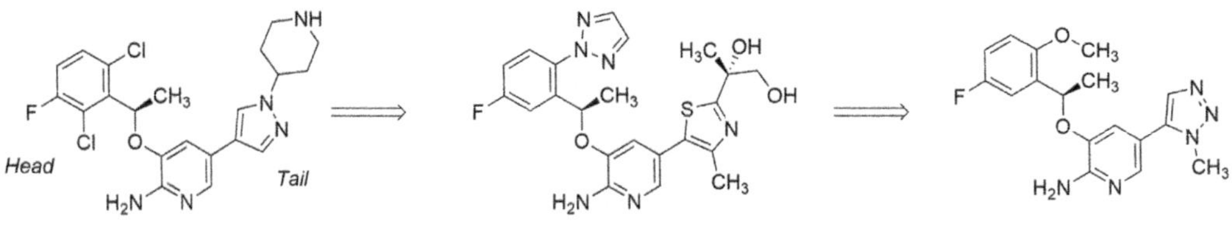

Crizotinib (**1**)
ALK-cell IC_{50} = 88 nM
ALK-L1196M cell IC_{50} = 800 nM
MDR (BA/AB) = 45

PF-06439015 (**5**)
ALK-cell IC_{50} = 0.76 nM
ALK-L1196M cell IC_{50} = 6.6 nM
MDR (BA/AB) = 10.9
LipE = 5.5

Compound (**10**)
ALK-cell IC_{50} = 95 nM
ALK-L1196M cell IC_{50} = 3200 nM
MDR (BA/AB) = 0.82
LipE = 3.1

FIGURE 10.6 Acyclic designs to modulate efflux ratio.

Given that we were aware of the factors that lead to the observed lack of permeability of PF-06439015 (**5**), and the fact that although this molecule is a potent inhibitor, it was not capable of penetrating the CNS, we sought to systematically carve away MW while removing polarity. This strategy proved to be successful in achieving the goal of reducing the MDR reflux ratio to below 2.5 (consistent with good predicted brain availability in humans) as exemplified by Compound (**10**), but unfortunately, this was at the expense of potency as both ALK and gatekeeper-mutant cellular IC_{50} values suffered (Figure 10.6).

This ongoing dichotomy of balancing potency with the requisite MDR ratio can be exemplified in a highly visual manner in the Spotfire plot presented in Figure 10.7. To orient the reader, this plot captures IC_{50} cell potency on the y-axis against metabolic stability (HLMs – human liver microsomes) on the x-axis. Furthermore, both the color (MDR) and the size (RRCK – Ralph Russ canine kidney) of the boxes are utilized to highlight the physicochemical properties related to permeability and CNS exposure. From the representation, we can see that we have identified several stable potent inhibitors, though have failed in the quest to couple these properties with the desired efflux ratio (a blue box in the highlighted square).

10.8 MOVING TO MACROCYCLIC INHIBITORS

10.8.1 Opportunities and Challenges

To this point, the advantages of having access to co-crystal structures within a program have been highlighted, and examination of several of these provided key structural insights to further enhance protein–ligand interactions and facilitated design of inhibitors with improved binding affinity and efficiency (Figure 10.8). During our discussion of the acyclic inhibitors, we emphasized that the

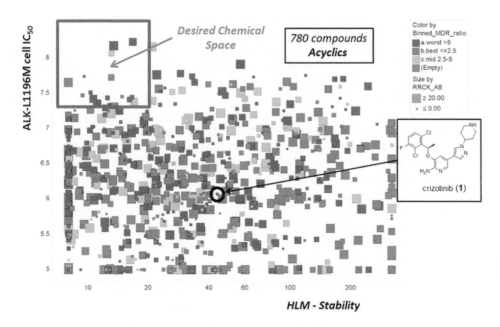

FIGURE 10.7   The challenge of balancing properties in acyclic space.

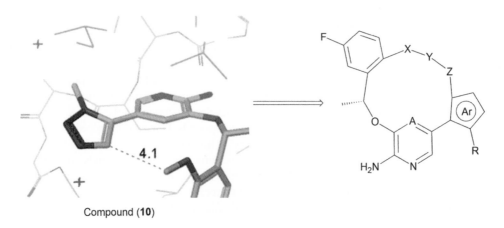

FIGURE 10.8   Acyclic co-crystal structures suggest macrocycle design.

G-loop was disordered when crizotinib was bound to the ALK protein, suggesting that the binding affinity was sub-optimal. A further trend emerged on examination of several co-crystals of acyclic inhibitors, which tended to adopt a U-shaped geometry in the bound conformation; this minimizes A-1,3-strain while placing the amide and the hydrogen on the chiral carbon in the same plane. Most notably, these structures highlighted the close proximity in the bound conformation of the "head" and "tail" portions of the molecule. This fueled the hypothesis that joining these to form a macrocycle would (a) reinforce the conformation without the introduction of additional binding strain and (b) provide the opportunity to pick up additional, enhanced specific protein–ligand interactions leading to an improvement in potency and thus LipE. We also felt that this structural change would improve ADME properties, increase the chances of moving into the desired MDR space through reducing the number of rotatable bonds and masking some of the molecule's polarity, and overall create a more compact, "globular" inhibitor with a lower surface area.

Despite the attractiveness of the macrocyclization approach, there are a number of risks associated with pursuing this. Firstly, when designing a rigid macrocycle, it is imperative for the molecule to adopt the right conformation to avoid locking the molecule in either an unproductive or sub-optimal binding conformation. There is also a degree of trepidation from a synthetic chemistry perspective in pursuing a series of macrocyclic inhibitors. In spite of their continuing recent renaissance in medicinal chemistry programs, such compounds still possess a stigma of being difficult to make in particular "on scale."[82–85] The challenges of ring-closure reactions are well established with 8- to 11-membered rings being the hardest to close (typically macrocycles are defined as molecules containing one large ring with at least 12 atoms) with rates and reaction efficiencies depending on enthalpic and entropic considerations as well as the conformation of the precursors.[86,87] Generally with regard to the reaction rates, there is a balance wherein for medium size rings, enthalpy outweighs entropy. This trend reverses for larger rings. Techniques used to facilitate cyclization such as slow addition and high dilution in particular present challenges when bulk quantities of a specific compound are required, though other strategies are emerging to alleviate this concern. This whole opportunities/challenges discussion presents an interesting contradiction as it is the inherent advantages of utilizing a macrocyclic template in a lead series that lead to the problems in accessing the compounds of interest. Often forming a ring is thought of solely improving potency through entropic improvements, though it is likely that conformational entropy is a minor factor, while enthalpy considerations are more substantial. In general, enthalpic impact more often outweighs any entropy contributions in the employment of both make-a-ring and break-a-ring strategies for improved binding efficiency.[88]

10.8.2 ETHER-LINKED MACROCYCLES

The most straightforward incursion into the macrocyclic space involved directly extending the methyl ether linker in Compound (**10**) to connect to the five-membered heterocycle in the tail. Despite creating potential complexities in the synthesis, this enabled us to provide a direct comparison between acyclic and cyclic inhibitors. In addition, we evaluated three discrete series (variations in the tail region) in the ether-linked macrocycles (variations in the tail region) to evaluate the optimal size among 12- to 14-membered ring compounds. During this process, we were very rapidly able to validate the "macrocyclization theory" while also able to establish that the smaller 12-membered rings consistently provided the most lipophilic efficient compounds, displaying both picomolar binding affinities and good cellular potency (Figure 10.9). Despite these improvements in both potency and LipE relative to the acyclic analogue, the macrocyclic ethers were generally too lipophilic (log D > 3) and still lacked the required efficiency for more facile overlap of potency, ADME, and CNS availability.

10.8.3 AMIDE-LINKED MACROCYCLES

A facile switch to lower the lipophilicity involved replacing the ether linker with an amide linker, and this also provided the added benefit of facilitating synthetic access to the compounds given that macrolactamization approaches are so well established. Staying within the optimal ring size established in our previous ether series studies, the first amide-based macrocycle (Compound (**13**)) we evaluated increased LipE by lowering log D (2.2), while retaining potency and exhibiting high metabolic stability. Evaluation of the co-crystal structure of the compound bound to the kinase domain of ALK also serves to highlight the desired enhanced interactions of the linker region of the inhibitor, specifically stabilizing and leading to ordering of the G-loop of the protein. To facilitate this, the *N*-methyl moiety of the amide was brought in closer contact with the carbonyl of Leu1122 (3.4 Å) and the nearby side chains of Leu1122 (4.1 Å), Gly1123 (4.7 Å), and Val1130 (4.7 Å), while the amide carbonyl interacts with both Lys1150 and His1124 through the formation of a water bridge to each residue (Figure 10.10).

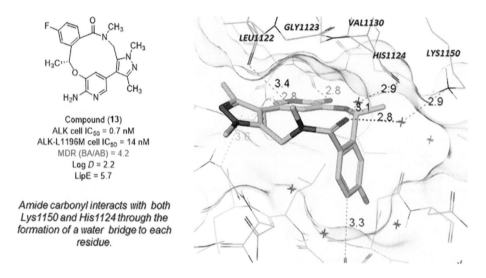

FIGURE 10.9 Ether-linked macrocycles.

Compound (**10**)
ALK cell IC$_{50}$ = 95 nM
ALK-L1196M cell IC$_{50}$ = 3200 nM
Log D = 2.4
LipE = 3.1

Compound (**11**)
ALK cell IC$_{50}$ = 0.9 nM
ALK-L1196M cell IC$_{50}$ = 21 nM
Log D = 3.8
LipE = 3.9

Compound (**12**)
ALK cell IC$_{50}$ = 1 nM
ALK-L1196M cell IC$_{50}$ = 20 nM
Log D = 3.3
LipE = 4.4

Compound (**13**)
ALK cell IC$_{50}$ = 0.7 nM
ALK-L1196M cell IC$_{50}$ = 14 nM
MDR (BA/AB) = 4.2
Log D = 2.2
LipE = 5.7

Amide carbonyl interacts with both Lys1150 and His1124 through the formation of a water bridge to each residue.

FIGURE 10.10 The initial amide-linked macrocycle (**13**).

The high LipE and reduced MW for Compound (**13**) clearly validated the macrocyclic amide as a potential series to navigate into the desired CNS-ADME space without the need to either sacrifice potency or compromise the safety profile of the molecule. In terms of the safety profile, achieving exquisite levels of kinase selectivity would be a key consideration. A brief overview of some of the SAR trends leading to the identification of lorlatinib (**18**) is provided in Figure 10.11, which enables the following observations to be made.

- Removal of the chiral methyl group from the ether linker (Compound (**14**)) led to an erosion of potency for the L1196M mutation, though this was accompanied by an unexpected boost in the desired selectivity. However, as we will see, this modification led to observed atropisomerism, which added an additional level of complexity to pursuing this series.
- The (*R*)-enantiomer (Compound (**15**)) is significantly more potent that the (*S*)-analogue (Compound (**16**)).
- Replacement of the aminopyridine core (Compound (**13**)) with an aminopyrazine (Compound (**15**)) was well tolerated, and the latter led to an improvement in potency. Somewhat surprisingly, this was accompanied by only a modest change in LipE as the aminopyrazines lead to an increase as opposed to an expected decrease in lipophilicity. This is believed to be due to the fact that the additional nitrogen atom is buried and thus poorly solvated, leading also to a decrease in the basicity of the aminopyrazine.

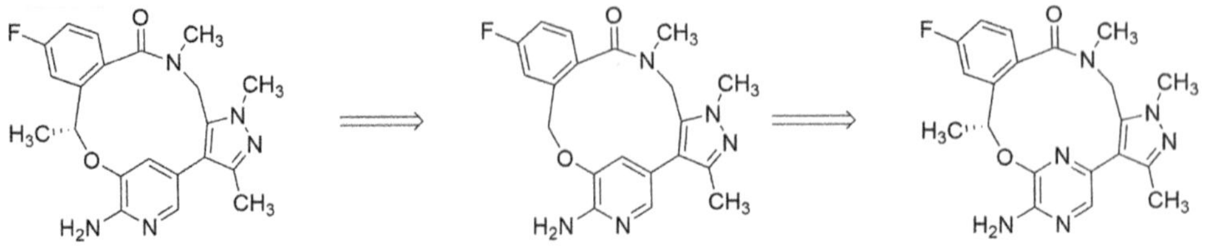

Compound (**13**)
ALK cell IC_{50} = 0.7 nM
ALK-L1196M cell IC_{50} = 14 nM
MDR (BA/AB) = 4.2
HLM = 8.6 mL/min/kg
Log *D* = 2.2, LipE = 5.7

Compound (**14**)
ALK cell IC_{50} = 6.4 nM
ALK-L1196M cell IC_{50} = 97 nM
MDR (BA/AB) = 5.5
HLM = 8.4 mL/min/kg
Log *D* = 2.0, LipE = 5.0

Compound (**15**)
ALK cell IC_{50} = 0.21 nM
ALK-L1196M cell IC_{50} = 1.4 nM
MDR (BA/AB) = 2.2
HLM = 8.0 mL/min/kg
Log *D* = 2.5, LipE = 6.4

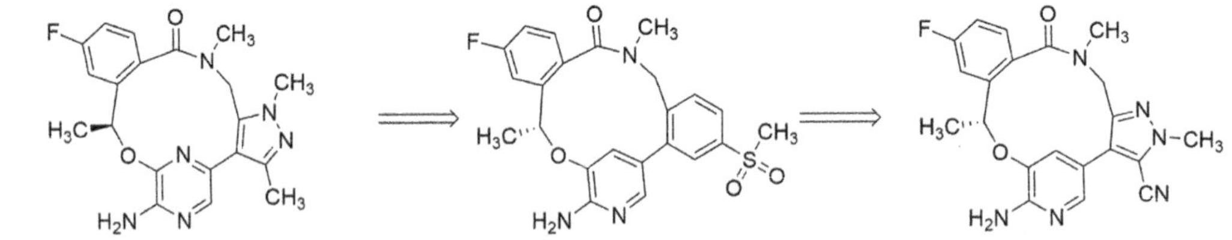

Compound (**16**)
ALK cell IC_{50} = 156 nM
ALK-L1196M cell IC_{50} = 1230 nM
MDR (BA/AB) = 3.1
HLM = 9.7 mL/min/kg
Log *D* = 2.5, LipE = 3.4

Compound (**17**)
ALK cell IC_{50} = 1.3 nM
ALK-L1196M cell IC_{50} =21 nM
MDR (BA/AB) = 9.2
HLM = 9.1 mL/min/kg
Log *D* = 1.8, LipE = 5.9

Lorlatinib (**18**)
ALK cell IC_{50} = 1.3 nM
ALK-L1196M cell IC_{50} = 21 nM
MDR (BA/AB) = 1.5
HLM ≤ 8 mL/min/kg
Log *D* = 2.3, LipE = 6.6

FIGURE 10.11 SAR highlights of the amide-linked macrocyclic series.

- One of the key learnings of the SAR analysis, which we were able to subsequently exploit in both our physicochemical property and selectivity optimization strategies, was the ability of the tailpiece of the molecule to accommodate a fair amount of steric bulk. In fact, five-membered, six-membered, and bicyclic (hetero)-aromatics were all well tolerated. For example, the polar sulfone (Compound (**17**)) led to a LipE increase as the polar sulfone was exposed to solvent without incurring any additional desolvation penalties. Although this compound maintained potency, based on the MDR ratio, it would be unlikely to penetrate the CNS. However, this compound presented an additional (in addition to PF-06439015) lead to the team as a non-CNS penetrant ALK inhibitor if challenges arose in achieving our selectivity objectives within the brain.

Lorlatinib (**18**) was selected as the lead molecule for the program based on it meeting all of our laboratory objectives specifically with regard to potency, log *D*, excellent PK properties, and a favorable MDR ratio.[89] The design of the tailpiece of the molecule with respect to the incorporation the nitrile at the C-3 position of the pyrazole will be elaborated upon in the discussion on achieving our selectivity goals.

10.9 DEVELOPMENT OF A SELECTIVITY STRATEGY FOR CNS PENETRATION

With a highly potent series of macrocyclic lactam-based inhibitors in hand, the goal for the program now changed gears to establish the high level of kinase selectivity required for a brain-penetrant compound. We have focused on minimization of the efflux ratio as a measure of the ability to access the CNS, though it is important to remain cognizant of the role that efflux (notably P-gp) plays in normal cells in acting as a barrier on the cell surface to prevent toxins from gaining access. Thus, if the molecule is to enter the cell, and more critically cross the BBB to enter the CNS, it is vital to consider selectivity to avoid off-target effects. Further, one must be aware in advance of families of anti-targets which are most likely to lead to these liabilities. In the current program, we were acutely concerned regarding inhibition of the Trk family (and in particular TrkB) of proteins, as inhibition thereof has been shown to have a negative effect on cognition. TrkB is expressed throughout the CNS and is involved in excitatory signaling, long-term potentiation, and feeding behavior. Previous clinical studies on CE-245677 (Pfizer), an oral pan-TrK/Tie2 kinase inhibitor under development at multiple doses for the treatment of certain cancers, were stopped in Phase 1 due to the development of significant CNS adverse events, including cognitive deficits, personality changes, and sleep disturbances. These effects were fully resolved upon cessation of dosing.

10.9.1 TARGETING H-5 IN THE ATP-BINDING POCKET

One of the more prevalent strategies for achieving kinase selectivity is through the targeting of steric interactions with specific amino-acid residues in the protein's ATP-binding pocket, which contains 27 amino acids.[90–92] The current project presents an additional conundrum, however. Generally achieving the desired "bumps" with proteins is achieved through building up of steric bulk, which typically involves additional MW. Herein with CNS access a requirement, it is important to try to achieve the desired steric interactions in as atom-economical a fashion as possible. The residue differences between the ALK and Trk family of proteins are highlighted in Figure 10.12. As can be seen at H_5, ALK (along with 26% of the kinome) features a relatively small leucine residue (Leu1198) wherein Trk features a larger tyrosine residue (Tyr635).[93] From a bigger picture perspective selectivity-wise, 60% of the kinome features a larger residue (either a tyrosine or phenylalanine at this position).

The rationale for targeting H_5 specifically can be explained by once again examining the co-crystal structures of our initial macrocyclic amide-based inhibitor (Compound (**13**)) with both ALK and TrkB (Figure 10.13). These clearly demonstrate the proximity of the H_5 residue to the methyl group at the C-3 *ortho*-position of the pyrazole. Further, given the unyielding framework that the macrocycle presents, this vector is positioned in a rigid fashion, thus necessitating protein movement if clashes through increased steric bulk occur. In the case of Compound (**13**), this is not necessary (the methyl group is within 3.1 Å of Tyr635), and this is reflected in the poor selectivity (1.7 fold) for ALK-1196M over TrkB. A similar low level of selectivity is observed when a cyclopropyl group is placed at this position. However, in the case of lorlatinib (**18**) where the C-3 vector is a nitrile group which extends rigidly towards H_5 in a linear fashion, the steric interaction corresponds to a commensurate boost in the desired selectivity (34-fold). Here, the nitrile group is within 1.0–1.8 Å of the terminal two atoms of the tyrosine, increasing the likelihood of an unfavorable steric interaction in TrkB, and it is believed that the nitrile is interacting unfavorably with the *meta*-carbon

Residue ID/Kinase	H1	H3	H5	H7	H8	H12	H13
ALK	I	L	L1198	A	G	K	S
TrkA	M	F	Y	R	H	N	R
TrkB	M	F	Y635	K	H	N	K
TrkC	M	F	Y	K	H	N	K

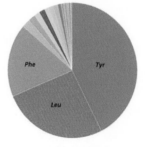

H5 by Residue

FIGURE 10.12 Targeting H-5 in the ATP-binding pocket.

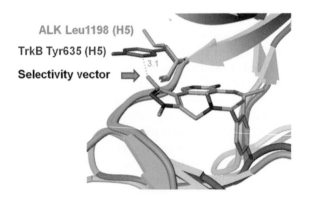

FIGURE 10.13   Co-crystal structure of Compound (**13**) ALK vs. Trk.

of the tyrosine residue while a further selectivity enhancement is being gained due to desolvation penalties or unfavorable electrostatic interactions between the electron-rich nitrile nitrogen and the tyrosine. A further increase in steric bulk in the case of the cyanoimidazopyridine (Compound (**19**)) leads to enhanced selectivity (93-fold), but this is at the cost of increased molecular weight, thus lowering the chance of the molecule accessing the CNS (Figure 10.14). Thus, lorlatinib (**18**) is believed to present the perfect balance in this scenario in terms of selectivity gain with a minimal (one atom) gain in MW.

10.9.2   The Benzylic Methyl Substituent: Encounters with Atropisomerism

We have already noted that absence of the chiral methyl group leads to an erosion in biochemical potency, and this is also seen to be the case for the direct des-methyl analogue of lorlatinib (**18**). However, interestingly the des-methyl compounds, which possess no stereogenic center, exist as atropisomers as initially revealed by analysis of the proton NMR spectra in which the methylenes present in the macrocyclic ring system display diastereotopic behavior (Figure 10.15). Atropisomers are stereoisomers which result from inhibited rotation about a single bond due to steric interaction or other factors rather than possession of a stereogenic center. Although it is well known that atropisomers are enantiomeric and as such energetically redundant, the transition barrier for conversion from one conformation to the other determines the racemization rates. (In the case of the macrocycles, a ring-flip or inversion occurs involving the passage of several atoms through the small space at the center of the ring, thus giving rise to an energy barrier.) There has been a growth in awareness of atropisomerism in drug development based on the facts that it is becoming more common given that

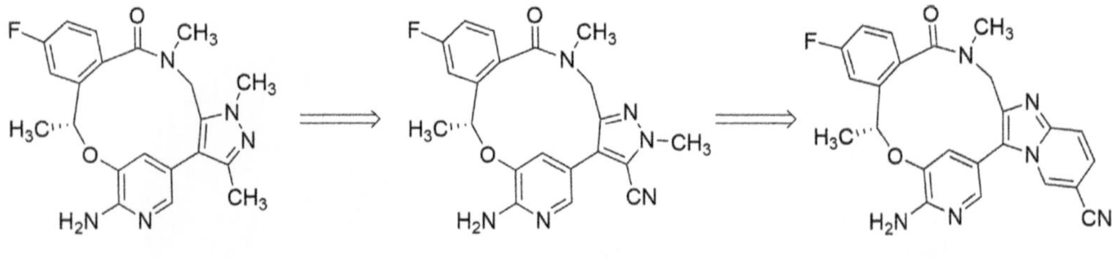

Compound (**13**)
ALK-L1196M K_i = 0.29 nM
TrkB K_i = 0.5 nM
Selectivity = 1.7

Lorlatinib (**18**)
ALK-L1196M K_i = 0.7 nM
TrkB K_i = 23 nM
Selectivity = 34

Compound (**19**)
ALK-L1196M K_i = 0.56 nM
TrkB K_i = 65 nM
Selectivity = 93

FIGURE 10.14   Effect of the C-3 pyrazole substituent on selectivity for ALK over Trk.

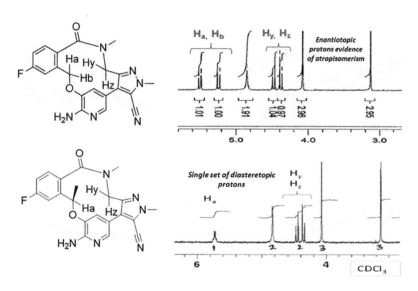

FIGURE 10.15   NMR of desmethyl compounds indicates atropisomerism.

emerging synthetic methodologies have made sterically encumbered bonds easier to form, as well as the important downstream implications due to possible differences in the compound's biological activity, pharmacokinetics, and toxicity.[94–99] A system has been proposed separating cases based on the energy barriers for interconversion with suggested strategies for compound development provided for each class.[100] It appears that the most difficult compounds to deal with are those that experience delayed axial interconversion (DE_{rot} between 20 and 30 kcal/mol) – that is, with $t_{1/2}$ values in the range of minutes, days, or months.

In the presence of the α-methyl stereocenter, a single set of diastereotopic protons is observed for the methylene and methine protons, reflecting a significant energy difference between the possible atrop-diastereomers. This energy difference is supported by MM calculations which provide a difference in ground-state energies of 8.5 kcal/mol. Examination of these atrop-diastereomers indicates that in the favored (and bioactive) conformation, the benzylic center minimizes A-1,3 strain; thus, a late-stage transition state in the reaction forming the macrocycle leads exclusively to this conformation (Figure 10.16).

The mixture of atropisomers was separated by chiral supercritical fluid chromatography (SFC),[101] and the discrete atropisomers, as well as lorlatinib and its enantiomer were submitted to our *in vitro* panel assessing ALK-L1196M potency, CNS exposure, ADME, and selectivity (Figure 10.17).[102] As we have previously observed, configuration at the chiral methyl is critical, and consequently, biochemical potency for Compound (**21**) lost several 100-fold relative to lorlatinib (**18**). The des-methyl compounds were slightly less lipophilic as expected and showed slightly diminished potencies, but the increased selectivity ratios for Compound (**22**) (831-fold) and Compound (**23**) (244-fold) relative to **18** (33-fold) were initially intriguing and worthy of further investigation. All the compounds evaluated showed low *in vitro* clearance in the HLM assay and were highly permeable with low efflux ratios consistent with P-gp substrate incompatibility.

However, there is a potential issue here, and closer inspection of the data presented herein in addition to further experimental work suggests that this will be a difficult problem to overcome. Although we have separated the atropisomers (Compounds (**22**) and (**23**)) by chiral SFC chromatography, it is highly probable that significant racemization has occurred prior to and/or during testing as the values obtained appear to be more consistent with those expected for racemic mixtures (Compound (**20**)). Modeling supports this hypothesis, demonstrating that the conformation of Compound (**23**) is completely inconsistent with optimal binding. As such, Compound (**23**) would be expected to result

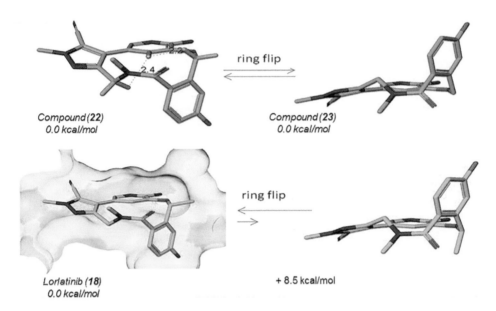

FIGURE 10.16 Conformations of atropisomers.

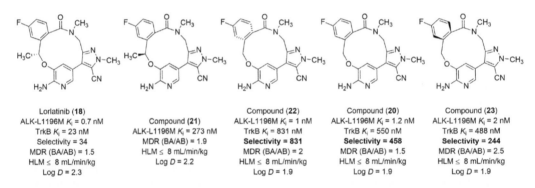

Lorlatinib (18)	Compound (21)	Compound (22)	Compound (20)	Compound (23)
ALK-L1196M K_i = 0.7 nM	ALK-L1196M K_i = 273 nM	ALK-L1196M K_i = 1 nM	ALK-L1196M K_i = 1.2 nM	ALK-L1196M K_i = 2 nM
TrkB K_i = 23 nM	MDR (BA/AB) = 1.9	TrkB K_i = 831 nM	TrkB K_i = 550 nM	TrkB K_i = 488 nM
Selectivity = 34	HLM ≤ 8 mL/min/kg	**Selectivity = 831**	**Selectivity = 458**	**Selectivity = 244**
MDR (BA/AB) = 1.5	Log D = 2.2	MDR (BA/AB) = 2	MDR (BA/AB) = 1.5	MDR (BA/AB) = 2.5
HLM ≤ 8 mL/min/kg		HLM ≤ 8 mL/min/kg	HLM ≤ 8 mL/min/kg	HLM ≤ 8 mL/min/kg
Log D = 2.3		Log D = 1.9	Log D = 1.9	Log D = 1.9

FIGURE 10.17 Selectivity data for lorlatinib analogs.

in significant losses in potency relative to Compound (**22**), which can adopt a similar binding active conformation to that observed for lorlatinib (**18**). This hypothesis can of course be experimentally confirmed, and it is also valuable to determine the $t_{1/2}$ for interconversion between the atropisomers. To achieve this, samples were evaluated at three temperatures (25°C, 45°C, and 65°C), with auto- mated sampling carried out at 5-min intervals to determine the Enantiomeric Excess (*ee*). Sampling was continued so that the reactions were monitored for at least two to three half-lives to ensure that the interconversion approximates a first-order reaction, and from this, the rate constant of racemiza- tion (k_{rac}) can be determined at each temperature. Analyzing this kinetic data with an Eyring plot, the $\Delta G^{\ddagger}$ (Gibbs energy of activation) at physiological temperature (37°C) was determined to be 24.56 kcal/mol, which corresponds to a $t_{1/2}$ of approximately 6 h.

Structure-based and modeling studies enable a hypothesis to be generated for the enhanced selectivity for ALK over TrkB as observed in the case of the des-methyl analogues. This indicates that there is more space between the (*R*)-methyl group and the Phe gatekeeper in the Trk protein, and removal of this substituent could potentially lead to high-energy water molecules populating this area, thus significantly decreasing the potency against TrkB.

The challenge to pursue this enhanced selectivity advantage presented by the des-methyl ana- logues in our macrocyclic amide series was to make subtle structural changes elsewhere in the

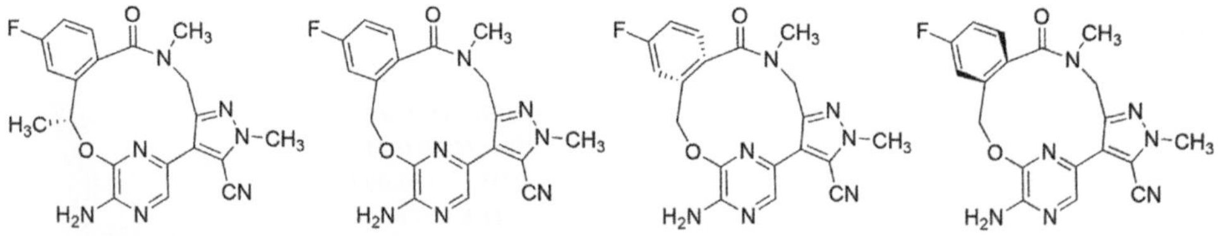

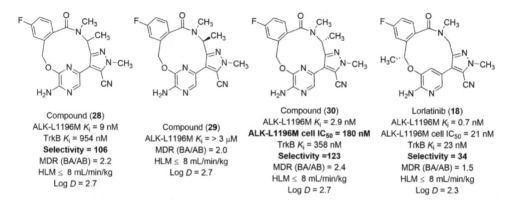

FIGURE 10.18   Des-methyl aminopyrazine analogs.

FIGURE 10.19   Addition of a new chiral center.

molecule with the hope of either decreasing the barrier of inversion and thus speeding up the "flipping" process or locking the inhibitor into a single conformation. In order to achieve the former, we evaluated the corresponding aminopyrazine compound believing that switching a CH for a N would allow a slightly larger gap in the center of the ring, thus enabling faster passage of the atoms through this to enable the flip to occur. The data for the aminopyrazine atropisomers (Compounds (**26**) and (**27**)) did indicate that $t_{1/2}$ was shorter, but again, it was not fast enough as the pair of compounds could be resolved on both the NMR and HPLC timescales (Figure 10.18).

A second strategy pursued was to add a second component of chirality elsewhere in the molecule. In our case, we bulked up the macrocyclic linker to make it closer to the pyrazole tail and thus prevent atropisomer interconversion. On the basis of modeling, it was predicted that a new "chiral methyl" next to the pyrazole tail region would play the role of a "chiral source" and lead the macrocyclization reaction to a single low-energy conformation incapable of interconversion. This would have the added benefit of simplifying both synthesis and purification. For this purpose, Compound (**28**) was made (separated into its enantiomers Compounds (**29**) and (**30**)) and indeed had improved ALK selectivity with no atropisomerism observed. Unfortunately, these compounds suffered from a significant loss of cell potency against the mutant protein (Figure 10.19).

Given this issue, we turned our attention back to lorlatinib (**18**) and carried out a more extensive biochemical kinase screening assay against a panel of 206 recombinant kinases (Invitrogen panel). The findings from these preliminary assays demonstrated that lorlatinib showed >100-fold selectivity for the ALK-L1196M gatekeeper mutation compared against 95% of the tested kinases with numerous key off-targets being spared including c-Met (>1,000×) and members of the IR family, including IGF1R (211×), IRR (600×), and INSR (1,000×) (Figure 10.20).

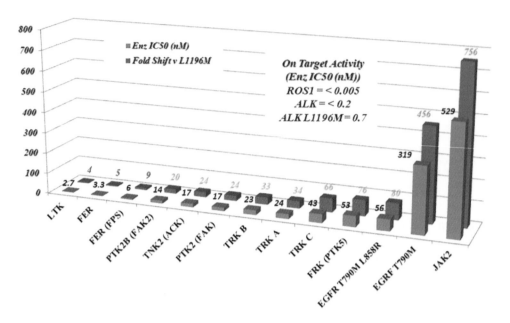

FIGURE 10.20   Kinase selectivity data for lorlatinib (**18**).

10.10   SYNTHESIS OF LORLATINIB (18)

Upon initial consideration for the synthesis of the amide-based macrocycles, several potential disconnections were evaluated with the key bond formation perceived to be the one leading to the formation of the ring (Figure 10.21). During progression of the program, these deliberations looked not only at expediting analogue synthesis but also on a long-term scalable approach to access a development candidate.

At an early stage, discovery of the formation of the ether bond as the ring-forming step (Disconnection "A") was obviated largely due to the fact that we had developed an approach to make this bond (with complete control of stereochemistry) for the synthesis of crizotinib (**1**). A second factor was that during the mainstream development of the SAR for the project, this fragment in the aminopyridine (**36**) (or aminopyrazine) tended to remain relatively constant. The initial chemistry to access lorlatinib utilized a macrolactamization reaction to close the ring mostly because this had the greatest precedence (Disconnection "B"); this approach is illustrated in Figure 10.22.[103]

Several factors are worthy of discussion regarding this chemistry. In our initial medicinal chemistry studies, formation of the boronate and Suzuki-coupling was achieved in a one-pot manner, though this was shown through screening to be counterproductive as different catalyst systems proved to be optimal for each step. Protection of the aminopyridine core during the sequence facilitated the reactions and prevented sequestration of palladium (Pd) to the API (Active Pharmaceutical Ingredient). The conditions for ester hydrolysis were carefully developed to avoid concomitant nitrile hydrolysis with trimethylsilanolate potassium (TMSOK), also enabling the penultimate macrolactamization precursor to be isolated as the potassium salt (Compound (**37**)). However, there were concerns regarding reagent quality as well as with long-term bulk supply of TMSOK. Screening of the coupling reagents determined that the reaction rate was in the order of HATU > COMU > T_3P > TPTU $\cong$ TSTU, and since a faster reaction would minimize the substrate concentration and, hence, lower the competitive intermolecular coupling impurities (which allowed a chromatography-free synthesis to be carried out), comprehensive safety evaluations were performed to enable the use of HATU in our scale-up facility.[104]

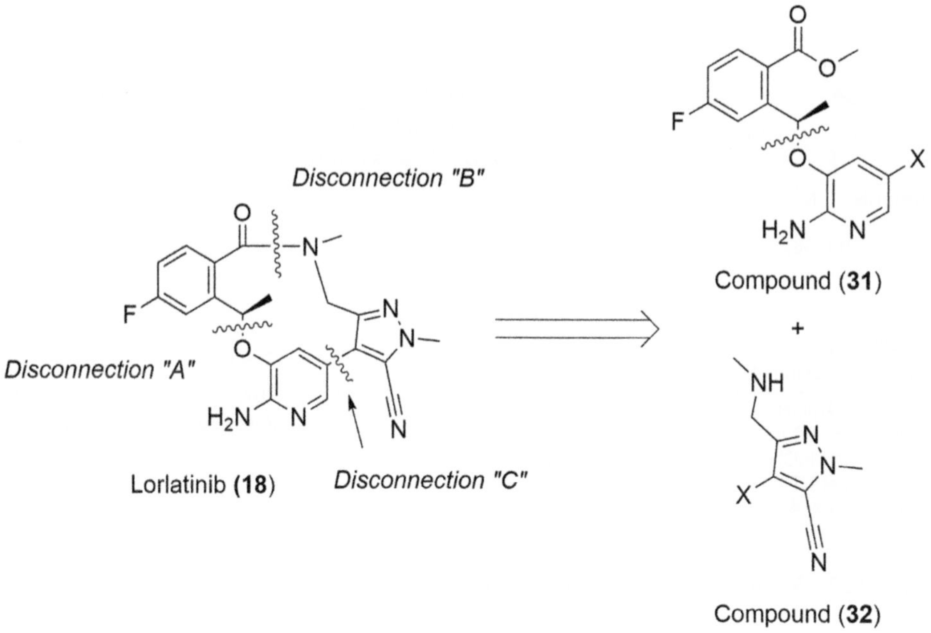

FIGURE 10.21   Synthetic disconnections for lorlatinib (**18**).

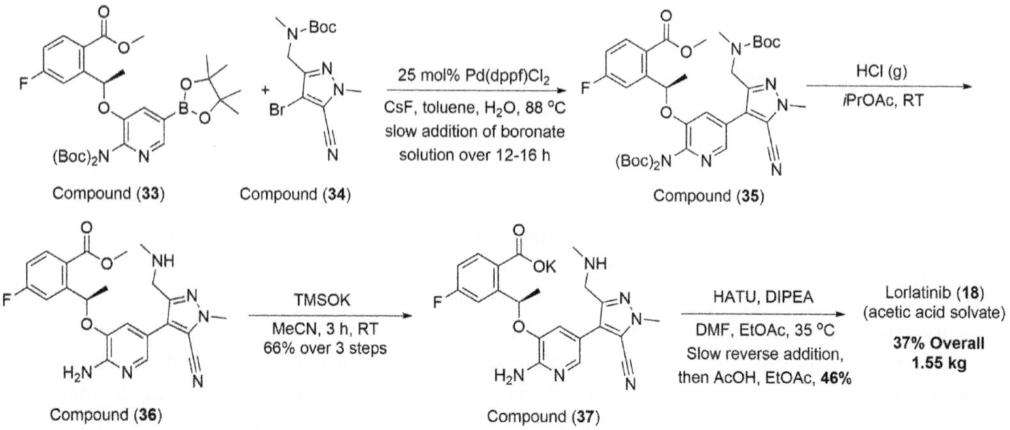

FIGURE 10.22   Synthesis of lorlatinib (**18**) through macrolactamization.

Although this synthetic approach represents a robust methodology to access our molecules of interest, several factors in the discovery space motivated us to investigate an alternative approach. Firstly, as is evidenced in the discussion on gaining selectivity through targeting the H-5 factor, the fragment of the molecule that underwent the most optimization was the (hetero)aromatic tail group. The monomers required to introduce these are in themselves highly engineered (see, for example, Compound (**34**)), and often these require a substantial amount of synthetic resource investment to access them. Therefore, minimization of the number of steps after which these have been incorporated into the target would be beneficial, especially if these steps lead to side reactions (*c.f.* nitrile hydrolysis). Secondly, there was a drive to access our analogues more quickly, specifically to develop a synthesis that would be amenable to a PMC (parallel medicinal chemistry) approach. Finally, we considered that an alternative synthesis might be better able to satisfy long-term API requirements,

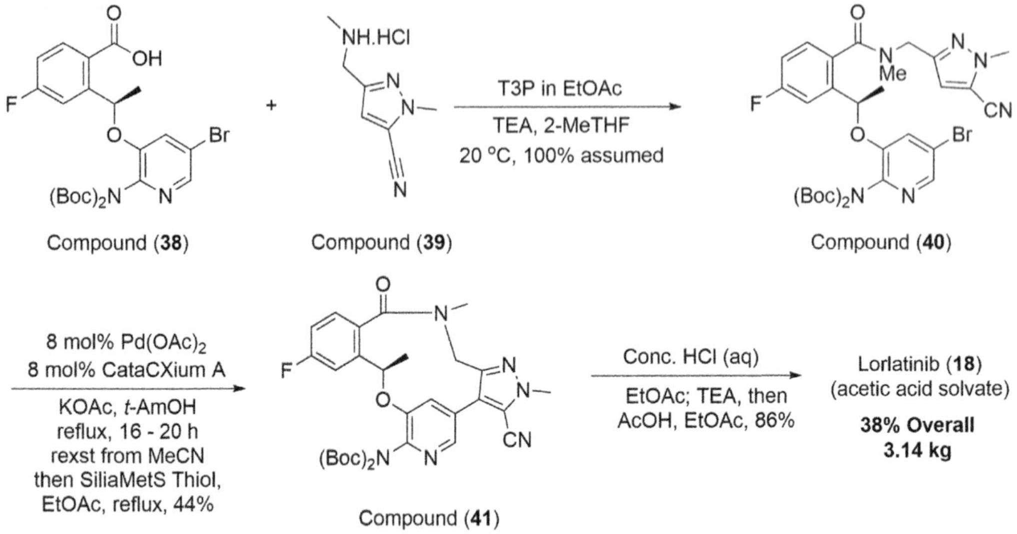

FIGURE 10.23 Synthesis of lorlatinib (**18**) through C–H functionalization.

particularly if the key ring-forming reaction can be run in a more concentrated manner. With these considerations in mind, it was recognized that it would be more sustainable to reverse the bond formation to shorten the number of steps and achieve both faster access to analogues and a potentially higher throughput. This alternative route employs a direct C–H functionalization for the C–C coupling and is one step shorter in the GMP endgame synthesis. In addition, it does not require the introduction of either the boronate ester (in Compound (**33**)) (thus avoiding a potential genotoxic impurity) or the bromo groups (in Compound (**34**)) on the respective coupling partners.

The second-generation synthesis was initially developed for the aminopyrazine series, but it could be directly applied to the aminopyridines provided the primary 2-amino group was protected. As noted previously, this has advantages regarding facilitating the removal of residual Pd after the C–C bond-forming step. Several rounds of high-throughput experimentation revealed promising conditions (Pd(OAc)$_2$, CataCXium A, and KOAc in t-amyl alcohol at reflux) for C–C cross-coupling, which results in lorlatinib (**18**) after deprotection and conversion to the final form. Issues with removal of an impurity arising from des-cyanation in the C–C bond formation were solved through crystallization of the penultimate intermediate (Compound (**41**)). This route was utilized in a development campaign to give the final API as an acetic acid solvate (3.14 kg, 38% overall, concentration for ring formation 35 L/kg) in 99.90% ultra pure liquid chromatography (UPLC) purity with only a single impurity (resulting from des-cyanation) detectable at 0.10%. Further development has established this direct arylation as the key ring-closing step in the current commercial synthesis of lorlatinib (**18**) (Figure 10.23).

10.11 ACCESSING THE CNS: PET STUDIES

One of the cornerstones of our project objectives was to develop an ALK inhibitor which can penetrate the CNS, and through our design cycles, we have utilized permeability measures to assay whether we are likely achieving this goal. Prior to entering the clinic, however, it is desirable to develop a tool in order to unequivocally demonstrate CNS penetrance. PET-imaging studies not only offered an opportunity to evaluate exposure at the target site of action (the brain) but also are able to confirm target engagement through binding to ALK is occurring (two of the so-called "Three Pillars of Survival").[105] In addition, PET imaging is an attractive technique as it involves no structural changes of the molecule of interest, though the caveat is that the radiolabel needs to be introduced late in the synthesis owing to the short half-lives of the radionuclides typically used.

We developed syntheses of both ^{11}C ($t_{1/2}$ = 20.4 min) and ^{18}F ($t_{1/2}$ = 109.7 min) isotopologs of lorlatinib (**18**) (Figure 10.24).[106] The former involved methylation of the amide on the *bis*-Boc-protected amino-pyridine (Compound (**42**)) with ^{11}CH$_3$I. Crucially, the methyl moiety on the amide was shown not to be a downstream metabolic liability, thus preserving the integrity of the radiolabel in the PET-imaging studies. HPLC purification followed by deprotection and reformulation resulted in the synthesis of [11C]-lorlatinib in 3% uncorrected radiochemical yield (RCY) (17% decay-corrected) at the end of synthesis (50 min total) and a high specific activity of 3Ci µ/mol with >95% radiochemical purity. Given the longer half-life of ^{18}F, access to this radiolabeled version of lorlatinib is attractive. This is particularly true for the prospect of advanced PET evaluations in rodent tumor models and normal non-human primates (NHPs) with the ultimate goal of clinical translation. However, we have seen that the fluorine moiety in lorlatinib is introduced at the outset of the synthesis, as it is present in one of the starting materials. Attempts to introduce this functionality through a late-stage fluorodenitration or S$_N$Ar process lead to only trace levels of the desired product with extensive decomposition observed because of the harsh reaction conditions employed. This moiety was successfully introduced through a spirocyclic iodonium ylide (SCIDY)-based radiofluorination on the *bis*-Boc-protected precursor (Compound (**43**)).[107,108] Purification followed by deprotection obtained [^{18}F]-lorlatinib (**18**) in 14% uncorrected RCY at the end of synthesis (40 min total) with >97% radiochemical purity.

Imaging of ^{11}C-radiolabeled lorlatinib (**18**) in nonhuman primates provided clear *in vivo* evidence for the desired BBB permeability with high initial uptake of the molecule to the brain observed.[109] Whole-body dynamic PET imaging revealed radioactivity primarily in the liver and kidneys with a progressive shift to the urinary bladder. At ~10 min post injection, peak-measured brain concentrations were locally high with the cerebellum exceeding a standardized uptake value (SUV) of 2 and the maximum measured percent injected dose in the brain shown to be 1.4%. Regional uptake exhibits modest heterogeneity but is generally concordant with expected ALK distribution, and these results support pillar 1 in that lorlatinib (**18**) crosses the BBB at sufficient concentrations to be potentially effective against BM.

With regard to target engagement (Pillar 2), tumor uptake of [^{11}C]-lorlatinib (**18**) was evaluated in mice bearing subcutaneous human H$_3$122 (EML4-ALK positive) xenografts by PET-CT in conjunction with blocking studies. These studies demonstrated that tumor uptake reached its peak 30–60 min post injection (>2% ID/g), and that co-injection with unlabeled lorlatinib led to a

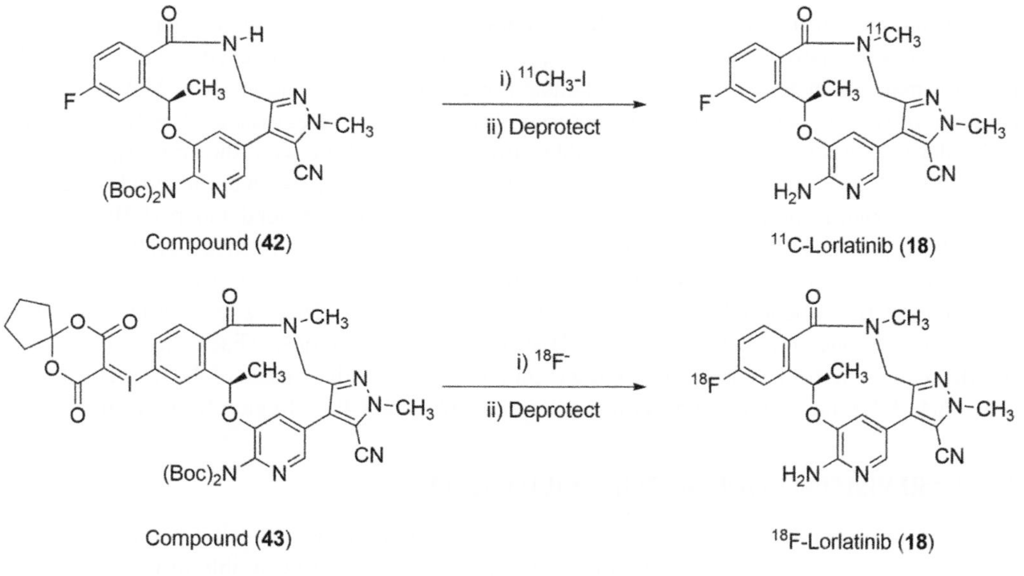

FIGURE 10.24   Synthesis of ^{11}C and ^{18}F isotopologs of lorlatinib (**18**).

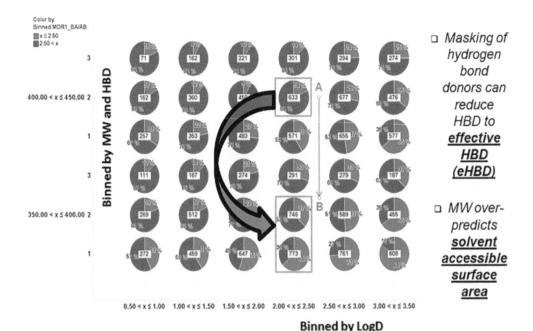

FIGURE 10.25 MDR analysis of Pfizer compound collection.

significant decrease (<0.4% ID/g) in tumor uptake during the entire 90 min of imaging. Given the short physical half-life, doses from C-11 compounds are generally low, and dosimetry estimates from NHPs are consistent with those observed for other ^{11}C-labeled compounds. Clinical translation certainly appears to be warranted.

We have previously reported that based on an analysis of the physical properties and MDR values of 115,000 compounds from the Pfizer file, lorlatinib (**18**) would only be expected to have an 18% chance of achieving a favorable MDR ratio (Region A on Figure 10.25). Critical to this analysis are the number of H-bond donors and the ***MW of the molecule. However, there are two factors inherently built into lorlatinib which we believe significantly increases its chances in respect to this analysis. Firstly, as we hypothesized at the outset, we believe the cyclic framework of the inhibitor makes it more compact (less rotatable bonds, more rigid), and as such, MW is an overpredictor of size. To support this claim, across a series of macrocycles evaluated, within a defined molecular weight range, the solvent accessible surface areas (SASAs) were shown to be on average approximately 10% smaller than those for non-macrocycles. Secondly, with aminopyridine selected as the hinge-binding motif, lorlatinib possesses two potential hydrogen-bond donors (HBDs). One of these HBDs provides the key interaction with the carbonyl of the hinge-residue Glu1197 within the protein–ligand complex, and the second NH_2 hydrogen has the opportunity to participate in an intramolecular hydrogen bond (IMHB) with the adjacent ether oxygen, leading to an improved chance of low efflux through lowering the effective HBD count from two to one.[110] Factoring in both a 10% size decrease with this lower effective HBD count, this increases the chance of the molecule having a favorable MDR ratio for brain penetration (between 37% and 64% – Region B) in the analysis.

10.12   REVISITING LABORATORY OBJECTIVES

Returning to our initial laboratory objectives with crizotinib (**1**) as a starting point, we can see that through a series of iterative structural modifications, we have been able to meet all of our goals for a next-generation ALK inhibitor and progress a compound with both excellent ADME

TABLE 10.4
Revisiting Laboratory Objectives

	Crizotinib (1)	Lorlatinib (18)
Dosing	500 mg/day BID	100 mg/day QD
EML4-ALK-WT cell IC_{50} (nM)	80	1.3
EML4-ALK-L1196M cell IC_{50} (nM)	841	21
ROS cell IC_{50} (nM)	45	0.15
TrkB cell IC_{50} (nM)	17	23
Efflux (MDR BA/AB)	45	1.5
LipE (cell IC_{50})	5.1	6.6
MW	450	406
Log D	2.0	2.3
Free brain: free plasma (rodent)	<0.03	0.23–0.33

and PK properties (Table 10.4). Our initial changes focused on the acyclic series and featured both removal of the piperidine and optimization of the head group. Although an attractive ALK inhibitor was identified in terms of potency and efficiency, the compound was not CNS penetrant. Macrocyclization proved to be the vital structural change to access desirable CNS-ADME space with enhanced selectivity being obtained through introduction of the nitrile at C-3 of the pyrazole.

At the conclusion of our discussion on the acyclic inhibitors, we utilized a Spotfire plot to visualize the complete cohort of compounds to demonstrate the difficulty in overlapping potency with the requisite ADME properties (Figure 10.26). Revisiting this plot now with the addition of the macrocyclic inhibitors (solids) highlights the dramatic impact that this structural modification had in terms of potency, stability, and favorable MDR ratio (colored green) for CNS penetration. Many macrocycles can access optimal log D (2–3) space as well as leading to improvements in potency specifically through being able to adopt a more rigid binding conformation as well as picking up

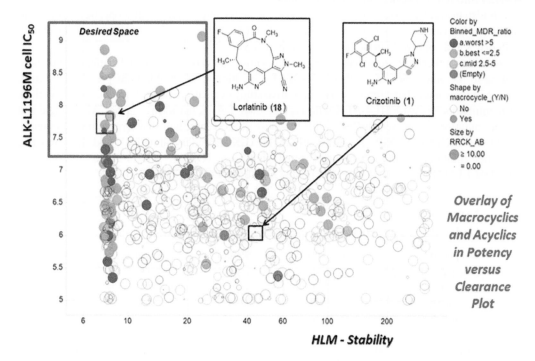

FIGURE 10.26 Spotfire overlay of macrocyclic and acyclic inhibitors.

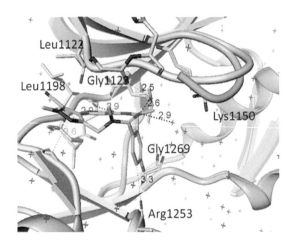

FIGURE 10.27   Co-crystal structure of lorlatinib (**18**) bound to L1196M ALK.

additional binding interactions with the protein through the linker region of the molecule. The macrocyclic series' economical use of each atom in exploiting complementarity to the binding pocket of the kinase with minimal ligand strain is further validated by analyses focusing on both lipophilic efficiencies and heavy atom count (HACNT).

The co-crystal structure of lorlatinib (**18**) bound to the kinase domain of the ALK (L1196M) mutant again highlights the key interactions to the protein from the linker region of the inhibitor, which lead to an ordering of the G-loop and, hence, enhanced binding (Figure 10.27). This can be clearly seen as the blue structure indicates the protein on binding with lorlatinib (**18**) whereas the pink structure is the same protein with crizotinib (**1**) bound. The amide methyl group of the macrocycle accesses a lipophilic pocket as previously seen with the carbonyl interacting with Lys1150 and His1124 through two structural water molecules.

It is also instructive to revisit Table 10.2 and evaluate the physicochemical properties of lorlatinib (**18**) in comparison to not only those of other ALK inhibitors but also the average properties of both CNS drugs and approved kinase inhibitors (Table 10.5).

Numerous other mutations have been noted within the crizotinib-treated patient population, leading to the emergence of a cohort of second-generation ALK inhibitors as previously shown. Shaw and co-workers have reported a comparative study evaluating the performance of the various ALK inhibitors against EML4-ALK wild-type or EML4-ALK harboring various ALK single and double mutations expressed in Ba/F$_3$ cells.[111] These studies indicated that lorlatinib (**18**) potently inhibited ALK phosphorylation against all single ALK secondary mutations, and in contrast to the other inhibitors, lorlatinib retained either significant (D1203N + E1210K) or intermediate (D1203N + F1174C) potency against the double mutations studied (Table 10.6).

TABLE 10.5
Updated Key Physicochemical Properties of ALK Inhibitors

Compound	HBD	TPSA (Å^2)	cLogP	MW
CNS drugs ($n = 119$)	1	45	2.8	305
Crizotinib (**1**)	3	77	4.2	450
Ceritinib (**2**)	3	105	6.5	558
Alectinib (**3**)	1	72	5.2	483
Brigatinib (**4**)	2	82	5.9	529
Kinase inhibitors ($n = 34$)	2	91	4.2	483
Lorlatinib (**18**)	2	110	2.7	406

TABLE 10.6

Reported Cellular ALK Phosphorylation Mean IC$_{50}$ (nmol/L) of ALK Inhibitors

Mutation Status	Crizotinib (1)	Ceritinib (2)	Alectinib (3)	Brigatinib (4)	Lorlatinib (18)
Parental Ba/F$_3$	763.9	885.7	890.1	2774.0	11293.8
EML4-ALK	38.6	4.9	11.4	10.7	2.3
C1156Y	61.9	5.3	11.6	4.5	4.6
I1171N	130.1	8.2	397.7	26.1	49.0
I1171S	94.1	3.8	177.0	17.8	30.4
I1171T	51.4	1.7	33.6	18.0	8.0
F1174C	115.0	38.0	27.0	18.0	8.0
L1196M	339.0	9.3	117.6	26.5	34.0
L1198F	0.4	196.2	42.3	13.9	14.8
G1202R	381.6	124.4	706.6	129.5	49.9
G1202del	58.4	50.1	58.8	95.8	5.2
D1203N	116.3	35.3	27.9	34.6	11.1
E1210K	42.8	5.8	31.6	24.0	1.7
G1269A	117.0	0.4	25.0	ND	10.0
D1203N + F1174C	338.8	237.8	75.1	123.4	69.8
D1203N + E1210K	153.0	97.8	82.8	136.0	26.6

10.13 CURRENT STATUS OF LORLATINIB (18)

Shaw and co-workers have reported the results from an international, multicenter, open-label, single-arm first-in-man Phase 1 trial evaluating the safety, efficacy, and pharmacokinetic properties of lorlatinib (**18**).[112] Between January 22, 2014 and July 10, 2015, 54 patients received at least one dose of lorlatinib, including 41 (77%) with ALK-positive and 12 (23%) with ROS1-positive NSCLC. Furthermore, among this cohort of patients, 28 (52%) had received two or more TKIs, while 39 (72%) had CNS metastases. The drug was generally well tolerated with the most common treatment-related adverse events among the 54 patients being hypercholesterolemia (72%), hypertriglyceridemia (39%), peripheral neuropathy (39%), and peripheral edema (39%). Though no maximum tolerated dose was identified (the recommended Phase 2 dose was selected as 100 mg once daily), one dose-limiting toxicity occurred at 200 mg, which was a grade 2 neurocognitive adverse event comprised of slowed speech, slowed mentation, and word-finding difficulty. From the study, it was demonstrated that lorlatinib showed both systemic and intracranial activity in patients with advanced ALK-positive and/or ROS1-positive NSCLC, most of whom had CNS metastases and had previously had two or more TKI treatments fail. (Statistics for ALK-1 positive NSLC was 46% ORR, 95% CI 31–63; ROS-1 positive NSLC statistics included 50% ORR, 95% CI 21–79.) The clinicians concluded that "lorlatinib might be an effective therapeutic strategy for patients with ALK-positive NSCLC who have become resistant to currently available TKIs, including second-generation ALK TKIs."[113] A Phase 3 randomized controlled trial comparing lorlatinib to crizotinib is presently ongoing (ClinicalTrials.gov, NCT03052608).[114]

On November 2, 2018, the FDA granted accelerated approval to lorlatinib (**18**) (Lorbrena®) for patients with ALK-positive metastatic NSCLC[115] whose disease has progressed on crizotinib and at least one other ALK inhibitor for metastatic disease or whose disease has progressed on alectinib or ceritinib as the first ALK inhibitor therapy for metastatic disease.[116]

ACKNOWLEDGMENTS

The author would like to thank Ted Johnson, Jennifer Lafontaine, Leslie Robinson as well as all the members of the ALK-project team involved with the discovery and development of lorlatinib.

REFERENCES

1. http://gco.iarc.fr/today/data/factsheets/cancers/39-All-cancers-fact-sheet.pdf (last accessed April 21st, 2019).
2. Ettinger, D. S., Akerley, W., Borghaei, H., Chang, A. C., Cheney, R. T., Chirieac, L. R., et al., Non-small cell lung cancer, *J. Natl. Compr. Canc. Netw.*, 8, 740–801, 2010.
3. Stella, G. M., Luisetti, M., Pozzi, E., and Comoglio, P. M., Oncogenes in non-small-cell lung cancer: Emerging connections and novel therapeutic dynamics, *Lancet Respir. Med.*, 1, 251–261, 2013.
4. Forde, P. M., Chaft, J. E., Smith, K. N., Anagnostou, V., Cottrell, T. R., et al., Neoadjuvant PD-1 blockade in resectable lung cancer, *N. Engl. J. Med.*, 378, 1976–1986, 2018.
5. Gettinger, S. N., Horn, L., Gandhi, L., Spigel, D. R., Antonia, S. J., et al., Overall survival and long-term safety of nivolumab (anti–programmed death 1 antibody, BMS-936558, ONO-4538) in patients with previously treated advanced non–small-cell lung cancer, *J. Clin. Oncol.*, 33, 2004–2012, 2015.
6. Garon, E. B., Rizvi, N. A., Hui, R., Leighl, N., Balmanoukian, A. S., et al., Pembrolizumab for the treatment of non–small-cell lung cancer, *N. Engl. J. Med.*, 372, 2018–2028, 2015.
7. Malhotra, J., Jabbour, S. K., and Aisner, J., Current state of immunotherapy for non-small cell lung cancer, *Transl. Lung Cancer Res.*, 6, 196–211, 2017.
8. Moya-Homo, I., Viteri, S., Karachaliou, N., and Rosell, R., Combination of immunotherapy with targeted therapies in advanced non-small cell lung cancer (NSCLC), *Ther. Adv. Med. Oncol.*, 10, 1–12, 2018.
9. Faehling, M., Schwenk, B., Kramberg, S., Eckert, R., Volckmar, et al., Oncogenic driver mutations, treatment, and EGFR-TKI resistance in a Caucasian population with non-small cell lung cancer: Survival in clinical practice, *Oncotarget*, 8, 77897–77914, 2017.
10. Nahar, R., Zhai, W., Zhang, T., Takano, A., Khng, A. J., et al., Elucidating the genomic architecture of Asian EGFR-mutant lung adenocarcinoma through multi-region exome sequencing, *Nat. Comm.*, 9, 216, 2018.
11. Olavarria, E., Craddock, C., Dazzi, F., Marin, D., Marktel, S., et al., Imatinib mesylate (STI571) in the treatment of relapse of chronic myeloid leukemia after allogeneic stem cell transplantation, *Blood*, 99, 3861–3862, 2002.
12. Soda, M., Choi, Y. L., Enomoto, M., Takada, S., Yamashita, Y., et al., Identification of the transforming EML4–ALK fusion gene in non-small-cell lung cancer, *Nature*, 448, 561–566, 2007.
13. Mano, H., Non-solid oncogenes in solid tumors: EML4-ALK fusion genes in lung cancer, *Cancer Sci.*, 99, 2349–2355, 2008; c) Shaw, A. T. and Solomon, B., Targeting anaplastic lymphoma kinase in lung cancer, *Clin. Cancer Res.*, 17, 2081–2086, 2011.
14. Morris, S. W., Naeve, C., Mathew, P., James, P. L., Kirstein, M. N., et al., ALK, the chromosome 2 gene locus altered by the t(2;5) in non-Hodgkin's lymphoma, encodes a novel neural receptor tyrosine kinase that is highly related to leukocyte tyrosine kinase (LTK), *Oncogene*, 14, 2175–2188, 1997.
15. www.ncbi.nlm.nih.gov/gene/238 (last accessed September 9th, 2018).
16. Morris, S. W., Kirstein, M. N., Valentine, M. B., Dittmer, K. G., Shapiro, D. N., et al., Fusion of a kinase gene, ALK, to a nucleolar protein gene, NPM, in non-Hodgkin's lymphoma, *Science*, 263, 1281–1284, 1994.
17. Bilsland, J. G., Wheeldon, A., Mead, A., Znamenskiy, P., Almond, S., et al., Behavioral and neurochemical alterations in mice deficient in anaplastic lymphoma kinase suggest therapeutic potential for psychiatric indications, *Neuropsychopharmacology* 33, 685–700, 2008.
18. Griffin, C. A., Hawkins, A. L., Dvorak, C., Henkle, C., Ellingham, T., et al., Recurrent involvement of 2p23 in inflammatory myofibroblastic tumors. *Cancer Res.*, 59, 2776–2780, 1999.
19. Yu, H., Huang, J. X., Wang, C. F., and Shi, D. R., ALK-positive large B-cell lymphoma: Report of a case, *Zhonghua Bing Li Xue Za Zhi*, 40, 561–562, 2011.
20. Bresler, S. C., Weiser, D. A., Huwe, P. J., Park, J. H., Krytska, K., et al., ALK mutations confer differential oncogenic activation and sensitivity to ALK inhibition therapy in neuroblastoma. *Cancer Cell*, 26, 682–694, 2014.
21. Combaret, V., Iacono, I., Bellini, A., Bréjon, S., Bernard, V., et al., Detection of tumor ALK status in neuroblastoma patients using peripheral blood, *Cancer Med.*, 4, 540–550, 2014.
22. Montavon, G., Jauquier, N., Coulon, A., Peuchmaur, M., Flahaut, M., et al., Wild-type ALK and activating ALK-R1275Q and ALK-F1174L mutations upregulate Myc and initiate tumor formation in murine neural crest progenitor cells, *Oncotarget*, 5, 4452–4466, 2014.
23. Martelli, M. P., Sozzi, G., Hernandez, L., Pettirossi, V., Navarro, A., et al., *EML4-ALK* rearrangement in non-small cell lung cancer and non-tumor lung tissues, *Amer. J. Path.*, 171, 661–670, 2009.

24. Zhang, N.-N., Liu, Y.-T., Ma, L., Wang, L., Hao, X.-Z., et al., The molecular detection and clinical significance of ALK rearrangement in selected advanced non-small cell lung cancer: ALK expression provides insights into ALK targeted therapy, *PLoS One*, 9, e84501, 2014.

25. https://diagnostics.roche.com/global/en/news-listing/2017/roche-receives-fda-approval-for-ventana-alk--d5f3--cdx-assay-to-.html.

26. Fujishima, S., Imai, K., Nakamura, R., Nanjo, H., et al., Novel method for rapid fluorescence in-situ hybridization of ALK rearrangement using non-contact alternating current electric field mixing, *Sci. Rep.*, 7, 155116, 2017.

27. Matsushime, H., Wang, L. H., and Shibuya, M., Human c-ros-1 gene homologous to the v-ros sequence of UR2 sarcoma virus encodes for a transmembrane receptor-like molecule, *Mol. Cell Biol.*, 6, 3000–3004, 1986.

28. Birchmeier, C., Sharma, S., and Wigler, M., Expression and rearrangement of the ROS1 gene in human glioblastoma cells, *Proc. Natl. Acad. Sci. U. S. A.*, 84, 9270–9274, 1987.

29. Davies, K. D., Le, A. T., Theodoro, M. F., Skokan, M. C., Aisner, D. L., et al., Identifying and targeting ROS1 gene fusions in non-small cell lung cancer, *Clin. Cancer Res.*, 18, 4570–4579, 2012.

30. Bergethon, K., Shaw, A. T., Ou, S.-H. I., Katayama, R., Lovly, C. M., et al., ROS1 rearrangements define a unique molecular class of lung cancers, *J. Clin. Oncol.*, 30, 863–870, 2012.

31. Pasquini, G. and Giaccone, G., C-MET inhibitors for advanced non-small cell lung cancer, *Expert Opin. Invest. Drugs*, 27, 363–375, 2018.

32. Cui, J. J., Targeting receptor tyrosine kinase MET in cancer: Small molecule inhibitors and clinical progress, *J. Med. Chem.*, 57, 4427–4453, 2014.

33. Cui, J. J., Tran-Dube, M., Shen, H., Nambu, M., Kung, P. P., et al., Structure based drug design of crizotinib (PF-02341066), a potent and selective dual inhibitor of mesenchymal-epithelial transition factor (c-MET) kinase and anaplastic lymphoma kinase (ALK), *J. Med. Chem.*, 54, 6342–6363, 2011.

34. Cui, J. J., McTigue, M., Kania, R., and Edwards, M., Case history: Xalkori™ (Crizotinib), a potent and selective dual inhibitor of mesenchymal epithelial transition (MET) and anaplastic lymphoma kinase (ALK) for cancer treatment, *Annu. Rep. Med. Chem.*, 48, 421–434, 2013.

35. Kung, P. P., Jones, R. A., and Richardson, P., Crizotinib (Xalkori): The First-in-Class ALK/ROS Inhibitor for Non-small Cell Lung Cancer, Innovative Drug Synthesis. In: J. J. Li and D. S. Johnson (Eds) *Innovative Drug Synthesis*. John Wiley & Sons: Hoboken, NJ, 119–156, 2016.

36. Christensen, J. G., Zou, H. Y., Arango, M. E., Li, Q., Lee, J. H., et al., Cytoreductive antitumor activity of PF-02341066, a novel inhibitor of anaplastic lymphoma kinase and c-Met, in experimental models of anaplastic large-cell lymphoma, *Mol. Cancer Ther.*, 6 (12, Part 1), 3314–3322, 2007.

37. Zou, H. Y., Li, Q., Lee, J. H., Arango, M. E., McDonnell, S. R., et al., An orally available small-molecule inhibitor of c-Met, PF-02341066, exhibits cytoreductive antitumor efficacy through antiproliferative and antiangiogenic mechanisms, *Cancer Res.*, 67, 4408–4417, 2007.

38. Drilon, E., Camidge, D. R., Ou, S.-H. I., Clarke, J. W., Socinski, M. A., et al., Efficacy and safety of crizotinib in patients with advanced MET exon-14 altered non-small cell lung cancer, *J. Clin. Oncol.*, 34 (suppl. 15, abstr. p. 108.), 2016.

39. Camidge, D. R., Ou, S.-H. I., Shapiro, G., Otterson, G. A., Cosca Villaruz, L. C., et al., Efficacy and safety of crizotinib in patients with advanced c-MET-amplified non-small cell lung cancer (NSCLC), *J. Clin. Oncol.*, 32 (suppl. 15, abstr. p. 8001.), 2014.

40. Ou, S.-H. I., Govindan, R., Eaton K. D., Otterson, G. A., Gutierrez, M. E., et al., Phase I results from a study of crizotinib in combination with erlotinib in patients with advanced nonsquamous non-small cell lung cancer, *J. Thorac. Oncol.*, 12, 145–151, 2017.

41. Jänne, P. A., Shaw A. T., Camidge, D. R., Giaccone, G., Shreeve, S. M., et al., Combined Pan-HER and ALK/ROS1/MET inhibition with dacomitinib and crizotinib in advanced non-small cell lung cancer: Results of a phase I study, *J. Thorac. Oncol.*, 11, 737–747, 2016.

42. Kwak, E. L., Camidge, D. R., Clark, J., Shapiro, G. I., Maki, R. G., et al., Clinical activity observed in a phase I dose escalation trial of an oral c-met and ALK inhibitor, PF-02341066, *Clin. Oncol.*, 27 (suppl, abstr.) e3509, 2009.

43. Tan, W., Wilner, K. D., Bang, Y. E., Kwak, L., Maki, R. G., et al., Pharmacokinetics (PK) of PF-02341066, a dual ALK/MET inhibitor after multiple oral doses to advanced cancer patients, *J. Clin. Oncol.*, 28 (suppl, abstr.) p. e2596, 2010.

44. Camidge, D. R., Bang, Y., Kwak, E. L., Iafrate, A. J., Varella-Garcia, M., et al., Activity and safety of crizotinib in patients with ALK-positive non-small-cell lung cancer: Updated results from a phase 1 study, *Lancet Oncol.*, 13, 1011–1019, 2012.

45. Shaw, A. T., Yeap, B. Y., Solomon, B. J., Riely, G. J., Gainor, J., et al., Effect of crizotinib on overall survival in patients with advanced non-small-cell lung cancer harbouring ALK gene rearrangement: A retrospective analysis, *Lancet Oncol.*, 12, 1004–1012, 2011.

46. Shaw, A. T., Ou, S.-H. I., Bang, Y.-J., Camidge, D. R., Solomon, B. J., et al., Crizotinib in ROS1-rearranged non–small-cell lung cancer, *N. Engl. J. Med.*, 371, 1963–1971, 2014.

47. Katayama, R., Khan, T. M., Benes, C., Lifshits, E., Ebi, H., et al., Therapeutic strategies to overcome crizotinib resistance in non-small cell lung cancers harboring the fusion oncogene EML4-ALK, *Proc. Natl. Acad. Sci. U. S. A.*, 108, 7535–7540, 2011.

48. Casaluce, F., Sgambato, A., Sacco, P. C., Palazzolo, G., Maione, P., et al., Resistance to crizotinib in advanced non-small cell lung cancer (NSCLC) with ALK rearrangement: mechanisms, treatment strategies and new targeted therapies, *Curr. Clin. Pharmacol.*, 11, 77–87, 2016.

49. Katayama, R., Shaw, A. T., Khan, T. M., Mino-Kenudson, M., Solomon, B. J., et al., Mechanisms of acquired crizotinib resistance in ALK-rearranged lung cancers, *Sci. Transl. Med.*, 120, 120ra17, 2012.

50. Chi, A. and Komaki, R., Treatment of brain metastasis from lung cancer, *Cancers*, 2, 2100–2137, 2010.

51. Weickhardt, A. J., Scheier, B., Burke, J. M., Gan, G., Bunn, P. A., et al., Local ablative therapy of oligoprogressive disease prolongs disease control by tyrosine kinase inhibitors in oncogene-addicted non-small-cell lung cancer, *J. Thorac. Oncol.*, 7, 1807–1814, 2012.

52. Steeg, P. S., Camphausen, K. A., and Smith, Q. R., Brain metastases as preventive and therapeutic targets, *Nat. Rev. Cancer*, 11, 352–363, 2011.

53. Heffron, T. P., Small molecule kinase inhibitors for the treatment of brain cancer, *J. Med. Chem.*, 59, 10030–10066, 2016.

54. Camidge, D. R., Taking aim at ALK across the blood–brain barrier, *J. Thorac. Oncol.*, 8, 389–390, 2013.

55. Rankovic, Z., CNS physicochemical property space shaped by a diverse set of molecules with experimentally determined exposure in the mouse brain, *J. Med. Chem.*, 60, 5943–5954, 2017.

56. Rankovic, Z., CNS drug design: Balancing physicochemical properties for optimal brain exposure, *J. Med. Chem.*, 58, 2584–2608, 2015.

57. Yamanaka, R., Management of refractory or relapsed primary central nervous system lymphoma, *Mol. Med. Rep.*, 2, 879–885, 2009.

58. Deeken, J. F. and Löscher, W., The blood-brain barrier and cancer: Transporters, treatment, and Trojan horses, *Clin. Cancer Res.*, 13, 1663–1674, 2007.

59. Wager, T. T., Hou, X., Verhoest, P. R., and Villalobos, A., Central nervous system multiparameter optimization desirability: Application in drug discovery, *ACS Chem. Neurosci.*, 7, 767–775, 2016.

60. Costa, D. B., Kobayashi, S., Pandya, S. S., Yeo, W.-L., Shen, Z., et al., CSF concentration of the anaplastic lymphoma kinase inhibitor crizotinib, *J. Clin. Oncol.*, 29, e443–e445, 2011.

61. Gainor, J. F., Tseng, D., Yoda, S., Dagogo-Jack, I., Friboulet, L., et al., Patterns of metastatic spread and mechanisms of resistance to crizotinib in ROS1-positive non-small-cell lung cancer, *JCO Precis. Oncol.*, 2017. doi: 10.1200/PO.17.00063.

62. Awad, M. M. and Shaw, A. T., ALK inhibitors in non-small cell lung cancer: Crizotinib and beyond, *Clin. Adv. Hematol. Oncol.*, 12, 429–439, 2014.

63. Galkin, A. V., Melnick, J. S., Kim, S., Hood, T. L., et al., Identification of NVPTAE684, a potent, selective, and efficacious inhibitor of NPM-ALK, *Proc. Natl. Acad. Sci. U. S. A.*, 104, 270–275, 2007.

64. Marsilje, T. H., Pei, W., Chen, B., Lu, W., Uno, T., et al., Synthesis, structure-activity relationships, and *in vivo* efficacy of the novel potent and selective anaplastic lymphoma kinase (ALK) inhibitor 5-chloro-N2-(2 isopropoxy-5-methyl-4-(piperidin-4-yl)phenyl)- N4-(2-(isopropylsulfonyl)phenyl)pyrimidine-2,4-diamine (LDK378) currently in phase 1 and phase 2 clinical trials, *J. Med. Chem.*, 56, 5675–5690, 2013.

65. Kinoshita, K., Kobayashi, T., Asoh, K., Furuichi, N., Ito, T., et al., 9-substituted 6,6-dimethyl-11-oxo-6,11-dihydro-5H-benzo[b]carbazoles as highly selective and potent anaplastic lymphoma kinase inhibitors, *J. Med. Chem.*, 54, 6286–6294, 2011.

66. Kinoshita, K., Asoh, K., Furuichi, N., Ito, T., Kawada, H., et al., Design and synthesis of a highly selective, orally active and potent anaplastic lymphoma kinaseinhibitor (CH5424802), *Bioorg. Med. Chem.*, 20, 1271–1280, 2012.

67. Huang, W. S., Liu, S., Zou, D., Thomas, M., Wang, Y., et al., Discovery of brigatinib (AP26113), a phosphine oxide-containing, potent, orally active inhibitor of analplastic lymphoma kinase, *J. Med. Chem.*, 59, 4948–4964, 2016.

68. Shaw, A. T., Kim, D. W., Mehra, R., Tan, D. S. W., Felip, E., et al., Ceritinib in ALK-rearranged non-small-cell lung cancer, *N. Engl. J. Med.*, 370, 1189–1197, 2014.

69. Kim, D.-W., Mehra, R., Tan, D. S.-W., Felip, E., Chow, L. Q. M., et al., Ceritinib in advanced anaplastic lymphoma (ALK)-rearranged (ALK+) non-small cell lung cancer (NSCLC): Results of the ASCEND-1 trial, *J. Clin. Oncol.*, 32 (suppl.), 8003, 2014.

70. Shaw, A., Mehra, R., Tan, D. S. W., Felip, E., Chow, L. Q., et al., Evaluation of ceritinib-treated patients (pts) with anaplastic lymphoma kinase rearranged (ALK+) non-small cell lung cancer (NSCLC) and brain metastases in the ASCEND-1 study, *Ann. Oncol.*, 25 (suppl. 4), iv455–iv456, 2014.

71. Kodama, T., Hasegawa, M., Takanashi, K., Sakurai, Y., Kondoh, O., et al., Antitumor activity of the selective ALK inhibitor alectinib in models of intracranial metastases, *Cancer Chemother. Pharmacol.*, 74, 1023–1028, 2014.

72. Zhang, S., Anjum, R., Squillace, R., Nadworny, S., Zhou, T., et al., The potent ALK inhibitor brigatinib (AP26113) overcomes mechanisms of resistance to first- and second-generation ALK inhibitors in preclinical models, *Clin. Cancer. Res.*, 22, 5527–5538, 2016.

73. Gadgeel, S. M., Gandhi, L., Riely, G. J., Chiappori, A. A., West, H. L., et al., Safety and activity of alectinib against systemic disease and brain metastases in patients with crizotinib-resistant ALK rearranged non-small-cell lung cancer (AF-002JG): Results from the dose-finding portion of a phase 1/2 study, *Lancet Oncol.*, 15, 1119–1128, 2014.

74. Ajimizu, H., Kim, Y. H., and Mishima, M., Rapid response of brain metastases to alectinib in a patient with non-small-cell lung cancer resistant to crizotinib, *Med. Oncol.*, 32, 3, 2015.

75. Gainor, J. F., Sherman, C. A., Willoughby, K., Logan, J., Kennedy, E., et al., Alectinib salvages CNS relapses in ALK positive lung cancer patients previously treated with crizotinib and ceritinib, *J. Thorac. Oncol.*, 10, 232–236, 2015.

76. Gettinger, S. N., Bazhenova, L., Salgia, R., Langer, C. J., Gold, K. A., et al., Updated efficacy and safety of the ALK inhibitor AP26113 in patients with advanced malignancies, including ALK+ non-small cell lung cancer, *J. Thorac. Oncol.*, 8 (suppl. 2), S296, 2013.

77. Camidge, D. R., Bazhenova, L., Salgia, R., Langer, C. J., Gold, K. A., et al., Safety and efficacy of brigatinib (AP26113) in advanced malignancies, including ALK+ non-small cell lung cancer (NSCLC), *J. Clin. Oncol.*, 33 (suppl.), 8062, 2015.

78. Edwards, M. P. and Price, D. A., Role of physicochemical properties and ligand lipophilicity efficiency in addressing drug safety risks, *Annu. Rep. Med. Chem.*, 45, 381–391, 2010.

79. Freeman-Cook, K. D., Hoffman, R. L., and Johnson, T. W., Lipophilic efficiency: The most important efficiency metric in medicinal chemistry, *Future Med. Chem.*, 5, 113–115, 2013.

80. Johnson, T. W., Gallego, R. A., and Edwards, M. P., Lipophilic efficiency as an important metric in drug design, *J. Med. Chem.*, 61, 6401–6420, 2018.

81. Huang, Q., Johnson, T. W., Bailey, S., Brooun, A., Bunker, K. D., et al., Design of potent and selective inhibitors to overcome clinical anaplastic lymphoma kinase mutationsresistant to crizotinib, *J. Med. Chem.*, 57, 1170–1187, 2014.

82. Driggers, E. M., Hale, S. P., Lee, J., and Terrett, N. K., The exploration of macrocycles for drug discovery: An underexploited structural class, *Nat. Rev. Drug Discovery*, 7, 608–624, 2008.

83. Giordanetto, F. and Kihlberg, J., Macrocyclic drugs and clinical candidates: What can medicinal chemists learn from their properties? *J. Med. Chem.*, 57, 278–295, 2014.

84. Marsault, E. and Peterson, M. L., Macrocycles are great cycles: Applications, opportunities, and challenges of synthetic macrocycles in drug discovery, *J. Med. Chem.*, 54 (7), 1961–2004, 2011.

85. Vendeville, S. and Cummings, M. D., Synthetic macrocycles in small-molecule drug discovery, *Annu. Rep. Med. Chem.*, 48, 371–386, 2013.

86. Lightstone, F. C. and Bruice, T. C., Enthalpy and entropy in ring closure reactions, *Bioorg. Chem.*, 26, 193–199, 1998.

87. Marti-Centelles, V., Pandey, M. D., Burguete, M. I., and Luis, S. V., Macrocyclization reactions: The importance of conformational, configurational and template-induced preorganization, *Chem. Rev.*, 115, 8736–8834, 2015.

88. Breslin, H. J., Lane, B. M., Ott, G. R., Ghoshe, A. K., Angeles, T. S., et al., Design, synthesis, and anaplastic lymphoma kinase (ALK) inhibitory activity for a novel series of 2,4,8,22-tetraazatetracyclo[14.3.1.13,7.19,13]docosa-1(20),3(22),4,6,9(21),10,12,16,18-nonaene macrocycles, *J. Med. Chem.*, 55, 449–464, 2011.

89. Johnson, T. W., Richardson, P. F., Bailey, S., Brooun, A., Burke, B. J., et al., Discovery of (10R)-7-amino-12-fluoro-2,10,16-trimethyl-15-oxo-10,15,16,17-tetrahydro-2H-8,4(metheno)pyrazolo[4,3-h][2,5,11]benzoxadiaza cyclotetradecine-3-carbonitrile (PF-06463922), a macrocyclic inhibitor of anaplastic lymphoma kinase (ALK) and cros oncogene 1 (ROS1) with preclinical brain exposure and broadspectrum potency against ALK-resistant mutations, *J. Med. Chem.*, 57, 4720–4744, 2014.

90. Henderson, J. L., Kormos, B. L., Hayward, M. M., Coffman, K. J., Jasti, J., et al., Discovery and preclinical profiling of 3-[4-(morpholin-4-yl)-7H-pyrrolo[2,3-d]pyrimidin-5-yl]benzonitrile (PF-06447475), a highly potent, selective, brain penetrant, and in vivo active LRRK2 kinase inhibitor, *J. Med. Chem.*, 58, 419–432, 2015.

91. Huang, D., Zhou, T., Lafleur, K., Nevado, C., and Caflisch, A., Kinase selectivity potential for inhibitors targeting the ATP binding site: A network analysis, *Bioinformatics*, 26, 198–204, 2010.

92. Xing, L., Rai, B., and Lunney E. A., Scaffold mining of kinase hinge binders in crystal structure database, *J. Comput-Aided Mol. Des.*, 28, 13–23, 2014.

93. Bertrand, T., Kothe, M., Liu, J., Dupuy, A., Rak, A., et al., The crystal structures of TrkA and TrkB suggest key regions for achieving selective inhibition, *J. Mol. Biol.*, 423, 439–453, 2012.

94. Toenjes, S. T. and Gustafson, J. L., Atropisomerism in medicinal chemistry: Challenges and opportunities, *Future Med. Chem.*, 10, 409–422, 2018.

95. Glunz, P. W., Recent encounters with atropisomerism in drug discovery, *Bioorg. Med. Chem. Lett.*, 28, 53–60, 2018.

96. Smyth, J. E., Butler, N. M., and Keller, P. A., A twist of nature: The significance of atropisomers in biological systems, *Nat. Prod. Rep.*, 32, 1562–1583, 2015.

97. Zask, A., Murphy, J., and Ellestad, G. A., Biological stereoselectivity of atropisomeric natural products and drugs, *Chirality*, 25, 265–274, 2013.

98. LaPlante, S. R., Fader, L. D., Fandrick, K. R., Fandrick, D. R., Hucke, O., et al., Assessing atropisomer axial chirality in drug discovery and development, *J. Med. Chem.*, 54, 7005–7022, 2011.

99. Clayden, J., Moran, W. J., Edwards, P. J., and LaPlante, S. R., The challenge of atropisomerism in drug discovery, *Angew. Chem. Int. Ed.*, 48, 6398–6401, 2009.

100. LaPlante, S. R., Edwards, P. J., Fader, L. D., Jakalian, A., and Hucke, O., Revealing atropisomer axial chirality in drug discovery, *ChemMedChem.*, 6, 505–513, 2011.

101. Yan, T. Q., Riley, F., Philippe, L., Davoren, J., Cox, L., et al., Chromatographic resolution of atropisomers for toxicity and biotransformation studies in pharmaceutical research, *J. Chromatogr. A*, 1398, 108–120, 2015.

102. Elleraas, J., Ewanicki, J., Johnson, T. W., Sach, N. W., Collins, M. R., and et al., Conformational studies and atropisomerism kinetics of the ALK clinical candidate lorlatinib (PF-06463922) and des-methyl congeners, *Angew. Chem. Int. Ed.*, 55, 3590–3595, 2016.

103. Li, B., Barnhart, R. W., Hoffman, J. E., Nematalla, A., Raggon, J., et al., Exploratory process development of lorlatinib, *Org. Process. Res. Dev.*, 2018. doi: 10.1021/acs.oprd.8b00210.

104. Sperry, J. B., Minteer, C. J., Tao, J. Y., Johnson, R., Duzguner, R., et al., Thermal stability assessment of peptide coupling reagents commonly used in pharmaceutical manufacturing, *Org. Process. Res. Dev.*, 2018. doi: 10.1021/acs.oprd.8b00193.

105. Morgan, P., Van Der Graaf, P. H., Arrowsmith, J., Feltner, D. E., Drummond, K. S., et al., Can the flow of medicines be improved? Fundamental pharmacokinetic and pharmacological principles toward improving Phase II survival, *Drug Disc. Today*, 17, 419–424, 2012.

106. Collier, T. L., Normandin, M. D., Stephenson, N. A., Livni, E., Liang, S. H., et al., Synthesis and preliminary PET imaging of ^{11}C and ^{18}F isotopologues of the ROS1/ALK inhibitor lorlatinib, *Nat. Commun.*, 8, 15761, 2017.

107. Rotstein, B. H., Stephenson, N. A., Vasdev, N., and Liang, S. H., Spirocyclic hypervalent iodine (III)-mediated radiofluorination of non-activated and hindered aromatics, *Nat. Commun.*, 5, 4365, 2014.

108. Rotstein, B. H., Wang, L., Liu, R. Y., Patteson, J., Kwan, E. E., et al., Mechanistic studies and radiofluorination of structurally diverse pharmaceuticals with spirocyclic iodonium (III) ylides, *Chem. Sci.*, 7, 4407–4417, 2016.

109. Holland, J. P., Cumming, P. and Vasdev, N., PET radiopharmaceuticals for probing enzymes in the brain, *Am. J. Nucl. Med. Mol. Imaging*, 3, 194–216, 2013.

110. Kuhn, B., Mohr, P., and Stahl, M., Intramolecular hydrogen bonding in medicinal chemistry, *J. Med. Chem.*, 53, 2601–2611, 2010.

111. Gainor, J. F., Dardaei, L., Yoda, S., Friboulet, L., Leshchiner, I., et al., Molecular mechanisms of resistance to first- and second-generation ALK inhibitors in ALK-rearranged lung cancer, *Cancer Discov.*, 6, 1119–1133, 2016.

112. Shaw, A. T., Felip, E., Bauer, T. M., Besse, B., Navarro, A., et al., Lorlatinib in non-small-cell lung cancer with ALK or ROS1 rearrangement: An international, multicenter, open-label, single-arm, first-in-man phase 1 trial, *Lancet Oncol.*, 18, 1590–1598, 2017.

113. http://clinicaltrials.gov/ct2/show/NCT01970865 (last accessed April 21st, 2019).

114. Akamine, T., Toyakawa, G., Tagawa, T., and Seto, T., Spotlight on lorlatinib and its potential in the treatment of NSCLC: The evidence to date, *Onco. Ther.*, 11, 5093–5011, 2018.
115. www.fda.gov/Drugs/InformationOnDrugs/ApprovedDrugs/ucm625027.htm (last accessed April 21st, 2019).
116. Richardson, P. F. and Johnson, T. W., Discovery of the ALK/ROS1 inhibitor lorlatinib (PF-06463922), *Med. Chem. Rev.*, 52, 45–66, 2017.

Section III

Drug Development

11 Integrated Drug Product Development

From Lead Candidate Selection to Life-Cycle Management

Madhu Pudipeddi
Parakam Pharma LLC

Abu T.M. Serajuddin
St. John's University

Ankita V. Shah
Freund-Vector Corporation

Daniel Mufson
Apotherx

CONTENTS

11.1 INTRODUCTION

Historically, medicines have been administered through the obvious portals following their preparation first by the shaman and then by the physician and later by the apothecary. These natural products were ingested, rubbed in, or smoked. For the past century, the person diagnosing the disease no longer prepares the potion, eliminating, no doubt, some of the power of the placebo, and as a consequence, drug discovery, development, and manufacturing have grown into a separate pharmaceutical industry. In particular, the last 75 years have been a period of astounding growth in our insight of the molecular function of the human body. This has led to discovery of medicines to treat diseases that were not even recognized before the mid-20th century. This chapter reflects the role of pharmaceutics and the diversity of the approaches taken to achieve these successes, including approaches that were introduced within recent years and describes how the role of the "industrial" pharmacist has evolved to become the technical bridge between discovery and development activities and, indeed, commercialization activities. No other discipline follows the progress of the new drug candidate as far with regard to the initial refinement of the chemical lead through preformulation evaluation to dosage-form design, clinical trial material (CTM) preparation, process scale-up, manufacturing, and then life-cycle management (LCM).

The pharmaceutical formulation was once solely the responsibility of the pharmacist, first in the drugstore and later in an industrial setting. Indeed, many of today's major drug companies, such as Merck, Lilly, and Pfizer components Searle, Warner-Lambert, Parke-Davis, and Wyeth, started in the backrooms of drugstores. During the second half of the 20th century, physicochemical and biopharmaceutical principles underlying pharmaceutical dosage forms were identified and refined, thanks to the pioneering works by Higuchi,[1] Nelson,[2] Levy,[3] Gibaldi,[4] and their coworkers. Wagner,[5] Wood,[6] and Kaplan[7] were among the earliest industrial scientists to systematically link formulation design activities and biology. Nevertheless, until the end of 20th century, formulations were developed somewhat in isolation with different disciplines involved in drug development operating independently. For example, during the identification and selection of new chemical entities (NCEs) for development, not much thought was given into how they would be formulated, and during dosage-form design, adequate considerations of *in vivo* performance of formulations were lacking. Wagner[5] first termed our evolving understanding of the relationship between the dosage form and its anatomical target "biopharmaceutics" in the early 1960s. Since then, it has been apparent that careful consideration of a molecule's physicochemical properties and those of its carrier, the dosage form, must be understood to enhance bioavailability, if given orally, and to enhance the ability of drug to reach the desired site of action, if given by other routes of administration. This knowledge allows for a rational stepwise approach in selecting new drug candidates, developing optimal dosage forms, and later, when it is necessary, making changes in the formulation or manufacturing processes. During the past two decades, the basic approach of dosage-form development in the pharmaceutical industry has changed dramatically. Dosage-form design is now an "integrated process" starting from identification of drug molecules for development to their ultimate commercialization as dosage forms. This is often performed by a multidisciplinary team consisting of pharmacists, medicinal chemists, physical chemists, analytical chemists, material scientists, pharmacokineticists, chemical engineers, and other individuals from related disciplines.

In its simplest terms, the dosage form is a carrier of the drug. It must further be reproducible, bioavailable, stable, readily scaleable, and elegant. The skill sets employed to design the first units of a dosage form, for example, a tablet, are quite different than those required to design a process to make hundreds of thousands of such units per hour, reproducibly, in ton quantities, almost anywhere in the world. Nevertheless, most decisions regarding dosage form design and manufacturability of drug products have to be made within a limited time period at the early stage of drug product development when the minimal drug substance supply may not favor extensive investigation. The manufacturability situation becomes understandably more complex as the dosage form becomes more sophisticated or if a drug-delivery system (DDS) is needed.

In recent years, there has been much progress in the level of sophistication in dosage-form design to keep pace with advances in drug discovery methods. New excipients, new materials, and combination products that consist of both a drug and a device have arisen to meet new delivery challenges. For example, many of the NCEs generated by high-throughput screening (HTS) are profoundly water-insoluble. What was considered a lower limit for adequate water solubility[7] (~0.1 mg/mL) in the 1970s has been surpassed by one to two orders of magnitude due to changes in the way drug discovery is performed. Traditional methods such as particle size reduction to improve the aqueous dissolution rate of these ever more insoluble molecules are not always sufficient to overcome the liability. New approaches have evolved to meet these challenges ranging from cosolvent systems[8] to the use of lipid-based DDSs[9] and amorphous solid dispersion.[10,11]

Many literature sources describing formulation and manufacture of different pharmaceutical dosage forms are available.[12,13] The primary objective of this chapter is to describe an integrated process of drug product development, demonstrating how all activities from lead selection to LCM are interrelated. Various scientific principles underlying these activities are described.

A survey of new drug approvals (NDAs) during the last 5 years (from 2013 to mid-2018) showed that nearly 50% of them are oral dosage forms. The percentage is higher if abbreviated NDAs for generics are included. Additionally, over 80% of dosage units dispensed by pharmacists at the retail level are oral solids (tablets, capsules, etc.) and liquids. Therefore, the primary focus of this chapter is the development of oral dosage forms with a few other dosage forms described only briefly. However, many of the principles described in this chapter are common to all dosage forms.

11.2 DEVELOPABILITY ASSESSMENT

The dosage-form design is guided by the properties of the drug candidate. If an NCE does not have suitable physical and chemical properties or pharmacokinetic attributes, the development of a dosage form (product) may be difficult and may sometimes be even impossible. Any heroic measures to resolve issues related to physicochemical and biopharmaceutical properties of drug candidates add to the time and cost of drug development. Therefore, the interaction between discovery and development scientists increased greatly to maximize the opportunity to succeed.[14]

11.2.1 DRUG DISCOVERY AND DEVELOPMENT INTERACTION

The drug discovery process typically involves[15]:

- Target identification.
- Target validation.
- Lead identification.
- Candidate(s) selection.

A drug target can be a receptor/ion channel, enzyme, hormone/factor, DNA, RNA, nuclear receptor, or other, unidentified, biological entity. Once drug targets are identified, they are exposed to numerous compounds in an *in vitro* or cell-based assay in an HTS mode. The advent of combinatorial chemistry has enabled synthesis of a very large number of compounds for testing against a particular target. Compounds that elicit a positive response in a particular assay are called "hits." Hits that continue to show positive response in more complex models rise to "leads" (lead identification). A selected few of the optimized leads are then advanced to preclinical testing. Despite extensive efforts in identifying and selecting leads, they suffer extensive attrition during preclinical and later in clinical testing due to lack of desirable physicochemical and biopharmaceutical attributes, unacceptable toxic effects, and inadequate efficacy during preclinical and clinical studies, thus adding to the drug development cost.[16–18] It has been estimated that poor pharmacokinetic factors, i.e., absorption, elimination, distribution, and metabolism (ADME) contributed to about 40% of failed candidates.[19] Some of the undesirable attributes of "leads" may be related to how new molecules are discovered and leads are identified. In particular, use of combinatorial chemistry and HTS technologies has resulted in the generation and selection of increasingly lipophilic drug molecules with potential biopharmaceutical hurdles in downstream development.[20] In HTS, stock solutions of compounds in dimethylsulfoxide (DMSO) are used for *in vitro* and even *in vivo* testing, which often lead to the selection water-insoluble drugs with potential dissolution and bioavailability issues. To reduce attrition of compounds during preclinical and clinical development, in recent years, pharmaceutical companies introduced developability assessment of new drug molecules and, especially, potential leads, as an integral part of the drug development process.[21–23]

11.2.2 SCREENING FOR DRUGGABILITY OR DEVELOPABILITY

Compounds with acceptable pharmaceutical properties, in addition to acceptable biological activity and safety profile, are considered "drug-like" or developable. Typical acceptable pharmaceutical properties for oral delivery of a drug-like molecule include sufficient aqueous solubility, permeability across biological membranes, satisfactory stability to metabolic enzymes, resistance to degradation in the gastrointestinal (GI) tract (pH and enzymatic stability), and adequate chemical stability for successful formulation into a stable dosage form. A number of additional barriers, such as efflux transporters[24] (i.e., export of drug from blood back to the gut) and first-pass metabolism by intestinal or liver cells, have been identified that may limit oral absorption. A number of computational and experimental methods are emerging for testing (or profiling) drug discovery compounds for acceptable pharmaceutical properties.

In this section, discussion of physicochemical profiling is limited to solubility, permeability, drug stability, and limited solid-state characterization (as we will see in Section 11.4, there are other physical–mechanical properties that must also be considered). For convenience, methods available for physicochemical profiling are discussed under the following categories: computational tools (sometimes referred to as *in silico* tools), HTS methods, and in-depth physicochemical profiling.[21,22,25]

11.2.2.1 Computational Tools

Medicinal chemists have always been adept in recognizing trends in physicochemical properties of molecules and relating them to molecular structure. With rapid increase in the number of hits and leads, computational tools have been proposed to calculate molecular properties that may predict potential absorption hurdles. For example, Lipinski's "Rule of 5"[20] states that poor absorption or permeation are likely when:

1. There are more than five H-bond donors (expressed as the sum of –NH and –OH groups).
2. The molecular weight is more than 500.
3. Log P > 5 (or c log P > 4.5).
4. There are more than ten H-bond acceptors (expressed as the sum of Ns and Os).

If a compound violates more than two of the four criteria, it is likely to encounter oral absorption issues. Compounds that are substrates for biological transporters and peptidomimetics are exempt from these rules. The Rule of 5 is a very useful computational tool for highlighting compounds with potential oral absorption issues. A number of additional reports on pharmaceutical profiling and developability of discovery compounds[25–28] have been published since the report of the Rule of 5. Polar surface area (PSA) and number of rotatable bonds have also been suggested as means to predict oral bioavailability. PSA is defined as the sum of surfaces of polar atoms in a molecule. A rotatable bond is defined as any single bond, not in a ring, bound to a nonterminal heavy (i.e., non-hydrogen) atom. Amide bonds are excluded from the count. It has been reported that molecules with the following characteristics will have acceptable oral bioavailability[29]:

1. Ten or fewer rotatable bonds.
2. PSA ≤140 Å2 (or 12 or fewer H-bond donors and acceptors).

Aqueous solubility is probably the single most important biopharmaceutical property that pharmaceutical scientists are concerned with. It has been the subject of computational prediction for several years.[30–33] The overall accuracy of the predicted values can be expected to be in the vicinity of 0.5–1.0 log units (a factor of 3–10) at best. Although a decision on acceptance or rejection of a particular compound cannot be made only on the basis of predicted parameters, these predictions may be helpful in identifying molecules with improved drug-like properties.[34]

11.2.2.2 High-Throughput Screening (HTS) Methods

High-throughput drug-like property profiling is increasingly used during lead identification and candidate selection. HTS pharmaceutical profiling may include:

- Compound purity or integrity testing using methods such as UV absorbance, evaporative light scattering, MS, and NMR.[35]
- Solubility.
- Lipophilicity (log P).
- Dissociation constant (pK_a).
- Permeability.
- Solution/solid-state stability determination.

Compound purity (or integrity testing) is important to ensure purity in the early stages because erroneous activity or toxicity results may be obtained by impure compounds. It is initiated during hit identification and continued into lead and candidate selection.

Solubility is measured to varying degrees of accuracy by HTS methods.[36] Typical methods in the lead identification stage include determination of "kinetic solubility" by precipitation of a drug solution in DMSO into the test medium. Since the solid-state form of the precipitate (crystalline or amorphous) is often not clearly known by this method, the measured solubility is approximate and generally higher than the true (equilibrium) solubility. Kinetic solubility, however, serves the purpose of identifying solubility limitations in activity or *in vitro* toxicity assays or in identifying highly insoluble compounds. Lipinski et al.[20] observed that, for compounds with a kinetic solubility of >65 µg/mL (in pH 7 non-chloride containing phosphate buffer at room temperature), poor oral absorption is usually due to factors unrelated to solubility. The acceptable solubility for a drug compound depends on its permeability and dose. This point will be further elaborated later. Methods to assess the need for solubility improvement and the development of suitable solubility-enhancing formulations in lead optimization have been reported.[37,38] Zheng et al.[39] reported a decision-making approach for selecting optimal formulation strategies for preclinical studies (PK parameters, toxicological evaluation, etc.). Solid dispersion and lipid-based formulations proved to be viable dosing approaches when the compounds were found to be relatively water-insoluble. Estimation or measurement of pK_a is important to understand the state of ionization of the drug under physiological conditions and to evaluate salt-forming ability.[40] Log P determines the partitioning of a drug between an aqueous phase and a lipid phase (i.e., lipid bilayer). Log P and acid pK_a can be theoretically estimated with reasonable accuracy.[20,41,42] High-throughput methods are also available for measurement of log P[43] and pK_a.[44]

Physical flux of a drug molecule across a biological membrane depends on the product of concentration (which is limited by solubility) and permeability. High-throughput artificial membrane permeability (also called "parallel artificial membrane permeability assay") has been used in early discovery to estimate compound permeability.[45–47] This method measures the flux of a compound in solution across an artificial lipid bilayer deposited on a microfilter. Artificial membrane permeability is a measure of the actual flux (rate) across an artificial membrane, whereas log P or log D, as mentioned earlier, represents equilibrium distribution between an aqueous and a lipid phase. Sometimes the term "intrinsic permeability" is used to specify the permeability of the unionized form. Artificial membrane permeability can be determined as a function of pH. The fluxes across the artificial membrane in the absence of active transport have been reported to relate to human absorption through a hyperbolic curve. The correlation of permeability through artificial membranes may depend on the specific experimental conditions such as the preparation of membranes and pH. Therefore, guidelines on what is considered acceptable or unacceptable permeability must be based on the individual assay conditions. For example, Hwang et al.[48] ranked compound permeation on the basis of the percent transport across the lipid bilayer in 2 h: <2% (low), 2%–5% (medium), and >5% (high), respectively.

Caco-2 monolayer, a model for human intestinal permeability, is commonly used in drug discovery to screen discovery compounds.[49–51] The method involves measurement of flux of the compound dissolved in a physiological buffer through a monolayer of human colonic cells deposited on a filter. Caco-2 monolayer permeability has gained considerable acceptance to assess human absorption. Compounds with a Caco-2 monolayer permeability (P_{app}) similar to or greater than that of propranolol ($\sim30 \times 10^{-6}$ cm/s) are considered highly permeable, while compounds with P_{app} similar to or lower than that of ranitidine ($<1 \times 10^{-6}$ cm/s) are considered poorly permeable. Hurdles associated with determination of permeability of poorly soluble compounds using Caco-2 method as well as potential laboratory to laboratory variability of results have been discussed in the literature.[52]

11.2.2.3 In-Depth Physicochemical Profiling

Once compounds enter the late lead selection or candidate selection phase, more in-depth physicochemical profiling is conducted. The extent of characterization may vary from company to company; however, it likely includes:

- Experimental pK_a and log P (as a function of pH, if necessary)
- Thermodynamic solubility (as a function of pH)

- Solution/suspension stability
- Solid-state characterization.

In the above list, solid-state characterization typically involves:

- Solid-state stability
- Feasibility of salt formation
- Polymorph characterization
- Particle size
- Hygroscopicity
- Dissolution rate.

In a more traditional pharmaceutical setting, the characterization physicochemical properties would be done during preformulation studies. With the availability of automation and the ability to conduct most of these experiments with small quantities of material, more preformulation activities are being shifted earlier into drug discovery.[22] Balbach and Korn[53] reported a "100 mg approach" to pharmaceutical evaluation of early development compounds. Additional absorption, metabolism, distribution, elimination, and toxicity[54] screens may also be conducted at this stage.

Overall, the scientific merit of physicochemical profiling is clear. It provides a better assessment of development risks of a compound early on. The important question is how can pharmaceutical companies utilize the vast amount of physicochemical information to advance the right drug candidates to preclinical and clinical testing. Scorecards or flags may be used to rank drug candidates for their physicochemical properties. These scores or flags, however, have to be appropriately weighted with biological activity, safety, and pharmacokinetic profiling of compounds. The relative weighting of various factors depends on the specific issues of a discovery program. However, the basic question of how does it help reduce attrition due to unacceptable physicochemical properties remains to be answered in a statistical sense. In 1997, Lipinski et al.[20] reported that a trend had been seen since the implementation of the Rule of 5 towards more drug-like properties in Pfizer's internal drug discovery program. Overall, the goal of a discovery program is to steer the leads in the right direction using computational and HTS approaches and then utilize the in-depth screening tools to select the most optimal compound without undue emphasis on a single parameter such as biological activity. The overall success of a compound is a function of its biological and biopharmaceutical properties.

11.3 OVERVIEW OF DOSAGE-FORM DEVELOPMENT AND PROCESS SCALE-UP

Once a compound is selected for development, the pharmaceutics group begins a series of studies to further evaluate the salient physical–chemical and physical–mechanical properties of NCEs to guide the actual design (formulation, recipe) of the dosage form that will carry the drug. These investigations consist of the following steps that span the years that the molecule undergoes clinical evaluation:

- Preformulation
- Consideration of biopharmaceutical aspects
- Dosage-form design
- CTMs manufacture
- Scale-up studies including technical transfer to manufacturing sites
- Initiation of long-term stability studies to guide the setting of the expiration date
- Production of "biobatches"
- Validation and commercial batches
- LCM.

While there is a natural desire to front-load these evaluations, this must be balanced against the sad fact that many molecules fail to survive clinical testing: sufficient characterization is performed to help select the "right" molecule to minimize losses at the later, and much more costly, clinical evaluation stages. Great thought and planning are required to optimize the level of effort expended on a single molecule when so many are known to fail during clinical evaluation.

There are many references on the design[12,13] and scale-up[55] of pharmaceutical dosage forms. It is appropriate to mention that the design of the dosage form should be well documented from its inception as this information is required at the NDA stage to explain the development approach. The FDA and its global counterparts are seeking documentation that the product quality and performance are achieved and assured by design of effective and efficient manufacturing processes. The product specifications should be based upon a mechanistic understanding of how the formulation and processing factors impact the product performance. As we describe later, it is important that an ability to effect continuous improvement to the production process with continuous real-time assessment of quality (quality by design; QbD) must be incorporated in drug development. Recent regulatory initiatives require that the product/process risks are assessed and mitigated.

11.4   PREFORMULATION

Although, in the earlier days, preformulation activities would start when a compound had been chosen as the lead candidate and advanced for preclinical development, in the current integrated discovery-development scenario, many of the classical preformulation activities are conducted while screening compounds are in the lead identification or compound selection stage.[22,23,39] Irrespective of the actual timing of preformulation studies, preformulation lays the foundation for robust formulation and process development. The purpose of preformulation is to understand the basic physicochemical properties of the drug compound so that the challenges in formulation are foreseen and appropriate strategies are designed. Some physicochemical properties are independent of the physical form (i.e., crystal form) but simply are a function of the chemical nature of the compound. They include chemical structure, pK_a, partition coefficient, log P, and solution-state stability. A change in the physical form does not affect these properties.

11.4.1   PREFORMULATION ACTIVITIES: INDEPENDENT OF SOLID FORM

11.4.1.1   Dissociation Constant

The pK_a, or dissociation constant, is a measure of the strength of an acid or a base. The dissociation constant of an organic acid or base is defined by the following equilibria:

$$HX + H_2O \leftrightarrow H_3O^+ + X^- \text{ (acid)}$$

$$BH^+ + H_2O \leftrightarrow H_3O^+ + B \text{ (base)}$$

$$K_a = \frac{\left[H_3O^+\right]\left[X^-\right]}{[HX]} \text{ (acid)} \tag{11.1}$$

$$K_a = \frac{\left[H_3O^+\right][B]}{\left[HB^+\right]} \text{ (base)}$$

whereby the $-\log K_a$ is defined as pK_a.

The pK_a of a base is actually that of its conjugate acid. As the numeric value of the dissociation constant increases (i.e., pK_a decreases), the acid strength increases. Conversely, as the acid dissociation constant of a base (that of its conjugate acid) increases, the strength of the base decreases.

For a more accurate definition of dissociation constants, each concentration term must be replaced by thermodynamic activity. In dilute solutions, concentration of each species is taken to be equal to activity. Activity-based dissociation constants are true equilibrium constants and depend only on temperature. Dissociation constants measured by spectroscopy are "concentration dissociation constants." Most pK_a values in the pharmaceutical literature are measured by ignoring activity effects and, therefore, are actually concentration dissociation constants or apparent dissociation constants. It is customary to report dissociation constant values at 25°C.

Drug dissociation constants are experimentally determined by manual or automated potentiometric titration or by spectrophotometric methods.[42,56] Current methods allow determination of pK_a values with drug concentrations as low as 10–100 μM. For highly insoluble compounds (concentration <1 to 10 μM), the Yesuda–Shedlovsky method[57] is commonly used where organic cosolvents (i.e., methanol) are employed to improve solubility. The method takes three or more titrations at different cosolvent concentrations, and the result is then extrapolated to pure aqueous system. The dissociation constant can also be determined with less accuracy from the pH–solubility profile using the following modification of the Henderson–Hasselbalch equation:

$$S = \begin{cases} S_{HA}\left(1 + 10^{pH-pK_a}\right) \text{for an acid} \\ S_B\left(1 + 10^{pK_a-pH}\right) \text{for a base} \end{cases} \tag{11.2}$$

where S_{HA} or S_B is the intrinsic solubility of the unionized form.

Some drugs exhibit concentration-dependent self-association in aqueous solutions. The dissociation constant of these compounds may change upon self-association. For example, pK_a of dexverapamil has been reported to shift from 8.90 to 7.99 in the micellar concentration range.[58]

11.4.1.2 Partition or Distribution Coefficient

The partition coefficient (P) is a measure of how a drug partitions between a water-immiscible lipid (or an organic phase) and water. It is defined as follows:

$$P = \frac{[\text{Neutral species}]_o}{[\text{Neutral species}]_w} \tag{11.3}$$

$$\log P = \log_{10} P$$

The distribution coefficient is the partition coefficient at a particular pH. The following equilibrium is often used to define D, with the assumption that only the unionized species partition into the oil or lipid phase:

$$D = \frac{[\text{Unionized species}]_o}{[\text{Unionized species}]_w + [\text{ionized species}]_w}$$

$$\log D_{at\,pH} = \begin{cases} \log P - \log\left[1 + 10^{(pH-pK)}\right] \text{for acids} \\ \log P - \log\left[1 + 10^{(pK-pH)}\right] \text{for bases} \end{cases} \tag{11.4}$$

The measurement of $\log P$ is important because it has been shown to correlate with biological activity and toxicity.[59] As discussed in the previous section, a range of $\log P$ (0–5) has been shown to be critical for satisfactory oral absorption of drug compounds.

11.4.1.3 Solution Stability Studies

Forced degradation studies provide information on drug degradation pathways, potential identification of degradation products in the drug product, structure determination of degradation products, and determination of intrinsic stability of a drug. Regulatory guidance and best practices for conducting forced degradation studies have been reviewed.[60] Typical conditions for forced degradation testing include strong acid/base, oxidative, photostability, thermal, and thermal/humidity conditions. Stress conditions are utilized that result in approximately 10% degradation.

Stability of the drug in solution over a pH range of 1–13 is assessed during preformulation. The purpose is twofold. In the short run, stability in the GI pH range of 1–7.5 is important for drug absorption. In the long run, knowledge of solution-state stability is important for overall drug product stability and possible stabilization strategy. Drug degradation in solution typically involves hydrolysis, oxidation, racemization, or photodegradation. The major routes of drug degradation have been thoroughly reviewed.[61,62]

Although determination of a complete pH-degradation rate profile is desired, it may not always be practical due to limitations of drug supply and time. Also, insufficient solubility in purely aqueous systems may limit determination of pH-degradation rate profiles. Organic cosolvents may be used to increase solubility; however, extrapolation to aqueous conditions must be done with caution. Stability of the drug in a suspended form in the desired buffer can be tested in lieu of solution stability. The stress test results must, however, be interpreted in relation to the solubility in the suspension medium. The test may provide an empirical indication of pH stability in the presence of excess water. Satisfactory stability in the GI pH range (1–7.5) is important for oral absorption. While there are examples of successful solid oral dosage forms of drug that are highly unstable at GI pH (didanosine, esomeprazole magnesium[63]), excessive degradation in the GI tract limits oral absorption and may require heroic efforts to address the problem via formulation. Although no hard and fast rule exists on acceptable GI stability, Balbach and Korn[53] considered <2% to 5% degradation under simulated *in vivo* conditions (37°C, pH 1.2–8, fed and fasted conditions) as acceptable. Higher degradation may require additional investigation. The effect of the GI enzymes on drug stability should also be evaluated.

11.4.2 Preformulation Activities: Dependent on Solid Form

A new drug substance may exist in a multitude of crystalline and salt forms with different physical properties such as shape, melting point, and solubility that can profoundly impact the manufacturing and performance of its dosage form. Different methods for screening and determining physicochemical for pharmaceutical solid forms have been reviewed.[64,65]

11.4.2.1 Solubility

Solubility is highly influenced by the solid-state form (e.g., crystalline or amorphous) of the drug. Rigorous solubility studies using the final solid form (i.e., salt form or crystal form) as a function of temperature (i.e., 25°C and 37°C) and pH (range 1–7.5) are conducted during preformulation. Solubility in nonaqueous solvents is also screened. Solubility in simulated GI fluids is also important.

For accurate determination of solubility:

- Attainment of equilibrium must be ensured by analyzing solution concentration at multiple time points until the concentration does not change considerably (i.e., <5% change in concentration).
- The pH of the saturated solution must be measured.
- The solid phase in equilibrium with the saturated solution must be analyzed by techniques such as hot stage microscopy, differential scanning calorimetry, or powder x-ray diffraction, to verify if the starting material has undergone a phase transformation.

11.4.2.2 Salt-Form Selection

The selection of an optimal chemical and physical form is an integral part of the development of an NCE.[66] If an NCE is neutral or if its pK_a value(s) is not conducive to salt formation, it has to be developed in the neutral form (unless a prodrug is synthesized) and the only form selection involves the selection of its physical (crystal) form. However, if it exists as a free acid or a free base, then the "form" selection involves the selection of both chemical and physical forms. A decision must be made whether a salt or its free acid or base form should be developed. As will be described in Section 11.5, a salt form may lead to a higher dissolution rate and higher bioavailability for a poorly water-soluble drug. For a drug with adequate aqueous solubility, a salt form may not be necessary, unless, of course, a salt provides an advantage with respect to its physical form. In the pharmaceutical industry, salt selection is usually performed by a multidisciplinary team comprising representatives from the drug discovery, chemical development, pharmaceutical development, ADME, and drug safety departments. Serajuddin and Pudipeddi[40] reported that the following questions need to be satisfactorily addressed by the team in the selection of an optimal salt form for a compound: "Is the acid or base form preferred because of biopharmaceutical considerations? Is the salt form more suitable? Is the preparation of stable salt forms feasible? Among various potential salt forms of a particular drug candidate, which has the most desirable physicochemical and biopharmaceutical properties?" With respect to physical properties, questions involve whether the compound exists in crystalline or amorphous form, and, if crystalline, whether it exhibits polymorphism.

At the outset of any salt selection program, it is important to determine whether the salt formation is feasible for the particular compound and, if yes, what counterions are to be used. Although it is generally agreed that a successful salt formation requires that the pK_a of a conjugate acid should be well below the pK_a of the conjugate base to ensure sufficient proton transfer from the acidic to the basic species, there is a tendency among certain quarters in the pharmaceutical industry to conduct a large number of experiments during salt selection. Hundreds of individual experiments for salt formation of a particular compound are not uncommon. Because of the availability of HTS screening techniques in recent years, there is no pressure to limit the number of such experiments. Serajuddin and Pudipeddi[40] reported that the number of feasibility experiments can be greatly reduced by studying the solubility vs. pH relationship of the drug and identifying the pH_{max} (the pH of maximum solubility). The nature of the pH–solubility profile and the position of pH_{max} depends on pK_a, intrinsic solubility (solubility of unionized species), and the solubility of any salt (K_{sp}) formed. For a basic drug, the pH must be decreased below the pH_{max} by using the counterion for a salt to be formed, and, for an acidic drug, the pH must be higher than the pH_{max}. Any counterion that is not capable of changing the pH in this manner may be removed from consideration. While salts may be formed from organic solvents by counterions that are not capable of changing the aqueous pH in this manner, such salts may readily dissociate in an aqueous environment. When the synthesis of multiple salts for a compound is feasible, the number may be narrowed down and the optimal salt may ultimately be selected by characterizing physicochemical properties of solids according to a multitier approach proposed by Morris et al.[67]

11.4.2.3 Co-Crystal Formation

Another class of solid form that has attracted much attention in recent years is co-crystal. According to the FDA guidelines,[68] "co-crystals are crystalline materials composed of two or more different molecules, typically drug and co-crystal formers ('coformers'), in the same crystal lattice." Unlike salts, no proton transfer between acids and bases is involved in co-crystal formation.[69] They usually form by electrostatic or hydrogen bonding between two species, and, thus, co-crystals may form even between neutral molecules. They also differ from polymorphs, which usually contain only single components with different arrangements or conformations of the molecules in crystal lattices. Co-crystals are expected to dissociate into individual species once they dissolve in aqueous media. Despite much research in this area, to date, one co-crystal has been developed as the drug product (sacubitril/valsartan or Entresto® Tablet; Novartis), where sodium salts of two drug molecules – one angiotensin receptor 1 (AT1) blocker and another neprilysin (NEP) inhibitor – formed a supramolecular complex.[70] In addition to being able to deliver two drug molecules as the singly crystalline package, the primary advantage of this product was that the co-crystal had superior crystallinity as compared to that of individual sodium salts. Much research has also been conducted to improve solubility, dissolution rate, and bioavailability of drugs by co-crystal formation. However, since crystals undergo dissociation into individual drugs and coformers in aqueous media, any definite advantage of co-crystals on bioavailability enhancement has not yet been established. Further research will be necessary before co-crystals may be used to develop bioavailability-enhancing formulations.

11.4.2.4 Polymorphism

Polymorphism of pharmaceutical solids has been extensively studied.[64] It is defined as the ability of a substance to exist as two or more crystalline phases that have different arrangements or conformations of the molecules in the crystal lattice. Many drug substances exhibit polymorphism. The definition of polymorphism according to the International Conference on Harmonization (ICH) guideline Q6A[71] includes polymorphs, solvates, and amorphous forms. Amorphous solids lack long-range order and do not possess a distinguishable crystal lattice. Solvates are crystal forms containing stoichiometric or nonstoichiometric amounts of solvent in the crystal. When the solvent is water, they are termed hydrates. A thorough screening of possible crystal forms is conducted during candidate lead selection or shortly thereafter.[72]

Typical methods for generation of polymorphs include sublimation, crystallization from different solvents, vapor diffusion, thermal treatment, melt crystallization, and rapid precipitation. HTS methods have been reported for polymorph screening.[73]

Methods for characterization of polymorphs include crystallographic techniques (single crystal and powder x-ray diffraction), microscopic characterization of morphology, thermal characterization (DSC/TGA – differential scanning calorimetry/thermogravimetric analysis), solution calorimetry, solid-state spectroscopic methods (IR, Raman, NMR), and solubility and intrinsic dissolution rate methods. Of these, the relative solubility or intrinsic dissolution rate is directly related to the free energy difference and, hence, the relative stability of polymorphs. Thermal data can also be used to assess relative stability of polymorphs. The form with the lowest solubility and, hence, free energy is the most stable form at a given temperature. Other forms would eventually transform to the stable form. The kinetics of crystal nucleation and growth determines the crystal form obtained during crystallization. Sometimes metastable forms are more readily crystallized than the most stable (and often desired) form. The kinetics of transformation of a metastable form to the stable form may be very slow and unpredictable. The unexpected appearance of a less soluble polymorphic form of a marketed antiviral drug, ritonavir, caused serious problems with manufacture of its oral formulation.[74]

The crystal form can have a profound influence on physicochemical properties of a drug substance. Melting point, solubility, stability, and mechanical properties may depend on the crystal form. A difference in solubility of crystal forms may manifest as a difference in bioavailability, but the impact would depend on the dose, solubility of each form, and permeability. Because of the much concern in the pharmaceutical field on the impact of polymorph solubility on dissolution and bioavailability of drugs, Pudipeddi and Serajuddin[75] conducted a trend analysis of polymorph solubility differences reported in the literature, where they analyzed solubility ratios of the polymorphs of 55 compounds (81 solubility ratios due to the existence of multiple forms for some compounds). The general trend revealed that the ratio of solubility between the most soluble crystal form and the least soluble crystal form of a compound is typically <2, while occasionally, higher ratios can be observed. Although potential effect of polymorphism on drug solubility and bioavailability is definitely a concern, other formulation factors such as the presence of amorphous material, particle size, and wettability often complicate the observed difference in the bioavailability of crystal forms. Hancock and Park[76] reported that amorphous solids may have tens and even hundreds of times higher solubility than their crystalline counterparts. Thus, the presence of even very small amounts of amorphous form generated by milling and other pharmaceutical operations may artificially increase solubility of certain polymorphs.

11.4.2.5   Solid-State Stability

The shelf life of a product is often predicted by the stability of the drug substance. A thorough investigation of the drug substance stability is, therefore, necessary during preformulation following the identification of the final salt or crystal form. Typical stability testing during preformulation includes accelerated testing for 2–4 weeks at high temperatures (50°C–80°C) in dry or moist (75% RH) conditions. Photostability is also conducted at this time.[53] The criteria for an acceptable solid-state stability are compound-specific. Balbach and Korn[53] recommended a degradation of <3% to 5% at 60°C (dry) and <10% to 20% at 60°C (100% RH) in 2 weeks as acceptable. Physical stability (i.e., change in crystal form) must also be investigated under accelerated temperature and humidity conditions.

11.4.2.6   Drug-Excipient Interactions

The stability of drug substance is often greater in its neat form than when it is formulated. Excipients may facilitate moisture transfer in the drug product or initiate solid-state reactions at the points of contact and adversely impact the stability of the drug substance. For the same reason, it is not uncommon for the drug product stability to decrease when the drug concentration in the formulation is reduced.[77] Pharmaceutical excipients can interact chemically with the active pharmaceutical ingredient (API). Drug-excipient compatibility studies are conducted during preformulation to select the most appropriate excipients for formulation. The following classes of excipients for oral dosage forms are commonly employed: diluents or fillers, binders, disintegrants, glidants, colors, compression aids, lubricants, sweeteners, preservatives, suspending agents, coatings, flavors, and printing inks.

A typical drug-excipient compatibility study includes preparation of binary mixtures of the drug and excipients in glass vials. The ratio of the drug to each excipient must be comparable to that in the formulation. It is prudent to set up two sets of samples to bracket the drug concentration foreseen in the final products. Multicomponent mixtures of drug and excipients mimicking prototype formulations can also be made and tested under accelerated conditions. The mixtures are subjected to accelerated stress conditions for a period of 2–4 weeks. The samples are analyzed by HPLC for the percent drug remaining and any degradation products formed. Samples are typically stored at a reference condition (−20°C or 5°C) and at elevated temperatures of 40°C–60°C under dry and moist conditions. Moist conditions are usually obtained by storing samples at 75% or 100% RH or by adding 10%–20% (w/w) water to the mixtures. Serajuddin et al.[78] described the principles and practice of drug-excipient compatibility studies for selection of solid dosage-form composition. Whether physical mixtures represent real drug-excipient(s) interactions in a capsule or tablet formulation has been debated.[79] The use of prototype formulations instead of physical mixtures has been suggested. However, drug substance availability and practicality may limit such an approach. According to Serajuddin et al.,[78] a well-conducted drug-excipient compatibility study using physical mixtures can lead to a good understanding of drug stability and robust formulations.[78] They observed that the addition of 20% water to drug-excipient mixtures in vials and then subjecting the closed vials to storage at 40°C or 50°C for up to 30 days provided stability-indicating results as opposed to storing samples at high humidity. This is because there could be variation in moisture sorption by different excipients that may influence drug-excipient interaction in binary mixtures during compatibility testing, while such variation may not exist in dosage forms and the moisture sorption could be higher due to the possible presence of multiple excipients. The added water may normalize the effects of any variation in moisture content.

One must also consider that pharmaceutical excipients can possess acidity (or basicity). Although the chemical nature of excipients (i.e., mannitol, starch, sucrose) may itself be neutral, trace levels of acidic or basic residuals from the manufacturing process often impart acidity or basicity to excipients. Depending on the relative proportion of the drug and its acid/base nature, the microenvironmental pH of the drug product can be influenced by the excipient acidity or basicity. For example, addition of 2% magnesium stearate to a low drug content formulation can result in a microenvironmental pH above 8. An approximate but simple way to measure the microenvironmental pH of a drug-excipient mixture is to measure the pH of a slurry prepared with minimal amount of water.[80] Excipients may also contain impurities that can accelerate drug degradation. For example, the following impurities may be present in the excipients listed with them: aldehydes and reducing sugars in lactose; peroxides and aldehydes in polyethylene glycol; heavy metals in talc, lignin, and hemicellulose in microcrystalline cellulose, formaldehyde in starch; and alkaline residues such as magnesium oxide in stearate lubricants.[81] The physical state of an excipient (i.e., particle size,[82] hydration state,[83] and crystallinity[84]) can also affect drug stability. It is, therefore, important to review the excipient literature data, such as *Handbook of Pharmaceutical Excipients*,[85] prior to its use in a formulation.

In assessing drug-excipient compatibility, the same guidelines as in solid-state stability can be used. The primary objectives of the study should be to determine whether any incompatibility occurs and, if so, what are the nature of degradation products formed. The excipients responsible for drug degradation should be avoided or, if that is not possible, those causing least degradation may be selected based on analysis of risks involved. Some of the interactions may be avoided by the application appropriate formulation development approaches. For example, if the microenvironmental pH is responsible for the instability of drugs, a pH adjustment may be adequate to stabilize the formulation.[86] If the presence of moisture or water is primarily responsible drug degradation (e.g., hydrolysis), it may be necessary to avoid wet granulation and the protection from moisture may stabilize the formulation during shelf life. It should also be noted whether any potentially toxic compounds are formed during drug-excipient interaction, since they may need to be qualified during the toxicological evaluation or be avoided, if the toxicity is unacceptable.

11.4.2.7   Powder Properties of Drug Substance

For solid oral dosage forms, powder properties such as powder flow, density, and compactibility are important. For products with low drug concentration (e.g., <5% to 10%), the powder properties of the drug substance are usually less influential than excipients on the overall mechanical properties of the formulation. For products with high drug concentration (i.e., >50%, w/w), powder properties of the drug substance (or API) may have significant influence on processability of the formulation. Since pharmaceutical scientists often manage to find engineering solutions to address poor powder properties of materials, there are no hard and fast rules for acceptable powder properties. However, consideration of powder properties from the early stages of development can result in more optimized and cost-saving processes.[87] Some of the powder properties of interest are:

- Particle morphology
- Particle size and particle size distribution (PSD), surface area
- True and relative densities
- Powder flow properties
- Blend uniformity of powders
- Compaction properties
- Powder sticking propensity with metal surfaces.

Hancock et al.[88] reported a comprehensive review of a wide variety of pharmaceutical powders. Investigation of electrostatic properties of the drug substance or drug-excipient mixtures during preformulation has been recommended.[89] Such studies may not only be relevant to solid oral dosage forms but also to dry powder inhalation systems.

Optimal compression or compaction properties of powders are critical for a robust solid dosage form. Although prediction of compaction properties of powders is not fully possible, tableting indexes[90] and other material classification methods[91] have been proposed to assess compaction properties of powders. Recently, Sun et al.[92,93] have undertaken systematic studies to determine tableting behavior of pharmaceutical crystals as well as their propensity for sticking with punch by using only small amounts of materials. It is hoped that such material-sparing methods will enable accurate predictions of mechanical properties and tabletability of new drug candidates at the early stage of drug development process, thus saving later time and efforts.

11.5 BIOPHARMACEUTICAL CONSIDERATIONS IN DOSAGE-FORM DESIGN

For systemic activity of a drug, it must be absorbed and reach the bloodstream or the site of action if given by the oral, topical, nasal, inhalation, or other route of administration where a barrier between the site of administration and the site of action exists. Even when a drug is injected intravenously or intramuscularly, one must ensure that it is not precipitated at the site of administration and that it reaches the site of action. In the development of dosage forms for a particular drug, a formulator must, therefore, carefully consider various physicochemical, biopharmaceutical, and physiological factors that may influence absorption and transport of drugs.[25] Appropriate formulation strategies must be undertaken to overcome the negative influences of any of these factors on the performance of dosage forms.

A large majority of pharmaceutical dosage forms are administered orally, and, in recent years, the drug solubility has become the most difficult challenge in the development of oral dosage forms. For example, in the 1970s and 1980s, when dissolution, bioavailability, and bioequivalence of drugs came under intense scrutiny and many of the related FDA guidelines were issued, a drug with solubility <20 µg/mL was practically unheard of. Presently, new drug candidates with intrinsic solubility <1 µg/mL are very common. The solubility of as low as 4 ng/mL for a marketed compound, itraconazole, has been reported.[94] In addition to solubility, physicochemical factors influencing oral absorption of drugs include dissolution rate, crystal form, particle size, surface area, ionization constant, partition coefficient, and so forth. Among the physiological factors, drug permeability through the GI membrane is of critical importance. Other physiological factors playing important roles in the performance of an oral dosage form are transit times in different regions of the GI tract, GI pH profile, and the presence of bile salts and other surfactants. Physiological differences such as the unfed vs. the fed state also need to be considered.

Since solubility and permeability are the two most important factors influencing oral absorption of drugs, the following Biopharmaceutical Classification System (BCS) for drug substances, based on the work by Amidon et al.,[95] has been recommended by the FDA:[96]

- Class I – Drug is highly soluble and highly permeable
- Class II – Drug is poorly soluble, but highly permeable
- Class III – Drug is highly soluble, but poorly permeable
- Class IV – Drug is both poorly soluble and poorly permeable.

For a BCS Class I compound, there are no rate-limiting steps in drug absorption, except gastric emptying, and therefore, no special drug-delivery consideration may be necessary to make the compound bioavailable. On the other hand, for a BCS Class II compound, appropriate formulation strategy is necessary to overcome the effect of low solubility. For a BCS Class III compound, formulation steps may be taken to enhance drug permeability through the GI membrane, although the options could be very limited. Often, the continued development of a BCS Class III compound depends on whether its low bioavailability from a dosage form because of poor permeability is clinically acceptable or not. A BCS Class IV compound presents the most challenging problem for oral delivery. Here, the formulation strategy is often related to enhancing the dissolution rate to deliver maximum drug concentration to the absorption site. However, similar to a Class III compound, formulation options to enhance drug permeability of Class IV compound are often limited. Because of the importance of BCS in the development of dosage-form design strategy, an early classification of new drug candidates is essential to identify their formulation hurdles.

Some of the physicochemical and physiological factors involved in the design of oral dosage forms are discussed below in more detail.

11.5.1 PHYSICOCHEMICAL FACTORS

11.5.1.1 Solubility

Low or poor aqueous solubility is a relative term. Therefore, the solubility of a compound must be considered together with its dose and permeability. A simple approach to assess oral absorption with a drug substance could be the calculation of its maximum absorbable dose (MAD)[97]:

$$MAD = S \times K_a SIWV \times SITT \tag{11.5}$$

where S is solubility (mg/mL) at pH 6.5, K_a the transintestinal absorption rate constant (per min) based on rat intestinal perfusion experiment, SIWV the small intestinal water volume (250 mL), and SITT the small intestinal transit time (4 h). One limitation of the MAD calculation is that only the aqueous solubility in pH 6.5 buffer is taken into consideration. There are many reports in the literature where the "*in vivo* solubility" of drugs in the GI tract in the presence of bile salts, lecithin, lipid digestion products, etc., was found to be much higher than that in the buffer alone. Therefore, MAD may be considered to be a conservative guide to potential solubility-limited absorption issues and whether any special dosage forms need to be considered to overcome such issues. Advanced software tools are available to estimate oral absorption.[98,99]

For an acidic and basic drug, the solubility over the GI pH range varies depending on the intrinsic solubility (S_o) of the compound (i.e., solubility of unionized or nonprotonated species), pK_a, and the solubility of the salt form.[100]

11.5.1.2 Dissolution

Dissolution rate, or simply dissolution, refers to the rate at which a compound dissolves in a medium. The dissolution rate may be expressed by the Nernst–Brunner diffusion layer form of the Noyes–Whitney equation:

$$J = \frac{dm}{Adt} = \frac{D}{h}(c_s - c_b) \tag{11.6}$$

where J is the flux, defined as the amount of material dissolved in unit time per unit surface area (A) of the dissolving solid, D the diffusion coefficient (diffusivity) of the solute, c_s the saturation solubility that exists at the interface of the dissolving solid and the dissolution medium, and c_b the concentration of drug at a particular time in the bulk dissolution medium. Under "sink" conditions ($c_b < 10\% \; c_s$), the above equation is reduced to

$$J \approx \frac{D}{h}c_s \tag{11.7}$$

where the dissolution rate is proportional to solubility. However, this equation is applicable only at the early stage of dissolution. If the dissolution continues, the dissolution rate decreases progressively as the concentration gradient between the solubility and the concentration in dissolution medium decreases, according to Equation 11.6, and no further dissolution occurs when the solubility reaches equilibrium.

The importance of dissolution on drug absorption and bioavailability may be described in relation to the concept of dissolution number, D_n, introduced by Amidon et al.[95] D_n may be defined as

$$D_n = \frac{t_{res}}{t_{diss}} \tag{11.8}$$

where t_{res} is the mean residence time of drug in the GI tract and t_{diss} the time required for a particle of the drug to dissolve. It is evident from this equation that the higher the D_n, the better the drug absorption, and a maximal drug absorption may be expected when $D_n > 1$, i.e., $t_{res} > t_{diss}$. However, for the most poorly water-soluble drugs, $D_n < 1$, and as a result, a dissolution rate-limited absorption is expected. To ensure bioavailability of poorly soluble drugs, formulation strategies must, therefore, involve increasing the dissolution rate such that the full dose is dissolved during the GI residence time, i.e., D_n becomes ≥ 1.

Some of the current methods of increasing dissolution rates of drugs are particle size reduction, salt formation, and development of the optimized delivery systems, such as solid dispersion and soft gelatin encapsulation.

The effect of particle size on drug absorption as a function of dose and drug solubility was analyzed by Johnson and Swindell.[101] They have shown that for dissolution rate-limited absorption ($D_n < 1$), the fraction of drug absorbed at any particle size will decrease with the increase in dose. On the other hand, if the dose is kept constant, the fraction of drug absorbed will increase with an increase in drug solubility. Salt formation increases the dissolution rate by modifying pH and increasing the drug solubility in the diffusion layer at the surface of the dissolving solid.[102] Depending on the pH in the GI fluid, a drug may precipitate out in its respective free acid or base form; however, if redissolution of the precipitated form is relatively rapid,[103] faster drug absorption is expected from the salt form. Although salt formation increases drug dissolution rate in most cases, exceptions exist. Under certain situations, a salt may convert into its free acid or base form during dissolution directly on the surface of the dissolving solid, thus coating the drug surface and preventing further dissolution at a higher rate. In such a case, the salt formation may not provide the desired advantage and a free acid or base form may be preferred.[104]

In recent years, advanced DDSs, such as solid dispersion, lipid-based formulations, and nanoparticles, have been extensively investigated to increase dissolution rate, absorption, and bioavailability of poorly water-soluble drugs. They will be discussed in greater detail later in Section 11.9.

11.5.2 Physiological Factors

In any formulation development, the GI physiology that can influence dosage-form performance must be kept in mind. The GI contents play important roles, since even in the fasted state in humans, the *in vivo* dissolution medium is a complex and highly variable milieu consisting of various bile salts, electrolytes, proteins, cholesterol, and other lipids. The GI pH is another factor that plays a critical role in the performance of the dosage form. The pH gradient in humans begins with a pH of 1–2 in the stomach, followed by a broader pH range of 5–8 in the small intestine, with the intermediate range of pH values of around 5–6 being found in the duodenum. Colonic absorption in the last segment of the GI tract occurs in an environment with a pH of 7–8. The average pH values significantly differ between the fed and the fasted states. Finally, gastric emptying time and intestinal transit time are very important for drug absorption. The majority of the liquid gets emptied from the stomach within 1 h of administration.[105] Food and other solid materials, on the other hand, take 2–3 h for half of the content to be emptied. The small intestinal transit time generally does not differ greatly, and it usually ranges from 3 to 4 h.[106]

11.5.2.1 Assessment of *In Vivo* Performance

Different formulation principles, dosage forms, and DDS are commonly evaluated in animal models, and attempts are made to predict human absorption on the basis of such studies.[107] Human studies are also conducted in some cases to confirm predictions from animal models. Chiou et al.[108,109] demonstrated that there is a highly significant correlation of absorption ($r^2 = 0.97$) between humans and rats with a slope near unity. In comparison, the correlation of absorption between dog and human was observed to be poor ($r^2 = 0.512$) as compared to that between rat and human ($r^2 = 0.97$). Therefore, although dog has been commonly employed as an animal model for studying oral absorption in drug discovery and development, one may need to exercise caution in the interpretation of data obtained.

In recent years, there has also been much interest in applying physiologically based pharmacokinetic (PBPK) modeling and simulation to predict the pharmacokinetic behavior of drugs, such as absorption, distribution, metabolism, and excretion (ADME) in humans using preclinical data.[110,111] Various *in vitro* data, such as physicochemical properties of drug, reversible inhibition (IC_{50}), permeability, protein binding, and enzymatic metabolism, and results from animal models and initial human dosing, when available, have been used in such modeling and simulation. Using five case studies in different pharmaceutical companies, Kesisoglou et al.[112] described how PBPK modeling may be used to address different biopharmaceutics or formulation issues, including (a) prediction of absorption prior to first-in-human (FIH) studies, (b) optimization of formulation and dissolution method post-FIH data, (c) early exploration of a modified-release formulation, (d) addressing bridging questions for late-stage formulation changes, and (e) prediction of pharmacokinetics in the fed state.

11.5.2.2 *In Vitro–In Vivo* Correlation

The various biopharmaceutical considerations discussed above can help in correlating *in vitro* properties of dosage forms with their *in vivo* performance.[25] *In vitro–in vivo* correlation (IV–IVC), when successfully developed, can serve as a surrogate of *in vivo* tests in evaluating prototype formulations or DDSs during development and reduce the number of animal experiments. This may also help in obtaining biowaivers from regulatory agencies, when applicable, and especially for BCS Class II compounds mentioned above. It can also support and validate the use of dissolution methods and specifications during drug development and commercialization.

The IV–IVC is generally established by comparing *in vitro* dissolution of drug with certain *in vivo* PK parameters. There are certain FDA guidelines for this purpose, where the correlations are categorized as Level A, Level B, Level C, and multiple Level C correlations.[113]

Level A correlation is generally linear and represents the correlation of *in vitro* dissolution with the drug fraction absorbed, which is obtained from the deconvoluted *in vivo* plasma levels. The Level B correlation is more limited in nature, and it does not uniquely reflect the actual *in vivo* plasma level curve; here, the mean *in vitro* dissolution time is compared with the mean *in vivo* residence or dissolution time. A Level C correlation establishes a single-point relationship between a dissolution parameter and a pharmacokinetic parameter. While a multiple Level C correlation relates one or several pharmacokinetic parameters of interest to the amount of drug dissolved at several time points of the dissolution profile, a Level C correlation can be useful in the early stages of formulation development when pilot formulations are being selected. For the purpose of IV–IVC, it is essential that the dissolution medium is biorelevant such that *in vitro* dissolution testing is indicative of *in vivo* dissolution of the dosage form. Dressman[114] published extensively on the identification and selection of biorelevant dissolution media.

11.6 BIOPHARMACEUTICS RISK ASSESSMENT ROADMAP (BIORAM): A PATIENT-CENTRIC DRUG DEVELOPMENT APPROACH

Despite extensive efforts in optimizing clinical leads, selection of drug candidates through drug discovery-development interaction (developability assessment), drug safety evaluation, and physicochemical/biopharmaceutical evaluation, many new drug candidates fail during clinical development as they cannot achieve the desired therapeutic outcome. Even if they are developed and marketed in certain cases, they have only limited patient benefits and poor commercial success due to such unfavorable attributes as inadequate therapeutic action, toxic effects, need for frequent dosing, etc.[115] In 2014, an international team of experts from industry, academia, and regulatory agency critically analyzed the situation and recommended that, to improve the success rate of drug product development, clinical relevance of drug must be directly built into early formulation development.[116] Currently, in a rush to promote new leads to human studies, formulation considerations relevant to clinical needs are often ignored. The team proposed a Biopharmaceutics Risk Assessment Roadmap (BioRAM) for identifying the robust and reliable link between the desired therapeutic outcome and the optimum drug-delivery approach. In this context, the biopharmaceutics risk was defined as "the risk of not achieving the intended in vivo drug product performance by delivering the drug substance in a manner such that it has the best chance of consistently meeting the patient/consumer needs." The team suggested that the development of most drug products may be classified into development scenarios as illustrated in Figure 11.1:

Scenario 1: This scenario depicts a situation where a rapid onset of drug action is desired, which, for example, may be necessary in cases of breakthrough pain, analgesia, acute angina pectoris, insomnia, etc. Rapid onset of action in this scenario is considered to take place in <1 h. For oral dosage form, the drug should not only dissolve rapidly, it should also be well absorbed.

Scenario 2: This scenario involves relatively rapid attainment of therapeutic concentration combined with the maintenance of such a therapeutic concentration, which may be attained by appropriately formulating a single drug or by the delivery of multiple drugs as a fixed combination product. When the compound has long half-life or at least one of the compounds in a fixed combination product has a long half-life, relatively simple formulations may be used to attain Scenario 2; otherwise, modified release (MR) dosage form containing a rapidly dissolving formulation component along with an extended release (ER) component may be necessary. This scenario may be required for combination antihistamines, antimigraine drugs, drugs for attention-deficit hyperactivity disorder (ADHD), and some sleep aids where rapid onset and sleep maintenance are desired.

Scenario 3: This scenario may be most suitable for chronotherapeutic administration of drug in which both the timing (onset of action) and maintenance (exposure) may be critical for therapeutic effects. Some of the drugs that may fall into this scenario are oncology products targeted to tumor growth cycles, blood pressure–lowering drugs intended to address the circadian rhythm of blood pressure, blood glucose–lowering agents to match physiological response to food effect, and so forth. The formulations for Scenario 3 may be relatively complex as they may require both immediate release (IR) and ER components and the adjustment for timing windows.

Scenario 4: This scenario is desired to maintain a target exposure over a time period using MR formulation or repeated dosing. It is appropriate for antibiotics, antihypertensives, antiepileptics, Alzheimer's disease (AD) drug products, chronic pain medications, and antispasmodics.

It is evident from Figure 11.1 that if the formulation strategies do not provide the necessary blood level profiles, there are risks that the products may not be able to attain desired therapeutic outcomes and may ultimately fail. For example, an antimigraine drug will not be effective if it does not dissolve and absorb rapidly after oral administration and a cephalosporin antibiotic will not treat infection if the drug concentration is not maintained above the microbial minimum inhibitory concentration (MIC). The implementation of BioRAM requires a multidisciplinary approach where asking critical questions, timely decision-making, and well-designed integrated studies are needed. The team responsible for the development of BioRAM also developed a companion scoring grid to guide implementation of the strategy.[117] It is expected that the BioRAM strategy will have a broad impact on how the drug products will be developed in the future. It may also influence the LCM of older products by reformulation to better address patients' needs.

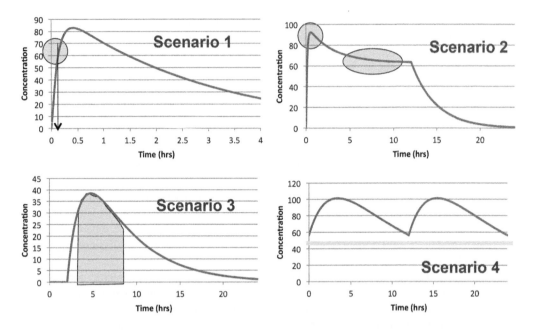

FIGURE 11.1 Four drug-delivery scenarios depicted as drug concentration–time profiles according to the four therapy-driven drug-delivery scenarios: rapid therapeutic onset (Scenario 1), multiphasic delivery (Scenario 2), delayed therapeutic onset (Scenario 3), and maintenance of target exposure (Scenario 4). (Reproduced with permission from Selen, A. et al., *J Pharm Sci.* 103(11), 2014, 3377–3397.)

11.7 CLINICAL FORMULATION DEVELOPMENT: CLINICAL TRIAL MATERIALS

A pharmaceutical dosage form is a means to deliver the active ingredient in an efficacious, stable, and elegant form to meet a medical need. The requirements and specifications of the product will depend on the scope and extent of its intended use in the various clinical phases and then through commercialization. Once the decision is made to move a drug candidate into clinical studies, the interaction of formulation scientists begins to shift from preclinical groups (medicinal chemistry, ADME) to those from development (analytical, process chemistry, clinical), marketing, and operations groups. The formulation, manufacture, packaging, and labeling of CTMs precede the individual phases of clinical development.

Clinical development of NCEs is broadly divided into Phase I, II, III, and then IV (post-marketing) studies.

Phase I studies evaluate the pharmacokinetics and safety of the drug in a small number (20–100) of healthy volunteers. Phase I studies are sometimes conducted in a small patient population (Proof of Concept studies) with a specific objective such as the validation of the relevance of preclinical models in man. The purpose of Phase I studies may be the rapid elimination of potential failures from the pipeline, definition of biological markers for efficacy or toxicity, or demonstration of early evidence of efficacy. These studies have a potential go/no-go decision criteria such as safety, tolerability, bioavailability/PK, pharmacodynamics, and efficacy. Dosage forms used in Phase I or Proof of Concept studies must be developed with objectives of the clinical study in mind.

Phase II studies encompass a detailed assessment of the compound's safety and efficacy in a larger patient population (usually 100–500 patients). It is important that any formulation selected for these studies must be based on sound biopharmaceutical and pharmaceutical technology principles. Phase III clinical studies, also referred to as pivotal studies, involve several thousands of patients (usually 1,000–5,000 or more) in multiple clinical centers, which are often in multiple countries. The aim of these studies is to demonstrate long-term efficacy and safety of the drug. Since these studies are vital in the approval of the drug, the dosage form plays a very critical role.

11.7.1 PHASE I CLINICAL TRIAL MATERIAL

A decision tree approach for reducing the time to develop and manufacture formulations for the first oral dose in humans has been described by Hariharan[118] and is reproduced in the scheme in Figure 11.2. The report summarized numerous approaches to the development and manufacture of Phase I formulations. Additional examples of rapid extemporaneous solution or suspension formulations for Phase I studies have been reported.[119,120]

In deciding the appropriate approach for early clinical studies, it is important to consider the biopharmaceutical properties of the drug substance and the goals of the clinical study. Practical considerations, such as the actual supply of the bulk drug and the time frame allotted for development, enter into the picture. The advantage of using extemporaneous formulations, including "powder-in-bottle", is the short development timelines (a few months) and the minimal drug substance requirements (a few hundred grams depending on the dose). Additional benefits include high-dose flexibility, minimal compatibility or formulation development, and minimal analytical work. The disadvantages include possible unpleasant taste, patient compliance issues, and dosing inconvenience for multiple-dose studies. For poorly soluble compounds, use of a nonaqueous solution may result in high systemic exposure that may be difficult to reproduce later on with conventional formulations.

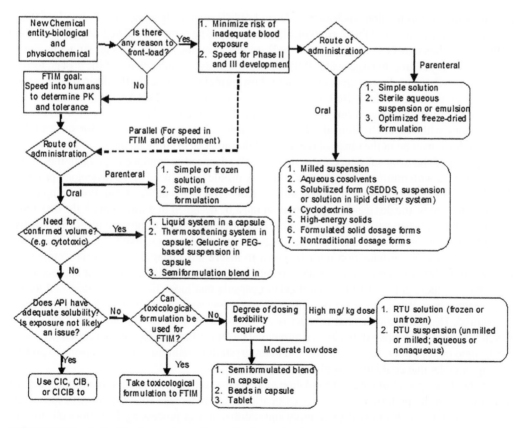

FIGURE 11.2 A decision tree approach for the development of first human dose. FTIM, first in man; CIC, chemical in capsule; CIB, chemical in bottle; CICIB, chemical in capsule in bottle; RTU, ready to use. (Reproduced with permission from Hariharan, M. et al., *Pharm Technol.* 2003, 68–84.)

The "drug-in-capsule" approach, where the neat drug substance is encapsulated in hard gelatin capsules, has similar time, material, and resource advantages but is limited to compounds that exhibit rapid dissolution.

An intermediate approach for Phase I CTM is a "formulated capsule" approach, where the preformulation and compatibility data are optimally utilized and a dosage form that has scale-up potential is developed. The pharmaceutical composition is chosen on the basis of excipient compatibility and preliminary *in vitro* dissolution data. Extensive design of experiments is not conducted at this stage. Processing conditions are derived from past experience or simple reasoning. Several hundred grams of the drug substance are typically necessary to develop a formulated capsule product. However, the advantage of a well-formulated capsule product is that it may be processed on a low- to medium-speed automatic equipment to meet the demands of a larger clinical study, if needed. A formulated tablet cannot be ruled out for Phase I; there is anecdotal information that a few drug companies use this approach. However, tablet development activities require more material and resources. While formulation design activities are in progress, the development of analytical methods must be initiated to develop an assay to separate the drug from its excipients, a stability-indicating method, and other tests (content uniformity, dissolution) that will be required to release a batch for human use. Limited drug substance and drug product stability testing are required to support a shelf life (typically a few months, preferably at room temperature or under refrigeration) sufficient to conduct the Phase I study.

11.7.2 PHASE II CLINICAL TRIAL MATERIAL

Extemporaneous formulations such as "powder-in-bottle" are unlikely to meet the demands of Phase II. Formulated capsules or tablets are typically necessary. The same formulated dosage form used in Phase I studies may be continued for Phase II studies, if all the required dosage strengths can be supported and medium- to large-scale (100,000 or more units) manufacturing of the dosage form is feasible. Alternatively, development of a more robust formulation closer to the desired commercial form may be undertaken. It is also the time to start considering the BioRAM approach mentioned earlier in formulating dosage forms since some of the Phase II clinical results may be used in critical decisions with respect to clinical efficacy. The drug substance requirements for such a Phase II dosage form may be in the range of the tens of kilograms, depending on the dose. The chemical development of the drug substance must also be well advanced to produce large quantities of the drug substance with minimal change in the impurity profile that is used in the toxicology program. The design of the clinical study may also influence the type of the dosage form. For example, in a double-blind study, the dosage-form presentation must be designed in a way to mask the difference in appearance between various dose strengths. It is often necessary to prepare placebos or active controls that consist of a marketed product disguised (masked) to look like the CTM. This is not a trivial exercise as consideration must be given to insure any changes made to the commercial product in the blinding process do not alter its bioavailability or stability. A popular approach is to place the commercial product into a hard gelatin capsule that looks like the capsules used for the investigative drug. When this is not possible (due to size issues) then the "double-dummy" approach is required. Here, the commercial product is encapsulated in a larger capsule, and a placebo to match the capsule is prepared. In case a commercial tablet is filled in a capsule, it is often necessary to fill some inert powder, such as lactose and microcrystalline cellulose, into the capsule to prevent rattling of tablet that could unmask blinding.

For the development of robust formulation and process, critical formulation issues must be first identified from the preformulation work. Critical issues may include solubility or dissolution rate for poorly water-soluble drugs, drug stability and stabilization, or processing difficulties due to poor powder properties of the drug substance (for high-dose formulations). In the initial stages of product development, *in vitro* dissolution in physiologically meaningful media must be utilized to guide the development of the prototype formulations.[96,114,121] If drug stability is identified as a potential issue, more thorough and careful drug-excipient compatibility studies may be necessary. On the basis of results of forced degradation and compatibility studies, a stabilization mechanism may be identified. Yoshioka and Stella[62] have thoroughly reviewed drug stability and stabilization. Stabilization strategies such as incorporation of antioxidants,[122] protection from moisture, or use of pH modifiers[123] may be considered. If a "functional" excipient such as an antioxidant or a pH modifier is required, then tests to monitor its performance (i.e., consumption of the antioxidant during storage) must be developed.

A pragmatic but critical factor in Phase II formulation development is the identification of the dosage strengths by the clinical development scientists. Since Phase II studies cover a large dose range, the number of dosage strengths may be quite large. The selection of minimal number of dosage strengths that can still support the needs of the clinical studies is best handled by a team comprising pharmaceutical scientists, pharmacokineticists, and the clinical study and clinical supply coordinators. Individual strategies must be developed for both the high and low dose strengths. The limitations on the high dose strength may include dissolution, processability, and size of the unit, whereas the low dose strengths may face content uniformity or stability issues. Once the critical formulation issues are identified, a systematic approach is necessary to develop the optimal composition and process. Statistical design of experiments provides a systematic basis for screening of the prototype formulations or variants. It is important to correctly identify the desired responses to be optimized (i.e., dissolution, chemical stability, processing of a unit operation). When the drug substance is in short supply, it is, however, not uncommon to use empirical or semi-statistical approaches to the formulation development. In developing Phase II formulations with scale-up potential, manufacturing equipment that operates on the same principle as the production equipment should be used as much as possible. A difference in the operating principle such as low- vs. high-shear granulation may yield different product characteristics.[124] About three to four variants are identified from the screening studies for stress stability testing. During Phase I and II clinical studies, the duration of the stability study depends on the extent of the clinical studies. One- to two-year stability studies under ICH[125] recommended storage conditions are desired with both accelerated and real-time storage conditions. Stability studies to monitor changes in drug product performance characteristics include potency, generation of degradation products, product appearance, and dissolution. Stability data from these studies are utilized to develop the final product and to select the most appropriate packaging conditions.

Hard gelatin capsules and tablets are among the most common dosage forms for Phase II clinical studies. Capsules are considered to be more flexible and forgiving than tablets for early clinical studies. Capsules may also offer an easy means of blinding as noted previously for the use of active controls. The filling capacity of capsules is limited (usually <400 mg fill weight of powders for a size 0 capsule), and large doses may require capsules that are not easy to swallow.

Production of tablets at medium-to-large scale requires more stringent control of powder properties due to the high-speed compression step. Processing of tablets and the physics of tablet compaction have been the subject of extensive investigation and voluminous literature exists on the topic.

Overall, the development of a robust formulation with scale-up potential for Phase II studies involves integration of physicochemical, biopharmaceutical, and technical considerations. Whether a rudimentary formulated capsule or a more robust formulation closer to the commercial form will be used in Phase II studies will depend on the company policy, material cost, the complexity of clinical design, and the development strategy.

11.7.3 PHASE III CLINICAL TRIAL MATERIAL

The purpose of Phase III clinical studies is to demonstrate the long-term safety and efficacy of the drug product. For optimal therapeutic outcome, it is important to consider the BioRAM approach in the development of Phase III CTM. To insure the same safety and efficacy in the commercial product, it is necessary that the Phase III CTM should be close to or identical to the final commercial form when possible. A brief overview of the regulatory requirements for the commercial dosage form may help contribute to the understanding of the strategies that can be used for Phase III CTM.[125] According to the ICH guidelines, data from stability studies should be provided on at least three primary batches of the drug product. The primary batches should be of the same formulation and packaged in the same container closure system as proposed for marketing. Two of the primary batches should be of pilot scale or larger (i.e., the greater of 1/10th of the production scale or 100,000 units) and the third batch may be of smaller (lab) scale. Production scale batches can replace pilot batches. The manufacturing processes for the product must simulate those to be applied to the production batches and should provide a product of the same quality. The stability study should typically cover a minimum period of 12 months at the long-term storage condition (i.e., 25°C/60% RH). As a result of these requirements, the formulation, manufacturing process, and packaging of the drug product must be finalized at about 1.5–2 or more years prior to the filing of a regulatory application for an NCE.

If the company policy is to use at least pilot-scale batches for Phase III clinical studies, the formulation and manufacturing process must be finalized prior to the initiation of the pivotal clinical studies. This activity may require large quantities of the drug substance (i.e., hundreds of kilograms depending on the dose and dosage strengths). Additionally, the synthesis of the drug substance must be established.[125] The development of Phase III CTM must occur in parallel to Phase II clinical studies. The advantage of this approach is that a robust and well-developed formulation is utilized in Phase III clinical studies and scale-up risk is reduced. Scale-up and process validation can occur subsequently as per regulatory guidelines.[55] The concepts of design for manufacturability and process analytics technologies[126] are best incorporated into the initial design phases to promote the generation of robust, cost-efficient processes. Alternatively, the dosage form may be scaled-up to production scale at the launch site, and the final commercial dosage form may be utilized for Phase III studies. This approach requires even more of the drug substance and a fully validated drug substance synthesis. Depending on the project strategy and the availability of resources, this approach may eliminate the risk of scale-up-related delays in launch. A detailed description of product scale-up and validation is beyond the scope of this chapter, but extensive literature is available on this topic.[127,128]

If the company resources, project timelines, or the complexity of the pivotal study design do not allow for the development of a pilot or the production scale formulation, the Phase III CTM may be significantly different from the final commercial form. The dosage form used for Phase III must, however, be robust enough to provide an uninterrupted drug supply for clinical studies. The Phase I or II CTM with the appropriate modifications (i.e., dosage strengths, blinding, manufacturability) may be used to meet the specific goals of Phase III. When significant composition or process changes are made to the CTM, a biopharmaceutical assessment must be made on the potential impact on systemic exposure. Clinical studies to establish bioequivalence between the previously used CTM and the new CTM or with the intended commercial dosage form may be required. The systemic exposure from early clinical formulations serves as a benchmark for later formulation changes. Demonstration of the bioequivalence of the Phase III CTM with the commercial dosage form, if different, may be necessary before regulatory approval and also in the case of certain post-approval changes.[129] In a bioequivalence study, a test product (such as an intended commercial dosage form) is compared with a reference formulation (i.e., a Phase III CTM material or an innovator's product in case of pharmaceutical equivalents) according to the guidelines established by the regulatory authorities. In the case of formulation changes for a CTM, the systemic exposure is compared to the earlier benchmark formulations, and substitution by the new product may be made on the basis of the product's efficacy and safety. Two products may be assessed to be equivalent even if the bioequivalence criteria are not strictly met, provided the new product does not compromise the safety and efficacy established by the previous product.[130] It may be possible to obtain a waiver for bioavailability or bioequivalence of formulations, depending on the BCS classification and the availability of established IV–IVC as discussed in the previous sections.

11.8   NON-ORAL ROUTES OF ADMINISTRATION

As noted earlier, this chapter is focused on the design and evaluation of oral dosage forms. Here, only a glimpse of some of the different biopharmaceutical issues that confront the design of a few non-oral dosage forms is provided.

11.8.1   PARENTERAL SYSTEMS

Many drugs are administered as parenterals for speed of action because the patient is unable to take oral medication or because the drug is a macromolecule such as a protein that is unable to be orally absorbed intact due to stability and permeability issues. The U.S. Pharmacopoeia defines parenteral articles as preparations intended for injection through the skin or other external boundary tissue, rather than through the alimentary canal. They include intravenous, intramuscular, or subcutaneous injections. Intravenous injections are classified as small volume (<100 mL per container) or large volume (100 mL or larger volume per container) injections. The majority of parenteral dosage forms are supplied as ready-to-use solutions or reconstituted into solutions prior to administration. Suspension formulations may also be used.[131]

The decision to develop a commercial parenteral dosage form must be made during drug discovery itself because the developability criteria are different than those of an oral dosage form. Additionally, extensive preformulation studies must be conducted to fully understand pH-solubility and pH-stability properties of the drug substance: equilibrium solubility and long-term solution stability are important for parenteral dosage-form development.[132,133] The drugs generated by the "biopharmaceutical" industry are typically macromolecules such as proteins. Macromolecular drugs are generated via genetic engineering/fermentation techniques and purified in an aqueous solution. The material at the final processing step is measured into a package for ultimate delivery to the patient. If further processing such as lyophilization is required to maintain adequate shelf life of the macromolecule, the composition (excipients) of the final product is carefully selected to maintain the physical and chemical stability of the bulk drug. This requires the coordination of the protein/processing specialist and the formulator to insure that the requisite stabilizers, pH control, and preservatives are added to the broth in the final processing steps.

Subcutaneous and intramuscular injections are administered in small volumes as bolus (1–4 mL). Intravenous injections can be given as bolus (typically 5 mL or less) with larger volumes administered by infusion. Due to the small volume of bolus injections, high drug concentrations (up to 100–200 mg/mL) may be required for administration of large doses. As a result, solubility enhancement is a major consideration for parenteral dosage forms. Solubilization principles for parenteral dosage forms have been reviewed.[134] Common approaches to enhance solubility include pH adjustment, addition of a cosolvent, addition of a surfactant, complexation, or a combination of these approaches. Ideally, parenteral dosage forms should have a neutral pH. Injection volume and potential for pain[135] at the site of injection must be carefully evaluated in the selection of the pH of the formulation.

Judicious selection of excipients for parenteral dosage forms is critical due to their systemic administration.[136–138] Excipients of a parenteral dosage form may have a significant effect on product safety including injection site irritation or pain. Permissible excipients for parenteral dosage forms are far less than those for oral dosage forms.

Parenteral products must be sterile, particulate-free, and should be isotonic. An osmolarity of 280–290 mOsmol/L is desirable. Slightly hypertonic solutions are permissible, but hypotonic solutions must be avoided to prevent hemolysis. It is essential that any product for injection be sterilized by a validated process starting with Phase 1 CTM. The method of sterilization (heat, filtration, high energy) of the product must be carefully considered as it can have a profound influence on the stability of the drug. This is also true for the choice of the packaging components as they too can influence stability by releasing materials (such as vulcanizing materials from rubber stoppers) capable of interacting with the drug.

Strategies for development of CTMs of parenteral dosage forms depend on their end use. For example, the requirements of an intravenous formulation intended only to determine the absolute bioavailability of a solid dosage form are more flexible than that being designed for commercial use. For example, if the stability of a solution formulation is limited, it may be supplied frozen to conduct a "one-time" bioavailability study. On the other hand, a parenteral formulation for long-term clinical use may require more extensive development to overcome such liabilities. If the stability of a solution formulation is not suitable for the desired storage condition (room temperature or refrigeration), a lyophilized formulation may be necessary. Principles and practice of lyophilization have been reviewed.[139]

11.8.2 INHALATION SYSTEMS

To treat diseases of the upper airways, drugs have been formulated as dry powders for inhalation, as solutions for nebulization, or in pressurized metered dose inhalers with the goal of delivering the drug topically in the bronchial region. The FDA considers inhalation dosage forms as unique as their performance is markedly influenced by the formulation, the design of the packaging (container and valve) as well as patient-controlled factors. The factors such as rate and extent of the breath can influence the delivered dose uniformity and the PSD. An additional challenge for the formulator is the limited number of excipients that have been demonstrated to be safe when delivered into the lung, which must be carefully sourced and controlled.[140]

The formulator's primary focus is on how to produce particles of the requisite size and how to maintain them as such prior to and during administration. For solid drugs, the particles are targeted to be produced in the range of 1–5 μm by milling, controlled precipitation, or spray drying. When a drug is sufficiently soluble in water, its solutions can be processed to create droplets in the appropriate particle size range by nebulizers (solutions must be sterile). Such fine particles are subject to agglomeration due to static charges or humidity and require exquisite control of the environment (atmospheric such as humidity, and local such as excipients) during production, packaging, and storage as changes in the PSD can profoundly influence the efficiency of delivery. These constraints can have a negative impact on the cost of goods. The pulmonary humidity may also result in changes to the PSD as the particles travel the length of the airways.[141] In addition to local lung delivery, pharmaceutical scientists have noted the extensive surface area of the lower, gas-transport region of the lung, the alveolar region, and have been seeking a means of reproducibly administering potent therapeutic agents into this region for rapid transport into the systemic circulation, thus bypassing the GI tract and liver. The pharmacokinetic profile from this noninvasive method mimics that of IV injection.[142] There is currently great interest to deliver macromolecules such as insulin in this manner employing the drug in solution or dry powders.[143] The ability of a vaporized (smoked) drug to exert a rapid central nervous system (CNS) effect was well known to our great ancestors, and such a rapid delivery is also currently under consideration for treatment of acute and episodic conditions such as migraine or breakthrough pain.[144]

11.9 DRUG-DELIVERY SYSTEMS

DDSs are essentially specialized dosage forms developed to overcome the limitations of conventional dosage forms, such as simple tablets, capsules, oral solutions, and injectable solutions. Depending on the type or relative novelty of such delivery systems, they are also referred to as new or novel DDS. In case of oral delivery of drugs, the DDS are often used to address various drug developability issues, such as the following:

- Poor solubility
- Low dissolution rate
- Unfavorable lipophilicity

- Poor permeability
- Poor stability
- Relatively short half-life
- First-pass metabolism
- GI pH effect
- Drug instability in GI fluid
- Effect of food and other GI contents
- Low absorption rate and bioavailability.

The DDS may also be necessary for the following:

- Increase patient convenience
- Develop once-a-day dosage form (compared to multiple-time dosing)
- Geriatric formulation
- Pediatric formulation
- Intellectual property
- New formulation principles
- Product and process improvement.

Many different technologies to address these drug-delivery issues are either available or are emerging through the efforts of Big Pharma or various smaller drug-delivery companies. For certain specific applications, multiple technologies may be available. For example, matrix tablets, coated tablets, osmotic release systems, and single unit vs. multiple units are available for prolonged-release dosage forms. Of course, each of these systems has its own advantages and disadvantages.

Currently, over two-thirds of NCEs in the drug discovery and development pipeline are extremely insoluble in aqueous media, and therefore, there is a great need for the solubility and dissolution-enhancing DDS in drug development. Williams et al.[145] reviewed extensively various DDS available to address solubility and dissolution. They include solid dispersion,[10,11] lipid-based DDS,[9,146] and nanoparticles.[147] Various *in vitro* models to predict *in vivo* performance of oral DDS have also been reported.[148]

Although solid dispersion as a DDS was first reported almost 60 years ago by Sekiguchi,[149] its application in the development of pharmaceutical products was very limited due to different manufacturing issues.[78] Due to the introduction of hot melt extrusion (HME) in the manufacture of solid dispersion,[150] its use in the pharmaceutical field has been reinvigorated as several products have already been marketed in recent years and others are under development.[151] Recently, Singh et al.[152] and Parikh et al.[153] developed a novel solid dispersion system by acid–base interaction in aqueous media and then drying. They observed extremely high solubility of drugs, which they called super-solubilization, when the solid dispersions were mixed with aqueous media. One issue with such a solid dispersion was that it was somewhat semisolid and viscous in nature, which was later resolved by adsorbing it onto silicates[154] or by HME along with polymers.[155]

The lipid-based DDS has also its own issues. Since lipids and surfactants commonly used for such a DDS are usually liquids, they result into liquid formulations that need to be encapsulated in soft gelatin capsules or filled into bottles. In recent years, there have been multiple reports on the successful conversion of liquid lipid-based DDS into solid dosage forms.[156–158]

The DDS has also been used extensively in the development of MR dosage forms. As mentioned earlier, the MR dosage forms are required to meet the needs of different BioRAM scenarios. Strategies to conduct feasibility assessment for the development oral MR formulations have been reported.[159]

The need for DDSs is possibly the greatest in the case of biotechnologically derived products that cannot be orally absorbed, such as peptides, proteins, and oligonucleotides. Some of the major formulation issues with such products are:

- Invasive dosage forms (injection)
- Short half-life
- Poor stability of drugs or dosage forms
- Lack of drug targeting.

Many different approaches are being applied to resolve these issues. Some of the techniques include biodegradable microspheres, PEGylation, liposomal delivery, electroporation, prodrug and conjugate formation, dry powder inhalers, supercritical fluid-based nanoparticles, viral and nonviral vectors, and so forth. Some of the successes in this area include first microsphere sustained release formulation of a peptide LHRH (Lupron Depot®; TAP/Takeda) in 1989; first PEGylated sustained release formulation of a protein adenosine deaminase (Adagen®; Enzon) in 1990; first microsphere sustained release formulation of a recombinant protein, human growth hormone (Nutropin Depot®; Genentech) in 2000; and first PEGylated sustained release formulation of a recombinant protein, interferon a-2b (PEG-Intron®; Schering-Plough) in 2000. Since then, many more biotech products utilizing DDS, including one for inhaled insulin,[160] have been marketed.

It is impossible to describe the breadth of drug-delivery activities in one section of a chapter. The general approach is to focus on a particular "anatomical niche," i.e., use of the lung as a portal for systemic delivery of drugs and to develop technology to overcome the barriers as they are unveiled. Physiologic barriers can be overcome with knowledge of their molecular basis: use of inhibitors to block CYP metabolism of a drug, use of specific permeability enhancers to facilitate transport across a membrane system, blockage of efflux transporters, and use of a prodrug to optimize the partition coefficient and transport mechanisms. Another approach is to take the notion that a particular technology may have DDS applicability, for example, nanoparticles, and then seek applications. Additionally, as "new" diseases are identified such as age-related macular disease, new noninvasive methods of delivery are required to reach the (retinal) targets at the back of the eye.[161]

The following are among the many novel delivery system approaches currently under development:

- *Tablets*: fast dissolving, lipid-carrier based, slow releasing, disintegrate without water, float in the stomach (gastric retentive), buccal
- *Dermal*: patches, iontophoresis
- *Solubility enhancement systems*: lipid-based systems, nanoparticles, surfactants, semi-solid formulations.

Oral methods for peptides

- *Inhalation*: peptide delivery, vaporization
- *Site-specific delivery*: liposomes, drug monoclonal antibody
- *Implants*: biodegradable polymers
- Drug-eluting stents.

A new trend in the delivery of medicines is to employ a device component. This may be an implantable pump for insulin, a metallic stent coated with a drug, or unit capable of rapidly vaporizing a discrete dose for inhalation. Such products are regulated by the FDA as "combination" products and may be reviewed by multiple centers within the agency, which may require additional levels of documentation to support the product design.

In selecting one DDS over another, technical feasibility must be considered. One consideration that is often overlooked during the development of the drug-delivery opportunities is the dose. For example, an oral tablet may be able to deliver as much as 1 g of drug, while the maximum limit for an inhaled dose could be 1 mg/dose and a transdermal patch 5–10 mg/day. Therefore, an inhalation product or transdermal patch could not be the substitute for the poor bioavailability or first-pass metabolism of a 500 mg tablet. Even the maximum dose in a tablet could be limited if a larger amount of excipients is needed for controlled release, taste masking, etc. The size limitation usually does not permit a larger than 300 mg dose in a capsule. For injectables, 1–2 µg/day could be the maximum dose for a monthly depot system, while 50–100 mg might be delivered intramuscularly or up to 750 mg by intravenous bolus administration, and as much as 1 g or even more may be administered by infusion. Another important consideration is the cost. The cost of manufacturing of a DDS should not be so high that the product is not commercially feasible.

11.10 PRODUCT LIFE-CYCLE MANAGEMENT

Because of the high attrition rate of new drug candidates during development, drug companies are often forced to conserve resources during initial development and bring a product to the market that may not have optimal pharmaceutical and clinical attributes. This approach also leads to faster availability of new therapies to patients. After the initial launch, further development activities leading to superior products, known as the LCM, continue. The LCM may lead to new dosage forms and delivery systems, new dosing regimen, new delivery routes, patient convenience, intellectual property, and so forth. An LCM program is considered successful only if it leads to better therapy and patient acceptance. There are numerous examples of successful LCM through the development of prolonged-release formulations. Development of nifedipine (Procardia® XL, Pfizer), diltiazem (Cardizem CD®, Aventis), and bupropion HCl (Wellbutrin SR® and XL®, GSK) prolonged-release products that not only provided more convenient and better therapy to the patients by reducing dosing frequency but at the same time greatly increased sales of the products are well-known examples. Even old compounds like morphine and oxycodone were turned into significant products by the development of more convenient prolonged-release formulations (MS Contin® and Oxycontin®, respectively; Purdue Pharma). Because of the widespread abuse of opioids, including morphine and oxycodone, the company has further managed the LCM of products by introducing newer abuse-proof versions (Oxycontin® ER).

Issues with the bioavailability of drugs due to their poor solubility or reduced permeability were discussed earlier. Many of the future LCM opportunities may come through bioavailability enhancement. Solid dispersion, microemulsion, soft gelatin capsule formation, solubilization, lipid-based DDSs, nanoparticle or nanocomposite formation, etc. are some of the common bioavailability approaches that can be utilized for LCM. Development of Lanoxicaps® by Burroughs Wellcome in 1970s by encapsulating digoxin solutions in soft gelatin capsules is a classic example of LCM by bioavailability enhancement and better pharmacokinetic properties. The development of a microemulsion preconcentrate formulation by Novartis (Neoral®), where the variability in plasma and the effect of food were reduced greatly, is another well-known example.[162]

LCM through the development of fixed combination products, where two or more drugs are developed or copackaged into a single entity, is gaining increased popularity. The fixed combination products often provide synergistic effects, better therapy, patient compliance, patient convenience, increased manufacturing efficiency, and reduced manufacturing cost. However, a clear risk/benefit advantage is essential for the successful LCM by combination products; mere patient convenience may not be sufficient. Common justifications for the development of fixed combination products include improvement of activity such as synergistic or additive effect, improved tolerance by reduced dose of individual ingredients, broadening of activity spectrum, improvement of pharmacokinetic properties, and simplification of therapy for the patient.

The development of oral dosage forms that disintegrate or dissolve in the mouth is providing LCM opportunities for pediatric, geriatric, or bedridden patients who have difficulty in swallowing. They are also being used by active adult patients who may not have ready access to water for swallowing tablets or capsules.

11.11 FDA INITIATIVES TO IMPROVE DRUG PRODUCT QUALITY AND MANUFACTURING

The pharmaceutical sector in the United States is a major, trillion-dollar industry that has great impacts on health and well-being of population. However, unlike other industries, such as electronics, semiconductors, food, chemical, automotive, and aerospace, where tremendous progress has been made during the past quarter century with respect to quality, process, and manufacturing technologies, the progress in the pharmaceutical industry has been limited.[163] The manufacturing processes in the pharmaceutical industry have remained static since drug products are still produced by batch manufacturing rather than continuous processing and the drug release is performed by conducting certain tests after manufacturing rather than based on the monitoring of product quality throughout the manufacturing process. As the focus has been on testing rather than building quality into drug products, product specifications have often been based on discrete or zero-tolerance criteria, resulting in inadequate understanding of the process and product quality. Any deviation from product specification results in batch failures although the overall quality of the product may not be compromised. To address the situation, and in particular to improve product quality, decrease product failures, and lower manufacturing costs, the FDA has undertaken several initiatives in collaboration with the pharmaceutical industry, which are described below.

11.11.1 QUALITY BY DESIGN

In 2002, the FDA introduced the initiative on pharmaceutical cGMPs of the 21st century, for which the final report was issued in 2004.[126] Prior to this initiative, only limited development information was included in the regulatory submission, and preapproval inspections were focused on systems as opposed to science. The process was essentially locked once the manufacturing process validation was completed, and no meaningful changes could be made after the approval of a product without prior approval from regulatory agencies. The objective of the FDA initiative was to address the situation, which Dr. Janet Woodcock, head of the Center for Drug Evaluation and Research (CDER) in the FDA summed up as follows: "To make [the] link [between measurement and risk], we must turn to the science of manufacturing and the concept of quality by design (QbD), which means that product and process performance characteristics are scientifically designed to meet specific objectives, not merely empirically derived from performance of test batches. To achieve QbD objectives, product and process characteristics important to desired performance must be derived from a combination of prior knowledge and experimental assessment during product development."[164]

FDA sought international collaboration to implement the pharmaceutical cGMPs of the 21st-century initiative. Several guidances were issued through ICH: Q8 for pharmaceutical development, Q9 for quality risk management, and Q10 for quality management. According to the ICH guidance Q8 (R2), the aim of pharmaceutical development is to design a quality product and its manufacturing process to consistently deliver the intended performance of the product.[165] This and other guidances formed the basis of pharmaceutical QbD. By definition, QbD is a systematic, scientific, risk-based, holistic, and proactive approach to pharmaceutical development that begins with predefined objectives and emphasizes product and processes understanding and process control.[166] It also means designing and developing formulations and manufacturing processes to ensure predefined product quality objectives.[164] The key elements of the QbD approach include: (a) define quality target product profile (QTPP); (b) design and develop product and manufacturing processes; (c) identify critical quality attributes (CQAs), critical process parameters (QPPs), critical material attributes (QMAs), and sources of variability; and (d) control manufacturing processes to produce consistent quality over time. Yu et al.[167] described comprehensively how QbD can be applied in pharmaceutical unit operations for manufacturing of solid dosage forms. They also identified various QTPP, CQAs, CMAs, and CPPs for various unit operations.

The QbD, along with the ICH guidance Q8, also establishes the concept of design space in drug product development, within which one can change all formulation and variables without impacting the final quality of the product desired or produced.[163] The extensive information and knowledge gained through years of pharmaceutical development and manufacturing studies, the underlying scientific principles, and the application of QbD framework form the basis of design space. It is expected that products manufactured anywhere within the design space would meet product specifications, thus giving flexibility for any changes during scale-up, technology transfer, and commercial manufacturing. In some cases when the design space is broad, for practical purposes, one can select a narrower "control" space, where the ranges of parameters are set for the final process.

Although the QbD firmly establishes that quality cannot be tested into products, rather it should be built by design, it has not yet been fully implemented for regulatory submission of new drug products. It has found general application only in the development of generic pharmaceutical products.[168] However, there is good appreciation for QbD in the pharmaceutical industry, and it is hoped that QbD will be applied to all pharmaceutical products in the future.

11.11.2 Process Analytical Technology (PAT)

PAT is a system for designing, analyzing, and controlling manufacturing through timely measurements (i.e., during processing) of critical quality and performance attributes of raw and in-process materials and processes with the goal of ensuring final product quality.[169] The term *analytical* in PAT includes chemical, physical, microbiological, mathematical, and risk analyses combined in an integrated manner. The on-, in-, and/or at-line measurements and control in PAT can accomplish the following[169,170]:

- Reduce production time
- Prevent rejects, scrap, and re-processing
- Offer the possibility of real-time release
- Increase automation to improve operator safety and reduce human error
- Facilitate continuous processing
- Improve process understanding
- Reduce cost.

Various tools applied for PAT may be categorized as[171]:

- Multivariate data acquisition and analysis tools
- Modern process analyzers or process analytical chemistry tools
- Endpoint monitoring and process control tools
- Knowledge management tools.

Some commonly used instruments in PAT are near-infrared (NIR), process Raman spectroscopy, process infrared (IR), online UV-visible analysis, NIR chemical imaging, process mass spectroscopy, etc.

11.11.3 Continuous Processing

As mentioned earlier, pharmaceutical manufacturing is essentially a batch process. Dr. Janet Woodcock and her associates at the FDA[172] opinioned that the batch process is not suitable for implementing the FDA's pharmaceutical quality for the 21st-century initiative,[126] since it does not "promote a maximally efficient, agile, flexible pharmaceutical sector that reliably produces high-quality drugs without extensive regulatory oversight." Another report from the FDA stated that "the pharmaceutical manufacturing sector remains relatively inefficient and less understood as compared with those in other chemical process industries."[173] For these reasons, there is great emphasis in continuous processing, where the material is fed from one end and product is obtained on the other end in a continuous manner.

In recent years, there has been considerable progress towards achieving the FDA goal of continuous manufacturing. Various attempts have been made to bring pharmaceutical unit operations from a batch process into a continuous one. One such attempt was made by Leuenberger[174] by integrating multi-cell fluidized bed dryers with high-shear granulator to produce dried granules for the manufacture of tablets. This approach, however, did not produce materials in a continuous fashion; rather, it produced small batches in a *semi-continuous* manner. Granulation is the critical and most difficult unit operation for manufacturing of tablets. Conventionally, wet granulation or even melt granulation has been a batch process due to the inherent design of the high-shear granulators. Therefore, to convert tablet manufacturing into a continuous process, the granulation process must be converted to a continuous one. By applying twin-screw granulators, both wet and melt granulation processes have been converted from batch processes into continuous ones.[175–177] In the twin-screw wet granulation process, material blend is fed at one end of the extruder barrel and water is added continuously using one of the ports in the extruder. Wet granules are obtained on the other end of the barrel. Usually, twin-screw wet granulation process has been carried out at room temperature, and therefore, wet granules require drying, milling, and sieving before further processing. The granules with suitable PSD are blended with lubricant and disintegrant before tablet compression. Although twin-screw extruders can produce wet granules continuously, integration of other unit operations like drying, milling, and sieving makes the process *non-continuous*. Recently, Meena et al.[175] developed a twin-screw wet granulation process at elevated temperature and optimized the process to produce the granules in a specific size range suitable for tablet processing. In addition, due to high temperature used in the extruder (around 70°C), the granules were completely dry and did not require further drying. In this process, a granulation step was combined with drying, milling, and sieving in a one-step process.[175,178] Similarly, twin-screw extruders were successfully adapted to develop a melt granulation process to produce granules of poorly compactible, high-dose drugs.[176,179,180] The granules obtained from extruders can be easily transported to the blender for mixing with disintegrants and lubricants before tablet compression. Such a process has the potential for converting the whole tablet manufacturing process into a continuous one. Overall, the continuous manufacturing of pharmaceuticals can lead to the ultimate application of QbD, where the process is controlled at each step of manufacturing and evaluated by process analytical tools to ensure the quality of the product without testing them at the end of the entire process.

11.12   SUMMARY

This chapter describes the many approaches employed to identify and then develop new pharmaceuticals. This chapter has endeavored to outline the careful multidisciplinary tasks that must take place in a timely fashion to insure that the inherent power of each new drug candidate is allowed to be fulfilled. As has been noted, the pressure befalling the product development scientists to make the right choices with regard to issues such as salt and polymorph selection at earlier time points has actually increased under this paradigm. Decisions regarding physical–chemical and physical–mechanical properties and their related impact on high-speed, efficient processing of the commercial dosage forms have to be made with little time and even less drug substance. This is remarkable in that a successful product will require a manufacturing process that is robust enough to consistently make millions of dosage units. The risk-based approach to pharmaceutical cGMP requires a complete understanding of pharmaceutical processes, which can only be achieved by the application of sound scientific principles throughout the discovery and development of a product. This chapter describes the underlying scientific principles of integrated product development, from lead selection to LCM. Some of the basic elements of QbD and continuous manufacturing as well as the application of PAT drug product development and manufacturing have also been discussed.

REFERENCES

1. Lemberger AP, Higuchi T, Busse LW, Swintosky JV, Wurster DE. Preparation of some steroids in microcrystalline form by rapid freeze-sublimation technique. *J Am Pharm Assoc.* 1954;43(6):338–341.
2. Nelson E, Schaldemose I. Urinary excretion kinetics for evaluation of drug absorption I. Solution rate limited and nonsolution rate limited absorption of aspirin and benzyl penicillin; absorption rate of sulfaethylthiadiazole. *J Am Pharm Assoc.* 1959;48(9):489–495.
3. Levy G, Nelson E. Pharmaceutical formulation and therapeutic efficacy. *JAMA.* 1961;177(10):689–691.
4. Gibaldi M, Kanig J. Absorption of drugs through the oral mucosa. *J Oral Ther Pharmacol.* 1965;1:440–450.
5. Wagner JG. Biopharmaceutics: Absorption aspects. *J Pharm Sci.* 1961;50(5):359–387.
6. Lieberman SV, Kraus SR, Murray J, Wood JH. Aspirin formulation and absorption rate I. Criteria for serum measurements with human panels. *J Pharm Sci.* 1964;53(12):1486–1491.
7. Kaplan SA. Biopharmaceutical considerations in drug formulation design and evaluation. *Drug Metab Rev.* 1972;1(1):15–33.
8. Yalkowsky SH, Rubino JT. Solubilization by cosolvents I: Organic solutes in propylene glycol–water mixtures. *J Pharm Sci.* 1985;74(4):416–421.
9. Prajapati HN, Dalrymple DM, Serajuddin ATM. A comparative evaluation of mono-, di- and triglyceride of medium chain fatty acids by lipid/surfactant/water phase diagram, solubility determination and dispersion testing for application in pharmaceutical dosage form development. *Pharm Res.* 2012;29(1):285–305.
10. Serajuddin ATM. Solid dispersion of poorly water-soluble drugs: Early promises, subsequent problems, and recent breakthroughs. *J Pharm Sci.* 1999;88(10):1058–1066.
11. Baghel S, Cathcart H, O'Reilly NJ. Polymeric amorphous solid dispersions: A review of amorphization, crystallization, stabilization, solid-state characterization, and aqueous solubilization of biopharmaceutical classification system class II Drugs. *J Pharm Sci.* 2016;105(9):2527–2544.
12. Augsburger LL, Hoag SW. *Pharmaceutical Dosage Forms: Tablets*, 3rd ed. Boca Raton, FL: CRC Press; 2016.
13. Gibson M. *Pharmaceutical Preformulation and Formulation: A Practical Guide from Candidate Drug Selection to Commercial Dosage Form*, 2nd ed. Boca Raton, FL: CRC Press; 2016.
14. Venkatesh S, Lipper RA. Role of the development scientist in compound lead selection and optimization. *J Pharm Sci.* 2000;89(2):145–154.
15. Drews J. Drug discovery: A historical perspective. *Science.* 2000;287(5460):1960–1964.
16. Kola I, Landis J. Can the pharmaceutical industry reduce attrition rates? *Nat Rev Drug Discov.* 2004;3:711.
17. Subramaniam S. Productivity and attrition: Key challenges for biotech and pharma. *Drug Discov Today.* 2003;12(8):513–515.
18. Wang J, Urban L. The impact of early ADME profiling on drug discovery and development strategy. *Drug Discov World.* 2004;5(4):73–86.
19. Prentis R, Lis Y, Walker S. Pharmaceutical innovation by the seven UK-owned pharmaceutical companies (1964–1985). *Br J Clin Pharmacol.* 1988;25(3):387–396.
20. Lipinski CA, Lombardo F, Dominy BW, Feeney PJ. Experimental and computational approaches to estimate solubility and permeability in drug discovery and development settings. *Adv Drug Delivery Rev.* 1997;23(1):3–25.
21. Kerns EH, Di L. Pharmaceutical profiling in drug discovery. *Drug Discov Today.* 2003;8(7):316–323.
22. Saxena V, Panicucci R, Joshi Y, Garad S. Developability assessment in pharmaceutical industry: An integrated group approach for selecting developable candidates. *J Pharm Sci.* 2009;98(6):1962–1979.
23. Landis MS, Bhattachar S, Yazdanian M, Morrison J. Commentary: Why pharmaceutical scientists in early drug discovery are critical for influencing the design and selection of optimal drug candidates. *AAPS PharmSciTech.* 2018;19(1):1–10.
24. Ho NFH, Raub TJ, Burton PS, Barsuhn CL, Adson A, Audus KL, Borchardt RT. Quantitative approaches to delineate passive transport mechanisms in cell culture monolayers. In: Amidon GL, Lee PI, Topp EM, eds. *Transport Processes in Pharmaceutical Systems.* New York: Marcel Dekker; 2000:219–316.
25. Li S, He H, Parthiban LJ, Yin H, Serajuddin ATM. IV-IVC considerations in the development of immediate-release oral dosage form. *J Pharm Sci.* 2005;94(7):1396–1417.
26. Wang J, Urban L, Bojanic D. Maximising use of in vitro ADMET tools to predict in vivo bioavailability and safety. *Exp Opin Drug Metab Toxicol.* 2007;3(5):641–665.

27. Butler JM, Dressman JB. The developability classification system: Application of biopharmaceutics concepts to formulation development. *J Pharm Sci.* 2010;99(12):4940–4954.
28. Clark DE. Computational methods for the prediction of ADME and Toxicity. *Adv Drug Delivey Rev.* 2002;54(3):253–254.
29. Veber DF, Johnson SR, Cheng H-Y, Smith BR, Ward KW, Kopple KD. Molecular properties that influence the oral bioavailability of drug candidates. *J Med Chem.* 2002;45(12):2615–2623.
30. Yalkowsky SH, Valvani SC. Solubility and partitioning I: Solubility of nonelectrolytes in water. *J Pharm Sci.* 1980;69(8):912–922.
31. Klopman G, Wang S, Balthasar DM. Estimation of aqueous solubility of organic molecules by the group contribution approach. Application to the study of biodegradation. *J Chem Inf Comput Sci.* 1992;32(5):474–482.
32. Huuskonen J, Salo M, Taskinen J. Aqueous solubility prediction of drugs based on molecular topology and neural network modeling. *J Chem Inf Comput Sci.* 1998;38(3):450–456.
33. Chen X-Q, Cho SJ, Li Y, Venkatesh S. Prediction of aqueous solubility of organic compounds using a quantitative structure–property relationship. *J Pharm Sci.* 2002;91(8):1838–1852.
34. McKenna JM, Halley F, Souness JE, et al. An algorithm-directed two-component library synthesized via solid-phase methodology yielding potent and orally bioavailable p38 MAP kinase inhibitors. *J Med Chem.* 2002;45(11):2173–2184.
35. Kenseth JR, Coldiron SJ. High-throughput characterization and quality control of small-molecule combinatorial libraries. *Curr Opin Chem Biol.* 2004;8(4):418–423.
36. Sugano K, Okazaki A, Sugimoto S, Tavornvipas S, Omura A, Mano T. Solubility and dissolution profile assessment in drug discovery. *Drug Metab Pharmacokinet.* 2007;22(4):225–254.
37. Faller B. Improving solubility in lead optimization. *Am Pharm Rev.* 2004;7:30–33.
38. Dai W-G, Pollock-Dove C, Dong LC, Li S. Advanced screening assays to rapidly identify solubility-enhancing formulations: High-throughput, miniaturization and automation. *Adv Drug Delivery Rev.* 2008;60(6):657–672.
39. Zheng W, Jain A, Papoutsakis D, Dannenfelser R-M, Panicucci R, Garad S. Selection of oral bioavailability enhancing formulations during drug discovery. *Drug Dev Ind Pharm.* 2012;38(2):235–247.
40. Serajuddin ATM, Pudipeddi M. Salt-selection strategies. In: Stahl P, Wermuth CG, eds. *IUPAC Handbook of Pharmaceutical Salts: Properties, Selection, and Use*, Chapter 6. Zurich: Wiley-VCH; 2002: 135–160.
41. Hilal SH, El-Shabrawy Y, Carreira LA, Karickhoff SW, Toubar SS, Rizk M. Estimation of the ionization pK$_a$ of pharmaceutical substances using the computer program SPARC. *Talanta.* 1996;43(4):607–619.
42. Albert A, Serjeant EP. *The Determination of Ionization Constants: A Laboratory Manual*, 3rd ed. New York: Chapman & Hall; 1984.
43. Donovan SF, Pescatore MC. Method for measuring the logarithm of the octanol–water partition coefficient by using short octadecyl–poly(vinyl alcohol) high-performance liquid chromatography columns. *J Chromatogr A.* 2002;952(1):47–61.
44. Comer J, Box K. High-throughput measurement of drug pK$_a$ values for ADME screening. *JALA.* 2003;8(1):55–59.
45. Kansy M, Senner F, Gubernator K. Physicochemical high throughput screening: Parallel artificial membrane permeation assay in the description of passive absorption processes. *J Med Chem.* 1998;41(7):1007–1010.
46. Avdeef A. The rise of PAMPA. *Expert Opin Drug Metab Toxicol.* 2005;1(2):325–342.
47. Liu X, Testa B, Fahr A. Lipophilicity and its relationship with passive drug permeation. *Pharm Res.* 2011;28(5):962–977.
48. Hwang K-K, Martin NE, Jiang L, Zhu C. Permeation prediction of M100240 using the parallel artificial membrane permeability assay. *J Pharm Pharm Sci.* 2003;6:315–320.
49. Artursson P, Palm K, Luthman K. Caco-2 monolayers in experimental and theoretical predictions of drug transport. *Adv Drug Delivery Rev.* 1996;22(1):67–84.
50. van Breemen RB, Li Y. Caco-2 cell permeability assays to measure drug absorption. *Expert Opin Drug Metab Toxicol.* 2005;1(2):175–185.
51. Hubatsch I, Ragnarsson EGE, Artursson P. Determination of drug permeability and prediction of drug absorption in Caco-2 monolayers. *Nat Protoc.* 2007;2:2111.
52. Hayeshi R, Hilgendorf C, Artursson P, et al. Comparison of drug transporter gene expression and functionality in Caco-2 cells from 10 different laboratories. *Eur J Pharm Sci.* 2008;35(5):383–396.
53. Balbach S, Korn C. Pharmaceutical evaluation of early development candidates "the 100 mg-approach." *Int J Pharm.* 2004;275(1):1–12.

54. Yu H, Adedoyin A. ADME–Tox in drug discovery: Integration of experimental and computational technologies. *Drug Discov Today*. 2003;8(18):852–861.
55. Berry IR, Nash RA, eds. *Pharmaceutical Process Validation*, 2nd ed. New York: Taylor & Francis; 1993.
56. Avdeef A. Physicochemical profiling (solubility, permeability and charge state). *Curr Top Med Chem*. 2001;1(4):277–351.
57. Avdeef A, Box KJ, Comer JEA, et al. PH-metric logP 11. pK$_a$ determination of water-insoluble drugs in organic solvent–water mixtures. *J Pharm Biomed Anal*. 1999;20(4):631–641.
58. Surakitbanharn Y, McCandless R, Krzyzaniak J, Dannenfelser R-M, Yalkowsky SH. Self-association of dexverapamil in aqueous solution. *J Pharm Sci*. 1995;84(6):720–723.
59. Hansch C, Leo A. *Substituent Constants for Correlation Analysis in Chemistry and Biology*. New York: Wiley-Interscience; 1979.
60. Reynolds DW, Facchine KL, Mullaney JF, Alsante KM, Hatajik TD, Motto MG. Conducting forced degradation studies. *Pharm Technol*. 2002;26(2):48–56.
61. Rhodes CT, Carstensen JT. *Drug Stability: Principles and Practices*. New York: Marcel Dekker; 2000.
62. Yoshioka S, Stella VJ. *Stability of Drugs and Dosage Forms*. New York: Kluwer Academic Publishers; 2007.
63. *Physicians' Desk Reference*. Montvale, NJ: Thomson PDR; 2004.
64. Brittain HG. *Polymorphism in Pharmaceutical Solids*, 2nd ed. New York: Informa Healthcare; 2016.
65. Aaltonen J, Allesø M, Mirza S, Koradia V, Gordon KC, Rantanen J. Solid form screening: A review. *Eur J Pharm Biopharm*. 2009;71(1):23–37.
66. Serajuddin ATM. Salt formation to improve drug solubility. *Adv Drug Delivery Rev*. 2007;59(7):603–616.
67. Morris KR, Fakes MG, Thakur AB, et al. An integrated approach to the selection of optimal salt form for a new drug candidate. *Int J Pharm*. 1994;105(3):209–217.
68. FDA. Regulatory classification of pharmaceutical co-crystals guidance for industry. www.fda.gov/downloads/Drugs/Guidances/UCM281764.pdf. Published February 2018. Accessed July 9, 2018.
69. Schultheiss N, Newman A. Pharmaceutical co-crystals and their physicochemical properties. *Cryst Growth Des*. 2009;9(6):2950–2967.
70. Feng L, Karpinski PH, Sutton P, et al. LCZ696: A dual-acting sodium supramolecular complex. *Tetrahedron Lett*. 2012;53(3):275–276.
71. ICH. Specifications: Test procedures and acceptance criteria for new drug substances and new drug products: Chemical substances. Q6A. *In International Conference on Harmonisation of Technical Requirements for Registration of Pharmaceuticals for Human Use*. 1999.
72. Newman A. Specialized solid form screening techniques. *Org Process Res Dev*. 2013;17(3):457–471.
73. Morissette SL, Almarsson Ö, Peterson ML, et al. High-throughput crystallization: Polymorphs, salts, co-crystals and solvates of pharmaceutical solids. *Adv Drug Delivery Rev*. 2004;56(3):275–300.
74. Bauer J, Spanton S, Henry R, et al. Ritonavir: An extraordinary example of conformational polymorphism. *Pharm Res*. 2001;18(6):859–866.
75. Pudipeddi M, Serajuddin A. Trends in solubility of polymorphs. *J Pharm Sci*. 2005;94(5):929–939.
76. Hancock BC, Parks M. What is the true solubility advantage for amorphous pharmaceuticals? *Pharm Res*. 2000;17(4):397–404.
77. Farag Badawy SI, Williams RC, Gilbert DL. Chemical stability of an ester prodrug of a glycoprotein IIb/IIIa receptor antagonist in solid dosage forms. *J Pharm Sci*. 1999;88(4):428–433.
78. Serajuddin ATM, Thakur AB, Ghoshal RN, et al. Selection of solid dosage form composition through drug–excipient compatibility testing. *J Pharm Sci*. 1999;88(7):696–704.
79. Monkhouse DC, Maderich A. Whither compatibility testing? *Drug Dev Ind Pharm*. 1989;15(13):2115–2130.
80. Pudipeddi M, Zannou EA, Vasanthavada M, et al. Measurement of surface pH of pharmaceutical solids: A critical evaluation of indicator dye-sorption method and its comparison with slurry pH method. *J Pharm Sci*. 2008;97(5):1831–1842.
81. Crowley P, Martini LG. Drug-excipient interactions. *Pharm Technol*. 2001;4:7–12.
82. Landin M, Casalderrey M, Martinez-Pacheco R, et al. Chemical stability of acetylsalicylic acid in tablets prepared with different particle size fractions of a commercial brand of dicalcium phosphate dihydrate. *Int J Pharm*. 1995;123(1):143–144.
83. Irwin WJ, Iqbal M. Solid-state stability: The effect of grinding solvated excipients. *Int J Pharm*. 1991;75(2):211–218.
84. Villalobos-Hernández JR, Villafuerte-Robles L. Effect of carrier excipient and processing on stability of indorenate hydrochloride/excipient mixtures. *Pharm Dev Technol*. 2001;6(4):551–561.

85. Rowe RC, Sheskey PJ, Owen S. *Handbook of Pharmaceutical Excipients*. London: Pharmaceutical Press; 2006.

86. Zannou EA, Ji Q, Joshi YM, Serajuddin A. Stabilization of the maleate salt of a basic drug by adjustment of microenvironmental pH in solid dosage form. *Int J Pharm*. 2007;337(1):210–218.

87. Sun CC. Materials science tetrahedron: A useful tool for pharmaceutical research and development. *J Pharm Sci*. 2009;98(5):1671–1687.

88. Hancock BC, Colvin JT, Mullarney MP, Zinchuk AV. Pharmaceutical powders, blends, dry granulations, and immediate-release tablets. *Pharm Technol*. 2003;64 - 80.

89. Rowley G. Quantifying electrostatic interactions in pharmaceutical solid systems. *Int J Pharm*. 2001;227(1):47–55.

90. Hiestand EN, Smith DP. Three indices for characterizing the tableting performance of materials. *Adv Ceram*. 1983;9:47.

91. Roberts RJ, Rowe RC. The compaction of pharmaceutical and other model materials: A pragmatic approach. *Chem Eng Sci*. 1987;42(4):903–911.

92. Wang C, Paul S, Wang K, Hu S, Sun CC. Relationships among crystal structures, mechanical properties, and tableting performance probed using four salts of diphenhydramine. *Cryst Growth Des*. 2017;17(11):6030–6040.

93. Paul S, Taylor LJ, Murphy B, et al. Powder properties and compaction parameters that influence punch sticking propensity of pharmaceuticals. *Int J Pharm*. 2017;521(1):374–383.

94. Janssens S, De Zeure A, Paudel A, Van Humbeeck J, Rombaut P, Van den Mooter G. Influence of preparation methods on solid state supersaturation of amorphous solid dispersions: A case study with itraconazole and Eudragit E100. *Pharm Res*. 2010;27(5):775–785.

95. Amidon GL, Lennernäs H, Shah VP, Crison JR. A theoretical basis for a biopharmaceutic drug classification: The correlation of in vitro drug product dissolution and in vivo bioavailability. *Pharm Res*. 1995;12(3):413–420.

96. FDA. Guidance for industry dissolution testing of immediate release solid oral dosage forms. www.fda.gov/downloads/drugs/guidances/ucm070237.pdf. Published August 1997. Accessed July 9, 2018.

97. Curatolo W. Physical chemical properties of oral drug candidates in the discovery and exploratory development settings. *Pharm Sci Technol Today*. 1998;1(9):387–393.

98. Agoram B, Woltosz WS, Bolger MB. Predicting the impact of physiological and biochemical processes on oral drug bioavailability. *Adv Drug Delivery Rev*. 2001;50:S41–S67.

99. De Buck SS, Sinha VK, Fenu LA, Nijsen MJ, Mackie CE, Gilissen RAHJ. Prediction of human pharmacokinetics using physiologically based modeling: A retrospective analysis of 26 clinically tested drugs. *Drug Metab Dispos Biol Fate Chem*. 2007;35(10):1766–1780.

100. Kramer S, Flynn G. Solubility of organic hydrochlorides. *J Pharm Sci*. 1972;61(12):1896–1904.

101. Johnson KC, Swindell AC. Guidance in the setting of drug particle size specifications to minimize variability in absorption. *Pharm Res*. 1996;13(12):1795–1798.

102. Pudipeddi M, Serajuddin ATM, Grant DJW, Stahl PH. Solubility and dissolution of weak acids, bases, and salts. In: Stahl P, Wermuth CG, eds. *IUPAC Handbook of Pharmaceutical Salts: Properties, Selection, and Use*, Chapter 2. Zurich: Wiley-VCH; 2002: 19–39.

103. Dill WA, Kazenko A, Wolf LM, Glazko AJ. Studies on 5,5'-diphenylhydantoin (Dilantin) in animals and man. *J Pharmacol Exp Ther*. 1956;118(3):270–279.

104. Serajuddin ATM, Sheen P-C, Mufson D, Bernstein DF, Augustine MA. Preformulation study of a poorly water-soluble drug, α-pentyl-3-(2-quinolinylmethoxy) benzenemethanol: Selection of the base for dosage form design. *J Pharm Sci*. 1986;75(5):492–496.

105. Washington N, Washington C, Wilson C. *Physiological Pharmaceutics: Barriers to Drug Absorption*, 1st ed. London: CRC Press; 2000.

106. Davis SS, Hardy JG, Fara JW. Transit of pharmaceutical dosage forms through the small intestine. *Gut*. 1986;27(8):886–892.

107. Smith CG, Poutsiaka JW, Schreiber EC. Problems in predicting drug effects across species lines. *J Int Med Res*. 1973;1(6):489–503.

108. Chiou W, Ma C, Chung S, Wu T, Jeong H. Similarity in the linear and non-linear oral absorption of drugs between human and rat. *Int J Clin Pharmacol Ther*. 2000;38(11):532–539.

109. Chiou WL, Jeong HY, Chung SM, Wu TC. Evaluation of using dog as an animal model to study the fraction of oral dose absorbed of 43 drugs in humans. *Pharm Res*. 2000;17(2):135–140.

110. Jones H, Chen Y, Gibson C, et al. Physiologically based pharmacokinetic modeling in drug discovery and development: A pharmaceutical industry perspective. *Clin Pharmacol Ther*. 2015;97(3):247–262.

111. Zhuang X, Lu C. PBPK modeling and simulation in drug research and development. *Acta Pharm Sin B*. 2016;6(5):430–440.

112. Kesisoglou F, Chung J, van Asperen J, Heimbach T. Physiologically based absorption modeling to impact biopharmaceutics and formulation strategies in drug development: Industry case studies. *J Pharm Sci*. 2016;105(9):2723–2734.

113. FDA. Guidance for industry extended release oral dosage forms: Development, evaluation, and application of in vitro/in vivo correlations. www.fda.gov/downloads/drugs/guidances/ucm070239.pdf. Published September 1997. Accessed July 9, 2018.

114. Jantratid E, Janssen N, Reppas C, Dressman JB. Dissolution media simulating conditions in the proximal human gastrointestinal tract: An update. *Pharm Res*. 2008;25(7):1663.

115. Gupta SK, Sathyan G. Pharmacokinetics of an oral once-a-day controlled-release oxybutynin formulation compared with immediate-release oxybutynin. *J Clin Pharmacol*. 1999;39(3):289–296.

116. Selen A, Dickinson PA, Müllertz A, et al. The biopharmaceutics risk assessment roadmap for optimizing clinical drug product performance. *J Pharm Sci*. 2014;103(11):3377–3397.

117. Dickinson PA, Kesisoglou F, Flanagan T, et al. Optimizing clinical drug product performance: Applying biopharmaceutics risk assessment roadmap (BioRAM) and the BioRAM scoring grid. *J Pharm Sci*. 2016;105(11):3243–3255.

118. Hariharan M. Reducing the time to develop and manufacture formulations for first oral dose in humans. *Pharm Technol*. 2003;2003:68–84.

119. Sistla A, Sunga A, Phung K, Koparkar A, Shenoy N. Powder-in-bottle formulation of SU011248. Enabling rapid progression into human clinical trials. *Drug Dev Ind Pharm*. 2004;30(1):19–25.

120. Aubry A-F, Sebastian D, Hobson T, et al. In-use testing of extemporaneously prepared suspensions of second generation non-nucleoside reversed transcriptase inhibitors in support of Phase I clinical studies. *J Pharm Biomed Anal*. 2000;23(2):535–542.

121. Shah VP, Noory A, Noory C, et al. In vitro dissolution of sparingly water-soluble drug dosage forms. *Int J Pharm*. 1995;125(1):99–106.

122. Waterman KC, Adami RC, Alsante KM, et al. Stabilization of pharmaceuticals to oxidative degradation. *Pharm Dev Technol*. 2002;7(1):1–32.

123. Al-Omari MM, Abdelah MK, Badwan AA, Jaber AMY. Effect of the drug-matrix on the stability of enalapril maleate in tablet formulations. *J Pharm Biomed Anal*. 2001;25(5):893–902.

124. Shiromani PK, Clair J. Statistical comparison of high-shear versus low-shear granulation using a common formulation. *Drug Dev Ind Pharm*. 2000;26(3):357–364.

125. ICH. Guidance for industry Q1A(R2) stability testing of new drug substances and products. www.fda.gov/downloads/drugs/guidances/ucm073369.pdf. Published November 2003. Accessed July 9, 2018.

126. FDA. Pharmaceutical cGMPs for the 21st century: A risk based approach: Final report. www.fda.gov/downloads/drugs/developmentapprovalprocess/manufacturing/questionsandanswersoncurrent-goodmanufacturingpracticescgmpfordrugs/ucm176374.pdf. Published September 2004. Accessed July 9, 2018.

127. Levin M. *Pharmaceutical Process Scale-Up*, vol 157, 2nd ed. Boca Raton, FL: CRC Press; 2005.

128. Carleton F, Agalloco J. *Validation of Pharmaceutical Processes*, 3rd ed. Boca Raton, FL: CRC Press; 2007.

129. FDA. Guidance for industry bioavailability and bioequivalence studies for orally administered drug products: General considerations. www.fda.gov/downloads/Drugs/Guidances/ucm154838. Published July 2002. Accessed July 9, 2018.

130. Cupissol D, Bressolle F, Adenis L, et al. Evaluation of the bioequivalence of tablet and capsule formulations of granisetron in patients undergoing cytotoxic chemotherapy for malignant disease. *J Pharm Sci*. 1993;82(12):1281–1284.

131. Akers MJ, Fites AL, Robison RL. Formulation design and development of parenteral suspensions. *J Parenter Sci Technol*. 1987;41(3): 88–96.

132. Broadhead J, Gibson M. Parenteral dosage forms. In: Gibson, M ed. *Pharmaceutical Preformulation and Formulation: A Practical Guide from Candidate Drug Selection to Commercial Dosage Form*, 2nd ed, Drugs and the Pharmaceutical Sciences. New York: CRC Press; 2016: 325–347.

133. Akers MK, Larrimore D, Guazzo D. *Parenteral Quality Control: Sterility, Pyrogen, Particulate, and Package Integrity Testing*, 3rd ed. New York: Marcel Dekker; 2002.

134. Sweetana S, Akers MJ. Solubility principles and practices for parenteral drug dosage form development. *PDA J Pharm Sci Technol*. 1995;50(5):330–342.

135. Gupta PK, Patel JP, Hahn KR. Evaluation of pain and irritation following local administration of parenteral formulations using the rat paw lick model. *PDA J Pharm Sci Technol*. 1994;48(3):159–166.

136. Nema S, Washkuhn RJ, Brendel RJ. Excipients and their use in injectable products. *PDA J Pharm Sci Technol.* 1997;51(4):166–171.
137. Strickley RG. Parenteral formulations of small molecules therapeutics marketed in the United States (1999). Part III. *PDA J Pharm Sci Technol.* 2000;54(2):152–169.
138. Strickley RG. Solubilizing excipients in oral and injectable formulations. *Pharm Res.* 2004;21(2):201–230.
139. Tang X (Charlie), Pikal MJ. Design of freeze-drying processes for pharmaceuticals: Practical advice. *Pharm Res.* 2004;21(2):191–200.
140. Poochikian G, Bertha CM. Regulatory view on current issues pertaining to inhalation drug products. *Respiratory Drug Delivery VIII Conference Davis Horwood International*, Raleigh, NC. 2002:159–164.
141. Hickey AJ, Martonen TB. Behavior of hygroscopic pharmaceutical aerosols and the influence of hydrophobic additives. *Pharm Res.* 1993;10(1):1–7.
142. Rabinowitz JD, Wensley M, Lloyd P, et al. Fast onset medications through thermally generated aerosols. *J Pharmacol Exp Ther.* 2004;309(2):769–775.
143. Amiel SA, Alberti KG. Inhaled insulin. *BMJ.* 2004;328(7450):1215–1216.
144. Hodges CC, Lloyd PM, Mufson D, Rogers DD, Wensley MJ. Delivery of aerosols containing small particles through an inhalation route. U.S. Patent 6,682,716, January 2004.
145. Williams HD, Trevaskis NL, Charman SA, et al. Strategies to address low drug solubility in discovery and development. *Pharmacol Rev.* 2013;65(1):315–499.
146. Feeney OM, Crum MF, McEvoy CL, et al. 50 years of oral lipid-based formulations: Provenance, progress and future perspectives. *Adv Drug Delivery Rev.* 2016;101:167–194.
147. Merisko-Liversidge E, Liversidge GG, Cooper ER. Nanosizing: A formulation approach for poorly-water-soluble compounds. *Eur J Pharm Sci.* 2003;18(2):113–120.
148. Kostewicz ES, Abrahamsson B, Brewster M, et al. In vitro models for the prediction of in vivo performance of oral dosage forms. *Eur J Pharm Sci.* 2014;57:342–366.
149. Sekiguchi NOK. Studies on absorption of eutectic mixture. I. A comparison of the behavior of eutectic mixture of sulfathiazole and that of ordinary sulfathiazole in man. *Chem Pharm Bull.* 1961;9(11):866–872.
150. Breitenbach J. Melt extrusion: From process to drug delivery technology. *Eur J Pharm Biopharm.* 2002;54(2):107–117.
151. Jermain SV, Brough C, Williams RO. Amorphous solid dispersions and nanocrystal technologies for poorly water-soluble drug delivery: An update. *Int J Pharm.* 2018;535(1):379–392.
152. Singh S, Parikh T, Sandhu HK, Shah NH, Malick AW, Singhal D, Serajuddin ATM. Supersolubilization and amorphization of a model basic drug, haloperidol, by interaction with weak acids. *Pharm Res.* 2013;30:1561–1573.
153. Parikh T, Sandhu HK, Talele TT, Serajuddin ATM. Characterization of solid dispersion of itraconazole prepared by solubilization in concentrated aqueous solutions of weak organic acids and drying. *Pharm Res.* 2016;33(6):1456–1471.
154. Shah A, Serajuddin ATM. Conversion of solid dispersion prepared by acid–base interaction into free-flowing and tabletable powder by using Neusilin® US2. *Int J Pharm.* 2015;484(1):172–180.
155. Parikh T, Serajuddin ATM. Development of fast-dissolving amorphous solid dispersion of itraconazole by melt extrusion of its mixture with weak organic carboxylic acid and polymer. *Pharm Res.* 2018;35(7):127.
156. Shah AV, Serajuddin ATM. Development of solid self-emulsifying drug delivery system (SEDDS) I: Use of poloxamer 188 as both solidifying and emulsifying agent for lipids. *Pharm Res.* 2012;29(10):2817–2832.
157. Gumaste SG, Dalrymple DM, Serajuddin ATM. Development of solid SEDDS, V: compaction and drug release properties of tablets prepared by adsorbing lipid-based formulations onto Neusilin® US2. *Pharm Res.* 2013;30(12):3186–3199.
158. Gumaste SG, Freire BOS, Serajuddin ATM. Development of solid SEDDS, VI: Effect of precoating of Neusilin® US2 with PVP on drug release from adsorbed self-emulsifying lipid-based formulations. *Eur J Pharm Sci.* 2017;110:124–133.
159. Thombre AG. Assessment of the feasibility of oral controlled release in an exploratory development setting. *Drug Discov Today.* 2005;10(17):1159–1166.
160. Klonoff DC. Afrezza inhaled insulin: The fastest-acting FDA-approved insulin on the market has favorable properties. *J Diabetes Sci Technol.* 2014;8(6):1071–1073.

161. Hastings MS, Li SK, Miller DJ, Bernstein PS, Mufson D. Visulex: Advancing iontophoresis for effective noninvasive back to-the-eye therapeutics. *Drug Delivery Technol.* 2004;4:53–57.
162. Meinzer A. Microemulsion-A suitable galenical approach for the absorption enhancement of poorly soluble compounds. *Bull Tech-Gattefosse.* 1995;88:21–26.
163. Joshi Y, LoBrutto R, Serajuddin ATM. Industry opinion of regulatory influence: The initiative on pharmaceutical cGMPs for the 21st century. *J Process Anal Technol.* 2006;3:6–14.
164. Woodcock J. The concept of pharmaceutical quality. *Am Pharm Rev.* 2004;7(6):10–15.
165. Food and Drug Administration. Guidance for industry: Q8(R2) pharmaceutical development. www.fda.gov/downloads/drugs/guidances/ucm073507.pdf. Published November 2009. Accessed April 2, 2018.
166. Yu LX. Pharmaceutical quality by design: Product and process development, understanding, and control. *Pharm Res.* 2008;25(4):781–791.
167. Yu LX, Amidon G, Khan MA, et al. Understanding pharmaceutical quality by design. *AAPS J.* 2014;16(4):771–783.
168. FDA. Quality by design for ANDAs: An example for immediate-release dosage forms. www.fda.gov/downloads/Drugs/DevelopmentApprovalProcess/HowDrugsareDevelopedandApproved/ApprovalApplications/AbbreviatedNewDrugApplicationANDAGenerics/UCM304305.pdf). Published April 2012.
169. FDA. Guidance for industry PAT: A framework for innovative pharmaceutical development, manufacturing, and quality assurance. www.fda.gov/downloads/drugs/guidances/ucm070305.pdf. Published September 2004. Accessed July 9, 2018.
170. Munson J, Freeman Stanfield C, Gujral B. A review of process analytical technology (PAT) in the US pharmaceutical industry. *Curr Pharm Anal.* 2006;2(4):405–414.
171. Yu LX, Lionberger RA, Raw AS, D'Costa R, Wu H, Hussain AS. Applications of process analytical technology to crystallization processes. *Adv Drug Delivery Rev.* 2004;56(3):349–369.
172. Lee SL, O'Connor T, Yang X, Cruz CN, Chatterjee S, Madurawe RD, Moore CMV, Lawrence XY, Woodcock J. Modernizing pharmaceutical manufacturing: From batch to continuous production. *J Pharm Innov.* 2015;10(3):191–199.
173. Myerson AS, Krumme M, Nasr M, Thomas H, Braatz RD. Control systems engineering in continuous pharmaceutical manufacturing May 20–21, 2014 continuous manufacturing symposium. *J Pharm Sci.* 2015;104(3):832–839.
174. Leuenberger H. New trends in the production of pharmaceutical granules: Batch versus continuous processing. *Eur J Pharm Biopharm.* 2001;52(3):289–296.
175. Meena AK, Desai D, Serajuddin ATM. Development and optimization of a wet granulation process at elevated temperature for a poorly compactible drug using twin screw extruder for continuous manufacturing. *J Pharm Sci.* 2017;106(2):589–600.
176. Batra A, Desai D, Serajuddin ATM. Investigating the use of polymeric binders in twin screw melt granulation process for improving compactibility of drugs. *J Pharm Sci.* 2017;106(1):140–150.
177. Djuric D, Kleinebudde P. Continuous granulation with a twin-screw extruder: Impact of material throughput. *Pharm Dev Technol.* 2010;15(5):518–525.
178. Shah AV, Serajuddin ATM. Twin screw continuous wet granulation. In: Narang, A, Badawy, S, eds. *Handbook of Pharmaceutical Wet Granulation: Theory and Practice in a Quality by Design Paradigm.* Philadelphia, PA: Elsevier; 2018.
179. Lakshman JP, Kowalski J, Vasanthavada M, Tong W-Q, Joshi YM, Serajuddin ATM. Application of melt granulation technology to enhance tabletting properties of poorly compactible high-dose drugs. *J Pharm Sci.* 2011;100(4):1553–1565.
180. Vasanthavada M, Wang Y, Haefele T, Lakshman JP, Mone M, Tong W, Joshi YM, Serajuddin, ATM. Application of melt granulation technology using twin-screw extruder in development of high-dose modified-release tablet formulation. *J Pharm Sci.* 2011;100(5):1923–1934.

12 New Trends in Pharmacological and Pharmaceutical Profiling

Lyn Rosenbrier Ribeiro
AstraZeneca

Duncan Armstrong
Novartis Institutes for Biomedical Research

Thierry Jean
Levitha

Jean-Pierre Valentin
UCB-Biopharma SPRL

CONTENTS

12.1 WHAT IS PHARMACOLOGICAL PROFILING?

In this chapter, *in vitro* secondary pharmacological profiling is a term used to describe the investigation of the pharmacological effects of a drug at molecular targets distinct from the intended therapeutic molecular target. It may also be described as selectivity screening, pharmacological profiling, or secondary pharmacology (Anon, 2001; Bowes et al., 2012; Papoian et al., 2015; Valentin & Hammond, 2008; Whitebread et al., 2005, 2016). Typically, simple *in vitro* biochemical and pharmacological assays are employed to determine whether a compound interacts with a range of G protein-coupled receptors (GPCRs), enzymes, kinases, nuclear hormone receptors, ion channels, and transporters. To mediate a biological effect, a drug binds specifically to the primary therapeutic target or to other molecular targets (Figure 12.1). These effects may be mediated by the parent drug or by metabolites of the drug. The molecular targets may be closely related to the therapeutic target (e.g., receptor subtypes) or very distantly related, for example, the therapeutic target may be an enzyme, and profiling investigates activity at other enzymes as well as receptors, ion channels, and transporters.

12.2 WHY DO PHARMACOLOGICAL PROFILING?

12.2.1 UNDERSTANDING THE CAUSES OF DRUG FAILURES

The first step to increase success in clinical development is to understand the causes of drug failure (Paul et al., 2010). Attrition can be caused by inadequate efficacy, unacceptable safety risks, poor ADME (absorption, distribution, metabolism, and excretion) characteristics, or pharmaco-economic considerations. Pharmaceutical industry surveys have revealed that over the last two decades the number of new medicines being launched failed to increase despite significant investments in Research and Development (R&D) (Anon, 2012a, 2017b; David et al., 2009; Munos, 2009; Pammolli et al., 2011; Ward et al., 2013; Waring et al., 2015). Over the same period, non-clinical and clinical safety has remained a major cause of drug attrition (Arrowsmith & Miller, 2013; Cook et al., 2014; Harrison, 2016; Kola & Landis, 2004; Munos, 2009). Such attrition can occur during preclinical or clinical development and during the post-approval stage, resulting in withdrawal of marketed

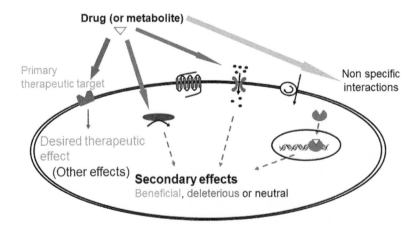

FIGURE 12.1 Mechanisms of drug action. To mediate a biological response, a drug may bind to the desired therapeutic target or to other molecular targets such as GPCRs, ion channels, or transporters on the cell membrane or to intracellular targets such as enzymes and nuclear hormone receptors. Side effects of the parent drug and/or metabolites may result from activity at primary or secondary targets or nonspecific interactions. The biological responses resulting from interactions at targets other than the primary target can be beneficial, deleterious, or neutral.

drugs accounting for approximately between one-quarter and one-third of all drug discontinuations (Arrowsmith & Miller, 2013; Cook et al., 2014; Harrison, 2016; Kola & Landis, 2004; Morgan et al., 2018; Redfern et al., 2010; Redfern & Valentin, 2017; Valentin et al., 2009). The relative importance of major causes of drug failures in clinical trials has evolved significantly over the last two decades (Kola & Landis, 2004). In the early 1990s, major causes of clinical attrition were poor pharmaco-kinetics (PKs) and bioavailability which together were responsible for over 40% of failures (Prentis et al., 1988). In 2000–2010, these causes contributed to ~10% to attrition in drug development in 2000 and were reduced to almost nothing at the end of the decade, (Arrowsmith, 2011; Arrowsmith & Miller, 2013; Kola & Landis, 2004) probably as a result of the efforts made by pharmaceutical firms during this period to address these issues in the early phases of drug discovery by incorporation of predictive screens. It is a common goal to improve the discovery process through constant efforts, to identify as early as possible, drug candidates that will have minimal chances to fail before or after they are approved for marketing. This can be achieved by the early identification of candidates that will have the best safety profiles, which means drugs that are active on their therapeutic target and have minimal adverse effects.

12.2.2 Economic Environment: The Cost of Drug Failure

Discovery and development of new drugs are costly and require time and luck (Anon, 2012a, 2017b). Using current methods, the average drug development time from target identification to New Drug Application (NDA) filing is 12 years (Dickson & Gagnon, 2004). R&D expenses have kept increasing since the 1970s, accounting for a significant proportion of revenues (Anon, 2012a, 2017b; Booth & Zemmel, 2004; Munos, 2009). Despite increasing effort, the number of new chemical entities (NCEs) reaching the market failed to increase (Anon, 2017b; Kola & Rafferty, 2002; Mullard, 2018). Bringing an NCE to market carries an average cost of more than US$ >1 billion (Anon, 2012a; DiMasi et al., 2003). Interestingly, 70% of this amount is spent on drugs that fail during discovery or development phases. Considering all therapeutic areas, the average success rate of a drug candidate entering clinical Phase I is less than 10% (Kola & Rafferty, 2002). Effort is currently focused on developing methods and approaches to decrease clinical attrition through early selection of drug candidates with the highest likelihood of success.

12.2.3 The Regulatory Drive for Safe Medicines

Adverse drug reactions (ADRs) in man account for 10% of the failure of drugs in clinical development (Kennedy, 1997). In addition, ADRs account for ~5% of all hospital admissions and around 0.15% of deaths following admission (Einarson, 1993; Pirmohamed et al., 2004). These ADRs may be linked to the activity of the drug at the primary molecular target, linked to the activity at other molecular targets, or mediated by nonspecific mechanisms. During the development of a drug, extensive pre-clinical safety studies are performed to assess the effects of the drug, and the selection and design of these studies are influenced strongly by worldwide regulatory requirements. The International Conference on Harmonization (ICH) of Technical Requirements for Registration of Pharmaceuticals for Human Use comprises a body of experts that develops and publishes guidelines and recommendations pertaining to the clinical use of NCEs. Major regulatory authorities worldwide adopt and implement these guidelines. The ICH S7A document outlines recommendations for safety pharmacology studies for human pharmaceuticals and defines three types of pharmacology studies that are applied during the drug discovery and development process (Anon, 2001). They are defined as follows:

- *Primary pharmacodynamic studies* investigate the mode of action or effects of a substance in relation to its desired therapeutic target.
- *Secondary pharmacodynamic studies* investigate the mode of action or effects of a substance not related to its desired therapeutic target.

- *Safety pharmacology studies* investigate the undesirable pharmacodynamic effects of a substance on physiological functions in relation to exposure in the therapeutic range and above.

Safety pharmacology studies are a regulatory requirement. Pharmacological profiling studies referred to as "secondary pharmacodynamic studies" by ICH S7A are recommended but not required. In contrast to most safety pharmacology studies, pharmacological profiling does not need to be conducted in compliance with Good Laboratory Practice (GLP) (Anon, 2001, 2009). The ICH S7A guidelines clearly indicate that pharmacological profiling should be considered and applied in two ways. The studies should be applied to provide information for the selection and design of safety pharmacology studies (proactive approach) and used to interpret the findings of *in vivo* safety pharmacology and toxicology studies or clinical findings (retrospective approach) (Valentin & Hammond, 2008).

A number of molecular targets have been associated with safety hazards, thus making those targets key culprits as part of the non-clinical safety evaluation of NCEs (44 and 70 targets in Bowes et al., 2012; Lynch et al., 2017, respectively). Although there is no specific non-clinical guidance regarding the assessment of off-target profiling, *in vitro* molecular target studies are to be considered as part of the ICH S7A (Anon, 2001) with a specific assessment of the human ether-a-go-go-related gene (hERG) channel in the ICH S7B (Anon, 2005) and the assessment of the abuse potential of drugs (Anon, 2017a). Although there are no specific non-clinical guidance on the assessment of seizure liability or suicidal ideation and behaviors, pharmacological targets associated with those safety risks have been described and should be evaluated early in drug discovery (Anon, 2012b; Easter et al., 2009; Hartmann et al., 2015; Muller et al., 2015; Stone et al., 2009; Urban et al., 2014). More recently, the assessment of drug-induced Torsade de Pointes, a potentially life-threatening arrhythmia, via a *Comprehensive in vitro Proarrhythmia Assessment* (CiPA), refers to the evaluation of drug candidates against seven cardiac ion channels responsible for the shape and duration of the ventricular action potential (Gintant et al., 2016; Sager et al., 2014). Ultimately, it is desirable for a drug to be devoid of ADRs when first administered to humans and when used in clinical practice; the end result is development of a medicine that has an acceptable benefit/risk profile for a given disease indication and patient population.

Although the true contribution of off-target pharmacology to drug attrition is not fully understood, there is evidence suggesting that highly promiscuous drugs are more likely to be discontinued during clinical development or withdrawn from market (Bowes et al., 2012). In a recent Phase 1 trial, a fatty acid amide hydrolase inhibitor (BIA 10-2474) led to the death of one volunteer and produced mild-to-severe neurological symptoms in four others (Moore, 2016a,b). Although the cause of the clinical neurotoxicity is unknown, it has been suggested that off-target activities at several lipase enzymes may have played a role potentially causing alterations in lipid networks and metabolic dysregulation in the nervous system (van Esbroeck et al., 2017).

12.2.4 BENEFIT TO DRUG DISCOVERY AND DEVELOPMENT

A major benefit of pharmacological profiling is the potential influence on decision-making in drug discovery (Redfern et al., 2002; Valentin & Hammond, 2008; Wakefield et al., 2002). If key decisions can be made earlier in discovery, this avoids complications during the latter, more resource intensive and expensive phases of drug development. One example of decision-making at early stages is where multiple chemical series are being evaluated and the primary pharmacology and physicochemical properties of the compounds make it difficult to distinguish between one chemical series and another. By generating an *in vitro* pharmacological profile, the information can influence the decision as to which chemical series is the best to take forward into the next phase of the drug discovery process. The structure–activity relationships can be explored within a chemical series and the unwanted activities designed out or additional beneficial activities that are identified can continue to be incorporated. Understanding the pharmacological profile of a drug is also useful in interpreting observed functional responses in *in vivo* efficacy, toxicology/safety studies, or, most importantly, in clinical trials or in patients. If the *in vitro* pharmacology data highlight activity at a target that may result in a deleterious effect, then specific *in vivo* studies can be designed to explore the functional effects in relation to the expected therapeutic plasma concentration (taking into account plasma protein binding) of the drug. Greater understanding of the profile of the compound lowers the risk of an ADR in volunteers in clinical trials and in patients in clinical practice to remain undetected and explained.

Pharmacological profiling can also reveal potential beneficial effects, i.e., drug discovery by serendipity. In this case, it may be that the drug has dual activity on two targets that may play a role in a disease. For example, a drug that is a dual antagonist at histamine H_1 receptors and platelet-activating factor (PAF) receptors (e.g., rupatadine) has added benefit over a selective histamine H_1 receptor antagonist (e.g., loratadine) for the treatment of allergic rhinitis because both PAF and histamine play key roles in mediating inflammatory responses (Izquierdo et al., 2003). Similarly, a drug that has dual activity at the dopamine D_2 receptor and β_2-adrenoceptor (e.g., sibenadet) has the potential for greater efficacy for the treatment of chronic obstructive pulmonary disease when compared with a selective β_2-adrenoceptor agonist (e.g., salbutamol) (Dougall et al., 2002; Ind et al., 2003).

Pharmacological profiling may also result in the discovery of novel drugs at known targets. For example, by profiling a drug that is being developed as a potent blocker of an ion channel, affinity at a GPCR may be identified. The drug may represent a novel chemical structure with activity at the GPCR, resulting in the generation of a new patent. If the drug had not been profiled, the pharmaceutical company would be unaware of this potentially novel drug for the therapy of, perhaps, another disease indication. For example, quetiapine and norquetiapine, its major metabolite, showed multiple *in vitro* pharmacological actions, and results from preclinical studies suggested that activity at noradrenaline transporter (NET) and 5-HT$_{1A}$ receptors contributed to the antidepressant and anxiolytic effects in patients treated with quetiapine, opening up additional disease indications (Cross et al., 2016).

Understanding the pharmacological profile of a drug contributes to the development of drugs that have improved safety and tolerability over existing therapies for a disease, and this provides the pharmaceutical company with a significant advantage over their competitors and benefit to the patients. For example, tricyclic antidepressants (TCAs) were discovered in the 1950s–1960s and were the therapy of choice for depression. Nonselective TCAs (e.g., amitriptyline) inhibit the reuptake of serotonin, noradrenaline, and dopamine but also have activity at a range of receptors such as histamine, muscarinic, and adrenoceptors. These nonselective TCSs have a range of side effects that are a consequence of the lack of selectivity for the monoamine uptake transporters (Holm & Markham, 1999). The therapy of choice for depression nowadays is the selective serotonin reuptake inhibitors (SSRIs) that have similar efficacy, but significantly fewer side effects (e.g., sertraline) (Doogan, 1991; Ferguson, 2001), and therefore increased patient compliance. Another example is the utilization of phosphodiesterase (PDE) 5 inhibitors for the therapy of erectile dysfunction (Goldstein et al., 1998). For example, tadalafil has equivalent efficacy but is more selective for PDE 5 over PDE 6 (780-fold selective for PDE 5) (Curran & Keating, 2003) when compared with sildenafil that has the potential to inhibit PDE 6 (approximately tenfold selective for PDE 5 over PDE 6) at higher doses. Inhibition of PDE 6 is associated with visual side effects in the form of disturbance in discrimination between the green and blue colors and increased perception of brightness mediated by inhibition of PDE 6 activity in the retinal rod cells of the eye (Stockman et al., 2007). The greater selectivity of tadalafil for PDE 5 over PDE 6 results in potentially a safer drug profile for the patient and a competitive edge for the pharmaceutical company that developed it.

12.3 STRATEGY AND METHODS APPLIED TO PHARMACOLOGICAL PROFILING

12.3.1 STRATEGY

It is likely that there is considerable diversity in *in vitro* pharmacological profiling strategies adopted by different pharmaceutical companies (Bowes et al., 2012; Lynch et al., 2017; Whitebread et al., 2005, 2016). The ultimate objective is to establish the selectivity of the drug and identify or predict pharmacological effects that may arise as a consequence of drug action at these targets. For a strategy to have impact on drug discovery and development, it has to be applied in a consistent way within a company to maximize the output and analysis of the data.

One strategy for pharmacological profiling is to front-load *in vitro* screening at key points in the drug discovery process. Pharmacological profiling is applied in a stepwise manner where the drug is assayed for activity at an increasing number of molecular targets as it progresses through the drug discovery process (Figure 12.2). The screens applied in the early stages of discovery consist of a small panel of key targets (e.g., ~20) that mediate functional effects of the vital physiological systems: cardiovascular, central nervous, and respiratory systems. During the optimization phases, the profile is expanded to a greater number of assays (e.g., ~50–150) adding targets that mediate effects on other organ systems such as the gastrointestinal and renal systems. As the drug evolves into a candidate for the development phase, a greater number of targets (e.g., several hundreds) covering all organ systems are screened. The result is the generation of a comprehensive pharmacological profile of the drug before it is first administered to humans.

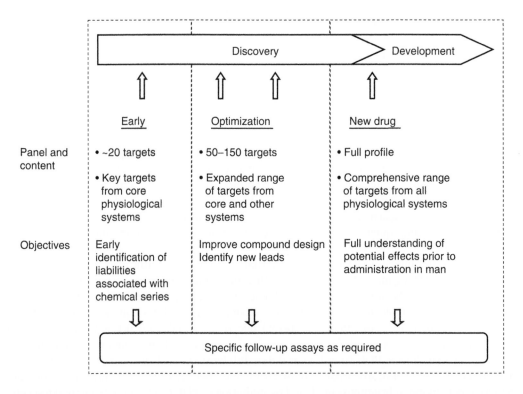

FIGURE 12.2 Pharmacological profiling in the drug discovery and development process. The number of molecular targets tested increases at each stage of the process and the information generated at each stage assists different types of decisions. The strategy is on the basis of a stepwise approach, focused on front loading of *in vitro* assays to generate a comprehensive pharmacological profile prior to the first administration of the drug to humans.

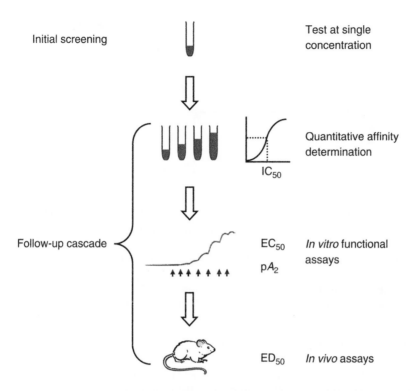

Initial screening   Test at single concentration

Quantitative affinity determination

IC_{50}

Follow-up cascade

EC_{50} *In vitro* functional
pA_2 assays

ED_{50} *In vivo* assays

FIGURE 12.3   The stepwise process applied to pharmacological profiling. Initially *in vitro* radioligand-binding assays or enzyme activity assays are used to profile the drug. If the drug has significant (e.g., >50% inhibition) activity, then *in vitro* functional assays are performed to understand the mode of action of the drug at the target and, if required, *in vivo* studies are subsequently performed.

The pharmacological action of drugs can be investigated *in vitro* or *in vivo* by a variety of methods, all of which are aimed at quantifying the drug action at a particular target so as to compare one drug with another. The parameters measured can be (a) the affinity of a drug to a target, (b) a functional response in a cell or tissue, and (c) a physiological response in an animal. A combination of these studies allows an integrated assessment of the biological effects (pharmacology) of a drug (Figure 12.3).

12.3.2   Methods

12.3.2.1   Radioligand-Binding Assays

Determination of the affinity of a drug for a target is commonly done by performing *in vitro* radioligand-binding assays, measuring the displacement of a radiolabeled tracer by the test compound from the receptor of interest. Concentration–response curves are constructed and activity quantified as IC_{50}: the molar concentration of drug that inhibits the binding of a radioligand by 50% (Figure 12.4) (Neubig et al., 2003). If the binding is competitive, the Cheng–Prusoff equation may be used to account for the concentration and affinity of the radiolabeled compound, which are used to determine the equilibrium dissociation constant, K_i, for the test molecule (Cheng & Prusoff, 1973). K_i is a representation of the affinity of the test molecule for the receptor of interest and is independent of the assay conditions used, so it should allow comparison of results between assays. IC_{50} is dependent on the assay conditions used, so it is suitable for ranking compounds tested within the same assay but may not be appropriate to compare with results from other assays or laboratories (Hulme & Trevethick, 2010).

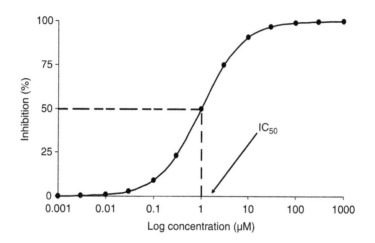

FIGURE 12.4 An example of a concentration–response curve for a drug in a radioligand-binding assay. Increasing concentration of drug (*x*-axis) causes an increase in the percentage inhibition of binding of the radiolabeled compound. The IC_{50} is defined as the concentration of drug that displaces 50% of the specific binding of the radioligand. In this example, the IC_{50} for the drug is 1 µM.

There are many advantages when applying the technique of radioligand-binding assays to pharmacological profiling; they are technically easy to perform, only small quantities of drug are required, and if validated correctly, they are extremely robust, reproducible, and suitable for high-throughput screening (HTS) formats. Importantly, the drug is incubated to reach equilibrium so quantitative determinations of affinity are attainable. Where the target has been cloned, the drug can be profiled against the human isoform of the target by screening using membranes from cells that express the recombinant form of the human receptor. Equilibrium radioligand-binding assays do not provide any information on the pharmacological consequence of the interaction of test compound with the receptor: they cannot distinguish between agonists, antagonists, partial agonists, or inverse agonists. While sodium or GTP-shift assays and kinetic radioligand-binding techniques are available (Hulme & Trevethick, 2010; Noel & do Monte, 2016; Uhlen et al., 2016), which allow access to more detailed understandings of ligand–receptor interactions, the additional complexity of such assays means that they are not typically employed in routine screening.

12.3.2.2 Functional *In Vitro* Assays

Pharmacological analysis of a drug in functional assays involves determining whether it is an agonist or antagonist at a particular receptor and the measurement of the potency of the drug in the assay system. An agonist is a drug that binds to a receptor and activates it to produce a biological response, and its effects depend on affinity (tendency to bind to a receptor) and efficacy (ability once bound, to initiate changes which result in the biological effect) (Neubig et al., 2003). An antagonist is a drug that binds to a receptor but does not result in a biological response (i.e., has no efficacy) but attenuates the response of an agonist. Potency is a measure of the concentration of a drug at which it is effective. For agonists, this is expressed as EC_{50}: the molar concentration of an agonist that produces 50% of the maximum possible effect of that agonist, conversely the concentration of an antagonist which inhibits 50% of a reference agonist response is the IC_{50}. Both parameters are dependent on the assay conditions used so, while suitable for ranking compounds tested in the same assay, IC_{50} and EC_{50} may not be appropriate parameters to allow comparison between assays or laboratories.

Functional *in vitro* assays can be subcellular, whole-cell, or animal tissue assays. For *in vitro* subcellular and cell-based assays, recombinant molecular targets can be expressed in cell lines and linked to second-messenger systems to measure functional responses. Examples of functional readouts in cell-based assay include second messengers such as measurement of changes in intra-cellular calcium (Chambers et al., 2003; Sullivan et al., 1999), cAMP (Williams, 2004), inositol phosphates, phosphorylation/dephosphorylation, or G protein activation. Cell-based assays are ame-nable to medium-throughput screening, and these assays provide key information as to whether the drug is an agonist or antagonist at the target. The main limitations are that cell culture on a rela-tively large scale is resource-intensive, and in some assays, true equilibrium may not be achieved (Chambers et al., 2003).

12.3.2.3 Comparative Assay Formats

It is clear that there are many different ways to assess the interaction of a test molecule with a tar-get. It is important to have an understanding of what information can be gained from a particular assay format and what may not be assessed. Perhaps the most commonly debated target family is the GPCRs, for which radioligand-binding assays and functional assays offer access to different aspects of test compound interaction with the receptor of interest (Hill Stephen, 2009; Kenakin, 2003). If we take the 5-HT_{2B} receptor as an example, radioligand binding offers an inexpensive and robust assessment of compound affinity for the target. However, in secondary pharmacologi-cal assessment, it is typically agonism at this receptor, associated with the development of cardiac valvulopathy (Papoian et al., 2017; Rothman et al., 2000), which is of greatest concern. Detection of binding at the receptor is insufficient to enable hazard identification – data from a functional assay are required to determine whether the binding interaction leads to an agonist or antagonist effect. If the latter, then it is likely that whatever the binding affinity, there is no known risk for developing cardiac valvulopathy. However, if an agonist effect is detected, then some form of quantitative risk assessment is desirable, and at the moment, this quantitation comes from the radioligand-binding data (Papoian et al., 2017). This is discussed in more detail in Section 12.4.

Other differences in GPCR assay format often considered include the assessment of allosteric (non-competitive) interactions of compounds with receptor. Not all such interactions are readily detected in radioligand-binding assays optimized for competitive equilibrium displacement; modifi-cations to the binding assay format may be required. Functional assays are sometimes more flexible and allosterism may more readily be detected in some functional assays without additional optimi-zation; however, it must always be remembered that the nature of a cooperative interaction of two (or more) molecules at a single protein is dependent on both ligands involved – such that use of the native agonist as a reference or stimulatory ligand is preferred; use of a non-native agonist may mis-represent the action of an allosteric ligand (Kenakin, 2007). Of course, there are very few marketed drugs which are known to act via an allosteric mechanism, and so, understanding the translation of *in vitro* allosteric activity to possible physiological consequence is more challenging.

Another well characterized example is the hERG (Kv11.1) encoded cardiac ion channel, with a strong link to cardiovascular adverse events involving lengthening of the QT interval on the car-diac electrocardiogram, leading to potentially life-threatening ventricular arrhythmias (Gintant et al., 2016). Radioligand binding using, for example, [^{3}H]dofetilide to label the hERG channel recombinantly expressed in membranes prepared from a host cell such as HEK293 or CHO may provide an inexpensive and high-capacity assay. Such an assay may be amenable to simple control of incubation time and temperature, but while it allows assessment of equilibrium affinity, it gives no information about the effect of test compound on the function of the channel. Alternate tech-nologies such as HTS electrophysiological assessment offer the ability to measure current passing through the channel in intact cells, giving information on the consequence of test compound inter-action with the channel. However, not all HTS electrophysiology platforms offer control of tem-perature, and the duration of the experiment may be limited by the ability to maintain an electrical seal leading to lack of equilibrium. It is also important to note that the majority of hERG blockers bind to the cytoplasmic side of the channel (Fermini & Fossa, 2003) and so compounds which are poorly membrane permeable may appear more potent in a binding assay using isolated mem-brane than whole-cell–based assays. Novartis has found that for the majority of compounds, data

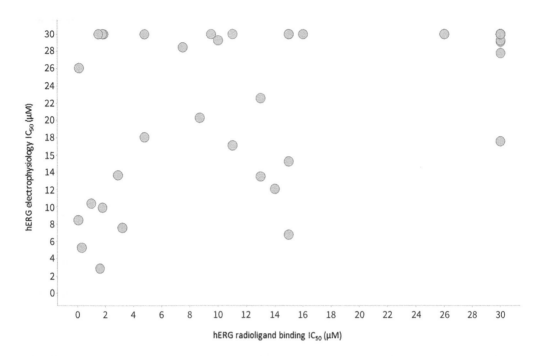

FIGURE 12.5 Correlation of hERG radioligand-binding data with QPatch IC_{50} data for compounds with poor membrane permeability. IC_{50} values for hERG inhibition measured by whole-cell electrophysiology using the QPatch platform are shown on the y-axis; IC_{50} values for displacement of [^{3}H]dofetilide from isolated membranes expressing the hERG channel are shown on the x-axis. Data from Novartis, 2017–2018.

generated in radioligand binding and the QPatch electrophysiological platform are well correlated; however, molecules with poor permeability characteristics may show notably greater potency in a membrane-based radioligand-binding assay than a whole-cell–based electrophysiological assay (Figure 12.5).

Another common example of the impact of assay format on compound characterization is the assessment of kinase activity. Although the technological details may vary, kinase assays may be broadly characterized as using isolated enzyme or using intact cells. As with the hERG example, differences exist between assay formats in the ability of test compound to access the kinase. There is also variation in the type and concentration of substrate, and of ATP, with cellular assays offering less control of these parameters. Differences in the apparent off-target kinase profile of R406 (the active metabolite of Fostamatinib) using isolated enzyme radioligand-binding assays compared to isolated enzyme functional assay were highlighted by Rolf et al. (2015).

These examples serve to highlight the importance of *in vitro* assay characterization in terms of standard performance measures (e.g., reproducibility, dynamic range), in terms of understanding the pharmacological context of the measures provided by the particular assay format, and in terms of validation of the translational value by examining the effects of compounds with known charac-teristics (discussed in Section 12.4).

12.3.2.4 Complex Assay Systems

While the characterization of the interaction of test compounds with individual molecular targets is generally well served by simple binding and functional assay as discussed above, it is sometimes desirable to gain an understanding of pharmacological effects in more integrated systems.

An alternative method used for *in vitro* functional assays is the use of animal tissue – the more traditional "organ-bath pharmacology" to measure physiological responses in an intact piece of tissue. Some examples include investigation of the muscarinic receptor–mediated contraction of guinea pig ileum, of β_1-adrenoceptors in mediating the force and rate of contraction of rat-isolated atria, and α-adrenoceptors in mediating vasoconstriction of rabbit aorta (Kenakin, 1984). The advantage of these assays compared with cell-based assays is that the drug can be incubated over time to ensure equilibrium, thereby enabling potency to be determined quantitatively. The main limitations of this type of functional assays are the use of tissue from animals, and that the drug is profiled with nonhuman targets. This may result in incorrect interpretation and prediction of functional effects of the drug in clinical studies if the potency and efficacy at the human and animal target are different. However, in some cases, functional assays can be performed using human tissue, although this may be limited by availability of human tissue for these experiments (Borman et al., 2002; Holmes et al., 2015; Jackson et al., 2018).

The pharmacology of a drug may also be investigated by administering the drug to animals and measuring biological responses. The responses can be pharmacologically quantified by expressing the response as an ED_{50}, the dose that gives 50% of the maximum response to that drug. This type of study may be applicable, for example, in the investigation of effects of the drug on a major physiological system, such as the cardiovascular system. In this case, the drug may be administered to either conscious or anesthetized animals and its effects on heart rate, blood pressure, and regional blood flows can be explored over a range of doses. To interpret the effects, it is useful in these studies to measure the concentration of drug in the plasma when the effect is observed. The advantage of this approach is that the effect of the drug is determined in an intact integrated physiological system. The major limitation in this methodology is that the receptor(s) mediating the response is/ are not known, as many different targets may mediate the same response, which adds complexity to the interpretation. Also, subtle effects may not be detected in *in vivo* studies where compensatory mechanisms may be present.

The interests of animal welfare; application of the 3R principles of replacement, reduction, and refinement; and the natural limitations of simple cellular models in approximating physiologically relevant cell-to-cell interactions all drive development of more complex cell-based systems. Such complex systems possess integrated pharmacology without recourse to whole animals or extracted tissues. One approach commonly applied is co-culturing of different cell types in three-dimensional structures, rather than as monolayers, offering the opportunity to explore more complex biological mechanisms. The current cutting edge of cell-based models is in the generation of complex multicellular 3D structures with integrated microscale structure and microfluidics, "organs-on-chips" (so called because the microengineering technology involved in the construction of the structures is analogous to that used in generating computer microchips). To date, models of bone, brain, breast, eye, gut, liver, lung, and kidney tissues have been generated (reviewed in Huh et al., 2011). Microengineered models generate the opportunity to explore multicell interactions in the context of organ-relevant spatial and structural constraints and in the presence of mechanical stresses. There is also the potential to connect multiple organs-on-chips giving the opportunity to study ADME of drugs across different organ systems. While the organ-on-a-chip technology is still in its early stages, these systems pose great potential to investigate pharmacological and safety effects of NCEs.

12.4   IMPACT OF PHARMACOLOGICAL PROFILING: DATA INTERPRETATION AND RISK ASSESSMENT

The impact of the *in vitro* screens applied to each stage in the drug discovery process will be different depending on the stage. The optimal set of assays at the early stages should assess drug activity against a small number of targets that play a key role in the function of core

TABLE 12.1

Examples of Functional Effects due to Activity at Pharmacological Targets

Target	Mechanism	Administration	Affect Compartment	Potential Functional Effect	Example
5-HT$_{2B}$ receptor	Agonism	Chronic	Cardiovascular (CVS)	Valvulopathy	Fenfluramine (Hutcheson et al., 2011; Papoian et al., 2017)
PDE4	Inhibitor	Acute	CNS	Emesis	Rolipram (Heaslip & Evans, 1995; Horowski & Sastre-Y-Hernandez, 1985)
Muscarinic M$_3$ receptor	Antagonist	Acute	GI	Reduced GI motility	Darifenacin (Chiba et al., 2002)
Histamine H$_1$ receptor	Antagonist	Acute	CNS	Sedation	Diphenhydramine (Kay et al., 1997)
β_1-adrenoceptor	Agonist	Acute	CVS	Tachycardia	Isoprenaline (Motomura et al., 1990)
Mu Opioid receptor	Agonist	Acute	GI	Constipation	Loperamide (Hanauer, 2008)
	Agonist	Acute	CNS	Respiratory depression	Morphine (Merel et al., 2012)
Cyclooxygenase-1	Inhibitor	Chronic	GI	Ulceration/bleeding in gastrointestinal tract	Naproxen (Lewis et al., 2002)
hERG potassium channel	Blocker	Acute	CVS	QT prolongation, cardiac arrhythmia	Cisapride (Bran et al., 1995)
		Chronic	Reproductive tract	Teratogenicity	Cisapride (Skold et al., 2002)
Adrenergic α_{1A} receptor	Antagonist	Acute	CVS	Orthostatic hypotension	Prazosin (Schoenberger, 1991)

physiological systems. A well-designed set of targets facilitates decision-making, and the interpretation relating to these targets is straightforward. For example, the panel may include key receptors that control the autonomic nervous system such as adrenoceptors and muscarinic receptors and other key targets known to cause ADRs in humans (some examples are given in Table 12.1). As a drug progresses further along the drug discovery process, the data may be used to identify unwanted activities, and this would allow the medicinal chemists to explore the structure–activity relationships at the molecular target and design out these undesirable activities from the chemical series. As the drug evolves into a potential new candidate for development, and when preclinical safety pharmacology and toxicology studies are being considered, pharmacological profiling contributes to the study design by identifying potential safety issues that will need to be monitored.

Dosing levels may also be set, taking into account the pharmacological profile, so that sufficient exposure is achieved in the studies to investigate the potential effects mediated via other molecular targets.

Data generated from *in vitro* pharmacological profiling therefore requires careful interpretation and contextualization to enable predictions of the potential functional consequences that pose a risk due to the activity presented in a molecule. There are a number of key factors that should be considered.

12.4.1 OCCUPANCY AND THERAPEUTIC INDEX

During early drug discovery, the affinity of investigative lead molecules at secondary targets is generally analyzed relative to the affinity at the desired therapeutic target. This is referred to as the selectivity of a molecule. However, based on the pharmacological principle that a level of occupancy of a target is required for a drug to elicit a biological response, a more informed approach is to analyze the response relative to the human-predicted (or measured) maximum free plasma concentration of the drug. Building on this idea, one can compare the therapeutic concentration required for therapeutic benefit to the concentration which elicits a toxicity. This ratio of toxic over therapeutic concentration is referred to the therapeutic index and is a measure of safety (Muller & Milton, 2012). During drug discovery both parameters, selectivity and therapeutic index, are used. At the very early stages of discovery, the selectivity approach tends to dominate as predicted drug concentrations expected in humans are unlikely to be accessible. However, a compound series may look selective but as progression refines chemical structure and PK/PD modeling establishes a predicted dose for man, the value of selectivity becomes less critical and the therapeutic index too is much more informative for the safety risk assessment.

To help facilitate understanding, the assessment data can be presented in graphical form as illustrated in Figure 12.6. The general rule of thumb is that, if the affinity at the off-target receptor is 100-fold lower than at the receptor mediating therapeutic benefit at the maximal free exposure concentration, then it is unlikely that there will be any observed functional effects in humans, as it is unlikely that the drug will reach levels high enough to have significant activity at the secondary target. However, this "margin" may differ depending on the target under analysis, and this is explored further in Section 12.4.3.

12.4.2 THERAPEUTIC PROFILE AND PATIENT POPULATION

Drugs can be dosed by multiple routes of administration and possess varying physicochemical and PK properties that affect their disposition within a whole animal or human. Given this, we

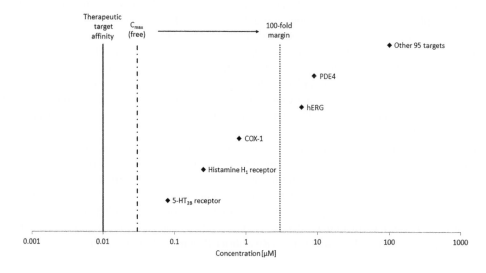

FIGURE 12.6 Graphical representation of therapeutic index to assess pharmacological profiling data. This diagram can be used to assess the therapeutic margins by plotting the K_I or IC_{50} potency values for the targets and comparing these to the therapeutic target affinity and predicted free plasma levels for therapeutic efficacy. In this example, the drug shows selectivity of less than 100-fold at three molecular targets when tested in a panel of 100 target assays. The potential functional effects of the drug because of the actions at these targets are described in Table 12.1.

need to consider the localization of expression of the target and the tissue(s) in which their function predominates. For example, targets primarily expressed in the brain will not be affected *in vivo* if the drug does not cross the blood–brain barrier, and the side effects potentially mediated by activity of the drug at these central receptors can therefore be ruled out under normal physiological conditions. Analysis of the physicochemical properties of the drug allows predictions to be made about the ability of the drug to penetrate the brain. Conversely, if a drug is administered by the oral route, the local concentration of drug in the gastrointestinal tract may be high relative to the plasma efficacious concentration, and thus, functional effects as a consequence of activity at a target expressed in the gastrointestinal tract may be revealed. Alternatively, the drug may be developed for inhaled administration, where the focus for interpretation will be on targets expressed in the respiratory system, and it is unlikely that there will be significant systemic exposure of the drug. Where a target plays a role within multiple tissues, understanding the biological role in each compartment can aid the risk assessment and influence chemical design. This is exemplified by the number of peripherally restricted opioid analgesics currently under clinical development that aim to maintain pain relief but reduce the risk of centrally driven respiratory depression through the Mu opioid receptor (Vadivelu et al., 2011) or dysphoria via the Kappa opioid receptor (Albert-Vartanian et al., 2016).

Another point to consider is the different interpretation that may be applied if a drug is administered chronically versus acutely. For example, drugs with activity at the β_2-adrenoceptor (e.g., isoprenaline) or the K_{ATP} channel (e.g., minoxidil) cause peripheral vasodilation, and this is associated with reflex tachycardia (Dogterom et al., 1992). However, there is clear evidence that chronic administration of these drugs results in necrosis of the left ventricle and papillary muscle of the heart in dogs (Dogterom et al., 1992; Herman et al., 1989). Interestingly, this effect does not seem to translate into humans (Sobota, 1989). The reason for this is that in humans, heart rate is closely controlled, whereas in dogs, these compounds elicit a tripling of resting heart rate to 180–200 beats/min, and it is this high heart rate that results in necrosis. Identification of vasodilator action from pharmacological profiling can be used to predict and explain this observed pathology but provide confidence that this is unlikely to translate into humans. Conversely, there are a number of target mechanisms that do manifest and translate to safety concerns in humans following chronic dosing, for example, myocardial hypertrophy through agonism of the vasopressin V_{1A} receptor (Lynch et al., 2017), cardiac valvulopathy through agonism of the 5-HT_{2B} receptor (Papoian et al., 2017), and thrombosis through inhibition of cyclooxygenase II enzyme (Lynch et al., 2017). Ultimately, applying pharmacological profiling early to detect these risks and design them away is the optimal approach to drug design. However, understanding these risks in context of the dosing regimen may enable a risk to be mitigated or a strategy such as application of intermittent dosing schedules to be applied to reduce chronic exposure. These need to be carefully considered given the patient context and benefit-risk profile.

Where possible and within the realms of information available, we must also try to consider any differences in the biological background of the intended disease indication or concomitant medications that could affect the manifestation or severity of the off-target–driven side effects. For example, it was recently proposed that the dyspnea and bronchospasm effects associated with ENT1 transporter inhibition in respiratory disease patient and not healthy volunteers were due to the superfluous increase in extracellular adenosine concentrations caused by the high levels of adenosine already present within the disease state (Rosenbrier Ribeiro & Storer, 2017).

12.4.3   Integrated Risk Assessment (*In Vitro* to *In Vivo*)

Understanding the translation of off-target *in vitro* pharmacology to an *in vivo* effect is important to enable contextualization and the likelihood of manifestation of side effects mediated by off-target activity. In preclinical toxicology studies' exposures are pushed to high level, which can result in off-target occupancy and concomitant toxicological effects. As such, understanding the relation of off-target activity to exposure can aid mechanistic understanding and help define the therapeutic index. In addition, for toxicities that cannot be detected preclinically due to limitations of naïve animal studies or species differences, translational understanding can aid early risk assessment for the intended patient population and can be informative to influence clinical plans with the addition of safety biomarkers, monitoring, or define patient inclusion/exclusion criteria.

In some cases, the target mediates a physiological response that will clearly be revealed *in vivo*. For example, inhibition of PDE4 activity is associated with emesis (rolipram; Heaslip & Evans, 1995; Horowski & Sastre-Y-Hernandez, 1985), antagonism of the histamine H_1 receptor in the central nervous system (CNS) causes sedation (diphenhydramine; Kay et al., 1997), and inhibition of PDE 6 activity causes visual disturbances (sildenafil; Goldstein et al., 1998). However, it is also possible that compensatory mechanisms may override the drug's action and a net functional effect may be undetectable. For example, if a drug has activity at the hERG potassium channel and also at the L-type calcium channel, although antagonism of the hERG potassium channel may result in the prolongation of the QT interval of the electrocardiogram, blockade of the L-type calcium channel may result in the shortening of the QT interval. In this case, it may be anticipated that overall there would be no effect on the QT interval *in vivo*. This is the case for verapamil, which blocks both L-type calcium channels and hERG potassium channels, and does not have any effect on the QT interval *in vivo* (Nademanee & Singh, 1988).

For target mechanisms that manifest a clear response *in vivo*, we can utilize data from known drugs that drive that mechanism to improve our translational understanding of the target occupancy versus effect and quantify the therapeutic index that poses an acceptable/unacceptable risk. Over the past 5 years, there have been numerous reports emerging to facilitate our translational understanding of target-related mechanisms of toxicity, highlighting the importance, complexity, and impact such knowledge can have on selecting the right molecules for progression. For the hERG potassium channel, it is clear that to be confident that proarrhythmia will not be observed in the clinic, the margin between the effective therapeutic plasma concentration (unbound) or free C_{max} target and potency at hERG should be at least 30-fold (Redfern et al., 2003). The severe dyspnea and bronchospasm caused through inhibition of the ENT1 transporter have been shown to be attributed to patients with underlying respiratory disorders only and particularly at risk when the effective free plasma concentration reaches a level 4× over the *in vitro* ENT1 IC_{50} potency (Rosenbrier Ribeiro & Storer, 2017). Unfortunately, quantitative translational data are not currently available in the public domain for many of the targets explored in pharmacological profiling. Therefore, the generalized "100-fold" margin is commonly utilized as a starting point for risk assessment and the considerations discussed in Section 12.4.2 become important to refine and focus in on key areas of safety liability. As more data emerge from preclinical and clinical studies with investigational products, we will be able to improve our translational understanding of off-target liabilities and use this to improve the accuracy of the risk assessment.

12.5 PHYSICOCHEMICAL PROPERTIES AND *IN SILICO* PHARMACOLOGICAL PROFILING

In vitro pharmacological profiling is mainly used to predict overall pharmacodynamic properties of a given drug, aiming to answer the question: "what does the drug do to the biological system?" It is also necessary to predict "what the biological system does to the drug", through the determination of PK properties, better known as ADME properties of the drug (Ekins et al., 2007b). Early prediction of bioavailability is a key step in the drug discovery process, not only because it gives an indication of the drugability of the compound but also because it helps interpreting *in vitro* pharmacological profiling data, taking into consideration the potential availability of the compound at its active site(s). Similarly, important efforts have been put on *in silico* prediction of pharmacological effects either on a single target or on several targets, in an attempt to predict safety issues or to develop promiscuous drugs with effects on specific targets (polypharmacology).

Finally, models to predict toxicity issues, such as cardiotoxicity, are described in the literature and made available for the community. This section gives a rapid overview of these different aspects of drug discovery.

12.5.1 Predicting Bioavailability

Bioavailability can be assessed early in the discovery process by combining data from a number of independent *in vitro* assays run as part of a pharmaceutical properties profile. These assays generally include some measurement of aqueous solubility, lipophilicity, cell or membrane permeability, and metabolic stability. In most cases, proper interpretation of data from a pharmaceutical properties profile will give reliable guidance for the identification of compounds with reasonable bioavailability. When testing large numbers of diverse compounds, one can be guided by a set of categorical rules on the basis of the knowledge derived from, or validated by, profiling data from large numbers of reference compounds with known *in vivo* properties. Examples for some commonly measured parameters and values associated with moderate to good bioavailability are listed in Table 12.2. It should be noted that there are examples of successful drugs falling outside of these guidelines, and information on the likely route of administration, dosage, formulation, serum protein binding, ionization effects, etc., need to be considered in addition to standard results from a pharmaceutical profile.

To bring a relevant context to pharmaceutical profile interpretation, it is important to understand how currently marketed drugs behave in each assay. Results from the profiling of approximately 300 marketed drugs with well-documented human % absorption and bioavailability are shown in Figures 12.7 and 12.8.

A well-designed apparent permeability assay can give guidance in the identification of compounds with desired human absorption (Figure 12.7). The low contributions made by solubility and lipophilicity data alone are not surprising as compounds with poor solubility or extreme log D values are likely to also show low apparent permeability. The situation with bioavailability is more complex (Figure 12.8). In this case, there are the expected independent contributions to human bioavailability of both *in vitro* permeability and metabolic stability assay results. In the case of bioavailability, the most significant guidance is given by the "one or more red flag" results. It is important not to overinterpret individual assay results. Note that nearly 30% of the drugs having less than 20% oral bioavailability did not have any red flag for solubility, log D values, permeability, or metabolic stability. At the other end of the spectrum, ~15% of the compounds with 80%–100% oral bioavailability did have one or more red flags (Figure 12.8).

To further strengthen decision-making, it is helpful to use statistical clustering to place each experimental compound in the context of a database created from the pharmaceutical profiling of compounds with well-characterized human PK. In the absence of additional guiding information,

---

TABLE 12.2

Bioavailability Assessment through the Use of *In Vitro* Assays and Values Associated with Moderate to Good Bioavailability

Pharmaceutical Properties	Quantified Results Obtained in Corresponding *In Vitro* Assay
Aqueous solubility (equilibrium)	>20 µM
Lipophilicity	Log D between −1 and 4
Apparent permeability	>5 × 10^{-6} cm/s
Metabolic stability	>40% remaining

Source: Data are taken from BioPrint® (Anon, 2018).

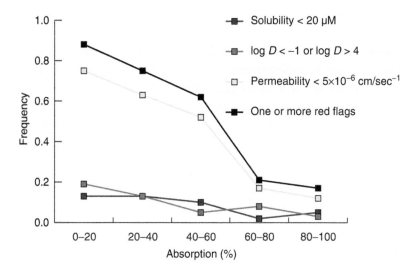

FIGURE 12.7 Approximately 300 drugs were tested in aqueous solubility, log D, and apparent permeability assays. Compounds having low values for solubility, or apparent permeability, or extreme log D values were flagged. The frequency of compounds with flags in each human absorption (%) bin is shown. Methods: **Solubility** was measured by adding dimethyl sulfoxide (DMSO)–solubilized drug to an isotonic pH 7.4 aqueous buffer to yield a final concentration of 2% DMSO and a theoretical 200 μM test compound. Ultrafiltration was used to remove precipitated drug, and final concentration was measured by liquid chromatography–ultraviolet (LC–UV) or liquid chromatography–mass spectrometry detection (LC–MS). **Lipophilicity** was measured as by n-octanol/PBS pH 7.4 partitioning (log D). **Apparent permeability** was measured using a subclone of the Caco2 cell line. A red flag is given to a compound having solubility less than 20 μM, log $D < -1$ or >4, or apparent permeability less than 5×10^{-6} cm/s. Data are taken from BioPrint® (Anon, 2018).

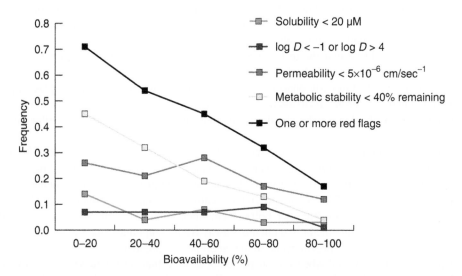

FIGURE 12.8 Approximately 300 drugs were tested in aqueous solubility, log D, apparent permeability, and metabolic stability assays. Compounds having low values for solubility, apparent permeability, or metabolic stability, or extreme log D values were flagged. The frequency of compounds with flags in each human bioavailability (%) bin is shown. Methods: see Figure 12.7 for **solubility, lipophilicity,** and **apparent permeability**. **Metabolic stability** was measured by incubating drug at 1 μM for 60 min with human liver microsomes. A red flag is given to a compound having solubility less than 20 μM, log $D < -1$ or >4, apparent permeability less than 5×10^{-6} cm/s, or metabolic stability less than 40% remaining. Data are taken from BioPrint® (Anon, 2018).

TABLE 12.3

Single Test Compound Clustering with Several Marketed Drugs on the Basis of the Similarity of *In Vitro* Profile Results Including Human Absorption, Oral Bioavailability, Permeability, Solubility, log *D*, and Metabolic Stability Characteristics

Compound	Solubility (μM)	Log *D*	Apparent Permeability (10^{-6} cm/s)	Metabolic Stability (% Remaining)	Oral Bioavailability (%)	Absorption (%)
Test compound	123	3.9	36	26	N/A	N/A
Drug A	210	3.3	117	22	15	93
Drug B	180	3.1	17	27	23	80
Drug C	128	3.6	32	51	15	N/A
Drug D	205	3.8	55	69	18	53
Drug E	113	3.3	25	68	75	75
Drug F	23	3.6	29	20	17	43
Drug G	33	3.5	33	28	60	60
Drug H	79	2.9	51	30	15	95

Source: Data are taken from BioPrint® (Anon, 2018).

it is reasonable to assume that compounds with overall similar pharmaceutical profiles will have similar *in vivo* PK properties. Table 12.3 shows a single test compound that clusters with several marketed drugs on the basis of the similarity of *in vitro* profile results. Human absorption and oral bioavailability are shown for the drugs. From this table, it can be seen that the group of selected drugs with favorable apparent permeability, solubility, and log *D* values all have good human absorption. It is reasonable to project that the test compound will likely have similar *in vivo* absorption. However, there is a distribution of 20%–69% remaining for the *in vitro* metabolic stability assay. As all compounds shown have good apparent permeability, it would be appropriate to focus attention on the compounds with metabolic stability most similar to the test compound. A reasonable assumption would be that the test compound should be flagged as having a potentially low bioavailability (15%–20%) (Table 12.3).

Additional assays used in early pharmaceutical property profiles usually include plasma protein binding, individual cytochrome P450 assays, stability in the presence of serum, production of metabolites likely to be involved in covalent binding to biomolecules, and interaction with efflux pumps:

- Serum protein binding is an important contributor to distribution and clearance *in vivo*. In addition, results from *in vitro* tests of compounds with high serum protein binding are expected to be sensitive to the presence of serum or other proteins in assay incubations.
- Panels of cytochrome P450 assays are run to identify compounds that interact as substrates or inhibitors of the major drug-metabolizing enzymes. P450 results in early drug discovery are often used to guide medicinal chemists away from compounds that interact primarily with the P450s that are polymorphic in the target patient population.
- Serum stability is tested when it is suspected or desired that serum enzymes may hydrolyze the parent drug. Some compounds or their metabolites are subject to glutathione conjugation, which may lead to covalent binding to protein and deoxyribonucleic acid (DNA). The potential for glutathione conjugation can be tested *in vitro* using human liver preparations.
- There are several efflux pumps which may affect absorption, blood–brain barrier penetration, and reabsorption from kidney microtubules. The most commonly tested efflux pump in early drug discovery is P-glycoprotein. Assays to identify P-glycoprotein substrates or inhibitors can be run using a variety of cell lines.

A variety of *in vitro* assays are available to assess important parameters that impact bioavailability, and these pharmaceutical profiling assays are commonly run to support drug discovery efforts. The major challenge is the interpretation of these results. Misinterpretation can lead to dropping promising compounds or alternatively taking unsuitable compounds into early development. In our experience, the interpretation of pharmaceutical profiling data is greatly strengthened by understanding how the active ingredients of all currently marketed drugs behave in these same *in vitro* assays.

There have been many attempts to use molecular properties to predict *in silico* bioavailability of drugs. In 1997, Lipinski (Lipinski, 2004; Lipinski, Lombardo, Dominy, & Feeney, 1997) proposed a general rule referred to as "the rule of five" which states that a compound will have a better chance to be bioavailable when its cLogP (calculated 1-octanol-water partition coefficient) is <5, its molecular mass is <500 Da, its number of hydrogen-bond donors is <5, and its number of hydrogen-bond acceptors is <10. However, as described above, bioavailability depends on many factors including absorption, solubility, binding to proteins, gut transit time, and first-pass metabolism (Leeson & Springthorpe, 2007). Consequently, simple molecular property-based models (Lu et al., 2004; Veber et al., 2002) fail to explain human bioavailability data (Hou et al., 2007; Tian et al., 2011). If each pharmaceutical company has its own computational model to predict bioavailability, none of these models has imposed itself as a standard.

In 2011, Tian et al. used a database of 1,014 molecules with human oral bioavailability data. They generated and evaluated a number of property-based rules for bioavailability classification. No rule was an effective predictor for oral bioavailability. However, using the genetic function approximation (GFA), they could construct multiple linear regression models for oral bioavailability with structural fingerprints as the basic parameters, together with important molecular properties. The best model was able to predict human oral bioavailability with an r of 7.9, a q of 0.72, and a root-mean-square error (RMSE) of 22.30% of the compound from the training set and an r_{test} of 0.71 and RMSE of 23.55% for a separate test set of 80 compounds.

It is well established that lipophilicity is a major determinant of pharmacodynamics and PK responses (Hansch & Dunn William, 1972; Hansch & Fujita, 1964). Consequently, recently synthesized drug candidates in leading drug discovery companies show a marked increase in lipophilicity (Leeson & Springthorpe, 2007). Interestingly, this increase in lipophilicity has proved to be associated with a lack of selectivity (Leeson & Springthorpe, 2007) and an increased potency on hERG potassium channel (Waring & Johnstone, 2007), leading to increased attrition in drug development.

12.5.2 *In Silico* Pharmacological Profiling

Computational models have been built for some molecules related to the action of drugs, such as the hERG potassium channel (see below), cytochrome P450 enzymes, and other promiscuous proteins (Ekins, 2004). Compared to the large number of potential molecular targets, the number of published models is still limited.

In silico pharmacological profiling can be applied in many ways to predict activity of a drug at pharmacological targets without the need to do actual experiments. A simple first step involves comparing the chemical structure of a new drug to structures of drugs that have already been profiled. Examination of the profiles of similar compounds may reveal common pharmacological activities, and this may aid in the identification of activities that may be detected *in vivo* quickly and easily. A second, more advanced strategy involves building detailed pharmacophore models for the targets contained within the database so that predictions can be made for each individual target. In theory, this could then allow a full pharmacological profile to be generated without actually needing to run laboratory assays. This would have significant benefits in terms of the number of compounds that could be screened in a given time using minimal resources. Another consideration is the impact on the 3Rs rule of animal welfare (reduction, refinement, and replacement), which is particularly valued in some countries. However, current computational pharmacophore models are generally not accurate enough to act as the sole source of pharmacological data. While a model may be sufficient to rank several compounds within a chemical series, absolute affinity or potency estimates are often not reliable, and predictions based on a single compound can be misleading. Therefore, *in silico* models are currently best used as a high-throughput filter allowing large numbers of chemical structures to be screened at low cost. One of the advantages of *in silico* prediction is then to reduce the number of molecules made and tested *in vitro* (Ekins et al., 2007a, 2007b). The most promising candidates would then be further profiled using the experimental pharmacology assays. Data on compounds with *in silico* predictions that have been validated by laboratory assays can then be used to further improve the *in silico* models. In the future, *in silico* models may replace *in vitro assays*, at least for early-stage screening of drugs prior to regulatory pharmacology and toxicology studies.

12.5.3   *In Silico* Prediction of hERG Inhibition and Cardiotoxicity

The hERG encoded voltage-gated potassium channel carries the delayed rectifying current (IKr) responsible for repolarization of the cardiac action potential (Tristani-Firouzi et al., 2001). Inhibition of the hERG encoded channel has led to many drug withdrawals from the market and represents a significant regulatory hurdle for new drugs in development (Shah, 2005). Consequently, *in vitro* screening of compounds against hERG is now regularly performed at an early stage in drug discovery programs. The potential effect of molecules on hERG channel activity is typically measured in whole-cell patch-clamp experiments. Higher throughputs may be reached using radioligand-binding assays but with a decreased physiological relevance than electrophysiological methods (see Section 12.3.2.3).

Compared to *in vivo* studies and even to *in vitro* screening, *in silico models* would require much less effort and time to predict affinity of drug candidates to hERG-binding site. There have been considerable efforts to develop such models in the past 20 years (Etkins, Crumb, Sarazan, Wikel, & Wrighton, 2002). A recent review (Jing et al., 2015) presents an extensive list of *in silico* models, classified as 2D and 3D quantitative structure–activity relationship (QSAR) models, pharmacophore models, classification models, and structure-based models using homology models of hERG. In summary of their review, Jing et al. noted that, as previously reported (Waring & Johnstone, 2007), lipophilic groups, especially aromatic groups, increase hERG-binding affinity of a compound. On the contrary, hydrophilic groups except secondary aliphatic amines were more likely to decrease hERG-binding affinity (Song & Clark, 2006) as does the addition of oxygen, the removal of carbon atoms, or the increase in the topological polar surface area by at least 60 Å (Braga et al., 2014). Many pharmacophore models for identifying hERG inhibitors have one positive ionizable group and at least one hydrophobic/aromatic group. Finally, two amino acids, Y652 and F656, in S6 helix play a vital role in forming π stacking with hERG inhibitors (Jing et al., 2015). Thus, *in silico* prediction of hERG inhibition is proven to aid molecular design of new drug candidates and allows for a significant reduction in the number of compounds to be synthesized and screened. However, as noted by the authors, *in vivo* electrocardiogram studies in preclinical mammalian species remain a requirement prior to testing in man (Jing et al., 2015).

Indeed, ICH standards require drug candidates to be tested for hERG inhibition in cultured cells. But not all drugs with hERG activity turn out to be torsadogenic, and more than a dozen of ion channels interact to affect the rhythm of heart contractions (Servick, 2016). Consequently, molecules that might in the end prove to be promising drug candidates are discarded for their hERG activity. This has prompted a panel of experts to launch a project called CiPA (see above, Section 12.2.1) to validate new and accurate safety tests to predict cardiac toxicity combining *in vitro* assays and *in silico* modeling. *In vitro* assays are patch-clamp measurements of seven human currents, whereas modeling efforts aim at *in silico* reconstruction of human ventricular cellular electrophysiology. As outlined in recent publications, "the new CIPA paradigm will be driven by a suite of mechanistically based in vitro assays coupled to *in silico* reconstructions of cellular cardiac electrophysiologic activity, with verification of completeness through comparison of predicted and observed responses in human-derived cardiac myocytes. To the extent that the ongoing validation program is successful, we envision petitioning for corresponding changes to regulatory requirements for proarrhythmia assessment and these might include eliminating or waiving the need for the TQT" (Gintant et al., 2016; Sager et al., 2014).

12.6   CHALLENGES FOR THE FUTURE

In this chapter, we have focused on small-molecule evaluation, as historically this is the predominant approach to discovering new therapeutic treatments for benefiting patients. Scientific advancements over the last decade have expanded our understanding of pharmacological space across the genome and driven novel approaches to design of therapeutic treatments. This diversification poses a challenge to the standard paradigm of secondary pharmacology *in vitro* testing described.

Approaches to small-molecule design are evolving to explore novel mechanisms of action (e.g., allosteric modulators, protein–protein interactions) (Conn et al., 2009; Fischer et al., 2015), novel modalities (e.g., peptides, antibody drug conjugates, PROTACS) (Valeur et al., 2017), or targeting new classes or proteins (e.g., epigenetics) (Prachayasittikul et al., 2017). With this, the remit of secondary pharmacology studies may require evolution. While it is certainly appropriate to continue to assess small molecules with these new modalities in traditional *in vitro* profiling assays, it may be appropriate to consider whether new technologies or target groups might also be included. Epigenetic targets, for example, are not common in general small-molecule *in vitro* profiling panels (Bowes et al., 2012; Lynch et al., 2017) but might be selectively included for molecules with primary targets involved in epigenetics. Similarly, for targeted protein degraders, it is appropriate to determine whether there are "standard" pharmacological off-target interactions but also to consider whether off-target proteins are degraded, which requires nonstandard approaches to be included in the secondary pharmacology strategy, such as proteomics (Sutton, 2012; van Esbroeck et al., 2017).

Indeed, it may be appropriate to consider nonstandard approaches to secondary pharmacological profiling, such as genomics, proteomics, or metabolomics (Sutton, 2012; van Esbroeck et al., 2017; Wilmes et al., 2015). While these approaches are perhaps not yet applicable as regular screening tools, they may be valuable in assessing specific questions, particularly in reverse pharmacology when attempting to ascribe molecular mechanisms to observed *in vivo* or clinical effects (Section 12.4.3; Moore, 2016a,b; van Esbroeck et al., 2017). They also offer advantages to broaden the breadth of assessment wider than a designed target panel, opening up exploration of the proteome and identify potentially new targets associated with toxicological mechanisms.

12.7   CONCLUDING REMARKS

ADRs are a major cause of failure for drugs entering clinical development. Marketed drugs with undesirable side effects are a major factor in reducing patient compliance. Every drug that enters clinical development represents an investment of millions of dollars that will only be recouped if the drug is approved for use and is frequently prescribed. Therefore, it is extremely important to ensure that new drugs are safe and tolerable as well as efficacious. By introducing pharmacological profiling early in the drug discovery process and increasing the scope of profiling as drugs progress toward the clinic, a pharmaceutical company can minimize the risks associated with unwanted side effects. Early and incremental *in vitro* profiling in parallel with a drug's progression through the stages of drug discovery represents an optimal approach, supplying high-quality data at key decision points. These activities may also result in the discovery of novel targets of known drugs to treat additional diseases – discovery by serendipity. Consistent application of this strategy within a pharmaceutical company's portfolio is key to success.

REFERENCES

Albert-Vartanian, A., Boyd, M. R., Hall, A. L., Morgado, S. J., Nguyen, E., Nguyen, V. P., … Raffa, R. B. (2016). Will peripherally restricted kappa-opioid receptor agonists (pKORAs) relieve pain with less opioid adverse effects and abuse potential? *Journal of Clinical Pharmacy and Therapeutics, 41*(4), 371–382.

Anon. (2001). ICH guildeline S7A Safety Pharmacology Studies for Human Pharmaceuticals. International Council for Harmonisation of Technical Requirements for Pharmaceuticals for Human Use (ICH).

Anon. (2005). ICH guideline S7B the Non-Clinical Evaluation of the Potential for Delayed Ventricular Repolarization (QT Interval Prolognation) by Human Pharmaceuticals. International Council for Harmonisation of Technical Requirements for Pharmaceuticals for Human Use (ICH).

Anon. (2009). ICH guideline M3(R2) on non-clinical safety studies for the conduct of human clinical trials and marketing authorisation for pharmaceuticals. International Council for Harmonisation of Technical Requirements for Pharmaceuticals for Human Use (ICH).

Anon. (2012a). Report to the President on Propelling Innovation in Drug Discovery, Development, and Evaluation. US Government. Retrieved from www.whitehouse.gov/sites/default/files/microsites/ostp/pcast-fda-final.pdf.

Anon. (2012b). Suicidal Ideation and Behavior: Prospective Assessment of Occurrence in Clinical Trials. U.S. Food and Drug Administration. Center for Drug Evaluation and Research (CDER). Retrieved from www.fda.gov/downloads/Drugs/GuidanceComplianceRegulatoryInformation/Guidances/UCM225130.pdf.

Anon. (2017a). Assessment of Abuse Potential of Drugs Guidance for Industry. U.S. Food and Drug Administration. Center for Drug Evaluation and Research (CDER). Retrieved from www.fda.gov/downloads/drugs/guidances/ucm198650.pdf.

Anon. (2017b). EvaluatePharma® World Preview 2017, Outlook to 2022. Retrieved from http://info.evaluategroup.com/rs/607-YGS-364/images/WP17.pdf: Evaluate Group.

Anon. (2018). Bioprint™ Database. Retrieved May 2018 www.cerep.fr/Cerep/Users/pages/ProductsServices/bioprintservices.asp.

Arrowsmith, J. (2011). Trial watch: Phase II failures: 2008–2010. *Nature Reviews: Drug Discovery, 10*(5), 328–329.

Arrowsmith, J., & Miller, P. (2013). Trial watch: Phase II and phase III attrition rates 2011–2012. *Nature Reviews: Drug Discovery, 12*(8), 569.

Booth, B., & Zemmel, R. (2004). Prospects for productivity. *Nature Reviews: Drug Discovery, 3*, 451–457.

Borman, R. A., Tilford, N. S., Harmer, D. W., Day, N., Ellis, E. S., Sheldrick, R. L., … Baxter, G. S. (2002). 5-HT(2B) receptors play a key role in mediating the excitatory effects of 5-HT in human colon in vitro. *British Journal of Pharmacology, 135*(5), 1144–1151.

Bowes, J., Brown, A. J., Hamon, J., Jarolimek, W., Sridhar, A., Waldron, G., & Whitebread, S. (2012). Reducing safety-related drug attrition: The use of in vitro pharmacological profiling. *Nature Reviews: Drug Discovery, 11*(12), 909–922.

Braga, R. C., Alves, V. M., Silva, M. F. B., Muratov, E., Fourches, D., Tropsha, A., & Andrade, C. H. (2014). Tuning hERG out: Antitarget QSAR models for drug development. *Current Topics in Medicinal Chemistry, 14*(11), 1399–1415.

Bran, S., Murray, W. A., Hirsch, I. B., & Palmer, J. P. (1995). Long QT syndrome during high-dose cisapride. *Archives of Internal Medicine, 155*(7), 765–768.

Chambers, C., Smith, F., Williams, C., Marcos, S., Liu, Z. H., Hayter, P., … Sewing, A. (2003). Measuring intracellular calcium fluxes in high throughput mode. *Combinatorial Chemistry & High Throughput Screening, 6*(4), 355–362.

Cheng, Y., & Prusoff, W. H. (1973). Relationship between the inhibition constant (K1) and the concentration of inhibitor which causes 50 per cent inhibition (I50) of an enzymatic reaction. *Biochemical Pharmacology, 22*(23), 3099–3108.

Chiba, T., Bharucha, A. E., Thomforde, G. M., Kost, L. J., & Phillips, S. F. (2002). Model of rapid gastrointestinal transit in dogs: Effects of muscarinic antagonists and a nitric oxide synthase inhibitor. *Neurogastroenterology and Motility, 14*(5), 535–541.

Conn, J. P., Christopoulos, A., & Lindsley, C. W. (2009). Allosteric modulators of GPCRs: A novel approach for the treatment of CNS disorders. *Nature Reviews: Drug Discovery, 8*, 41.

Cook, D., Brown, D., Alexander, R., March, R., Morgan, P., Satterthwaite, G., & Pangalos, M. N. (2014). Lessons learned from the fate of AstraZeneca's drug pipeline: A five-dimensional framework. *Nature Reviews: Drug Discovery, 13*(6), 419–431.

Cross, A. J., Widzowski, D., Maciag, C., Zacco, A., Hudzik, T., Liu, J., ... Wood, M. W. (2016). Quetiapine and its metabolite norquetiapine: Translation from in vitro pharmacology to in vivo efficacy in rodent models. *British Journal of Pharmacology, 173*(1), 155–166.

Curran, M. P., & Keating, G. M. (2003). Tadalafil. *Drugs, 63*(20), 2203–2212.

David, E., Tramontin, T., & Zemmel, R. (2009). Pharmaceutical R&D: The road to positive returns. *Nature Reviews: Drug Discovery, 8*(8), 609–610.

Dickson, M., & Gagnon, J. P. (2004). Key factors in the rising cost of new drug discovery and development. *Nature Reviews: Drug Discovery, 3*(5), 417–429.

DiMasi, J. A., Hansen, R. W., & Grabowski, H. G. (2003). The price of innovation: New estimates of drug development costs. *Journal of Health Economics, 22*(2), 151–185.

Dogterom, P., Zbinden, G., & Reznik, G. K. (1992). Cardiotoxicity of vasodilators and positive inotropic/vasodilating drugs in dogs: An overview. *Critical Reviews in Toxicology, 22*(3–4), 203–241.

Doogan, D. P. (1991). Toleration and safety of sertraline: Experience worldwide. *International Clinical Psychopharmacology, 6*(Suppl 2), 47–56.

Dougall, I. G., Young, I. A., Ince, I. E., & Jackson, D. M. (2002). Dual dopamine D_2 receptor and β_2-adrenoceptor agonists for the treatment of chronic obstructive pulmonary disease: The pre-clinical rationale. *Respiratory Medicine, 97*(Suppl A), S3–S7.

Easter, A., Bell, M. E., Damewood, J. R., Redfern, W. S., Valentin, J.-P., Winter, M. J., ... Bialecki, R. A. (2009). Approaches to seizure risk assessment in preclinical drug discovery. *Drug Discovery Today, 14*(17), 876–884.

Einarson, T. R. (1993). Drug-related hospital admissions. *Annals of Pharmacotherapy, 27*(7–8), 832–840.

Ekins, S. (2004). Predicting undesirable drug interactions with promiscuous proteins in silico. *Drug Discovery Today, 9*(6), 276–285.

Etkins, S., Crumb, W. J., Sarazan, R. D., Wikel, J. H., & Wrighton, S. A. (2002). Three-dimensional quantitative structure-activity relationship for inhibition of human ether-a-go-go-related gene potassium channel. *Journal of Pharmacology and Experimental Therapeutics, 301*(2), 427–434.

Ekins, S., Mestres, J., & Testa, B. (2007a). In silico pharmacology for drug discovery: Applications to targets and beyond. *British Journal of Pharmacology, 152*(1), 21–37.

Ekins, S., Mestres, J., & Testa, B. (2007b). In silico pharmacology for drug discovery: Methods for virtual ligand screening and profiling. *British Journal of Pharmacology, 152*(1), 9–20.

Ferguson, J. M. (2001). SSRI antidepressant medications: Adverse effects and tolerability. *Primary Care Companion to the Journal of Clinical Psychiatry, 3*(1), 22–27.

Fermini, B., Fossa, A. A. (2003). The impact of drug-induced QT interval prolongation on drug discovery and development. *Nature Reviews, 2*, 439–447.

Fischer, G., Rossmann, M., & Hyvönen, M. (2015). Alternative modulation of protein–protein interactions by small molecules. *Current Opinion in Biotechnology, 35*, 78–85.

Gintant, G., Sager, P. T., & Stockbridge, N. (2016). Evolution of strategies to improve preclinical cardiac safety testing. *Nature Reviews: Drug Discovery, 15*(7), 457–471.

Goldstein, I., Lue, T. F., Padma-Nathan, H., Rosen, R. C., Steers, W. D., & Wicker, P. A. (1998). Oral sildenafil in the treatment of erectile dysfunction. Sildenafil Study Group. *New England Journal of Medicine, 338*(20), 1397–1404.

Hanauer, S. B. (2008). The role of loperamide in gastrointestinal disorders. *Reviews in Gastroenterological Disorders, 8*(1), 15–20.

Hansch, C., & Dunn William, J. (1972). Linear relationships between lipophilic character and biological activity of drugs. *Journal of Pharmaceutical Sciences, 61*(1), 1–19.

Hansch, C., & Fujita, T. (1964). p-σ-π Analysis. A method for the correlation of biological activity and chemical structure. *Journal of the American Chemical Society, 86*(8), 1616–1626.

Harrison, R. K. (2016). Phase II and phase III failures: 2013–2015. *Nature Reviews: Drug Discovery, 15*(12), 817–818.

Hartmann, A., Whitebread, S., Hamon, J., Fekete, A., Trendelenburg, C., Müller, P. Y., & Urban, L. (2015). Targets associated with drug-related suicidal ideation and behavior. In L. Urban, V. F. Patel, & R. J. Vaz (Eds.), *Antitargets and Drug Safety* (pp. 457–478). Weinheim: Wiley-VCH Verlag GmbH & Co. KGaA.

Heaslip, R. J., & Evans, D. Y. (1995). Emetic, central nervous system and pulmonary activities of rolipram in the dog. *European Journal of Pharmacology, 286*, 281–290.

Herman, E. H., Ferrans, V. J., Young, R. S., & Balazs, T. (1989). A comparative study of minoxidil-induced myocardial lesions in beagle dogs and miniature swine. *Toxicologic Pathology, 17*(1 Pt 2), 182–192.

Hill Stephen, J. (2009). G-protein-coupled receptors: Past, present and future. *British Journal of Pharmacology, 147*(S1), S27–S37.

Holm, K. J., & Markham, A. (1999). Mirtazapine: A review of its use in major depression. *Drugs, 57*(4), 607–631.

Holmes, A., Bonner, F., & Jones, D. (2015). Assessing drug safety in human tissues—What are the barriers? *Nature Reviews: Drug Discovery, 14*, 585.

Horowski, R., & Sastre-Y-Hernandez, M. (1985). Clinical effects of the neurotropic selective cAMP phosphodiesterase inhibitor rolipram in depressed patients: Global evaluation of the preliminary reports. *Current Therapeutic Research, 38*(1), 23–29.

Hou, T., Wang, J., Zhang, W., & Xu, X. (2007). ADME evaluation in drug discovery. 6. Can oral bioavailability in humans be effectively predicted by simple molecular property-based rules? *Journal of Chemical Information and Modeling, 47*(2), 460–463.

Huh, D., Hamilton, G. A., & Ingber, D. E. (2011). From 3D cell culture to organs-on-chips. *Trends in Cell Biology, 21*(12), 745–754.

Hulme, E. C., & Trevethick, M. A. (2010). Ligand binding assays at equilibrium: Validation and interpretation. *British Journal of Pharmacology, 161*(6), 1219–1237.

Hutcheson, J. D., Setola, V., Roth, B. L., & Merryman, W. D. (2011). Serotonin receptors and heart valve disease—It was meant 2B. *Pharmacology & Therapeutics, 132*(2), 146–157.

Ind, P. W., Laitinen, L., Laursen, L., Wenzel, S., Wouters, E., Deamer, L., & Nystrom, P. (2003). Early clinical investigation of Viozan™ (sibenadet HCI), a novel D_2 dopamine receptor, β_2-adrenoceptor agonist for the treatment of chronic obstructive pulmonary disease symptoms. *Respiratory Medicine, 97*, S9–S21.

Izquierdo, I., Merlos, M., & Garcia-Rafanell, J. (2003). Rupatadine: A new selective histamine H1 receptor and platelet-activating factor (PAF) antagonist. A review of pharmacological profile and clinical management of allergic rhinitis. *Drugs of Today, 39*(6), 451–468.

Jackson, S. J., Prior, H., & Holmes, A. (2018). The use of human tissue in safety assessment. *Journal of Pharmacological and Toxicological Methods, 93*, 29–34.

Jing, Y., Easter, A., Peters, D., Kim, N., & Enyedy, I. J. (2015). In silico prediction of hERG inhibition. *Future Medicinal Chemistry, 7*(5), 571–586.

Kay, G. G., Berman, B., Mockoviak, S. H., Morris, C. E., Reeves, D., Starbuck, V., … Harris, A. G. (1997). Initial and steady-state effects of diphenhydramine and loratadine on sedation, cognition, mood, and psychomotor performance. *Archives of Internal Medicine, 157*(20), 2350–2356.

Kenakin, T. (1984). The classification of drugs and drug receptors in isolated tissues. *Pharmacological Reviews, 36*(3), 165–222.

Kenakin, T. (2003). Predicting therapeutic value in the lead optimization phase of drug discovery. *Nature Reviews: Drug Discovery, 2*, 429.

Kenakin, T. (2007). Allosteric theory: Taking therapeutic advantage of the malleable nature of GPCRs. *Current Neuropharmacology, 5*, 149–156.

Kennedy, T. (1997). Managing the drug discovery/development interface. *Drug Discovery Today, 2*(10), 436–444.

Kola, I., & Landis, J. (2004). Can the pharmaceutical industry reduce attrition rates? *Nature Reviews: Drug Discovery, 3*(8), 711–716.

Kola, I., & Rafferty, M. (2002). New Technologies that May Impact Drug Discovery in the 5–10 Year Timeframe Workshop. Data derived from Prous Science, Drugs News Prospect (2002). Paper presented at the Biomed Expo, Ann Arbor, MI.

Leeson, P. D., & Springthorpe, B. (2007). The influence of drug-like concepts on decision-making in medicinal chemistry. *Nature Reviews: Drug Discovery, 6*(11), 881–890.

Lewis, S. C., Langman, M. J., Laporte, J. R., Matthews, J. N., Rawlins, M. D., & Wiholm, B. E. (2002). Dose-response relationships between individual nonaspirin nonsteroidal anti-inflammatory drugs (NANSAIDs) and serious upper gastrointestinal bleeding: A meta-analysis based on individual patient data. *British Journal of Clinical Pharmacology, 54*(3), 320–326.

Lipinski, C. A. (2004). Lead- and drug-like compounds: The rule-of-five revolution. *Drug Discovery Today Technology, 1*(4), 337–341.

Lipinski, C. A., Lombardo, F., Dominy, B. W., & Feeney, P. J. (1997). Experimental and computational approaches to estimate solubility and permeability in drug discovery and development settings. *Advanced Drug Delivery Reviews, 23*(1), 3–25.

Lu, J. J., Crimin, K., Goodwin, J. T., Crivori, P., Orrenius, C., Xing, L., … Burton, P. S. (2004). Influence of molecular flexibility and polar surface area metrics on oral bioavailability in the rat. *Journal of Medicinal Chemistry, 47*(24), 6104–6107.

Lynch, J. J., Vleet, T. R. V., Mittelstadt, S. W., & Blomme, A. G. (2017). Potential functional and pathological side effects related to off-target pharmacological activity. *Journal of Pharmacological and Toxicological Methods, 87*, 108–126.

Merel, B., Marieke, N., Elise, S., Leon, A., Terry, W. S., & Albert, D. (2012). Non-analgesic effects of opioids: Opioid-induced respiratory depression. *Current Pharmaceutical Design, 18*(37), 5994–6004.

Moore, N. (2016a). ERRATUM: Lessons from the fatal French study BIA-10-2474. *British Medical Journal, 353,* i2956.

Moore, N. (2016b). Lessons from the fatal French study BIA-10-2474. *British Medical Journal, 353,* i2727.

Morgan, P., Brown, D. G., Lennard, S., Anderton, M. J., Barrett, J. C., Eriksson, U., … Pangalos, M. N. (2018). Impact of a five-dimensional framework on R&D productivity at AstraZeneca. *Nature Reviews: Drug Discovery, 17,* 167–181.

Motomura, S., Reinhard-Zerkowski, H., Daul, A., & Brodde, O. E. (1990). On the physiologic role of beta-2 adrenoceptors in the human heart: In vitro and in vivo studies. *American Heart Journal, 119*(3 Pt 1), 608–619.

Mullard, A. (2018). 2017 FDA drug approvals. *Nature Reviews: Drug Discovery, 17*(2), 81–85.

Muller, P. Y., Dambach, D., Gemzik, B., Hartmann, A., Ratcliffe, S., Trendelenburg, C., & Urban, L. (2015). Integrated risk assessment of suicidal ideation and behavior in drug development. *Drug Discovery Today, 20*(9), 1135–1142.

Muller, P. Y., & Milton, M. N. (2012). The determination and interpretation of the therapeutic index in drug development. *Nature Reviews: Drug Discovery, 11*(10), 751–761.

Munos, B. (2009). Lessons from 60 years of pharmaceutical innovation. *Nature Reviews: Drug Discovery, 8*(12), 959–968.

Nademanee, K., & Singh, B. N. (1988). Control of cardiac arrhythmias by calcium antagonism. *Annals of the New York Academy of Sciences, 522,* 536–552.

Neubig, R. R., Spedding, M., Kenakin, T., & Christopoulos, A. (2003). International Union of Pharmacology Committee on Receptor Nomenclature and Drug Classification. XXXVIII. Update on terms and symbols in quantitative pharmacology. *Pharmacological Reviews, 55*(4), 597–606.

Noel, F., & do Monte, F. M. (2016). Validation of a Na^+-shift binding assay for estimation of the intrinsic efficacy of ligands at the A_{2A} adenosine receptor. *Journal of Pharmacological and Toxicological Methods, 84,* 51–56.

Pammolli, F., Magazzini, L., & Riccaboni, M. (2011). The productivity crisis in pharmaceutical R&D. *Nature Reviews: Drug Discovery, 10*(6), 428–438.

Papoian, T., Chiu, H. J., Elayan, I., Jagadeesh, G., Khan, I., Laniyonu, A. A., … Yang, B. (2015). Secondary pharmacology data to assess potential off-target activity of new drugs: A regulatory perspective. *Nature Reviews: Drug Discovery, 14*(4), 294.

Papoian, T., Jagadeesh, G., Saulnier, M., Simpson, N., Ravindran, A., Yang, B., … Szarfman, A. (2017). Utility of in vitro secondary pharmacology data to assess risk of drug-induced valvular heart disease in humans: Regulatory considerations. *Toxicologic Pathology, 45,* 1–8.

Paul, S. M., Mytelka, D. S., Dunwiddie, C. T., Persinger, C. C., Munos, B. H., Lindborg, S. R., & Schacht, A. L. (2010). How to improve R&D productivity: The pharmaceutical industry's grand challenge. *Nature Reviews: Drug Discovery, 9*(3), 203–214.

Pirmohamed, M., James, S., Meakin, S., Green, C., Scott, A. K., Walley, T. J., … Breckenridge, A. M. (2004). Adverse drug reactions as cause of admission to hospital: Prospective analysis of 18 820 patients. *British Medical Journal, 329,* 15–19.

Prachayasittikul, V., Prathipati, P., Pratiwi, R., Phanus-umporn, C., Malik, A. A., Schaduangrat, N., … Nantasenamat, C. (2017). Exploring the epigenetic drug discovery landscape. *Expert Opinion on Drug Discovery, 12*(4), 345–362.

Prentis, R. A., Lis, Y., & Walker, S. R. (1988). Pharmaceutical innovation by the seven UK-owned pharmaceutical companies (1964–1985). *British Journal of Clinical Pharmacology, 25*(3), 387–396.

Redfern, W., Carlsson, L., Davis, A., Lynch, W., Mackenzie, I., Palethorpe, S., … Wallis, R. (2003). Relationships between preclinical cardiac electrophysiology, clinical QT interval prolongation and torsade de pointes for a broad range of drugs: Evidence for a provisional safety margin in drug development. *Cardiovascular Research, 58*(1), 32–45.

Redfern, W., Ewart, L., Hammond, T. G., Bialecki, R., Kinter, L., Lindgren, S., … Valentin, J. P. (2010). Impact and frequency of different toxicities throughout the pharmaceutical life cycle. *The Toxicologist: Supplement to Toxicological Sciences, 114*(1), 231.

Redfern, W., & Valentin, J. P. (2017). Prevalence, frequency and impact of safety related issues throughout the pharmaceutical life cycle. *The Toxicologist, 150*(1), 170.

Redfern, W., Wakefield, I. D., Prior, H., Pollard, C. E., Hammond, T. G., & Valentin, J. P. (2002). Safety pharmacology—A progressive approach. *Fundamental & Clinical Pharmacology, 16*(3), 161–173.

Rolf, M. G., Curwen, J. O., Veldman-Jones, M., Eberlein, C., Wang, J., Harmer, A., … Braddock, M. (2015). In vitro pharmacological profiling of R406 identifies molecular targets underlying the clinical effects of fostamatinib. *Pharmacology Research & Perspectives, 3*(5), e00175.

Rosenbrier Ribeiro, L., & Storer, I. R. (2017). A semi-quantitative translational pharmacology analysis to understand the relationship between in vitro ENT1 inhibition and the clinical incidence of dyspnoea and bronchospasm. *Toxicology and Applied Pharmacology, 317*, 41–50.

Rothman, R. B., Baumann, M. H., Savage, J. E., Rauser, L., McBride, A., Hufeisen, S. J., & Roth, B. L. (2000). Evidence for possible involvement of 5-HT2B receptors in the cardiac valvulopathy associated with fenfluramine and other serotonergic medications. *Circulation, 102*(23), 2836–2841.

Sager, P. T., Gintant, G., Turner, J. R., Pettit, S., & Stockbridge, N. (2014). Rechanneling the cardiac proarrhythmia safety paradigm: A meeting report from the Cardiac Safety Research Consortium. *American Heart Journal, 167*(3), 292–300.

Schoenberger, J. A. (1991). Drug-induced orthostatic hypotension. *Drug Safety, 6*(6), 402–407.

Servick, K. (2016). A painstaking overhaul for cardiac safety testing. *Science, 353*, 976–977.

Shah, R. R. (2005). Drugs, QT interval prolongation and ICH E14. *Drug Safety, 28*(2), 115–125.

Skold, A. C., Danielsson, C., Linder, B., & Danielsson, B. R. (2002). Teratogenicity of the I(Kr)-blocker cisapride: Relation to embryonic cardiac arrhythmia. *Reproductive Toxicology, 16*(4), 333–342.

Sobota, J. T. (1989). Review of cardiovascular findings in humans treated with minoxidil. *Toxicologic Pathology, 17*(1), 193–202.

Song, M., & Clark, M. (2006). Development and evaluation of an in silico model for hERG binding. *Journal of Chemical Information and Modeling, 46*(1), 392–400.

Stockman, A., Sharpe, L. T., Tufail, A., Kell, P. D., Ripamonti, C., & Jeffery, G. (2007). The effect of sildenafil citrate (Viagra®) on visual sensitivity. *Journal of Vision, 7*(8), 4–4.

Stone, M., Laughren, T., Jones, M. L., Levenson, M., Holland, P. C., Hughes, A., … Rochester, G. (2009). Risk of suicidality in clinical trials of antidepressants in adults: Analysis of proprietary data submitted to US Food and Drug Administration. *British Medical Journal, 339*, b2880.

Sullivan, E., Tucker, E. M., & Dale, I. L. (1999). Measurement of [Ca^{2+}] using the fluorometric imaging plate reader (FLIPR). *Methods in Molecular Biology, 114*, 125–133.

Sutton, C. W. (2012). The role of targeted chemical proteomics in pharmacology. *British Journal of Pharmacology, 166*(2), 457–475.

Tian, S., Li, Y., Wang, J., Zhang, J., & Hou, T. (2011). ADME evaluation in drug discovery. 9. Prediction of oral bioavailability in humans based on molecular properties and structural fingerprints. *Molecular Pharmaceutics, 8*(3), 841–851.

Tristani-Firouzi, M., Chen, J., Mitcheson, J. S., & Sanguinetti, M. C. (2001). Molecular biology of K$^+$ channels and their role in cardiac arrhythmias. *The American Journal of Medicine, 110*(1), 50–59.

Uhlen, S., Schioth, H. B., & Jahnsen, J. A. (2016). A new, simple and robust radioligand binding method used to determine kinetic off-rate constants for unlabeled ligands. Application at alpha2A- and alpha2C-adrenoceptors. *European Journal of Pharmacology, 788*, 113–121.

Urban, L., Maciejewski, M., Lounkine, E., Whitebread, S., Jenkins, J. L., Hamon, J., … Muller, P. Y. (2014). Translation of off-target effects: Prediction of ADRs by integrated experimental and computational approach. *Toxicological Research, 3*(6), 433–444.

Vadivelu, N., Mitra, S., & Hines, R. L. (2011). Peripheral opioid receptor agonists for analgesia: A comprehensive review. *Journal of Opioid Management, 7*(1), 55–68 (1551–7489 (Print)).

Valentin, J. P., & Hammond, T. (2008). Safety and secondary pharmacology: Successes, threats, challenges and opportunities. *Journal of Pharmacological and Toxicological Methods, 58*(2), 77–87.

Valentin, J. P., Keisu, M., & Hammond, T. G. (2009). Predicting human adverse drug reactions from nonclinical safety studies. In S. C. Gad (Ed.), *Clinical Trials Handbook* (pp. 87–113). Hoboken, NJ: John Wiley & Sons, Inc.

Valeur, E., Guéret Stéphanie, M., Adihou, H., Gopalakrishnan, R., Lemurell, M., Waldmann, H., … Plowright Alleyn, T. (2017). New modalities for challenging targets in drug discovery. *Angewandte Chemie International Edition, 56*(35), 10294–10323.

van Esbroeck, A. C. M., Janssen, A. P. A., Cognetta, A. B., Ogasawara, D., Shpak, G., van der Kroeg, M.,…van der Stelt, M. (2017). Activity-based protein profiling reveals off-target proteins of the FAAH inhibitor BIA 10-2474. *Science, 356*, 1084–1087.

Veber, D. F., Johnson, S. R., Cheng, H.-Y., Smith, B. R., Ward, K. W., & Kopple, K. D. (2002). Molecular properties that influence the oral bioavailability of drug candidates. *Journal of Medicinal Chemistry, 45*(12), 2615–2623.

Wakefield, I. D., Pollard, C., Redfern, W., Hammond, T. G., & Valentin, J. P. (2002). The application of in vitro methods to safety pharmacology. *Fundamental & Clinical Pharmacology, 16*(3), 209–218.

Ward, D. J., Martino, O. I., Simpson, S., & Stevens, A. J. (2013). Decline in new drug launches: Myth or reality? Retrospective observational study using 30 years of data from the UK. *BMJ Open, 3*(2), e002088. doi:10.1136/bmjopen-2012-002088.

Waring, M. J., Arrowsmith, J., Leach, A. R., Leeson, P. D., Mandrell, S., Owen, R. M., … Weir, A. (2015). An analysis of the attrition of drug candidates from four major pharmaceutical companies. *Nature Reviews: Drug Discovery, 14*(7), 475–486.

Waring, M. J., & Johnstone, C. (2007). A quantitative assessment of hERG liability as a function of lipophilicity. *Bioorganic & Medicinal Chemistry Letters, 17*(6), 1759–1764.

Whitebread, S., Dumotier, B., Armstrong, D., Fekete, A., Chen, S., Hartmann, A., … Urban, L. (2016). Secondary pharmacology: Screening and interpretation of off-target activities-focus on translation. *Drug Discovery Today, 21*(8), 1232–1242.

Whitebread, S., Hamon, J., Bojanic, D., & Urban, L. (2005). Keynote review: In vitro safety pharmacology profiling: An essential tool for successful drug development. *Drug Discovery Today, 10*(21), 1421–1433.

Williams, C. (2004). cAMP detection methods in HTS: Selecting the best from the rest. *Nature Reviews: Drug Discovery, 3*(2), 125–135.

Wilmes, P., Heintz-Buschart, A., & Bond, P. L. (2015). A decade of metaproteomics: Where we stand and what the future holds. *Proteomics, 15*(20), 3409–3417.

13 Pharmacokinetics–Pharmacodynamics in New Drug Development

Sarfaraz K. Niazi

University of Illinois
University of Houston
Pharmaceutical Scientist, LLC

CONTENTS

13.1   HISTORY AND BASIC CONCEPTS OF PHARMACOKINETICS AND PHARMACODYNAMICS

13.1.1   BACKGROUND

This history of using drugs to treat ailments dates back to our foraging days, when we had, at first, accidentally consumed a botanical or animal product that resulted in alleviation of body disorders; then, we started experimenting, and surely not all experiments worked. The era of natural therapies lasted for thousands of years, getting more sophisticated with our ability to write and publish our observations. The Sumerian archaic (pre-cuneiform) writing and the Egyptian hieroglyphs are generally considered the earliest true writing systems, both emerging out of their ancestral proto-literate symbol systems from 3400 to 3100 BC, with earliest coherent texts from about 2600 BC. The history of printing starts as early as 3500 BCE, when the Persian and Mesopotamian civilizations used cylinder seals to certify documents written in clay. Other early forms include block seals, pottery imprints, and cloth printing. Woodblock printing on paper originated in China around 200 CE. It led to the development of movable type in the 11th century and the spread of book production in East Asia. Woodblock printing was also used in Europe, but it was in the 15th century that European printers combined movable type and alphabetic scripts to create an economical book publishing industry.

New drugs were not required to undergo premarket safety testing in the United States until 1938, when a therapeutic disaster – the Elixir Sulfanilamide tragedy – prompted Congress to pass a bill mandating this now-routine process. History repeated itself nearly 25 years later, when the thalidomide tragedy led to passage of new amendments in 1962 to ensure drug efficacy and greater drug safety. As is typical with major tragedies, critical information was gained, in this case leading to novel approaches for understanding, predicting, diagnosing, and managing drug-induced toxicities.

As we improved our understanding of the time course of drug action, disciplines like pharmacokinetics and pharmacodynamics came into existence. The term "pharmacokinetics" was first introduced by F. H. Dost in 1953 in his text, *Der Blutspiegel-Kinetic der Konzentrationsablaufe in der Frieslaujjlussigkeit*. However, some of the subject matter was published before the word was coined. It is also of interest that the first English language review of the subject matter, published in 1961, was entitled *Kinetics of Drug Absorption, Distribution, Metabolism and Excretion* and did not include the word *pharmacokinetics*. Another early book on the subject matter, *Textbook of Biopharmaceutics and Clinical Pharmacokinetics*, was published by this author in 1979 and remains in print.

Pharmacokinetics has been defined in a number of ways. Literally, the word means the application of kinetics to *pharmakon*, the Greek word for drugs and poisons. *Kinetics* is the branch of knowledge which involves the change of one or more variables as a function of time. The purpose of pharmacokinetics is to study the time course of drug and metabolite concentrations or amounts in biological fluids, tissues, and excreta; and to construct suitable models to interpret such data. In pharmacokinetics, the data are analyzed using a mathematical representation of a part or the whole of an organism. Broadly, the purposes of pharmacokinetics are to reduce data to a number of meaningful parameter values and to use the reduced data to predict either the results of future experiments or the results of a host of studies which would be too costly and time-consuming to complete. A similar definition was given by Gibaldi and Levy as follows: "Pharmacokinetics is concerned with the study and characterization of the time course of drug absorption, distribution, metabolism and excretion, and with the relationship of these processes to the intensity and time course of therapeutic and adverse effects of drugs. It involves the application of mathematical and biochemical techniques in a physiologic and pharmacologic context."

Pharmacokinetics is currently defined as the study of the time course of drug absorption, distribution, metabolism, and excretion. Clinical pharmacokinetics is the application of pharmacokinetic principles to the safe and effective therapeutic management of drugs in an individual patient.

The maxim often used to describe pharmacokinetics is "what the body does to the drug" – namely, move it around the various compartments of the body until it is metabolized or excreted.

Pharmacodynamics (PD), in contrast, refers to the relationship between drug concentration at the site of action and the resulting effect. Elucidating this relationship enables quantitative prediction of drug effects, a major goal in clinical pharmacology. "Dynamics" in the term "pharmacodynamics" refers to the Greek term "dynamos" or "power"; hence, pharmacodynamics is the study of how the presence of the drug powers a bodily reaction. This includes the time course and intensity of therapeutic and adverse effects and how these effects vary as a function of drug concentration. Thus, the maxim used to describe PD is "what the drug does to the body" – the converse of the maxim for PK.

Pharmacodynamics history is the history of medicine that shows how societies have changed in their approach to illness and disease from ancient times to the present. Early medical traditions include those of Babylon, China, Egypt, and India. The Greeks introduced the concepts of medical diagnosis, prognosis, and advanced medical ethics. The Hippocratic Oath was written in ancient Greece in the 5th century BCE and is a direct inspiration for oaths of office that physicians swear upon entry into the profession today. Universities began systematic training of physicians around the years 1,220 in Italy. During the Renaissance, understanding of anatomy improved, and the microscope was invented. The germ theory of disease in the 19th century led to cures for many infectious diseases. The mid-20th century was characterized by new biological treatments, such as antibiotics. These advancements, along with developments in chemistry, genetics, and radiography led to modern medicine.

The majority of drugs either mimic or inhibit normal physiological/biochemical processes or inhibit pathological processes in animals or inhibit vital processes of endo- or ectoparasites and microbial organisms. The main drug actions include

1. Stimulating action through direct receptor agonism and downstream effects.
2. Depressing action through direct receptor agonism and downstream effects (e.g., inverse agonist).
3. Blocking/antagonizing action (as with silent antagonists), where the drug binds the receptor but does not activate it.
4. Stabilizing action where the drug seems to act either as a stimulant or as a depressant (e.g., some drugs possess receptor activity that allows them to stabilize general receptor activation, like buprenorphine in opioid-dependent individuals or aripiprazole in schizophrenia, all depending on the dose and the recipient).
5. Exchanging/replacing substances or accumulating them to form a reserve (e.g., glycogen storage); direct beneficial chemical reaction as in free radical scavenging.
6. Direct harmful chemical reaction which might result in damage or destruction of the cells, through induced toxic or lethal damage (cytotoxicity or irritation).

The undesirable effects of a drug include increased probability of cell mutation (carcinogenic activity), a multitude of simultaneous assorted actions which may be deleterious, drug–drug interaction (additive, multiplicative, or metabolic), induced physiological damage, or abnormal chronic conditions.

Simple models relating drug dose or concentration in plasma to effect date back to the 1960s,[1] and these theories have become more sophisticated over time to include a space called the biophase which surrounds the site of action and is in equilibrium with the circulating blood, distribution models, models for indirect mechanism of action that involve primarily the modulation of endogenous factors, and models for cell trafficking and transduction systems.

Combining pharmacokinetics with pharmacodynamics allowed us the ability to study the time course of drug action, a major breakthrough that expedited development of new drugs by reducing the studies required to establish a safe dosing range. The field of pharmacokinetic/

pharmacodynamic (PK/PD) modeling has made many advances from the basic concept of the dose–response relationship to extended mechanism-based models. Remarkable evolution has taken place in tolerance and time-variant models, non- and semi-parametric models, and population PK/PD modeling; the future possible directions for PK/PD modeling include equations for general classes of novel semi-parametric models, as well as describing additive or set-point, of regulatory, additive feedback models in their direct and indirect action variants. The recent availability of highly sophisticated mathematical and statistical modeling software has brought this complex modeling to the reach of the bench scientist, assuring that PK/PD modeling will remain, perhaps the most important tool in the development of new drugs.

PK/PD modeling, an integral component of the drug development process, is a mathematical technique for predicting the effect and efficacy of drug dosing over time. Broadly speaking, pharmacokinetic models describe how the body manipulates a drug in terms of absorption, distribution, metabolism, and excretion. Pharmacodynamic (PD) models describe how a drug affects the body by linking the drug concentration to an efficacy (or safety) metric. A well-characterized PK/PD model is an important tool in guiding the design of future experiments and trials.

The PK/PD modeling process includes the following steps:

- Import, process, and visualize time-course data.
- Select a pharmacokinetic model from a library or create mechanism-based PK/PD models using the interactive block-diagram editor.
- Estimate model parameters using nonlinear regression or NLME methods.
- Explore system dynamics, using parameter sweeps and sensitivity analysis.
- Simulate dosing strategies and what-if scenarios.

The regulatory submissions have, during the past 40 years, become more demanding as a result of the greater emphasis placed on the safety of new drug molecules. A significant component of the safety profile of a drug and its recommended dosing regimen is based on the PK characteristics of the drug in typical Phase I studies. With our improving understanding of human physiology, the use of biological markers (biomarkers), and the availability of better computing tools, highly sophisticated mathematical models can be invoked to support the safety profile of new drug molecules. One goal of the clinical pharmacology section of a regulatory filing is to describe and predict the relationship between drug dose and drug effect. Pharmacokinetic profiles of a new molecule describe the factors affecting the dose-active site concentration process, and PD describes the activity site concentration–effect process.

13.1.2 THE EXPOSURE–RESPONSE RELATIONSHIP

The term "exposure" refers to the product of the amount of drug entering and staying in the body and the time it is in the body. Various measures of observed or integrated drug concentrations in the plasma and other biological fluids are used to describe exposure, such as C_{max}, C_{min}, C_{ss}, and most notably AUC. *Response* refers to a direct measure of the pharmacologic effect of the drug and includes many endpoints or biomarkers: clinically remote biomarkers (such as receptor occupancy), a presumed mechanistic effect (such as angiotensin-converting enzyme or ACE inhibition), a potential or accepted surrogate (such as effects on blood pressure, lipids, or cardiac output), or the full range of short-term or long-term clinical effects related to either efficacy or safety.

The exposure of a drug is determined by a set of pharmacokinetic processes – often referred to as absorption–distribution–metabolism–excretion (ADME) – which ultimately control the systemic exposure of the body to a drug and its metabolites after drug administration. This relationship is often extended to the release or liberation (L) of drug molecules from the dosage form: LADME. PK parameters which describe the ADME behavior include the area under the curve (AUC) and the concentration at maximum (C_{max}) as well as other parameters calculated from those measured parameters, such as clearance, half-life, and volume of distribution. The AUC is a measure reflecting both the amount of drug in the body as well as the persistence of the drug, which can be critical for therapeutic or toxic effect. In contrast, the peak rather than the duration of the drug may be critical, and this is encompassed by C_{max}.

A drug can be eliminated either by metabolism (M), as one or more active or inactive metabolites, or by excretion of the unchanged drug through various routes, notably the kidneys. This systemic exposure, reflected in plasma drugs and metabolite concentrations, is generally used to relate the drug dose to both its beneficial and adverse effects. All drugs show inter- and intra-individual variances in PK measures and parameters, compounding the mathematical difficulties in making projections about the time course of drug concentration and activity in the body.

The exposure–response relationship is critical for regulatory review because it determines the safety and effectiveness of drugs. A drug can be determined to be safe and effective only when the relationship between the beneficial and adverse effects *vis-à-vis* an exposure to the body is known. Drug molecules can range from the well-tolerated drugs with little dose-related toxicity, in which the PK properties from a single dose to multiple doses are linear, to toxic molecules, in which clinical use must be determined on the basis of a nonlinearity in their PK or PD properties as a function of dose. Sometimes with the toxic drugs, the doses can be titrated for tolerability, but it is always important to develop information on population exposure–response relationships for favorable and unfavorable effects and to adjust the exposure for various subsets of the population. Pharmacokinetic characterization in pediatric cases is of particular importance to FDA.

The choice of dose, dosage form, type of drug release, dosing interval, and how the drug is monitored depends on the magnitude of an effect and the time course. Exposure–response and PK data further clarify this relationship *vis-à-vis* intrinsic and extrinsic patient factors, such as genetic phenotypes and age effects, for instance. Additionally, exposure–response data can be used as well-controlled clinical studies to support effectiveness; to emphasize the efficacy, especially where surrogates are used; and to support dosing in populations other than those studied. It is important to know that in several situations, the relationship between dose and plasma concentration is rather poor, primarily because of nonlinear PK results or inter-individual variations; in such situations, the systemic exposure studies offer the most valuable supporting tool for validating the efficacy profile of a new drug molecule. For example, it might be reassuring to observe that even patients with increased plasma concentrations (metabolic outliers or patients on other drugs in a study, for instance) do not have increased toxicity in general or with respect to a particular concern (such as QT prolongation on electrocardiograms). However, determining blood concentration remains indispensable. This is particularly true when both the drug and its metabolites are active, when different exposure measurements (such as C_{max} and AUC) provide different relationships between exposure and efficacy or safety, when the number of fixed doses in the dose–response studies is limited, and when responses are highly variable and the data are intended for an exploration of the underlying causes of the response variability.

Exposure–response data, using short-term biomarkers or surrogate endpoints, can sometimes make further exposure–response studies from clinical endpoints unnecessary. For example, if it can be shown that the short-term effect does not increase beyond a particular dose or concentration, there may be no reason to explore higher doses or concentrations in the clinical trials. Similarly, short-term exposure–response studies with biomarkers might be used to evaluate early (i.e., first dose) responses seen in clinical trials.

Exposure–response information for a new target population can sometimes be used to eliminate the need for clinical studies by showing comparable concentration–response relationships using short-term clinical or PD endpoint. The same holds true for dose and dosing-interval changes, formulation changes, and the like. In *some* cases, if there is a change in route of administration, dosage form, or a difference in the proportion of parent and active metabolites between two populations. In these cases, additional exposure–response data with short-term endpoints can support such a change without further clinical trials. For instance, a change in drug-input rate would be an example of a change in dosage form whose introduction may be supported by additional exposure-response data alone. Likewise, an example of two populations where the ratio of parent and active molecules may differ would be the pediatric and adult populations.

In new formulations, exposure–response data are used to isolate unintentional PK differences (e.g., in the release profile and, thus possibly, the metabolic profile through the saturable metabolic system). With these data in hand, a company can use *in vitro* or *in vivo* bioavailability studies as sufficient proof of equivalence, particularly between the formulations used in establishing the initial efficacy proof and the formulation intended for marketing. In biological drugs, changes in the manufacturing process often lead to subtle, unintentional changes in the product, resulting in altered PKs. When a change in the product can be determined as not having any pharmacologic effects (e.g., no effect of unwanted immunogenicity), exposure–response information may allow appropriate use of the new product. The comparability protocols suggested by FDA, and recently by European authorities, heavily rely on these studies for biological drugs.

Exposure–response data, however, are not likely to obviate the need for clinical data when formulation or manufacturing changes result in altered PKs, unless the relationships between measured responses and relevant clinical outcomes are well understood. The sponsors of studies need to make a financial evaluation of these studies as well. Often, while working with regulatory authorities, the choice of studies submitted depends on what would be the least-expensive alternative. In my experience, many such cost-saving measures end up actually costing more; the key is to provide the most robust data supported by multiple studies to account for most of the plausible variables.

In measuring exposure–response relationships, the choice of active moiety is important; ideally, it should include the active parent and active metabolites, particularly in those instances where the metabolism is route-dependent (e.g., first-pass metabolism). In such instances, hepatic and renal impairments become important considerations. There are many important considerations in selecting one or more active moieties to measure in plasma and in choosing specific measures of systemic exposure. When drugs are optically active and administered as the racemate, the differences in the PK and PD properties of the enantiomers should be elucidated to ascertain whether replacing the racemate with one of the pure enantiomers as the final drug product might be valuable.[2]

Complex drug substances can include drugs derived from animal or plant materials and drugs derived from traditional fermentation processes (yeasts, molds, bacteria, or other microorganisms). For some of these drug substances, identification of individual active moieties or ingredients is difficult or impossible. In such a circumstance, measurement of only one or more of the major, active moieties can be used as a "marker of exposure" in understanding exposure–response relationships and can even be used to identify the magnitude of contribution from individual, active moieties. Protein binding must be considered in the special case of the development of antibodies to a drug. Antibodies can alter the PK of a drug and can also affect PK/PD relationships by neutralizing the activity of the drug or preventing its access to the active site. This is of major concern in the development of biosimilar drugs because many biological drugs are coming off patent. The fact that antibodies produced by these molecules can result in sensitizing the body toward its endogenous production of the product (e.g., insulin or erythropoietin) and that there is no good way to measure this sensitization had kept the regulatory authorities from allowing biosimilar drugs. However, now with greater understanding of the safety of these products, their use has become widely accepted.

Pharmacokinetic concentration–time curves for a drug and its metabolites are used to identify primary exposure metrics such as AUC, C_{max}, or C_{min}, which are time-independent in contrast to the sequential concentration measurements over time. A peak plasma concentration of a drug is often associated with a PD response, especially with an adverse event. There can be large inter-individual variability in the time-to-peak concentration, and closely spaced sampling times are often critical to determining the peak plasma concentration accurately in individual patients because of potential differences in demographics, disease states, and food effects. All these elements are clearly spelled out in the protocols written to conduct these studies.

During chronic therapy, collection of multiple plasma samples over a dosing interval is often not practical. As a substitute, a trough plasma sample can be collected just before administration of the next dose at scheduled study visits. Trough concentrations are often proportional to AUC because they do not reflect drug-absorption processes as peak concentrations do in most cases. For many of the drugs that act slowly relative to the rates of their ADME, trough concentration and AUC can often be equally well correlated with drug effects.

Although collection of serial plasma samples and measurements of the corresponding response is the most desirable method, in some instances this process may not be ethical or possible, such as in pediatric or geriatric populations. An alternative method is to obtain plasma samples at randomly selected times during the study, or at prespecified but different times, to measure drug concentration and, in some cases, response. Although this type of sampling will not allow the usual PK data analysis for making precise estimates of individual PK parameters, the use of a specialized technique – population PK analysis combined with the Bayesian estimation method – can be used to approximate population and individual PK parameters. This will allow evaluation of exposure variables that can be more readily correlated to response than the few plasma concentrations measured.

The efficacy and safety of a drug can be characterized using a variety of measurements or response endpoints. These include clinical outcomes (clinical benefit or toxicity), effects on a well-established surrogate (e.g., change in blood pressure or QT interval), and effects on a more remote biomarker (e.g., change in ACE inhibition or bradykinin levels). In many cases, multiple response endpoints are desirable especially when less-persuasive clinical endpoints (biomarkers, surrogates) are used in choosing doses for the larger and more complex clinical endpoint trials. Greater problems arise in studies conducted across many sites or where multiple laboratories are used to measure the outcome.

The regulatory authorities place great emphasis on exposure–response studies. Poorly designed studies are likely to confound essential properties of drug molecules. As a result, the protocols must be written with precision and clarity to prevent misinterpretations.

13.1.3 BIOMARKERS

Biomarkers can provide great predictive value in early drug development if they reflect the mechanism of action for the drug, even if they do not become surrogate endpoints. Data from genomic and proteomics that differentiate healthy from disease states lead to biomarker discovery and identification. Multiple genes control complex diseases via hosts of gene products in biometabolic pathways and cell and organ signal transduction. Pilot exploratory studies are required to identify key biomarkers for predictive clinical assessment of disease progression and the effect of drug intervention.

Most biomarkers are endogenous macromolecules, which can be measured in biological fluids. Many biomarkers exist in heterogeneous forms with varying activity and immunoreactivity, posing challenges for bioanalysis. Reliable and selective assays can be validated under a good laboratory practice (GLP) environment for quantitative methods. Whereas the need for consistent reference standards and quality control monitoring during sample analysis for biomarker assays is similar to that for drug molecules, many biomarkers have special requirements for sample collection that demand well-coordinated team management. Bioanalytical methods should be validated to meet study objectives at various drug development stages and should perform adequately at quantifying biochemical responses specific to the target disease progression and drug intervention. Protocol design for producing sufficient data for PK/PD modeling would be more complex than that for PK alone.

Knowledge of the mechanism of action is helpful in planning clinical studies using cascade, sequential, crossover, or replicate study designs. Biomarkers associated with a clinical endpoint must be first identified and practical bioanalytical methods to quantify the biomarker must be developed and validated. Using mechanism-based PK/PD models, the resultant biomarker concentration data in response to drug therapy can be used to predict clinical endpoints. In addition to accelerating the testing of safety and efficacy for new drug candidates, biomarker data can be utilized for personalized therapeutic treatment. The use of biomarkers and surrogate endpoints of efficacy, hazard, and exposure in preclinical studies has evolved rapidly in validating projections to drug response on the basis of PK studies.

Biomarkers consist of a variety of physiologic, pathologic, or anatomic measurements that relate to normal or pathological biologic processes.[3,4] These include characteristics – such as physical signs, blood analytes, physiological measurements, and the like – that are objectively measured and validated as indicators of normal biological processes, pathogenic processes, or pharmacological responses to the use of a drug. The use of biomarkers presupposes that the cause or the susceptibility or the disease progression is related to a biomarker measurement and that this correlation is applicable at the clinical level (from pharmacologic effect to clinical response). Biomarkers differ in how close they come to serving the intended purpose:

- Some biomarkers are valid surrogates for a clinical benefit (e.g., blood pressure, cholesterol, and viral load).
- Some reflect a pathologic process and are, at least, candidate surrogates (e.g., brain appearance in Alzheimer's disease, brain infarct size, and various radiographic or isotopic function tests).
- Some reflect a drug action but have an uncertain relationship with the clinical outcome, such as the inhibition of platelet aggregation dependent on adenosine diphosphate or ACE inhibition.
- Some biomarkers are more remote from the clinical benefit endpoint (e.g., the degree of binding to a receptor or inhibition of an agonist).

Regulatory authorities do not accept a biomarker as an acceptable, surrogate endpoint for the determination of a new drug's efficacy unless it has been empirically shown to function as a valid indicator of clinical benefit (i.e., it is a valid surrogate). This relationship cannot be built on a theoretical justification for leaving biomarkers to support but not replace surrogate endpoints; this is because many biomarkers will never undergo the rigorous statistical evaluation that would establish their value as a surrogate endpoint. As a result, biomarkers that are not validated as surrogate endpoints cannot be used as supportive data. The time course of changes in biomarkers is often different from changes in clinical endpoints as they more closely approximate the time course of plasma drug concentrations. Therefore, the arguments presented against using plasma concentration as a measure of efficacy apply to biomarkers as well. However, biomarkers do offer an alternative in establishing the exposure–response relationships over a range of doses in clinical trials, which can provide insight into efficacy and toxicity. Biomarkers can also be useful during the drug discovery and development stages, where they can help link preclinical and early clinical exposure–response relationships and better establish dose ranges for clinical testing.

Often, drug responses are difficult to obtain (such as that of a bacterial cure from an antibiotic), are difficult to measure quantitatively (such as the mood elevation produced by an antidepressant), have variable effect on longevity (for instance, in the survival time from a cancer therapy), or are unethical to measure (a necropsy score for a safety evaluation, for instance). Therefore, the effect of ultimate interest in a PK/PD trial may be replaced by a surrogate endpoint. The clinical validity or relevance of a surrogate is determined by its statistical association and mechanistic links with a clinical outcome (surrogate accuracy). In addition, the surrogate should have desirable metrological properties, namely, reproducibility (of measurement), continuity (for a graded quantitative measurement), objectivity, specificity, and linearity.

Surrogate endpoints, a subset of biomarkers, are laboratory measurements or physical signs used in therapeutic trials as a substitute for clinical endpoints expected to predict the effect of the therapy.[3] A fully validated, surrogate endpoint consistently predicts the clinically meaningful endpoint of a therapy.[4]

The FDA is able to rely on less well-established surrogates for accelerated approval of drugs that provide meaningful benefits over existing therapies for serious or life-threatening illnesses, such as AIDS, rare carcinomas, or breaking infectious diseases. In these cases, the surrogates are likely to predict clinical benefit on the basis of epidemiologic, therapeutic, pathophysiologic, or other scientific evidence. Surrogates generally would not be used to evaluate clinical relevance or side effects when these effects are mechanistically unrelated to the surrogate.[3] Examples in which prospective and retrospective trials have demonstrated statistical correlations between the surrogate markers and either clinical success or prevention of resistance include antibacterial drug development, where some indices are mechanistically related to clinical outcome because they are all constructed using the minimum inhibitory concentration (MIC) values. For ACE inhibitors that help prevent heart failure, PK/PD relationships have been investigated using plasma and tissue ACE inhibition. On the basis of these relationships, dosage regimens have been designed.[5] As is with the case of AUC and MIC or C_{max} and MIC, ACE inhibition is only a surrogate endpoint – survival time and quality of life are the ultimate goals. Nevertheless, survival time and quality of life are difficult to monitor in ACE-inhibition therapy.

13.1.4 Adding Modeling to New Drug Development

There is broad recognition within the pharmaceutical industry that the drug development process, especially the clinical component of it, needs considerable improvement to keep pace with changes in the requirements and needs of the health care environment.[6] Modeling and simulation are mathematically validated techniques that have long been used extensively in disciplines other than the pharmaceutical industry to develop products more efficiently. Examples of other industries using extensive modeling and simulation include the automobile, aerospace, and computer design industries. Both modeling and simulation rely on the use of mathematical and statistical models that are simplified descriptions of the complex systems under investigation. Other types of models that are becoming increasingly important are population models, stochastic simulations, and ongoing efforts to integrate models for disease progression and patient behavior (such as compliance).

For both modeling and simulation, a guidance on "good pharmacokinetic practice" – analogous to GLP, good clinical practices, and good manufacturing practices – is currently evolving through the joint efforts of academia, regulatory agencies, and industry. These guidelines will be a great impetus for model-based analysis and simulation, contributing to the streamlining of the pharmaceutical drug development processes.[7]

13.2 PHARMACOKINETIC/PHARMACODYNAMIC MODELING ELEMENTS

13.2.1 Introduction

Drug development identifies the right molecule for the right target using a dosage regimen that yields the optimal therapeutic outcome. Safety information and adequate and well-controlled clinical studies that establish a drug's effectiveness are the basis for the approval of new drugs. The risk inherent in the administration of any new drug entity depends on the nature and extent of exposure to the body. This information is pivotal in determining the safety and effectiveness of drugs. A drug is determined to be safe and effective only when the relationship of beneficial and adverse effects with a defined exposure is known. Though rarer as more potent molecules are developed, there have been few instances in which a drug is well tolerated with little dose-related toxicity by all patients. In these cases, single-dose studies can be used to represent the full nature of exposure, taking into account PK and idiosyncratic parameters. More often though, the drugs are very toxic, and their clinical use can only be based on weighing the favorable and unfavorable effects at a particular dose or doses. While in some instances, the doses can be titrated to effect or to tolerability, often one must develop information on population exposure–response relationships and on how and whether exposure can be adjusted for various subpopulations. Exposure may vary from the general population in pediatric, elderly, or genetically compromised patients.

The critical exposure–response data generally derived from the preclinical and clinical studies provide a basis for integrated model-based analysis and simulation.[8,9] Simulation allows us to predict an expected relationship between exposure and response in situations where real data are absent or difficult to obtain. Models often do not establish a causal relationship or provide an explanation for the mechanism of action for a drug, and as a result, they may not be used as a basis for approval of a new drug. They can, however, help in analyzing data from well-controlled clinical trials. Predicting an exposure–response relationship, well-controlled clinical studies that investigate several fixed doses or that measure systemic exposure levels are required. This exposure and the associated response can be analyzed using scientifically reasonable causal models, validating hypotheses linking exposure to response which can be used to predict the effects of alternative doses and dosage regimens that are not actually tested. This information is useful in optimizing the dosing regimen and individualizing treatment in specific patient subsets for which limited data are provided.

In a clinical response, there are two major sources of variation: the PK and the PD of the drug. Applying computer-based methods to the population PK/PD approach helps to separate these sources of variability.[10] Mixed-effect models analyzing population PK/PD data explain variation between subjects (or groups of subjects). These models are capable of handling pooled (often sparse) data while allowing for fixed or random effects. Improving the power of the estimation process for the PK and PD parameters improves the suggested optimal dosing regimens.

All reasonable dosing regimens cannot be studied using the current gold standard of a 48-week controlled study of efficacy and safety. Instead, an optimized dosing regimen is best achieved by integrating PK (describing the relationship between dose and concentration vs. time) and PD parameters (describing the relationship between concentration and effect vs. time). This integration is often performed though a link model, which bridges the PK and PD models, and a statistical model describing the intra- and inter-individual variability. When a PK/PD model is employed, both the time course and the variability in the effect vs. time relationship are predicted for different dosage-regimen scenarios. However, mechanistic PK/PD models can be extrapolated to predict behavior rather than merely describing it. A prior determination of a safe and effective dosage regimen for use in pivotal clinical trials always proves beneficial, in terms of both cost and risk in clinical trials.[11] Several elements are crucial for successful PK/PD modeling and simulation for rational drug development, including

- Mechanism-based biomarker selection and correlation to clinical endpoints.
- Quantification of drug and metabolites in biological fluids, under GLP.
- GLP-like validation of the biomarker method and measurements.
- Mechanism-based PK/PD modeling and validation. This involves the four distinct steps of building PK model, building PD model, linking PK and PD models, and simulation of treatment regimens or trials for useful prediction.

13.2.1.1 Assumptions in Pharmacokinetic/Pharmacodynamic Modeling

In the process of PK/PD modeling, it is important to prospectively describe the objectives of the modeling, the study design, and the available PK and PD data. The assumptions of the model can be related to dose–response, PK, PD, or one or more of the assumptions listed in Table 13.1.

The assumptions can be based on previous data or on the results of any available current analysis. What constitutes an appropriate model depends on the mechanism of the drug's action, the assumptions made, and the intended use of the model in decision-making. If the assumptions do not lead to a mechanistic model, an empirical model can be selected, in which case validating the model's predictability becomes especially important. (Note that non-mechanistic models do not get good reviews from the FDA.) The model-selection process comprises a series of trial-and-error steps,

---

TABLE 13.1

Assumptions in Pharmacokinetic/Pharmacodynamic Modeling

The mechanism of the drug actions for efficacy and adverse effects

Development of tolerance or absence of tolerance

Disease state progression

Circadian variations in basal conditions

Absence or presence of an effect compartment

The PK model of absorption and disposition and the parameters to be estimated

Inclusion or exclusion of specific patient data

Distribution of PK and PD measures and parameters

Presence or absence of active metabolites and their contribution to clinical effects

Immediate or cumulative clinical effects

Drug-induced inhibition or induction of PK processes

Response in a placebo group

Influential covariates

The PD model of effect and the parameters to be estimated

Distributions of intra- and inter-individual variability in parameters

---

in which different model structures or newly added or dropped components to an existing model can be assessed by visual inspection and can be tested using one of several objective criteria. New assumptions can be added when emerging data justifies it.

13.2.1.2 Model Validation

Model validation is a process that involves establishing the predictive power of a model during the study design as well as in the data analysis stages. The predictive power is estimated through simulation that considers distributions of PK, PD, and study-design variables. A robust study design will provide accurate and precise model-parameter estimations that are insensitive to model assumptions.

During the analysis stage of a study, models can be validated on the basis of internal or external data. A common method for estimating predictability is to split the data set into two parts: build the model on the basis of one set of data and test the predictability of the resulting model on the second set of data. The predictability is especially important when the model is intended to provide supportive evidence for primary efficacy studies, to address safety issues, and to support new doses and dosing regimens in new target populations or subpopulations defined by intrinsic and extrinsic factors, or when there is a change in dosage form or route of administration. A PK/PD model can be related to dose–response, to PK, to PD, or to one or more of the following assumptions (Table 13.1).

13.2.2 PHARMACOKINETIC MODELING STUDIES

13.2.2.1 Compartment Pharmacokinetic Modeling

Pharmacokinetics is the study of the movement of drug molecules in the body, requiring appropriate differential calculus equations to study various rates and processes. The rate of elimination of a drug is described as being dependent on, or proportional to, the amount of drug remaining to be eliminated. By definition, such a process obeys first-order kinetics. The rate of elimination can, therefore, be described as

$$\frac{dX}{dt} = -kX \tag{13.1}$$

where k is a mere proportionality constant or a rate constant, and X is the amount remaining to be eliminated (and therefore, X_0 is the initial amount or the dose administered). Integration allows converting Equation 13.1 to

$$X = X_0 e^{k_{el} t} \tag{13.2}$$

Because the amount X is proportional to the concentration, a similar equation describes the time-decay profile of the drug concentration instead of the amount:

$$Cp_t = Cp_o e^{-k_{el} t} \tag{13.3}$$

This simple, first-order relationship allows a linear association between the log (more appropriately, the natural logarithm) of concentration and time. It is noteworthy that this concentration is the "effective" concentration and not necessarily the measured concentration. *Effective* refers to a thermodynamic activity rather than the physical concentration. Drugs decay in proportion to the concentration of "free" drug molecules, and whatever is bound to proteins may not be available for disposition. This extrapolation becomes more complex when we take into account other factors that might alter the "activity" (in a thermodynamic sense) of the drug in a biological fluid. For example, structuring of water inside protoplasm imparts lipophilic characteristics, which create significant differences in the available concentration gradients. This is a primary reason as to why it is not always possible to correlate measured concentrations with pharmacologic responses because the level of drug at the site of action or at the receptors depends highly on the thermodynamic activity of the drug, which is difficult to assess.

The relationship between the amount of drug and its concentration is classically represented by the following equation, which functions as if there were a physical space (called distribution volume) throughout, which the drug distributes evenly:

$$V_d = \frac{\text{Amount of drug in the body}}{\text{Concentration measured in plasma}} \tag{13.4}$$

This relationship is an oversimplification of the distribution characteristics of drug molecules in the body and can provide results in volumes often much larger than the body weight. For example, if a drug were selectively stored in different parts of the body, such as digoxin or diazepam are, the apparent distribution volumes using Equation 13.4 would be several multiples of the body's volume. Because the distribution of a drug is a time-dependent process, even within the same "compartment," it is suggested that this parameter be treated as a time-dependent variable[12] (as we will see later); treating a "bolus" dose as a short-term infusion improves the results of the deconvolution of integrated equations. This assumption allows a more accurate physical representation of the PK models because an "instantaneous" intravenous (IV) injection is treated as a very short duration, zero-order, input function. As we shall see, this consideration is more important as we integrate PD models where the action and effect of the drug are delayed for several reasons, including the input and distribution variables.

The area under the plasma-concentration–time curve, the AUC, is a useful parameter in defining the overall body exposure to a drug; this parameter integrates the concentration-over-time function:

$$\text{AUC} = \int_{t=0}^{t=\infty} Cp_t \, dt \tag{13.5}$$

Because the time function of drug concentration is dependent on the rate at which the drug is cleared from the hypothetical "volume," the AUC function is inversely proportional to the rate constant of elimination k_{el}: total body clearance, CL:

$$\text{AUC} = \frac{\text{Dose}}{V_d * k_{el}} \qquad (13.6)$$

Rearranging Equation 13.6 results in an expression for V_d.

$$V_d = \frac{\text{Dose}}{\text{AUC} * k_{el}} \qquad (13.7a)$$

The rate constant of elimination, in turn, is related to clearance (CL). Clearance is the product of volume of distribution, V_d, and the elimination-rate constant, k_{el}, describing drug elimination through renal excretion, metabolism, or removal from the sampled compartment by another means.

$$\text{CL} = k_{el} * V_d \qquad (13.7b)$$

The definition of clearance often confuses students of PK. Clearance is an inherent phenomenon, in which distribution volumes are high and rate constants are small to compensate for the distribution. Both the volume of distribution and the rate constant are derived phenomena and do not determine clearance. Note that total body clearance is a composite of all pathways that clear or remove the drug from the sampled compartment or the compartment from which the drug is cleared; this is based on the mathematical relationship between the observed elimination-rate constant and its components. Each of the pathways is involved in the turnover of the drug within the body. Using the parameters described above, it is possible to "simulate" concentration of a sampled compartment (of fluid) as a function of time in a single- or multiple-dose application using simple, iterative programs. Numerous computer programs are now available, which are drug- and model-specific, and allow simulations of steady-state blood levels that depend on various body functions and body characteristics that affect the clearance of the drug. Mixed models, involving a zero-order infusion, a bolus, or other similar combinations, can be made to estimate blood concentrations under different circumstances related to drug administration.

When drugs are received by routes other than IV injection, input is not "instantaneous" or zero order, and the function must often be represented as a mixed-order, primarily a first-order, process. This mixed-order model must then be taken into account in simulating drug concentration. Drug clearance, however, is not always a constant parameter, especially when an organ such as the liver is involved in the removal of the drug from the body:

$$\text{Organ clearance} = \frac{Q(C_a - C_v)}{C_a} = Q * E \qquad (13.7c)$$

where Q is the blood-flow rate to the organ, C_a is the concentration of the drug in the blood when entering the organ (in the arterial blood), and C_v is the concentration of drug in the blood when leaving the organ (in the venous blood). The term E is the steady-state extraction ratio. High E values mean high clearance by the liver and thus extensive metabolism. The liver blood-flow rate is a physiological parameter that can be altered in disease states. The extraction ratio depends not only on the function of liver but also on the nature of the drug. Both the hepatic clearance and the extraction ratio are empirical parameters and depend on the total hepatic blood flow, the unbound fraction of the drug, and the intrinsic clearance rate. Intrinsic clearance is differentiated from total clearance; the former is the ability to process the drug when other factors are not present. In other words, intrinsic clearance is the property of a body organ that clears the drug such as liver or kidney; for example, the maximum clearance in kidneys cannot exceed the total blood-flow rate to the kidneys and the hepatic clearance cannot exceed the total blood flow to the liver. The actual clearance of a drug from the body depends on the intrinsic clearance as well as how much drug is presented to the organ responsible for elimination – a lower amount of exposure of the drug to the eliminating organ resulting from distribution to body tissues will reduce the total clearance but will have no effect on the intrinsic clearance:

$$CL = Q * \frac{fu\ CL_{int}}{Q + \left(fu\ Cl_{int}\right)} = \frac{Q\ CL_{int}^{total}}{Q + CL_{int}^{total}} \tag{13.8}$$

where fu = unbound fraction of the drug, CL_{int}^{total} = the intrinsic clearance of unbound and bound drug, CL_{int} = the intrinsic clearance of unbound drug only.

This makes the extraction ratio:

$$E = \frac{fu\ CL_{int}}{Q + \left(fu\ CL_{int}\right)} \tag{13.9}$$

High-clearance drugs are those for which there is no saturation of the reaction that converts the drug, and therefore, the clearance rate approaches the blood-flow rate. In contrast, flow rate is irrelevant for capacity-limited drugs, and clearance is a simple product of the unbound fraction and the intrinsic clearance.

The traditional method of PK data analysis uses a two-stage approach: estimation of PK parameters through nonlinear regression using extensive concentration–time data from an individual, and using these data parameters as input data for the second-stage calculation of descriptive summary statistics on the sample. These statistics typically include the mean parameter estimates, and the variance and the covariance of the individual parameter estimates. Analysis of dependencies between parameters and covariates using classical statistical approaches (linear stepwise regression, covariance analysis, cluster analysis) can be included in the second stage. Covariates include patient characteristics such as age and body weight. The two-stage approach yields adequate estimates of population characteristics. Mean estimates of parameters are usually unbiased, but the random effects (variance and covariance) are likely to be overestimated in all realistic situations. Refinements such as the global two-stage approach have been proposed to improve the traditional approach through bias correction for the random effects of covariance and differential weighting of individual data according to the data's quality and quantity.

13.2.2.2   Physiologically Based Pharmacokinetic (PBPK) Studies

Physiologically based PK studies take a different perspective in modeling drug disposition in human body – a mechanistic physiologic distribution model. This approach had been in use in other disciplines long before the compartment kinetic modeling was applied to the study of drugs. In 1937, the mathematical basis for physiological PK modeling was established by Teorell,[13,14] but the solution to the equations was too difficult to obtain before the invention of the digital computer. An automatic solution of a physiologically realistic, mathematical description of the uptake, distribution, and clearance of a chemical agent was proposed by Bischoff in the early 1960s.[15] At that time, computation limitations forced several simplifications to the models, including the assumption that the distribution of the drug between tissues and blood is instantaneously at equilibrium, which led to physiological models with blood-flow–limited delivery of chemicals to tissues. Inhalation PK models using instantaneous distribution are well known. Physiological PK studies progressed no further until the early 1970s, when the physiological parameters of human organ system became better known and digital computers became more widely available. Today, physiological PK modeling is critical to understanding the behavior of a drug at the site of action.

Exposure modeling studies are often based on the physiologic functions that determine uptake, distribution, and elimination of drugs from the body. This approach was pioneered using anesthetics, in which physical distribution determines both the onset and termination of action.[16] Similar results have been reported for other compounds such as D_2O and ethanol,[17] propranolol,[18] and inulin and protein-bound antibiotics.[19] The modeling is based on a quantitative description of the

TABLE 13.2

Perfusion Rates of Organs of the Human Body

Organ	% Body Weight	Perfusion Rate (L/min/kg)
Fat	20	0.056
Muscles	48	0.03
Brain	2.0	0.56
Skin	3.7	0.12

distribution process using standardized organ weights and blood-flow rates (Table 13.2). A simpler model used for propranolol assumes no solute binding and a tissue/plasma equilibrium coefficient of 10 for all tissues, except for muscles where this value is 3.62 and for fats where this value is 2.42.[20] Also in this simple model, there is no first-pass effect and kidney excretion is the only mechanism of drug removal from the body; thus, the input function is equal to systemic availability. In more complex models, tissue binding and other factors that produce nonequilibrium of the tissue/plasma ratios are introduced. The simple model, when used to determine bolus-response function, is well described by a simple two-exponential function; in the more complicated models, three exponents generally provide good fit, and often, going to higher exponents does not improve the predictability. More important is the timing of the first data point obtained in the bolus-response function. This should, ideally, be obtained at or before the end of the constant infusion. (Note that better estimates are obtained from infusion studies than from single-bolus doses because there is always an inevitable delay in the dispersion of drug in the bolus dosing, even though the model assumes no delay.) When a deconvolution method is used (see below), the robustness of analysis depends on the accuracy of venous-concentration data because the response function $r(t)$ is established from these data. Therefore, any errors in this function reduce the reliability of the analysis, particularly when a later time sample, such as 10 min, is used as the first data point.

In a systems pharmacology context, the PBPK model parameters should be divided into three categories: parameters relating to the system or species, the drug, or the study design. Relevant parameters for the system or species can include age, weight, height, and genetic make-up of human or animal subjects. Parameters related to the drug include physicochemical characteristics determining permeability through membranes, partitioning to tissues, binding to plasma proteins, or affinities toward certain enzymes and transporter proteins. Study design parameters may include dose, route and frequency of administration, and the effect of concomitant drugs and food. Separation of pharmacokinetic parameters into these three categories is vital to allow development of generic drug-independent models that can be used for a wide range of compounds. Further, it facilitates independent development of various databases of anatomical, biological, physiological, and genetic characteristics of healthy and disease populations that can be used to simulate clinical studies.

The following factors have significantly expanded our ability to combine and integrate various prior data sets into PBPK models:

- The availability of *in vitro* systems which act as surrogates for *in vivo* reactions relevant to the absorption, distribution, metabolism, and excretion (ADME) processes
- Recent development and refinement of *in vitro–in vivo* extrapolation (IVIVE) techniques
- Methods to predict tissue partition coefficients using physicochemical properties and protein binding data.

This systems approach expedites prediction and investigation of important intrinsic (e.g., organ dysfunction, race, genetics, and disease) and extrinsic (e.g., drug–drug interactions, smoking, diet, and environmental) factors on drug exposure and response. This in turn helps with designing and optimizing clinical studies and selecting the optimal dosing regimens. Such capabilities become even more important when drugs are to be dosed in very young children or disease populations where running clinical studies is not commonly feasible or is very challenging.

Over the past decade, PBPK modeling has had a significant impact on regulatory science and decisions. The US FDA has identified innovation in clinical evaluations as a major scientific priority area, and especially innovation through modeling and simulation (M&S). Accordingly, they have used M&S strategies to address various drug development, regulatory, and therapeutic questions over the past decade. Parekh and co-workers have highlighted some examples where M&S has served as a useful predictive tool which include dose selection for pivotal trials, including dosing in select populations such as pediatrics, optimization of dose and dosing regimen in a subset patient population, and prediction of efficacy and dosing in an unstudied patient population.

The PBPK models using nonclinical and clinical data to predict drug PK/PD properties in healthy and patient subjects are increasingly used at various stages of drug development and regulatory interactions. However, initially the regulatory applications of PBPK models were mainly focused on predicting drug–drug interactions (DDI), and the areas of application are gradually expanding. These have expanded in other areas such as drug formulation and/or absorption modeling, age, and ethnic-related changes in PK and disposition and the assessment of PK changes in case of different physio-pathological conditions (e.g., renal and/or hepatic deficiencies).

The current PBPK modeling techniques will improve with time as there currently is a lack of adequate and reliable systems data. As the models become more complex, they demand more detailed knowledge of the system, and while there has been a significant improvement in identifying and generating missing information, we still have a long way to go. Systems data which are currently insufficient include the abundance and activity of non-CYP enzymes and transporters in various tissues, and absorption-related data and how these are changing by age/disease status. Such challenges are even bigger when developing mechanistic PBPK models for biologics and/or models that aim at providing mechanistic insights into drug efficacy and safety. However, since these data are related to the biological system, the burden and benefit of generating these data can be shared through collaboration between various stakeholders in a pre-competitive manner.

Developing bottom-up system pharmacology models requires inputs from numerous experts including experimentalists, biologists, epidemiologists, pharmacists, pharmacologists, and mathematical modelers. Therefore, by its nature, it is a multi-disciplinary endeavor requiring adequate education and communication among various stakeholders who have traditionally been working in isolation. These experts usually work independent of each other and need to learn how to interact and communicate with other disciples. Furthermore, the best practice in developing and accessing such models is evolving. While there are few articles in literature that provide guidelines, more rigorous standards are lacking as highlighted in Regulatory agencies around the world have started addressing these needs and providing guidelines. A PBPK concept paper has been published by EMA which may lead to a specific European guideline on qualification and reporting of PBPK modeling and analysis.

The applications of PBPK models will expand even further if they are adequately equipped to connect and interact with other tools/platforms (interoperability), such as quantitative system pharmacology models of various disease progressions. Innovative applications of PBPK have started appearing in the literature, such as introduction of time-varying physiology into pediatric PBPK models; modeling drug disposition in kidney, brain, and lung; virtual bioequivalent studies; and modeling antibody-drug conjugates (ADC). In addition, given the integrative nature of PBPK models, which allows incorporation of characteristics of patients, it will not be surprising to see applications of these models in implementing precision dosing at the point-of-care in the near future. In particular, when relevant and affordable biomarkers for enzyme/transporters activities are developed/identified, such information can be incorporated within PBPK models to determine and optimize doses in, for instance, DDI cases.

Table 13.3 shows the products where the PBPK simulations have been included in the drug label (FDA, EMA, and PMDA). For detailed information about specific drug products, visit the FDA-approved drug product database (www.accessdata.fda.gov/).

TABLE 13.3
Labels of Recently Approved Drugs Listing PBPK Studies (Company Drug Brand)

Actelion Macitentan Opsumit: www.accessdata.fda.gov/drugsatfda_docs/nda/2013/204410Orig1s000ClinPharmR.pdf; www.accessdata.fda.gov/drugsatfda_docs/label/2013/204410s000lbl.pdf; www.ema.europa.eu/docs/en_GB/document_library/EPAR_-_Product_Information/;human/002697/WC500160899.pdf (EMA)

Alkermes Aripiprazole lauroxil Aristada: www.accessdata.fda.gov/drugsatfda_docs/nda/2015/207533Orig1s000ClinPharmR.pdf; www.accessdata.fda.gov/drugsatfda_docs/label/2015/207533s000lbl.pdf

Ariad Ponatinib hydrochloride Iclusig: www.accessdata.fda.gov/drugsatfda_docs/nda/2012/203469Orig1s000ClinPharmR.pdf; www.accessdata.fda.gov/drugsatfda_docs/label/2012/203469lbl.pdf

Astrazeneca Naloxegol oxalate Movantik: www.accessdata.fda.gov/drugsatfda_docs/nda/2014/204760Orig1s000ClinPharm.pdf; www.accessdata.fda.gov/drugsatfda_docs/label/2014/204760s000lbl.pdf

Astrazeneca Olaparib Lynparza: www.accessdata.fda.gov/drugsatfda_docs/nda/2014/206162Orig1s000ClinPharmR.pdf; www.accessdata.fda.gov/drugsatfda_docs/label/2014/206162lbl.pdf

Astrazeneca Osimertinib mesylate Tagrisso: www.accessdata.fda.gov/drugsatfda_docs/nda/2015/208065Orig1s000ClinPharmR.pdf; www.accessdata.fda.gov/drugsatfda_docs/label/2015/208065s000lbl.pdf

Eisai Lenvatinib mesylate Lenvima: www.accessdata.fda.gov/drugsatfda_docs/nda/2015/206947Orig1s000ClinPharmR.pdf; www.accessdata.fda.gov/drugsatfda_docs/label/2015/206947s000lbl.pdf

Genentech Alectinib hydrochloride Alecensa: www.accessdata.fda.gov/drugsatfda_docs/nda/2015/208434Orig1s000ClinPharmR.pdf; www.accessdata.fda.gov/drugsatfda_docs/label/2015/208434s000lbl.pdf

Genentech Cobimetinib fumarate Cotellic ttp: //www.accessdata.fda.gov/drugsatfda_docs/nda/2015/206192Orig1s000ClinPharmR.pdf; www.accessdata.fda.gov/drugsatfda_docs/label/2015/206192s000lbl.pdf

Genzyme Eliglustat tartrate Cerdelga: www.accessdata.fda.gov/drugsatfda_docs/nda/2014/205494Orig1s000ClinPharmR.pdf; www.accessdata.fda.gov/drugsatfda_docs/label/2014/205494Orig1s000lbl.pdf; www.info.pmda.go.jp/downfiles/ph/PDF/340531_3999037M1023_1_02.pdf (PMDA); www.ema.europa.eu/docs/en_GB/document_library/EPAR_-_Product_Information/;human/003724/WC500182387.pdf (EMA)

Incyte Ruxolitinib phosphate Jakafi: www.accessdata.fda.gov/drugsatfda_docs/nda/2011/202192Orig1s000PharmR.pdf; www.accessdata.fda.gov/drugsatfda_docs/label/2014/202192s006lbl.pdf

Janssen Rilpivirine hydrochloride Edurant: www.accessdata.fda.gov/drugsatfda_docs/nda/2011/202022Orig1s000ClinPharmR.pdf; www.accessdata.fda.gov/drugsatfda_docs/label/2011/202022s000lbl.pdf

Janssen Rivaroxaban Xarelto: www.accessdata.fda.gov/drugsatfda_docs/nda/2011/022406Orig1s000ClinPharmR.pdf; www.accessdata.fda.gov/drugsatfda_docs/label/2011/022406s000lbl.pdf

Janssen Simeprevir sodium, Olysio: www.accessdata.fda.gov/drugsatfda_docs/nda/2013/205123Orig1s000ClinPharmR.pdf; www.accessdata.fda.gov/drugsatfda_docs/label/2013/205123s001lbl.pdf

Novartis Ceritinib Zykadia: www.accessdata.fda.gov/drugsatfda_docs/nda/2014/205755Orig1s000ClinPharmR.pdf; www.accessdata.fda.gov/drugsatfda_docs/label/2014/205755s000lbl.pdf; www.ema.europa.eu/docs/en_GB/document_library/EPAR_-_Product_Information/;human/003819/WC500187504.pdf (EMA)

Novartis Panobinostat lactate Farydak: www.accessdata.fda.gov/drugsatfda_docs/nda/2015/205353Orig1s000ClinPharmR.pdf; www.accessdata.fda.gov/drugsatfda_docs/label/2015/205353s000lbl.pdf; www.pmda.go.jp/PmdaSearch/iyakuDetail/ResultDataSetPDF/300242_

Novartis Sonidegib phosphate Odomzo: www.accessdata.fda.gov/drugsatfda_docs/nda/2015/205266Orig1s000ClinPharmR.pdf; www.accessdata.fda.gov/drugsatfda_docs/label/2015/205266s000lbl.pdf; www.ema.europa.eu/docs/en_GB/document_library/EPAR_-_Product_Information/;human/002839/WC500192970.pdf (EMA)

Pfizer Sildenafil citrate, Revatio: www.accessdata.fda.gov/drugsatfda_docs/nda/2009/022473s000_ClinPharmR.pdf; www.accessdata.fda.gov/drugsatfda_docs/label/2012/022473s003lbl.pdf

Pharmacyclics Ibrutinib Imbruvica: www.accessdata.fda.gov/drugsatfda_docs/nda/2013/205552Orig1s000ClinPharmR.pdf; www.accessdata.fda.gov/drugsatfda_docs/label/2013/205552s000lbl.pdf; www.ema.europa.eu/docs/en_GB/document_library/EPAR_-_Product_Information/;human/003791/WC500177775.pdf (EMA)

13.2.2.3 Bioequivalence and Systemic Exposure Models

Screening of drug molecules for suitability for use in humans is often subjected to certain basic toxicity or workability solutions to reduce the cost. The human body must be able to remove the drug in a reasonable time. Drug clearance is an intrinsic parameter; however, body clearance (extent of drug removal) is dependent on cardiac output and the overall extraction ratio:

$$ER = \frac{[(\text{Entrance concentration}) - (\text{Exit concentration})]}{\text{Entrance concentration}} \tag{13.10}$$

The extraction ratio (ER) ranges from 0 to 1, and the cardiac output is proportional to body size:

$$\dot{Q}\left(\frac{\text{mL}}{\text{kg} * \text{min}}\right) = 180 \ [\text{BW (kg)}]^{-0.19} \tag{13.11}$$

Cross-species comparisons can be made for crude estimates, and generally drugs that have clearance of less than 4 mL/min/kg would be evaluated only if there are special reasons related to their particular mechanisms of actions.

In addition to the "removal potential" of a drug (or ability to be eliminated quickly), the "entry potential" is also a good screening parameter. For drugs that are poorly bioavailable, further development should proceed only if proper modification to the molecular structure or to the drug delivery system is made to provide a reasonable possibility of entry. When evaluating bioavailability, it is important to first establish a PK basis because of the large variation in bioavailability as a result of the differences in population PK. Population models are most appropriate for this type of evaluation. Obviously, the consideration of bioequivalence in establishing compliance of generic products is important, and the guidelines for these measurements are defined in the *United States Pharmacopoeia* and other guidelines provided by the FDA. It should be noted that the purpose of these studies is to compare the systemic exposure of the body to the drug molecules. This requires measurement of both the extent of absorption and the rate of absorption. Traditionally, parameters such as AUC, T_{max}, and C_{max} are studied using specified statistical models. For drugs given orally, these studies cannot be substituted with PD studies, which may be required for some drugs in which the plasma or sample tissue concentration is not available.

13.2.2.4 Deconvolution Techniques

The bolus-response function $r(t)$ is defined as the "systemic (e.g., venous) concentration that is produced by a bolus systemic input of unit amount."[24] This is generally described using a multiexponential function:

$$r(t) = \sum_{i=1}^{\rho} a_i \left(e^{-\frac{t}{T_i}} \right) \tag{13.12}$$

The optimized values for a_i and T_i are determined by using mathematical approach without any physiological significance.[21–23] Generally, the resorting required to use a three-exponential term takes the estimates out of the population parameters or global minimum.

The three parameters in γ-distribution are chosen by minimizing the error function:

$$\text{Error function} = \sum_i \frac{(y_{gam})_i - (y_{dat})_i}{(y_{dat})_i + \text{noise}} \tag{13.13}$$

where $(y_{dat})_i$ is the sum of overall data points for the experimental venous concentration, and $(y_{gam})_i$ is the venous concentration-determined y convoluting the γ-distribution input using a polyexponential equation as described above for $r(t)$. The *noise* factor in Equation 13.13 determines the weighting of each data point used. When there is no error of *noise*, then the error is simplified for each point. When the error is large, the term $(y_{dat})_i$ drops out in the denominator and the error is proportional to the numerator of the error term. Because the γ-distribution function is highly nonlinear, it is important to use a global annealing procedure such as that used in PKQuest (Minneapolis, MN; www.pkquest.com, requiring Maple software, Maplesoft, Ontario, Canada, www.maplesoft.com),[24] and then follow it with nonlinear minimization. The venous concentration is fitted by using interpolation, meaning that it goes through each data point or uses a smoothing cubic-spline function, followed by the deconvolution. The B-spline function defines the number and position of "breakpoints" and the order of the spline function. Highly sophisticated models have been used for this purpose.[25]

The course of systemic exposure to a drug is studied by comparing IV administration studies using a deconvolution approach, in which $r(t)$ is the systemic concentration produced from IV administration (also called bolus function) and $I(t)$ is the systemic input rate (in units such as g/min) from the non-intravenous (non-IV) route:

$$c(t) = \int_0^t r(t-\tau)[I(\tau)]d\tau \tag{13.14}$$

If there is no first-pass effect involved, then $I(\tau)$ is equal to the rate of intestinal absorption upon administration of equal doses (in IV and non-IV forms). In first-pass metabolism, $I(\tau)$ is the systemic availability of the drug upon oral (or sublingual, rectal, buccal, etc.) dosing. The function $r(\tau)$ is obtained by fitting the data upon IV administration to a variety of exponential equations and selecting the best fit through residual mean error of fit. The duration of infusion can be instantaneous (a few seconds for bolus input) but, more realistically, is usually a few minutes. Whereas it is desirable to obtain the sample as early as possible, sampling earlier than 2 min after injection is not advised, so as to allow time for venous mixing. Long-term IV infusions are also used to obtain the $r(\tau)$ function. Mathematical solutions of the deconvolution are easily obtained by using such validated software as PKQuest requiring Maple[24] software. Several methods are used for deconvolution; γ-distribution input is a parametric-fitting technique. Whereas polyexponential-fitting techniques are widely used, better fits are obtained by using a parametric approach for simulating $I(\tau)$, where A is the amount of drug reaching the circulation, Γ the γ-function, a the γ-number that ranges from 1 to 6, and b a constant that has inverse time units:

$$I(t) = \frac{(Ab)^a \, t^{a-1}e^{-bt}}{\Gamma(a)} \tag{13.15}$$

This approach offers a superior simulation, particularly in situations where there is a delay in the input function such as in intestinal absorption and gastric emptying variations.[26] The three parameters (A, a, and b) given above are estimated by global (also called *simulated annealing*) and local (also called *Powell*) nonlinear optimization.[27] The fitting of data using γ-deconvolution method smooths data noise, and with no user-adjustable parameters, the bias is removed. If it is not possible to define the input using a single γ-distribution, then other deconvolution approaches, such as analytical, spline, or uniform approaches, which remove the "roughness" of the input rate are used. The choice of parameters is additionally improved by experimental Akaike criterion[28] and the "generalized cross-validation."[29]

The analytical deconvolution involves approximation of $C(t)$ by an interpolating or smoothed spline function and subsequent deconvolution.[27,30] The analytical deconvolution method is most commonly used for the advantage of being fast, and also where data are exact, excellent results are obtained. However, the robustness of this approach depends on the value chosen for the smoothing parameter, which is poorly estimated even when standard deviation is available (very rare). Where there is noisy data, it adds more error in analytical deconvolution compared to spline and uniform methods. Also, analytical deconvolution does not allow use of negative values for input. In spline function input consideration, the input $I(t)$ is parameterized using a general B-spline function and then obtaining deconvolution by a constrained regression.[31,32] In using uniform input, $I(t)$ is estimated on dense uniform sequence of time points using stochastic regularization procedure for deconvolution.[25,33]

13.2.2.5 Population Pharmacokinetics

The main purpose of generating PK data is to simulate various disease conditions, physiologic variables, and patient characteristics in arriving at effective dosing regimen. This exercise is important because it is not always possible to collect data in all eventual situations or to obviate the need for extensive data whose collection may not be possible. Planned population PK studies (see below) to test the design factors prove very useful. A simulation scheme requires repetitive simulations, followed by analysis of data sets, to account for the effect of sampling variability on parameter estimates. A permutation of study designs can be simulated to determine the most useful design, yielding the best information.

There are generally two types of protocols used for simulations: the add-on and the standalone. Both protocols contain clear statements of the population-analysis objectives and the proposed sampling design and data collection procedures. The specific PK parameters to be investigated are identified in advance. If the population PK study is added on to a clinical trial (an add-on study), as is done frequently, then the PK protocol should not compromise the primary objectives of the clinical study. Investigators should be made aware of the value of including a population PK study in a clinical trial.

The population model defines at least two levels of hierarchy. At the first level, PK observations in an individual (such as concentrations of drug species in biological fluids) are viewed as component of the individual probability model. In this model of individual observations, the mean distribution is represented by a fitted model such as a biexponential model where mean is given by a PK model. At the second level, the individual parameters are regarded as random variables, and the probability distribution of these (often the mean and the variance, i.e., the inter-subject variance) is modeled as a function of individual-specific covariates. These models – their parameter values and the use of study designs and data analysis methods designed to elucidate population PK models and their parameter values – all fit under the umbrella of *population PKs*.

Defining it more explicitly, population PK is the study of the sources and correlates of variability in drug concentrations among individuals who are the target patient population receiving clinically relevant doses of a drug of interest.

Population PK analysis is now pivotal during drug development for understanding the quantitative relationships among the drug-dosing profile, patient characteristics, and drug disposition. It provides a better understanding of the variability within a target population. Nonlinear, mixed-effects modeling proves very useful in dose-ranging studies (such as so-called titration or effect-controlled designs), in which there is a reasonable expectation that inter-subject kinetic variations may require changing dosing regimens for some subgroups within the target population. An example of such a scenario is when the population is heterogeneous or when the therapeutic window is narrow.

Pharmacokinetic characterization through estimation of PK parameters returns greater value if the parameters of disposition and their statistical distribution in larger populations can be ascertained (as shown above in the two-stage modeling practice). Patients even within a narrow demographic group are different in how their body acts on drug molecules (the PK part of the equation); this includes drug distribution, biotransformation, and excretion. Obviously, disease state and use of other medications can have a significant impact on these disposition profiles and alter the dose–concentration response. Variations become larger when we compare age, sex, genetic, and even environmental factors. For example, steady-state concentrations of drugs that are mostly eliminated by the kidney are usually higher in patients suffering from renal failure. For drugs with uncomplicated mechanisms of excretion, the kidney function would be a good measure to project half-lives of drugs in any population – the goal of population PK studies. However, a difference in the plasma concentration alone cannot be a predictor of drug response or drug toxicity; there must exist a relationship between concentration (or another surrogate) and the response to allow determining the window of variation allowed. Population PK seeks to identify the measurable pathophysiologic factors that alter the dose–concentration relationship and the extent of those changes. This allows one to appropriately modify the dose if such changes are associated with clinically significant shifts in the therapeutic index. These projections can be made (when subjected to appropriate statistical models) even when the amount of data is limited. Additionally, the analysis of data from a variety of unbalanced designs as well as from studies that are normally excluded because they do not lend themselves to the usual forms of PK analysis can also be used in a population PK modeling; examples of more challenging data sets to analyze include concentration data obtained from pediatric and elderly patients or data obtained during the early development of drugs when the relationships between dose or concentration and efficacy or safety is established.

The magnitude of the unexplained or random variability in PK evaluation is important because the efficacy and safety of a drug can decrease as unexplainable variability increases. In addition to inter-individual variability, the degree to which steady-state drug concentrations in individuals typically vary about their long-term average is also important. Concentrations may vary as a result of inexplicable, day-to-day or week-to-week kinetic variability or may be the result of errors in concentration measurements. Estimates of this kind of variability (residual intra-subject, inter-occasion variability) are important for therapeutic drug monitoring. Knowledge of the relationship among concentration, response, and physiology is essential to design the dosing strategies for rational therapeutics that may not necessarily require therapeutic drug monitoring.

Whereas the variability in PD parameters is likely to be higher than what the PK would predict, given the highly variable inter- and intra-individual effects, cross-species comparisons are more likely to vary because of PK effects. For example, the binding of drug can be different between species, resulting in different disposition and PD characteristics; in such instances, comprehensive PK/PD modeling proves extremely useful in projecting effects and side effects in humans as a method of inter-species extrapolation.

Another reason for using population PK approach is to obviate the inherent weaknesses built into pure PK studies. These PK studies are designed to limit the variation between and among individuals through strict acceptance criteria of healthy subjects; historically, kineticists have tried to define PK parameters within a very narrow range as a measure of goodness of a study. The reality is that when a drug is used by consumers, it will be subject to all types of variations possible, including many that could not have been envisioned in the development phases. Therefore, focusing on a single PK parameter (such as half-life or clearance) on determining appropriate dose is not always the best choice. In contrast to traditional PK evaluation, the population PK approach encompasses the following information:

- The collection of relevant PK information in patients who are representative of the target population to be treated with the drug.
- The identification and measurement of variability during drug development and evaluation.
- The explanation of variability by identifying factors such as demographic, pathophysiologic, environmental, or concomitant drug-related origin that may influence the PK behavior of a drug.
- The quantitative estimation of the magnitude of the unexplained variability in the patient population.

Population PK modeling is also used to estimate population parameters of a "response surface model" in Phase I and late Phase IIb clinical drug development. It is this stage when the developer decides on what is the most appropriate application of the drug. Since the number of samples at this stage can be extensive, there is little need to resort to complex data analysis systems and the use of the two-stage method often suffices. The more complex nonlinear, mixed-effects model can be used when it is desired to pool the data from several studies. Population PK modeling is also used in early Phase IIa and Phase III drug development to gain information on drug safety (*vis-à-vis* efficacy) and to gather additional information on drug PK in special populations, such as the elderly or pediatric populations. This approach is also useful in postmarketing surveillance (Phase IV) studies. The studies performed during Phase III and Phase IV clinical drug development allow use of a full-population sampling PK study design (few blood samples drawn from several subjects at various time points). This sampling design can provide important information during new drug evaluation, regulatory decision-making, and drug labeling.

There are two common methods for obtaining estimates of the fixed effects (the mean) and the variability: the two-stage approach and the nonlinear, mixed-effects modeling approach. The two-stage approach involves multiple measurements on each subject. The nonlinear, mixed-effects model can be used in situations where extensive measurements cannot or will not be made on all or any of the subjects.

When properly applied, population PK studies in patients combined with suitable mathematical and statistical analyses (e.g., using nonlinear, mixed-effects modeling) are valid and, on some occasions, are preferred alternatives to extensive studies. This applies to both PK and PD studies. When large data are not available, the traditional two-stage approach is not applicable because estimates of individual parameters cannot be obtained; in such instances, nonlinear, mixed-effects modeling proves valuable. This modeling approach is developed from the realization that where PK and PD are to be investigated in patients, practical considerations should prevail in determining how much data can be gathered. This would inevitably have to be less stringent because the patients subjected to dose-determining studies are less likely to be tolerant than the healthy subjects. Other approaches such as naive averaged-data modeling can also be used which are also simpler systems; in this model, the study of a population sample rather than that of an individual is used as a unit of analysis for the estimation of the distribution of parameters and their relationships with covariates within the population. This approach uses individual PK data of an observational (experimental) type, which can be sporadic, unbalanced, or fragmentary, compared to extensive uniform data in well-designed PK studies. Analysis according to the nonlinear, mixed-effects model in such situations provides estimates of population characteristics that define the population distribution of the PK (or PD) parameter. In the mixed-effects modeling context, the collection of population characteristics is composed of population mean values (derived from fixed-effects parameters) and their variability within the population (generally the variance and covariance values derived from random-effects parameters). A nonlinear, mixed-effects modeling approach to the population analysis of PK data, therefore, consists of directly estimating the parameters of the population from the full set of individual concentration values. The individuality of each subject is maintained and accounted for, even when data are not extensive.

13.2.2.6 Pharmacokinetics in Disease States

13.2.2.6.1 Hepatic Impairment

The time course of drug concentration (and hence, in most instances, its PD) is determined by the mechanisms that remove the drug from the body, which are primarily mediated by the liver and the kidney. The liver clears drugs through a variety of oxidative and conjugative metabolic pathways or through biliary excretion of unchanged drugs or metabolites. Alterations of these excretory and metabolic activities by hepatic impairment can lead to drug accumulation or, less often, failure to form an active metabolite. Kidneys remove drugs by filtration and secretion, and renal impairment can also lead to drug accumulation (see Section 13.2.2.6.2).

It is well established that hepatic disease alters the absorption and disposition of drugs (as described in PK studies) as well as their efficacy and safety (as described in PD studies). Disease states shown to alter PKPD properties include common hepatic diseases, such as alcoholic liver disease and chronic infections with hepatitis viruses B and C; and less common diseases, such as acute hepatitis D or E, primary biliary cirrhosis, primary sclerosing cholangitis, and α-antitrypsin deficiency. Liver disease can also alter kidney function, which can lead to accumulation of a drug and its metabolites, even when the liver is not primarily responsible for elimination. Liver disease may also alter PD effects, such as the increased encephalopathy with certain drugs in patients with hepatic failure. The specific effect of any disease on hepatic function is often poorly described and highly variable, particularly with regard to effects on the PK and PD of a drug.

Hepatic function is assessed using markers such as endogenous substances affected by the liver (e.g., bilirubin and albumin); by using functional measures such as prothrombin time; or by using the ability of the liver to eliminate marker substrates, such as antipyrine,[34] indocyanine green,[34] monoethylglycine-xylidide,[35] or galactose.[36] Clinical variables and combinations of variables have also been studied. Clinical variables utilized to assess hepatic function include ascites or encephalopathy, nutritional status, peripheral edema, and histologic evidence of fibrosis. Combinations of variables to determine hepatic function include the Child–Pugh classification for alcoholic cirrhosis and portal hypertension,[37,38] the Mayo risk scores for primary biliary cirrhosis and primary sclerosing cholangitis,[39] and the Maddrey–Carithers discriminant function for acute alcoholic hepatitis.[40,41] Despite extensive efforts, no single measure or group of measures is adequate to describe the impact of renal impairment universally. In almost all instances, the sponsor will need to develop a detailed set of parameters to adequately describe the PK and PD in hepatic impairment. The FDA and other regulatory agencies recommend PK studies in patients with impaired hepatic function if hepatic metabolism or excretion accounts for a substantial portion (greater than 20% of the absorbed drug) of the elimination of a parent drug or active metabolite. Even if this 20% rule of thumb does not apply, the FDA requires liver impairment studies for drugs (or their metabolites) with narrow therapeutic or toxic index. This is necessary to take into account population variation in hepatic function. It is noteworthy that unless the sponsor shows that the drug is not extensively metabolized, it is assumed to be removed primarily from the liver, thus requiring liver impairment studies.

For some drugs, hepatic functional impairment is not likely to alter PK sufficiently to require dosage adjustment. In such cases, a study to confirm the prediction is generally not important. Drug properties that support this conclusion include drugs excreted entirely via renal routes, with no involvement of the liver, drugs metabolized in the liver to a small extent (less than 20%), and drugs in which the therapeutic range is wide, so that modest impairment of hepatic clearance will not lead to drug toxicity directly or by increasing its interaction with other drugs. This also applies to gaseous or volatile drugs that are primarily eliminated via the lungs. For drugs intended for single-dose administration, hepatic-impairment studies are not generally useful, unless clinical concerns suggest otherwise.

Population PK screening in Phases II and III is useful in assessing the impact of altered hepatic function (as a covariate) in PKs if those patients are not excluded from Phase II and III trials and if there is sufficient PK information collected about the patients to characterize them reasonably well. If a population PK approach is used, patients in Phase II and III studies are assessed for encephalopathy, ascites, serum bilirubin, serum albumin, and prothrombin time (which are components of the Child–Pugh score) or a similar group of measures of hepatic function. The population PK study, then, would include the following features:

- Preplanned analysis of the effect of hepatic impairment
- Appropriate evaluation of the severity of liver disease
- A sufficient number of patients and a sufficient representation of the entire range of hepatic functions to allow the study to detect PK differences large enough to warrant dosage adjustment
- Measurement of the unbound concentrations of the drug, when appropriate
- Measurement of the parent drug and the active metabolites.

These features are important if the sponsor intends to use the results to support a conclusion that no dosage adjustment is required for patients with impaired hepatic function. PD assessments may be useful in studies designed to assess the effect of altered liver function, especially if concentration–response data are not available or if there is a concern that an altered hepatic function could alter the PD response.

Plasma-concentration data (and urine-concentration data, if collected) are analyzed to estimate the measures or parameters that describe the PKs of the drug and its active metabolites. Such measures or parameters include AUC, peak concentration (C_{max}), apparent clearance (CL/F), renal and nonrenal clearance, apparent volume of distribution (*Vdz* or *Vdss*), and terminal half-life (*t*1/2). (*F* in the expression CL/*F* is bioavailability) Where relevant, measures or parameters can be expressed in terms of unbound drug concentrations, such as the apparent clearance relative to the unbound drug concentration (CL_u/F = dose/AUC_u, where subscript *u* indicates unbound drug). Noncompartmental or compartmental modeling approaches to parameter estimates can be used.

Relationships between hepatic functional abnormalities (e.g., hepatic blood flow, serum albumin concentration, or prothrombin time), overall impairment scores (such as the Child–Pugh), and selected PK parameters are sought using linear and nonlinear models. Examples of selected PK parameters studied in the context of hepatic impairment include total body clearance, oral clearance, apparent volume of distribution, unbound clearance, or dose-normalized area under the unbound, concentration–time curve. A regression approach for continuous variables describing hepatic impairment and PK parameters is appropriate with the understanding that some correlations will rely on categorical variables such as the Child–Pugh. Typically, modeling results include parameter estimates of the chosen model and measures of their precision such as standard errors or confidence intervals. Prediction-error estimates are also desirable to assess the appropriateness of the model.

13.2.2.6.2   *Renal Impairment*

Similar to the recommendations made above in adjusting the dosing in hepatic function modification, a PK study in patients with impaired renal function is recommended if a significant portion of the drug is removed from the body through kidneys. In such a case, PK parameters are likely to be different in patients with impaired renal function. Generally, drugs given in a single dose, drugs with high therapeutic indices and eliminated via hepatic clearance, or gaseous or volatile drugs eliminated through the lungs may not require studies in renal impairment. This rule-of-thumb changes, however, where renal impairment affects hepatic metabolism or where the impact of dialysis on the PK of a drug should be considered. Note that this study of changes in distribution and elimination in renal impairment affects both the active drug and its metabolites, and it is particularly important where a narrow therapeutic or toxic index is observed. In some instances, a study in patients with renal impairment is required even though the drug is not significantly removed by the kidneys but where compromised renal function affects plasma protein binding of drugs. In such instances, a drug highly cleared by liver (but dependent on protein binding) may show a significant difference in its disposition characteristics in renal impairment. Patients on dialysis often require larger doses; dialysis often significantly affects the PK requiring dose adjustment. As a result, PK should be studied under both dialysis and nondialysis for patients with end-stage renal disease requiring dialysis. If the drug and metabolites have a large unbound volume of distribution (V_d), only a small fraction of the amount in the body will be removed by dialysis. Measurements, such as plasma creatinine or creatinine clearance, have been used successfully to adjust dosing regimens for drugs eliminated primarily by the kidneys.

In some cases, renal impairment studies are conducted to establish the thesis that a dosage adjustment is not necessary; these studies certainly make an impressive presentation to regulatory authorities and, more importantly, allow the sponsor to make certain additional claims on the label. Note that renal impairment is a major pathologic change that induces many physiologic changes. Consequently, unrelated effects resulting from renal impairment are often observed unexpectedly such as altered absorption and distribution of drugs in renal impairment. This may occur even when the renal route does not represent a significant elimination mechanism. The kidneys also participate in the metabolism of drugs and produce hormones that can indirectly affect pathophysiology of disease and drug disposition. For example, in renal impairment, secretion of erythropoietin is reduced, leading to anemia; where a drug extensively bound to components of red blood cells, its disposition is likely to be altered in renal impairment.

13.2.2.6.3 Pharmacokinetic Studies in Special Populations

The FDA places special importance on PK studies in pediatric patients because of the essential differences in physiological characteristics between adults and children; variance can sometimes be substantial. In pediatric populations, growth and developmental changes affect the rates of adsorption, distribution, metabolism, and excretion (ADME) and lead to changes in PK measures and parameters. To achieve AUC and C_{max} values in children similar to values associated with effectiveness and safety in adults, it may be important to evaluate the PK of a drug over the entire pediatric age range in which the drug will be used. Where growth and development are rapid, adjustment in dose within a single patient over time may be important to maintain stable systemic exposure.

The PK in the elderly is another subject of great interest to regulatory authorities. Unlike pediatric PKs, in most instances some linear extrapolations can be drawn on the basis of such parameters as lean body mass and age. Nevertheless, these extrapolations should be studied and validated.

13.2.2.7 Computational Support

During the past few years, computational efficiency has advanced, and many software programs have become available – often on free domains – that are useful in the simulation of PK models. The following "Software for Use in Pharmacokinetic/Pharmacodynamic Modeling" sidebar lists some of these programs:

General-Purpose High-Level Scientific Computing Software

Berkeley Madonna, University of California at Berkeley, www.berkeleymadonna.com
MATLAB®-Simulink®, The MathWorks, Inc., www.mathworks.com/
MLAB, Civilized Software, Inc., www.civilized.com
GNU Octave, University of Wisconsin, www.octave.org/
SOURCEFORGE: https://sourceforge.net/. **OPEN SOURCE**
MONOLIX SUITE: http://lixoft.com/

Biomathematical Modeling Software

ADAPT II, Biomedical Simulations Resource, USC http://bmsr.usc.edu/
ModelMaker, ModelKinetix, www.modelkinetix.com/
NONMEM, University of California at San Francisco and Globomax Service Group, www.globomaxservice.com/
Stella, High Performance Systems, Inc., www.hps-inc.com/
WinNonlin, Pharsight Corp., www.pharsight.com
SAAM II, SAAM Institute, Inc., www.saam.com

Toxicokinetic Software

ACSL Toxicology Toolkit, AEgis Technologies Group, Inc., www.aegistg.com/
CERTARA PHOENIX WINNONLIN, www.certara.com

Physiologically Based Custom-Designed Software

GastroPlus, Simulations Plus, Inc., www.simulations-plus.com
Pathway Prism, Physiome Sciences, Inc., www.physiome.com
Physiolab, Entelos, Inc., www.entelos.com/

One should definitely examine them to appreciate how easy it has become to create and analyze PK and PD models. If you are interested in further information, literature searches of electronic databases for material on PK/PD modeling should use both the spellings "modeling" and "modelling" because they are used interchangeably.

13.2.2.8 Concentration–Response Relationships

The basic model for ascertaining dose–effect relationships is derived from the sigmoidal effect model, which correlates maximal response (E_{max}), placebo (E_0), and the dose producing 50% of effect (ED_{50}). This is called the Hill model:

$$\text{Effect} = E_0 + \frac{E_{max} \times dose}{ED_{50} + dose} \tag{13.16}$$

The effect is related to PK characteristics through the clearance model wherein the ED_{50} is correlated to the plasma concentration, CL, and bioavailability. Note that EC_{50} is one-half of the concentration found at E_{max} and the ED_{50} results in one-half of the maximum effect (assuming $E_0 = 0$ and that dose is the amount of drug that reaches systemic circulation." ED_{50} is expressed in terms of EC_{50} in the following equation:

$$ED_{50} = \frac{Cl \times EC_{50}}{Bioavailability} \tag{13.17}$$

The significance of the term EC_{50}, a mathematical parameter only, lies in its independence of any PK parameters (other than V_d) or dosage-form factors, and depends only on the type of drug used. On the other hand, ED_{50} is highly dependent on the delivery system. During the development phase, a drug company may study several dosage forms which require development of ED_{50} values for each; on the other hand, EC_{50} is a more fundamental parameter related to potency and toxicity of drug that can be readily obtained from single-dose studies wherein the complete plasma level profile provides essential information. This is in sharp contrast to dose-range studies where at least three doses are administered.

The usual definitions of maximal effect (E_{max}) and potency (EC_{50} or IC_{50}) require another look at this stage before proceeding to developing mathematical relationship between drug concentration, dose, and body functions using a PK/PD approach.[42] Note that while potency is often referred to as EC_{50} or IC_{50}, potency is actually inversely proportional to these quantities.

Potency comprises both action and inhibition of action and is predicted by the Hill model based on dose–response curve; though a 50% level is chosen, it is an arbitrary percentage and other values such as 60 or 40% action can also be calculated and used. Potency is not a relevant factor unless it is so low that the dose requirement is very high (to a level where nonlinear binding with albumin can be observed, resulting in nonlinear kinetics) or where the serious side effects are dose-dependent and make an effective dose unacceptably toxic. The potency, EC_{50}, is expressed in a mechanistic equilibrium model where the action is direct:

$$EC_{50} = \frac{K_d}{(1 + \tau)} \tag{13.18}$$

where K_d is drug affinity, and τ is a transducer constant described as

$$\tau = \frac{R_{total}}{KE} \tag{13.19}$$

Here, R_{total} is the size of receptor pool, and KE is the dissociation constant for the agonist–receptor complex formed; it is a measure of drug-binding properties and physiologic-system response.

Note that low KE means a highly efficient stimulus–response relationship. Mathematically, low KE yields a high value of τ; thus, the term EC_{50} or potency turns out to be dependent on the size of receptor pool, affinity, and the efficiency of binding. How EC_{50} compared to K_d determines whether a drug is efficacy- or affinity-driven agonist. When EC_{50} is much lower than K_d, the drug is efficacy-driven, since in this case $\tau > 1$; an opposite case applies where τ is close to 1 when EC_{50} is lower than K_d (though not too low). Efficacy-driven drugs are little affected when the density of receptor is altered, meaning that there is a change in potency but not in the maximal effect. The affinity-driven drugs, on the other hand, are affected by changes in receptor density that result in changes in maximal response. The affinity-driven drugs have complicated PK/PD relationships and are often difficult to model since highly efficacious agonists are likely to produce maximal response regardless of tissue or species studied, and low-efficacy drugs of this type may interact with only specific tissues and are more likely to show interspecies differences, something that is of great significance in scaling of pharmacologic and clinical effect from one species to another. A lot of work is done on developing good animal species models in the early drug development; an early appreciation of whether a drug is likely to have larger interspecies differences is crucial in reducing the cost of this stage of drug development.

When studying the maximal efficacy (E_{max}), one should not confuse it with potency or intrinsic effects; two drugs with entirely different potency can elicit the same maximal response, except that the dose will be different. This parameter is often difficult or impossible to study experimentally as increasing the dose invariably introduces safety issues. Also, the maximal effect is what is observed and not necessarily what an organ is capable of reaching. The maximal effect (E_{max}) of system is of great importance in PK/PD modeling:

$$E_{max,drug} = \frac{E_{max,system}\tau}{\tau+1} \tag{13.20}$$

Toutain[42] provides an excellent review of these equations (described below). The classical approach in dose selection is dose ranging. This is based on a parallel dose–response design, where subjects are randomly assigned to dose levels, and the response is analyzed using a statistical model where the observed response (Y_{ij}) is in ith subject and jth dose, θ_{ij} is the mean response from jth dose, and ε_{ij} is the error in the observed vs. the expected response from the jth dose:

$$Y_{ij} = \theta_j + \varepsilon_{ij} \tag{13.21}$$

In the crossover design, the drawback is that it is unable to provide information about the shape of the individual dose–response curve. In a crossover design, the "effective dose" is determined by statistical analysis (by testing the null hypothesis) and, therefore, is highly dependent on the power of the study. This results in crossover studies in which sample sizes are small and the errors are large, and which end up with recommendations for higher doses – something the drug companies actually prefer because it is always easy to reduce dosage later without additional, expensive trials. A combination of these two approaches, i.e., dose range and crossover, provides greater robustness to the data obtained.

Besides the continuous effect models, drugs are often fitted to all-or-none models where the responses such as disappearance of arrhythmia in response to a drug dose cannot be graded; the same results hold when we classify the response ending in cure or not, or presence or absence of a given side effect or effects. In such instances, EC_{50} is the median concentration for which half of the subject population is above the threshold and the slope of the curve becomes the variance of the threshold in that population.

Graphic representation of cumulative frequency distribution of selected effects as a function of concentration is also prepared where median effective concentration is considered for drug selectivity using different endpoints.

13.2.3 PHARMACODYNAMIC MODELS

The PD models fall under two categories: graded or quantal of fixed-effect model. Graded refers to a continuous response at different concentrations, whereas the quantal model would evaluate discrete response such as dead or alive, desired or undesired and are almost invariably clinical endpoints.

13.2.3.1 The Hill Model

For a graded response, one can resort to using surrogates, and the classic Hill model (or sigmoid model as described in Equation 13.16) is used to correlate observed effect ($E(t)$) with the concentration modified by an exponent that is called the Hill coefficient; in classic sigmoidal model, the Hill coefficient (h) would be equal to 1:

$$E(t) = E_0 + \frac{E_{max} \times C^h(t)}{EC_{50}^h + \left[C^k(t)\right]} \quad (13.22)$$

The shape of curve is modulated by varying the Hill coefficient to fit the observed behavior. Sensitivity of a drug determines how concentration translates into effect as described by the shape coefficient or the Hill coefficient; at a value of 1, it is a classical parabola. E_0 is generally the baseline or placebo effect; in the event when the response involves an inhibitory effect (IC_{50}), the response is subtracted from observed response and thus the above equation reads as follows:

$$E(t) = E_0 - \frac{E_{max} \times C^h(t)}{IC_{50}^h + C^h(t)} \quad (13.23)$$

where there is full inhibition possible, the term E_0 is replaced by E_{max}, reducing the above equation to a fractional equation:

$$E(t) = E_{max}\left[1 - \frac{c^H(t)}{IC_{50}^h + C^h(t)}\right] \quad (13.24)$$

where the term $\dfrac{c^H(t)}{IC_{50}^h + C^h(t)}$ in the parenthesis is the fraction of maximum effect lost at a given time at any given time and ranges from 1 to 0; the fraction of E_{max} remaining at a given time is the term subtracted from 1 as shown in the parenthesis. This is, therefore, labeled as the fractional Hill model.

The models presented in Equations 13.23 and 13.24 have three types of parameters: the independent parameter of concentration, the dependent parameter of effect (E), and those parameters that are obtained from the observed relationship between E and C: E_0, E_{max}, EC_{50} or IC_{50}, or h.[43,44] The variation of these parameters is pivotal in establishing the relationship between PD and PK properties. When evaluating this relationship of parameters, it is important to differentiate drug action from drug response. The action pertains to mechanism of action and the response or effect – that is, the outcome. For example, the action of aspirin is to block cyclooxygenase, and the effect is reduction in atheroma. Where the intent is to determine a dosing regimen, the response or effect is more important. However, there are instances where it is more relevant to study drug levels at the site of action because a direct relationship is not always established between these levels and the response. As an example, the levels of NSAIDs at the site of action can be a useful parameter since the response is highly variable. The term $C(t)$ in Equations 13.22–13.24 is an independent parameter that represents blood or plasma levels; however, this can well be urinary concentration or concentration in other biological tissue samples.[45,46] For instances where an instantaneous equilibrium between plasma concentration and the biophase or the surrounding tissue is possible, Equation 13.23 in mono-compartment situation would be converted to Equation 13.25 (instantaneous equilibrium model).

An example of instantaneous equilibration between plasma and biophase includes the case of ACE, where equilibration is quickly achieved between plasma and surface of blood vessels:

$$E(t) = E_0 + \frac{E_{max} \left[\dfrac{Dose}{V_c} \exp\left(-\dfrac{CL}{V_c} \times time \right) \right]^h}{EC_{50}^h + \left[\dfrac{Dose}{V_c} \exp\left(-\dfrac{CL}{V_c} \times time \right) \right]^h} \tag{13.25}$$

Notice how Equation 13.25 correlates with Equation 13.16, which represents PK/PD parameters, while Equation 13.25 is a PK parameter-dependent equation. Equation 13.25 introduces another independent parameter, time, allowing determination of both the optimal dose and optimal dosing interval for drugs intended for multiple dosing. Examples of drugs intended for multiple dosing include antibiotics, antihypertensive drugs, and antiepileptic drugs.

Equations 13.22 and 13.25 are intended for drugs where a relationship between drug concentration and effect exists and for these drugs, the profiles are meaningful. However, when concentration is of lesser importance such as where a drug induces a long-term toxicity, a better approach is to use AUC or the time when a threshold concentration is achieved.[47] This can be studied by integrating Equations 13.22 and 13.25 to give a relationship between E and AUC in a Hill model as shown in Equation 13.26:

$$E = E_0 + \frac{E_{max} \times (AUC)^h}{AUC_{50}^h + (AUC)^h} \tag{13.26}$$

where the independent variable, determined by $F \times Dose \times CL$; AUC_{50} is the exposure that yields $E_{max}/2$.

In those instances of time-dependent drugs where concentration is related to a direct but unobserved action, A_{direct}, the observed response is related to the cumulative action of drug through this direct (and primary) action:

$$E_{observed} = \frac{(E_{observed} \times AUC) - A_{direct}}{AUC - A_{direct,50} + AUC - A_{direct}} \tag{13.27}$$

The term $AUC - A_{direct,50}$ is the duration of maximum direct action that produces half the maximal observed response. Where the concentration is much smaller than EC_{50}, dividing the above equation by EC_{50} gives

$$E_{observed} = \frac{E_{observed,max} \times \left(\dfrac{AUC}{EC_{50}} \right)}{\left(\dfrac{AUC}{EC_{50}} \right)_{50} + \left(\dfrac{AUC}{EC_{50}} \right)} \tag{13.28}$$

Now the term AUC/EC_{50} becomes the independent potency parameter that will predict efficacy of concentration-dependent drugs such as antibiotics (ala MIC_{50} or MIC_{90} that predict ordinal effects). The term $(AUC/EC_{50})_{50}$ has dimensions of time.

13.2.3.2 The Receptor Theory Model

The Hill coefficient measures system cooperativity in receptor theory models. When $h > 1$, this means that agonist molecules facilitate binding of subsequent molecules arriving at the receptor. For drugs with low h value, the drug effects are only moderate over a wide range of dosing. This means that the curve is shallow giving measurable responses at low plasma levels – in such cases of low h value, the terminal half-life proves very useful in predicting the duration of effect.[48] On the other hand, where the slope is steep, variations in concentration become more important and around EC_{50} these variations can produce a wide range of effects, from no effect to maximal response. This consideration is more important where the drug has low therapeutic index. Whereas antibiotics can be classified into high and low h using the Hill model, this does not necessarily mean that one class is time-dependent and the other concentration-dependent. More appropriately, PK properties and the concentration–effect profile are integrated where the effect becomes proportionate to time above a critical concentration or AUC; at a high h value of above 5, the concentration range exceeds a threshold, and the graded PD model turns into a quantal model.[49]

The dose–response relationship in a quantal model can be analyzed with the help of a logistic model where we calculate probability of an event at a given concentration, AUC, or dose:

$$\pi_{outcome} = \frac{e^{\alpha+\beta x}}{1+e^{\alpha+\beta x}} \tag{13.29}$$

The exponent α fixes location while β is the slope; the probability of an event *not* happening is simply

$$1-\pi_{outcome} = \frac{1}{1+e^{\alpha+\beta x}} \tag{13.30}$$

A ratio of odds whether an event is likely or not is simply the ratio of odds of happening and odds of not happening:

$$\frac{\pi_{outcome}}{1-\pi_{outcome}} = e^{\alpha+\beta x} \tag{13.31}$$

The logit (L) is the natural logarithm given by

$$L = \alpha + \beta x \tag{13.32}$$

The logistic model and the Hill model differ in that the dependent variable is categorical in the logistic model, though both models use the same parameters and the same underlying model. In the Hill model, the probability of outcome can be calculated from the known concentration C and the unknown median concentration EC_{50}. When EC_{50} (the concentration at which the probability of effect is 50%) is very low, the probability of action is always certain and almost equal to 1, but when EC_{50} is very high compared to C, then it is simply the ratio of C and EC_{50}:

$$\pi_{outcome} = \frac{C_h}{EC_{h,50} + C_h} \tag{13.33}$$

Comparison of Equation 13.33 with Equation 13.29 shows that α is $\ln(EC_{50})^h$, x is $\ln C$, and β corresponds to the Hill coefficient, h, which now represents interpatient variability.

13.2.4 THE HYSTERESIS LOOP PHENOMENON

If the effect is proportional to concentration at all ranges and times, then we will have a simple relationship with effect rising with concentration until it reaches a plateau. The same type of relationship is always observed if we were to sample the drug at the site of action. In real situations, this seldom happens for two reasons. First, the PK profile need not be similar to the PD model due to a lag in drug distribution, for example. Second, the drug response when concentration is rising may not be the same as when drug concentration is falling. These two factors lead to a plot if effect is plotted against concentration that appears like a loop, a hysteresis loop. If an inverse relationship exists in the suppression of an effect, the loop will go clockwise and is termed a proteresis. A proteresis will occur in the rare case of arterial route of administration, or if the site of action equilibrates faster than the plasma concentration, or where there is an accumulation of antagonist.[48] Several interesting correlations are drawn from the appearance of the hysteresis loop phenomenon. If this phenomenon is observed, then an attempt should be made to consider the delay in action. This can be done through mathematical intervention such as volume of distribution estimation since biophase is poorly perfused or protected by a barrier.[49] Also, it can be done by developing understanding of the underlying mechanism such as induction of an enzyme or conversion of drug to an active form. Hysteresis is also observed when the tissue becomes sensitized over time, thereby giving a higher response. When there are changes in the neurotransmitters (increased or decreased) or other such actions where the drug action is indirect, it is better to employ indirect response models.

Therefore, we can conclude that hysteresis can be of PK origin or of PD origin. In the latter case, it is desirable to sample the site of action, if possible. Otherwise, one can resort to hypothetical effect-compartment modeling (see below). Running numerous steady-state experiments does not prove very useful or practical in such situations.

13.2.5 The Effect-Compartment Model

The milieu of effect can be represented by a hypothetical compartment, as if this were a continuous space or a composite of similar spaces; the concentration within the effect compartment can then be related to the observed effect:

$$E(t) = \frac{E_{max} \times Ce(t)}{EC_{50} + [Ce(t)]} \tag{13.34}$$

It is important to understand that $Ce(t)$ is not a measurable concentration, but it is dependent on the distribution of drug from the central compartment (where the drug is administered) and any delays in reaching a certain level of $Ce(t)$ are attributed to this distribution delay. Similarly, the terms E_{max} and EC_{50} can only be obtained indirectly. (If delays in response are not PD effects, this would be more suitably modeled by an indirect action model as in Section 13.2.6). Since the rise in $Ce(t)$ is dependent on the amount of drug entering the effect compartment, the equilibration constant between the effect compartment and the central compartment controls this function:

$$\frac{dA_e}{dt} = \left(K_{1e} \times A_1\right) - \left(K_{e0} \times A_e\right) \tag{13.35}$$

The input function is the product of amount in the central compartment A_1 and the entry rate constant K_{1e}; the output function is given by the amount in the effect compartment A_e and the outward rate constant K_{e0}. Both rate constants are of first order. In most instances, $Ce(t)$ or the amount in the effect compartment is much smaller compared to the amount in the central compartment, and thus, a small error assumption will assume that the drug is eliminated directly from the effect compartment rather than first returning to the central compartment, which, in reality, is the case. This assumption allows simplification of the above equation. Generally, the concentration in the effect compartment is easily described in the traditional PK modeling situations. For example, in a classic single-compartment model,

$$Ce(t) = \frac{K_{1e} \times Dose}{V_e \left(K_{e0} - K_{10}\right)} \left(e^{-K_{1e}t} - e^{K_{e0}t}\right) \tag{13.36}$$

The term V_e refers to the distribution volume of the effect compartment, and thus, the effect compartment becomes synonymous with the central compartment. The above equation contains three unknown parameters, making it impossible to predict the concentration or the effect ($E(t)$) as a function of time; if we define the ratio of concentration in the effect compartment and the central compartment, or the partition coefficient, at equilibrium,

$$K_p = \frac{K_{1e}V_1}{K_{e0}V_e} \tag{13.37}$$

then, this will allow rewriting the above equation as

$$Ce(t) = \frac{Ke_0 \times Dose \times K_p}{V_1(Ke_0 - K_{10})}\left(e^{-K_{10}t} - e^{-Ke_0t}\right) \tag{13.38}$$

In this equation, we now have the term for the volume of central compartment, V_1, instead of the volume term in the effect compartment. If we use the term $Ce(t)/K_p$ rather than $Ce(t)$ in the effect model, we can rewrite the effect as a function of time:

$$E(t) = \frac{E_{max} \times \left[Ce(t)/K_p\right]}{\left(EC_{50}/K_p\right) + \left[Ce(t)/K_p\right]} \tag{13.39}$$

By substituting the value of $Ce(t)$, we get

$$E(t) = \frac{E_{max} \times \dfrac{Ke_0 \times Dose}{V_1(Ke_0 - K_{10})}\left(e^{-K_{10}t} - e^{-Ke_0t}\right)}{\left(EC_{50}/K_p\right) + \left[\dfrac{Ke_0 \times Dose}{V_1(Ke_0 - K_{10})}\left(e^{-K_{10}t} - e^{-Ke_0t}\right)\right]} \tag{13.40}$$

Now we have an equation that can be resolved from PD data over time and having on hand the basic PK parameters. It should be understood that the term EC_{50}/K_p is the EC_{50} that produces $E_{max}/2$ under equilibrium condition; this parameter is of practical use in adjusting dosing regimen. The K_p factor can only be estimated by sampling the biophase, which is generally not possible; in such instance, the plasma concentration serves a clinical endpoint.

The equilibration of drug between plasma and biophase is determined by the parameter K_{e0} and the equilibration half-time $\ln 2/K_{e0}$, and $T_1/2K_{e0}$ is the length of time it takes for plasma concentration and effective concentration to reach equilibrium; this time can range from a few minutes to several hours. For drugs with short half-time, the value of K_{e0} is large, and an equilibration is reached fast, and thus, the plasma concentration is a good indicator of biophase levels; for drugs with low K_{e0} (lower than the terminal elimination-rate constant), an equilibrium will never be reached. The time at which peak concentration is achieved (Te_{max}) is expressed as

$$Te_{max} = \frac{\ln\left(\dfrac{Ke_0}{K_{10}}\right)}{Ke_0 - K_{10}} \tag{13.41}$$

The plasma concentration at Te_{max}, when divided by the dose, gives an estimate of distribution volume that can be used to calculate dosing for fast equilibrating drugs such as anesthetics. This distribution volume term often proves more useful than the traditional central volume of distribution or the steady-state distribution volume.

The term K_{e0} determines how a hysteresis loop closes and thus represents a steady-state–effect relationship.[50,51] Using this parameter, there is no need to predefine a PD model as the nature of the model is decided by the hysteresis loop encountered. One of the most common mistakes encountered in PD modeling is to set E_{max} to 100%. At those instances where the concentration profiles are not well defined or cannot be defined by traditional PKs, one can model the observed concentration into PD models by smoothing the function using traditional linear or cubic-spline functions.[52] Such a case where this would be necessary would be when modeling an endogenous compound such as insulin or erythropoietin, which is administered in a dosage form.

13.2.6 THE INDIRECT ACTION MODELS

When the concentration of a drug cannot be directly related to effect, indirect modeling is suggested. This will most aptly apply to situations where, for example, the response is not directly a result of action (e.g., binding a receptor site) as there may be several steps involved in between the action and response with their own specific mathematical relationships. Unlike the direct action models where any delay in the response is likely a result of PK phenomenon, the delay where indirect models are used is a result of the intrinsic nature of drug action and response relationship. Several models can be used in such instances:

$$\frac{dR}{dt} = K_{in} - \left(K_{out} R \right) \tag{13.42}$$

The rate equation describes the variation in response variable R (with initial value of R_0); the measured response appears at a constant rate (zero order) of K_{in} and is eliminated by K_{out}, the first-order constant. The indirect response models will generally fall into two categories: inhibition or stimulation function. The inhibition response is classically described in terms of IC_{50}, the drug concentration that produces 50% of maximal inhibition, and I_{max}, which is a number from 0 to 1 where 1 represents total inhibition:

$$I(t) = -\frac{I_{max} \times C(t)}{IC_{50} + [C(t)]} \tag{13.43}$$

For example, the equation would be applicable to the action of synthetic glucocorticoid on adrenal glands or effect of furosemide on sodium absorption in the loop of Henley.

Applying this equation in the above response model (Equation 13.42) gives

$$\frac{dR}{dt} = K_{in} - K_{in} \times I(t) - \left(K_{out} \times R \right) \tag{13.44a}$$

Substituting for $I(t)$ yields

$$\frac{dR}{dt} = K_{in} \left\{ 1 - \frac{I_{max} \times C(t)}{IC_{50} + [C(t)]} \right\} - \left(K_{out} \times R \right) \tag{13.44b}$$

Similarly, where a stimulation action (e.g., antipyretic effect of NSAIDS through thermolysis or production of cAMP by bronchodilator beta-2 agonist[53]) is involved, the stimulation response

$$S(t) = \frac{S_{max} \times C(t)}{SC_{50} + [C(t)]} \tag{13.45}$$

when applied to the PD model gives

$$\frac{dR}{dt} = K_{in} \left\{ 1 + \frac{S_{max} \times C(t)}{SC_{50} + [C(t)]} \right\} - \left(K_{out} \times R \right) \tag{13.46}$$

The maximal effect in the models given above occurs later than the time when C_{max} is reached[54] since the drug produces an incremental effect, either inhibition of stimulation, provided the concentration remains above IC_{50} or SC_{50}. How the response returns to its baseline after concentration decay is governed by K_{in} the zero-order input rate, K_{out} the first-order output rate, and the rate of drug elimination from the body. Thus, it is possible to see a persistent drug effect without any detectable drug concentration. The maximum response in indirect model drugs is linearly proportional to the logarithm of dose, contrasting the direct action models where the response E_{max} is a dose-independent parameter[55] as shown in the effect-compartment model. Studies on the shape of concentration–effect response in drugs that act by stimulation show that the rate of drug delivery is crucial and thus sustained or targeted delivery of these drugs would maximize efficacy and reduce the side effects.

13.3   EXAMPLES

The science and art of pharmacokinetic modeling is evolving very fast, as evidenced by novel applications to establish safety and efficacy of new drugs, particularly the biologic drugs.[56] Table 13.4 lists some of the most recent attempts to project PK and PD in combined studies (Table 13.5).

TABLE 13.4
Examples in Literature Where PK/PD Studies Have Been Reported

Apomorphine[57]	Ivabradine[67]
Erythropoietin recombinant[58]	Lorazepam[68]
Digoxin[59]	Cetrizine[69]
Mefloquine[60]	Isepamicin[70]
Telithromycin[61]	Formoterol[71]
Insulin[62]	Furesimide[72]
Angiotensin II receptor antagonist class[63]	Piperacillin-tazobactam[73]
Insulin[64]	Diltiazem[74]
Octreotide acetate[65]	Oxprenolol[75]
Thalidomide[66]	Trimazosin[76,77]

TABLE 13.5
Recent Applications of PK/PD Modeling

Pharmacokinetic-pharmacodynamic modeling in pediatric drug development, and the importance of standardized scaling of clearance.[78]

Precision medicine with imprecise therapy: computational modeling for chemotherapy in breast cancer.[79]

Regulatory perspectives in pharmacometric models of osteoporosis.[80]

Leveraging model-informed approaches for drug discovery and development in the cardiovascular space.[81]

Assessment and modeling of antibacterial combination regimens.[82]

Imipenem-relebactam and meropenem-vaborbactam: two novel carbapenem-β-lactamase inhibitor combinations.[83]

Semi-mechanistic pharmacokinetic-pharmacodynamic modeling of antibiotic drug combinations.[84]

Organ-on-a-chip technology for reproducing multiorgan physiology.[85]

Experimental design and modeling approach to evaluate efficacy of β-lactam/β-lactamase inhibitor combinations.[86]

Levodopa in Parkinson's disease: a review of population pharmacokinetics/pharmacodynamics analysis.[87]

Pharmacokinetic/pharmacodynamic modeling for drug development in oncology.[88]

Population pharmacokinetic analyses of lithium: a systematic review.[89]

Population pharmacokinetic modeling, Monte Carlo simulation and semi-mechanistic pharmacodynamic modeling as tools to personalize gentamicin therapy.[90]

(Continued)

TABLE 13.5 (*Continued*)

Recent Applications of PK/PD Modeling

Animal models in the pharmacokinetic/pharmacodynamic evaluation of antimicrobial agents.[91]

From lead optimization to NDA approval for a new antimicrobial: use of preclinical effect models and pharmacokinetic/pharmacodynamic mathematical modeling.[92]

Understanding and applying pharmacometric modeling and simulation in clinical practice and research.[93]

Clinical pharmacokinetics and pharmacodynamics of biologic therapeutics for treatment of systemic lupus erythematosus.[94]

Pediatric clinical pharmacology of voriconazole: role of pharmacokinetic/pharmacodynamic modeling in pharmacotherapy.[95]

Pharmacology-based toxicity assessment: toward quantitative risk prediction in humans.[96]

Challenges in the clinical assessment of novel tuberculosis drugs.[97]

A comprehensive review of novel drug-disease models in diabetes drug development.[98]

Chronobiology and pharmacologic modulation of the renin-angiotensin-aldosterone system in dogs: what have we learned?[99]

Dosing of rivaroxaban by indication: getting the right dose for the patient.[100]

Clinical development of galunisertib (LY2157299 monohydrate), a small molecule inhibitor of transforming growth factor-beta signaling pathway.[101]

Propofol: a review of its role in pediatric anesthesia and sedation.[102]

The role of extracellular binding proteins in the cellular uptake of drugs: impact on quantitative *in vitro*-to-*in vivo* extrapolations of toxicity and efficacy in physiologically based pharmacokinetic-pharmacodynamic research.[103]

Pharmacokinetic-pharmacodynamic and dose-response relationships of antituberculosis drugs: recommendations and standards for industry and academia.[104]

Integration of PKPD relationships into benefit-risk analysis.[105]

Application of pharmacokinetic-pharmacodynamic modeling and simulation for antibody-drug conjugate development.[106]

Vancomycin pharmacokinetic models: informing the clinical management of drug-resistant bacterial infections.[107]

Making the most of clinical data: reviewing the role of pharmacokinetic-pharmacodynamic models of anti-malarial drugs.[108]

Cancer and comparative imaging.[109]

Clinical pharmacology in neonates: small size, huge variability.[110]

A review of the studies using buprenorphine in cats.[111]

Pharmacokinetic/pharmacodynamic modeling approaches in pediatric infectious diseases and immunology.[112]

Design of optimized hypoxia-activated prodrugs using pharmacokinetic/pharmacodynamic modeling.[113]

Ceftolozane/tazobactam: a novel cephalosporin/β-lactamase inhibitor combination with activity against multidrug-resistant gram-negative bacilli.[114]

Use of pharmacokinetic/pharmacodynamic systems analyses to inform dose selection of tedizolid phosphate.[115]

Pediatric models in motion: requirements for model-based decision support at the bedside.[116]

Pharmacokinetic-pharmacodynamic modeling in anesthesia.[117]

Population pharmacokinetic-pharmacodynamic modeling in oncology: a tool for predicting clinical response.[118]

The p53 protein and its molecular network: modeling a missing link between DNA damage and cell fate.[119]

Animal models of human disease: challenges in enabling translation.[120]

Translation of central nervous system occupancy from animal models: application of pharmacokinetic/pharmacodynamic modeling.[121]

A genetic algorithm based global search strategy for population pharmacokinetic/pharmacodynamic model selection.[122]

Pharmacokinetics and pharmacokinetic-pharmacodynamic correlations of therapeutic peptides.[123]

Moving from basic toward systems pharmacodynamic models.[124]

13.4 CONCLUSIONS AND RECOMMENDATIONS

The integration of PK and PD principles into the drug development process to make it more rational and efficient is highly laudable. This integration relies extensively on PK/PD models to describe the relationships among dose; concentration (and, more generally, exposure); and the responses such as surrogate markers, efficacy measures, and adverse events. Well-documented empirical and physiological PK/PD models are increasingly becoming available.

In addition to the characterization of PK and PD, population PK/PD models involve relationships between covariates (i.e., patient characteristics) and PK/PD parameters, and allow us to assess and quantify potential sources of variability in exposure and response in specific target population, even under erratic and limited sampling conditions. Often implications of significant covariate effects can be evaluated by computer simulations using the population PK/PD model. Stochastic simulation is widely used as a tool for evaluating statistical methodology, including, for example, the evaluation of performance measures for bioequivalence assessment. Recent trends indicate expanded use of simulations to support drug development for predicting the outcome of planned clinical trials. The methodological basis for this approach is provided by population PK/PD models and random sampling techniques. Models for disease progression and behavioral features such as compliance, drop-out rates, adverse event-dependent dose reductions, and the like, have to be added to population PK/PD models to simulate real situations. Use of computer simulation helps to evaluate the consequences of design features on the safety and efficacy assessment of a drug, enabling scientists and regulators to identify statistically valid and practical study designs.

REFERENCES

1. Csajka, C. and Verotta, D., Pharmacokinetic-pharmacodynamic modelling: History and perspectives, *J. Pharmacokinet. Pharmacodyn.*, 33, 227, 2006.
2. Food and Drug Administration, Development of new stereoisomeric drugs, United States Food and Drug Administration, Washington, DC, 1992, Available at https://www.fda.gov/regulatory-information/search-fda-guidance-documents/development-new-stereoisomeric-drugs
3. Temple, R.J., A regulatory authority's opinion about surrogate endpoints, in *Clinical Measurement in Drug Evaluation*, Nimmo, W.S. and Tucker, G.T., Eds., Wiley, Indianapolis, IN, 1995, pp. 1–22.
4. Lesko, L.J. and Atkinson, A.J., Jr., Biomarkers and surrogate endpoints: Use in drug development and regulatory decision making: Criteria, validation, strategies, *Ann. Rev. Pharmacol. Toxicol.*, 41, 347–366, 2001.
5. Toutain, P.L., Pharmacokinetic/pharmacodynamic integration in drug development and dosage-regimen optimization for veterinary medicine, *AAPS PharmSci.*, 4, E38, 2002.
6. Gieschke, R. and Steimer, J.L., Pharmacometrics: Modelling and simulation tools to improve decision making in clinical drug development, *Eur. J. Drug Metab. Pharmacokinet.*, 25, 49–58, 2000.
7. Wakefield, J.C. and Bennett, J.E., The Bayesian modeling of covariates for population pharmacokinetic models, *J. Am. Statist. Ass.*, 91, 917–927, 1996.
8. Machado, S.G., Miller, R., and Hu, C., A regulatory perspective on pharmacokinetic/pharmacodynamic modeling, *Stat. Methods Med. Res.*, 8, 217–245, 1999.
9. Sheiner, L.B. and Steimer, J.L., Pharmacokinetic/pharmacodynamic modeling in drug development, *Ann. Rev. Pharmacol. Toxicol.*, 40, 67–95, 2000.
10. Aarons, L., Population pharmacokinetics: Theory and practice, *Br. J. Clin. Pharmacol.*, 32, 669–670, 1991.
11. Peck, C.C., Barr, W.H., Benet, L.Z., Collins, J., Desjardins, R.E., Furst, D.E., Harter, J.G., Levy, G., Ludden, T., and Rodman, J.H., Opportunities for integration of pharmacokinetics, pharmacodynamics, and toxicokinetics in rational drug development, *Int. J. Pharm.*, 82, 9–19, 1992.
12. Niazi, S., Volume of distribution and tissue level errors in instantaneous intravenous input assumptions, *J. Pharm. Sci.*, 65, 1539–1540, 1976.
13. Teorell, T., Kinetics of distribution of substances administered to the body. I: The extravascular modes of administration, *Arch. Intern. Pharmacodyn.*, 57, 205–225, 1937.
14. Teorell, T., Kinetics of distribution of substances administered to the body. II: The intravascular mode of administration, *Arch. Intern. Pharmacodyn.*, 57, 226–240, 1937.
15. Bischoff, K.B. and Brown, R.G., Drug distribution in mammals, *Chem. Eng. Prog. Symp.*, 62, 33–45, 1966.
16. Levitt, D.G., PKQuest: Volatile solutes—Application to enflurane, nitrous oxide, halothane, methoxyflurane and toluene pharmacokinetics, *BMC Anesthesiol.*, 2, 5, 2002.
17. Levitt, D.G., PKQuest: Measurement of intestinal absorption and first pass metabolism application to human ethanol pharmacokinetics, *BMC Clin. Pharmacol.*, 2, 4, 2002.
18. Levitt, D.G., PKQuest: A general physiologically based pharmacokinetic model—Introduction and application to propranolol, *BMC Clin. Pharmacol.*, 2, 5, 2002.

19. Levitt, D.G., PKQuest: Capillary permeability limitation and plasma protein binding—Application to human inulin, dicloxacillin and ceftriaxone pharmacokinetics, *BMC Clin. Pharmacol.*, 2, 7, 2002.

20. Cai, W. and Shao, X., A fast annealing evolutionary algorithm for global optimization, *J. Comput. Chem.*, 23, 427–435, 2002.

21. Department of Mathematical Sciences, Maple System, Villanova University, Villanova, PA, 2003, Available at https://www1.villanova.edu/villanova/artsci/mathematics/resources-and-opportunities/maple.html

22. Levenberg, K.A., Method for the solution of certain problems in least squares, *Q. Appl. Math.*, 2, 164–168, 1944.

23. Marquardt, D., An algorithm for least-squares estimation of non-linear parameters, *SIAM J. Appl. Math.*, 11, 431–441, 1963.

24. Levitt, D.G., The use of a physiologically based pharmacokinetic model to evaluate deconvolution measurements of systemic absorption, *BMC Clin. Pharmacol.*, 3, 1, 2003. doi:10.1186/1472-6904-3-1.

25. De Nicolao, G., Sparacino, G., and Cobelli, C., Nonparametric input estimation in physiological systems, problems, methods, and case studies, *Automatica*, 33, 851–870, 1997.

26. Debord, J., Risco, E., Harel, M., Le Meur, Y., Buchler, M., Lachatre, G., Le Guellec, C., and Marquet, P., Application of a gamma model of absorption to oral cyclosporin, *Clin. Pharmacokinet.*, 40, 375–382, 2001.

27. Veng-Pedersen, P., Linear and nonlinear system approaches in pharmacokinetics: How much do they have to offer? I. General considerations, *J. Pharmacokinet. Biopharm.*, 16, 413–472, 1988.

28. Akaike, H., A new look at the statistical model identification, *IEEE Trans. Automat. Contr.*, 19, 716–723, 1974.

29. Craven, P. and Wahba, G., Smoothing noisy data with spline functions, *Num. Math.*, 31, 377–403, 1979.

30. Gillespie, W.R. and Veng-Pedersen, P., A polyexponential deconvolution method: Evaluation of the gastrointestinal bioavailability and mean in vivo dissolution time of some ibuprofen dosage forms, *J. Pharmacokinet. Biopharm.*, 13, 289–307, 1985.

31. Verotta, D., Concepts, properties, and applications of linear systems to describe distribution, identify input, and control endogenous substances and drugs in biological systems, *Crit. Rev. Biomed. Eng.*, 24, 73–139, 1996.

32. Verotta, D., Estimation and model selection in constrained deconvolution, *Ann. Biomed. Eng.*, 21, 605–620, 1993.

33. Sparacino, G. and Cobelli, C., Deconvolution of physiological and pharmacokinetic data: Comparison of algorithms on benchmark problems, in *Modeling and Control in Biomedical Systems*, Linkens, D.A. and Carson, E., Eds., Elsevier, Oxford, 1997, pp. 151–153.

34. Figg, W.D., Dukes, G.E., Lesesne, H.R., Carson, S.W., Songer, S.S., Pritchard, J.F., Hermann, D.J., Powell, J.R., and Hak, L.J., Comparison of quantitative methods to assess hepatic function: Pugh's classification, indocyanine green, antipyrine, and dextromethorphan, *Pharmacotherapy*, 15, 693–700, 1995.

35. Testa, R., Caglieris, S., Risso, D., Arzani, L., Campo, N., Alvarez, S., Giannini, E., Lantieri, P.B., and Celle, G., Monoethylglycinexylidide formation measurement as a hepatic function test to assess severity of chronic liver disease, *Am. J. Gastroenterol.*, 92, 2268–2273, 1997.

36. Tang, H.-S. and Hu, O.Y.-P., Assessment of liver function using a novel galactose single point method, *Digestion*, 52, 222–231, 1992.

37. Zakim, D. and Boyer, T.D., *Hepatology: A Textbook of Liver Disease*, W.B. Saunders Co., Philadelphia, PA, 1996.

38. Pugh, R.N., Murray-Lyon, I.M., Dawson, J.L., Pietroni, M.C., and Williams, R., Transection of the oesophagus for bleeding oesophageal varices, *Br. J. Surg.*, 60, 646–649, 1973.

39. Wiesner, R.H., Grambsch, P.M., Dickson, E.R., Ludwig, J., MacCarty, R.L., Hunter, E.B., Fleming, T.R., Fisher, L.D., Beaver, S.J., and LaRusso, N.F., Primary sclerosing cholangitis: Natural history, prognostic factors and survival analysis, *Hepatology*, 10, 430–436, 1989.

40. Carithers, R.L., Jr., Herlong, H.F., Diehl, A.M., Shaw, E.W., Combes, B., Fallon, H.J., and Maddrey, W.C., Methylprednisolone therapy in patients with severe alcoholic hepatitis, *Ann. Intern. Med.*, 110, 685–690, 1989.

41. Maddrey, W.C., Boitnott, J.K., Bedine, M.S., Weber, F.L., Jr., Mezey, E., and White, R.I., Jr., Corticosteroid therapy of alcoholic hepatitis, *Gastroenterology*, 75, 193–199, 1978.

42. Toutain, P.L., Pharmacokinetic/pharmacodynamic integration in drug development and dosage-regimen optimization for veterinary medicine, *AAPS PharmSci.*, 4, 1–29, 2002.

43. Holford, N.H. and Sheiner, L.B., Pharmacokinetic and pharmacodynamic modeling in vivo, *Crit. Rev. Bioeng.*, 5, 273–322, 1981.

44. Holford, N.H. and Sheiner, L.B., Kinetics of pharmacologic response, *Pharmacol Ther.*, 16, 143–166, 1982.
45. Hammarlund-Udenaes, M. and Benet, L.Z., Furosemide pharmacokinetics and pharmacodynamics in health and disease: An update, *J. Pharmacokinet. Biopharm.*, 17, 1–46, 1989.
46. Landoni, M.F. and Lees, P., Pharmacokinetic/pharmacodynamic modeling of non-steroidal anti-inflammatory drugs, *J. Vet. Pharmacol. Ther.*, 20, 118–120, 1997.
47. Karlsson, M.O., Molnar, V., Bergh, J., Freijs, A., and Larsson, R., A general model for time–dissociated pharmacokinetic–pharmacodynamic relationships exemplified by paclitaxel myelosuppression, *Clin. Pharmacol. Ther.*, 63, 11–25, 1998.
48. Campbell, D.B., The use of kinetic-dynamic interactions in the evaluation of drugs, *Psychopharmacology*, 100, 433–450, 1990.
49. Mattie, H., Antibiotic efficacy in vivo predicted by in vitro activity, *Int. J. Antimicrob. Agents*, 14, 91–98, 2000.
50. Mager, D.E. and Jusko, W.J., Pharmacodynamic modeling of time-dependent transduction systems, *Clin. Pharmacol. Ther.*, 70, 210–216, 2001.
51. Fuseau, E. and Sheiner, L.B., Simultaneous modeling of pharmacokinetics and pharmacodynamics with a nonparametric pharmacodynamic model, *Clin. Pharmacol. Ther.*, 35, 733–741, 1984.
52. Sheiner, L.B. and Beal, S.L., Bayesian individualisation of pharmacokinetics: Simple implementation and comparison with non-Bayesian methods, *J. Pharm. Sci.*, 71, 1344–1348, 1982.
53. Jusko, W.J. and Ko, H.C., Physiologic indirect response models characterize diverse types of pharmacodynamic effects, *Clin. Pharmacol. Ther.*, 56, 406–419, 1994.
54. Sharma, A. and Jusko, W.J., Characterization of four basic models of indirect pharmacodynamic responses, *J. Pharmacokinet. Biopharm.*, 24, 611–635, 1996.
55. Gobburu, J.V. and Jusko, W.J., Role of dosage regimen in controlling indirect pharmacodynamic responses, *Adv. Drug Deliv. Rev.*, 46, 45–57, 2001.
56. Levy, G., Mechanism-based pharmacodynamic modeling, *Clin. Pharmacol. Ther.*, 56, 356–358, 1994.
57. Aymard, G., Berlin, I., de Brettes, B., and Diquet, B., Pharmacokinetic–pharmacodynamic study of apomorphine's effect on growth hormone secretion in healthy subjects, *Fundam. Clin. Pharmacol.*, 17, 473–481, 2003.
58. Varlet-Marie, E., Gaudard, A., Audran, M., Gomeni, R., and Bressolle, F., Pharmacokinetic–pharmacodynamic modeling of recombinant human erythropoietin in athletes, *Int. J. Sports Med.*, 24, 252–257, 2003.
59. Hornestam, B., Jerling, M., Karlsson, M.O., and Held, P.; DAAf Trial Group, Intravenously administered digoxin in patients with acute atrial fibrillation: A population pharmacokinetic/pharmacodynamic analysis based on the digitalis in acute atrial fibrillation trial, *Eur. J. Clin. Pharmacol.*, 58, 747–755, 2003.
60. Svensson, U.S., Alin, H., Karlsson, M.O., Bergqvist, Y., and Ashton, M., Population pharmacokinetic and pharmacodynamic modeling of artemisinin and mefloquine enantiomers in patients with falciparum malaria, *Eur. J. Clin. Pharmacol.*, 58, 339–351, 2002.
61. Nicolau, D.P., Pharmacodynamic rationale for short-duration antibacterial therapy, *J. Infect.*, 44, 17–23, 2002.
62. Lin, S. and Chien, Y.W., Pharmacokinetic–pharmacodynamic modeling of insulin: Comparison of indirect pharmacodynamic response with effect-compartment link models, *J. Pharm. Pharmacol.*, 54, 791–800, 2002.
63. Csajka, C., Buclin, T., Fattinger, K., Brunner, H.R., and Biollaz, J., Population pharmacokinetic–pharmacodynamic modelling of angiotensin receptor blockade in healthy volunteers, *Clin. Pharmacokinet.*, 41, 137–152, 2002.
64. Miyazaki, M., Mukai, H., Iwanaga, K., Morimoto, K., and Kakemi, M., Pharmacokinetic–pharmacodynamic modeling of human insulin: Validity of pharmacological availability as a substitute for extent of bioavailability, *J. Pharm. Pharmacol.*, 53, 1235–1246, 2001.
65. Zhou, H., Chen., T.L., Marino, M., Lau, H., Miller, T., Kalafsky, G., and McLeod, J.F., Population PK and PK/PD modeling of microencapsulated octreotide acetate in healthy subjects, *Br. J. Clin. Pharmacol.*, 50, 543–552, 2000.
66. Eriksson, T., Bjorkman, S., Roth, B., and Hoglund, P., Intravenous formulations of the enantiomers of thalidomide: Pharmacokinetic and initial pharmacodynamic characterization in man, *J. Pharm. Pharmacol.*, 52, 807–817, 2000.
67. Duffull, S.B. and Aarons, L., Development of a sequential linked pharmacokinetic and pharmacodynamic simulation model for ivabradine in healthy volunteers, *Eur. J. Pharm. Sci.*, 10, 275–284, 2000.

68. Blin, O., Jacquet, A., Callamand, S., Jouve, E., Habib, M., Gayraud, D., Durand, A., Bruguerolle, B., and Pisano, P., Pharmacokinetic–pharmacodynamic analysis of mnesic effects of lorazepam in healthy volunteers, *Br. J. Clin. Pharmacol.*, 48, 510–512, 1999.

69. Urien, S., Tillement, J.P., Ganem, B., and Kuch, M.D., A pharmacokinetic–pharmacodynamic modeling of the antihistaminic (H_1) effects of cetirizine, *Int. J. Clin. Pharmacol. Ther.*, 37, 499–502, 1999.

70. Tod, M., Minozzi, C., Beaucaire, G., Ponsonnet, D., Cougnard, J., and Petitjean, O., Isepamicin in intensive care unit patients with nosocomial pneumonia: Population pharmacokinetic–pharmacodynamic study, *J. Antimicrob. Chemother.*, 44, 99–108, 1999.

71. van den Berg, B.T., Derks, M.G., Koolen, M.G., Braat, M.C., Butter, J.J., and van Boxtel, C.J., Pharmacokinetic/pharmacodynamic modeling of the eosinopenic and hypokalemic effects of formoterol and theophylline combination in healthy men, *Pulm. Pharmacol. Ther.*, 12, 185–192, 1999.

72. Wakelkamp, M., Alvan, G., and Paintaud, G., The time of maximum effect for model selection in pharmacokinetic–pharmacodynamic analysis applied to frusemide, *Br. J. Clin. Pharmacol.*, 45, 63–70, 1998.

73. Dalla Costa, T., Nolting, A., Rand, K., and Derendorf, H., Pharmacokinetic–pharmacodynamic modeling of the *in vitro* antiinfective effect of piperacillin-tazobactam combinations, *Int. J. Clin. Pharmacol. Ther.*, 35, 426–433, 1997.

74. Luckow, V. and Della Paschoa, O., PK/PD modeling of high-dose diltiazem—absorption-rate dependency of the hysteresis loop, *Int. J. Clin. Pharmacol. Ther.*, 35, 418–425, 1997.

75. Koopmans, R., Oosterhuis, B., Karemaker, J.M., Wemer, J., and van Boxtel, C.J., Pharmacokinetic–pharmacodynamic modeling of oxprenolol in man using continuous noninvasive blood pressure monitoring, *Eur. J. Clin. Pharmacol.*, 34, 395–400, 1988.

76. Kelman, A.W., Meredith, P.A., Elliott, H.L., and Reid, J.L., Modeling the pharmacokinetics and pharmacodynamics of trimazosin, *Biopharm. Drug Dispos.*, 7, 373–388, 1986.

77. Meredith, P.A., Kelman, A.W., Elliott, H.L., and Reid, J.L., Pharmacokinetic and pharmacodynamic modeling of trimazosin and its major metabolite, *J. Pharmacokinet. Biopharm.*, 11, 323–325, 1983.

78. Germovsek, E., Barker, C.I.S., Sharland, M., and Standing, J.F., Pharmacokinetic-pharmacodynamic modeling in pediatric drug development, and the importance of standardized scaling of clearance. *Clin. Pharmacokinet.*, 58, 39, 2019.

79. McKenna, M.T., Weis, J.A., Brock, A., Quaranta, V., and Yankeelov, T.E., Precision medicine with imprecise therapy: Computational modeling for chemotherapy in breast cancer, *Transl. Oncol.*, 11(3), 732–742, 2018 April 16.

80. Madrasi, K., Li, F., Kim, M.J., Samant, S., Voss, S., Kehoe, T., Bashaw, E.D., Ahn, H.Y., Wang, Y., Florian, J., Schmidt, S., Lesko, L.J., and Li, L., Regulatory perspectives in pharmacometric models of osteoporosis, *J. Clin. Pharmacol.*, 58(5), 572–585, 2018 May.

81. Dockendorf, M.F., Vargo, R.C., Gheyas, F., Chain, A.S.Y., Chatterjee, M.S., and Wenning, L.A., Leveraging model-informed approaches for drug discovery and development in the cardiovascular space, *J. Pharmacokinet. Pharmacodyn.*, 45(3), 355–364, 2018.

82. Rao, G.G., Li, J., Garonzik, S.M., Nation, R.L., and Forrest, A., Assessment and modelling of antibacterial combination regimens, *Clin. Microbiol. Infect.*, 2017 December 18, pii:S1198-743X(17)30677-8.

83. Zhanel, G.G., Lawrence, C.K., Adam, H., Schweizer, F., Zelenitsky, S., Zhanel, M., Lagacé-Wiens, P.R.S., Walkty, A., Denisuik, A., Golden, A., Gin, A.S., Hoban, D.J., Lynch, J.P., III, and Karlowsky, J.A., Imipenem-relebactam and meropenem-vaborbactam: Two novel carbapenem-β-lactamase inhibitor combinations, *Drugs*, 78(1), 65–98, 2018 January.

84. Brill, M.J.E., Kristoffersson, A.N., Zhao, C., Nielsen, E.I., and Friberg, L.E., Semi-mechanistic pharmacokinetic-pharmacodynamic modelling of antibiotic drug combinations. *Clin. Microbiol. Infect.*, 2017 December 8, pii:S1198-743X(17)30670-5.

85. Lee, S.H. and Sung, J.H., Organ-on-a-chip technology for reproducing multiorgan physiology. *Adv. Healthc. Mater.*, 7(2), 2018 January, doi:10.1002/adhm.201700419.

86. Sy, S.K.B. and Derendorf, H., Experimental design and modelling approach to evaluate efficacy of β-lactam/β-lactamase inhibitor combinations, *Clin. Microbiol. Infect.*, 2017 July 29, pii:S1198-743X(17)30404-4.

87. Marsot, A., Guilhaumou, R., Azulay, J.P., and Blin, O., Levodopa in Parkinson's disease: A review of population pharmacokinetics/pharmacodynamics analysis, *J. Pharm. Pharm. Sci.*, 20(0), 226–238, 2017.

88. Garralda, E., Dienstmann, R., and Tabernero, J., Pharmacokinetic/pharmacodynamic modeling for drug development in oncology, *Am. Soc. Clin. Oncol. Educ. Book*, 37, 210–215, 2017.

89. Methaneethorn, J., Population pharmacokinetic analyses of lithium: A systematic review, *Eur. J. Drug Metab. Pharmacokinet.*, 43(1), 25–34, 2018 February.

90. Llanos-Paez, C.C., Hennig, S., and Staatz, C.E., Population pharmacokinetic modelling, Monte Carlo simulation and semi-mechanistic pharmacodynamic modelling as tools to personalize gentamicin therapy, *J. Antimicrob. Chemother.*, 72(3), 639–667, 2017 March 1.

91. Zhao, M., Lepak, A.J., and Andes, D.R., Animal models in the pharmacokinetic/pharmacodynamic evaluation of antimicrobial agents, *Bioorg. Med. Chem.*, 24(24), 6390–6400, 2016 December 15.

92. Drusano, G.L., From lead optimization to NDA approval for a new antimicrobial: Use of pre-clinical effect models and pharmacokinetic/pharmacodynamic mathematical modeling, *Bioorg. Med. Chem.*, 24(24), 6401–6408, 2016 December 15.

93. Standing, J.F., Understanding and applying pharmacometric modelling and simulation in clinical practice and research, *Br. J. Clin. Pharmacol.*, 83(2), 247–254, 2017 February.

94. Yu, T., Enioutina, E.Y., Brunner, H.I., Vinks, A.A., and Sherwin, C.M., Clinical pharmacokinetics and pharmacodynamics of biologic therapeutics for treatment of systemic lupus erythematosus, *Clin. Pharmacokinet.*, 56(2), 107–125, 2017 February.

95. Kadam, R.S. and Van Den Anker, J.N., Pediatric clinical pharmacology of voriconazole: Role of pharmacokinetic/pharmacodynamic modeling in pharmacotherapy, *Clin. Pharmacokinet.*, 55(9), 1031–1043, 2016 September.

96. Sahota, T., Danhof, M., and Della Pasqua, O., Pharmacology-based toxicity assessment: Towards quantitative risk prediction in humans, *Mutagenesis*, 31(3), 359–374, 2016 May.

97. Dooley, K.E., Phillips, P.P., Nahid, P., and Hoelscher, M., Challenges in the clinical assessment of novel tuberculosis drugs. *Adv. Drug Deliv. Rev.*, 102, 116–122, 2016 July 1.

98. Gaitonde, P., Garhyan, P., Link, C., Chien, J.Y., Trame, M.N., and Schmidt, S., A comprehensive review of novel drug-disease models in diabetes drug development, *Clin. Pharmacokinet.*, 55(7), 769–788, 2016 July.

99. Mochel, J.P. and Danhof, M., Chronobiology and pharmacologic modulation of the renin-angiotensin-aldosterone system in dogs: What have we learned? *Rev. Physiol. Biochem. Pharmacol.*, 169, 43–69, 2015.

100. Escolar, G., Carne, X., and Arellano-Rodrigo, E., Dosing of rivaroxaban by indication: Getting the right dose for the patient, *Expert Opin. Drug Metab. Toxicol.*, 11(10), 1665–1677, 2015.

101. Herbertz, S., Sawyer, J.S., Stauber, A.J., Gueorguieva, I., Driscoll, K.E., Estrem, S.T., Cleverly, A.L., Desaiah, D., Guba, S.C., Benhadji, K.A., Slapak, C.A., and Lahn, M.M., Clinical development of galunisertib (LY2157299 monohydrate), a small molecule inhibitor of transforming growth factor-beta signaling pathway, *Drug Des. Devel. Ther.*, 9, 4479–4499, 2015 August 10.

102. Chidambaran, V., Costandi, A., and D'Mello, A., Propofol: A review of its role in pediatric anesthesia and sedation, *CNS Drugs*, 29(7), 543–563, 2015 July.

103. Poulin, P., Burczynski, F.J., and Haddad, S., The role of extracellular binding proteins in the cellular uptake of drugs: Impact on quantitative in vitro-to-in vivo extrapolations of toxicity and efficacy in physiologically based pharmacokinetic-pharmacodynamic research, *J. Pharm. Sci.*, 105(2), 497–508, 2016 February.

104. Gumbo, T., Angulo-Barturen, I., and Ferrer-Bazaga, S., Pharmacokinetic-pharmacodynamic and dose-response relationships of antituberculosis drugs: Recommendations and standards for industry and academia, *J. Infect. Dis.*, 211(Suppl. 3), S96–S106, 2015 June 15.

105. Bellanti, F., van Wijk, R.C., Danhof, M., and Della Pasqua, O., Integration of PKPD relationships into benefit-risk analysis, *Br. J. Clin. Pharmacol.*, 80(5), 979–991, 2015 November.

106. Singh, A.P., Shin, Y.G., and Shah, D.K., Application of pharmacokinetic-pharmacodynamic modeling and simulation for antibody-drug conjugate development, *Pharm. Res.*, 32(11), 3508–3525, 2015 November.

107. Stockmann, C., Roberts, J.K., Yu, T., Constance, J.E., Knibbe, C.A., Spigarelli, M.G., and Sherwin, C.M., Vancomycin pharmacokinetic models: Informing the clinical management of drug-resistant bacterial infections, *Expert Rev. Anti. Infect. Ther.*, 12(11), 1371–1388, 2014 November.

108. Simpson, J.A., Zaloumis, S., DeLivera, A.M., Price, R.N., and McCaw, J.M., Making the most of clinical data: Reviewing the role of pharmacokinetic-pharmacodynamic models of anti-malarial drugs, *AAPS J.*, 16(5), 962–974, 2014 September.

109. LeBlanc, A.K., Cancer and comparative imaging, *ILAR J.*, 55(1), 164–168, 2014.

110. Allegaert, K. and van den Anker, J.N., Clinical pharmacology in neonates: Small size, huge variability, *Neonatology*, 105(4), 344–349, 2014.

111. Steagall, P.V., Monteiro-Steagall, B.P., and Taylor, P.M., A review of the studies using buprenorphine in cats, *J. Vet. Intern. Med.*, 28(3), 762–770, 2014 May–June.

112. Barker, C.I., Germovsek, E., Hoare, R.L., Lestner, J.M., Lewis, J., and Standing, J.F., Pharmacokinetic/pharmacodynamic modelling approaches in paediatric infectious diseases and immunology, *Adv. Drug Deliv. Rev.*, 73, 127–139, 2014 June.

113. Foehrenbacher, A., Secomb, T.W., Wilson, W.R., and Hicks, K.O., Design of optimized hypoxia-activated prodrugs using pharmacokinetic/pharmacodynamic modeling, *Front Oncol.*, 3, 314, 2013 December 27.

114. Zhanel, G.G., Chung, P., Adam, H., Zelenitsky, S., Denisuik, A., Schweizer, F., Lagacé-Wiens, P.R., Rubinstein, E., Gin, A.S., Walkty, A., Hoban, D.J., Lynch, J.P., III, and Karlowsky, J.A., Ceftolozane/tazobactam: A novel cephalosporin/β-lactamase inhibitor combination with activity against multidrug-resistant gram-negative bacilli, *Drugs*, 74(1), 31–51, 2014 January.

115. Lodise, T.P. and Drusano, G.L., Use of pharmacokinetic/pharmacodynamic systems analyses to inform dose selection of tedizolid phosphate, *Clin. Infect. Dis.*, 58(Suppl. 1), S28–S34, 2014 January.

116. Barrett, J.S., Paediatric models in motion: Requirements for model-based decision support at the bedside, *Br. J. Clin. Pharmacol.*, 79(1), 85–96, 2015 January.

117. Gambús, P.L. and Trocóniz, I.F., Pharmacokinetic-pharmacodynamic modelling in anaesthesia, *Br. J. Clin. Pharmacol.*, 79(1), 72–84, 2015 January.

118. Bender, B.C., Schindler, E., and Friberg, L.E., Population pharmacokinetic-pharmacodynamic modelling in oncology: A tool for predicting clinical response, *Br. J. Clin. Pharmacol.*, 79(1), 56–71, 2015 January.

119. Eliaš, J., Dimitrio, L., Clairambault, J., and Natalini, R., The p53 protein and its molecular network: Modelling a missing link between DNA damage and cell fate, *Biochim. Biophys. Acta*, 1844(1 Pt. B), 232–247, 2014 January.

120. McGonigle, P. and Ruggeri, B., Animal models of human disease: Challenges in enabling translation, *Biochem. Pharmacol.*, 87(1), 162–171, 2014 January 1.

121. Melhem, M., Translation of central nervous system occupancy from animal models: Application of pharmacokinetic/pharmacodynamic modeling, *J. Pharmacol. Exp. Ther.*, 347(1), 2–6, 2013 October.

122. Sale, M. and Sherer, E.A., A genetic algorithm based global search strategy for population pharmacokinetic/pharmacodynamic model selection, *Br. J. Clin. Pharmacol.*, 79(1), 28–39, 2015 January.

123. Dio, L. and Meibohm, B., Pharmacokinetics and pharmacokinetic-pharmacodynamic correlations of therapeutic peptides, *Clin. Pharmacokinet.*, 52(10), 855–868, 2013 October.

124. Jusko, W.J. Moving from basic toward systems pharmacodynamic models, *J. Pharm. Sci.*, 102(9), 2930–2940, 2013 September.

14 The Evolving Role of the Caco-2 Cell Model to Estimate Intestinal Absorption Potential and Elucidate Transport Mechanisms

Jibin Li and Ismael J. Hidalgo
Absorption Systems

CONTENTS

14.1 INTRODUCTION

For most therapeutic indications, oral ingestion represents the preferred method of drug administration because of its convenience and patient compliance. The intestinal epithelium constitutes a selective permeability barrier between the blood and the environment. It is composed of a single layer of heterogeneous cells, which include enterocytes or absorptive cells, undifferentiated (crypt) cells, endocrine cells, and goblet cells.

The absorptive cells originate in the villus crypt as undifferentiated cells. As undifferentiated cells move along the crypt–villus axis, they undergo enterocytic differentiation, developing a tall, columnar appearance, microvilli, and tight junctions that connect adjacent cells to form a cell monolayer, which behaves both as a physical and biochemical barrier to intestinal drug absorption. The tight junctions, an important component of the physical barrier, prevent free transepithelial drug diffusion via the paracellular route. Since paracellular permeation represents a minor transepithelial transport pathway, drugs that undergo predominantly paracellular transport will exhibit a low-to-moderate extent of absorption.[1,2] The transcellular route of permeability is accessible to compounds that display suitable physicochemical properties. Since this type of compound can diffuse passively across the intestinal epithelial cell membrane, they often achieve a high extent of absorption.[3,4]

Passive transcellular transport across the intestinal epithelium involves three discrete steps: (a) uptake across the apical membrane, (b) diffusion through the cytoplasm, and (c) efflux across the basolateral membrane. Occasionally, drug molecules without favorable physicochemical properties to undergo passive transcellular diffusion can traverse the intestinal epithelium aided by membrane transporters.[5–7] In addition to the permeability resistance of the intestinal mucosa, the numerous drug-metabolizing enzymes (e.g., CYP3A4, CYP2C9) and efflux transporters (e.g., Pgp and BCRP) represent a formidable biochemical barrier.[8]

In the recent past, technological advances in combinatorial chemistry and high-throughput screening (HTS) permitted medicinal chemists to synthesize and test large numbers of chemicals in a short time in an attempt to accelerate the identification of lead drug candidates. During the late 1980s and early 1990s, pharmaceutical scientists began to use both primary cell cultures and cell lines for drug-transport studies. Among the various cell lines established, Caco-2, derived from human colon adenocarcinoma, gained the most acceptance because it undergoes spontaneous differentiation under normal culture conditions.[9,10] Caco-2 cells differentiation is characterized by the development of enterocyte-like morphological structures such as tight junctions and microvilli.[9–11] In addition to expressing drug-metabolizing enzymes (e.g., aminopeptidases, esterases, sulfatases, and cytochrome P450), they also express several uptakes (e.g., ASBT, PEPT1, MCT1, OATP2B1) and efflux (e.g., Pgp, BCRP, MRP2, and MRP3, and OSTαβ) transporters.[12–23]

14.2   CACO-2 CELLS AS AN INTESTINAL PERMEABILITY MODEL

The main objective of earlier studies with Caco-2 cells was to evaluate the utility of this system in the prediction of the absorption potential of new molecular entities. First, there was a great deal of interest in validating Caco-2 cells as an in vitro model of intestinal permeability. A number of studies determined the permeability coefficients in Caco-2 monolayers of a relatively large number of drugs and evaluated the correlation between in vitro apparent permeability coefficient (P_{app}) values and extent of absorption (F_{abs}) in humans.[24–26] A strong correlation between in vitro P_{app} and F_{abs} in humans is extremely valuable because it demonstrates the utility of in vitro P_{app} values to predict the absorption potential of drug candidates. Beyond an interest in proving the ability of this in vitro model to predict in vivo absorption, some efforts have sought to compare various aspects of Caco-2 cells and intestinal enterocytes to assess its suitability for studying the specific processes that impact intestinal drug absorption. As the pharmaceutical research community became more comfortable with the use of cellular models to assess drug absorption, the role of Caco-2 cells continued to evolve. From its initial application in the drug development arena, where Caco-2 permeability data helped estimate the absorption potential of drug development candidates, this model system has been shifted upstream, to assess absorption potential of compound analogs even before a lead compound has been selected and, downstream, to support more developmental work such as formulation development/optimization and to obtain permeability data that can be submitted to regulatory bodies to make decisions on the need, or not, to conduct clinical studies (e.g., bioequivalence or transporter-mediated drug–drug interactions or DDIs).[27,28]

14.2.1 Caco-2 Cell Culture

For permeability experiments, Caco-2 cells are generally seeded on polycarbonate filters (Transwell™) uncoated or coated with extracellular attachment factors (e.g., rat tail collagen, type I). The cells are cultured in high-glucose Dulbecco's modified Eagle's medium (DMEM), supplemented with 10% fetal bovine serum (FBS), 1% nonessential amino acids (NEAA), 1% L-glutamine, penicillin (100 U/mL), and streptomycin (100 µg/mL) at 37°C under a humidified air–5% CO_2 atmosphere for about 3 weeks.[11] Prior to starting transport studies, the integrity of cell monolayers is assessed by measuring their transepithelial electrical resistance (TEER) and determining the transepithelial fluxes of markers of passive paracellular diffusion. TEER measurements, which are quick, easy, and nondestructive, can be obtained using an EVOM (Epithelial Voltohmmeter) (World Precision Instrument, New Haven, CT, USA). In our laboratory, Caco-2 monolayers used in most studies have TEER values of at least 450 Ω cm^2 and for studies conducted for permeability classification according to the biopharmaceutics classification system (BCS), the acceptable TEER values range is restricted to 450–650 Ω cm^2. Monolayers that do not meet the TEER value requirements are discarded.

The permeability coefficients P_{app} of low-molecular-weight passive paracellular flux markers (e.g., mannitol and Lucifer yellow) are more sensitive than TEER values in indicating monolayer integrity.[11,29] Although larger permeability markers, such as inulin and PEG 4000, have also been used, due to their molecular size they are less sensitive and less relevant than mannitol or Lucifer yellow as indicators of monolayer integrity in drug-permeability studies.[11,30] Owing to the difficulty of detecting mannitol by common analytical techniques, the use of mannitol usually requires radioactive material. Thus, when the use of radioactive material is undesirable, Lucifer yellow, a fluorescent compound, is a good alternative marker of monolayer integrity. The acceptable transepithelial permeability coefficient (P_{app}) value of Lucifer yellow is less than 4×10^{-7} cm/s in our laboratory. Because the P_{app} value of Lucifer yellow is more sensitive than TEER in detecting monolayer imperfections, monolayers with Lucifer yellow P_{app} values greater than 4×10^{-7} cm/s are rejected, even if their TEER values are in the acceptable range.

Unlike drug permeability, where, after having demonstrated a great deal of utility, Caco-2 cells are widely used, this is not the case in intestinal metabolism studies. Although Caco-2 cells express several intestinal drug-metabolizing enzymes such as peptidases, CYP1A1, CYP1A2, CYP3A5, UDP glucuronyltransferases, and phenol sulfotransferases,[22,23,31–34] their utility in the study of intestinal drug metabolism has not been fully established. One difficulty is that the level of expression of CYPs, including CYP3A4, the most abundant CYP isozyme in the intestinal mucosa, appears low and variable.[31,32,35] Two approaches have been used to increase the utility of Caco-2 cells in the study of CYP3A4 metabolism. The first consists of inducing the enzymes by culturing the cells with Di OH vitamin D$_3$.[36] Based on the scarce number of studies using this system, it appears that the application of Di OH vitamin D$_3$-induced CYP3A4 to drug metabolism studies is very limited. As is the case with transporters, the expression of enzymes is very variable, and the response to Di OH vitamin D$_3$ treatment is likely to show even greater variability in enzyme activity.

In another attempt to create Caco-2 cells expressing CYP3A4 activity, Caco-2 cells were transfected with human CYP3A4 cDNA.[37] Some success was reported regarding the expression of CYP3A4 activity, and the transfected cells failed to develop stable barrier properties.[37] Although the mentioned disadvantages that resulted from the transfection are probably responsible, at least in part, for the lack of acceptance of these transfected cells, it is possible that an interest in simplifying the cells may have played a role, because the integration of metabolism and transport processes, especially when membrane transporters are involved, would further complicate the interpretation of permeability data. This same mindset has resulted in a migration towards simpler permeability models (e.g., parallel artificial membrane permeation assay or PAMPA) and to the use of alternative cell lines such as Madin–Darby canine kidney (MDCK).[38,39] The lack of widespread acceptance of Caco-2 in intestinal metabolism studies in recent years, judging by the limited number of

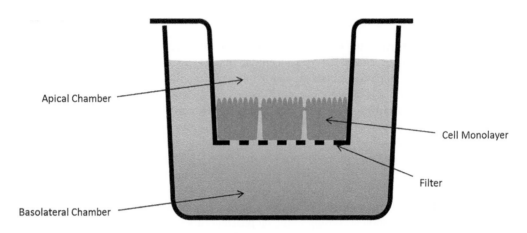

Apical Chamber

Cell Monolayer

Filter

Basolateral Chamber

FIGURE 14.1 Schematic representation of the Transwell™ system.

studies utilizing Caco-2 cells for this type of studies, is likely to continue in the near future and that Caco-2 cells will be preferentially applied to permeability and transporter studies with little involvement in drug metabolism.

Since 1995, there has been a growing interest in the interplay between metabolism (i.e., CYP3A4) and efflux transport (i.e., Pgp).[40–42] Conceptually, Pgp controls intestinal CYP3A4-catalyzed metabolism by increasing the exposure to CYP3A4 through repeated recycling (i.e., uptake followed by efflux).[42,43] However, the value of Caco-2 monolayers in these studies has been limited because although these cells exhibit a robust expression of Pgp, they express very low levels of CYP3A.[44] Although one study successfully induced CYP3A4 with Di OH vitamin D3 and was able to identify the relative roles of Pgp and CYP3A4 on the apical secretion of indinavir metabolites in Caco-2 monolayers,[43] in general, other systems are being increasingly used to evaluate the functional CYP3A–Pgp relationship.[45,46]

14.2.2 CACO-2 PERMEABILITY ASSAY

Permeability screening experiments are commonly carried out in 12- or 24-well Transwell plates. Cell monolayers grown on porous polycarbonate filters are washed and bathed in a transport buffer such as Hanks' balanced salt solution (HBSS), containing 2 mM glucose. For pH dependence experiments, the transport buffer solution can be adjusted using 10 mM MES for pH 5.5–6.5 or using 10 mM HEPES for pH 6.5–7.4. For low-solubility compounds, 1%–4% BSA can be added to the transport solution to enhance the apparent solubility. Although organic solvents can be used for this purpose, the concentration of these solvents needed to produce an effective increase in the solubility of most poorly soluble compounds is likely to cause cellular damage. At time zero, compounds dissolved in transport solution are dosed to either the apical chamber (apical-to-basolateral) transport or the basolateral chamber (basolateral-to-apical) transport (Figure 14.1). The Transwell plates are incubated at 37°C with appropriate agitation to reduce the thickness of the unstirred water layer. Samples are collected from the receiver chambers at preselected time points (e.g., 30, 60, 90, and 120 min) and replenished with drug-free transport buffer. To determine the percent recovery (or mass balance), samples should be collected from the donor chamber at the end of the experiment. Drug transport is usually expressed by apparent permeability coefficient P_{app} as

$$P_{app} = \frac{\Delta Q}{\Delta t A C_0} \tag{14.1}$$

where ΔQ is the amount of drug transported during the time interval Δt, A the surface area, and C_0 the initial concentration in the donor chamber. P_{app} values are used to determine the absorption potential of the test compounds. In our laboratory, the absorption potential is categorized as follows: high, $P_{appA-B} > 1 \times 10^{-6}$ cm/s; medium, 0.5×10^{-6} cm/s $< P_{appA-B} < 1 \times 10^{-6}$ cm/s; and low, $P_{appA-B} < 0.5 \times 10^{-6}$ cm/s. For passive diffusion and non-saturated transporter-mediated transport, P_{app} values are independent of the initial drug concentration. A decrease in the absorptive flux, as indicated by a decrease in P_{appA-B} value, with increasing initial drug concentration indicates saturation of transporter-mediated influx. The ratio B-A P_{app} (indicator of secretory flux) divided by the A-B P_{app} (i.e., P_{appB-A}/P_{appA-B}) can be used to assess the involvement of efflux transport mechanisms (e.g., P-glycoprotein).

Compared to in vivo studies, the Caco-2 permeability model substantially increases the speed at which absorption potential can be estimated and requires less amount of drug substance; however, manually performed assays are still too slow compared to biological HTS assays. Caco-2 cells take about 3 weeks to form monolayers of fully differentiated cells. At this point, Caco-2 monolayers are used to evaluate absorption potential under a variety of permeability protocols, depending on the specific application. When permeability assays are applied to drug development activities such as lead optimization and prediction of in vivo drug absorption, there is a greater emphasis on data quality than on speed because of the smaller number of compounds involved and the type of decisions made based on these data.

14.2.3 ESTIMATION OF ABSORPTION POTENTIAL

The correlation between drug absorption in rats and humans that has been repeatedly demonstrated[24–26,47,48] permits the use of absorption in rats to predict drug absorption in humans. Ideally, the assessment of the absorption potential of new molecular entities should be an integral part of the lead-optimization phase because this information would constitute one of the criteria used for selecting lead compounds with overall favorable characteristics for drug development. However, the routine use of animals in screening absorption for large numbers of compounds is not feasible for several reasons. First, if the compound undergoes pre-systemic metabolism, determination of extent of drug absorption requires the administration of radioactively labeled compound, which is rarely available during the lead-optimization phase. Second, absorption studies in animals require relatively large amounts of material, drug quantification in plasma can be difficult for some compounds, and experiments are too laborious and time-consuming to be effective as a permeability screening tool. Third, in vivo studies provide little insight into intestinal absorption mechanisms (e.g., passive diffusion vs. transporter-mediated transport and paracellular vs. transcellular diffusion). Although in vitro systems do not comprise all the factors responsible for intestinal drug absorption, they have greater utility than animals in the study of mechanisms of drug absorption. Thus, these limitations have made cell culture models such as Caco-2 a viable alternative for screening the absorption potential of lead compounds. The Caco-2 permeability assay allows the determination of the absorption potential of large numbers of compounds in experiments of short duration (1–2 h), utilizing a simple buffer solution and only sub-milligram quantities of test material.

14.2.4 INTER-LABORATORY VARIABILITY IN CACO-2 PERMEABILITY MEASUREMENTS

Reports from several laboratories showing a good correlation between Caco-2 P_{app} and fraction absorbed in humans (F_{abs})[24–26,47,48] support the attractiveness of Caco-2 permeability measurements for the estimation of oral-absorption potential. However, the increasing use of Caco-2 cells in drug-permeability measurements has put in evidence a large inter-laboratory variability in permeability values. This inter-laboratory variability is influenced by biological factors such as culture conditions, passage number, days in culture, and cell viability; and experimental design factors such as composition and pH of transport buffer, drug concentration, co-solvents, sampling times, stirring conditions, and calculation method.

A large inter-laboratory variability in the two most common indexes of monolayer integrity, TEER and P_{app} value of mannitol, has been reported.[24–26,47,48] This variability in P_{app} values does not apply only to mannitol because when Caco-2 P_{app} data from five laboratories were plotted against known F_{abs} in humans, the inter-laboratory variability was very large.[49] The variability in P_{app} values makes it extremely difficult to combine, or even compare, permeability data from different laboratories. After conducting a thorough examination of the complex issues associated with permeability measurements in Caco-2 and MDCK monolayers, Volpe found that, as a result of the heterogeneity of Caco-2 and MDCK cells and differences in culturing conditions, researchers unintentionally end selecting the cell population(s) that thrive under the respective conditions used in the different laboratories. Since these populations of Caco-2 or MDCK cells may have different phenotypes, they may exhibit differences in biochemical and/or physical barrier properties, which would contribute to inter-laboratory variability. After concluding that standardization of the assay would not only be difficult but could also hinder further improvements in the area, she recommended a more general approach to permeability measurements.[50] Volpe's conclusions that inter-laboratory variability can be reduced with continuous use of well-defined controls agree with an earlier review by Hidalgo, who suggested the measurement of transporter and enzyme activity at regular intervals to detect phenotypic drift in the cultures.[49] Most laboratories attempt to control phenotypic drift by maintaining the cells under a certain number of passages; however, ultimately, the most effective way to avoid long-term changes in cell phenotype by periodic monitoring enzyme/transporter functionality of Caco-2 cells because these are likely to reflect changes in the underlying composition of the cultures. Thus, the stability of the cell monolayers should be based on whether the monolayers meet the acceptance criteria established by the laboratory. Moreover, the use of alternative cell lines does not necessarily avoid these issues and may introduce unique drawbacks associated with these cells, which have not been noticed simply due to their limited use. In any case, performing all the experiments in a single laboratory using robust protocols eliminates inter-laboratory variability and should increase the reliability of the permeability coefficients values.

14.3   USE OF CACO-2 CELLS IN DRUG DISCOVERY

The Caco-2 cell model is considered the most common in vitro model for prediction of intestinal drug absorption. The model is widely used in pharmaceutical industry for screening of new chemical entities (NCEs) regarding their absorption potential and also for identifying substrates and/or inhibitors of drug transporters. Increasing recognition that early consideration of the permeability characteristics of hit compounds would enhance the drug-like quality, and ultimately the probability of success, of selected lead candidates caused a quick upstream migration of drug-permeability measurements. This adaptation makes it possible to incorporate permeability information into the hit-to-lead phase of the drug discovery process, but for these data to be useful, permeability measurements must be rapid and feasible with the small amounts of material usually available in early discovery phase. Thus, efforts have been made to automate and miniaturize the Caco-2 permeability assay.

Caco-2 cells require 3 weeks of culturing to reach full maturity, during this period sterility, and several other factors have to be carefully controlled to ensure viable and consistent cell monolayers for permeability assays. Additionally, advances in modern chemistry enable rapid generation of chemical libraries of NCEs; the number of compounds to be screened is ever increasing. These demands have compelled pharmaceutical companies to develop robotic systems in cell culturing and/or high-throughput permeability screening assays. Some tedious manual cell culture procedures, such as plate coating, and cell seeding and feeding, can be performed using robotic systems. High-throughput permeability screening assays can be conducted by liquid-handlers to automate compound dosing and sample collections, thus reducing potential human errors.

The implementation of automated permeability assays required a continuous change in the type and size of transport plates. While the early work with Caco-2 cells was done in 6-well Transwell trays, the need for higher throughput was quickly accompanied by a shift towards 12- and 24-well plates.[15,17,21–24,34,42] One important pitfall of the single-piece 24-well HTS is the difficulty in dealing with bad monolayers, because it consists of a single piece with 24 inserts. In standard transport plates, inserts containing monolayers with low TEER can be excluded from the experiment and replaced with suitable monolayers from other plates; however, in the single-piece 24-well HTS system, bad inserts cannot be discarded. Often, all the monolayers are dosed, but only the samples from the good monolayers are processed, a procedure that creates a lot of unnecessary and potentially confusing data.

To further exploit the benefits of automation, permeability plates have undergone noticeable miniaturization, as evidenced by the recent development of a single-piece 96-well (HTS) Transwell plate. Implementation of 96-well HTS plates in permeability assays is not necessarily trivial because of the difficulty associated with achieving homogeneous seeding of such small filters. Nonhomogeneous seeding could lead to cell stacking, and the difficulty related to the treatment of bad monolayers, as mentioned in connection with the single-piece 24-well HTS plates, is expected to be greater in the 96-well HTS plates.[51] In the 1990s, combinatorial chemistry promised to generate huge numbers of chemical leads in record time,[52,53] and this anticipation provided impetus for the development of the 96-well permeability assay; failure to deliver appreciable numbers of drug candidates with this approach has tempered expectations in this regard and, in turn, the urgency to implement the assay in this scale.

Despite this development, some effort to establish the use of the 96-well HTS plates will continue driven by the desired to potentially achieve a fourfold throughput increase in the permeability assay compared with 24-well HTS plates.[54,55] As with the PAMPA, the tiny surface area of the filters of the 96-well HTS presents an analytical challenge for compounds with low-to-moderate permeability.[56] Miniaturization and automation have resulted in a dramatic increase in the throughput of permeability measurements. This increase would be of little value without a concurrent increase in the analytical methodology used to quantify the samples generated in permeability assays. Liquid chromatography/mass spectrometry (triple quadrupole) (LC-MS/MS) technology, together with automation and miniaturization, has enabled the overall increase in throughput of the permeability assays.[57] LC-MS/MS not only makes method development faster and easier compared to HPLC but also accelerates sample analysis.

14.4 USE OF CACO-2 CELLS IN DRUG DEVELOPMENT

14.4.1 GENERAL

Initially, Caco-2 cells were used mainly by pharmaceutical scientists trying to predict the absorption of NCEs as they entered the development phase. However, as the acceptance of the system expanded, the application of Caco-2 progressively migrated upstream towards earlier phases of drug discovery, as medicinal chemists realized that permeability information obtained earlier could help inform decisions regarding the design of subsequent chemical analogs, thus making sure the potential success of selected drug candidates would not be derailed by poor intestinal absorption. The application of the Caco-2 permeability assay also migrated downstream towards more development-oriented tasks such as formulation development/optimization,[58,59] study of permeation enhancers,[60,61] and permeability classification according to the BCS.[27] In addition, the purpose of the US FDA guidance for enzymes and transporters-mediated DDIs is to outline a framework for the execution of in vitro experiments, mostly in cellular systems, to assess the potential of NCEs to interact with co-administered drugs. According to these guidelines, if an NCE fails to interact with a given transporter in an in vitro assay, it can be exempted from clinical transporter-mediated DDIs studies.[28] Although in vitro experiments are done with several cell lines (e.g., HEK293, CHO) containing a variety of uptake transporters,[62,63] Caco-2 cells, with their full complement of basolateral transporters, are widely accepted in the evaluation of drug interactions with the efflux transporters Pgp and BCRP.[28,64,65] LLC-PK1 cells have been transfected with human, monkey, canine, rat, and mouse P-glycoprotein for inter-species comparison of Pgp-mediated drug efflux.[66] The MDCK transfected with BCRP (i.e., BCRP-MDCK) or Pgp (MDR1-MDCK) is used to study drug interactions with BCRP and Pgp, respectively; however, they lack some basolateral transporters, which may be necessary for some compounds to enter the cells and to create the possibility that these compounds yield false-negative interaction results in these cells. A remarkable milestone in the evolution of the impact of Caco-2 cells in drug development was the application of the cell line in the context of the BCS. The novelty of this strategy was that it makes it possible to waive in vivo bioequivalence studies for BCS Class 1 and Class 3 drugs classified using in vitro systems, as long as the suitability of these systems is previously demonstrated. Likewise, Caco-2 cells have been widely used to assess the potential involvement of NCEs in transporter-mediated DDIs. As per FDA guidelines, it is possible to make decisions regarding the need to conduct P-glycoprotein (Pgp) and breast cancer resistance protein (BCRP)–mediated DDI studies in vivo[28] based on in vitro drug-transporter interaction and DDI data (obtainable in Caco-2 cells).

14.4.2 DRUG DELIVERY

Two factors that often limit intestinal absorption are poor aqueous solubility and low membrane permeability. If low intestinal absorption is due to insufficient membrane permeability, formulation approaches to enhance permeability can be undertaken, some involving the use of Caco-2 cells. For example, a popular strategy has been the use of prodrugs to increase hydrophobicity and drug permeation across the lipid bilayer of the cell membrane. Prodrugs have also been used to target intestinal mucosal transporters such as PepT1, thus capitalizing on a transporter-mediated route, that would not be accessible to the intact drug, to achieve systemic drug absorption.[67–69] Drug permeability can also be enhanced by modifying the barrier properties of the intestinal epithelial layer through opening of intercellular junctions or increasing membrane fluidity. Although these approaches to increase drug absorption have been tested in animals and intestinal tissue in vitro,[70–72] Caco-2 cells have clear advantages in this type of studies. These advantages include the following: (a) greater experimental flexibility (e.g., pH, directionality); (b) ease of data interpretation; (c) samples are cleaner, and thus, drug analysis is easier compared to plasma; (d) relatively small amounts of material are required; (e) higher sensitivity permits to screen large numbers of potential enhancers; and (f) experiments are simple and rapid.

In addition to being useful in the evaluation of formulation strategies, the flexibility of the Caco-2 cell permeability model makes it possible to evaluate the relative contribution of specific permeability pathways. The low permeability of clodronate and other bisphosphonates, which permeate the intestinal epithelium through passive paracellular diffusion,[73] is restricted by the intercellular junctions and the formation of complexes with calcium. A potentially useful strategy to increase the permeation of this type of compounds is to shift the permeability pathway from paracellular diffusion to transcellular diffusion. After synthesizing and determining the permeability coefficients in Caco-2 cells of a series of ester prodrugs of clodronate having varying degrees of lipophilicity, Raiman et al. found that, when log D reached 1, the predominant permeability pathway of the esters of clodronate shifted from paracellular to transcellular.[73] Obtaining this detailed information on intestinal permeability mechanisms in more complex systems is much more difficult.

In general, when the low absorption is due to low membrane permeability, available options to increase absorption are very limited. To increase the absorption of compounds with low permeability, it is necessary to alter the barrier properties of the intestinal mucosa through permeability enhancers. Permeability enhancers can act either by opening tight junctions or by modifying the cell membrane. However, lowering the diffusional resistance of the intestinal mucosa might potentially allow the penetration of foreign substances into the systemic circulation, which could result in infections and/or toxicity. Therefore, a thorough evaluation of this strategy should include the conduct of studies to understand the interaction of any permeability enhancer with the intestinal mucosa. Questions, such as onset and reversibility of permeability enhancement effect, size limit for "bystander" molecules permeation, mucosal damage, and fate of the enhancer, should be addressed. Very few systems have the necessary versatility to be useful in answering this type of questions. In one study with Caco-2 cells, in which the role of nine common excipients on the transepithelial permeation of seven low-permeability compounds was evaluated,[74] most excipients did not influence the permeation of low-permeability drugs across Caco-2 cell monolayers. The only exception was sodium lauryl sulfate (SLS), which increased the permeation of all the drugs tested,[74] probably by impacting the physical integrity of the cells. Other excipients had limited effect on Caco-2 permeability. For example, Docusate Sodium caused a moderate increase in cimetidine permeability, and Tween 80 had marked increase on the permeability of furosemide and cimetidine, two drugs with intrinsically low permeability.[74] Although these studies provide useful insight into the effect of excipients on different transcellular transport pathways, the evaluation of potential detrimental effect on the cells was very limited for two reasons. First, in general, excipients were tested at only one (usually low) concentration, which makes it difficult to evaluate the true potential for a detrimental effect on the cell monolayers. Also, because the concentrations of excipients were different, it is not possible to make a direct comparison of the relative potential of different excipients to cause cellular damage. Second, evaluation of monolayer damage was limited to monitoring the impact on the passive paracellular transport pathway, but did not assess the effect on cellular biochemistry or viability.

Thus, systematic studies focusing on the concentration dependence of excipient effect on barrier properties, biochemical integrity, and cell viability are necessary for a better understanding of the mechanisms through which excipients can influence drug absorption.

14.4.3 Formulation Development

A successful application of cellular systems, particularly, Caco-2 cells, in formulation development was slow to begin, partly due to the belief that these cells are hypersensitive to excipients. Despite the sensitivity of Caco-2 cells to pharmaceutical excipients, they may be utilized in support of certain aspects of formulation development. For example, a device (i.e., distribution/partition or D/P chamber) developed by Kataoka et al.[75] provided a much-needed impetus to the use of Caco-2 cells in this context. While drug solubility and permeability are the most important determinants of drug absorption, they are often evaluated under experimental conditions that do not take into consideration the environment in which the dosage form (e.g., tablet) disintegrates/dissolves and the drug is absorbed. In other words, drug solubility and permeability and drug product dissolution are normally tested in separate experiments under conditions that are not physiologically relevant. The D/P chamber consists of two compartments separated by a filter containing a Caco-2 monolayer.[75] In this chamber, following the addition of a small amount of powder form material (e.g., macerated tablet and drug-excipient mix) to the apical compartment, samples are taken from the apical compartment to determine drug dissolution and from the basolateral (receiver) compartment to determine permeation.[75] By measuring drug dissolution and permeation concurrently, it is possible to characterize the interplay between these two processes, variable information for formulation scientists. To enhance the physiological relevance of these experiments, the composition of the donor and receiver compartment solutions mimics, at least partially, the composition of the intestinal fluid and blood plasma, respectively. As part of the D/P chamber, Caco-2 cells have been very useful in the investigation of various aspects of drug formulation for BCS Class 2 drugs. More recently, we introduced an innovative dissolution–permeability chamber (In-vitro Dissolution Absorption System, IDAS2) that can be used in a standard drug dissolution apparatus.[76] As is the case with the D/P chamber, IDAS2 allows the simultaneous evaluation of dissolution and permeability; however, IDAS2 has the unique advantage that it can be used to test intact tablets. Caco-2 cells, an integral part of IDAS2, should be of great utility to formulation efforts because by allowing the evaluation of drug permeability following the application of intact dosage forms, it is possible to assess the role of dosage form disintegration/dissolution on drug permeation. This information is extremely useful for formulation scientists as they engage in the process of optimizing formulations for clinical studies. The experimental conditions used with IDAS2 include (a) a Caco-2 monolayer that can measure drug permeation, (b) dissolution (apical) fluid that mimics the gastrointestinal (GI) fluid, (c) permeation (basolateral) solution that mimics blood plasma, and (d) two-stage method that mimics stomach-to-intestine transfer. In addition, the dissolution volume (250 mL) approaches the volume used in clinical studies (i.e., one tablet swallowed with 8 oz of water). Thus, dissolution and permeation data obtained in this system is expected to be of greater physiological relevance than data obtained using the routine practice of conducting dissolution and permeability tests separately; therefore, in the end, IDAS2 data should be more useful to predict in vivo drug product performance.

14.5   MECHANISTIC STUDIES WITH CACO-2 CELLS

Not long after the characterization of Caco-2 monolayers as a model system of intestinal permeability and its subsequent application to the screening of new molecular entities to assess intestinal absorption potential in humans, some scientists started to explore the utility of this system in the elucidation of the cellular mechanism underlying drug transport across the intestinal epithelial layer. This initiative was well justified because, for many compounds, intestinal permeation involves transporters that either facilitate or limit transepithelial transport and Caco-2 cells express most membrane transporters involved in intestinal drug transport.[13,15–19,77,78] However, when interpreting results of studies that involve carrier-mediated transport, discretion, and scaling factors may be required because of the difference in expression level of transporters between in vitro and in vivo systems.[79] Another important consideration when using Caco-2 cells to study carrier-mediated transport is that not all transport systems in Caco-2 cells may achieve maximal expression level at the same days in culture.[13,17,47] Thus, for greater utility of Caco-2 cells in mechanistic transport studies, the culture-time window for optimal transporter expression determined and a qualitative evaluation of the transporters, demonstrating that they are representative of the native intestinal transporters, should be performed. A great deal of the work with Caco-2 monolayers was focused on trying to identify the specific pathways through which compounds were able to traverse epithelial cell layers. Interest in using Caco-2 monolayers to study transport pathways is well founded because Caco-2 monolayers exhibit morphological characteristics similar to enterocytes such as tight intercellular junctions and highly developed microvilli.[9–11,80] Drug transport across Caco-2 monolayers is limited by the action of biochemical and physical barriers (Figure 14.2). The biochemical barrier comprises drug-metabolizing enzymes, uptake transporters, and efflux transporters, and the physical barrier consists of the cell membrane and intercellular junctions.[13,15–19,78,80] Beyond morphological similarity, these components bestow on Caco-2 cells monolayers permeability resistance characteristics reminiscent of those found in the intestinal epithelium.

This permeability barrier shows selectivity in that small hydrophobic molecules can partition into and diffuse across the lipid bilayer of the cell membrane, whereas small hydrophilic molecules can only diffuse between cells (i.e., through the intercellular junctions). However, it is difficult to predict intestinal permeability based on physicochemical properties alone because uptake and efflux transporters may interact with drugs to increase or decrease their absorptive flux. Furthermore, the complexity of the permeability process makes it difficult to elucidate permeability pathways in animals and perfused intestinal tissues. For this reason, Caco-2 cells in particular have been used extensively to investigate the role of specific permeability pathways in drug absorption.

Paracellular transport generally takes place via passive diffusion and is accessible to small (i.e., molecular weight [MW] < 200) hydrophilic molecules.[81–83] The barrier restricting the passive movement of solutes through the paracellular pathway is the tight junction, and the driving force for passive paracellular diffusion is the electrochemical potential gradient resulting from the

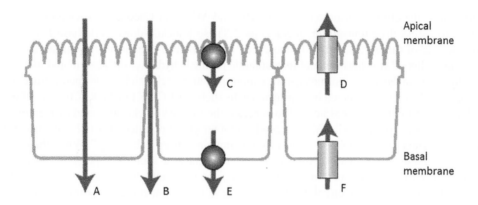

FIGURE 14.2   Major transport pathways in Caco-2 monolayers. A, Passive transcellular; B, passive paracellular; C, transporter-mediated apical uptake; D, transporter-mediated apical efflux; E, transporter-mediated basolateral efflux; F, transporter-mediated basolateral uptake.

differences in concentration, electrical potential, and hydrostatic pressure between the two sides of the epithelial membrane. Although the paracellular pathway plays a minor role in the intestinal permeability of high-absorption compound, the converse is true for low-absorption compounds. To determine the contribution to permeation across Caco-2 monolayers of the paracellular route, the permeability resistance of this pathway can be modulated by the addition of calcium-chelating agents such as EGTA or EDTA, which causes opening of tight junctions.[84] Usually, the extracellular concentrations of EDTA or EGTA must be at least 0.5 mM to open the paracellular transport pathway, and a 1.26 mM EDTA was found to increase the permeation across Caco-2 monolayers of ranitidine by 5–20-fold without affecting the permeation of ondansetron, indicating that ranitidine permeates the cell layer paracellularly and ondansetron transcellularly route.[84–87] However, caution must be exercised in the use of Caco-2 monolayers to study paracellular transport. Tanaka et al.[88] found that the TEER values of Caco-2 monolayers (468 Ω cm^2) were higher than those of rat jejunum (37 Ω cm^2) and colon (83 Ω cm^2). Consistent with higher TEER values, the P_{app} of the paracellular permeability marker FITC–dextran (MW = 4,000) in Caco-2 monolayers, rat jejunum, and rat colon was 0.22, 2, and 0.95 × 10^{-5} cm/min, respectively. Based on the observation that Caco-2 monolayers exhibited the greatest decrease in TEER in the presence of 10 mM EDTA, it is suggested that they are more sensitive than rat jejunum and rat colon to tight junction.[88]

Passive transcellular drug transport involves movement of drug molecules across the enterocyte membrane. For hydrophobic compounds, transcellular transport is the main route of intestinal epithelial permeation due to its magnitude. The surface area of the cell membrane, the transcellular route, accounts for 99.9% of the total intestinal epithelial surface area, whereas the surface area of the tight junctions, the paracellular route, accounts for only 0.01%.[89,90] Generally, the extent of absorption of compounds whose permeability is limited to the paracellular pathway is low, and the extent of absorption of compounds that readily traverse the cell membrane is high. However, the involvement of membrane transporters (uptake or efflux) and metabolic enzymes can change this pattern, making it virtually impossible to predict drug absorption in the absence of specific information on drug–transporter interaction. For example, the absorption of some hydrophilic molecules such as cephalosporins and angiotensin-converting enzyme (ACE) inhibitors is much higher than expected from their intrinsic membrane permeation characteristics.[91,92] Even when drug-transporter interactions are characterized in in vitro assays, the difference in transporter level of expression and presence of other transporters in vivo makes it difficult to translate in vitro data to in vivo absorption. Caco-2 cells have been useful in the development of in vitro tools to predict the role of drug transporters in vivo. For example, to complement the utility of transporters chemical inhibitors, which can interact with multiple transporters, Zhang et al.[93] used shRNA technology to silence the expression of the efflux transporters Pgp, BCRP, and MRP2 in Caco-2 to provide an alternative strategy to assess drug-transporter interactions.[94] These knocked-down cells have been used to study the role of these transporters in the efflux of statin drugs[95] and to demonstrate the role of Pgp in the efflux of ximelagatran and its metabolic products hydroxyl-melagatran and melagatran. Since hydroxyl-melagatran and melagatran cannot penetrate cells and are formed intracellularly, this study could not be conducted in transfected cells such as LLC-PK1 or MDCK because they lack the drug-metabolizing enzymes needed to generate the hydroxyl-melagatran and melagatran prior to their extrusion from the cells. More recently, another group produced Caco-2 cells where these same transporters were completely knocked out using zinc finger technology.[96] While the complete silencing of these transporters seems advantageous because it avoids the potential noise due to residual transporter function in knocked-down cells, one important drawback of this approach was that the cells tended to compensate for the silencing of these transporters by overexpressing other transporters.[96]

14.6   RECENT APPLICATIONS OF CACO-2 CELLS

Caco-2 cells have been valuable in the estimation of drug absorption potential, transport mechanisms, and effect of permeation enhancers on transepithelial transport.[35,39,53,67–69,78–81] Owing to the sensitivity of the cells and the limited solubility of new molecular entities, Caco-2 permeability studies are routinely done with relatively low concentration of compounds. One way to increase the solubility of these compounds is to use organic solvents. The low tolerability of Caco-2 cells to organic solvents limits the use of this approach in permeability studies.

14.6.1   Biopharmaceutics Classification System

With the Hatch–Waxman Act of 1984 the US government sought to control the rocketing prices of drug products by increasing the participation of generic drugs. However, realizing that the use of lower cost generic drugs would not be helpful unless their quality could be guaranteed, the FDA quickly introduced the bioequivalence requirement and issued guidelines for the conduct of these studies. To ensure that generic products were comparable in quality, safety, and efficacy to innovator products, they must demonstrate bioequivalence in studies in humans. In addition, during the development of drug products containing NCEs, bioequivalence tests are required whenever a substantial change (e.g., change in formulation, manufacturing site, or production scale) was implemented. As a result, an average of 3–6 BE studies were conducted before a new product was launched.[97] These BE studies not only increase the time and cost of drug discovery and development, but also expose large number of healthy subjects to the potential side effects of these drugs. With the intention of reducing the regulatory burden weighing down on efforts to develop new drugs, the US FDA introduced new guidelines describing the possibility of waiving BE studies for compounds belonging to BCS Class 1 (high aqueous solubility and high intestinal permeability) and Class 3 (high aqueous solubility and low intestinal permeability.[27] While the publication of the US FDA guidelines allowing the waiver of bioequivalence study for compounds classified as Class 1 based on the BCS bolstered the role of Caco-2 cells in permeability measurements,[27] the well-characterized Caco-2 cells helped with the implementation and widespread acceptance of the BCS. While the first two drafts of the guidelines permitted to avoid costly and time-consuming bioequivalence studies for BCS Class 1 drugs, the final version of the document also makes Class 3 drugs eligible for biowaivers.[27] Although the FDA guidelines did not single out either Caco-2 or any other cell line for BCS permeability classification, and in principle, any cell line can be used as long as it is validated according to criteria described in the guidelines, Caco-2 cells have become the preferred cell line for this type of test. The primary reason for this preference is that many pharmaceutical scientists are familiar with Caco-2 cells, which are available in many pharmaceutical laboratories. In addition, given that the validation of this model for BCS permeability classification involves a fair amount of work, it does make sense to use a cell line that is widely used throughout the industry rather than undertaking the validation of a lesser-known cell line for this limited application.

Before Caco-2 or other cellular model can be used in the classification of drug permeability according to the BCS, the model must demonstrate suitability for this application. Validation can be accomplished by demonstrating a correlation between in vitro permeation and fraction absorbed (F_{abs}) in humans for a set of about 20 compounds that contain compounds whose extent of absorption is low (<50%), moderate (50%–89%), and high (at least 90%).[27] Since the reliability of the BCS classification of a drug is critical, before the in vitro–in vivo correlation is undertaken, the stability of cellular model and the robustness of the assay in each laboratory should be demonstrated. As validation requires correct rank ordering but not an absolute value for any of the reference compounds, the inter-laboratory variability problem is likely to persist in the near future. Among other things, the model needs to show (a) a correlation between in vitro permeability and in vivo absorption for at least 20 compounds, preferentially chosen from the list provided in the guidance;[27] (b) functionality of known membrane transporters; and (c) physical integrity to eliminate the possibility overestimating the permeability of test drugs. After method suitability has been established, the permeability class of a test drug is determined by comparing its permeability coefficient to that of the high-permeability reference compound, which must be a drug whose extent of absorption in human is at least 85% (e.g., antipyrine or minoxidil). If the permeability coefficient of the test drug is higher than that of the high-permeability reference compound, the permeability of the test compound is classified as high; otherwise, it is classified as low. Together with the high-permeability reference control, the permeability assay must include a low-permeability control (e.g., mannitol or atenolol), to monitor cell monolayer integrity, thus avoiding the possibility of artificially high permeability of the test drug as a result of leaky cell monolayers.

The utilization of in vitro models in general, and Caco-2 cells in particular, to determine the BCS class of compounds is becoming increasingly popular as scientists become more familiar with the FDA guidelines and start to benefit from their application. For example, in the last few years we have witnessed the implementation of BCS-based biowaivers for Class 1 and Class 3 drugs not only in the US and Europe but also in Canada, Australia, and numerous developing countries.[98]

A recent study that assessed the impact of the US FDA BCS on drug development found that more than 160 applications were approved based on this approach and that the applications included new and generic drugs and were related to several therapeutic areas.[99] The authors of the study concluded that the impact of the BCS has been substantial as evidenced by savings in excess of $100 million, the adoption by other regulatory agencies of some version of the BCS, and the large number of healthy human subjects who were not exposed to the risks associated with *in vivo* studies.[99] And, in the spirit of further facilitating drug registration in different regions of the world, the US FDA has incorporated important changes into its original BCS guidelines to increase harmonization with the EMA's BCS guidelines.[100] It is interesting that although the FDA guidelines for BCS-based biowaivers describe in great detail the necessary steps to demonstrate the suitability of the Caco-2 model in any laboratory, one of the main reasons for the rejection of applications for BCS Class 1 classification filed between 2004 and 2017 time was insufficient information demonstrating the suitability of the Caco-2 permeability model prior to its utilization in permeability classification studies.[99]

Although the last BCS Biowaiver guidance document does not specify the epithelial cells that can be used for in vitro permeability classification, due to the widespread utilization of Caco-2 cells as an in vitro permeability model Caco-2 cells have and will continue to facilitate the implementation of the BCS in many laboratories in the near future. Thus, the abundance of Caco-2 cell permeability data will most likely cement the role of this cell line as the in vitro "gold standard" for permeability classification.

14.6.2 MEMBRANE TRANSPORTERS AND TRANSPORTER-MEDIATED DRUG–DRUG INTERACTIONS (IH)

Caco-2 cells express a number of intestinal epithelial transporters (Figure 14.3). Some of these transporters, which are referred to as uptake transporters, mediate solute transport from the intestinal lumen into the enterocytes and thus help the drug absorption process. Uptake transporters found in Caco-2 cells include the following: large neutral amino acid transporter, bile acid transporter, oligopeptide transporter (PepT1), monocarboxylic acid transporter, nucleoside transporter, and fatty-acid transporter.[17–19,101,102] Caco-2 cells also express a number of transporters that mediate the apical efflux of solutes, thus countering the drug absorption process. These transporters include members of the ABC superfamily of transporters such as P-glycoprotein (Pgp), MRP, and BCRP. The expression of uptake and efflux transporters by Caco-2 cells has increased the utility of this cell line in the study of mechanistic aspects of intestinal drug transport.

Before engaging in the use of Caco-2 cells to investigate the role of transporters in intestinal permeability, an important consideration that must be kept in mind is the variability of transporter expression. The optimal time to evaluate transporter function is not necessarily correlated with the optimal time for monolayer barrier functions. Although some studies have shown that based on paracellular permeability and TEER values, two indicators of monolayer integrity, the barrier properties are achieved within 2 weeks post-seeding; however, the maximal expression of several transporters is not necessarily optimal at this time.[17,20,103] Moreover, the level of transporter expression is influenced by the culture conditions used. Thus, it is unlikely that the level of transporter expression can be standardized based on passage number or post-seeding time alone. The reason for this situation is that Caco-2 cell cultures consist of a series of cell subpopulations rather than a single cell population. The existence of subpopulations has made it possible to isolate clones that differ from the general cell culture in the level of expression of characteristics of interest. For example, clones that express high levels of the brush-border marker enzyme, sucrase–isomaltase, bile acid transporter, or Pgp have been isolated.[104–106] The experimental conditions used to culture the cells can affect transporter expression in an undefined manner.

Intestinal transporters are responsible for the high absorption of some drugs whose passive absorption is expected to be low. PepT1 is involved in the absorption of the ACE inhibitors (e.g., enalapril and captopril), penicillins (e.g., amoxicillin), and cephalosporins (e.g., cephalexin, cephradine, and cefaclor). Attempt to exploit the apparently broad substrate specificity of PepT1 conjugates of amino acids and drugs has been made to target this transporter. Ideally, the conjugate which needs to be stable in the GI tract would reach PepT1, undergo translocation across the enterocyte, and

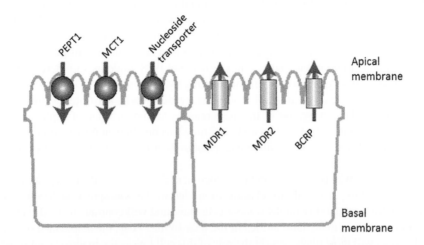

FIGURE 14.3 Partial representation of uptake and efflux transporters expressed in Caco-2 cells.

would then release the drug in the plasma. If the conjugate is biologically active, the release of the drug in plasma is not a requirement. For example, the bioavailability of L-Dopa, which is transported by the large neutral amino acid transporter, is low due to poor membrane permeability and metabolic instability. Following conjugation to L-Phe, the L-Dopa–L-Phe conjugate was recognized by PepT1 resulting in a much higher amount of L-Dopa being transported across Caco-2 monolayers as part of the L-Dopa–L-Phe conjugate.[107] Another example of a successful application of this approach was the conjugation of the antiviral agent, acyclovir. The oral bioavailability of acyclovir is limited (~20%) due to low intestinal mucosal permeation.[67,108] Synthesis of the valine–acyclovir (valacyclovir) prodrug caused a three- to fourfold increase in acyclovir bioavailability compared to acyclovir.[109]

Evidence of the expression of several efflux transporters in Caco-2 cells is varied. It ranges from quantification of mRNA using real-time PCR to assessment of functionality by determining the vectorial transport of substrates. Although mRNA evidence for MDR1, MRP1, MRP2, MRP3, MRP4, MRP5, BCRP, and LRP has been found in Caco-2 cells, functionality data have been extensively demonstrated only for MDR1.[16,35] Data showing the functionality in Caco-2 cells of MRP2 and BCRP is very limited and for other efflux transporters is practically nonexistent.

14.6.3 P/D Chamber, IDAS1, and IDAS2 Systems for Simultaneous Assessment of Drug Dissolution and Permeation In Vitro

The potential use of Caco-2 cells to screen drug formulations appears attractive; however, experiments have to be conducted under carefully controlled conditions. The Caco-2 model has been applied in conjunction with a dissolution system to evaluate the dissolution–absorption relationship.[75,110,111] Kataoka et al.[75,111] designed an in vitro system, namely dissolution/permeation system (D/P system), for simultaneous assessment of dissolution and permeation. The D/P system is composed of two half-cambers with a Caco-2 cell monolayer mounted in between; the apical chamber is filled with 8 mL of biorelevant dissolution medium and the basolateral chamber with 5.5 mL of transport buffer. The dissolution medium is a modified fasted or fed state simulated intestinal fluid (FaSSIF$_{mod}$ or FeSSIF$_{mod}$) with isotonic osmolality and has no detrimental effect on Caco-2 cell monolayer; the media in both chambers are consistently stirred by magnetic stirrers. The system has been demonstrated to be a useful tool for (a) predicting in vivo drug absorption in the discovery stage because it requires a small amount of compound, (b) identifying factors that affect drug absorption such as food effect, and (c) comparative evaluation of drug formulations.

The in vitro dissolution absorption system 1, IDAS1, is the product of modifications made to the D/P system). IDAS1 comprises four components: dissolution–permeation chambers, magnetic stirring bar motor, motor controller, and temperature heating block. While the internal design of the IDAS1 chamber is identical to the D/P chamber, they differ in the chamber locking mechanism, external size and shape, type of motor (i.e., step vs. brushless), ability of the controller to display the stirring speed (i.e., rpm), and the use of a heating block. The potential utility of IDAS1 has been shown in numerous studies performed with the D/P system.[75,111,112–115]

The in vitro dissolution absorption system 2, IDAS2 (Figure 14.4), is a more recent invention consisting of two permeability chambers placed inside a dissolution vessel. In this system, the Caco-2 cell monolayers are mounted on the permeation chambers with the apical membrane of cell monolayers in contact with the dissolution media. The dissolution vessel is filled with 500 mL of biorelevant medium and the permeation chamber with 8 mL of transport buffer. While the dissolution of a drug product is measured in dissolution vessel, the drug permeation across Caco-2 cell monolayers is assessed by the appearance of the drug in the permeability chambers. A unique feature of IDAS2 is that it permits the assessment of intact, clinical-size dosage forms, i.e., whole tablets or capsules.

Caco-2 cells have been valuable in the estimation of drug absorption potential, transport mechanisms, and effect of permeation enhancers on transepithelial transport. Owing to the sensitivity of

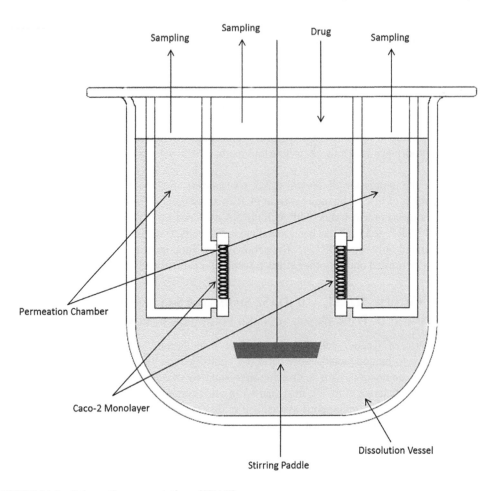

FIGURE 14.4 Schematic representation of IDAS2.

the cells and the limited solubility of many new molecular entities, Caco-2 permeability studies are routinely done with concentration of compounds below the anticipated luminal concentration, which may result in overestimation of the contribution of active transport processes. When using D/P or IDAS systems, because the modified biorelevant media used are not harmful to Caco-2 cells, drug products can be dosed at a dose/volume ratio equivalent to the administration of a clinical dose with 240 mL (8 fluid ounces) of water, which is consistent with the FDA recommendations for drug administration during clinical trials[27] and allows the evaluation of a wide range of drug concentrations in these experiments. In addition, experiments with IDAS2 can be performed in two consecutive phases to mimic transition from gastric to intestinal environment. In this experimental setting, a drug product is introduced into acidic medium (e.g., simulated gastric fluid, SGF) and allowed to undergo dissolution for 15–30 min (Phase 1). After Phase 1, a concentrated buffer solution is added to increase the pH to 6.5 and introduce bile salts and lecithin at appropriate concentrations to mimic the intestinal contents under fed or fasted conditions. After medium transition, the permeation chambers mounted with Caco-2 cell monolayers are submerged into the dissolution medium, and Phase 2 (intestinal dissolution–permeation phase) is initiated. This two-phase IDAS2 procedure has potential for assessing supersaturation phenomena, particularly for basic drugs which have higher solubility in acidic gastric condition while low solubility in neutral intestinal environment.

One constraint of the existing in vitro models that combine dissolution and permeability measurements is the experimental scale of the setup. Since, typically, dissolution experiments are performed in large media volumes (i.e., hundreds of milliliters), whereas permeability experiments are conducted in small volumes (i.e., few milliliters) of assay buffer, the resulting ratio of permeation surface area to dissolution volume is much lower than the ratio of intestinal surface area to luminal volume found in vivo. Thus, one concern is that in these models, drug permeation may not correlate with drug dissolution; however, we believe that this concern is unwarranted because permeation is a function of drug concentration at the diffusional surface, and independent of the volume of dissolution media. The diffusional area of Caco-2 cell monolayers in D/P and IDAS models is equal or larger than the value (1.13 cm2) available in a 12-well Transwell plate, which is typically used in Caco-2 assays. Coupled with sensitive LC-MS/MS instruments, the majority of Caco-2 assays can be performed at relative low dosing concentrations (1–100 µM) in 12-well or even 48-well Transwell plates. In fact, because drug products can be dosed at clinical level in IDAS models, drug concentrations in dissolution media are most likely higher than those in traditional Caco-2 assays in plate format; thus, the detection of permeated drug in permeation chamber is less of an issue.

On the other hand, the proportion of drug permeation to dissolution in the in vitro dissolution–permeation models does present a difference from in vivo system where the large surface area of the intestinal membrane permits efficient drug absorption. If a drug is highly permeable, the drug may effectively penetrate the intestinal wall and dissipate into the blood circulation. This rapid absorption process creates sink conditions in the lumen which leads to greater dissolution and in turn higher absorption, even if the drug is poorly water soluble. To create sink conditions in in vitro dissolution testing, a biphasic dissolution test was developed by including a water-immiscible organic phase (e.g., octanol) intended to mimic drug partitioning into the intestinal membrane.[116] The biphasic dissolution test under sink condition was able to discriminate among different formulations of a poorly soluble model drug, celecoxib, whereas those formulations were undistinguishable in the single-phase dissolution test under non-sink conditions.[116] The biphasic test system offers significant advantages for drug product development, but concerns remain on how well the organic phase can represent intestinal membrane permeability and partitioning of dissolution media components into the organic phase.[117] One way to reconcile in vitro dissolution–permeation results with in vivo observation is to use PBPK modeling approaches. The rate constants of drug dissolution and permeation obtained from in vitro experiments are applied into a set of differential equations incorporated with physiological parameters of the GI tract to describe the dynamic dissolution–permeation processes in vivo. The potential of this approach is illustrated by the oral-absorption PBPK model built with STELLA® software.[118] Despite the physical constraints of in vitro dissolution–permeation models, the systems constitute practical tools to study the dynamic interplay between dissolution and permeation processes in an integrated experimental setting. It has been demonstrated that these models permit the comparative evaluation of different formulation technologies and the determination of various factors affecting drug absorption. It is likely that these in vitro systems, by (a) allowing the conduct of experiments under conditions that mimic both the GI environment and intestinal drug concentrations and (b) capturing the interplay between dissolution and permeation, will produce experimental data more useful for developing in silico models truly predictive of in vivo product performance than standard in vitro data from separate dissolution/permeability experiments.

14.7 ROLE OF CACO-2 CELLS IN *IN SILICO* PREDICTION OF DRUG ABSORPTION

Oral dosing is the preferred route of drug administration, but the labor involved in generating experimental data to estimate drug absorption potential is substantial. Therefore, the possibility of developing computer programs capable of predicting drug absorption is very attractive. The simplest approach consists of applying Lipinski's rule of 5, which is not to predict drug absorption but as a filter to flag compounds with potential absorption problems.[119] Other computational strategies consist of predicting the absorption or permeability of molecules based only on molecular descriptors.[120,121] The success of this strategy is limited because permeability and absorption are not basic properties of the molecules determined by their chemical structures, and therefore, predictable by using chemical descriptors alone. Despite this intrinsic difficulty, researchers have predicted permeability coefficients (P_{app}) from the chemical structure of the compounds with varying degrees of success.[121–124] Molecular descriptors found to be useful in predicting P_{app} are H bonds and dynamic polar surface area (PSA).[125,126] Drug absorption from the GI tract comprises a series of concurrent processes involving dissolution, partitioning into bile salt micelles, epithelial cellular permeation, and intestinal drug transport and metabolism. Oral drug absorption is influenced by multiple factors including physicochemical properties of drug molecule, characteristics of the formulation, and interplay with the underlying physiological properties of the GI tract[117]. The extent of oral drug absorption is dependent on compound properties such as solubility, permeability, ionization, crystallinity, and dissolution-related processes, as well as GI physiological parameters such as luminal pH, transit time, effective absorptive surface area, and bile salt and phospholipid contents. Solubility/dissolution of solid dosage forms in GI fluids and the permeability of a drug through the gut wall have been shown as the primary factors governing the intestinal drug absorption, and thus were selected as the two primary parameters for classifying drug substances in the BCS.[127,128] Dissolution of drugs from solid dosage forms in the GI tract is a prerequisite for oral drug absorption because in most cases, drug molecules must exist in free form dissolved in GI fluid before traversing the intestinal epithelium, the major barrier to intestinal absorption. The first dissolution model was published by Noyes and Whitney[129] and subsequently improved by considering the exposed drug surface area and diffusion layer concept.[130] The refined Noyes–Whitney equation is most often used in computational simulation for describing dissolution behavior.[131,132]

$$\frac{dW_d}{dt} = z \cdot W^{2/3} \cdot (C_s - C_t) \tag{14.2}$$

where W_d is the amount of drug dissolved at time t; W, the amount of drug to be dissolved; C_s, the drug solubility; C_t, the concentration of dissolved drug at time t; z, an apparent parameter determined for describing the initial dissolution rate.

Dissolution and permeability are extremely difficult to determine in vivo; thus, these parameters are commonly measured from in vitro experiments. One of the most important and yet very difficult tasks in drug development is to establish correlation between drug in vitro profiles (e.g., dissolution and permeability) and in vivo pharmacokinetics from in vitro drug profiles.[116] As knowledge in GI physiology grew, in vitro and in vivo data accumulated, and computational power continuously increased, it was possible to create sophisticated physiologically based pharmacokinetic (PBPK) models to simulate the plasma concentration–time profiles and extent of drug absorption in vivo. Many pharmaceutical companies are building in-house customized PBPK models for evaluating drug candidates in various discovery and development stages. Dedicated absorption simulation software packages are also widely used in industry and academia, such as GastroPlus™ 9 distributed by Simulations Plus (Simulation Plus, Inc., Lancaster, CA, USA), Simcyp® Simulator 17 distributed by Simcyp Limited (Certara, Sheffield, United Kingdom), PK-Sim® 7 distributed by Bayer Technology Services GmhH (Leverkusen, Germany), and gCOAS distributed by Process Systems Enterprise (London, United Kingdom).

Modern simulations use multi-compartment models to describe drug absorption in different segments along the GI tract,[131,133] e.g., the Advanced Compartmental Absorption Transit (ACAT), model implemented in GastroPlus™,[134] and the Advanced Dissolution Absorption and Metabolism (ADAM) model in Simcyp® Simulator[135]. The compartment models consist of a set of differential equations that describe simultaneous movement of a drug (dissolved and undissolved) through the GI tract and absorption of dissolved material from each compartment into the portal vein. Drug dissolution and absorption are kinetic processes and therefore can be described by two time-dependent differential equations. Simulation of dissolution process is basically based on Noyes–Whitney theory and described by modified Noyes–Whitney equation, Equation 14.1. Drug absorption through the gut wall is described by a second differential equation, as given in Equation 14.2:

$$\frac{dW_a}{dt} = P_{eff} \cdot A \cdot C_t \tag{14.3}$$

where W_a is the absorbed amount of drug; A, the effective intestinal membrane surface area for absorption; C_t, the concentration of dissolved drug.

To describe the permeability of the gut wall, in vitro permeability values determined from Caco-2 cell or PAMPA assays are commonly used in the simulations.[118] The effective permeability (P_{eff}) is the result of both cell membrane permeability (P_{Caco-2}) and unstirred water layer permeability (P_{UWL}), in the following relation:

$$\frac{1}{P_{eff}} = \frac{1}{P_{Caco-2}} + \frac{1}{P_{UWL}} \tag{14.4}$$

The in vivo human intestinal P_{eff} may be one or two order of magnitude higher than the permeability values obtained from in vitro Caco-2 assays, because of increased surface area by villous structure in the intestine,[136] thus scaling factors may be needed when applying in vitro permeability values in computational model to simulate in vivo absorption.

14.8 FINAL REMARKS

Originally, the perceived advantages of Caco-2 cells were their relevance (e.g., enterocytic morphology and epithelial barrier properties, expression of numerous transporters, and drug-metabolizing enzymes) and experimental convenience (e.g., required smaller amount of material, short duration of experiments, and simplicity of sample matrices). Although the initial impetus to explore the used of Caco-2 cells stemmed from the lack of suitable experimental models to screen the increasing numbers of drug candidates being produced, the utility of the model in the estimation of absorption potential drove its adaptation to other applications such as drug-transporter interactions, permeation enhancers, mechanisms of intestinal permeation, and formulation development. The Caco-2 cell system is largely responsible for the radical increase in awareness of the importance of permeability on the ultimate success of drug candidates among medicinal chemists and drug metabolism scientists. After having played a major role in permeability studies for almost three decades, Caco-2 cells have cemented their role as a tool for investigating myriad aspects of intestinal drug absorption. As a result of its widespread use in the academic, industrial, and regulatory environments across the world, it has become, by default, the gold standard for in vitro permeability studies. In recent years, the value of the Caco-2 cell model in pharmaceutical research has been cemented by its application in the context of the BCS and transporter-mediated DDI guidelines. While the idea of predicting drug absorption and permeability constitutes a highly desirable goal, the processes that control drug permeation in vivo are complex and the lack of relevance of most in vitro experimental assays generates data whose value in predicting in vivo performance is very limited. Fortunately, the recently developed D/P and In-vitro Dissolution Absorption Systems, that permit to evaluate drug dissolution and permeation simultaneously, reveal an incipient interest in developing systems of greater physiological grounding, likely to yield data of greater translatability to the in vivo situation.

ABBREVIATIONS

ABC: ATP-binding cassette
ACE: angiotensin-converting enzyme inhibitor
BBMV: Brush–Border membrane vesicles

BSA: bovine serum albumin
cMOAT: canalicular multispecific organic anion transporter
CYP: cytochrome P450
DMSO: dimethylsulfoxide
HPLC: high-performance liquid chromatography
HTS: high-throughput screening
LC/MS: liquid chromatography/mass spectrometry (single quadrupole)
LC-MS/MS: liquid chromatography/mass spectrometry (triple quadrupole)
Log D/P: logarithm of the octanol/water (or buffer) distribution/partition coefficient
MDCK: Madin–Darby canine kidney
MDR: multidrug resistance
MRP: multidrug-resistance associated protein
PAMPA: parallel artificial membrane permeation assay
PEPT1:   peptide transporter 1
Pgp: P-glycoprotein
TEER: transepithelial electrical resistance
UDPGT: UDP glucuronyltransferase
UV: ultraviolet

REFERENCES

1. Pade, V., and S. Stavchansky. 1997. Estimation of the relative contribution of the transcellular and paracellular pathway to the transport of passively absorbed drugs in the Caco-2 cell culture model. *Pharm. Res.* 14:1210–5.
2. Sakai, M., A. B. Noach, M. C. Blom-Roosemalen, et al. 1994. Absorption enhancement of hydrophilic compounds by verapamil in Caco-2 cell monolayers. *Biochem. Pharmacol.* 48:1199–210.
3. Ranaldi, G., K. Islam, and Y. Sambuy. 1994. D-cycloserine uses an active transport mechanism in the human intestinal cell line Caco 2. *Antimicrob. Agents Chemother.* 38:1239–45.
4. Smith, P.L., D.A. Wall, C.H. Gochoco, et al. 1922. (D) Routes of delivery: Case studies. (5) Oral absorption of peptides and proteins. *Adv. Drug Deliv. Rev.* 8:253–90.
5. Hu, M., and G. L. Amidon. 1988. Passive and carrier-mediated intestinal absorption components of captopril. *J. Pharm. Sci.* 77:1007–11.
6. Fisher, R. B. 1981. Active transport of salicylate by rat jejunum. *Q. J. Exp. Physiol.* 66:91–8.
7. Hidalgo, I. J., F. M. Ryan, G. J. Marks, et al. 1993. pH-dependent transepithelial transport of cephalexin in rabbit intestinal mucosa. *Int. J. Pharmaceut.* 98:83–92.
8. Hidalgo, I. J., and J. Li. 1996. Carrier-mediated transport and efflux mechanisms in Caco-2 cells. *Adv. Drug Deliv. Rev.* 22:53–66.
9. Pinto, M., S. Robine-Leon, M. D. Appay, et al. 1983. Enterocytic-like differentiation and polarization of the human colon adenocarcinoma cell line Caco-2 in culture. *Biol. Cell* 47:323–30.
10. Grasset, E., M. Pinto, E. Dussaulx, et al. 1984. Epithelial properties of human colonic carcinoma cell line Caco-2: Electrical parameters. *Am. J. Physiol.* 247:C260–7.
11. Hidalgo, I. J., T. J. Raub, and R. T. Borchardt. 1989. Characterization of the human colon carcinoma cell line (Caco-2) as a model system for intestinal epithelial permeability. *Gastroenterology* 96:736–49.
12. Hunter, J., M. A. Jepson, T. Tsuruo, et al. 1993. Functional expression of P-glycoprotein in apical membranes of human intestinal Caco-2 cells. Kinetics of vinblastine secretion and interaction with modulators. *J. Biol. Chem.* 268:14991–7.
13. Dix, C. J., I. F. Hassan, H. Y. Obray, et al. 1990. The transport of vitamin B_{12} through polarized monolayers of Caco-2 cells. *Gastroenterology* 98:1272–9.
14. Takanaga, H., I. Tamai, and A. Tsuji. 1994. pH-dependent and carrier-mediated transport of salicylic acid across Caco-2 cells. *J. Pharm. Pharmacol.* 46:567–70.
15. Said, H. M., R. Redha, and W. Nylander. 1987. A carrier-mediated, Na^+ gradient-dependent transport for biotin in human intestinal brush-border membrane vesicles. *Am. J. Physiol.* 253:G631–6.
16. Kool, M., M. de Haas, G. L. Scheffer, et al. 1997. Analysis of expression of cMOAT (MRP2), MRP3, MRP4, and MRP5, homologues of the multidrug resistance-associated protein gene (MRP1), in human cancer cell lines. *Cancer Res.* 57:3537–47.

17. Hidalgo, I. J., and R. T. Borchardt. 1990. Transport of bile acids in a human intestinal epithelial cell line, Caco-2. *Biochim. Biophys. Acta* 1035:97–103.

18. Hidalgo, I. J., and R. T. Borchardt. 1990. Transport of a large neutral amino acid (phenylalanine) in a human intestinal epithelial cell line: Caco-2. *Biochim. Biophys. Acta* 1028:25–30.

19. Dantzig, A. H., and L. Bergin. 1990. Uptake of the cephalosporin, cephalexin, by a dipeptide transport carrier in the human intestinal cell line, Caco-2. *Biochim. Biophys. Acta* 1027:211–7.

20. Ng, K. Y., and R. T. Borchardt. 1993. Biotin transport in a human intestinal epithelial cell line (Caco-2). *Life Sci.* 53:1121–7.

21. Nicklin, P. L., W. J. Irwin, I. F. Hassan, et al. 1995. The transport of acidic amino acids and their analogues across monolayers of human intestinal absorptive (Caco-2) cells in vitro. *Biochim. Biophys. Acta* 1269:176–86.

22. Baranczyk-Kuzma, A., J. A. Garren, I. J. Hidalgo, et al. 1991. Substrate specificity and some properties of phenol sulfotransferase from human intestinal Caco-2 cells. *Life Sci.* 49:1197–206.

23. Carriere, V., T. Lesuffleur, A. Barbat, et al. 1994. Expression of cytochrome P-450 3A in HT29-MTX cells and Caco-2 clone TC7. *FEBS Lett.* 355:247–50.

24. Artursson, P., and J. Karlsson. 1991. Correlation between oral drug absorption in humans and apparent drug permeability coefficients in human intestinal epithelial (Caco-2) cells. *Biochem. Biophys. Res. Commun.* 175:880–5.

25. Chong, S., S. A. Dando, K. M. Soucek, et al. 1996. In vitro permeability through Caco-2 cells is not quantitatively predictive of *in vivo* absorption for peptide-like drugs absorbed via the dipeptide transporter system. *Pharm. Res.* 13:120–3.

26. Yee, S. 1997. In vitro permeability across Caco-2 cells (colonic) can predict *in vivo* (small intestinal) absorption in man-fact or myth. *Pharm. Res.* 14:763–6.

27. U.S. Food and Drug Administration (FDA), Center for Drug Evaluation and Research (CDER). December 2017. Guidance for industry, waiver of *in vivo* bioavailability and bioequivalence studies for immediate-release solid oral dosage forms based on a biopharmaceutics classification system. www.fda.gov/downloads/Drugs/Guidances/ucm070246.pdf (accessed April 11, 2018).

28. U.S. Food and Drug Administration (FDA), Center for Drug Evaluation and Research (CDER). October 2017. Guidance for industry, in vitro metabolism- and transporter-mediated drug-drug interaction studies. www.fda.gov/downloads/Drugs/GuidanceComplianceRegulatoryInformation/Guidances/UCM581965.pdf (accessed April 11, 2018).

29. Walter, E., and T. Kissel. 1995. Heterogeneity in the human intestinal cell line Caco-2 leads to differences in transepithelial transport. *Eur. J. Pharm. Sci.* 3:215–30.

30. Cogburn, J. N., M. G. Donovan, and C. S. Schasteen. 1991. A model of human small intestinal absorptive cells. 1. Transport barrier. *Pharm. Res.* 8:210–6.

31. Boulenc, X., M. Bourrie, I. Fabre, et al. 1992. Regulation of cytochrome P450IA1 gene expression in a human intestinal cell line, Caco-2. *J. Pharmacol. Exp. Ther.* 263:1471–8.

32. Raeissi, S. D., Z. Guo, G. L. Dobson, et al. 1997. Comparison of CYP3A activities in a subclone of Caco-2 cells (TC7) and human intestine. *Pharm. Res.* 14:1019–25.

33. Yoshioka, M., R. H. Erickson, H. Matsumoto, et al. 1991. Expression of dipeptidyl aminopeptidase IV during enterocytic differentiation of human colon cancer (Caco-2) cells. *Int. J. Cancer* 47:916–21.

34. Munzel, P. A., S. Schmohl, H. Heel, et al. 1999. Induction of human UDP glucuronosyltransferases (UGT1A6, UGT1A9, and UGT2B7) by t-butylhydroquinone and 2,3,7,8-tetrachlorodibenzo-p-dioxin in Caco-2 cells. *Drug Metab. Dispos.* 27:569–73.

35. Prueksaritanont, T., L. M. Gorham, J. H. Hochman, et al. 1996. Comparative studies of drug-metabolizing enzymes in dog, monkey, and human small intestines, and in Caco-2 cells. *Drug Metab. Dispos.* 24:634–42.

36. Schmiedlin-Ren, P., K. E. Thummel, J. M. Fisher, et al. 1997. Expression of enzymatically active CYP_3A_4 by Caco-2 cells grown on extracellular matrix-coated permeable supports in the presence of 1α,25-dihydroxyvitamin D_3. *Mol. Pharmacol.* 51:741–54.

37. Crespi, C. L., B. W. Penman, and M. Hu. 1996. Development of Caco-2 cells expressing high levels of cDNA-derived cytochrome P4503A4. *Pharm. Res.* 13:1635–41.

38. Ranaldi, G., K. Islam, and Y. Sambuy. 1992. Epithelial cells in culture as a model for the intestinal transport of antimicrobial agents. *Antimicrob. Agents Chemother.* 36:1374–81.

39. Irvine, J. D., L. Takahashi, K. Lockhart, et al. 1999. MDCK (Madin-Darby canine kidney) cells: A tool for membrane permeability screening. *J. Pharm. Sci.* 88:28–33.

40. Wacher, V. J., C. Y. Wu, and L. Z. Benet. 1995. Overlapping substrate specificities and tissue distribution of cytochrome P450 3A and P-glycoprotein: Implications for drug delivery and activity in cancer chemotherapy. *Mol. Carcinog.* 13:129–34.

41. Tran, C. D., P. Timmins, B. R. Conway, et al. 2002. Investigation of the coordinated functional activities of cytochrome P450 3A4 and P-glycoprotein in limiting the absorption of xenobiotics in Caco-2 cells. *J. Pharm. Sci.* 91:117–28.

42. Cummins, C. L., W. Jacobsen, and L. Z. Benet. 2002. Unmasking the dynamic interplay between intestinal P-glycoprotein and CYP3A4. *J. Pharmacol. Exp. Ther.* 300:1036–45.

43. Hochman, J. H., M. Chiba, M. Yamazaki, et al. 2001. P-glycoprotein-mediated efflux of indinavir metabolites in Caco-2 cells expressing cytochrome P450 3A4. *J. Pharmacol. Exp. Ther.* 298:323–30.

44. Vaessen, S. F., M. M. van Lipzig, R. H. Pieters, et al. 2017. Regional expression levels of drug transporters and metabolizing enzymes along the pig and human intestinal tract and comparison with Caco-2 cells. *Drug Metab. Dispos.* 45:353–60.

45. Cummins, C. L., L. Salphati, M. J. Reid, et al. 2003. *In vivo* modulation of intestinal CYP3A metabolism by P-glycoprotein: Studies using the rat single-pass intestinal perfusion model. *J. Pharmacol. Exp. Ther.* 305:306–14.

46. Tamura, S., Y. Tokunaga, R. Ibuki, et al. 2003. The site-specific transport and metabolism of tacrolimus in rat small intestine. *J. Pharmacol. Exp. Ther.* 306:310–6.

47. Gres, M. C., B. Julian, M. Bourrie, et al. 1998. Correlation between oral drug absorption in humans, and apparent drug permeability in TC-7 cells, a human epithelial intestinal cell line: Comparison with the parental Caco-2 cell line. *Pharm. Res.* 15:726–33.

48. Yazdanian, M., S. L. Glynn, J. L. Wright, et al. 1998. Correlating partitioning and Caco-2 cell permeability of structurally diverse small molecular weight compounds. *Pharm. Res.* 15:1490–4.

49. Hidalgo, I. J. 2001. Assessing the absorption of new pharmaceuticals. *Curr. Top. Med. Chem.* 1:385–401.

50. Volpe, D. A. 2008. Variability in Caco-2 and MDCK cell-based intestinal permeability assays. *J. Pharm. Sci.* 97:712–25.

51. Rothen-Rutishauser, B., A. Braun, M. Gunthert, et al. 2000. Formation of multilayers in the Caco-2 cell culture model: A confocal laser scanning microscopy study. *Pharm. Res.* 17:460–5.

52. Gallop, M. A., R. W. Barrett, W. J. Dower, et al. 1994. Applications of combinatorial technologies to drug discovery. 1. Background and peptide combinatorial libraries. *J. Med. Chem.* 37:1233–51.

53. Gordon, E. M., R. W. Barrett, W. J. Dower, et al. 1994. Applications of combinatorial technologies to drug discovery. 2. Combinatorial organic synthesis, library screening strategies, and future directions. *J. Med. Chem.* 37:1385–401.

54. Alsenz, J., and E. Haenel. 2003. Development of a 7-day, 96-well Caco-2 permeability assay with high-throughput direct UV compound analysis. *Pharm. Res.* 20:1961–9.

55. Balimane, P. V., K. Patel, A. Marino, et al. 2004. Utility of 96 well Caco-2 cell system for increased throughput of P-gp screening in drug discovery. *Eur. J. Pharm. Biopharm.* 58:99–105.

56. Kansy, M., F. Senner, and K. Gubernator. 1998. Physicochemical high throughput screening: Parallel artificial membrane permeation assay in the description of passive absorption processes. *J. Med. Chem.* 41:1007–10.

57. Caldwell, G. W., S. M. Easlick, J. Gunnet, et al. 1998. In vitro permeability of eight beta-blockers through Caco-2 monolayers utilizing liquid chromatography/electrospray ionization mass spectrometry. *J. Mass Spectrom.* 33:607–14.

58. Whitehead, K., and S. Mitragotri. 2008. Mechanistic analysis of chemical permeation enhancers for oral drug delivery. *Pharm. Res.* 25:1412–9.

59. Frank, K. J., U. Westedt, K. M. Rosenblatt, et al. 2014. What is the mechanism behind increased permeation rate of a poorly soluble drug from aqueous dispersions of an amorphous solid dispersion? *J. Pharm. Sci.* 103:1779–86.

60. Lohani, S., H. Cooper, X. Jin, et al. 2014. Physicochemical properties, form, and formulation selection strategy for a biopharmaceutical classification system class II preclinical drug candidate. *J. Pharm. Sci.* 103:3007–21.

61. Forner, K., I. Hidalgo, J. Lin, et al. 2017. Dissolution/permeation: The importance of the experimental setup for the prediction of formulation effects on fenofibrate *in vivo* performance. *Pharmazie* 72:581–586.

62. Noe, J., R. Portmann, M. E. Brun, et al. 2007. Substrate-dependent drug-drug interactions between gemfibrozil, fluvastatin and other organic anion-transporting peptide (OATP) substrates on OATP1B1, OATP2B1, and OATP1B3. *Drug Metab. Dispos.* 35:1308–14.

63. Bachmakov, I., H. Glaeser, M. F. Fromm, et al. 2008. Interaction of oral antidiabetic drugs with hepatic uptake transporters: Focus on organic anion transporting polypeptides and organic cation transporter 1. *Diabetes* 57:1463–9.

64. Hodin, S., T. Basset, E. Jacqueroux, et al. 2018. In vitro comparison of the role of P-glycoprotein and breast cancer resistance protein on direct oral anticoagulants disposition. *Eur. J. Drug Metab. Pharmacokinet.* 43:183–191.

65. Fenner, K. S., M. D. Troutman, S. Kempshall, et al. 2009. Drug-drug interactions mediated through P-glycoprotein: Clinical relevance and in vitro-*in vivo* correlation using digoxin as a probe drug. *Clin. Pharmacol. Ther.* 85:173–81.

66. Takeuchi, T., S. Yoshitomi, T. Higuchi, et al. 2006. Establishment and characterization of the transformants stably-expressing MDR1 derived from various animal species in LLC-PK1. *Pharm. Res.* 23:1460–72.

67. Beauchamp, L.M., G.F. Orr, P. de Miranda, et al. 1992. Amino acid ester prodrugs of acyclovir. *Antivir. Chem. Chemother.* 3:157–164.

68. Ganapathy, M. E., W. Huang, H. Wang, et al. 1998. Valacyclovir: A substrate for the intestinal and renal peptide transporters PEPT1 and PEPT2. *Biochem. Biophys. Res. Commun.* 246:470–5.

69. Tamai, I., T. Nakanishi, H. Nakahara, et al. 1998. Improvement of L-dopa absorption by dipeptidyl derivation, utilizing peptide transporter PepT1. *J. Pharm. Sci.* 87:1542–6.

70. Tomita, M., T. Sawada, T. Ogawa, et al. 1992. Differences in the enhancing effects of sodium caprate on colonic and jejunal drug absorption. *Pharm. Res.* 9:648–53.

71. Sancho-Chust, V., M. Bengochea, S. Fabra-Campos, et al. 1995. Experimental studies on the influence of surfactants on intestinal absorption of drugs. Cefadroxil as model drug and sodium lauryl sulfate as model surfactant: Studies in rat duodenum. *Arzneimittelforschung* 45:1013–7.

72. Constantinides, P., C. M. Lancaster, J. Marcello, et al. 1995. Enhanced intestinal absorption of an RGD peptide from water-in-oil microemulsions of different composition and particle size. *J. Control. Rel.* 34:109–16.

73. Raiman, J., R. Niemi, J. Vepsalainen, et al. 2001. Effects of calcium and lipophilicity on transport of clodronate and its esters through Caco-2 cells. *Int. J. Pharm.* 213:135–42.

74. Rege, B. D., L. X. Yu, A. S. Hussain, et al. 2001. Effect of common excipients on Caco-2 transport of low-permeability drugs. *J. Pharm. Sci.* 90:1776–86.

75. Kataoka, M., Y. Masaoka, Y. Yamazaki, et al. 2003. In vitro system to evaluate oral absorption of poorly water-soluble drugs: Simultaneous analysis on dissolution and permeation of drugs. *Pharm. Res.* 20:1674–80.

76. Li, J., and I. J. Hidalgo. 2017. System for the concomitant assessment of drug dissolution, absorption and permeation and methods of using the same. U.S. Patent No. 9,546,991B2.

77. Pfrunder, A., H. Gutmann, C. Beglinger, et al. 2003. Gene expression of CYP3A4, ABC-transporters (MDR1 and MRP1-MRP5) and hPXR in three different human colon carcinoma cell lines. *J. Pharm. Pharmacol.* 55:59–66.

78. Gutmann, H., G. Fricker, M. Torok, et al. 1999. Evidence for different ABC-transporters in Caco-2 cells modulating drug uptake. *Pharm. Res.* 16:402–7.

79. Taipalensuu, J., H. Tornblom, G. Lindberg, et al. 2001. Correlation of gene expression of ten drug efflux proteins of the ATP-binding cassette transporter family in normal human jejunum and in human intestinal epithelial Caco-2 cell monolayers. *J. Pharmacol. Exp. Ther.* 299:164–70.

80. Artursson, P. 1990. Epithelial transport of drugs in cell culture. I: A model for studying the passive diffusion of drugs over intestinal absorptive (Caco-2) cells. *J. Pharm. Sci.* 79:476–82.

81. Lennernas, H. 1995. Does fluid flow across the intestinal mucosa affect quantitative oral drug absorption? Is it time for a reevaluation? *Pharm. Res.* 12:1573–82.

82. Fine, K. D., C. A. Santa Ana, J. L. Porter, et al. 1993. Effect of D-glucose on intestinal permeability and its passive absorption in human small intestine *in vivo*. *Gastroenterology* 105:1117–25.

83. Soergel, K. H. 1993. Showdown at the tight junction. *Gastroenterology* 105:1247–50.

84. Knipp, G. T., N. F. Ho, C. L. Barsuhn, et al. 1997. Paracellular diffusion in Caco-2 cell monolayers: Effect of perturbation on the transport of hydrophilic compounds that vary in charge and size. *J. Pharm. Sci.* 86:1105–10.

85. Artursson, P., and C. Magnusson. 1990. Epithelial transport of drugs in cell culture. II: Effect of extracellular calcium concentration on the paracellular transport of drugs of different lipophilicities across monolayers of intestinal epithelial (Caco-2) cells. *J. Pharm. Sci.* 79:595–600.

86. Boulenc, X., C. Roques, H. Joyeux, et al. 1995. Biophosphonates increase tight junction permeability in the human intestinal epithelial (Caco-2) model. *Int. J. Pharm.* 123:13–24.

87. Gan, L. S., P. H. Hsyu, J. F. Pritchard, et al. 1993. Mechanism of intestinal absorption of ranitidine and ondansetron: Transport across Caco-2 cell monolayers. *Pharm. Res.* 10:1722–5.

88. Tanaka, Y., Y. Taki, T. Sakane, et al. 1995. Characterization of drug transport through tight-junctional pathway in Caco-2 monolayer: Comparison with isolated rat jejunum and colon. *Pharm. Res.* 12:523–8.

89. Pappenheimer, J. R., and K. Z. Reiss. 1987. Contribution of solvent drag through intercellular junctions to absorption of nutrients by the small intestine of the rat. *J. Membr. Biol.* 100:123–36.

90. Madara, J. L., and J. R. Pappenheimer. 1987. Structural basis for physiological regulation of paracellular pathways in intestinal epithelia. *J. Membr. Biol.* 100:149–64.

91. Tsuji, A., and I. Tamai. 1996. Carrier-mediated intestinal transport of drugs. *Pharm. Res.* 13:963–77.

92. Bai, J. P., M. Hu, P. Subramanian, et al. 1992. Utilization of peptide carrier system to improve intestinal absorption: Targeting prolidase as a prodrug-converting enzyme. *J. Pharm. Sci.* 81:113–6.

93. Zhang, W., J. Li, S. M. Allen, et al. 2009. Silencing the breast cancer resistance protein expression and function in caco-2 cells using lentiviral vector-based short hairpin RNA. *Drug Metab. Dispos.* 37:737–44.

94. Li, J., D. A. Volpe, Y. Wang, et al. 2011. Use of transporter knockdown Caco-2 cells to investigate the in vitro efflux of statin drugs. *Drug Metab. Dispos.* 39:1196–202.

95. Darnell, M., J. E. Karlsson, A. Owen, et al. 2010. Investigation of the involvement of P-glycoprotein and multidrug resistance-associated protein 2 in the efflux of ximelagatran and its metabolites by using short hairpin RNA knockdown in Caco-2 cells. *Drug Metab. Dispos.* 38:491–7.

96. Sampson, K. E., A. Brinker, J. Pratt, et al. 2015. Zinc finger nuclease-mediated gene knockout results in loss of transport activity for P-glycoprotein, BCRP, and MRP2 in Caco-2 cells. *Drug Metab. Dispos.* 43:199–207.

97. Cook, J.A., and H. N. Bockbrader. 2002. An industrial implementation of the biopharmaceutics classification system. *Dissolution Technol.* 9:6–8.

98. Cardot, J. M., A. Garcia Arieta, P. Paixao, et al. 2016. Implementing the biopharmaceutics classification system in drug development: Reconciling similarities, differences, and shared challenges in the EMA and US-FDA-recommended approaches. *AAPS J.* 18:1039–46.

99. Mehta, M. U., R. S. Uppoor, D. P. Conner, et al. 2017. Impact of the US FDA "Biopharmaceutics Classification System" (BCS) Guidance on Global Drug Development. *Mol. Pharm.* 14:4334–8.

100. Davit, B. M., I. Kanfer, Y. C. Tsang, et al. 2016. BCS biowaivers: Similarities and differences among EMA, FDA, and WHO requirements. *AAPS J.* 18:612–8.

101. Hu, M., and R. T. Borchardt. 1992. Transport of a large neutral amino acid in a human intestinal epithelial cell line (Caco-2): Uptake and efflux of phenylalanine. *Biochim. Biophys. Acta* 1135:233–44.

102. Trotter, P. J., and J. Storch. 1991. Fatty acid uptake and metabolism in a human intestinal cell line (Caco-2): Comparison of apical and basolateral incubation. *J. Lipid Res.* 32:293–304.

103. Bravo, S. A., C. U. Nielsen, J. Amstrup, et al. 2004. In-depth evaluation of Gly-Sar transport parameters as a function of culture time in the Caco-2 cell model. *Eur. J. Pharm. Sci.* 21:77–86.

104. Woodcock, S., I. Williamson, I. Hassan, et al. 1991. Isolation and characterization of clones from the Caco-2 cell line displaying increased taurocholic acid transport. *J. Cell Sci.* 98:323–32.

105. Peterson, M. D., and M. S. Mooseker. 1992. Characterization of the enterocyte-like brush border cytoskeleton of the C2BBe clones of the human intestinal cell line, Caco-2. *J. Cell Sci.* 102:581–600.

106. Horie, K., F. Tang, and R. T. Borchardt. 2003. Isolation and characterization of Caco-2 subclones expressing high levels of multidrug resistance protein efflux transporter. *Pharm. Res.* 20:161–8.

107. Hu, M., P. Subramanian, H. I. Mosberg, et al. 1989. Use of the peptide carrier system to improve the intestinal absorption of L-alpha-methyldopa: Carrier kinetics, intestinal permeabilities, and in vitro hydrolysis of dipeptidyl derivatives of L-alpha-methyldopa. *Pharm. Res.* 6:66–70.

108. Han, H., R. L. de Vrueh, J. K. Rhie, et al. 1998. 5′-Amino acid esters of antiviral nucleosides, acyclovir, and AZT are absorbed by the intestinal PEPT1 peptide transporter. *Pharm. Res.* 15:1154–9.

109. Steingrimsdottir, H., A. Gruber, C. Palm, et al. 2000. Bioavailability of aciclovir after oral administration of aciclovir and its prodrug valaciclovir to patients with leukopenia after chemotherapy. *Antimicrob. Agents Chemother.* 44:207–9.

110. Ginski, M. J., and J. E. Polli. 1999. Prediction of dissolution-absorption relationships from a dissolution/Caco-2 system. *Int. J. Pharm.* 177:117–25.

111. Kataoka, M., Y. Masaoka, S. Sakuma et al. 2006. Effect of food intake on the oral absorption of poorly water-soluble drugs: In vitro assessment of drug dissolution and permeation assay system. *J. Pharm. Sci.* 95:2051–61.

112. Buch, P., P. Langguth, M. Kataoka, et al. 2009. IVIVC in oral absorption for fenofibrate immediate release tablets using a dissolution/permeation system. *J. Pharm. Sci.* 98:2001–9.

113. Buch, P., P. Holm, J. Q. Thomassen, et al. 2010. IVIVR in oral absorption for fenofibrate immediate release tablets using dissolution and dissolution permeation methods. *Pharmazie* 65:723–8.

114. Kataoka, M., S. Itsubata, Y. Masaoka, et al. 2011. In vitro dissolution/permeation system to predict the oral absorption of poorly water-soluble drugs: Effect of food and dose strength on it. *Biol. Pharm. Bull.* 34:401–7.

115. Kataoka, M., K. Sugano, C. da Costa Mathews, et al. 2012. Application of dissolution/permeation system for evaluation of formulation effect on oral absorption of poorly water-soluble drugs in drug development. *Pharm. Res.* 29:1485–94.

116. Shi, Y., P. Gao, Y. Gong, et al. 2010. Application of a biphasic test for characterization of in vitro drug release of immediate release formulations of celecoxib and its relevance to *in vivo* absorption. *Mol. Pharm.* 7:1458–65.

117. Kostewicz, E. S., L. Aarons, M. Bergstrand, et al. 2014. PBPK models for the prediction of *in vivo* performance of oral dosage forms. *Eur. J. Pharm. Sci.* 57:300–21.

118. Fei, Y., E. S. Kostewicz, M. T. Sheu, et al. 2013. Analysis of the enhanced oral bioavailability of fenofibrate lipid formulations in fasted humans using an in vitro-in silico-*in vivo* approach. *Eur. J. Pharm. Biopharm.* 85:1274–84.

119. Lipinski, C.A., F. Lombardo, B.W. Dominy, et al. 1997. Experimental and computational approaches to estimate solubility and permeability in drug discovery and development settings. *Adv. Drug Deliv. Rev.* 23:3–25.

120. Palm, K., K. Luthman, A. L. Ungell, et al. 1996. Correlation of drug absorption with molecular surface properties. *J. Pharm. Sci.* 85:32–9.

121. Klopman, G., L. R. Stefan, and R. D. Saiakhov. 2002. ADME evaluation. 2. A computer model for the prediction of intestinal absorption in humans. *Eur. J. Pharm. Sci.* 17:253–63.

122. Burton, P. S., J. T. Goodwin, T. J. Vidmar, et al. 2002. Predicting drug absorption: How nature made it a difficult problem. *J. Pharmacol. Exp. Ther.* 303:889–95.

123. Parrott, N., and T. Lave. 2002. Prediction of intestinal absorption: Comparative assessment of GASTROPLUS and IDEA. *Eur. J. Pharm. Sci.* 17:51–61.

124. Stenberg, P., U. Norinder, K. Luthman, et al. 2001. Experimental and computational screening models for the prediction of intestinal drug absorption. *J. Med. Chem.* 44:1927–37.

125. Palm, K., P. Stenberg, K. Luthman, et al. 1997. Polar molecular surface properties predict the intestinal absorption of drugs in humans. *Pharm. Res.* 14:568–71.

126. Oprea, T. I., and J. Gottfries. 1999. Toward minimalistic modeling of oral drug absorption. *J. Mol. Graph. Model.* 17:261–74.

127. Amidon, G. L., H. Lennernas, V. P. Shah, et al. 1995. A theoretical basis for a biopharmaceutic drug classification: The correlation of in vitro drug product dissolution and *in vivo* bioavailability. *Pharm. Res.* 12:413–20.

128. da Silva Junior, J. B., T. M. Dezani, A. B. Dezani, et al. 2015. Evaluating potential Pgp substrates: Main aspects to choose the adequate permeability model for assessing gastrointestinal drug absorption. *Mini Rev. Med. Chem.* 15:858–71.

129. Noyes, A. A., and W. R. Whitney. 1897. The rate of solution of solid substance in their own solutions. *J. Am. Chem. Soc.* 19:930–4.

130. Dokoumetzidis, A., and P. Macheras. 2006. A century of dissolution research: From Noyes and Whitney to the biopharmaceutics classification system. *Int. J. Pharm.* 321:1–11.

131. Nicolaides, E., M. Symillides, J. B. Dressman, et al. 2001. Biorelevant dissolution testing to predict the plasma profile of lipophilic drugs after oral administration. *Pharm. Res.* 18:380–8.

132. Sugano, K. 2009. Introduction to computational oral absorption simulation. *Expert Opin. Drug Metab. Toxicol.* 5:259–93.

133. Yu, L. X., E. Lipka, J. R. Crison, et al. 1996. Transport approaches to the biopharmaceutical design of oral drug delivery systems: Prediction of intestinal absorption. *Adv. Drug Deliv. Rev.* 19:359–76.

134. Agoram, B., W. S. Woltosz, and M. B. Bolger. 2001. Predicting the impact of physiological and biochemical processes on oral drug bioavailability. *Adv. Drug Deliv. Rev.* 50 Suppl 1:S41–67.

135. Jamei, M., D. Turner, J. Yang, et al. 2009. Population-based mechanistic prediction of oral drug absorption. *AAPS J.* 11:225–37.

136. Lennernas, H. 1998. Human intestinal permeability. *J. Pharm. Sci.* 87:403–10.

15 Preclinical Toxicology

Damani Parran
Nouryon Chemicals

CONTENTS

15.1 INTRODUCTION

Developing new therapies to treat disease and to improve quality of life is a long complex and costly process. The Pharmaceutical Research and Manufacturers of America (PhRMA) estimates that to launch a new pharmaceutical product to the market, the costs can exceed $1.2 billion and take 10–15 years of discovery and development (McVean, 2014). For every 5,000–10,000 compounds that enter the pipeline, only one will receive approval. Even medicines that reach clinical trials have only a 16% chance of being approved. The development of a pharmaceutical is a stepwise process involving the evaluation of both animal and human efficacy and safety information. The ultimate safety assessment of any new pharmaceutical is derived from extensive clinical studies. However, before a new pharmaceutical is administered to human subjects, an extensive series of preclinical studies are conducted to support the safe-usage in clinical trials. The goals of the preclinical safety evaluation generally include a characterization of toxic effects with respect to target organs, dose dependence, relationship to exposure, and, when appropriate, potential reversibility. Preclinical studies are conducted to define pharmacological and toxicological effects not only prior to initiation of human studies but throughout clinical development. This information is used to estimate an initial safe starting dose and dose range for the human trials and to identify parameters for clinical monitoring for potential adverse effects. The preclinical safety studies, although usually limited at the beginning of clinical development, help to define the conditions of the clinical trial to be supported. Both *in vitro* and *in vivo* can contribute to this characterization. Preclinical safety testing should consider the following: (a) selection of the relevant animal species; (b) age; (c) physiological sate; (d) the manner of delivery, including dose, route of administration, and treatment regimen; and (e) stability of the test material under the conditions of use. This chapter will focus on the regulatory preclinical toxicology requirements during pharmaceutical development. In addition, this chapter will focus on preclinical toxicology models and their role in this very complex process.

15.2   OVERVIEW OF DRUG REGULATIONS

Regulation of the development, production, marketing, and sales of pharmaceuticals entails contradictory objectives. Regulations are designed to ensure that new and effective medical treatments reach the public rapidly while, simultaneously, provide protection from ineffective or even unsafe products. In addition, the regulations must protect from predatory marketing practices that promote unsubstantiated claims to vulnerable consumers/patients. Balancing these goals falls globally in large measure to the Food and Drug Administration (FDA) in the United States and to regional and centralized regulatory bodies in the European Union (EU) (Kashyap et al., 2013). Both the FDA and the regulatory bodies in the EU have implemented proactive approaches for safety surveillance and risk assessment. This was in part due to the withdrawal of a number of high-profile branded pharmaceuticals including Seldane (Friedman et al., 1999), Vioxx (FDA, 2005), and Rezulin. Such debacles caused regulatory authorities to change their emphasis from the reactive collection of safety data to a more proactive risk management approach. In addition, public scrutiny of regulatory authorities has increased the focus on drug safety surveillance with the downstream impact of increased regulatory requirements for post-marketing pharmacovigilance.

15.2.1   HISTORY AND EVOLUTION OF THE US FOOD AND DRUG ADMINISTRATION

The FDA is the first consumer protection agency in the United States, originated in the US Patent Office in 1848 and later inherited by the Department of Agriculture in 1862 (FDA, 2018). The modern function of the agency in oversight of drug and medical device marketing was ultimately codified in the Pure Food and Drug Act of 1906, which was passed in response to a pressing need to curb interstate markets for adulterated and mishandled food and pharmaceuticals. The Pure Food and Drug Act of 1906 was the first of a series of significant consumer protection laws which was enacted by Congress in the 20th century. The primary objective of this Act was to ban foreign and interstate traffic in adulterated or mislabeled food and drug products. In addition, it directed the US Bureau of Chemistry in the US Department of Agriculture (which was renamed US Food and Drug Administration in 1930) to inspect products and refer offenders to prosecutors. The basis of the law rested on the regulation of product labeling rather than pre-market approval. The Act of 1906 required that active ingredients be placed on the label of a drug's packaging and that drugs could not fall below purity levels established by the United States Pharmacopeia or the National Formulary. A major point that the Act of 1906 did not regulate was drug efficacy.

To address the lack of drug efficacy oversight, the Food, Drug and Cosmetic (FD&C) Act passed and enacted in 1938. The FD&C Act required all drugs to be approved for safety by the FDA. The FD&C Act would replace the Act of 1906 was ultimately enhanced and passed in the wake of a therapeutic disaster in 1937. A Tennessee drug company marketed a form of the new sulfa wonder drug that would appeal to pediatric patients, Elixir Sulfanilamide. However, the solvent in this untested product was a highly toxic chemical analogue of antifreeze; over 100 people died, many of whom were children. The public outcry not only reshaped the drug provisions of the new law to prevent such an event from happening again but also propelled the bill itself through Congress. FDR signed the Food, Drug, and Cosmetic Act on 25 June 1938. The new law brought cosmetics and medical devices under control, and it required that drugs be labeled with adequate directions for safe use. Moreover, it mandated pre-market approval of all new drugs, such that a manufacturer would have to prove to FDA that drugs were safe before it could be sold. It irrefutably prohibited false therapeutic claims for drugs, although a separate law granted the Federal Trade Commission jurisdiction over drug advertising. The FD&C Act formally authorized factory inspections, and it added injunctions to the enforcement tools at the agency's disposal. Within 2 months of the passage of the Act, the FDA began to identify drugs such as the sulfas that simply could not be labeled for safe use directly by the patient – they would require a prescription from a physician. The ensuing debate by the FDA, industry, and health practitioners over what constituted a prescription and an over-the-counter drug was resolved in the Durham–Humphrey Amendment of 1951. From the 1940s to the 1960s, the abuse of amphetamines and barbiturates required more regulatory effort by FDA than all other drug problems combined. Furthermore, the new law ushered in a flood of new drugs applications, over 6,000 in the first 9 years, and 13,000 by 1962.

The FDA's mission for safe drug approval was expanded in 1962 by the Kefauver–Harris Amendments that added the requirement that drugs be proven "effective" as well as safe. These amendments placed strict controls on the use of investigational drugs. Regulations regarding drug safety oversight were expanded in 1976 to include medical devices. As with the FD&C Act of 1938, a therapeutic disaster compelled passage of the Kefauver–Harris Amendments; in this case, the disaster was narrowly averted. Thalidomide, a sedative that was never approved in this country, produced thousands of grossly deformed newborns outside of the United States. Three years later, Congress gave the FDA enhanced control over amphetamines, barbiturates, hallucinogens, and other drugs of considerable abuse potential in the Drug Abuse Control Amendments of 1965. That function was consolidated with similar responsibilities in 1968 under an organization that gave rise to the Drug Enforcement Administration.

Throughout the rest of the 20th century, FDA's role has undergone significant changes due to expanding federal regulations, increasing complexity of drugs and devices and the growth of the pharmaceutical industry into a major economic force in the United States. Currently, the United States has among the most stringent drug and medical device regulations for development and marketing. These strict regulations may have served the public with enhanced assurance of therapeutic safety; however, it has faced criticism. There are concerns that such strict regulations have caused a "drug lag" resulting in the decline in the number of drugs approved by the FDA from an average of 50 per year in the late 1950s to approximately 17 per after 1965. It is unclear whether FDA regulations were entirely responsible for the deceleration, because foreign countries also experienced a lag. However, it was nevertheless obvious that new drugs were reaching the market in other countries months to years before achieving FDA approval in the United States. Others are concerned that drug approval in the European Union (EU) is too quick and may be a detriment to patient safety.

15.2.2   REGULATION OF DRUGS BY THE **FDA** IN THE UNITED STATES

The United States historically has been the world's largest market for pharmaceuticals. However, this lead has been narrowing as managed care and generic substitution in the United States have slowed the growth of the drug industry by some economic measures (Vilas-Boas and Tharp, 1997). Access to the United States market demands compliance with the FDA's requirements. The FDA has a standard procedure for drug approval, and most drugs must go through this process. The first step is the completion of a large number of preclinical studies. Data from these studies are used in the completion and submission of an Investigative New Drug (IND) application. This application is submitted to the FDA for permission to conduct clinical studies and the transport of drugs across states. The clinical evaluation of a drug agent involves several phases where Phases 0–1 involve a small number of healthy subjects to clarify the pharmacology and dose range; Phase 2 includes several hundred patients with the targeted condition and to determine a dose–response relationship; and Phase 3 where several hundred to several thousand patients are administered the drug to show safety and efficacy. A list of many of the required preclinical and nonclinical studies is presented in Figure 15.1. After the completion of controlled clinical trials under an IND, pharmaceutical companies seeking marketing approval in the United States must now submit a

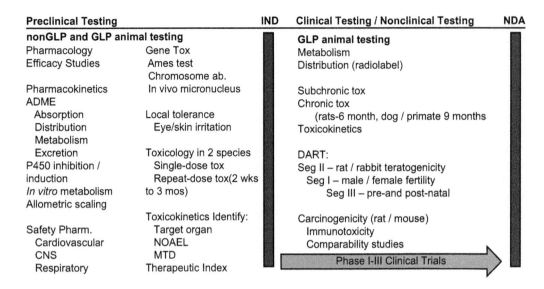

Preclinical Testing		IND	Clinical Testing / Nonclinical Testing	NDA
nonGLP and GLP animal testing			**GLP animal testing**	

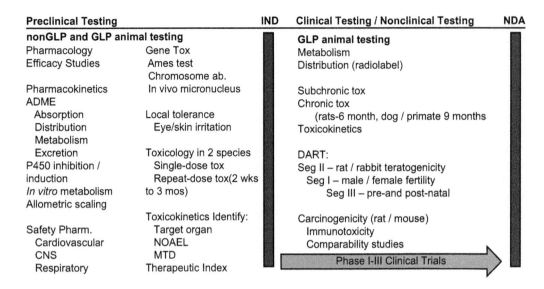

FIGURE 15.1 Preclinical and nonclinical studies: List of required preclinical and nonclinical studies during drug development.

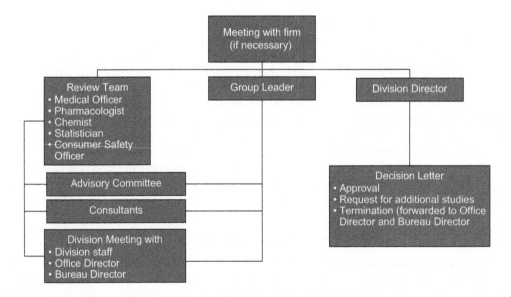

FIGURE 15.2 FDA review process for new drug applications: FDA in-house review teams.

new drug application (NDA) to the FDA. The NDA review process is conducted in-house by teams within the FDA, as indicated in Figure 15.2, with each team responsible for reviewing drugs in a particular therapeutic class (Grabowski and Vernon, 1983). At the end of the lengthy review process, the company is issued a decision letter, which may be an approval letter, a request for additional studies, or a termination letter.

15.2.3 REGULATION OF DRUGS IN THE EUROPEAN UNION

The evolution of the European regulations of drugs is much more recent than the FDA with significant changes after the formation of the European Union (EU) in 1993. Prior to the EU, regulation and marketing approval for drugs was enforced by the member states. Difference in regulations between the European member states often delayed the approval of drugs and even caused "protectionist" legislation to shield against revival state market competition. Since 1993, the EU member states have been reorganized and standardized, including interstate agencies. Efforts to standardize the European drug approval regulations came to fruition before the formation of the EU with the passage of EC Directive 75/65/EEC in 1965 (Council Directive, 1965). The Directive defined a medical product as "any substance or combination of substances which be administered to human beings or animals with a view to making a medical diagnosis or to restoring, correcting or modifying physiological functions in human beings or in animals." Under the Directive, any medicinal product marketed in the member states would first pass approval in the originating state. The Directive established consistent guidelines throughout the member states regarding the information that must be submitted for approval. This process is parallel to the regulations of the FDA regarding investigation of new drug applications and new drug approval applications. Many of the processes to approve drug in the EU are similar to those of the FDA. Figure 15.3 shows the comparison of the drug approval processes in the United States and the EU (Van Norman, 2016).

Similar to the drug approval process in the United States, the first step is for the investigator to obtain pre-authorization for use of the drug in clinical trials. All European clinical trials were regulated under the Clinical Trials Directive of the European Commission (2001/20/EC), later repealed and replaced in 2014 by Regulation No. 536/2014 of the European Parliament (European Commission, 2014). The next step of the drug approval process in the EU is also similar to the United States with the drug proceeding through the sequence of analogous studies: Phase I trials conducted in a small number of healthy subjects to determine the pharmacology and dose range, Phase II trials conducted in several hundred patients with the target condition to investigate the dose–response relationship, and Phase III confirmatory trials conducted in several hundred to several thousand patients to substantiate safety and efficacy. As in the United States, the EC provides means for approving "orphan drugs," or drugs that treat conditions affecting a very low number of patients, making it difficult to complete a randomized controlled trial. There are also methods for obtaining conditional approval for drugs to be used in emergency conditions or other conditional approvals (European Medicines Agency, 2004).

The final steps in the drug approval process in the EU differ from the United States. Unlike the FDA, the European Medicines Agency (EMA) does not oversee all drug approvals. In Europe, there are

United States	European Union

1. Application
   • Application to the FDA for permission to conduct clinical studies and transport drugs across states
	1. Application
	   • Application within one or more states of the EU for approval to conduct clinical studies; each state designates its own regulatory body that will carry out approvals

2. Clinical Trials Phase
   • Phase 0 and I: small number of healthy subjects, clarify pharmacology and dose range
   • Phase II: several hundred patients with the target condition, to determine dose / response relationship
   • Phase III: several hundred to several thousand patients to show safety and efficacy
	2. Clinical Trials Phase
	   • Phase 0 and I: small number of healthy subjects, clarify pharmacology and dose range
	   • Phase II: several hundred patients with the target condition, to determine dose / response relationship
	   • Phase III: several hundred to several thousand patients to show safety and efficacy

3. Emergency Use and Orphan Drugs
   • "Orphan drug" applications: special approval process for drugs showing promise in treating illnesses that affect fewer than 200,000 patients
   • EIND (Emergency drug application) process: for life-threatening situations; shorter process to IND approval; full IND approval application process must be initiated, but treatment can proceed after EIND approval
   • Treatment IND process: drug must be in clinical trials and show promise for treatment for life-threatening or serious condition
	3. Emergency Use and Orphan Drugs
	   • "Orphan drug" applications: special consideration for drugs to treat conditions experiences by < 1/50,000 patients annually
	   • Emergency drug use provided for in life-threatening situations: drug must be already engages in clinical trials

4. New Drug Application to the FDA	4. 4 Pathways to Drug Approval
	   1. Centralized process through the EMA for designated drugs
	   2. Applications to the designated national body within a single EU state
	   3. Mutual recognition: after approval in a single state, application for mutual recognition in all states via the EMA Decentralized process: simultaneous application in multiple EU states

FIGURE 15.3 Comparison of drug approval process in the United States and the EU: similarities and contrasts between the FDA and EU approval processes.

four routes by which a drug can be approved, depending on the drug class and manufacturer preference (Parvisi and Woods, 2014). The four routes are (a) centralized process, (b) national process, (c) mutual recognition, and (d) decentralized procedure.

The centralized process is controlled through the EMA. Every member state of the EU is represented on the EMA Committee for Medicinal Products, which issues a single license valid in all EU member states. This route of approval is mandatory for some classes of drugs, such as treatments for cancers, HIV/AIDS, and neurodegenerative disorders. For the national process, each EU state may have its own procedures for approving drugs that fall outside of those required to undergo the centralized process. For mutual recognition, the drug approvals in one EU state via that state's national process can obtain marketing authorization in another EU member state. The decentralized procedure allows manufacturers to apply for simultaneous approval in more than one EU state for products that have not yet been authorized in any EU state. This assumes that the drug does not fall under the mandatory centralized process. This is the most common route for drug approval (Parvisi and Woods, 2014).

15.2.4 Drug Regulatory Process in Japan

The Japan Ministry of Health and Welfare (the Koseisho) was established as a governmental administrative body under the Japanese constitution that was framed after World War II (Vilas-Boas and Tharp, 1997). The Ministry is responsible for the operation of the National Health System of Japan, national regulation of hospitals and other health facilities, registration and regulation of many of the health professionals, and approval and surveillance of drug products and other regulatory matters. The Koseisho also administers the reimbursement of providers, including those who prescribe and dispense in the mostly physician-controlled health delivery system.

In Japan, pharmaceutical companies are required to do clinical effectiveness and safety studies on a population of Japanese patients. This policy makes it a great deal more difficult for overseas companies to complete their Japanese new drug applications. On the other hand, the Koseisho recently had issued statements suggesting that it will be more permissive about allowing the submission of foreign clinical studies in support of new drug applications. The registration procedure for foreign companies may become less time-consuming in the future, especially if such companies form a joint venture, co-market or co-develop, or cross-license with a Japanese company.

Japanese regulatory authority has been confronted with the new challenge of regulating biosimilar/follow-on biologic products. To ensure the quality, safety, and efficacy of biosimilar/follow-on biologic products, Japan has published a guideline for quality, safety, and efficacy of biosimilar/follow-on biologics based on the similarity concept outlined by the EMA. Following the adoption of the guideline, two follow-on biologic products have been approved or marketed, and more than ten products are in development in Japan (Yamaguchi and Arato, 2011). WHO and Canada have also published guidelines for similar biotherapeutic products and subsequent entry biologics for products termed biosimilar/follow-on biologics. These guidelines have generally similar fundamental concepts and similar regulatory framework for the licensure of biosimilar/follow-on biologics. However, several regulatory requirements for these products seem to be different from each other. Therefore, it is very interesting to compare the requirements of each guideline and help in the harmonization of regulatory pathway for biosimilar/follow-on biologic products.

15.2.5 International Harmonization Conference

The International Harmonization Conference (ICH) is an effort by the governments and drug industries of the United States, the European Union, and Japan to integrate the global regulation of pharmaceuticals, to provide greater uniformity of drug approval procedures and manufacturing standards, and regulation between countries (Vilas-Boas and Tharp, 1997). This global gold standard undoubtedly will influence the outcome of the negotiations within the ICH, thus having a profound effect on the global market. Furthermore, many nations model their pharmaceutical legislation after this standard.

International minimum expectations for preclinical toxicology studies to support clinical trials are best summarized in ICH M3(R2) "Nonclinical Safety Studies for the Conduct of Human Clinical Trials and Marketing Authorization for Pharmaceuticals" (M3(R2)). The basic principle of this guidance is that preclinical studies should identify potential toxicities that may be associated with multiples of the anticipated clinical dose, formulation concentration, dosing schedule, route of administration, and duration of the proposed clinical studies. In addition to ICH M3(R2), local and regional regulatory agencies tend to adopt guidance documents that more clearly delineate specific expectations for a given region (Denny and Stewart, 2017). The legal requirements for the registration and approval of a pharmaceutical product can and do differ in individual countries. The ICH guidelines recommend international standards and promote harmonization of preclinical safety studies to support human clinical trials.

Preclinical toxicology studies from single dose to life-time exposure studies are required to be conducted under the good laboratory practice (GLP) regulations, as disseminated by the laws of the region in which a submission is planned. In the EU, there are two directives that address the requirements for conducting preclinical studies under GLP; specifically, directive 2004/10/EC and directive 2004/09/EC. These directives define the requirements of the EU Member States to designate the authorities responsible for GLP inspections in their given regions. The directives also define the requirements for reporting and for the mutual acceptance of the data generated from these studies. GLP directive 2004/09/EC includes the OECD revised guidelines for Compliance Monitoring Procedures for GLP. In addition, directive 2004/09/EC covers the OECD guidelines for the Conduct of Test Facility Inspections and Study Audits covers laboratory inspections and study audits for laboratories certified to conduct GLP studies in the EU. Both of the above directives help to assure regulatory authorities that the data submitted are an accurate and trustworthy reflection of the results obtained during the preclinical studies and can therefore be relied upon for making decisions in the human clinical studies.

15.3 PRECLINICAL TOXICOLOGY TESTING

As mentioned earlier, drug products are required to undergo a series of preclinical toxicology studies to gather information regarding the safety of a new drug candidate before it is allowed to enter clinical trials in humans. This is ultimately required for the final approval of the drug in the market. Preclinical toxicology data are generally required by regulatory agencies throughout the world; however, the specifics of the studies may vary. In an effort to unify the regulatory requirements for preclinical toxicology data, the United States, Europe, and Japan developed the International Conference for Harmonization (ICH) compendium of guidance. The ICH increased the efficiency of the development process by staggering the clinical study phases in accordance with the availability of the preclinical toxicology data. These studies are categorized based on duration of exposure, including (a) acute (single dose), subacute (14–28 days), subchronic (90 days), and chronic (180 days, rodents; 270 days, nonrodent). These exposure durations are standard study periods often addressed in regulatory requirements, along with categorization of the length of the studies into subsets based on the length of dosing or frequency of exposures to the drug product. The practical length of a general toxicology study, along with the number of doses administered to animals, is based on specific characteristics of the drug as well as on the expected human exposure during each phase of human clinical trials (Denny and Stewart, 2017).

A basic principle of general toxicology evaluations is that toxic effects are categorized as either an acute or a chronic effect due to drug exposure (Eaton and Klaassen, 2001). An acute effect is defined as an effect that occurs rapidly after a single or only a few exposures to a toxicant. Chronic effects are defined as those that occur after repeated exposures and are further subcategorized into (a) noncarcinogenic and (b) carcinogenic effects (Eaton and Klaassen, 2001; Denny and Stewart, 2017).

Each step in the preclinical toxicology evaluation process is designed to characterize potential dose-related adverse effects with respect to tissues, organs, organ systems, and their potential reversibility (McVean, 2014). Therefore, data from preclinical studies are valuable for estimating a safe initial starting dose for human clinical trials and to identify parameters for clinical monitoring of the potential adverse effects that may occur in humans (McVean, 2014). Preclinical toxicology studies are inherently designed to focus on finding potential adverse effects by testing the drug at exposure levels that are multiple times higher doses than and exceed the expected duration in human clinical trials (McVean, 2014; Denny and Stewart, 2017). The initial steps of the design of preclinical toxicology studies are the selection of the animal models, and the dose selection and routes of administration need to be established.

15.3.1 Animal Test Model Selection

Safety regulations require definitive toxicity studies in at least one animal species, which depends on the type of drug under development. It is much more common to use up to four different animal species in a preclinical toxicity assessment (Denny and Stewart, 2017). The use of this number of animal species is usually not apparent during the initial stages of development; however, it may become clear when planning the conduct of developmental and reproductive studies and the conduct of carcinogenicity studies. An example is the initial selection of the rat and dog during the early phases of drug development, but the selection of the rabbit model to assess the potential for teratogenicity and the mouse model to assess the potential for carcinogenicity (Eaton and Klaassen, 2001).

Regional and harmonized guidance documents specify that definitive toxicity studies should be conducted in at least one and possibly two laboratory animal species (one rodent, e.g., rat or mouse, and one nonrodent, e.g., rabbit, dog, pig, monkey, or other suitable species) (Denny and Stewart, 2017). However, there are many cases where only one species (rodent) may be acceptable, and a nonrodent species is not used or limited to the use of the rabbit (vaccines) or the monkey (antibodies) (Bussiere et al., 2009; Cavagnaro, 2008). The criteria for species selection are typically based on metabolic profiles, pharmacokinetic profiles, species tolerance, and pharmacological activity of the molecule in the species under consideration for safety evaluation. In order to obtain such information during the discovery phases of drug development as well as to assess structure–activity relationships, in vitro and in silico models can be used. Later on in development, in vitro testing may continue to be used, and programs may start to conduct screening evaluations in the in vivo systems as well. Within these in vivo systems, more predictive data for species selection can be generated for determination of metabolic profiles and for determination of systemic bioavailability, pharmacokinetics, and toxicity, which will allow the preclinical development team to make informed decisions regarding the species to be used in the definitive toxicology studies.

15.3.2 Dose Selection and Routes of Administration

The regulatory expectation for general toxicology studies is that the dose selected for the high dose will produce observable toxicity, as defined by decrements either in weight gain or in food consumption, and/or alterations in clinical observations, clinical chemistry, and hematology parameters as well as organ-related alterations. The five general criteria for defining the high dose are maximum tolerated dose (MTD), maximum feasible dose (MFD), 50-fold margin dose, limit dose, and saturation of exposure (M3(R2)). Guidance provided by the Laboratory Animal Science Association is also available regarding dose level selection for regulatory general toxicology studies for pharmaceuticals. The MTD is considered to be a dose where target-organ toxicity is expected to occur but should not cause mortality (Eaton and Klaassen, 2001). Conversely, the MFD is based on the fact that a dose cannot be given at a high enough level to demonstrate a definitive dose-limiting toxicity profile. This MFD determination is influenced by the technical capability to deliver the dose and/or by the physiochemical properties of the drug candidate. The 50-fold margin dose and the limit dose are similar, in practical application to the MFD. High doses that are determined based on these two criteria are not influenced by technical or physiochemical issues but are based on the opinion that any higher dose would not provide any additional relevance for the assessment of clinical safety.

The presumption within the explanations of the four criteria mentioned thus far is that systemic exposure increases with dose. However, the last criterion for high-dose selection evaluates this presumption by direct comparison. Saturation of exposure describes the systemic exposure achieved in animals and correlates that exposure to the dose, sex, species, and the time course of the toxicity study. The data obtained can indicate the absorption limitation of exposure to the drug candidate. In such, a case, the lowest dose that achieves high dose (in the absence of any other dose-limiting constraints).

The process of dose selection for general toxicology studies involves generating and compiling data from early in vitro and in vivo (non-GLP studies) studies to derive the doses proposed in the initial GLP animal studies (Eaton and Klaassen, 2001). The selection of doses for definitive general toxicity studies should cover the dose–response continuum from pharmacologically relevant doses through the MTD. By following such a plan, the doses selected will cover exposures that form relevant multiples of the proposed human therapeutic doses. The clinical indication(s) and the expected route of human exposure should also be considered. There are numerous regulatory positions regarding high-dose selection in toxicity studies (see Table 15.1). These studies are generally designed to maximize the potential for identifying adverse effects, whether "on-target" (associated with the intended pharmacology) or "off-target" (other than the intended biologic target) (Shankar et al., 2006). Typically, use of the term "high dose" reflects a consideration of an MTD or an MFD, when applied to the design of general toxicology studies conducted to support human clinical trials for new potential pharmaceuticals. In some cases, the pharmacological properties of the investigational drug can limit the number of doses that can be administered, or can alter the dosing frequency in a particular species. It is also reasonable to determine a severely toxic dose (STD, i.e., dose that causes death or irreversible severe toxicity) in 10% of rodents. Similar studies are also conducted in nonrodents as a precaution because the target patient population is highly susceptible to unanticipated adverse events. Thus, it becomes quite important to test the upper levels of toxicity and potential recovery from those toxicities. For other pharmaceutical targets, it may be reasonable and justifiable to choose a high dose in the range of the lowest-observed-effect level (LOEL) or lowest-observed-adverse-effect level (LOAEL) based on the premise that such doses would be more applicable to human exposure (Buckley et al., 2008). The basic principles of toxicology indicate that the dose selection in shorter-term studies should be on the higher side, maximizing the potential

TABLE 15.1

Relevant Regulations for Dosing

Regulation	Description
ICH S1C(R2)	Predicted to produce a minimum toxic effect over the course of the toxicity study … may be predicted from a 90-day range-finding study in which minimal toxicity is observed … factors to consider include: ≤10% decrease in body weight gain relative to controls; target organ toxicity; significant alterations in clinical pathology. Maximum doses for general toxicity studies may be based on margins of exposure or limit doses
US IGS	Highest dose in chronic study just high enough to elicit signs of minimal toxicity without significantly altering normal lifespan due to effects other than carcinogenicity. Determined in a 90-day study. Considers alteration in body and organ weight, clinical pathology and more definitive toxic, pathologic, or histopathologic endpoints
CHMP (2008)	Should enable identification of target organ toxicity or other nonspecific toxicity, or until limited by volume of dose … Ideally, systemic exposure to the drug and/or principal metabolites should be a significant multiple of the anticipated clinical systemic exposure … Need for adjustment if unexpected toxicity or lack thereof

for identification of "system failure" using an MTD. As exposure duration increases and additional perspectives are gained from clinical trials, the high dose in the preclinical general toxicology study could correctly move toward the more clinically relevant LOEL/LOAEL. Doses in the range of LOEL/LOAEL allow for development of potential toxicity over longer exposure duration, which is an important part of a toxicology profile (Buckley et al., 2008). The low dose in a general toxicology study should ideally demonstrate a no-observable-effect level (NOEL) while maintaining at least a 1× equivalent of the anticipated human therapeutic dose. However, depending on the level of target engagement, it is uncommon to find effects at the low dose that are related to the pharmacology of the drug candidate but that are deemed to be nonadverse. This would be an example of a dose that can be considered as the no-observable-adverse-effect level (NOAEL), because the effects are either considered not to be a side effect, not biologically relevant, and/or are within estimations of background historical data. An adverse effect is generally defined as a side effect or unintended effect of the drug candidate; however, not all side effects are biologically or physiologically relevant. Also, consideration must be made to the fact that many therapeutic effects (or intended effects) can be considered adverse because the effect can negatively impact biological or physiological process. Thus, the trained toxicologist needs to understand and utilize a number of contributing factors in the toxicological evaluation process. The intermediate dose(s) in a general toxicology study is needed to demonstrate a dose–response relationship, which is critical in defining the optimum pharmacological dose and important in establishing the margins of safety. Acute studies generally are conducted to find the MTD, but this MTD level cannot be administered repeatedly. Thus, the duration of administration will continuously influence the high-dose selection. As the highest dose level is necessarily decreased, the intermediate doses are also decreased to maintain an extrapolation for dose response. For example, it has been observed that the acute MTD might have been ≥1,000 mg/kg, but repeated dosing up to 28 days in duration may only allow for a high dose of ≤100 mg/kg because of toxicity. Because such situations can occur, it is prudent to always consider conducting short duration repeated-dose studies (e.g., 5–14 days) to evaluate tolerance under subacute conditions before conducting the definitive IND enabling repeated-dose studies. Unlike the selection of dose levels, the selection of the route of administration is more straightforward. The most common routes of clinical exposure are oral, parenteral, and dermal. Thus, the historical delivery route for a given class of compound will dictate the route used in the preclinical studies. When multiple routes of exposure are used clinically for a particular drug class, the preclinical administration plan for a drug candidate will still be based on the planned clinical route for the specific drug candidate.

15.4 STUDY TYPES USED IN THE ASSESSMENT OF GENERAL TOXICOLOGY

15.4.1 Acute/Dose-Range–Finding Toxicity Studies

One of the first toxicity tests performed during the drug development process is acute toxicity. Although basic in their nature, the specificity of the data obtained from such a study has specialized utility in regulatory safety assessments (Denny and Stewart, 2017). The objective of these studies is to determine a maximum tolerable dose (MTD) and the maximum feasible dose (MFD) in the initial toxicity characterization. These effects are determined after one or more routes of administration (one route being oral or the intended route of exposure) in one or more species. The species most often used are the mouse and rat, but sometimes, the rabbit and dog are employed (Eaton and Klaassen, 2001). Studies are performed in both adult male and female animals. To determine LD_{50}, the number of animals that die in a 14-day period after a single dosage is tabulated. In addition to mortality and weight, daily examination of test animals is conducted for signs of intoxication, lethargy, behavioral modifications, morbidity, food consumption, and other clinical observations. Gross necropsy is performed to evaluate tissues and organs systems for gross changes. However, in most cases, further histological evaluation of these tissues is not conducted. Acute toxicity tests (a) give a quantitative estimate of acute toxicity for comparison with other substances, (b) identify target organs and other clinical manifestations of acute toxicity, (c) establish the reversibility of the toxic response, and (d) provide dose-ranging guidance for other studies (Eaton and Klaassen, 2001). If there is a reasonable likelihood of substantial exposure to the material by dermal or inhalation exposure, acute dermal and acute inhalation studies are performed.

The data obtained from these studies are the foundation for dose selection in the safety pharmacology studies (i.e., cardiovascular, central nervous system, and pulmonary assessments) and in the definitive IND enabling general toxicity studies (Denny and Stewart, 2017). These early phase acute studies are used to select candidate compounds for potential further development, sometimes together with an abbreviated multiple-dose phase. It should be noted that for any acute study to provide quality information, some knowledge of bioavailability and systemic exposure must be available depending on route of exposure and clinical indication.

15.4.2 ACUTE/REPEATED-DOSE SCREENING STUDIES

The screening acute study can have many formats, but the primary purpose is to determine whether the drug candidate is tolerable at multiples of a potential clinically relevant dose. The added benefit of the screening study is the ability to compare multiple drug candidates in the same study (Denny and Stewart, 2017). When this is done, tolerability indicators, such as alterations in body weight and/or food consumption, as well as observations of abnormal clinical signs, can provide some perspective on which species may be the better lead candidate. At this stage, species selection is typically not complete, but at least a rodent and possibly a nonrodent species should be evaluated. With drug candidates that are intended for daily use, it is also reasonable to assess repeated exposure over a period of 5–7 days' duration. This can facilitate the initial assessment of potential alterations in clinical pathology endpoints and can be useful in the selection of dose levels in a regulatory program, but should not be used as a replacement for any of the acute studies that are intended for submission. This is because the study conditions may lack certain quality standards (e.g., drug-batch purity, environmental controls, and certified food source) necessary to minimize confounding factors in the interpretation of the data.

15.5 REPEATED-DOSE TOXICITY STUDIES

Repeated-dose toxicity studies are performed to obtain information on the toxicity of a drug agent after repeated administration and as an aid to establish doses. Although there is much harmony in the requirements of repeated-dose general toxicity studies intended to support human clinical trials, some regional differences in the suggested minimum duration of repeated-dose general toxicity studies do exist (M3(R2)). For example, in the United States and European Union, 2-week rodent and nonrodent studies can support a single-dose or short-term repeated-dose human clinical trial, while in Japan, a 4-week rodent and 2-week nonrodent study may be needed to support the Phase I clinical trials. In the repeated-dose preclinical studies, the goal is to determine the "no-observed-adverse-effect level" (NOAEL). Due to the fact that limited preclinical data are available, it is important to understand some basic "milestones" that can allow the investigator to make decisions about the progression or termination of the development of a species.

15.5.1 SUBACUTE TOXICITY STUDIES (2–4 WEEKS)

The duration of the repeated-dose toxicity studies is usually related to the duration, therapeutic indication, and proposed dose period of the Phase I clinical trial. For some clinical indications, a 2-week study may be sufficient to support the Phase I clinical trial. However, for most products, a 4-week study is necessary to provide sufficient safety information for trials where more than single doses are planned. The duration of the animal toxicity studies should equal, or exceed, the duration of the human clinical trials, up to the maximum recommended duration of the preclinical repeated-dose toxicity studies. The 4-week repeated-dose toxicity study is usually conducted as a "follow-on" study after the acute (single dose) and 1–2-week (dose-range finding) studies have been conducted. Thus, the 4-week study is conducted to strengthen the toxicity data for the potential pharmaceutical product. These studies should further develop the toxicity profile by elucidating the MTD and by establishing the first NOAEL when the test article is administered over the longer dose period. Repeated-dose studies evaluate additional parameters in the toxicokinetic profile because the number of doses (i.e., the dose period) is extended. Two- and four-week studies generally identify toxicity as evidenced by alterations in clinical observations, body weight, and food consumption, but also evaluate clinical pathology, hematology, and histopathology parameters (including the potential for target organs of toxicity), and these are the first studies in which a recovery period should be considered to evaluate the reversibility of any adverse effects (Pandher et al., 2012). As noted in previous sections, the route of administration should be the same as the intended route of human dosing, and the dosing regimen will depend on the pharmacodynamic properties of the product under development. In general, the 4-week rodent study utilizes the standard study design information given in previous sections, and the number of animals for the main toxicity evaluation is typically 10 animals per sex, per group, with additional animals included for toxicokinetic evaluations (Eaton and Klaassen, 2001). If toxicokinetic assessments are included, then the number of additional animals will depend on the acceptable collection volumes attainable from the rodent species of choice, the number of blood collections planned, and the timing of sample collections for analysis. There are guidelines covering blood volumes that can be safely collected over a 24-h period, which should be considered to prevent physiological complications that can potentially distort the bioanalytical data (Diehl et al., 2001). To provide adequate statistical power for the bioanalytical analysis, at least three individual animal samples per time point are needed. There is guidance to assist in determining appropriate timing and sampling for rodents (FELASA, 1994; Laboratory Animal Management, 1996). Both the quantity and frequency of blood sampling depend on the circulating blood volume of the animal. The approximate blood volume of a mouse is 80 and 70 mL/kg for a rat (Diehl et al., 2001). If the mouse is the rodent species, then more animals may be required for the toxicokinetic analyses because the blood volume available from each mouse is limited by physiological constraints, and at this early stage of development, the bioanalytical method typically requires a volume of serum or plasma that only allows a single collection time point in the mouse. As a result, individual mice are required for blood collections. For studies that include a recovery period, additional animals should be included in the main study groups and subjected to the same doses and conditions as the other study animals. If potential target-organ toxicities or evidence of other toxicities as evidenced by body weight changes or clinical observations is identified in previous studies, the inclusion of a recovery period is recommended to evaluate whether observed toxicities are partial to completely reversible over the designated recovery interval.

15.5.2 SUBCHRONIC TOXICITY STUDIES (90 DAYS/13 WEEKS)

The toxicity of a drug after subchronic exposure is then determined. Subchronic exposure can last for different periods of time, but 90 days/13 weeks is the most common test duration. The principal goals of the subchronic study are to establish a NOAEL and to further identify and characterize the specific organ or organs affected by the test compound after repeated administration. In addition, a LOAEL is determined (Eaton and Klaassen, 2001). The numbers obtained for NOAEL and LOAEL will depend on how closely the dosages are spaced and the number of animals examined. Determinations of NOAELs and LOAELs have numerous regulatory implications. A subchronic study is usually conducted in two species (rat and dog) by the route of intended exposure (usually oral). For rodent studies, 20–25 animals per sex, per group, are commonly assigned to the 13-week toxicology study. Most studies also include additional animals for toxicokinetic evaluation because of the size constraints previously discussed. Toxicokinetic sampling is done at the initiation, midway point, and end of the dosing period. Additional animals may be included for assignment to a recovery period to evaluate the reversibility of any adverse finding. A common study design is to assign 25 animals per sex, per group, necropsy 20 animals at the completion of the dose period, and place surviving animals into the recovery phase. In rodents, the 13-week study has one additional caveat. Animals should be observed once or twice daily for signs of toxicity, including changes in body weight, diet consumption, changes in fur color or texture, respiratory or cardiovascular distress, motor and behavioral abnormalities, and palpable masses. All premature deaths should be recorded and necropsied as soon as possible. Severely moribund animals should be terminated immediately to preserve tissues and reduce unnecessary suffering. At the end of the 90-day study, all the remaining animals should be terminated, and blood and tissues should be collected for further analysis. The gross and microscopic condition of the organs and tissues (about 15–20) and the weight of the major organs (about 12) are recorded and evaluated. Hematology and blood chemistry measurements are usually done before, in the middle of, and at the termination of exposure. Hematology measurements usually include hemoglobin concentration, hematocrit, erythrocyte counts, total and differential leukocyte counts, platelet count, clotting time, and prothrombin time. Clinical chemistry determinations commonly made include glucose, calcium, potassium, urea nitrogen, alanine aminotransferase (ALT), serum aspartate aminotransferase (AST), gamma-glutamyltranspeptidase (GGT), sorbitol dehydrogenase, lactic dehydrogenase, alkaline phosphatase, creatinine, bilirubin, triglycerides, cholesterol, albumin, globulin, and total protein. Urinalysis is usually performed in the middle of and at the termination of the testing period and often includes determination of specific gravity or osmolarity, pH, proteins, glucose, ketones, bilirubin, and urobilinogen as well as microscopic examination of formed elements. If humans are likely to have significant exposure to the chemical by dermal contact or inhalation, subchronic dermal and/or inhalation experiments may also be required. Subchronic toxicity studies not only characterize the dose–response relationship of a test substance after repeated administration but also provide data for a more reasonable prediction of appropriate doses for chronic exposure studies.

During the final stages of the preclinical program, it may be necessary to conduct carcinogenicity evaluations in a second rodent species. This is normally the case for small molecules. Because either the rat or the mouse was the primary species used in support of toxicology characterization, data will need to be collected on the second rodent species in preparation for the carcinogenicity evaluation. This necessitates an additional 13-week study (now termed DRF) exclusively for selection of doses to be used in the carcinogenicity evaluation. The data collected are the same as those in a standard repeated-dose general toxicity study, but the objective is not to fully characterize toxicity, but to find a high dose that is clinically relevant and that does not produce more than a 10% weight change (in comparison with respective controls) over the 13-week exposure period. For nonrodents, the canine and nonhuman primates are the two most commonly used species in 13-week general toxicology studies. The study design of the 13-week nonrodent study differs from the rodent study primarily in the number of animals assigned per group and the toxicokinetic sampling. Nonrodent 13-week studies typically have four to five animals per sex, per group. If a recovery phase is planned, then one to two animals per sex, per group are generally carried into the recovery phase of the study. In the event that mortality occurs in any group, the number of animals assigned to the recovery group is reduced and the animal will be evaluated with the main study animals. The 13-week studies are used not only to support extended dosing in human clinical trials but also to establish dose levels in chronic studies. These studies build on the toxicology data established so far and provide important toxicological and toxicokinetic data from longer exposure periods. They are therefore critical in assessing potential human toxicities from increased dose periods.

15.5.3 Chronic Toxicity Studies (6–12 Months)

Long-term or chronic studies are performed similarly to subchronic studies except that the period of exposure is longer than 3 months. In rodents, chronic exposures are usually for 6 months to 2 years. Chronic studies in nonrodent species are usually for 1 year but may be longer. The length of exposure is somewhat dependent on the intended period of exposure in humans. If the drug agent is planned to be used for short periods, such as an antimicrobial agent, a chronic exposure of 6 months may be sufficient, whereas the agent has the potential for life-time exposure in humans, a chronic study up to 2 years in duration is likely to be required.

Prior to the formation of the ICH, regulatory expectations for chronic toxicity testing differed between the three regions (the United States, Japan, and the EU). The United States and Japan required 12-month and the European agencies 6-month studies to cover the interim period between the 3-month and 2-year carcinogenicity studies. This resulted in many pharmaceutical companies conducting two chronic repeated-dose studies in rodents and in nonrodents, one of 6 months' duration in each species to support clinical trials, and one of 12 months' duration to support marketing in the United States and Japan. After lengthy evaluation of the databases of these duplicative studies, ICH S4 recommended that the maximum duration of long-term repeated-dose toxicity studies should be reduced from 12 to 6 months in rodents and from 12 to 9 months in nonrodents (ICH S4). The US FDA agreed to this change because the data showed that any late-emerging toxicity in the rat would be identified in the 2-year carcinogenicity study, which was required anyway. The situation with regard to the nonrodent was less straightforward, as the chronic study is the longest test in this species. The European and Japanese regulatory agencies agreed that nonrodent studies should be limited to a maximum of 6 months, based on the findings of the Center for Medicines Research and JPMA databases (Lumley and McAuslane, 1999). However, the US FDA continued to require a 12-month nonrodent study. Following reevaluation by US FDA, it was determined that 9 months may be sufficient to capture most of the new toxicity findings that emerged after 6 months. Thus, the 9-month study in nonrodents was proposed as the basis for harmonization within all ICH regions ICH S4. This conclusion was arrived at by an expert working group, which was assembled to evaluate the data from several chronic studies. Of the 18 cases evaluated, 11 supported a study-duration of 9–12 months, 4 supported a duration of 12 months, and the 3 remaining cases indicated that a 6-month study would be adequate. The expert working group recommended that there was sufficient evidence to support harmonized 9-month duration for nonrodent toxicity studies, which would be applicable for most categories of pharmaceuticals (ICH S4). In rodent and nonrodent studies, the number of animals used is generally the same as the number used in the 13-week studies, and the same toxicological parameters are evaluated. These extended repeated-dose studies are conducted for the following reasons:

- 6-month studies may be acceptable for indications of human exposure in chronic conditions with short-term or intermittent human exposure.
- 6-month studies are also acceptable for pharmaceuticals intended to treat life-threatening diseases (e.g., cancer or HIV).
- 12-month studies are more appropriate for chronically used drugs for which human clinical trial data are limited to short-term exposure.
- 12-month studies may also be more applicable for new molecular entities acting on new molecular targets where there is limited human data, especially for longer-term exposure. Since there are differences between the US FDA and the other ICH signatories regarding the appropriate length of repeated-dose studies greater than 13 weeks, the sponsor must assess the appropriate length of the studies to be conducted. These decisions must take into consideration the regulatory requirements for both the conduct of clinical trials and marketing applications based on regional differences (Lumley and McAuslane, 1999).

Dose selection is critical in these studies to ensure that premature mortality from chronic toxicity does not limit the number of animals that survive to a normal life expectancy. Most regulatory guidelines require that the highest dose administered be the estimated MTD. This is generally derived from subchronic studies, but additional longer studies (e.g., 6 months) may be necessary if delayed effects or extensive cumulative toxicity is indicated in the 90-day subchronic study.

15.6 CONCLUSION

The preclinical development of pharmaceuticals has evolved over the past 30–40 years. A major driver of this evolution was the promulgation of harmonized international guidelines. These guidelines describe the proper basic design and management of studies to produce appropriate data to support clinical (human) testing. Because of the effort put into harmonization of guidelines, various matters of concern have undergone focused scrutiny and discussion. This includes topics around animal welfare, types of studies, study methods, and data interpretation. The production of internationally acceptable consensus guidelines has helped to streamline preclinical development and has provided a level of confidence that appropriate data will be produced to answer the questions of the pharmaceutical's, or biologic's, safety and effectiveness, and will be accepted for regulatory review worldwide.

To put these guidelines into practice, a preclinical scientific program requires the attention of the toxicologist. A trained toxicologist ensures collection of appropriate preclinical data (especially reliable safety data) to support innovative pharmaceuticals. In general, acute toxicity evaluations are the first preclinical tests performed on a drug candidate. Therefore, prudent characterization of the acute toxicity is critical. Data obtained during this stage provide the basis for dose level, route, frequency of exposure, and other parameters for the next stages of drug development (subacute, subchronic, and chronic testing). In addition, the toxicologist must select the appropriate species (advantages/disadvantages of each), dose extrapolation from acute to chronic toxicity testing, applicable guidelines, the GLP requirements, and appropriate endpoints. Finally, if they understand the intricacies of the general toxicology plan, the toxicologist will be able to progress smoothly as the potential pharmaceutical moves forward in development.

REFERENCES

Buckley LA, Benson K, Davis-Bruno K, Dempster M, Finch GL, Harlow P, et al. (2008) Nonclinical aspects of biopharmaceutical development: Discussion of case studies at a PhRMA-FDA workshop. *Int J Toxicol* 27(4), 303–312.

Bussiere JL, Martin P, Horner M, Couch J, Flaherty M, Andrews L, et al. (2009) Alternative strategies for toxicity testing of species-specific biopharmaceuticals. *Int J Toxicol* 28(3), 230–253.

Cavagnaro JA. (2008) Preclinical safety evaluation of biopharmaceuticals. In: *A Science-Based Approach to Facilitating Clinical Trials* (Cavagnaro JA, ed.) John Wiley and Sons, Hoboken, NJ.

CHMP (Committee for Medicinal Products for Human Use). EMEA, London EMEA/CHMP/ EWP/692702. Available from: https://www.ema.europa.eu/en/documents/scientific-guideline/ reflection-paper-extrapolation-results-clinical-studies-conducted-outside-european-union-eu-eu_en.pdf

Council Directive 65/65/EEC of 26 January 1965 on the Approximation of Provisions Laid Down by Law, Regulation or Administrative Action Relating to Proprietary Medicinal Products. Available from: http:// eur-lex.europa.eu/legal-content/EN/TXT/?uri=CELEX:31965L0065. Accessed August 17, 2018.

Denny KH, Stewart CW. (2017) Acute, subacute, subchronic, and chronic general toxicity testing for preclinical drug development. In: *A Comprehensive Guide to Toxicology in Nonclinical Drug Development*, 2nd Edition (Faqi AS, ed.) Academic Press, London. pp. 109–127.

Diehl KH, Hull R, Morton D, Pfister R, Rabemampianina Y, Smith D, et al. (2001) A good practice guide to the administration of substances and removal of blood, including routes and volumes. *J Appl Toxicol* 21, 15–23.

Eaton DL, Klaassen CD. (2001) Principles of toxicology. In: *Casarett & Doull's Toxicology: The Basic Science of Poisons* (Klaassen CD, ed.).

European Commission Regulation (EU) No 536/2014 of the European Parliament and of the Council of April 16, 2014 on Clinical trials on medicinal products for human use, and repealing Directive 2001/20/EC.

European Commission. Clinical trials – Directive 2001/20/EC.

European Medicines Agency. (2004). Guideline on the scientific application and the practical arrangements necessary to implement commission regulation (EC) No. 507/2006 on the conditional marketing authorization for medical products for human use falling within the scope of regulation (EC) No. 726/2004.

FDA. (2005) U.S. Department of Health and Human Services, FDA public health advisory: safety of Vioxx. Available from: www.fda.gov/Drugs/Drugsafety/. Accessed August 30, 2018.

FDA. (2018) The history of FDA's fight for consumer protection and public health. Available from: www.fda. gov/AboutFDA/History/default.htm. Accessed August 31, 2018.

FELASA. (1994) Guidelines: Pain and distress in laboratory rodents and lagomorphs. 28, 97–112.

Friedman MA, Woodcock J, Lumpkin MM, Shuren JE, Hass AE, Thompson LJ. (1999) The safety of newly approved medicines: Do recent market removals mean there is a problem? *JAMA* 281, 1728–1734.

Grabowski HG, Vernon JM. (1983) *The Regulation of Pharmaceuticals*. Washington, DC, American Enterprise Institute.

ICH S4. Duration of chronic toxicity testing in animals (rodent and nonrodent toxicity testing). https://www.ema. europa.eu/en/documents/scientific-guideline/ich-s-4-duration-chronic-toxicity-testing-animals-rodent-non-rodent-toxicity-testing-step-5_en.pd.

Kashyap UN, Gupta V, Raghunandan HV. (2013) Comparison of drug approval process in United States and Europe. *J Pharm Sci Res* 5, 131–136.

Laboratory Animal Management. (1996) Rodents. National Research Council (Adopted September 1998).

Lumley CE, McAuslane JAN. (1999) Center for Medical Research: Assessment of pharmaceuticals for potential human carcinogenic risk [CMR Workshop, CMR International].

M3(R2) Nonclinical safety studies for the conduct of human clinical trials and marketing authorization for pharmaceuticals. In *International Conference on Harmonization (ICH)*. https://www.fda.gov/regulatory-information/search-fda-guidance-documents/m3r2-nonclinical-safety-studies-conduct-human-clinical-trials-and-marketing-authorization.

McVean M. (2014) Preclinical development to IND: Drugs, biologics, cellular/gene therapies and vaccines, presented at the BioBoot Camp on April 18, 2014.

Pandher K, Leach MW, Burns-Naas LA. (2012) Appropriate use of recovery groups in nonclinical toxicity studies: Value in a science-driven case-by-case approach. *Vet Pathol* 49(2), 357–361.

Parvisi N, Woods K. (2014) Regulation of medicines and medical devices: Contrasts and similarities. *Clin Med* 14, 6–12.

Shankar G, Shores E, Wagner C, Mire-Sluis A. (2006) Scientific and regulatory considerations on the immunogenicity of biologics. *Trends Biotechnol* 24(6), 274–280.

Van Norman GA. (2016) Drugs and devices: Comparison of European and U.S. approval processes. *J Am Coll Cardiol Basic Trans Sci* 1, 399–412.

Vilas-Boas IM, Tharp CP. (1997) The drug approval process in the U.S., Europe, and Japan. *J Manag Care Pharm* 3(4), 459–465.

Yamaguchi T, Arato T. (2011) Quality, safety and efficacy of follow-on biologics in Japan. *Biologicals* 39, 328–332.

16 Safety Pharmacology
Past, Present, and Future

Jean-Pierre Valentin, Annie Delaunois,
Marie-Luce Rosseels, and Vitalina Gryshkova
UCB Pharmaceuticals

Tim G. Hammond
Preclinical Safety Consulting Limited

CONTENTS

16.1 BACKGROUND

16.1.1 Safety-Related Drug Attrition

The reasons for drug attrition have evolved over the years; over the last 25 years, lack of safety (both non-clinical and clinical) remains the major cause of attrition during clinical development, which accounts for approximately 20%–30% of all drug discontinuation (see Figure 16.1).[1–4] More worrying is the fact that there is no clear trend toward a reduction of the attrition owing to safety reasons.

In this section, a brief summary of the nature, frequency, and consequences of adverse drug reactions (ADRs) in two clinical situations is presented. There are ADRs experienced by healthy volunteers and patients participating in clinical studies with potential new medicines and those experienced by patients who are prescribed licensed medicines. A review of these two situations points to areas of success with the current practices for non-clinical safety pharmacology testing but also identifies some areas where further research might lead to new or better safety pharmacology tests. Prior to reviewing the literature, some definitions are worth considering (see Valentin, J.P., Keisu, M., and Hammond, T.G., 2009 for details).[5]

An *adverse event* (AE) is defined as an unintended injury caused by medical management rather than the disease process. ADRs are a subset of AEs, which are thought to be causally related to the use of a medicine. A serious AE or ADR results in death, is life-threatening, requires hospitalization, results in persistent disability, and is a congenital abnormality. The severity can be classified as (a) mild – slightly bothersome; (b) relieved with symptomatic treatment; (c) moderate – bothersome, interferes with activities, only partially relieved with symptomatic treatment; or (d) severe – prevents regular activities, not relieved with symptomatic treatment. So, a serious ADR is always significant and has a high impact – it can lead to the discontinuation of a

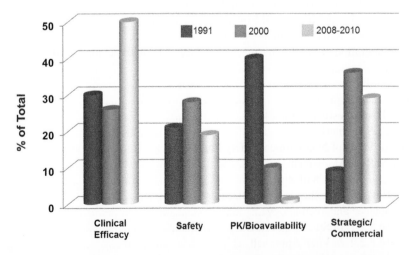

FIGURE 16.1 Relative contribution to drug attrition during clinical development over a 20-year period. Non-clinical safety and clinical safety remain a main contributor to drug attrition. PK, pharmacokinetic. (Data source from Kola and Landis[1] and Anon[2]; data further supported by Cook et al.[3] and Waring et al.[4])

drug in development, a significant limitation in the use of a drug (precaution, contra-indication), or even to the withdrawal of the drug from the market place. A nonserious ADR can be more or less severe in its intensity, and its impact will depend upon its frequency and intensity. The impact of serious and nonserious ADRs on a drug's commercial success will be titrated against the overall clinical benefit the drug brings to the patient.

ADRs in humans fall into five types. Of these, acute safety pharmacology studies can reasonably be expected to predict Type A ADRs (see Table 16.1).[6] This means that ~75% of clinical ADRs are potentially predictable on the basis of non-clinical safety pharmacology studies.

During early development, the first clinical studies (Phases I and IIA) are generally very safe.[6,7] In fact, molecules with a significant potential to generate serious ADRs are probably never given to healthy volunteers and are given to patients (e.g., refractory cancer patients) only with great care. These studies are to be conducted diligently with careful monitoring for the emergence of potentially worrisome ADRs. While AEs do occur, they are generally more related to the experimental procedures (e.g., needle puncture) than to the drugs. Safety pharmacology probably contributes significantly to the maintenance of this good track record. This is supported by published reports showing that single-dose non-clinical safety studies could overall accurately predict the clinical outcome.[8–10] The common ADRs observed with a high incidence (10%–30%) during these phases are linked to the gastrointestinal (GI) and central nervous systems (CNSs). In addition, ADRs that occurred with a low incidence are also detected. They are often specific to the new chemical entity (NCE) under investigation and are often pharmacologically mediated.

During Phase II/III clinical development, a large number (often the majority) of patients report AEs, with a wide variation in the type, frequency, and severity of events detected. Nonserious ADRs are often mechanism-, or drug class-, or disease-related (Table 16.2). Such ADRs limit the utility of a new medicine by restricting its use to those patients who either do not experience or can tolerate the ADRs, and they do not usually pose a safety issue. Serious ADRs tend to be present only at low frequencies. Pharmacological mechanism related to serious ADRs can occur in sensitive individuals, those with unusual kinetics, and in the presence of kinetic or occasionally dynamic drug interactions. In principle, such ADRs might be predictable from safety pharmacology testing, although it should be acknowledged that safety pharmacology testing is usually conducted in young adult healthy animals, conditions that may be suboptimal to detect such effects.

TABLE 16.1

Classification of ADRs in Humans

Type A	Dose-dependent; predictable from primary, secondary, and safety pharmacology	Main cause of ADRs (~75%), rarely lethal
Type B	Idiosyncratic response, not predictable, not dose-related	Responsible for ~25% of ADRs, but majority of lethal ones
Type C	Long-term adaptive changes	Commonly occurs with some classes of drug
Type D	Delayed effects (e.g., carcinogenicity, teratogenicity)	Low incidence
Type E	Rebound effects following discontinuation of therapy	Commonly occurs with some classes of drug

Note: Conventional safety pharmacology studies can reasonably be expected to predict "Type A" ADRs. Functional toxicological measurements may predict "Type C" ADRs. Conventional toxicology studies address "Type D" ADRs. Prediction of "Type B" responses requires a more extensive non-clinical and clinical evaluation, often only addressing risk factors for the idiosyncratic response. "Type E" ADRs are rarely investigated non-clinically using functional measurements unless there is cause for concern.

Source: Adapted from Redfern, W.S. et al., *Fundam. Clin. Pharmacol.*, 16, 161–173, 2002.

TABLE 16.2

Major Causes of Acute Functional ADRs

Acute ADR	Example
Augmented ("supratherapeutic") effect of interaction with the primary molecular target	Pronounced bradycardia with a beta-blocker; pronounced hypotension with an angiotensin II receptor antagonist
Interaction with the primary molecular target present in nontarget tissues	Sedation caused by antihistamines
Interactions with secondary molecular targets	Interactions with the hERG cardiac channel leading to QT interval prolongation (e.g., some antipsychotics and antihistamines drugs)
Non-specific effects	
Pharmacologically active metabolites	

Occasional non-pharmacological serious ADRs occur; these can be induced by direct chemical toxicity, hypersensitivity, or immunological mechanisms. Serious ADRs always limit the use of a new medicine by requiring warnings, precautions, and contra-indications; they can even preclude regulatory approval. Apart from preventing the development of NCEs likely to induce serious ADRs in larger clinical studies, a key contribution that can be made by non-clinical safety pharmacology is in the elucidation of the mechanisms responsible for these ADRs. Once the mechanism responsible for the ADR is known, it becomes possible to prepare soundly argued precautions and contra-indications.

When medicines are on the market, the actual incidence of serious ADRs is difficult to judge, but clearly they occur with sufficient frequency to be a serious concern. An authoritative review[11] concludes that between 1 in 30 and 1 in 60 physician consultations result from ADRs (representing 1 in 30–40 patients). The same review concludes that 4%–6% of hospital admissions could be the result of ADRs. Although there is debate over the number of deaths caused by ADRs – the figure of around 106,000 deaths per year in the United States is often quoted,[6,12] this has been suggested to be a gross overestimate and, for example, the US FDA MedWatch system recorded 6,894 deaths in 2000 (www.fda.gov/medwatch/index.html).

The frequency of serious ADRs can be very low (e.g., 0.25–1.0 cases of rhabdomyolysis per 10,000 patients treated with a statin[13]); however, when millions of patients are under treatment, this can generate substantial morbidity. Furthermore, ADRs may be due to (a) clinical error (e.g., misprescribing contra-indicated drugs) or (b) patient self-medication error – especially in the era of mass media communication and information. Although the frequency of these events can be very low, it is still necessary to investigate the pharmacological mechanisms driving these events. For example, the elucidation of the connection between drug-induced Torsades de Pointes (TdP), QT interval prolongation, and hERG potassium channel blockade has been considered as a major advance in this area and led to the rapid development of non-clinical *in vitro* screening assays of medium- to high-throughput capabilities. To better understand the main causes for ADR-related drug withdrawals, medicines withdrawn from either the USA or worldwide market were reviewed.[11,14] The review highlighted the fact that several of these toxicities fall into the remit of safety pharmacology such as cardiovascular, GI, and CNS–associated ADRs. The prominence of arrhythmias in Stephens' review probably reflects the interest in TdP-type arrhythmias over the two last decades.[11]

The main safety reasons of drug attrition, ADRs, and withdrawal throughout the pharmaceutical life cycle are presented in Figure 16.2. Toxicities associated with the cardiovascular, hepatic, and CNSs account for the majority of drug discontinuation, ADRs, and drug withdrawal although the profile varies from one target organ to another. For example, there is a significant attrition related to the cardiovascular system in drug discovery (pre-first in human (FIH) trials), that may reflect the availability of predictive assays, thus resulting in limited cardiovascular-related attrition in Phase I.

Phase	'Nonclinical'		Phase I		Phase I-III		Registration		Post-Marketing	
Information:	CD stopped		Severe ADRs		CD stopped		Approval delays/ denied	ADRs on label	Serious ADRs	Withdrawal from sale
Sample size:	88 CDs stopped	33 CDs stopped	1,015 participants	11,028 participants	82 CDs stopped	61 CDs stopped	80 CDs delayed/ denied	1,138 drugs	21,298 patients	53 drugs withdrawn
Time period:	1993-2006	2005-2010	1986-1995	2004-2011	Unknown-1999	2005-2010	2000-2012	(All)	2004-2005	2000-2014
Cardiovascular:	27%	17%	9%	Not stated	21%	24%	18%	36%	15%	49%
Nervous system:	14%	7%	28%	40%	21%	34%	9%	67%	39%	21%
Gastrointestinal:	3%	3%	23%	42%	5%	9%	6%	67%	14%	4%
Respiratory:	2%	8%	0%	Not stated	0%	3%	0%	32%	8%	2%
Renal:	2%	8%	0%	Not stated	9%	9%	5%	19%	2%	0%
Other:	47%	57%	25%	43%	42%	21%	56%	109%	49%	92%

0%   1-9%   10-19%   >20%

FIGURE 16.2 Main safety reasons of drug attrition, ADRs, and withdrawal throughout the pharmaceutical life cycle. The various toxicity domains have been ranked first by contribution to products withdrawn from sale and then by attrition during clinical development. The table focuses on the five organ systems relevant to safety pharmacology; All the "other organ systems" have been combined together. The color code represents the level of incidence. (Data extracted from Valentin and Redfern.[15])

However, the picture changes when drugs are administered for long period of time to patients suffering from chronic diseases, where cardiovascular-related attrition and ADRs raised again reflecting our current limitation to identify risk and to form optimal risk assessment, management, and mitigation. In contrast, there is limited attrition in drug discovery for neurotoxicity which may reflect the limited availability of highly predictive models or the inability to predict some toxicities (e.g., headache, suicidal ideation), thus resulting in significant CNS-related AEs and attrition in early Phase I clinical trials. The late-stage attrition, ADRs, and withdrawal related to CNS effects remain elevated reflecting again our current inability to identify certain risk and to form optimal risk assessment, management, and mitigation (e.g., cognitive dysfunction, suicidal ideation, auditory dysfunction).

In preclinical phases, the safety-related attrition results predominantly (75%) from the compound (i.e., to the chemistry) and to a lesser extent from the target (25%); the situation changes when it comes to the clinical phases where compounds and chemistry account equally (52% and 48%, respectively) for the safety-related attrition.[3] When considering the five organ systems of interest to safety pharmacology (namely cardiovascular, respiratory, GI, renal, and nervous systems), the safety-related attrition is equally distributed between structural (e.g., necrosis, apoptosis) and functional effects. In terms of failure modes, failure of hazard identification, suboptimal risk assessment, and inappropriate governance account, respectively, for ~40%, 40%, and 20% of the safety-related attrition.[15]

16.2 ORIGIN AND EVOLUTION OF SAFETY PHARMACOLOGY

16.2.1 REGULATORY REQUIREMENTS

Prior to 1990, regulatory guidance on non-clinical organ function testing was limited. The US and European regulations provided only general references to the evaluation of drug effects on organ system functions.[16–19] Organ function assessments included within investigational new drug (IND) applications and registrations (NDAs) were inconsistent and often viewed as unimportant.[20,21] However, in Japan, the Ministry of Health and Welfare (MHW), now referred to as the Ministry of Health, Labor, and Welfare, had promulgated comprehensive guidance for organ function testing as early as in 1975 (see Table 16.3). These guidances described which organ systems would be evaluated as a first-tier evaluation (List A studies) and made specific recommendations regarding study designs (including description of models, criteria for dose selection, and which endpoints would be included in the investigation). A second tier of studies (List B) to be conducted based on the significant findings in List A investigations (Table 16.3) was also presented.[22] Because the Japanese guidances were the most comprehensive of their time, they became the *de facto* foundation for organ function testing throughout the pharmaceutical industry.[6,23,24] The organ function studies included in Lists A and B were intertwined with studies whose aim was to catalog additional pharmacological functions and activities (i.e., secondary pharmacology) in addition to the primary pharmacological function/activity. Kinter et al.[17] distinguished two subgroups of objectives embedded in the Japanese studies as safety and pharmacological profiling. This concept was enlarged upon the International Conference on Harmonization (ICH) safety pharmacology Expert Working Group (EWG) to define three categories of pharmacology studies: primary and secondary pharmacodynamic (PD) studies, and safety pharmacology studies (see Section 16.3[24,25] and Table 16.3). During the same period, European, USA, and Japanese regulatory agencies prepared positions on general pharmacology/safety pharmacology in the form of guidance and concept papers.[24–27] Any complacency surrounding safety pharmacology was shattered in 1996 with the appearance of the first draft of a "Points to Consider" document on QT prolongation by the European Medicines Agency's Committee for Proprietary Medicinal Products (CPMP) and issued as an official document the following year.[28] One of the more controversial aspects of this document was the recommendation to incorporate screening of all non-cardiac drugs for effects on cardiac action

TABLE 16.3

International Guidances Referring to Physiological Functions as Relevant to Safety Pharmacology Assessment

Document and Source	Comments	Reference
JMHW notes on applications for approval to manufacture new drugs requested the evaluation of the effects of physiological functions	To assess the effects of the test substance on the CNS, peripheral nervous system, sensory organs, respiratory and cardiovascular systems, smooth muscles including uterus, peripheral organs, renal function, and adverse effects observed in clinical studies	33
Guidelines for General Pharmacology Studies – Japanese Guidelines for Non-clinical Studies of Drugs	Defines a list of studies to be conducted on all NCE (List A) and a list of studies to be conducted on a case-by-case basis (list B). Guidance has been superseded by ICH S7A and S7B	22
ICH M3: Timing of Non-clinical Safety Studies for the Conduct of Human Clinical Trials for Pharmaceuticals	Provides guidance on the timing of safety pharmacology studies in relation to clinical development. Establishes that safety pharmacology studies should be conducted prior to first administration to humans	30
ICH S6: Preclinical Safety Evaluation of Biotechnology-Derived Pharmaceuticals	Defines the objective of safety pharmacology studies to reveal functional effects on major physiological systems (e.g., cardiovascular, respiratory, renal, and CNSs)[28,29]	
ICH S7A: Note for Guidance on Safety Pharmacology Studies for Human Pharmaceuticals	Provides the general framework for *in vitro* and *in vivo* safety pharmacology studies, including studies addressing the risk for a drug to slow cardiac repolarization[31]	
ICH S7B: Note for Guidance on the Non-clinical Evaluation for Delayed Ventricular Repolarization (QT Interval Prolongation) by Human Pharmaceuticals	Describes a non-clinical testing strategy for assessing the potential of a test substance to slow ventricular repolarization. Includes information concerning non-clinical assays and integrated risk assessment[34]	
ICH E14: Note for Guidance on the Clinical Evaluation of QT/QT$_c$ Interval Prolongation and Proarrhythmic Potential for Nonantiarrhythmic Drugs	Provides recommendations concerning the design, conduct, analysis, and interpretation of clinical studies to assess the potential of a drug to slow cardiac repolarization[35,36]	
Guideline on the Non-Clinical Investigation of the Dependence Potential of Medicinal Products	Provides guidance on the need for testing of dependence potential in animals during the development of medicinal products and indicate what type of information is expected as part of a MAA[37]	
Guidance for Industry on the Assessment of Abuse Potential of Drugs	Provides guidance on the need for testing of dependence potential in animals and humans during the development and post-approval of medicinal products[38]	
Guidance for Industry, Investigators, and Reviewers: Exploratory IND Studies	Provides guidance on non-clinical and clinical approaches that should be considered when planning exploratory IND studies in humans[40]	
ICH M3-R2: Guidance on Non-clinical Safety Studies for the Conduct of Human Clinical Trials and Marketing Authorisation for Pharmaceuticals	Recommends international standards for the non-clinical safety studies recommended to support human clinical trials as well as marketing authorization for pharmaceuticals.[41]	
ICH S9-R1 (step 2): Guideline on Non-clinical Evaluation for Anticancer Pharmaceuticals – Questions and Answers[39]	Assists sponsors in prospectively assessing the occurrence of treatment-emergent suicidal ideation and behavior in clinical trials of drug and biological products[42]	
Guidance for Industry: Suicidal Ideation and Behavior: Prospective Assessment of Occurrence in Clinical Trials		

Note: GLP, Good laboratory practice; ICH, International Conference on Harmonization; IND, investigational new drug; JMHW, Japanese Ministry of Health and Welfare; MAA, marketing authorization application; NCE, new chemical entity.

potential *in vitro*. The opposition to this document from the pharmaceutical industry arose partly because this recommendation wrong-footed the industry. The positive impact of the CPMP document was that it resuscitated safety pharmacology as a rigorous scientific discipline. In 1998, the MHW and the Japanese Pharmaceutical Manufacturers' Association proposed to the ICH Steering Committee the adoption of an initiative on safety pharmacology. This proposal was accepted and given the designation of Topic S7.

The origin of the term "safety pharmacology" is obscure. It first appeared in the draft guidances of the ICH M3 and S6 (see Table 16.3).[29,30] ICH S6 stated that "...the aim of the safety pharmacology studies should be to reveal functional effects on major physiological systems (e.g., cardiovascular, respiratory, renal, and CNSs)..." The ICH S7 EWG began its work in the first quarter of 1999, and a harmonized safety pharmacology guidance was finalized and adopted by the regional regulatory authorities over 2000–2001.[31] The ICH S7A guidance describes the objectives and principles of safety pharmacology, differentiates tiers of investigations, establishes the timing of these investigations in relationship to the clinical development program, and introduces the requirement for good laboratory practice (GLP) where applicable.[31,32]

The ICH S7 EWG extensively debated how to evaluate the potential of new drugs to produce a rare but potentially life-threatening ventricular tachyarrhythmia (TdP) in susceptible individuals.[43–47] The incidence of TdP with drugs that are targeted at non-cardiac indications can be very low; hence, the imperative need to find non-clinical surrogates is to identify those drugs with the potential to elicit TdP.[48–55] The controversial issue has been the accuracy of the non-clinical models to identify problematic drugs and how the generated data may be assimilated into an assessment of human risk.[23]

Recognizing that the resolution would not be easily forthcoming, the ICH S7 EWG proposed to the ICH Steering Committee a new initiative to generate guidance for the assessment of the effects of drugs on cardiac ventricular repolarization. This proposal was accepted in November 2000 and was designated ICH S7B (see Table 16.3).[34] The guidance on safety pharmacology was finalized at the same meeting and was redesignated S7A (see Table 16.3).[31] Surveys of the pharmaceutical industry, regulatory agencies, and members of the audience of the 4th Safety Pharmacology Society meeting, conducted 3 years after the implementation of the ICH S7A, concluded that the guidance has been successfully implemented in which GLP-compliant safety pharmacology "core battery" studies are usually performed prior to first administration to humans.[56] The approach is science-driven and specifies the use of robust and sophisticated *in vitro* and *in vivo* assays as subsequently confirmed.[57]

There are, however, some areas that require further refinement/clarification such as the specifics of study design including the selection of dose–concentration, choice of species, modeling of the temporal PD changes in relation to pharmacokinetic (PK) profile of parent drug and major metabolites, use of an appropriate sample size, statistical power analysis, testing of human-specific metabolites, and demonstrating not only the model's sensitivity but also its specificity for predicting AEs in humans.[56] A year after the adoption of the ICH S7B, the US Food and Drug Administration (FDA) and the Pharmaceutical Research and Manufacturers of America proposed to the ICH Steering Committee the adoption of a parallel initiative to prepare guidance on clinical testing of NCEs for their potential to prolong ventricular repolarization. This proposal was accepted as ICH E14 (Table 16.3). Following a recommendation from the ICH Steering Committee, the activities of the ICH S7B and E14 EWGs were aligned from 2003 onward. In May 2005, both guidances were finalized and due for implementation in November of the same year (see Table 16.3).[44,55] In a field where the scientific understanding and technological advances evolve rapidly, a formal review and update of these guidance documents is happening.[35,58]

16.2.2 Drivers Influencing the Approach to Safety Pharmacology

In addition to regulatory requirements, several factors influence the safety pharmacology strategy in any given pharmaceutical organization. The main factors are presented in Table 16.4. Thus, the pharmaceutical industry besets with a number of significant challenges to achieve high-quality,

TABLE 16.4

Non-Exhaustive List of Factors Influencing the Approach to Safety Pharmacology

Increase number, complexity, and stringency of regulatory requirements

Increase number and novelty of chemical space and new chemical entities

Increase number and novelty of molecular targets (e.g., kinases, intracellular, intranuclear, protein–protein interaction)

Increase number and novelty of approaches (e.g., different format of antibodies (mono-, bi-, tri-specifics), ADC, gene, and cell therapies)

Increase throughput of *in silico*, and *in vitro* vs. *in vivo* assays

Increase "front loading"/de-risking initiatives

Increase awareness and application of the "3Rs" rule of animal usage and welfare

Increase patient awareness and expectations

Reduce reimbursement of medicines by payers

Limited supply of compound during the early discovery stages

Limited availability of scientific and technical expertise in key areas (e.g., integrative physiology, pharmacology, and toxicology)

Reduce late-stage attrition

Reduce discovery and development cycle timelines

Increase quality of candidate drugs

Reduce funding of non-clinical safety functions

Predictive value of *in silico*, *in vivo*, and *in vitro* non-clinical assays with respect to human safety

high-throughput, cost-effective, timely, efficient, and predictive safety pharmacology studies, during the early stages of the discovery process. Along with satisfying project demands, scientific safety questions, international regulatory guidances, and increased patient awareness, safety pharmacology is increasingly being used to enable informed decision-making. One of the key factors influencing the approach to safety pharmacology is the rapidly evolving scientific and technological knowledge. Examples include the evolution (a) from labor-intensive manual patch-clamp electrophysiology to high-throughput electrophysiology-based platforms; (b) from snap-shot manual recordings and measurements of the electrocardiogram (ECG) and blood pressure in conscious animals to real-time monitoring devices and semi-automated analytical software; (c) from the manual counting of respiration rate to the direct quantification using whole-body plethysmography chambers; (d) from the assessment of GI function (gastric emptying, intestinal transit) using charcoal meals to the utilization of scintigraphy techniques; and (e) from invasive measurement of GFR to measurements in conscious laboratory animals without the need for blood or urine sampling or laboratory assays.[59,60]

16.3 DEFINITION AND OBJECTIVES OF SAFETY PHARMACOLOGY STUDIES

During the course of the discovery and development of a drug, three types of pharmacology studies are to be conducted, namely primary, secondary, and safety pharmacology.[31] Primary pharmacology studies are defined as those that "investigate the mode of action and effects of a substance in relation to its desired therapeutic target." On the other hand, secondary PD studies are defined as those that "investigate the mode of action and effects of a substance not related to its desired therapeutic target," whereas safety pharmacology studies "investigate the potential undesirable PD effects of a substance on physiological functions in relation to exposure in the therapeutic range and above."[26,31] Therefore, safety pharmacology studies are designed to investigate functional effects as opposed to morphological changes induced by an NCE.[61] Although pharmacology studies have been divided into subcategories, it is recognized that undesired functional effects (the domain of safety pharmacology) may be mediated via the primary or secondary pharmacological targets. Therefore, it is important to take a holistic approach while assessing adverse functional effects.

The objectives of safety pharmacology studies are threefold: first "to identify undesirable PD properties of a substance that may have relevance to its human safety"; second "to evaluate adverse PD and pathophysiological effects of a substance observed in toxicology and clinical studies"; and third "to investigate the mechanism of the adverse PD effects observed and suspected."[31] Thus, the ICH S7A guidance objectives are primarily concerned with protecting clinical trial participants (volunteers and patients) and also patients that are receiving marketed products from any potential AEs of NCEs. This view is further supported by the clinician's perspective who sees the objectives of safety pharmacology as enabling (a) adapting the design (including parameters) of clinical studies; (b) preventing serious ADRs in early clinical trials; (c) providing guidance for setting up the doses/exposures in ascending dose tolerance clinical studies; (d) predicting the likelihood of unwelcome pharmacologically mediated ADRs that need monitoring during early clinical studies; and (e) reducing or eliminating serious ADRs in large-scale clinical trials and in clinical practice. Moreover, an unwritten objective may be to support business decisions mainly based on predictions of likely human safety profile or by identifying the risk in the early phases of drug discovery in order to design out, wherever possible, unwanted pharmacological activities.

16.4 CURRENT PRACTICES

16.4.1 ICH S7A/B

16.4.1.1 Core Battery, Follow-Up, and Supplemental Studies: Definitions and Expectations

Safety pharmacology studies have been subdivided into "core battery," "follow-up," and "supplemental" studies.[31] The "core battery" studies are aimed to investigate the effects of NCEs on the cardiovascular, respiratory, and CNSs that are considered as vital organ systems based on the fact that acute failure of these systems would pose an immediate hazard to human life. In some instances, based on scientific rationale, the "core battery" may or may not be supplemented.[31] Additionally, ADRs may be either (a) suspected based on the pharmacological class, or the chemical class, or (b) identified based on outcome from other non-clinical or clinical studies, pharmacovigilance, or from literature reports. When such potential ADRs raise concern for human safety, these should be explored in "follow-up" or "supplemental" studies. "Follow-up" studies are meant to provide a greater depth of understanding or additional knowledge to that provided by the "core battery" on vital functions. Moreover, "supplemental" studies are meant to evaluate potential adverse PD effects on organ system functions that have not been addressed in either the "core battery" or repeated-dose toxicity studies when there is a cause for concern. The organ systems falling into this category may include, but are not limited to, the GI, renal, urinary, immune, endocrine, or autonomic nervous systems.

16.4.1.2 General Considerations and Principles

Since the pharmacological effects of an NCE depend on its intrinsic properties, the studies should be selected and designed accordingly. General considerations in the selection and design of these studies that assess potential ADRs are to include effects associated (a) with the therapeutic class, (b) with members of a chemical class, (c) with nontarget-mediated activities, and (d) with effects observed in previous non-clinical/clinical studies that warrant further investigation.

16.4.1.2.1 Species, Gender Selection, and Animal Status

The selection of an appropriate species is crucial in having an understanding of the molecular and biochemical comparison of the underlying controls on the physiological system(s) in the test species with those that are operative in humans. As an example, the dog and monkey are considered appropriate species for evaluation of drug effects on cardiac ventricular repolarization because of their dependence on potassium rectifying currents, IKr and IKs, to repolarize the ventricular myocardium; a similar dependence is known to exist in the human myocardium. On the other hand, the primary cardiac repolarizing currents in mice and rats rely on the outward potassium current, Ito. Thus, these species would not be appropriate for assessing the potential human risk posed by drugs that can affect cardiac IKr and IKs repolarizing currents. Inversely, the guinea pig, which is a rodent species considered as suitable to evaluate IKr-related effects, does not express $K_V4.3$, the ion channel responsible for Ito current.[62] As another example, the rat does not possess a gall bladder; this raises important questions about the relevance of this species for assessing GI function. On the basis of GI functional homology for humans, especially motility, gastric emptying, and pH value, particularly in the fasted state which is analogous to the conditions prevailing in many Phase I trials, the dog is perhaps a more relevant species.[63] Moreover, the dog appeared as a better predictor of clinical GI ADRs than the monkey for 25 anticancer drugs.[64] Although physiological similarity is an important requirement, it is only one of many factors that must be considered during species selection. In the absence of specific scientific reasons, the species in safety pharmacology studies are selected to maintain their consistency across study types (i.e., toxicology studies where information on metabolism and toxicokinetic is available).

Ideally, the gender selected for a study should include both male and females. However, it could be argued that non-clinical safety pharmacology studies should focus primarily on the gender that will be included in Phase I (i.e., usually male) or on the most sensitive gender for a given assay (e.g., female gender to assess the proarrhythmogenic potential).[65] It is important to note that these animals are most often healthy, which may be an important distinction from the patient population for which the drug is being developed. In many cases, the patient population may have an enhanced risk of demonstrating ADRs. To compensate for this possible limitation, the exposure of animals to the NCE and any major metabolites should explore large multiples (e.g., 100-fold) of the anticipated therapeutic concentration wherever feasible.

16.4.1.2.2   Dose or Concentration Selection, Route of Administration, and Duration of Studies

Doses selected for safety pharmacology studies are typically based on the criteria established in the ICH S7A guidance.[31] Doses should exceed those projected for clinical efficacy and at the upper limit be bound by (a) adverse PD effects in the safety pharmacology study, (b) moderately adverse effects in other non-clinical studies that follow a similar route and duration of dosing, or (c) limit of solubility/toxicity. In the absence of adverse effects, the maximum administrable dose can be used. Most importantly, the doses/concentrations should establish the dose/concentration–response relationship of the adverse effect.

The route of administration of an NCE is typically the intended clinical route of administration. However, an alternative route may be used if this leads to an increase in systemic exposure of parent drug or major metabolites or if this alternative route satisfies another important objective of the study. For example, it is common to increase/boost the exposure following inhalation administration by associating a subcutaneous administration of the NCE. The intravenous route is commonly used in the anesthetized guinea pig cardiovascular model, as it allows building a PK/PD relationship by collecting both cardiovascular parameters and PK samples on the same animals along the infusion of the test compound.[66]

Data should be collected for a sufficient period of time to identify the onset, time course, and recovery of effects should they be seen. In absence of knowing whether effects of the NCE will be observed in the non-clinical study, the timing of measurements will be based on available PK or toxicokinetic data collected in the same species, using the formulation and route that have been selected for the safety pharmacology investigation. On the basis of PK/toxicokinetic profile of the NCE, measurements will be made for a period that encompasses the maximal blood/plasma concentration (i.e., C_{max}) of parent drug and major metabolites, with recovery encompassing a period of at least five half-lives beyond the C_{max}. Moreover, a guidance document suggests that for all pivotal non-clinical safety studies that include a toxicokinetic evaluation, control blood samples should be collected and analyzed irrespective of the route of administration to confirm the validity of the study.[67]

For *in vivo* cardiovascular studies, a best practice article has been recently published.[68] It is reported that when using telemetered animals, the Latin-square crossover design can be adopted. A relatively small pool of instrumented animals can be reused to study vehicle and two to three dose levels of drug as long as an appropriate washout period (at least five half-lives) is permitted between each dosing day. Most often, the studies follow single-dose administration, with an adequate washout period between doses when animals are reused. However, an NCE that exhibits a prolonged half-life (e.g., antibody vs. small molecule) may not allow for execution of this type of a Latin-square crossover study design within a practical timeframe. In this case, if this type of study design is used, it is recommended to allow up to 1 or 2 weeks between doses, accepting that some accumulation of drug will occur with each subsequent dose. An alternative design is to use separate groups of animals for each dose, treated in parallel. However, such design might dramatically increase the total number of animals used. In both cases, the use of a negative control group (i.e., vehicle or placebo treatment) will be important to assess any temporal changes associated with the environment, human intervention, accommodation of animals, or the handling procedures.

The use of a negative control group (i.e., placebo or vehicle) is a requirement. In particular, effects of novel vehicles should be assessed since it is recognized that pharmacological properties of some vehicles may interfere with the activity of the NCEs and compromise with the interpretation of the results. Hydroxypropyl-beta-cyclodextrins, for example, commonly used to solubilize test items, are well known to induce hypertensive effects when given by systemic route at concentrations as low as 10%.[69] The use of a positive control in each study may not be practical based on the need to demonstrate multidirectional changes in a multitude of parameters as possible outcomes of administration of the NCE. This concern has been addressed in the ICH S7A guidelines that does not require the use of a positive control in each *in vivo* study, but rather by demonstrating that the *in vivo* model has been fully characterized.[31] One situation, however, where a positive control may be warranted is to rule out a suspected activity of the NCE by demonstrating the sensitivity and specificity of the experimental model. Additionally, the use of positive control is recommended for *in vitro* studies, where results of the NCE might be normalized to the effects of the positive control.[34]

16.4.1.2.3 Application of Good Laboratory Practice

Similar to other non-clinical safety studies that are required for registration of human pharmaceuticals, the GLP standards[32] apply to safety pharmacology studies,[31,34] with some exceptions as noted below. GLP adherence is expected for the safety pharmacology "core battery" studies, but the guidance acknowledges that aspects of the "follow-up" and "supplemental" studies, if they use unique methodologies, may not necessarily comply with GLP standards.[31,34] In these cases, however, the guidance recommends that the reconstruction of the study must be insured through documentation and archiving, and adequate justification is required for those aspects of the study not complying with GLP standards. Additionally, the impact of the noncompliance on the generated results and their interpretation should be acknowledged.[31,34,56]

16.4.1.2.4 Application of Statistical Analysis

The application of statistical methodology should be appropriate to optimize the design of the study to detect changes that are biologically significant for human safety,[31,70] while avoiding the unnecessary use of animals. The use of statistical power analysis allows the determination of optimal sample size for the detection of biologically relevant changes. For example, group sizes of four to eight dogs are sufficient to detect, with an 80% chance, a 10%–15% change in cardiovascular parameters (e.g., blood pressure, left ventricular (LV) pressure, QT interval, QT_c interval).[71,72,73]

16.4.1.2.5 Testing of Isomers, Metabolites, and Finished Products

Another challenge is how and when to consider isomers, metabolites, and the actual finished product. In general, any parent compound and its major metabolite(s) that achieve, or are suspected to achieve, systemic exposure in humans should be evaluated. Assessment of the effects of major (i.e., >25% of the parent) human-specific metabolite(s), if absent or present only at relatively low concentrations in animals, should be considered.[31,74,75] This is of particular importance if the metabolite(s) is known to substantially contribute to the pharmacological actions of the NCE. *In vitro* or *in vivo* testing of the individual isomers should also be considered. Moreover, studies on the finished product are only necessary if the PK/PD is substantially altered in comparison with the active NCE tested previously.

16.4.1.2.6 Conditions under Which Safety Pharmacology Studies Are Not Necessary

The ICH S7A guidance makes provision for conditions under which safety pharmacology studies may not be necessary.[31] The conditions include (a) locally applied agents where the systemic exposure or distribution to other organs or tissues is low, (b) cytotoxic agents of known mechanisms of action for treatment of end-stage cancer patients, (c) new salts having similar PK and PD properties to the original NCE, and (d) biotechnology-derived products that achieve highly specific receptor targeting. In the latter example, the evaluation of safety pharmacology endpoints may be considered as part of the toxicology studies. However, if the biotechnology-derived products represent a novel therapeutic class or are not achieving high selectivity, a more extensive evaluation of safety pharmacology studies should be considered.

16.4.1.2.7 Timing of Safety Pharmacology Studies in Relation to Clinical Development

The safety pharmacology "core battery" studies should be available prior to first administration in humans. Furthermore, "follow-up" and "supplemental" studies should also be available prior to first administration in humans if there are specific causes for concerns. During clinical development, additional studies may be warranted to clarify observed or suspected adverse effects in animals or humans. Finally, prior to product approval, effects on organ systems that are defined as part of "follow-up" and "supplemental" studies should be assessed, unless not warranted. Available information from toxicology or clinical studies can support this assessment and replace the need for stand-alone safety pharmacology studies.

16.4.2 Assessment of Vital Organ Functions

The following sections provide examples of approaches to assess drug effects on the cardiovascular, respiratory, and CNSs in compliance with the current and emerging regulatory guidance documents.[31,34,37,40]

16.4.2.1 Cardiovascular System

A comprehensive evaluation of the cardiovascular system includes an assessment of heart and vascular function and an evaluation of alterations in blood components. The heart functions as a "pumping unit" and may be impacted by effects of NCEs on the contractile elements of the myocardium or a loss in synchrony of the depolarizing wavefront that transits the myocardium (the conduction of an electrical impulse from the atrium to the ventricle) or repolarization of the myocardium (recovery of the myocardial cell to allow the propagation of the next depolarizing impulse). Effects of NCEs on vasculature may be manifested by a redistribution of flow to specific vascular beds generated by changes in vascular resistance. Effects of an NCE on the heart and vasculature may be mediated through direct actions on receptors, ion channels, transporters, enzymes or intracellular second-messenger systems or indirect effects on neurons, hormones, or normal physiological reflex mechanisms.

Measurements of cardiac function may include an evaluation of heart rate (HR), output, and contractility. These indices of heart function are reflected in the HR (derived from either the pressure pulse or ECG), systemic blood flows, myocardial shortening, intra-cardiac pressures, cardiac wall thickness, and cardiac chamber size; the latter four parameters were measured during both systole and diastole. Electrical conduction from the atrium to the ventricle during depolarization and recovery of the myocardium during repolarization are evident in the ECG recorded from electrodes placed either on the body surface, just below the surface (subcutaneous), or in close proximity to the heart within the thoracic chamber. The durations and amplitudes of PR and QRS intervals of the ECG represent conduction through the atrium and ventricle, respectively, and QT interval encompasses both phases of ventricular depolarization and repolarization. In addition to measurements of the duration of these specific intervals, an investigator will also interrogate the ECG to identify any changes in the morphology that may be indicative of PD effects on discrete areas of the heart or drug-related pathology that may have resulted from acute, subchronic, or chronic exposure.

The integrated function of the vasculature and heart, as a closed circulatory system, supplies nutrients and oxygen to critical organs and removes metabolic wastes and carbon dioxide. This integrated system results from the careful control of cardiac output (CO), arterial blood pressure (systolic and diastolic pressures; integrated to derive mean arterial pressure), and systemic vascular resistance, thereby maintaining blood perfusion through organs that are critical for sustaining life. Arterial blood pressure is equivalent to CO (times) systemic vasculature resistance. Thus, an NCE's effect on either CO or systemic vasculature resistance may increase or decrease arterial blood pressure. For example, dilation of the systemic vasculature (reflected as a decrease in systemic vasculature resistance) without a change in CO will result in a decline in arterial blood pressure. In contrast, an increase in CO without a change in systemic vascular resistance will result in an increase in arterial blood pressure. In practice, CO and arterial blood pressure are measured parameters, and systemic vascular resistance is derived.

Effects of NCEs on the cardiovascular system may also be secondary to alterations in the cardiac or vascular microstructure resulting from NCE-induced cellular toxicity. In the same light, changes in the components that constitute the blood following acute, subchronic, or chronic exposure to an NCE may also be manifested. These important pathologic endpoints are routinely assessed in multiple-dose toxicology studies in rodent and non-rodent species. Cardiovascular PD endpoints that are incorporated in such studies may serve two roles: (a) to identify any underlying NCE-induced changes in systemic hemodynamic or cardiac function that could have produced pathologic findings (i.e., an enlarged heart and myocardial necrosis that is the result of a sustained increase in cardiac afterload) or (b) to identify whether functional changes accompany or precede the pathologic lesion (i.e., alterations in the ECG that is associated with, but may precede the onset of a myocardial lesion). Depending on the nature of the lesion and the time course over which it develops, the results may provide a biomarker that can be monitored in the clinic to indicate the eventual emergence of the pathologic lesion with continued dosing.

Cardiovascular ADRs are one of the most prominent issues of the pharmaceutical industry; in the last decade, the single most common cause of the withdrawal or restriction of the use of drugs that have been already marketed has been the prolongation of the QT interval associated with polymorphic ventricular tachycardia or TdP (see Table 16.4).[53,76–78] TdP is typically not seen in clinical trials prior to registration of the drug. For terfenadine, the recognition of this rare event required extensive use and detailed monitoring from 1985 until 1998 before the drug was finally recalled because of 125 suspected drug-related deaths in the United States alone. Recent epidemiological retrospective analysis suggests that the incidence of drug-induced TdP might have been underestimated.[79] TdP has been linked to delayed cardiac repolarization, as manifested by a prolongation of the QT interval on the ECG. As a consequence, NCE-induced QT prolongation is generally considered as a surrogate marker for drug-induced TdP. Almost all compounds that prolong the QT interval and produce TdP in humans do so via inhibition of the rapid form of the delayed rectifier potassium current, IKr. During the last few years, significant advances have been made in our ability to test for effects of NCEs on the IKr current and a range of other cardiac ion channels. These *in vitro* assays, in conjunction with the *in vivo* cardiovascular/cardiac models (see Table 16.5), are able to detect the vast majority of compounds capable of prolonging the QT interval in humans. Thus, the risk of NCE-induced changes in cardiac repolarization in humans have been greatly reduced but not eliminated.[76] Other notable recent examples of cardiovascular ADRs include the increased incidence of heart valve regurgitation of the anorectic fenfluramine, in association with phentermine[80] and the withdrawal from the market place of the cyclooxygenase type 2 inhibitor (Vioxx) for an unacceptable increased risk of myocardial infarction and stroke.[81]

Regulatory guidance for non-clinical cardiovascular safety pharmacology testing is given in the ICH S7A and B.[31,34] The effects of an NCE on blood pressure, HR, and the ECG should be evaluated. Furthermore, *in vivo*, *in vitro*, and *ex vivo* evaluations, including methods for assessing repolarization and conductance abnormalities, should also be considered. The evaluation of drug

TABLE 16.5
Examples of Commonly Used Cardiovascular "Follow-Up" Studies

Physiological Endpoint	Methodology/Test	Reference(s)
	In Vitro	
Disaggregated Cells		
Repolarizing currents (e.g., IKs, IK1, Ito), depolarizing currents (e.g., INa) currents, ICa (whole-cell patch-clamp)	Disaggregated cells ventricular myocytes, mouse atrial tumor cells (AT-1), immortalized cardiac muscle cells (HL-1)	Jost et al.[82]; Liu and Antzelevitch[83]; Jurkiewicz and Sanguinetti[84]; Li et al.[85]; Yang and Roden[86]; Banyasz et al.[87]; Xia et al.[88]
APD (whole-cell patch-clamp)	Disaggregated cells	Davie et al.[89]
Myocyte contraction, Ca transients with fluorescent dyes	Disaggregated cells	Suetake et al.[90]; Cordeiro et al.[91]; Graham et al.[92]; Hamilton et al.[93]
Heterologous expression systems hERG current (whole-cell patch-clamp)	Mammalian cell expression (CHO, HEK-293, mouse L-cells, COS-7) and *Xenopus* oocytes expression system	Witchel et al.[94]; Zou et al.[95]; Cavero et al.[96]; Martin et al.[97]; McDonald et al.[98]
KvLQT1-minK current (whole-cell patch-clamp)	Mammalian cell expression (CHO, HEK293)	Sanguinetti et al.[99]
Affinities/functional activities at relevant molecular targets	Pharmacological profiling – binding or functional assays at receptors, enzymes, ion channels, transporters	Wakefield et al.[100]; Bowes et al.[26]
Isolated Tissues		
Vasoconstrictor/vasodilator assessment	Isolated tissues (e.g., arterial/venous rings)	Krasner et al.[101]; Lefer et al.[102]
Chronotropism and inotropism	Isolated atrium/papillary muscle/isolated hearts	Sugiyama et al.[103]; Voss et al.[104]
APD (microelectrode)	Purkinje fibers, papillary muscle, mid-myocardial (M-cell) wedge preparation	Abi-Gerges et al.[105]; Gintant et al.[106]; Lu et al.[107]; Antzelevitch et al.[108]
MAP, ECG parameters, proarrhythmia	Isolated intact hearts (Langendorff)	Hondeghem et al.[109]; Eckardt et al.[110,111]
Flows, pressures	Isolated intact hearts (working)	Miyoshi et al.[112]; Hill et al.[113]
Aggregation	Platelets or whole blood cell assays	

(Continued)

TABLE 16.5 (*Continued*)
Examples of Commonly Used Cardiovascular "Follow-Up" Studies

Physiological Endpoint	Methodology/Test	Reference(s)
	In Vivo	
Hemodynamic and cardiac parameters: pressure (arterial, venous, ventricular, e.g., ventricular contractility, CO), HR, peripheral resistance, ECG parameters, body temperature, flow	Conscious (restrained or telemetry) and anesthetized	Takahara et al.[114]; Sato et al.[115]; Nekooeian and Tabrizchi[116]
Ejection fraction (% of blood expelled from the ventricle during contraction), % fractional shortening (contraction of left ventricle), stroke volume (volume of blood ejected per contraction), *dp/dt* (an index of contractility), valve insufficiency, or regurgitation	Echocardiography	Ozkanlar et al.[117]; Tsusaki et al.[118]; Hanton et al.[119]
Regional blood flow	Flow probes, Doppler, microspheres	Wakefield et al.[120]
Proarrhythmia	Conscious and anesthetized	Eckardt et al.[110]; van der Linde et al.[121]; Sugiyama et al.[122]; Chiba et al.[123]
Denervated autonomic nervous system	Denervated autonomic nervous system; anesthetized pithed rat	Collister and Osborn[124]; AbdelRahman[125]
Autonomic nervous system (baro-, mechano-, and chemoreflex)	HR variability, nerve recordings, bar-mechano reflex curves, tilt table	Verwaerde et al.[126]; Harada et al.[127]; Mangin et al.[128]
Orthostatic maneuvers	Tilt table	Bedford and Dormer[129]; Humphrey and McCall[130]
Coronary reperfusion	Transient coronary artery ligature (total vs. partial occlusion)	Ytrehus[131]
Myocardial infarct size	Permanent coronary artery ligature	Wu and Lima[132]
Thrombosis	Models of arterial or venous thrombosis	Shebuski et al.[133]

effects on the cardiovascular system can be conducted using a range of *in vivo* or *in vitro* techniques, and species (Table 16.5). The ICH S7B, however, specifically requires an *in vitro* IKr assay and an *in vivo* QT assay to be made available to the regulators (Figure 16.3).

The recently proposed new paradigm of CiPA (for Comprehensive in vitro Proarrhythmia Assessment) is now considering proarrhythmia risk beyond delayed ventricular repolarization.

16.4.2.1.1 In Vivo QT Assay

The aim of an *in vivo* QT assay is to measure indices of ventricular repolarization such as the QT interval. This assay can be designed to meet the objectives of both ICH S7A and S7B.[31,34] The development of telemetry techniques in conscious animals has had a major impact on the conduct of *in vivo* cardiovascular safety pharmacology studies. The telemetry technique permits a continuous collection of a range of physiological parameters, including HR, pressures (e.g., arterial, venous, pleural, LV), ECG (including the QT interval), and body temperature over longer periods of time in undisturbed animals. Thus, drug effects can be studied under physiological conditions using the clinical route of administration. Alternatively, anesthetized animals can be used under conditions

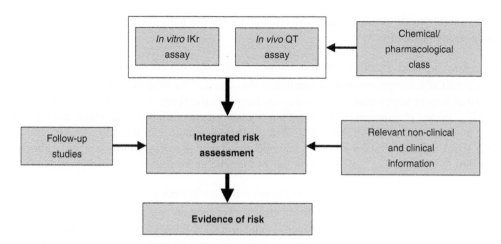

FIGURE 16.3 Component elements of the testing strategy risk for delayed ventricular repolarization and QT interval prolongation. (Adapted from Anon, CPMP/ICH/423/02/2005.)

where (a) the compound is poorly tolerated in the chosen species (e.g., due to emesis or tremor), (b) its bioavailability/exposure is expected to be low, or (c) insufficient information is known about the NCE at the time of the evaluation. Although the effect of anesthesia on the NCE under study may be unknown (e.g., drug–drug interaction), it is generally accepted that effects seen in anesthetized animals are qualitatively comparable to those detected in conscious animals.

16.4.2.1.2 In Vitro IKr Assay

The aim of an *in vitro* IKr assay is to evaluate the effects of NCE on this ionic current. This evaluation can be performed in cardiac myocytes or cell lines expressing hERG, the α-subunit of the IKr channel protein. The most commonly used cell lines are Chinese hamster ovary (CHO) and human embryonic kidney (HEK293) cells. Screening of NCEs for hERG inhibitory activity early in the drug discovery process may help to make business decisions on hERG data alone that ultimately may reduce attrition of NCEs in later non-clinical or clinical development stages. Screening during the early phases of the discovery process implies the ability to test a large number of compounds in a short period of time with minimal effort in terms of resources. Several approaches are currently being used. In binding assays, the displacement of a radioactively labeled channel antagonist (e.g., [3H]dofetilide) by the compound under investigation is measured.[134] Rubidium (Rb) flux assays rely on the high permeability of Rb^+ through voltage-sensitive K^+ channels.[135] Fluorescence assays make use of voltage-sensitive dyes, which measure the membrane potential of a living cell.[136] All these tests suffer from measuring the effects on the hERG channel indirectly and are therefore prone to artifacts. For the last two decades, the patch-clamp technique has been regarded as the gold standard, although the manual handling of the patch-clamp systems is labor-intensive. The introduction of new medium-/high-throughput patch-clamping technologies has significantly increased the ability to detect and design hERG-like properties.[137]

16.4.2.1.3 In Vitro Assays for Other Cardiac Ion Channels

In addition to hERG, many other ion channels, transporters, or exchangers play a role in cardiac electrophysiology. This is the reason why the first CiPA component consists of assessing the drug effects not on one single current but on seven ionic currents chosen based on their prominent impact on cardiac action potential. The currents selected are, in addition to IKr, ICaL generated through $Ca_V1.2$ channels, fast and late sodium currents both generated through $Na_V1.5$ channels, as well as IKs, Ito, IK1 outward repolarizing currents. The challenge associated with this component is the lack of standardization of protocols across laboratories conducting patch-clamp assays, leading to high variability in the results. Efforts are ongoing to develop harmonized experimental conditions and operating procedures.[58]

16.4.2.1.4   In Silico Reconstruction of Human Ventricular Repolarization Changes

The second CiPA component aims at simulating the effects of the drug on action potential generated by virtual human ventricular cardiomyocytes, based on the Rudy–O'Hara model.[138] The *in vitro* data (IC_{50} values) obtained in the different ion channel assays of the first component are introduced in the mathematical model of AP, which then predicts the changes in the different AP parameters (APD, EAD, etc.). This model allows to integrate the effects of a drug on multiple ion channels and takes into account the fact that effects on some channels can be counterbalanced by others.[139]

16.4.2.1.5   Human Stem-Cell–Derived Ventricular Cardiomyocytes

The third component is used to confirm effects predicted by the two first components in an *in vitro* cellular model, and possibly to identify changes that might not be related to the ion channels assessed. Different technology platforms are available, such as microelectrode arrays or voltage-sensitive dyes. The model usually combines multiple endpoints linked to electrophysiology (action or field potential), contractility, and beating rate. The challenges and questions associated with this component are the relative immaturity of the cells and its potential impact on the predictivity of the model, and the determination of clear go/no go criteria for each endpoint, which would allow to take decision on the progression of the test drug.

In addition to the "core battery" assays described above, the ICH S7A guidance recommends "follow-up" studies to investigate the effects on CO, ventricular contractility, vascular resistance as well as the effects of endogenous and exogenous substances on the cardiovascular responses. A non-exhaustive list of *in vitro*, *ex vivo*, and *in vivo* cardiovascular models is presented in Table 16.6 that can be used to investigate known or suspected issues. With respect to assessing the potential for

TABLE 16.6
Examples of Commonly Used Respiratory "Follow-Up" Studies

Physiological Endpoint	Methodology/Test	Reference(s)
Ventilatory disorders Hypo-/ hyperventilation	$PaCO_2$ *in vivo*, respiration rate, tidal volume, minute volume	Murphy et al.[140]
Central or peripheral mechanism	Responses to CO_2 and NaCN *in vivo*	Murphy et al.[142]
	Neuronal activity *in vitro/in vivo*	Jackson et al.[146]; Widdicombe[147]
Obstructive Disorders		
Airway narrowing	Forced maneuvers (FVC, FEV, IC) *in vivo*, airway resistance *in vivo*, airway resistance *in vitro* (tracheal rings)	Diamond et al.[148]; Mauderly[149,150]; O'Neil et al.[151]; Douglas et al.[152]; Kenakin[153]; Murphy et al.[154]
Restrictive Disorders		
Interstitial thickening	Lung compliance *in vivo*	Diamond et al.[148]; Mauderly[149,150]
Hemorrhage	Ciliary beat assays *in vitro*	Chilvers et al.[155]
Cellular infiltration	Mucus level assays *in vitro*	Khawaja et al.[156]
Abnormal surfactant production	Alveolar type II cell surfactant secretion assay *in vitro*	Dietl et al.[157]
Affinities/functional activities at relevant molecular targets	Pharmacological profiling — binding or functional assays at receptors, enzymes, ion channels, transporters	Bowes et al.[158]

Lynch et al.,[159] Wakefield et al.[100]

an NCE to slow ventricular repolarization and prolong the QT interval, "follow-up" studies can be used to understand the basis of discrepancies among non-clinical studies and between non-clinical and clinical studies.

16.4.2.2 Respiratory System

The respiratory system can be divided functionally into a pumping apparatus and a gas exchange unit.[140] The pumping apparatus includes those components of the nervous and muscular systems that are responsible for generating and regulating breathing patterns, whereas the gas exchange unit consists of the lung with its associated airways, alveoli, and interstitial area that contains blood and lymph vessels and an elastic fibrous network. The ICH S7A guidance recommends that respiratory rate and other measures of respiratory function (e.g., tidal volume or hemoglobin oxygen saturation) should be quantified using appropriate methodologies as part of the "core battery" studies. The assessment of lung mechanics (e.g., airway resistance, lung compliance) has been relegated to the rank of "follow-up" investigations and therefore is not required prior to first administration to humans (unless there is a cause for concern).[31]

In some respects, part of the content of the ICH S7A guidance regarding respiratory function monitoring and respiratory testing is inaccurate and misleading.[31,56,141] Therefore, to effectively protect clinical trials participant from respiratory ADRs while avoiding unnecessary animal usage, it is recommended that the effects of NCEs on respiratory function should include both ventilatory and mechanical function assessments.[56,141–143]

The primary test for evaluating the pumping apparatus is the measurement of ventilatory changes in conscious animals. To characterize ventilatory patterns, the measurement should include at least the respiratory rate, tidal volume, and minute volume. Additionally, to investigate for potential mechanisms, the measurement should also include the inspiratory flow, expiratory flow, and fractional inspiratory flow. The most common method used to collect these parameters is the plethysmography.[144] The rat is usually used as first choice species, as it can easily stay in a plethysmography chamber for several hours in freely moving conditions, thereby with reduced stress level. Non-invasive telemetry model has been developed although the robustness of such approaches has been questioned.[145] "Follow-up" evaluations should include tests for detecting the occurrence of hypo- or hyperventilation syndromes and for distinguishing central from peripheral effects. Functional changes in the gas unit, or lung are evaluated by measuring changes in the mechanical properties of the lung. The primary test for evaluating the function of the lung should include measurements of the lung resistance and compliance. Dynamic measurements of the resistance to lung airflow and lung compliance are the preferred tests, as they can be used to simultaneously measure both the ventilation and lung function parameters repeatedly in conscious animals.[141] "Follow-up" evaluations should be performed to determine the site and extent of an obstructive disorder, to confirm the presence of a restrictive disorder, and to characterize effects on forced expiratory airflows and lung capacities (see Table 16.6). The use of respiratory stimulation models allows to increase the sensitivity of detection or to quantify loss of functional reserve.[142] Although the ICH S7A guidance refers primarily to *in vivo* studies, *in vitro* approaches can be deployed to address specific respiratory endpoints.[100]

16.4.2.3 Central Nervous System

Most of the ADRs relating to the nervous system impact on the quality of life rather than the risk to life (e.g., lethargy, anorexia, insomnia, personality changes, and nausea). There are, however, some serious life-threatening adverse effects involving the nervous system (e.g., loss of consciousness and convulsions). Some of these reflect the fact that the nervous system controls the other two vital organ systems for that CNS impairment could be fatal (e.g., decreased respiratory drive leading to respiratory arrest, and decreased sympathetic outflow leading to cardiovascular collapse). The nervous system adjusts the function of the other acutely vital organ systems according to current and long-term requirements of the organism. Therefore, drug effects on cardiovascular and respiratory functions can be mediated via a direct action within the CNS, or via sensory nerve endings located in the cardiovascular and pulmonary systems. Some CNS adverse effects can be indirectly life-threatening. For example, drowsiness, cognitive impairment, motor coordination, dizziness, involuntary movement, and visual disturbances can all affect driving performance; moreover, depression and personality changes can lead to suicidal tendencies. As an illustration, the number of deaths in the United States between 1984 and 1996 in patients receiving terfenadine was 396, a proportion of which were attributed to sudden death resulting from TdP. This overall low incidence of fatalities is nevertheless a significant improvement over the first generation of antihistamine drugs that have been suspected to be responsible for significant fatalities in car accidents resulting from their sedative effects.[160,161]

The ICH S7A guidance states that effects of an NCE on "motor activity, behavioral changes, coordination, sensory/motor reflex responses, and body temperature should be evaluated. For example, a functional observational battery (FOB), modified Irwin's test, or other appropriate test can be used."[31,162–164] Whereas the Irwin's test was introduced in the pharmaceutical industry initially as a rapid psychotropic screening procedure for use in mice, the FOB arose from neurotoxicity testing in rats in the chemical and agrochemical industries and has been adapted and adopted for use in safety pharmacology testing of NCEs.[165–167] The FOB is a systematic evaluation of nervous system function in the rat, comprising more than 30 parameters and covering autonomic, neuromuscular, sensorimotor, and behavioral domains.[166]

"Follow-up" studies can include behavioral pharmacology, learning and memory, ligand-specific binding, neurochemistry, visual, auditory and electrophysiology examinations, etc. The ICH S7A guidance does not distinguish between CNS-targeted vs. non-CNS–targeted drugs, although a higher incidence of CNS side effects would be expected for CNS-targeted drugs compared to non-CNS-targeted drugs; hence, the number of "follow-up" studies is likely to be greater in the former case. In a survey, it was recommended that there would be a more extensive non-clinical assessment of CNS function.[56] The recommendations were based on the fact that some CNS AEs could not necessarily be predicted from an Irwin's or FOB test (e.g., pro-/anticonvulsive potential, abuse potential, and headache).[56] Table 16.7 summarizes some *in vivo* and *in vitro* physiological functions that can be evaluated as "follow-up" studies on a cause-for-concern basis and the associated methodologies. Although the ICH S7A guidance refers primarily to *in vivo* studies, *in vitro* approaches can be deployed to address specific nervous system endpoints (for review, see Wakefield et al.[100]). Of interest are the guidances on the non-clinical and clinical investigations that might be required to assess the dependence and suicidal ideation potential of drugs.[36,37,39] The proposed generic approach is presented in Figure 16.4.

16.4.3 Assessment of Non-Vital Organ Functions

The ICH S7A guidance states that "supplemental" studies are meant to evaluate potential adverse PD effects on organ systems functions that are not acutely essential for the maintenance of human life and not addressed by the "core battery" or repeated-dose toxicity studies when there is a cause for concern.[31] Examples of physiological functions that fall into that category include, but are not limited to, the renal/urinary, immune, GI, endocrine, and autonomic nervous systems. This section focuses on the renal and GI systems based on their potential impact on the clinical development program.

16.4.3.1 Gastrointestinal System

The GI tract, essentially a 10-m-long muscular tube extending from mouth to anus, is target to many clinical ADRs ranging from minor non–life-threatening (e.g., tooth discoloration) to severe and life-threatening (e.g., perforated ulcer).[190,191] Between the two ends of this severity spectrum range, a multitude of common, yet poorly understood, drug-induced disturbances of GI function that negatively impact patient safety, compliance, quality of life, and clinical benefit. The impact of GI ADRs

TABLE 16.7
Examples of Commonly Used CNS "Follow-Up" Studies

Physiological Endpoint	Methodology/Test	Reference
	In Vivo	
Nociception	Tail flick, hot plate, plantar test	Eddy and Leimbach[168]
Convulsion	Electric or chemically induced convulsion	Krall[142]
Locomotor activity	Interruptions of photoelectric beams, activity wheels, changes in electromagnetic fields, Doppler effects video image analysis, telemetry	Reiter and McPhail[169]
Motor coordination	Rotarod, beam walking	Dunham and Miya[170]; Abou-Donia et al.[171]; Jolkkonen et al.[172]
Auditory function	Startle reflex, brainstem auditory evoked response	Redfern et al.[6]; Wakefield et al.[100]; Herr et al.[173]
Visual function	Ophtalmoscopy, optokinetic	Prusky et al.[174]
Anxiety	Elevated plus maze	Redfern and Williams[175]
Sleep induction	Loss of righting reflex, electroencephalogram	Porsolt et al.[167]; Jouvet[176]
Learning and memory	Passive avoidance task, Morris water maze, radial maze, operant behavior tasks	Bammer[177]; Glick and Zimmerberg[178]; Morris[179]; Olton[180]; Dunnett et al.[181]
Electrophysiological examinations	Electroencephalogram *in vivo*	Itil[182]
Drug dependence	Physical withdrawal phenomena, drug preference tests	Meert[183]; Goudie et al.[184]
Drug abuse	Place preference tests, drug discrimination tests, self-administration procedures	Brady and Fischman[185]; Lal[186]; Schechter and Calcagnetti[187]
	In Vitro	
Electrophysiological examinations	Evoked potentials *in vitro* from brain slices	Fountain and Teyler[188]
Affinities/functional activities at relevant molecular targets	Pharmacological profiling – binding or functional assays at receptors, enzymes, ion channels, transporters	Bowes et al.[26]; Wakefield et al.[100]; Fliri et al.[189]

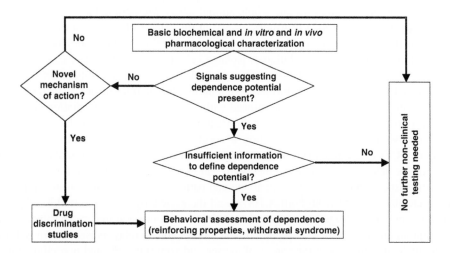

FIGURE 16.4 Proposed non-clinical testing strategy for assessing dependence potential of CNS-active NCEs. (Adapted from Anon, EMEA/CHMP/SWP/94227/2004, 2005.)

in terms of drug withdrawals from the market has been minimal, with only one example, pirprofen, in the period from 1960 to 1999[14]; however, drugs with labeling restrictions are commonplace, for example, Lotronex.[192]

GI AEs are one of the most common ADRs observed in humans, spanning across multiple modalities, and pharmacological modes of action.[193,194] According to the former author,[193] GI AEs, such as diarrhea/loose stool, constipation, nausea, vomiting (emesis), and decreased appetite, accounted for 21% and 14% of small and large molecule drug reactions, respectively, and were ranked as the top AEs in humans, followed next by neurological and hepatobiliary AEs, and similar incidences of GI AEs have been reported elsewhere with other datasets,[158,195,196] indicating a consistent theme regarding drug-induced effects on the on the GI system with marketed drugs. Given, however, that symptoms of disturbed GI function are encountered in everyday life, the actual incidence of drug-related effects is most likely extensively underreported. Such effects can significantly impact patient compliance and quality of life[197] as well as complicate drug development and impede progression of valuable new medicines.[198]

In addition to safety concerns, drug-related inhibition or enhancement of GI motor function can also lead to the alteration of a drug's PK profile owing to alterations in residency time of a drug in its site of absorption. Changes in plasma exposure of orally administered drugs, for example, aminophylline following application of an inhibitor of gastric motor function, for example, propantheline, are well documented.[199]

The GI system is responsible at its most basic level for providing a continual supply of water, electrolytes, minerals, and nutrients. This is achieved by a myriad of specialized cells and coordinated interplay of motility, secretion, digestion, absorption, blood flow, and lymph flow. These components are under elaborate control of the central and enteric nervous systems, endocrine, and paracrine regulation of hormones. The highly complex nature of GI function is clearly illustrated by the estimate that 80–100 million neurons exist within the enteric nervous system, a number comparable to that found within the spinal column, hence described as a "second brain."[200]

ICH S7A lists examples of GI parameters that can be measured; for example, gastric secretion, GI injury potential, bile secretion, transit time *in vivo*, ileal contraction *in vitro*, gastric pH, and pooling. The list of parameters is by no means comprehensive toward assessing function rather than histopathology; this inclination reflects the distribution of clinical GI ADRs, but functional assessment should not predominate at the expense of histopathology.

Harrison et al.[201] reviewed the methods and techniques capable of assessing specific changes in GI function at the membrane, cell, and whole animal levels. Membrane-based studies record the uptake of solutes and electrolyte transport, assessing the effects of NCEs on trans-epithelial GI transport and flux. Such methods lend themselves to permeability, immunocytochemistry, morphology, and molecular biology techniques. Isolated cells from the GI tract or cultured cell lines provide knowledge of regulation and function at a cellular level, while motility pattern, taken from *in vivo* or from biopsies, provides information at a more integrated level. In anesthetized animals, ligated segments of the intestine can be infused with NCEs, providing information about absorptive and secretory processes important for the treatment of diarrhea. Computer simulations and modeling are used to simulate the dissolution, absorption, distribution, metabolism, and excretion properties of NCEs in the human GI tract, thereby replacing to some extent animal testing. Finally, advances in the field of imaging, combined with endoscopy, have resulted in a wireless capsule, allowing the inspection of the GI tract anatomy and pathology without surgical intervention. One important element omitted from both ICHS7A and the review by Harrison et al.[201] of particular relevance to anticancer drugs is nausea and vomiting. Such omission may reflect the complexity of the symptom and that there are no non-clinical models able to relay nausea. There is, however, a growing understanding about the causes of nausea and vomiting[202,203] and non-clinical dog and ferret models amenable for safety pharmacology testing.[204,205] It is becoming increasingly evident that in addition to traditional *in vivo* techniques, safety pharmacologists could in the foreseeable future routinely incorporate membrane, isolated tissue, and endoscopy techniques for GI tract testing of NCEs.

16.4.3.2 Renal System

The kidney is a privileged target for toxic agents because of its physiological and PK properties. It receives the largest amount of blood per gram of tissue among any other organ, and therefore, it is more exposed to exogenous circulating NCE than many other organs. Moreover, tubular mechanisms of ion transport act to facilitate drug entry into renal tubular cells. From a PK perspective, the kidney is involved in filtration, excretion, and reabsorption of xenobiotics. The kidney concentrates urine so that intratubular drug concentration may be much higher than plasma concentration, and finally, the kidney has a high metabolic rate. Several drugs/drug classes are associated with nephrotoxicity (e.g., antibiotics, nonsteroidal anti-inflammatory drugs, immunosuppressors, angiotensin-converting enzyme inhibitors, chemotherapeutic drugs, and fluorinated anesthetics).[206,207] Deterioration of renal function over a period of hours to days results in the failure to excrete nitrogenous waste products and the inability to maintain fluid and electrolyte balance. Acute renal failure owing to toxic or ischemic injury is a clinical syndrome referred to as acute tubular necrosis and a common disease with high overall mortality (~50%).[16,208,209] There are a number of risk factors for acute renal failure that should be kept in mind during the development of an xenobiotics, especially in clinical trials: (a) patient-related risk factors such as age, sex, race, preexisting renal insufficiency, specific diseases, sodium retaining states; dehydration and volume depletion, sepsis, shock; (b) drug-related risk factors such as dose, duration, frequency, form of administration, repeated exposure; and (c) drug–drug interaction such as the associated use of NCEs with added or synergistic nephrotoxic potential. The challenge while assessing renal function results from the ability of the kidney to adapt to increases in single nephron glomerular filtration rate that consequently tend to mask renal injury until a considerable amount of kidney parenchyma is irreversibly lost. Such impairments, although not detectable in normal populations, can be quite significant in susceptible individuals. As a consequence, the use of animals that have been made more sensitive to functional effects should be considered (e.g., salt depletion, dehydration, coadministration of pharmacological agents, and unilateral nephrectomized animals).[16]

The ICH S7A states that the effects of the NCE on renal parameters should be assessed; for example, urinary volume, specific gravity, osmolality, pH value, fluid/electrolyte balance, proteins, cytology, and blood chemistry determinations such as blood urea nitrogen, creatinine, and plasma proteins can be used.[31] Most of these parameters are measured in rodent toxicology studies, although usually not after the first day of administration. In addition, seven novel nephrotoxicity biomarkers were approved by the FDA and EMA for non-clinical development in the rat.[210]

From a scientific point of view, glomerular filtration rate (inulin or creatinine clearance) is the best global estimate of renal function. Other parameters of interest include but are not limited to

1. Fractional excretion of electrolytes (i.e., sodium, potassium, and chloride)
2. Renal blood (or plasma) flow (*p*-aminohippuric acid clearance or ultrasonic transit time flowmetry)
3. Enzymuria (which could allow differential location of NCE-induced injuries, for example, alanine-aminopeptidase; γ-glutamyl-transferase; trehalase originating from the brush border; β-glucuronidase, *N*-acetyl-β-D-glucosaminidase, acid phosphatase, β-galactosidase originating from the lysosomes and lactate dehydrogenase, leucine aminopeptidase, β-glucosidase, fructose-1,6 biphosphatase, and pyruvate kinase originating from the cytosol)
4. Proteinuria (including albuminuria indicative of increases in the permeability of the glomerular capillary wall; β-2 microglobulin indicative of an impairment in tubular reabsorption)

5. Glucosuria (which may be indicative of proximal tubular damage if the serum glucose concentration is within the normal physiological range)
6. Diuresis
7. Concentrating ability of the kidney (measurement of urine osmolality assessed following withdrawal of food and water for 24 h; free water clearance).

Assessment of renal function is not mandatory prior to first administration to man or even during clinical development. Drug-induced kidney injury (DIKI) remains a significant cause of attrition in drug development, but safety attrition related to renal system in the pre-/non-clinical development and early clinical trial is small.[211] However, based on the potential implications of acute renal failure and the challenges in assessing it in normal healthy animals or humans, it would make sense to consider a proper assessment of renal function prior to first administration to humans. An industry survey conducted in 2012 concluded that most companies measured urinary excretion parameters in toxicology studies, generally in metabolism cages and conducted dedicated studies only when there is a cause of concern.[212]

As a whole, renal/urinary studies have little impact by themselves but are combined with other target organ system data to form a larger decision-making set to potentially stop projects.

16.4.4 Integration of Safety Pharmacology Endpoints in Toxicology Studies

Back in the late 1970s, Zbinden wrote that the ADRs, which the standard toxicological test procedures do not aspire to recognize, include most of the functional side effects. Clinical experience indicates, however, that these are much more frequent than the toxic reactions owing to morphological and biochemical lesions.[213] The commonly heard arguments against inclusion of functional tests in toxicology studies are the facts that (a) it is not a regulatory requirement; (b) the design, conduct, reporting, and interpretation of the tests require specialized scientific and technical expertise; (c) the increased handling may cause changes in stress hormones, which may influence the pathology; and (d) the tests are cost and labor-intensive for a limited value in return. However, several elements have argued in favor of the inclusion of functional endpoints in toxicology studies.[214] For example, some toxic responses are of a purely functional nature and are not accompanied by morphological lesions (e.g., arrhythmia and seizures); functional toxicity often occurs much earlier and at lower doses than those necessary to induce pathological organ damage; most pathological lesions are secondary to a functional disturbance (e.g., vasoconstriction, tachycardia, and endocrine responses).[213]

Table 16.8 illustrates examples of functional endpoints that have been successfully incorporated into toxicology studies. The examples cover primarily, but are not limited to, the vital organ systems (i.e., cardiovascular, respiratory, and CNSs). The repeat-dose toxicology studies offer an ideal opportunity to compare the pharmacological effects of repeated dosing with single-dose responses. Some assessments are incorporated routinely as per regulatory requirements (e.g., assessment of renal function using metabolism cage and assessment of cardiac electrical activity; Table 16.8). Some functional measures can be "bolted-on" without affecting the animals (e.g., home cage locomotor activity; Table 16.8) or with minimal impact (e.g., neurobehavioral assessment using a FOB test; Table 16.8). Some functional endpoints can be incorporated on a case-by-case basis if a cause for concern arises from the primary, secondary, or safety pharmacology studies, toxicology studies, or clinical observations, for example, the measurement of brainstem auditory evoked responses (BAER) in conscious-restrained beagle dogs for a drug suspected to affect the auditory function (Figure 16.5); or the measurement of ventilatory parameters for a drug suspected or known to affect the respiratory function or lung morphology. Finally, under some circumstances where the functional measurements would interfere with the primary goal of the toxicology study, satellite groups of animals can be included (e.g., telemetered animals to measure blood pressure).

TABLE 16.8

Examples of Measurement of Safety Pharmacology Endpoints in Toxicology Studies

Physiological Function	Parameters	Technique/Method	Species	Reference(s)
Respiratory	Respiration rate, inspiratory and expiratory times, tidal volume, minute volume, peak inspiratory, and expiratory flows	Whole-body plethysmography	Rat	McMahon et al.[215]
Cardiovascular	ECG intervals durations, amplitudes, and morphology	Surface electrodes using noninvasive telemetry in freely moving animals or conventional recordings in restrained animals	Dog	Hunter et al.[216]; Tattersall et al.[73]; Detweiller[217]; Derakhchan[218]
	Blood pressure		Dog Non-human primate	Ward G. et al[219]; MacMahon C. et al.[220]; Mitchell et al.[221]
Cardiovascular	Heart weight, wall thickness, LV and right ventricular (RV) end-diastolic and end-systolic lumen volumes (EDV and ESV, respectively), CO, HR, and LV diastolic filling pressure	Magnetic resonance imaging	Dog	Opie[222]
Cardiovascular	As above	Echocardiography	Dog, rat	Hamlin[223]
Nervous system	20–30 neurobehavioral endpoints	FOB	Rat (or mice) or dog	LeBel et al.[224]; Horner et al.[225]
	Behavior, seizure liability, and sleep pattern	EEG combined with videorecording	Dog and NHP	Authier et al.[226]
Motor coordination	Time spend on the rotarod	Rotarod	Rat	Chapillon et al.[227]
Auditory	Amplitudes and durations of waveforms	Brain auditory evoked response	Dog	Holliday et al.[228]
Auditory	Startle reflex	Prepulse modulation	Rat	Colley et al.[229]
Visual	Ophthalmoscopy	Ophthalmoscope	Dog	
Visual	Pupil diameter	Pupil diameter to dark/light stimuli	Rat	Redfern et al.[230]
Nociception	Time to withdraw tail	Tail flick	Rat	Horner et al.[225]
Muscle strength	Force	Grip strength	Rat	Horner et al.[225]
General activity	Distance traveled	Locomotor activity	Rat	Horner et al.[225]
Nervous system	Conduction time	Nerve conduction	Rat	Horner et al.[225]
Renal	Water intake, food intake, body weight, urine volume, urinary excretion of Na, K, Cl	Metabolism cage	Rat	Benjamin et al.[212]
GI	Feces weight, aspect	Clinical observation or metabolism cage	Rat, dog	

It is important to highlight that functional toxicology is not an alternative to acute dose safety pharmacology studies, but it is complementary. The main reasons are as follows: (a) functional measurements during repeat-dose studies are assessing the "effects" rather than "responses" to drugs in animals that may have some degree of multiple organ impairment owing to repeated drug exposure; (b) the experimental conditions may be difficult to optimize; for the variable being

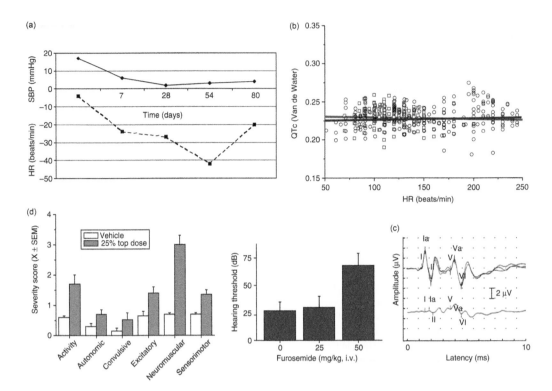

FIGURE 16.5 Examples illustrating the measurement of safety pharmacology endpoints in toxicology studies. (a) Monitoring of systolic blood pressure (SBP) and HR in rats over a 3-month period following once daily treatment. Values are vehicle subtracted mean absolute changes in SBP and HR. SBP was measured using the tail-cuff method, and HR was derived from pulsatile signal of BP. (b) Drug/vehicle effect on Van de Water-corrected QT interval in dogs treated once daily for up to 1 year. Data are from 177 dogs. Baseline (predose) values or vehicle data are presented as open squares, whereas postdose or drug-treated data are plotted as open circles. The regression lines were not statistically, significantly different from each other, and the slopes of the regression lines were not different from zero. (c) Brainstem auditory evoked responses (BAER) induced by auditory stimuli via an earpiece were recorded from three conscious restrained beagle dogs using subdermal needle electrodes via a Medelec Sapphire IIA System. Repeated determinations were made of background auditory threshold in decibels (dB) and the amplitudes and latencies of waveforms I to V at 80dB. Furosemide (25 or 50 mg/kg) was then administered intravenously, and the effects on these parameters were assessed over a 60-min period. Preliminary results demonstrated that Frusemide at 50 mg/kg but not 25 mg/kg increased the hearing threshold in dogs by approximately 40dB (left-hand side). In addition, the amplitudes of waves I to V were markedly reduced (right-hand side). These effects were transient returning to normal by 6-min postdose. (d) Neurobehavioral assessment using a FOB and an automated assessment of motor activity. Dose- and time-related effects of seven prototypic chemicals following both single and 4-week repeated exposures were tested in four laboratories in the United States and four in Europe. The results indicated that neurotoxicants could be detected and characterized, despite some differences on specific endpoints. This study also provides extensive data regarding the use of neurobehavioral screening methods over a range of laboratory conditions as well as the reliability, sensitivity, and robustness of the tests to detect neurotoxic potential of chemicals.[234] Domain scores represent the neurotoxicity of acrylamide after 4 weeks of dosing (5 days/ week). The scores for each functional domain are averaged across all laboratories, and data are presented as means ±SEM. For clarity, only the vehicle and the 25% of the top dose groups are shown.

measured, there may be a large number of animals to measure from in a short period of time; and other measurements are also being made; and (c) a tolerance may develop to the drug response; for example, several classes of drugs/drug classes undergo a diminution of their initial response following repeated dosing (e.g., benzodiazepines, ethanol, nicotine, morphine, and β_2-adrenoceptors). This is an adaptive mechanism, occurring at molecular, cellular, and system levels related to the pharmacological target. It can also involve the induction of P450 enzymes. As summarized by Haefely, "Some form of adaptive syndrome is the inevitable consequence of the reciprocal interaction between most or all classes of drugs and the organism."[231] Functional endpoints can be integrated into regulatory toxicology. Over the past decade, technical progress in noninvasive methodology and refined measurements for pharmacological parameters and standardization of study design allow the incorporation into regulatory toxicology studies today and probably even more in the future. In addition, revolution in telemetry provides now relevant and reliable *in vivo* data from a variety of systems that can be integrated within a single assay.[232] The limitation of conducting pharmacological measurements in regulatory toxicology studies should be acknowledged. Safety pharmacology studies should complement toxicity studies in terms of choice of species and dose regimen. Ethical consideration of animal usage, especially dogs and monkeys, can only be justified in the future when more clinically relevant data can be gained from fewer *in vivo* studies. Multidisciplinary cooperation between pharmacology, PK, and toxicology will lead to the refinement and reduction of *in vivo* studies when functional parameters are integrated into regulatory studies and may provide valuable functional explanations for toxicology findings. An industry survey, conducted in 2012, highlighted a trend toward the integration of safety pharmacology parameters within toxicology studies instead of standalone studies in about 40% of the respondents for new chemical entities (especially for non-rodent studies), and this proportion reached 75% when it refers to biological agents. For this class of compounds, non-human primate is the most frequent species used in safety assessment, and both ethical and regulatory (ICHS6 guideline) considerations favor the inclusion of safety pharmacology parameters in toxicology studies.[233]

16.5 BIOMARKERS

The term "biomarker" generally refers to a measurable indicator of a biological state or condition, either physiological or pathological. Another definition is "any substance, structure, or process that can be measured in the body or its products and influence or predict the incidence of outcome or disease."[235] In toxicology or safety pharmacology, the measured response may be functional (blood pressure, ECG, respiratory rate, etc.) or biochemical (glycemia, liver enzymes, urinary proteins, etc.). Fluidic biomarkers usually refer to biomarkers that can be detected in body fluids such as blood, urine, and cerebrospinal fluid. Biomarkers have gained more and more importance over the last decades and play a critical role in drug development. They are expected to be early predictors of safety issues, and when measured in animal species as part of the preclinical studies, they should be translatable to humans.

This section does not pretend to give an exhaustive list of all biomarkers used in safety pharmacology but provides a short overview of the most common biomarkers applied to cardiovascular, renal, GI, and CNSs.

16.5.1 Cardiovascular Biomarkers

In addition to the functional cardiovascular biomarkers already covered earlier in this chapter, such as QT_c interval, blood pressure, and HR, a number of biochemical biomarkers have been developed to predict drug-induced cardiac injury (DICI). In 2008, a Cardiac Safety Biomarkers Working Group was initiated by HESI to review potential biomarkers of cardiac injury. This group advocated the use of cardiac troponins (cTns), biomarkers already used by physicians and cardiologists, in preclinical toxicity studies.[236] The validation studies conducted by this group led to the qualification by FDA of cTnT and cTnI as biomarkers in safety assessment studies.[237] The kinetics of the cTn response depends on the dose and frequency of drug administration as well as the mechanism of the cardiac injury induced by the compound.[238] High sensitivity assays are needed to allow detection of low levels of cTn as early sign of possibly reversible damage to the heart.

Besides cTns, natriuretic peptides, in association with HR, have been suggested as interesting safety biomarkers in rats to detect cardiac hypertrophy in short-term studies.[239] A lot of other potential novel biomarkers of cardiac injury have been explored in medicine, such as C-reactive protein, fibrinogen, and uric acid,[240] but their application in preclinical drug development still remains uncertain.

MicroRNAs (mRNAs), small noncoding RNAs, also appear as promising biomarkers, as they are easily detectable in the circulation, stable, and evolutionary conserved. Several miRNAs (miR-1-3p, miR-208a-3p, miR-499-5p) are enriched in the myocardium across species and can be detected in biofluids upon acute cardiac injury injury.[241–243] In rat studies, a significant upregulation of miR-208a-3p in plasma was demonstrated in acute cardiac injury models which correlated with troponin level.[242,244] The link between miRNAs expressed in the heart upon DICI and circulating miRNAs was further explored in rats.[245] Dysregulation of miRNAs not only serves as a good biomarker but also points to different mechanisms of cardiac injury. The clear advantages of miRNAs over conventional biomarkers of cardiac injury promote their further investigation and validation for clinical and preclinical use.

In addition to DICI, medication can cause drug-induced vascular injury (DIVI). Efforts are ongoing to evaluate sensitive and specific biomarkers of vascular injury which include the propeptide von Willebrand factor (vWGpp), caveolin-1, vascular endothelial growth factor (VEGF), tissue inhibitor of metalloproteinase 1 (TIMP-1), TSP-1, smooth muscle alpha actin (SMA), calponin, and transgelin.[246,247]

16.5.2 RENAL BIOMARKERS

DIKI remains a significant cause of candidate drug attrition during drug development. However, the incidence of renal toxicities in preclinical studies is low, and the mechanisms by which drugs induce kidney injury are still poorly understood. Renal function is not part of the ICH S7A core battery; therefore, stand-alone safety pharmacology studies are not routinely conducted by pharmaceutical companies. When done, they include assessment of glomerular and hemodynamic functions. Moreover, in addition to the standard urinary analyses and histopathological examination of kidneys integrated in general toxicity studies, novel promising urinary biomarkers have emerged over the last decade, offering greater sensitivity and specificity than traditional renal parameters.

Seven of these biomarkers have been qualified by regulatory agencies for use in rat toxicity studies: kidney injury molecule (KIM-1), clusterin (CLU), trefoil factor 3 (TFF-3), β-2microglobulin (B2M), cystatin C (Cys C), albumin, and total protein.[248] Novel exploratory preclinical biomarkers of DIKI, urinary osteopontin (OPN), osteoactivin, neutrophil gelatinase-associated lipocalin (NGAL), α -glutathione S-transferase (α-GST), urinary retinol-binding protein 4 (RBP-4), and serum Cys C are being further investigated for their mechanistic specificity and ability to detect early kidney injury as well as their utility in monitoring injury reversibility.[249]

Developing a single biomarker for DIKI detection and monitoring is practically impossible due to a very complex kidney structure and function. A panel of biomarkers is more likely to detect and characterize renal injury at early stages. Omics approaches have been undertaken to study changes in the kidney upon nephrotoxicity at the level of mRNA, miRNA, protein, and metabolites.[250–253]

16.5.3 GASTROINTESTINAL BIOMARKERS

Unlike for DIKI biomarkers, the discovery and use of biomarkers for GI injury is still in its infancy. The challenges of developing biomarkers are even greater when the complexity of GI injuries is looked upon more carefully. The drug-induced GI side effects include but are not limited to, nausea, vomiting, ulceration, inflammation of the intestine, mucositis, altered fecal output, abdominal pain/discomfort, GI bleeding, and/or perforation which can be inflicted by various drug classes, chemical classes, or therapeutic areas. Several potential blood, stool, and breath biomarkers have been studied for GI injuries, including stool and serum inflammatory biomarkers and biomarkers of small intestinal enterocyte mass and function. Blood biomarkers of GI toxicity include but are not limited to citrulline, C-reactive protein, diamine oxidase, gastrins, and CD64. Fecal biomarkers of GI toxicity include calprotectin, lactoferrin, polymorphonuclear neutrophil elastase (PMN-e), fecal S100A12, and miRNAs.[254]

16.5.4 CNS BIOMARKERS

In contrast to DICI, DIVI, or DIKI, biomarkers for detecting the risk of drug-induced neurological disorders have been poorly explored so far. Although various functional biomarkers are used in preclinical studies through EEG recordings, activity measurements, or behavioral tests, fluidic biomarkers for CNS injury are not yet in place. The main challenge is that these biomarkers are usually not detectable in blood but requires invasive collection of CSF. However, recent technological advances in neurosciences allowed to identify biomarker signatures as potential diagnostic tools for various neurological diseases and conditions (AD, PD, MS, autism…). miRNAs have been also found to be potential targets or effectors of neurotoxicants which could lead to discovery of novel CNS biomarkers in the future. Several public–private consortia (HESI, Neutox, etc.) are currently working on this research domain, so that considerable progress can be expected in the coming decade with regard to CNS biomarkers.

16.6 PREDICTIVE VALUE OF NON-CLINICAL SAFETY PHARMACOLOGY TESTING TO HUMANS

The value of animal testing to ensure safety of human volunteers and patients during drug development has been widely debated. Recent publications have challenged the utility of conducting toxicology studies in animal species.[255,256] However, analysis of the value of non-clinical in silico, *in vitro*, and *in vivo* studies to predict adverse PD events in humans during clinical trials has been robustly and extensively studied over recent years focusing on understanding the confidence in the biology of the test systems to reflect human biology, the confidence in the assays themselves, and the confidence in the translational value to human subjects and patients.[257,258] Reports published over the last decade showed promises with specificity, sensitivity, and accuracy in the 75%–85%,[259–261] although the level of confidence is largely depending on the endpoints.[194] Since safety pharmacology is well positioned to assess the predictive capacity of non-clinical assays to predict human safety in that similar endpoints can be measured preclinically and clinically (e.g., blood pressure, ECG, EEG, EMG[262]), it is of the utmost importance to focus collaborative research efforts to understand our confidence of non-clinical models to predict risk to human subjects and patients.

It is difficult to answer the question "Are non-clinical safety pharmacology tests predictive of side effects in humans?" Primarily because hard evidence is not readily available in the public domain, if an NCE has an effect in a non-clinical test, there may be limited information in the public domain, as the result may have precluded clinical development and therefore no apparent value in communicating this information. If there has been no effect in a non-clinical test and likewise no effect in the corresponding variable in humans, these negative data may have been deemed not to be of publishable interest. What we are then left with are the high profile examples of side effects in humans that are apparently not detected non-clinically. One publication attempted to explore the predictive values of safety pharmacology assays to humans.[10] Some significant correlations were reported. For example, a decreased locomotor activity in rodents was positively correlated with dizziness and sleepiness in humans, a decreased intestinal transit in rodents was correlated with constipation and "anorexia" in humans, and a decreased urinary and sodium excretion in the rat was correlated with edema in humans. Rather more bizarrely, the findings of analgesia, decreased body temperature, and anticonvulsive activity in rodents were each correlated with "thirst" in humans. This indicates the limitations of such surveys.

There are, however, numerous examples of drugs that cause adverse effects in humans, which would be detectable in safety pharmacology studies conducted as per ICH S7A and S7B guidances.

Some notable examples of individual drugs showing untoward effects in non-clinical studies that are correlated in a quantitative sense with adverse effects in humans have been reported,[6] for example, (a) the sedative effects of clonidine in various animal species and man, (b) the propensity of cisapride to prolong ventricular repolarization, (c) the respiratory depressant effects of morphine, (d) the nephrotoxic effect of cyclosporine, and (e) the GI effects of erythromycin. These examples illustrated the very good agreement of effects across all species tested and across a narrow range of doses/concentrations. Over the last few years, data have been generated to assess the value of non-clinical tests to predict the potential of NCEs to prolong the QT interval of the ECG and ultimately the proarrhythmic potential of these drugs. The published data converged in that an integrated risk assessment (see below) based upon data on the potency against hERG, an *in vivo* repolarization assay, and if necessary an *in vitro* repolarization assay are in a qualitative sense predictive of the clinical outcome.[255,263–265] These data have been further supported by publications suggesting that a 30-fold margin between the highest free plasma concentration of a drug in clinical use (C_{max}), and the concentration inhibited by 50% of the hERG current (IC_{50}) could be adequate to insure an acceptable degree of safety from arrhythmogenesis with a low risk of obtaining false positives.[52,53,266] Similar conclusions have been reached in relation to sodium channel block and risk of QRS prolongation.[267]

In the same way as in efficacy and kinetic models, there are, however, various reasons why non-clinical safety pharmacology assays may not predict human adverse effects: (a) species differences in the expression or functionality of the molecular target mediating the adverse effects, (b) differences in PK properties between test species and man, (c) sensitivity of the test system (e.g., observations of a qualitative nature should be followed-up with specific quantitative assessment), (d) poor optimization of the test conditions (the baseline level has to be set correctly to detect drug-induced changes), (e) study designs that are statistically underpowered, (f) inappropriate timing of functional measurements in relation to the time of maximal effect (i.e., T_{max}), (g) delayed effects (safety pharmacology studies generally involved a single-dose administration with time points covering the PK profile of the parent drug), (h) difficulty of detection in animals (adverse effects such as arrhythmia, headache, disorientation, and hallucinations are quite a challenge to detect in safety pharmacology studies), and (i) assessment of a suboptimal surrogate endpoint that predicts with some degree of confidence the clinical outcome (e.g., QT/QT_c interval prolongation as a surrogate of TdP). Some framework has been established to facilitate the assessment of the predictive capacity of non-clinical models to humans focusing on understanding of (a) the confidence in the biology, (b) the robustness of the model (or assay or screen), and (c) the confidence in the translational using pharmacological tools.[257,258]

16.7   INTEGRATED RISK ASSESSMENT

The integrated risk assessment is the stepwise and holistic evaluation of non-clinical study results in conjunction with any other relevant information and should be scientifically based and individualized for an NCE. Such an assessment can contribute to the design of clinical investigations and the interpretation of their findings.

Risk assessment in terms of protecting Phase I clinical trial participants is relatively straightforward, as it does not take into account any consideration of the therapeutic target, patient population, disease indication (unless when the Phase I trials include patients), or the degree of unmet medical need. Therefore, the assessment of the safety pharmacology data has to take into consideration the severity of the outcome in any given safety pharmacology test (see below – bearing in mind the sensitivity and specificity of the assays), and the plasma concentration at which it occurred relative to the expected exposure in the trial. Depending on the stage of drug development, the integrated risk assessment should consider contribution of metabolites as well as metabolic differences between humans and animals.

In terms of early risk assessment of NCE viability, the situation is more complex. In an attempt to simplify and standardize how safety pharmacology data can contribute to early risk assessment of project viability, Redfern et al.[6] proposed a matrix-type approach described below. This requires a grading process; each of the factors in the risk assessment can be graded into, for example, three categories – low, medium, and high. Starting with the safety pharmacology tests themselves, they can be categorized as follows: (a) minor – predictive of nonserious, reversible side effects (e.g., certain GI or renal effects); (b) moderate – predictive of impairment of quality of life (e.g., sedation, motor coordination); and (c) major – predictive of potentially life-threatening effects (e.g., prolonged the QT interval, pronounced hypotension, or bronchoconstriction). The next step consists in grading the therapeutic target according to disease severity: (a) minor/moderate disease (e.g., eczema, rhinitis, and Raynaud's syndrome), (b) debilitating disease (e.g., asthma, epilepsy, Parkinson's disease, stroke, and angina), and (c) life-threatening disease (e.g., cancer, AIDS, and myocardial infarction). The third component that can be considered is the existing therapy; the potential new drug must be anticipated to be superior to existing therapy. Therefore, the existing therapy cannot be classified as excellent; instead, it can be rated as (a) good, (b) partially effective with side effects, and (c) poor/inexistent. Once collected, the set of information can be put together in a matrix that also takes into account the dose level (or concentration) at which the effects were observed in the safety pharmacology test, in comparison with the expected clinical exposure (total or free plasma concentration), as shown in Figure 16.6. Effects well to the left of the line of crosses are acceptable without further debate, those well to the right are unacceptable, and those on or near the line require further discussion.

Hypothetical examples are presented to illustrate the usefulness of such matrix. In the first example, an NCE targeted at Raynaud's syndrome – a minor/moderate; disease[6] under classification, for which existing therapy is poor. The NCE is found to block the hERG channel and prolong the QT interval at a relatively low multiple (e.g., ~30-fold) above the expected therapeutic plasma concentration (Figure 16.7 as a schematic illustration). According to the matrix analysis, the decision is either to discontinue the progression of the NCE into development or to accept embarking on an extensive and expensive clinical program in compliance with the ICH E14 guidance.[36] At the

Therapeutic target	Existing therapy	Minor 100 x TD	Minor 10 x TD	Minor 1 x TD	Moderate 100 x TD	Moderate 10 x TD	Moderate 1 x TD	Major 100 x TD	Major 10 x TD	Major 1 x TD
Minor disease	Good	x								
	Partially-effective/side-effects		x							
	Poor/none			x					⊕	
Debilitating disease	Good				x					
	Partially-effective/side-effects					x				
	Poor/none						x			
Life-threatening disease	Good							x		
	Partially-effective/side-effects								x	
	Poor/none									x

Column header spanning: "Maximum acceptability of SP effect (TD = therapeutic dose)" over Minor / Moderate / Major, each subdivided into 100 x TD, 10 x TD, 1 x TD.

FIGURE 16.6 Safety Pharmacology Integrated Risk Assessment Matrix. Outcomes to the left of the line of the red crosses are acceptable, those to the right are unacceptable, whereas outcomes on or near the line require further discussion and possibly further investigations. 100X, 10X, 1X represents 100-, 10- or 1-fold the TD. A hypothetical example, highlighted in gray, is presented to illustrate the usefulness of such a matrix; an NCE targeted at Raynaud's syndrome – a "minor/moderate" disease under classification, for which existing therapy is poor. The NCE is found to block the hERG channel and prolong the QT interval (both considered ADRs of major severity) at relatively low concentrations (e.g., ~10-fold) above the expected therapeutic plasma concentration. According to the matrix analysis, the decision is either to discontinue the progression of the NCE into development or to accept embarking on an extensive and expensive clinical program in compliance with the ICH E14 guidance. (Adapted from Redfern, W.S. et al., *Fundam. Clin. Pharmacol.*, 16, 161–173, 2002.)

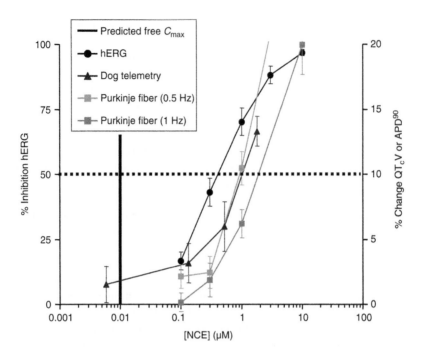

FIGURE 16.7 Schematic illustration of an integrated risk assessment aimed at assessing the liability for an NCE to prolong the QT interval in man. APD, action potential duration; NCE, new chemical entity; QT$_c$V, Van de Water-corrected QT interval duration.

other end of the spectrum, an NCE is targeted at leukemia, where the existing therapy is partially effective and has marked side effects. The NCE was found to increase intestinal motility at the therapeutic dose (TD); this would not be a major issue to prevent progression into clinical development. However, this could become an issue if the NCE eventually reaches the market and has to compete with a new rival drug of similar efficacy but lacking this side effect. Other factors should be considered such as (a) the target population, for example, cognitive impairment as a side effect may be more problematic in elderly and pediatric patients; and (b) the ultimate project objective, for example, QT interval prolongation as a side effect may be manageable and acceptable for an NCE aimed at demonstrating proof of mechanism or proof of principle. Overall, the evidence of risk, as part of an integrated risk assessment, can support the planning and interpretation of subsequent clinical studies.

16.8 CURRENT AND FUTURE CHALLENGES

The future of safety pharmacology will depend, in part, upon the scientific and technological advances as well as the regulatory challenges that envelop the constantly evolving pharmaceutical drug development.

16.8.1 NOVEL APPROACHES TO TREAT DISEASES

With advances in genetic and molecular biology and biotechnology, which allows for the identification of new clinical molecular targets, newer pharmaceutical agents are being identified that act at these novel molecular sites in an attempt to ameliorate the disease conditions. Moreover, new therapeutic approaches are being developed that encompass the broad class of biologic, e.g., vaccines, recombinant proteins, fusion proteins, monoclonal antibodies, RNA- and DNA-based, as well as cell-based therapies; these novel approaches present new challenges to the safety pharmacologist. Inherent in the novelty of new targets and new approaches is the risk of unwanted effects that may or may not be detected based on current scientific knowledge and techniques. These novel modalities have brought into focus a need to modify the way safety pharmacologists think about the profiling of such therapies. A rational tailored design, considering appropriate experimental models, study protocols, and overall risk assessment strategies to evaluate these novel treatment modalities, requires careful consideration. In addition, the duration of therapy, complementary nature encompassed within the continuum of preclinical studies leading into clinical studies, and the integration of data are in most cases applied where clinical experience with mode of therapy is lacking. For example, the complexities of advancing into the clinic with gene therapies illustrate the challenges and critical questions that must be addressed. Recognizing the limitations of what can be accomplished in preclinical models is also critical to establishing a strategy to mitigate AEs. Scientific consultation and alignment with regulatory authorities is an important aspect of advancing such novel therapeutics.

16.8.2 Science and Technology

The scientific challenge facing safety pharmacology is to keep pace, to adapt, and to incorporate new technologies in the evaluation of new drugs in non-clinical models, and to identify the effects that pose a risk to human volunteers and patients. In addition, the 3Rs (reduction, replacement, and refinement of laboratory animals) are now integrated in the current regulatory requirements and expectations.[268] Considerable efforts have been developed over the last decade to reduce, replace, and/or refine the use of animals in safety pharmacology. Another challenge is linked to the fact that non-clinical safety studies are typically conducted in normal, healthy, young adult, or adult animals. Therefore, these tests may not appropriately detect specific responses in humans at other ages (e.g., neonates, adolescents, and geriatrics) or those with underlying chronic diseases conditions (e.g., heart failure, renal failure, and type II diabetes), which may alter the PD response to an NCE. The challenge is to identify non-clinical models that reflect the overall human pathological condition and to incorporate these disease models along with traditional safety models/assays into safety pharmacology paradigms to produce integrated and more accurate assessment of possible human risk. Below are a few examples of promising future areas for development of safety pharmacology that further illustrate the technological and regulatory challenges in front of us.

16.8.2.1 In Silico Approach

There is a considerable interest in computational models to predict the safety of NCEs in the drug discovery and development phases. Insight into the secondary and safety pharmacological potential of a chemical scaffold early in the drug discovery process could help the medicinal chemists to prioritize particular chemical series or alternatively can contribute halting the discovery process for a given research project. It also allows building structure-activity relationships and modifies a chemical structure to remove or reduce a particular liability. The main safety pharmacology area that has been explored in terms of computational evaluation and prediction is cardiotoxicity. For example, there are now several models, filters, and decision trees based solely on physicochemical descriptors of hERG blockers. In silico hERG models that accurately predict molecules with a low potency at hERG (e.g., $IC_{50} < 10\,\mu M$) are valuable, as they contribute to prompt a project team to follow up with a relevant approach (either stop the compound or test it in priority to confirm the effect). Ideally, such models should have a low false-positive rate. Although it is recognized that hERG is probably a key contributor to drug-induced QT prolongation and TdP, some other molecular mechanisms are likely to be involved including other ion channels. As part of CiPA, mathematical models of the human ventricular action potential have been developed.[269] Using the effects of the test drug on different cardiac ion channels, they predict proarrhythmia risk associated with this molecule.

Some models are able to take into account the interindividual variability by simulating the response in a wide population of cardiomyocytes, mimicking a population of patients.[138,270] It is important to mention that molecular modeling in safety pharmacology is not restricted to the cardiovascular system, and valuable models predicting blood–brain barrier partitioning of NCEs have already been described.[271] Ultimately, such models should be key in defining the overall CNS program that a given NCE should go through.

Safety pharmacology is not the only discipline beneficiating from such in silico approach. Various in silico tools are available in the field of secondary pharmacology, also named off-target toxicity. Some consists of database collecting public information on the off-target profile of marketed or withdrawn drugs and the associated adverse reactions seen in the preclinical or clinical studies, while others are able to predict, based on the chemical structure, the expected secondary pharmacology profile and side effects of a new molecule.[272]

16.8.2.2 *In Vitro* Approach

An emerging approach is the use of human tissue and/or 3D models such as organ-on-chips, spheroid models, and human-induced pluripotent stem cells for safety assessments.[273] Human-induced pluripotent stem cell–derived cardiomyocytes have been successfully obtained. These cells present characteristics of differentiated ventricular cardiomyocytes, including a typical action potential. The ability to expand human stem cells *in vitro* provides unlimited potential for producing quantities of cardiomyocyte – progenitors, which can be employed to assess the effects of NCEs on cardiac ion channels, electrical activity, and contractility.[273] Human stem-cell–derived neurons are also being developed, which can be used to assess the anti- or pro-convulsant properties of new drugs or the potential for neurotoxic effects.[274,275] Another area of increasing interest is the microfluidic "organs-on-chips," in which living cells in continuously perfused chambers are arranged to mimic different tissues interconnected together. In this system, functional, metabolic, or genetic activities can be assessed.[276] Ultimately, increase in scientific knowledge and development of new technologies should lead to the development of new, robust, and predictive assays/models for ADRs that are currently poorly predicted such as nausea, headache, and arrhythmia.

16.8.2.3 *In Vivo* Approach

Besides the classical animal species used in safety pharmacology *in vivo* studies, namely rodents, dogs, monkeys or mini-pigs, the use of less sentient, non-mammalian species has raised increasing interest and applications for screening or mechanistic purposes.

The zebrafish (*Danio rerio*) has gained in popularity in the early 2000s, because of its small size and transparent body, allowing its use in 96-well plates.[277–280] Models have been developed to assess drug effects primarily on cardiovascular, nervous (including sensory functions), GI, and immune systems. As any other *in vivo* model, the zebrafish shows some limitations that might complicate its use for safety pharmacology testing, in particular the understanding and characterization of the kinetics properties and metabolism capabilities of the zebrafish larvae, and their proper exposure to the NCEs that are poorly soluble in water. Invertebrate species such as *Drosophila* or nematodes (*C. elegans* – Caenorhabditis) are increasingly used as alternative species for early safety assessment.[281]

Although, currently, the use of mammalian species cannot be completely avoided, a lot of refinements have been introduced in safety studies, such as the integration of multiple endpoints (including safety pharmacology parameters) in general toxicology studies, the use of non-invasive techniques like the jacketed external telemetry, the microsampling methods avoiding the use of satellite animals for toxicokinetics, and the social housing of both rodents and non-rodents even during telemetry studies. All these refinement procedures contribute to an improved animal welfare, which ultimately triggers optimized scientific data.[268]

16.8.3 REGULATORY REQUIREMENTS

The future of safety pharmacology is intertwined with international regulatory guidance. The ICH S7B and E14 guidances have been finalized and are due for implementation in November 2005.[34,36] An emerging draft guidance from the European Medicines Agency for the Evaluation of Medicines for Human Use on the non-clinical investigation of the dependence potential of medicinal products has been released for consultation and once finalized will influence the overall approach to safety pharmacology especially for CNS-targeted NCEs.[37] Safety pharmacology is also considered an important component to newly emerging regulatory guidance from the US FDA such as the safety evaluation of pediatric drug products and non-clinical studies for the development of pharmaceutical excipients. The discipline is considered integral to the evolving regulatory strategies for safety that accelerates the introduction of NCE into clinical phases (cf., Position paper on non-clinical safety studies to support clinical trials with a single microdose[282]) and the US FDA, screening IND application.[40,283] It is worth elaborating on this topic since it is likely to influence the approach to safety pharmacology. In its March 2004 Critical Path Report, the FDA explained that to reduce the time and resources expended during the early drug development on NCEs that are unlikely to succeed, tools are needed to distinguish earlier in the process those candidates that hold promise from those that do not.[284] The FDA suggests that limited exploratory IND investigations in humans can be initiated with less, or different, non-clinical support that is required for traditional IND studies because exploratory IND studies present fewer potential risks than do traditional Phase I studies that look for dose-limiting toxicities.[40] The draft guidance describes some early Phase I exploratory approaches that are consistent with regulatory requirements, but that will enable sponsors to move ahead more efficiently with the development of promising NCEs while maintaining needed human subject protections. Exploratory IND studies, which usually involve very limited human exposure and have no therapeutic intent, can serve a number of useful goals such as to (a) gain an understanding of the relationship between a specific mechanism of action and the treatment of a disease, (b) provide important information on PK properties of an NCE, (c) select the most promising lead product from a group of NCEs designed to interact with a particular therapeutic target in humans, and (d) explore the characteristics of the bio-distribution of a product using various imaging technologies. The non-clinical safety evaluation recommended for an exploratory IND application is more limited than for a traditional IND application[30] and may be tailored to the intended clinical study design and objective. For example, if the aim is to administer a microdose (i.e., 1/100th of the pharmacological dose and <100 gg), safety pharmacology endpoints may be incorporated into the single-dose toxicity studies (e.g., evaluation of respiratory and CNSs using clinical observations or FOB/Irwin's tests usually in rodent species) although we would strongly recommend an appropriate quantitative assessment of these vital organ functions. Furthermore, a comprehensive *in vitro* pharmacological profiling, aimed at determining the affinity and activity of the NCE at molecular targets known to be associated with unwanted activities, should be recommended (e.g., hERG, α-adrenoceptor). If the clinical trial is designed to study pharmacological effects or to investigate the mechanism(s) of action, then the effects of the NCE should be evaluated in safety pharmacology studies. Evaluation of the central nervous and respiratory systems can be performed as part of the rodent toxicology studies while evaluation of the cardiovascular system can be assessed in the non-rodent species, generally the dog. The replacement of stand-alone safety pharmacology studies by the inclusion of safety pharmacology endpoints into toxicology studies has clearly some benefit but also some potential drawbacks that one has to be aware, accept, and recognize the impact on quality and therefore the value of the data (see Section 16.4.3.3).

The introduction of pharmaceuticals into the environment is gaining growing attention of both regulators and pharmaceutical industry.[285–287] While this is not currently the subject of any international environmental guideline, the use of organ function endpoints may become an important component in bridging safety data collected from mammalian vertebrates (including humans) to aquatic species for purposes of the identification of relevant target and organ functions and the design of specific environmental toxicology studies.

16.8.4 TRAINING AND EDUCATION

Safety pharmacology also faces significant challenges of attracting, training, and certifying investigators in integrative approaches to physiology, pharmacology, and toxicology to insure the development of its promising new future.[288,289] The paucity of training in integrative biomedical sciences has had detrimental long-lasting effects such as (a) an impact on the development of intact animal models of human function and disease, (b) an impact on skills to conceptualize biomedical hypothesis and experiments at the level of the intact animal, and (c) an impact on the process of non-clinical and clinical drug discovery and development.[289] Over the recent years, several training programs in safety pharmacology have emerged either as stand-alone or as integrated part of broader modules.[290] Education remains a primary emphasis for the Safety Pharmacology Society (SPS; www.safetypharmacology.org/) through content derived from regional and annual meetings, webinars, and publication of its works it seeks to inform the general scientific and regulatory community.[291] The SPS has developed the Diplomate in Safety Pharmacology (DSP) certification process providing many benefits including authentication of the discipline within the overall pharmaceutical community and with regulatory authorities.[292] DSP provides an opportunity for candidates to strengthen their fundamental scientific knowledge and stimulates the sharing of data, methods, and model development in the form of publications and presentations on relevant topics in safety pharmacology.

16.8.5 THE SAFETY PHARMACOLOGY SOCIETIES

Prior to 2001, many of the same scientific experts of the "General Pharmacology/Safety Pharmacology Discussion Group" met once a year in Philadelphia. The historical birth of safety pharmacology is described by Bass, Kinter, and Williams.[24] The SPS was incorporated in 2001, at the time when the ICH was issuing ICH S7A.[31] The SPS is a global scientific society fostering best practice around the discipline of safety pharmacology. The society has a mission statement which declares it is a nonprofit organization that promotes knowledge, development, application, and training in safety pharmacology – a distinct scientific discipline that integrates the best practices of pharmacology, physiology, and toxicology. The objective of safety pharmacology studies is to further the discovery, development, and safe use of biologically active chemical entities and large molecules by the identification, monitoring, and characterization of potentially undesirable PD activities in non-clinical studies. The SPS also supports the human safety of drugs and biologicals by fostering scientific research, education, and dissemination of scientific information through meetings and other scientific interactions. The SPS is composed of members from across the globe and reaches out internationally through meetings, webinars, and publications. The society provides a number of resources which are available to members and nonmembers; members clearly enjoy the widest benefits. The SPS organizes an annual meeting in fall, comprising invited speakers, posters, continuing education courses, and exhibits.[293–296]

More recently, the Japanese and Chinese Safety Pharmacology Societies were incorporated in 2009 and 2015, respectively. From the "General Pharmacology/Safety Pharmacology Discussion Group" and its low 50 contributors, the Safety Pharmacology Societies have grown significantly with over 1,000 members, confirming the sustained strength and need for such impactful discipline.

16.9 CONCLUSION

Safety pharmacology is a discipline that has rapidly grown and evolved during the last 20 years, and is now facing new challenges on scientific, technological, regulatory, and human fronts. Thus, the answer to the future challenges of safety pharmacology will be contained within the vision of its current and future leaders, the issues and concerns that they will face, and the solutions that they will bring to them.

REFERENCES

1. Kola, I. and Landis, J., Can the pharmaceutical industry reduce attrition rates?, *Nat. Rev. Drug Discov.*, 3, 711–715, 2004.
2. Anon, Phase II failures: 2008–2010. *Nat. Rev. Drug Discov.*, 10, 1, 2011.
3. Cook, D., Brown, D., Alexander, R., March, R., Morgan, P., Satterthwaite, G., and Pangalos, M.N., Lessons learned from the fate of AstraZeneca's drug pipeline: A five-dimensional framework?, *Nat. Rev. Drug Discov.*, 13(6), 419–431, 2014.
4. Waring, M.J., Arrowsmith, J., Leach, A.R., Leeson, P.D., Mandrell, S., Owen, R.M., Pairaudeau, G., Pennie, W.D., Pickett, S.D., Wang, J., Wallace, O., and Weir, A., An analysis of the attrition of drug candidates from four major pharmaceutical companies, *Nat. Rev. Drug Discov.*, 14(7), 475–486, 2015.
5. Valentin, J.P., Keisu, M., and Hammond, T.G., Chapter 4: Predicting human adverse drug reactions from non-clinical safety studies, in *Clinical Trials Handbook*, Gad, S.C., Ed., John Wiley & Sons, Inc., Hoboken, NJ, 2009, pp. 87–113.
6. Redfern, W.S., Wakefield, I.D., Prior, H., Hammond, T.G., and Valentin, J.P., Safety pharmacology—A progressive approach, *Fundam. Clin. Pharmacol.*, 16, 161–173, 2002.
7. Sibille, M., Deigat, N., Janin, A., Kirkesseli, S., and Durand, D.V., Adverse events in phase-I studies: A report in 1015 healthy volunteers, *Eur. J. Clin. Pharmacol.*, 54, 13–20, 1998.
8. Olson, H., Betton, G., Robinson, D., Thomas, K., Monro, A., Kolaja, G., Lilly, P., Sanders, J., Sipes, G., Bracken, W., Dorato, M., Van Deun, K., Smith, P., Berger, B., and Heller, A., Concordance of the toxicity of pharmaceuticals in humans and in animals, *Regul. Toxicol. Pharmacol.*, 32, 56–67, 2000.
9. Greaves, P., Williams, A., and Eve, M., First dose of potential new medicines to humans: How do animals help?, *Nat. Rev. Drug Discov.*, 3, 226–236, 2004.
10. Igarashi, T., Nakane, S., and Kitagawa, T., Predictability of clinical adverse reactions of drugs by general pharmacology studies, *J. Toxicol. Sci.*, 20, 77–92, 1995.
11. Stephens, M.D.B., Chapter 1: Introduction, in *Stephens' Detection of New Adverse Drug Reactions*, 5th ed., Talbot, J. and Waller, P., Eds., John Wiley & Sons, Inc., Chichester, 2004, pp. 1–91.
12. Lazarou, J., Pomeranz, B.H., and Corey, P.N., Incidence of adverse drug reactions in hospitalized patients: A meta-analysis of prospective studies, *JAMA*, 279, 1200–1205, 1998.
13. McKenney, J.M., Pharmacologic options for aggressive low-density lipoprotein cholesterol lowering: Benefits versus risks, Am. J. Cardiol., 96, 60E–66E, 2005.
14. Fung, M., Thornton, A., Mybeck, K., Wu, J.H., Hornbuckle, K., and Muniz, E., Evaluation of the characteristics of safety withdrawal of prescription drugs from worldwide pharmaceuticals markets—1960 to 1999, *Drug Inform. J.*, 35, 293–317, 2001.
15. Valentin, J.P. and Redfern, W.S., Prevalence, frequency and impact of safety related issues throughout the pharmaceutical life cycle, *Toxicologist*, 150(1), 170, Abstract 1722, 2017.
16. Gad, S.C., Ed., *Safety Pharmacology in Pharmaceutical Development and Approval*, CRC Press, Boca Raton, FL, 2004, pp. 95–108.
17. Kinter, L.B., Gossett, K.A., and Kerns, W.D., Status of safety pharmacology in the pharmaceutical industry—1993, *Drug Dev. Res.*, 32, 208–216, 1994.
18. Lumley, C.E., General pharmacology, the international regulatory environment, and harmonization guidelines, *Drug Dev. Res.*, 32, 223–232, 1994.
19. Bass, A.S., Hombo, T., Kasai, C., Kinter, L.B., and Valentin, J.P., A historical view and vision into the future of the field of safety pharmacology, *Handb. Exp. Pharmacol.*, 229, 3–45, 2015.
20. Green, M.D., An FDA perspective on general pharmacology studies to determine activity and safety, *Drug Dev. Res.*, 35, 158–160, 1995.
21. Proakis, A.G., Regulatory considerations on the role of general pharmacology studies in the development of therapeutic agents, *Drug Dev. Res.*, 32, 233–236, 1994.
22. Anon, Guidelines for general pharmacology studies–Japanese guidelines for non-clinical studies of drugs manual, Nippo Yakuji, Ltd., Ed., 1995, pp. 71–80.
23. Kinter, L.B. and Valentin, J.P., Safety pharmacology and risk assessment, *Fundam. Clin. Pharmacol.*, 16, 175–182, 2002.
24. Bass, A.S., Kinter, L.B., and Williams, P., Origins, practices and future of safety pharmacology, *J. Pharmacol. Toxicol. Methods*, 49, 145–151, 2004.
25. Bass, A.S. and Williams, P.D., Status of international regulatory guidelines on safety pharmacology, in *Safety Pharmacology*, Williams, P.D. and Bass, A.S., Eds., Springer Publishing, New York, 2003, pp. 9–20.

26. Bowes, J., Rolf, M., and Valentin, J.P., Chapter 6: Pharmacological profiling: The early assessment of pharmacological properties, in *The Process of New Drug Discovery and Development*, 2nd ed., Smith, C.G. and O'Donnell, J., Eds., Informa, New York, 2006, pp. 5-16.

27. Kurata, M., Kanai, K., Mizuguchi, K., Yoshida, M., Nakamura, K., Furuga, Y., Kinoue, A., Totsuka, K., and Igarashi, S., Trends in safety pharmacology in the US and Europe, *J. Toxicol. Sci.*, 22, 237–248, 1997.

28. Anon, ICH guideline S6 (R1)—Preclinical safety evaluation of biotechnology-derived pharmaceuticals, Step 5, Committee for medicinal products for human use (CHMP), EMA/CHMP/ICH/731268/1998, www.ema.europa.eu/docs/en_GB/document_library/Scientific_guideline/2009/09/WC500002828.pdf, 2011.

29. Anon, ICH S6: Preclinical safety evaluation of biotechnology-derived pharmaceuticals, CPMP/ICH/302/95, London, 1997.

30. Anon, ICH M3: Timing of non-clinical safety studies for the conduct of human clinical trials for pharmaceuticals, CPMP/ICH/286/95, 1997, Modified version released November 2000.

31. Anon, ICHS7A: Safety pharmacology studies for human pharmaceuticals, CPMP/ICH/539/00, London, www.emea.eu.int/pdfs/human/ich/053900en.pdf, 2000.

32. Anon, *The GLP Pocket Book: The Good Laboratory Practice Regulations 1999 and Guide to UK GLP Regulations*, MCA Publications, London, 1999, pp. 1–75.

33. Anon, Japanese Ministry of Health and Welfare (MHW), in *Notes on Applications for Approval to Manufacture New Drugs*, 1975.

34. Anon, ICH S7B: The nonclinical evaluation of the potential for delayed ventricular repolarization (QT interval prolongation) by human pharmaceuticals, CPMP/ICH/423/02, London, www.emea.eu.int/pdfs/human/ich/042302en.pdf, 2005.

35. Anon, E14 implementation working group ICH E14 guideline: The clinical evaluation of QT/QTc interval prolongation and proarrhythmic potential for non-antiarrhythmic drugs questions & answers (R3), Current version dated 10 December 2015, www.ich.org/fileadmin/Public_Web_Site/ICH_Products/Guidelines/Efficacy/E14/E14_Q_As_R3__Step4.pdf.

36. Anon, ICH E14: The clinical evaluation of QT/QTc interval prolongation and proarrhythmic potential for non-antiarrhythmic drugs, CPMP/ICH/2/04, London, www.emea.eu.int/pdfs/human/ich/000204en.pdf, 2005.

37. Anon, Guideline on the non-clinical investigation of the dependence potential of medicinal products, www.ema.europa.eu/docs/en_GB/document_library/Scientific_guideline/2009/09/WC500003360.pdf, 2006.

38. Anon, Guidance for industry on the assessment of abuse potential of drugs, www.fda.gov/downloads/drugs/guidancecomplianceregulatoryinformation/guidances/ucm198650.pdf, 2017.

39. Anon, ICH S9-R1 (step 2): Guideline on nonclinical evaluation for anticancer 4 pharmaceuticals - questions and answers, EMA/CHMP/ICH/453684/2016, www.ema.europa.eu/docs/en_GB/document_library/Scientific_guideline/2016/07/WC500211310.pdf, 2016.

40. Anon, Guidance for industry, investigators, and reviewers: Exploratory IND studies, www.fda.gov/downloads/drugs/guidancecomplianceregulatoryinformation/guidances/ucm078933.pdf, 2006.

41. Anon, ICH M3 (R2) non-clinical safety studies for the conduct of human clinical trials for pharmaceuticals, EMA/CPMP/ICH/286/1995, www.ema.europa.eu/docs/en_GB/document_library/Scientific_guideline/2009/09/WC500002720.pdf, 2009.

42. Anon, Guidance for industry: Suicidal ideation and behavior: Prospective assessment of occurrence in clinical trials, www.fda.gov/drugs/guidancecomplianceregulatoryinformation/guidances/ucm315156.htm, 2012.

43. Ackerman, M.J., The long QT syndrome: Ion channel diseases of the heart, *Mayo Clin. Proc.*, 73, 250–269, 1998.

44. Anderson, M.E., Al-Khatib, S.M., Roden, D.M., and Califf, R.M., Cardiac repolarization: Current knowledge, critical gaps and new approaches to drug development and patient management, *Am. Heart J.*, 144, 769–781, 2002.

45. De Ponti, F., Poluzzi, E., and Montanaro, N., Organizing evidence on QT prolongation and occurrence of Torsades de pointes with non-anti-arrhythmic drugs: A call for consensus, *Eur. J. Clin. Pharmacol.*, 57, 185–209, 2001.

46. Haverkamp, E., Breithart, G., Camm, A.J., Janse, M.J., Rosen, M.R., Antzelevitch, C., Escande, D., Franz, M., Malik, M., Moss, A., and Shah, R., The potential for QT prolongation and proarrhythmia by non-antiarrhythmic drugs: Clinical and regulatory implications, *Cardiovasc. Res.*, 47, 219–233, 2000.

47. Fenichel, R.R., Malik, M., Antzelevitch, C., Sanguinetti, M., Roden, D.M., Priori, S.G., Ruskin, J.N., Lipicky, R.J., and Cantilena, L.R., Drug induced torsades de pointes and implication for drug development, *J. Cardiovasc. Electrophysiol.*, 15, 475–495, 2004.

48. Malik, M. and Camm, A.J., Evaluation of drug-induced QT interval prolongation. Implications for drug approval and labeling, *Drug Saf.*, 24, 323–351, 2001.

49. Moss, A.J., The QT interval and torsade de pointes, *Drug Saf.*, 21, 5–10, 1999.

50. Thomas, S.H.L., Drugs, QT interval abnormalities and ventricular arrhythmias, *Adverse Drug React. Toxicol. Rev.*, 13, 77–102, 1994.

51. Viskin, S., Long QT syndromes and torsade de pointes, *Lancet*, 354, 1625–1633, 1999.

52. Webster, R., Leischmann, D., and Walker, D., Towards a drug concentration effect relationship for QT prolongation and torsades de pointes, *Curr. Opin. Drug Discov. Dev.*, 5, 116–126, 2002.

53. Redfern, W.S., Carlsson, L., Davis, A.S., Lynch, W.G., MacKenzie, I., Palethorpe, S., Siegl, P.K.S., Strang, I., Sullivan, A.T., Wallis, R., Camm, A.J., and Hammond, T.G., Relationship between preclinical cardiac electrophysiology, clinical QT interval prolongation and torsade de pointes for a broad range of drugs: Evidence for a provisional safety margin in drug development, *Cardiovasc. Res.*, 58, 32–45, 2003.

54. Valentin, J.P., Hoffmann, P., De Clerck, F., Hammond, T.G., and Hondeghem, L., Review of the predictive value of the Langendorff heart model (Screenit system) in assessing the proarrhythmic potential of drugs, *J. Pharmacol. Toxicol. Methods*, 49, 171–181, 2004.

55. Lawrence, C.L., Pollard, C.E., Hammond, T.G., and Valentin, J.P., Nonclinical proarrhythmia models: Predicting Torsades de Pointes, *J. Pharmacol. Toxicol. Methods*, 52, 46–59, 2005.

56. Valentin, J.P., Bass, A.S., Atrakchi, A., Olejniczak, K., and Kannosuke, F., Challenges and lessons learned since implementation of the safety pharmacology guidance ICH S7A, *J. Pharmacol. Toxicol. Methods*, 52, 22–29, 2005.

57. Lindgren, S., Bass, A.S., Briscoe, R., Bruse, K., Friedrichs, G.S., Kallman, M.J., Markgraf, C., Patmore, L., and Pugsley, M.K., Benchmarking safety pharmacology regulatory packages and best practice, *J. Pharmacol. Toxicol. Methods*, 58(2), 99–109, 2008.

58. Gintant, G., Sager, P.T., and Stockbridge, N., Evolution of strategies to improve preclinical cardiac safety testing, *Nat. Rev. Drug Discov.*, 7, 1–22, 2016.

59. Schock-Kusch, D., Xie, Q., Shulhevich, Y., Hesser, J., Stsepankou, D., Sadick, M., Koenig, S., Hoecklin, F., Pill, J., and Gretz, N., Transcutaneous assessment of renal function in conscious rats with a device for measuring FITC-sinistrin disappearance curves, *Kidney Int.*, 79, 1254–1258, 2011.

60. Schreiber, A., Shulhevich, Y., Geraci, S., Hesser, J., Stsepankou, D., Neudecker, S., Koenig, S., Heinrich, R., Hoecklin, F., Pill, J., Friedemann, J., Schweda, F., Gretz, N., and Schock-Kusch, D., Transcutaneous measurement of renal function in conscious mice, *Am. J. Physiol. Renal Physiol.*, 303, F783–F788, 2012.

61. Zbinden, G., Neglect of function and obsession with structure in toxicity testing, in *Proceedings of the 9th International Congress of Pharmacology*, London, 1984, 1, pp. 43–49.

62. Huo, R., Sheng, Y., Guo, W.-T., and Dong, D.-L., The potential role of $K_V4.3$ K^+ channel in heart hypertrophy, *Channels*, 8(3), 203–209, 2014.

63. Dressman, J.B., Comparison of canine and human gastrointestinal physiology, *Pharm. Res.*, 3, 123–131, 1986.

64. Schein, P.S., Davis, R.D., Carter, S., Newman, J., Schein, D.R., and Rall, D.P., The evaluation of anticancer drugs in dogs and monkeys for the prediction of qualitative toxicities in man, *Clin. Pharmacol. Ther.*, 11, 3–40, 1970.

65. Abi-Gerges, N., Philp, K., Pollard, C., Wakefield, I., Hammond, T., and Valentin, J.P., Sex differences in ventricular repolarization: From cardiac electrophysiology to Torsades de Pointes, *Fund. Clin. Pharmacol.*, 18, 139–151, 2004.

66. Marks, L., Borland, S., Philp, K., Ewart, L., Lainée, P., Skinner, M., Kirk, S., and Valentin, J.-P., The role of the anaesthetised guinea-pig in the preclinical cardiac safety evaluation of drug candidate compounds, *Toxicol. Appl. Pharmacol.*, 263(2), 171–83, 2012.

67. Anon, Guideline on the evaluation of control samples in nonclinical safety studies: Checking for contamination with the test substance, CPMP/SWP/1094/04, London, 17 March 2005.

68. Leishman, D.J., Beck, T.W., Dybdal, N., Gallacher, D.J., Guth, B.D., Holbrook, M., Roche, B., and Wallis, R.M., Best practice in the conduct of key nonclinical cardiovascular assessments in drug development: Current recommendations from the Safety Pharmacology Society, *J. Pharmacol. Toxicol. Methods*, 65(3), 93–101, 2012.

69. Rosseels, M.-L., Delaunois, A.G., Hanon, E., Guillaume, P.J., Martin, F.D., and van den Dobbelsteen, D.J., Hydroxypropyl-β-cyclodextrin impacts renal and systemic hemodynamics in the anesthetized dog, *Regul. Toxicol. Pharmacol.*, 67(3), 351–359, 2013.

70. Festing, M.F.W., Overend, P., Gaines Das, R., Cortina Borja, M., and Berdoy, M., The design of animal experiments, in *Laboratory Animal Handbooks*, Laboratory Animals Ltd. by Royal Society of Medicine Press Ltd., London, 2003, 14, pp. 71–83.

71. Chiang, A.Y., Smith, W.C., Main, B.W., and Sarazan, R.D., Statistical power analysis for hemodynamic cardiovascular safety pharmacology studies in beagle dogs, *J. Pharmacol. Toxicol. Methods*, 50, 121–130, 2004.

72. Guth, B.D., Bass, A.S., Briscoe, R., Chivers, S., Markert, M., Siegl, P.K., and Valentin, J.-P., Comparison of electrocardiographic analysis for risk of QT interval prolongation using safety pharmacology and toxicological studies, *J. Pharmacol. Toxicol. Methods*, 60(2), 107–116, 2009.

73. Tattersall, M.L., Dymond, M., Hammond, T., and Valentin, J.P., Correction of QT values to allow for increases in heart rate in conscious beagle dogs in toxicology assessment, *J. Pharmacol. Toxicol. Methods*, 53, 11–19, 2006.

74. Baillie, T.A., Cayen, M.N., Fouda, H., Gerson, R.J., Green, J.D., Grossman, S.J., Klunk, L.J., LeBlanc, B., Perkins, D.G., and Shipley, L.A., Drug metabolites in safety testing, *Toxicol. Appl. Pharmacol.*, 182, 188–196, 2002.

75. Luffer-Atlas, D. and Atrakchi, A., A decade of drug metabolite safety testing: Industry and regulatory shared learning. *Expert Opin. Drug Metab. Toxicol.*, 13(9), 897–900, 2017.

76. Roden, D.M., Drug induced prolongation of the QT interval, *N. Engl. J. Med.*, 350, 1013–1022, 2004.

77. Shah, R.R. and Hondeghem, L.M., Refining detection of drug-induced proarrhythmia: QT interval and TRIaD, *Heart Rhythm*, 2, 758–772, 2005.

78. Belardinelli, L., Antzelevitch, C., and Vos, M.A., Assessing predictors of drug-induced torsade de pointes, *Trends Pharmacol. Sci.*, 24, 619–625, 2003.

79. Straus, S.M.J.M., Sturkenboom, M.C.J.M., Bleumink, G.S., Dieleman, J.P., van der Lei, J., de Graeff, P.A., Kingma, J.H., and Stricker, B.H.C., Non-cardiac QTc-prolonging drugs and the risk of sudden cardiac death, *Eur. Heart J.*, 26, 2007–2012, 2005.

80. Volmar, K.E. and Hutchins, G.M., Aortic and mitral fenfluramine-phentermine valvulopathy in 64 patients treated with anorectic agents, *Arch. Pathol. Lab. Med.*, 125, 1555–1561, 2001.

81. Edwards, I.R., What are the real lessons from Vioxx?, *Drug Saf.*, 28, 651–658, 2005.

82. Jost, N., Virag, L., Bitay, M., Takacs, J., Lengyel, C., Biliczki, P., Nagy, Z., Bogats, G., Lathrop, D.A., Papp, J.G., and Varro, A., Restricting excessive cardiac action potential and QT prolongation: A vital role for IKs in human ventricular muscle, *Circulation*, 112, 1392–1399, 2005.

83. Liu, D.W. and Antzelevitch, C., Characteristics of the delayed rectifier current (IKr and IKs) in canine ventricular epicardial, midmyocardial, and endocardial myocytes. A weaker IKs contributes to the longer action potential of the M cell, *Circ. Res.*, 76, 351–365, 1995.

84. Jurkiewicz, N.K. and Sanguinetti, M.C., Rate-dependent prolongation of cardiac action potentials by a methanesulfonanilide class III antiarrhythmic agent. Specific block of rapidly activating delayed rectifier K$^+$ current by dofetilide, *Circ. Res.*, 72, 75–83, 1993.

85. Li, G.R., Feng, J., Yue, L., and Carrier, M., Transmural heterogeneity of action potentials and Ito1 in myocytes isolated from the human right ventricle, *Am. J. Physiol.*, 275, H369–H377, 1998.

86. Yang, T. and Roden, D.M., Regulation of sodium current development in cultured atrial tumor myocytes (AT-1 cells), *Am. J. Physiol.*, 271, H541–H547, 1996.

87. Banyasz, T., Fulop, L., Magyar, J., Szentandrassy, N., Varro, A., and Nanasi, P.P., Endocardial versus epicardial differences in L-type calcium current in canine ventricular myocytes studied by action potential voltage clamp, *Cardiovasc. Res.*, 58, 66–75, 2003.

88. Xia, M., Salata, J.J., Figueroa, D.J., Lawlor, A.M., Liang, H.A., Liu, Y., and Connolly, T.M., Functional expression of L- and T-type Ca^{2+} channels in murine HL-1 cells, *J. Mol. Cell. Cardiol.*, 36, 111–119, 2004.

89. Davie, C., Valentin, J.P., Pollard, C., Standen, N., Mitcheson, J., Alexander, P., and Thong, B., Comparative pharmacology of guinea pig cardiac myocyte and cloned hERG (I(Kr)) channel, *J. Cardiovasc. Electrophysiol.*, 15, 1302–1309, 2004.

90. Suetake, I., Takisawa, H., and Nakamura, T., Related contractile activity and fluorescence changes in fluo-3-loaded isolated ventricular myocytes, *Jpn. J. Physiol.*, 42, 815–821, 1992.

91. Cordeiro, J.M., Greene, L., Heilmann, C., Antzelevitch, D., and Antzelevitch, C., Transmural heterogeneity of calcium activity and mechanical function in the canine left ventricle, *Am. J. Physiol. Heart Circ. Physiol.*, 286, H1471–H1479, 2004.

92. Graham, M.D., Bru-Mercier, G., Hopkins, P.M., and Harrison, S.M., Transient and sustained changes in myofilament sensitivity to Ca^{2+} contribute to the inotropic effects of sevoflurane in rat ventricle, *Br. J. Anaesth.*, 94, 279–286, 2005.

93. Hamilton, D.L., Boyett, M.R., Harrison, S.M., Davies, L.A., and Hopkins, P.M., The concentration-dependent effects of propofol on rat ventricular myocytes, *Anesth. Analg.*, 91, 276–282, 2000.

94. Witchel, H.J., Milnes, J.T., Mitcheson, J.S., and Hancox, J.C., Troubleshooting problems with in vitro screening of drugs for QT interval prolongation using HERG K^+ channels expressed in mammalian cell lines and *Xenopus* oocytes, *J. Pharmacol. Toxicol. Methods*, 48, 65–80, 2002.

95. Zou, A., Curran, M.E., Keating, M.T., and Sanguinetti, M.C., Single HERG delayed rectifier K^+ channels expressed in *Xenopus* oocytes, *Am. J. Physiol.*, 272, H1309–H1314, 1997.

96. Cavero, I., Mestre, M., Guillon, J.M., and Crumb, W., Drugs that prolong QT interval as an unwanted effect: Assessing their likelihood of inducing hazardous cardiac dysrhythmias, *Expert Opin. Pharmacother.*, 1, 947–973, 2000.

97. Martin, R.L., McDermott, J.S., Salmen, H.J., Palmatier, J., Cox, B.F., and Gintant, G.A., The utility of hERG and repolarization assays in evaluating delayed cardiac repolarization: Influence of multi-channel block, *J. Cardiovasc. Pharmacol.*, 43, 369–379, 2004.

98. McDonald, T.V., Yu, Z., Ming, Z., Palma, E., Meyers, M.B., Wang, K.W., Goldstein, S.A., and Fishman, G.I., A minK-HERG complex regulates the cardiac potassium current I(Kr), *Nature*, 388, 289–292, 1997.

99. Sanguinetti, M.C., Curran, M.E., Zou, A., Shen, J., Spector, P.S., Atkinson, D.L., and Keating, M.T., Coassembly of K(V)LQT1 and minK (IsK) proteins to form cardiac I(Ks) potassium channel, *Nature*, 384, 80–83, 1996.

100. Wakefield, I.D., Pollard, C., Redfern, W.S., Hammond, T.G., and Valentin, J.P., The application of in vitro methods to safety pharmacology, *Fundam. Clin. Pharmacol.*, 16, 209–218, 2002.

101. Krasner, L.J., Wendling, W.W., Cooper, S.C., Chen, D., Hellmann, S.K., Eldridge, C.J., McClurken, J.B., Jeevanandam, V., and Carlsson, C., Direct effects of triiodothyronine on human internal mammary artery and saphenous veins, *J. Cardiothorac. Vasc. Anesth.*, 11, 463–466, 1997.

102. Lefer, D.J., Nakanishi, K., Vinten-Johansen, J., Ma, X.L., and Lefer, A.M., Cardiac venous endothelial dysfunction after myocardial ischemia and reperfusion in dogs, *Am. J. Physiol.*, 263, H850–H856, 1992.

103. Sugiyama, A., Kobayashi, M., Tsujimoto, G., Motomura, S., and Hashimoto, K., The first demonstration of CGRP-immunoreactive fibers in canine hearts: Coronary vasodilator, inotropic and chronotropic effects of CGRP in canine isolated, blood-perfused heart preparations, *Jpn. J. Pharmacol.*, 50, 421–427, 1989.

104. Voss, H.P., Shukrula, S., Wu, T.S., Donnell, D., and Bast, A., A functional beta-2 adrenoceptor-mediated chronotropic response in isolated guinea pig heart tissue: Selectivity of the potent beta-2 adrenoceptor agonist TA 2005, *J. Pharmacol. Exp. Ther.*, 271, 386–389, 1994.

105. Abi-Gerges, N., Small, B.G., Lawrence, C.L., Hammond, T.G., Valentin, J.P., and Pollard, C.E., Evidence for gender differences in electrophysiological properties of canine Purkinje fibres, *Br. J. Pharmacol.*, 142, 1255–1264, 2004.

106. Gintant, G.A., Limberis, J.T., McDermott, J.S., Wegner, C.D., and Cox, B.F., The canine Purkinje fiber: An in vitro model system for acquired long QT syndrome and drug-induced arrhythmogenesis, *J. Cardiovasc. Pharmacol.*, 37, 607–618, 2001.

107. Lu, H.R., Vlaminckx, E., Teisman, A., and Gallacher, D.J., Choice of cardiac tissue plays an important role in the evaluation of drug-induced prolongation of the QT interval in vitro in rabbit, *J. Pharmacol. Toxicol. Methods*, 52, 90–105, 2005.

108. Antzelevitch, C., Shimizu, W., Yan, G.X., and Sicouri, S., Cellular basis for QT dispersion, *J. Electrocardiol.*, 30, 168–175, 1998.

109. Hondeghem, L.M., Carlsson, L., and Duker, G., Instability and triangulation of the action potential predict serious proarrhythmia, but action potential duration prolongation is antiarrhythmic, *Circulation*, 103, 2004–2013, 2001.

110. Eckardt, L., Haverkamp, W., Mertens, H., Johna, R., Clague, J.R., Borggrefe, M., and Breithardt, G., Drug-related torsades de pointes in the isolated rabbit heart: Comparison of clofilium, d,l-sotalol, and erythromycin, *J. Cardiovasc. Pharmacol.*, 32, 425–434, 1998.

111. Eckardt, L., Haverkamp, W., Borggrefe, M., and Breithardt, G., Experimental models of torsade de pointes, *Cardiovasc. Res.*, 39, 178–193, 1998.

112. Miyoshi, K., Taniguchi, M., Seki, S., and Mochizuki, S., Effects of magnesium and its mechanism on the incidence of reperfusion arrhythmias following severe ischemia in isolated rat hearts, *Cardiovasc. Drugs Ther.*, 14, 625–633, 2000.

113. Hill, A.J., Laske, T.G., Coles, J.A. Jr., Sigg, D.C., Skadsberg, N.D., Vincent, S.A., Soule, C.L., Gallagher, W.J., and Iaizzo, P.A., In vitro studies of human hearts, *Ann. Thorac. Surg.*, 79, 168–177, 2005.

114. Takahara, A., Dohmoto, H., Yoshimoto, R., Sugiyama, A., and Hashimoto, K., Utilization of telemetry system to assess the cardiovascular profile of AH-1058, a new cardioselective Ca^{2+} channel blocker, in conscious dogs, *Jpn. J. Pharmacol.*, 85, 331–334, 2001.

115. Sato, K., Kandori, H., and Sato, S., Evaluation of a new method using telemetry for monitoring the left ventricular pressure in free-moving rats, *J. Pharmacol. Toxicol. Methods*, 31, 191–198, 1994.

116. Nekooeian, A.A. and Tabrizchi, R., Effects of adenosine A2A receptor agonist, CGS 21680, on blood pressure, cardiac index and arterial conductance in anaesthetized rats, *Eur. J. Pharmacol.*, 307, 163–169, 1996.

117. Ozkanlar, Y., Nishijima, Y., da Cunha, D., and Hamlin, R.L., Acute effects of tacrolimus (FK506) on left ventricular mechanics, *Pharmacol. Res.*, 52, 307–312, 2005.

118. Tsusaki, H., Yonamine, H., Tamai, A., Shimomoto, M., Iwao, H., Nagata, R., and Kito, G., Evaluation of cardiac function in primates using real-time three-dimensional echocardiography as applications to safety assessment, *J. Pharmacol. Toxicol. Methods*, 52, 182–187, 2005.

119. Hanton, G., Gautier, M., Bonnet, P., and Herbet, A., Effect of milrinone on echocardiographic parameters after single dose in Beagle dogs and relationship with drug-induced cardiotoxicity, *Toxicol. Lett.*, 155, 307–317, 2005.

120. Wakefield, I.D., March, J.E., Kemp, P.A., Valentin, J.P., Bennett, T., and Gardiner, S.M., Comparative regional haemodynamic effects of the nitric oxide synthase inhibitors, S-methyl-L-thiocitrulline and L-NAME, in conscious rats, *Br. J. Pharmacol.*, 39, 1235–1243, 2003.

121. van der Linde, H., Van de Water, A., Loots, W., Van Deuren, B., Lu, H.R., Van Ammel, K., Peeters, M., and Gallacher, D.J., A new method to calculate the beat-to-beat instability of QT duration in drug-induced long QT in anesthetized dogs, *J. Pharmacol. Toxicol. Methods*, 52, 168–177, 2005.

122. Sugiyam, A., Ishida, Y., Satoh, Y., Aoki, S., Hori, M., Akie, Y., Kobayashi, Y., and Hashimoto, K., Electrophysiological, anatomical and histological remodeling of the heart to AV block enhances susceptibility to arrhythmogenic effects of QT-prolonging drugs, *Jpn. J. Pharmacol.*, 88, 341–350, 2002.

123. Chiba, K., Sugiyama, A., Hagiwara, T., Takahashi, S.I., Takasuna, K., and Hashimoto, K., In vivo experimental approach for the risk assessment of fluoroquinolone antibacterial agents-induced long QT syndrome, *Eur. J. Pharmacol.*, 486, 189–200, 2004.

124. Collister, J.P. and Osborn, J.W., The chronic infusion of hexamethonium and phenylephrine to effectively clamp sympathetic vasomotor tone: A novel approach, *J. Pharmacol. Toxicol. Methods*, 42, 135–147, 1999.

125. AbdelRahman, A.R., Inadequate blockade by hexamethonium of the baroreceptor heart rate response in anesthetized and conscious rats, *Arch. Int. Pharmacodyn. Ther.*, 297, 68–85, 1989.

126. Verwaerde, P., Senard, J.M., Mazerolles, M., Tran, M.A., DamaseMichel, C., Montastruc, J.L., and Montastruc, P., Spectral analysis of blood pressure and heart rate, catecholamine and neuropeptide Y plasma levels in a new model of neurogenic orthostatic hypotension in dog, *Clin. Auton. Res.*, 6, 75–82, 1996.

127. Harada, T., Abe, J., Shiotani, M., Hamada, Y., and Horri, I., Effect of autonomic nervous function on QT interval in dogs, *J. Toxicol. Sci.*, 30, 229–237, 2005.

128. Mangin, L., Swynghedauw, B., Benis, A., Thibault, N., Lerebours, G., and Carre, F., Relationships between heart rate and heart rate variability: Study in conscious rats, *J. Cardiovasc. Pharmacol.*, 32, 601–607, 1998.

129. Bedford, T.G. and Dormer, K.J., Arterial hemodynamics during head-up tilt in conscious dogs, *J. Appl. Physiol.*, 65, 1556–1562, 1988.

130. Humphrey, S.J. and McCall, R.B., A rat model for predicting orthostatic hypotension during acute and chronic antihypertensive drug therapy, *J. Pharmacol. Methods*, 7, 25–34, 1982.

131. Ytrehus, K., The ischemic heart—Experimental models, *Pharmacol. Res.*, 42, 193–203, 2000.

132. Wu, K.C. and Lima, J.A.C., Non-invasive imaging of myocardial viability: Current techniques and future developments, *Circ. Res.*, 93, 1146–1158, 2003.

133. Shebuski, R.J., Bush, L.R., Gagnon, A., Chi, L., and Leadley, R.J. Jr., Development and applications of animal models of thrombosis, *Methods Mol. Med.*, 93, 175–219, 2004.

134. Tang, W., Kang, J., Wu, X., Rampe, D., Wang, L., Shen, H., Li, Z., Dunnington, D., and Garyantes, T., Development and evaluation of high throughput functional assay methods for hERG potassium channel, *J. Biomol. Screen.*, 6, 325–331, 2001.

135. Rezazadeh, S., Hesketh, J.C., and Fedida, D., Rb^+ flux through hERG channels affects the potency of channel blocking drugs: Correlation with data obtained using a high-throughput Rb^+ efflux assay, *J. Biomol. Screen.*, 9, 588–597, 2004.

136. Cheng, C.S., Alderman, D., Kwash, J., Dessaint, J., Patel, R., Lescoe, M.K., Kimrade, M.B., and Yu, W., A high throughput hERG potassium channel function assay: An old assay with a new look, *Drug Dev. Ind. Pharm.*, 28, 177–191, 2002.

137. Sorota, S., Zhang, X.S., Margulis, M., Tucker, K., and Priestley, T., Characterization of a hERG screen using the IonWorks HT: Comparison to a hERG rubidium efflux screen, *Assay Drug Dev. Technol.*, 3, 45–57, 2005.

138. Dutta, S., Chang, K.C., Beattie, K.A., Sheng, J., Tran, P.N., Wu, W.W., Wu, M., Strauss, D.G., Colatsky, T., and Li, Z., Optimization of an in silico cardiac cell model for proarrhythmia risk assessment, *Front. Physiol.*, 8, 616, doi:10.3389/fphys.2017.00616, 2017.

139. Kramer, J., Obejero-Paz, C.A., Myatt, G., Kuryshev, Y.A., Bruening-Wright, A., Verducci, J.S., and Brown, A.M., MICE models: Superior to the HERG model in predicting Torsade de Pointes, *Sci. Rep.*, 3, 2100, doi:10.1038/srep02100, 2013.

140. Murphy, D.J., Safety pharmacology of the respiratory system: Techniques and study design, *Drug Dev. Res.*, 32, 237–246, 1994.

141. Murphy, D.J., Assessment of respiratory function in safety pharmacology, *Fund. Clin. Pharmacol.*, 16, 183–196, 2002.

142. Murphy, D.J., Respiratory safety pharmacology—Current practice and future directions, *Regul. Toxicol. Pharmacol.*, 69(1), 135–140, 2014.

143. Murphy, D.J., Optimizing the use of methods and measurement endpoints in respiratory safety pharmacology, *J. Pharmacol. Toxicol. Methods*, 70(3), 204–209, 2014.

144. Hoymann, H.G., Lung function measurements in rodents in safety pharmacology studies, *Front. Pharmacol.*, 3, 156, doi:10.3389/fphar.2012.00156, 2012.

145. Ewart, L.C., Haley, M., Bickerton, S., Bright, J., Elliott, K., McCarthy, A., Williams, L., Ricketts, S.A., Holland, T., and Valentin, J.-P., Pharmacological validation of a telemetric model for the measurement of bronchoconstriction in conscious rats, *J. Pharmacol. Toxicol. Methods*, 61(2), 219–229, 2010.

146. Jackson, D.M., Pollard, C.E., and Roberts, S.M., The effect of nedocromil sodium on the isolated rabbit vagus nerve, *Eur. J. Pharmacol.*, 221, 175–177, 1992.

147. Widdicombe, J.G., Clinical significance of reflexes from the respiratory system, *Anesthesiology*, 23, 434–444, 1962.

148. Diamond, L. and O'Donnell, M., Pulmonary mechanics in normal rats, *J. Appl. Physiol. Respir. Environ. Exerc. Physiol.*, 43, 942–948, 1977.

149. Mauderly, J.L., Effect of inhaled toxicants on pulmonary function, in *Concepts in Inhalation Toxicology*, McClellan, R.O. and Henderson, R.F., Eds., Hemisphere Publishing Corp., New York, 1989, pp. 347–401.

150. Mauderly, J.L., The influence of sex and age on the pulmonary function of the beagle dog, *J. Gerontol.*, 29, 282–289, 1974.

151. O'Neil, J.J. and Raub, J.A., Pulmonary function testing in small laboratory animals, *Environ. Health Perspect.*, 56, 11–22, 1984.

152. Douglas, J.S., Dennis, M.W., Ridgeway, P., and Bouhuys, A., Airway dilation and constriction in spontaneously breathing guinea-pigs, *J. Pharmacol. Exp. Ther.*, 180, 98–109, 1971.

153. Kenakin, T.P., The classification of drugs and drug receptors in isolated tissues, *Pharmacol. Rev.*, 36, 165–222, 1984.

154. Murphy, D.J., Renninger, J.P., and Gossett, K.A., A novel method for chronic measurement of pleural pressure in conscious rats, *J. Pharmacol. Toxicol. Methods*, 39, 137–141, 1998.

155. Chilvers, M.A. and O'Callaghan, C., Analysis of ciliary beat pattern and beat frequency using digital high speed imaging: Comparison with the photomultiplier and photodiode methods, *Thorax*, 55, 314–317, 2000.

156. Khawaja, A.M., Liu, Y.C., and Rogers, D.F., Effect of fenspiride, a non-steroidal anti-inflammatory agent, on neurogenic mucus secretion in ferret trachea in vitro, *Pulm. Pharmacol. Ther.*, 12, 363–368, 1999.

157. Dietl, P., Haller, T., Mair, N., and Frick, M., Mechanisms of surfactantexocytosis in alveolar type II cells in vitro and in vivo, *News Physiol. Sci.*, 16, 239–243, 2001.

158. Bowes, J., Brown, A.J., Hamon, J., Jarolimek, W., Sridhar, A., Waldron, G., and Whitebread, S., Reducing safety-related drug attrition: The use of in vitro pharmacological profiling, *Nat. Rev. Drug Discov.*, 11(12), 909–922, 2012.

159. Lynch, J.J., Vleet, T.R.V., Mittelstadt, S.W., and Blomme, A.G., Potential functional and pathological side effects related to off-target pharmacological activity, *J. Pharmacol. Toxicol. Methods*, 87, 108–126, 2017.

160. Cimbura, G., Lucas, D.M., Bennett, R.C., Warren, R.A., and Simpson, H.M., Incidence and toxicological aspects of drugs detected in 484 fatally injured drivers and pedestrians in Ontario, *J. Forensenic Sci.*, 27, 855–867, 1982.

161. Weiler, J.M., Bloomfield, J.R., Woodworth, G.G., Grant, A.R., Layton, T.A., Brown, T.L., McKenzie, D.R., Baker, T.W., and Watson, G.S., Effects of fexofenadine, diphenhydramine, and alcohol on driving performance. A randomised, placebo-controlled trial in the Iowa driving simulator, *Ann. Intern. Med.*, 132, 354–363, 2000.

162. Mattsson, J.L., Spencer, P.J., and Albee, R.R., A performance standard for clinical and functional observation battery examination of rats, *J. Am. Coll. Toxicol.*, 15, 239–254, 1996.

163. Irwin, S., Comprehensive observational assessment: 1a. A systematic, quantitative procedure for assessing the behavioural and physiologic state of the mouse, *Psychopharmacologia*, 13, 222–257, 1968.

164. Haggerty, G.C., Strategies for and experience with neurotoxicity testing of new pharmaceuticals, *J. Am. Coll. Toxicol.*, 10, 677–687, 1991.

165. Trabace, L., Cassano, T., Steardo, L., Pietra, C., Villetti, G., and Kendrick, K.M., Biochemical and neurobehavioural profile of CHF2819, a novel, orally active acetylcholinesterase inhibitor for Alzheimer's disease, *J. Pharmacol. Exp. Ther.*, 294, 187–194, 2000.

166. Redfern, W.S., Strang, I., Storey, S., Heys, C., Barnard, C., Lawton, K., Hammond, T.G., and Valentin, J.P., Spectrum of effects detected in the rate functional observational battery following oral administration of non-CNS targeted compounds, *J. Pharmacol. Toxicol. Methods*, 52, 77–82, 2005.

167. Porsolt, R.D., Lemaire, M., Durmuller, N., and Roux, S., New perspectives in CNS safety pharmacology, *Fundam. Clin. Pharmacol.*, 16, 197–207, 2002.

168. Eddy, N.B. and Leimbach, D., Synthetic analgesics: II-Dithienylbutenyl and dithienylbutylamines, *J. Pharmacol. Exp. Ther.*, 107, 385–393, 1953.

169. Reiter, L.R. and McPhail, R.C., Motor activity: A survey of methods with potential use in toxicity testing, *Neurobehav. Toxicol.*, 1, 53–66, 1979.

170. Dunham, N.W. and Miya, T.S., A note on a simple apparatus for detecting neurological deficit in mice and rats, *J. Am. Pharm. Assoc.*, 46, 208–209, 1957.

171. Abou-Donia, M.B., Dechkovskaia, A.M., Goldstein, L.B., Shah, D.U., Bullman, S.L., and Khan, W.A., Uranyl acetate-induced sensorimotor deficit and increased nitric oxide generation in the central nervous system in rats, *Pharmacol. Biochem. Behav.*, 72, 881–890, 2002.

172. Jolkkonen, J., Puurunen, K., Rantakomi, S., Harkonen, A., Haapalinna, A., and Sivenius, J., Behavioural effects of the alpha(2)-adrenoceptor antagonist, atipamezole, after focal cerebral ischemia in rats, *Eur. J. Pharmacol.*, 400, 211–219, 2000.

173. Herr, D.W., Graff, J.E., Derr-Yellin, E.C., Crofton, K.M., and Kodavanti, P.R., Flash-, somatosensory-, and peripheral nerve-evoked potentials in rats perinatally exposed to Aroclor 1254, *Neurotoxicol. Teratol.*, 23, 591–601, 2001.

174. Prusky, G.T., Alam, N.M., Beekman, S., and Douglas, R.M., Rapid quantification of adult and developing mouse spatial vision using a virtual optomotor system, *Investig. Ophthalmol. Visual Sci.*, 45, 4611–4616, 2004.

175. Redfern, W.S. and Williams, A., A re-evaluation of the role of α_2-adrenoceptors in the anxiogenic effects of yohimbine, using the selective antagonist delequamine in the rat, *Br. J. Pharmacol.*, 116, 2081–2089, 1995.

176. Jouvet, M., Biogenic amines and the states of sleep, *Science*, 163, 32–41, 1969.

177. Bammer, C., Pharmacological investigations of neurotransmitter involvement in passive avoidance responding: A review and some new results, *Neurosci. Biobehav. Rev.*, 6, 247–296, 1982.

178. Glick, S.D. and Zimmerberg, D., Amnesic effects of scopolamine, *Behav. Biol.*, 7, 2445–2454, 1972.

179. Morris, R.G.M., Spatial localization does not require the presence of local cues, *Learn. Motiv.*, 12, 239–260, 1981.

180. Olton, D.S., The radial maze as a tool in behavioral pharmacology, *Physiol. Behav.*, 40, 793–797, 1986.

181. Dunnett, S.B., Evenden, J.L., and Iversen, S.D., Delay-dependent short term memory deficits in aged rats, *Psychopharmacology*, 96, 174–180, 1982.

182. Itil, T.M., The discovery of psychotropic drugs by computer-analyzed cerebral bioelectrical potentials (CEEG), *Drug Dev. Res.*, 1, 373–407, 1981.

183. Meert, T.F., Effects of various serotonergic agents on alcohol intake and alcohol preference in Wistar rats selected at two different levels of alcohol preference, *Alcohol Alcohol.*, 28, 157–170, 1993.

184. Goudie, A.J., Harrison, A.A., and Leathley, M.J., Evidence for a dissociation between benzodiazepine withdrawal signs, *Neuroreport*, 4, 295–299, 1993.
185. Brady, J.V. and Fischman, M.W., Assessment of drugs for dependence potential and abuse liability: An overview, in *Behavioral Pharmacology: The Current Status*, Seiden, L.S. and Balster, R.L., Eds., Alan R. Liss, New York, 1985, pp. 361–382.
186. Lal, H., Ed., *Discriminative Stimulus Properties of Drugs: Advances in Behavioral Biology*, Plenum Press, New York, 1977, 22.
187. Schechter, M.D. and Calcagnetti, D.J., Trends in place preference conditioning with cross-indexed bibliography 1957–91, *Neurosci. Biobehav. Rev.*, 17, 21–41, 1993.
188. Fountain, S.B. and Teyler, T.J., Suppression of hippocampal slice excitability by 2-, 3- and 4-methylpyridine, *Ecotoxicol. Environ. Safety*, 48, 301–305, 1995.
189. Fliri, A.F., Loging, W.T., Thadeio, P.F., and Volkmann, R.A., Biological spectra analysis: Linking biological activity profiles to molecular structure, *Proc. Natl. Acad. Sci. USA*, 102, 261–266, 2005.
190. Henry, D.A., Ostapowicz, G., and Robertson, J., Drugs as a cause of gastrointestinal disease, *Baillieres Clin. Gastroenterol.*, 8, 271–300, 1994.
191. Makins, R. and Ballinger, A., Gastrointestinal side effects of drugs, *Expert Opin. Drug Saf.*, 2, 421–429, 2003.
192. Mayer, E.A. and Bradesi, S., Alosetron and irritable bowel syndrome, *Expert Opin. Pharmacother.*, 4, 2089–2098, 2003.
193. Tamaki, C., Nagayama, T., Hashiba, M., Fujiyoshi, M., Hizue, M., Kodaira, H., Nishida, M., Suzuki, K., Takashima, Y., Ogino, Y., Yasugi, D., Yoneta, Y., Hisada, S., Ohkura, T., and Nakamura, K., Potentials and limitations of nonclinical safety assessment for predicting clinical adverse drug reactions: Correlation analysis of 142 approved drugs in Japan, *J. Toxicol. Sci.*, 38(4), 581–598, 2013.
194. Mead, A.N., Amouzadeh, H.R., Chapman, K., Ewart, L., Giarola, A., Jackson, S.J., Jarvis, P., Jordaan, P., Redfern, W., Traebert, M., Valentin, J.-P., and Vargas, H.M., Assessing the predictive value of the rodent neurofunctional assessment for commonly reported adverse events in phase I clinical trials, *Regul. Toxicol. Pharmacol.*, 80, 348–357, 2016.
195. Ghahremani, G.G., Gastrointestinal complications of drug therapy, *Abdom. Imaging*, 24, 1–2, 1999.
196. Valentin, J.-P., Al-Saffar, A., Ewart, L., Glab, J., Mark, S.L., Redfern, W.S., Roberts, S., and Hammond, T., A retrospective analysis of gastrointestinal adverse events impacting on drug development, *J. Pharmacol. Toxicol. Methods*, 66, 179, 2012.
197. De Jonghe, B.C. and Horn, C.C., The importance of systematic approaches in the study of emesis, *Temperature*, 2(3), 322–323, 2015.
198. Holmes, A.M., Rudd, J.A., Tattersall, F.D., Aziz, Q., and Andrews, P.L., Opportunities for the replacement of animals in the study of nausea and vomiting, *Br. J. Pharmacol.*, 157, 865–880, 2009.
199. Kimura, T. and Higaki, K., Gastrointestinal transit and drug absorption, *Biol. Pharm. Bull.*, 25, 149–164, 2002.
200. Gershon, M.D., The enteric nervous system: A second brain, *Hosp. Pract.*, 34, 31–32, 1999.
201. Harrison, A.P., Erlwanger, K.H., Elbrond, V.S., Andersen, N.K., and Unmack, M.A., Gastrointestinal-tract models and techniques for use in safety pharmacology, *J. Pharmacol. Toxicol. Methods*, 49, 187–199, 2004.
202. Andrews, P.L. and Hawthorn, J., The neurophysiology of vomiting, *Baillieres Clin. Gastroenterol.*, 2, 141–168, 1998.
203. Hornby, P.J., Central neurocircuitry associated with emesis, *Am. J. Med.*, 111, 106–112, 2001.
204. Rudd, J.A. and Naylor, R.J., The actions of ondansetron and dexamethasone to antagonise cisplatin-induced emesis in the ferret, *Eur. J. Pharmacol.*, 322, 79–82, 1997.
205. Yamashita, M., Yamashita, M., Tanaka, J., Chagi, K., Takeda, S., Kurihara, T., Takeda, Y., and Fujii, Y., Vomiting induction by ipecac syrup in dogs and ferrets, *J. Toxicol. Sci.*, 22, 409–412, 1997.
206. Fillastre, J.P., Détection de la néphrotoxicité médicamenteuse, *J. Pharmacol. (Paris)*, 17(Suppl. I), 41–50, 1986.
207. Thatte, L. and Vaamonde, C.A., Drug-induced nephrotoxicity, *Postgrad. Med.*, 100, 83–100, 1996.
208. Thadani, R., Pascual, M., and Boventre, J.V., Acute renal failure, *N. Engl. J. Med.*, 334, 1448–1460, 1996.
209. Lieberthal, W. and Nigam, S.K., Acute renal failure II. Experimental models of acute renal failure: Imperfect but indispensable, *Am. J. Physiol.*, 278, F1–F12, 2000.
210. Dieterle, F., Sistare, F., Goodsaid, F., Papaluca, M., Ozer, J.S., Webb, C.P., Baer, W., Senagore, A., Schipper, M.J., Vonderscher, J., Sultana, S., Gerhold, D.L., Phillip, J.A., Maurer, G., Carl, K., Laurie, D., Harpur, E., Sonee, M., Ennulat, D., Holder, D., Andrews-Cleavenger, D., Gu, Y.Z., Thompson, K.L.,

Goering, P.L., Vidal, J.M., Abadie, E., Maciulaitis, R., Jacobson-Kram, D., Defelice, A.F., Hausner, E.A., Blank, M., Thompson, A., Harlow, P., Throckmorton, D., Xiao, S., Xu, N., Taylor, W., Vamvakas, S., Flamion, B., Lima, B.S., Kasper, P., Pasanen, M., Prasad, K., Troth, S., Bounous, D., Robinson-Gravatt, D., Betton, G., Davis, M.A., Akunda, J., McDuffie, J.E., Suter, L., Obert, L., Guffroy, M., Pinches, M., Jayadev, S., Blomme, E.A., Beushausen, S.A., Barlow, V.G., Collins, N., Waring, J., Honor, D., Snook, S., Lee, J., Rossi, P., Walker, E., and Mattes, W., Renal biomarker qualification submission: A dialog between the FDA-EMEA and Predictive Safety Testing Consortium, *Nat. Biotechnol.*, 28, 455–462, 2010.

211. Redfern, W.S., Ewart, L., Hammond, T.G., Bialecki, R., Kinter, L., Lindgren, S., Pollard, C.E., Roberts, R., Rolf, M., and Valentin, J.-P., Impact and frequency of different toxicities throughout the pharmaceutical life cycle, *Toxicologist*, 114(S1), 1081, 2010.

212. Benjamin, A., Gallacher, D.J., Greiter-Wilke, A., Guillon, J.M., Kasai, C., Ledieu, D., Levesque, P., Prelle, K., Ratcliffe, S., Sannajust, F., and Valentin, J.-P., Renal studies in safety pharmacology and toxicology: A survey conducted in the top 15 pharmaceutical companies, *J. Pharmacol. Toxicol. Methods*, 75, 101–110, 2015.

213. Zbinden, G., Ed., *Pharmacological Methods in Toxicology*, Pergamon, Elmsford, New York, 1997, pp. 3–6.

214. Luft, J. and Bode, G., Integration of safety pharmacology endpoints into toxicology studies, *Fundam. Clin. Pharmacol.*, 16, 91–103, 2002.

215. McMahon, N., Robinson, N.A., Martel, E., and Valentin, J.P., A method for the long term monitoring of cardiovascular and respiratory function in the Wistar rat, *Toxicology*, 202, 33–127, 2004.

216. Hunter, D., Schofield, J., Gracie, K., Moors, J., Philp, K., Carter, P., Prior, H., Valentin, J.P., and Hammond, T.G., Use of a non-invasive telemetry system (EMKA) for functional cardiovascular endpoints in toxicology studies, *J. Pharmacol. Toxicol. Methods*, 49, 234, 2004.

217. Detweiller, D.K., Electrocardiography in toxicological studies, in *Comprehensive Toxicology*, Snipes, I.G., McQueen, C.A., and Gandolfi, A.J., Eds., Pergamon Press, New York, 1997, pp. 209–234.

218. Derakhchan, K., Chui, R.W., Stevens, D., Gu, W., and Vargas, H.M., Detection of QTc interval prolongation using jacket telemetry in conscious non-human primates: Comparison with implanted telemetry, *Br. J. Pharmacol.*, 171(2), 509–522, 2014.

219. Ward, G., Miliken, P., Patel, B., and McMahon, N., Comparison of non-invasive and implanted telemetric measurement of blood pressure and electrocardiogram in conscious Beagle dogs, *J. Pharmacol. Toxicol. Methods*, 66(2), 106–113, 2012.

220. MacMahon, C., Mitchell, A.Z., Klein, J.L., Jenkins, A.C., and Sarazan, R.D., Evaluation of blood pressure measurement using a miniature blood pressure transmitter with jacketed external telemetry in cynomolgus monkeys, *J. Pharmacol. Toxicol. Methods*, 62(2), 127–135, 2010.

221. Mitchell, A.Z., McMahon, C., Beck, T.W., and Sarazan, R.D., Sensitivity of two noninvasive blood pressure measurement techniques compared to telemetry in cynomolgus monkeys and beagle dogs, *J. Pharmacol. Toxicol. Methods*, 62(1), 54–63, 2010.

222. Opie, L.H., *The Heart Physiology, From Cell to Circulation*, Lippincott-Raven, Philadelphia, PA, 1998.

223. Hamlin, R.L., Non-drug-related electrocardiographic features in animal models in safety pharmacology, *J. Pharmacol. Toxicol. Methods*, 52, 60–76, 2005.

224. LeBel, C.P. and Foss, J.A., Use of a rodent neurotoxicity screening battery in the preclinical safety assessment of recombinant-methionyl human brain-derived neurotrophic factor, *Neurotoxicology*, 17, 851–863, 1996.

225. Horner, S.A., Gould, S., Noakes, J.P., Rattray, N.J., Allen, S.L., Zotova, E., and Arezzo, J.C., Lack of neurotoxicity of the vascular targeting agent ZD6126 following repeated i.v. dosing in the rate, *Mol. Cancer Ther.*, 3, 783–791, 2004.

226. Authier, S., Delatte, M.S., Kallman, M.J., Stevens, J., and Markgraf, C., EEG in non-clinical drug safety assessments: Current and emerging considerations, *J. Pharmacol. Toxicol. Methods*, 81, 274–285, 2016.

227. Chapillon, P., Lalonde, R., Jones, N., and Caston, J., Early development of synchronised walking on the roda rod in rats—Effects of training and handling, *Behav. Brain Res.*, 93, 77–81, 1998.

228. Holliday, T.A., Nelson, H.J., Williams, D.C., and Willits, N., Unilateral and bilateral brainstem auditory-evoked response abnormalities in 900 Dalmatian dogs, *J. Vet. Intern. Med.*, 6, 166–174, 1992.

229. Colley, J.C., Edwards, J.A., Heywood, R., and Purser, D., Toxicity studies with quinine hydrochloride, *Toxicology*, 54, 219–226, 1989.

230. Redfern, W.S., Unpublished observations, 2005.

231. Haefely, W., Biological basis of drug-induced tolerance, rebound, and dependence. Contribution of recent research on benzodiazepines, *Pharmacopsychiatry*, 19, 353–361, 1986.

232. Hamdam, J., Sethu, S., Smith, T., Alfirevic, A., Alhaidari, M., Atkinson, J., Ayala, M., Box, H., Cross, M., Delaunois, A., Dermody, A., Govindappa, K., Guillon, J.M., Jenkins, R., Kenna, G., Lemmer, B., Meecham, K., Olayanju, A., Pestel, S., Rothfuss, A., Sidaway, J., Sison-Young, R., Smith, E., Stebbings, R., Tingle, Y., Valentin, J.P., Williams, A., Williams, D., Park, K., and Goldring, C., Safety pharmacology—Current and emerging concepts, *Toxicol. Appl. Pharmacol.*, 273(2), 229–241, 2013.

233. Authier, S., Arezzo, J., Delatte, M.S., Kallman, M.J., Markgraf, C., Paquette, D., Pugsley, M.K., Ratcliffe, S., Redfern, W.S., Stevens, J., Valentin, J.P., Vargas, H.M., and Curtis, M.J., Safety pharmacology investigations in toxicology studies: An industry survey, *J. Pharmacol. Toxicol. Methods*, 68, 44–51, 2013.

234. Moser, V.C., Becking, G.C., MacPhail, R.C., and Kulig, B.M., The IPCS collaborative study on neurobehavioral screening methods, *Fundam. Appl. Toxicol.*, 35, 143–151, 1997.

235. Strimbu, K. and Tavel, J.A., What are biomarkers?, *Curr. Opin. HIV AIDS*, 5(6), 463–466, 2010.

236. Pierson, J.B., Berridge, B.R., Brooks, M.B., Dreher, K., Koerner, J., Schultze, E., Sarazan, R.D., Valentin, J.-P., Vargas, H.M., and Pettit, S.D., A public-private consortium advances cardiac safety evaluation: Achievements of the HESI Cardiac Safety Technical Committee, *J. Pharmacol. Toxicol. Methods*, 68, 7–12, 2013.

237. Hausner, E.A., Hicks, K.A., Leighton, J., Szarfman, A., Thompson, M.A., and Harlow, P., Qualification of cardiac troponins for nonclinical use: A regulatory perspective, *Regul. Toxicol. Pharmacol.*, 67(1), 108–114, 2013.

238. Reagan, W.J., Troponin as biomarker of cardiac toxicity: Past, present and future, *Toxicol. Pathol.*, 38, 1134–1137, 2010.

239. Engle, S.K. and Watson, D.E., Natriuretic peptides as cardiovascular safety biomarkers in rats: Comparison with blood pressure, heart rate, and heart weight, *Toxicol. Sci.*, 149(2), 458–472, 2016.

240. Wang, J., Tan, G.J., Han, L.N., Bai, Y.Y., He, M., and Lui, H.B., Novel biomarkers for cardiovascular risk prediction, *J. Geriatr. Cardiol.*, 14, 135–150, 2017.

241. Devaux, Y., Vausort, M., Goretti, E., Nazarov, P.V., Azuaje, F., Gilson, G., Corsten, M.F., Schroen, B., Lair, M.L., Heymans, S., and Wagner, D.R., Use of circulating microRNAs to diagnose acute myocardial infarction, *Clin. Chem.*, 58(3), 559–567, 2012.

242. Thompson, K.L., Boitier, E., Chen, T., Couttet, P., Ellinger-Ziegelbauer, H., Goetschy, M., Guillemain, G., Kanki, M., Kelsall, J., Mariet, C., de La Moureyre-Spire, C., Mouritzen, P., Nassirpour, R., O'Lone, R., Pine, P.S., Rosenzweig, B.A., Sharapova, T., Smith, A., Uchiyama, H., Yan, J., Yuen, P.S., and Wolfinger, R., Absolute measurement of cardiac injury-induced microRNAs in biofluids across multiple test sites, *Toxicol. Sci.*, 154(1), 115–125, 2016.

243. Vacchi-Suzzi, C., Hahne, F., Scheubel, P., Marcellin, M., Dubost, V., Westphal, M., Boeglen, C., Büchmann-Moller, S., Cheung, M.S., Cordier, A., De Benedetto, C., Deurinck, M., Frei, M., Moulin, M., Oakeley, E., Grenet, O., Grevot, A., Stull, R., Theil, D., Moggs, J.G., Marrer, E., and Couttet, P., Heart structure-specific transcriptomic atlas reveals conserved microRNA-mRNA interactions, *PLoS One*, 8(1), e52442, 2013.

244. Glineur, S.F., De Ron, P., Hanon, E., Valentin, J.P., Dremier, S., and Nogueira da Costa, A., Paving the route to plasma miR-208a-3p as an acute cardiac injury biomarker: Preclinical rat data supports its use in drug safety assessment, *Toxicol. Sci.*, 149(1), 89–97, 2016.

245. Gryshkova, V., Fleming, A., McGhan, P., De Ron, P., Fleurance, R., Valentin, J.P., and Nogueira da Costa, A., miR-21-5p as a potential biomarker of inflammatory infiltration in the heart upon acute drug-induced cardiac injury in rats, *Toxicol. Lett.*, 286, 31–38, 2018.

246. Mikaelian, I., Cameron, M., Deidre Dalmas, A., Enerson Bradley, E., Gonzalez, R.J., Guionaud, S., Hoffmann, P.K., King, N.M., Lawton, M.P., Scicchitano, M.S., Smith, H.W., Thomas, R.A., Weaver, J.L., and Zabka, T.S.; Vascular Injury Working Group of the Predictive Safety Consortium, Nonclinical safety biomarkers of drug-induced vascular injury: Current status and blueprint for the future, *Toxicol. Pathol.*, 42(4), 635–657, 2014.

247. Dong, L.H., Lv, P., and Han, M., Roles of SM22alpha in cellular plasticity and vascular diseases, *Cardiovasc. Hematol. Disord. Drug Targets*, 12(2), 119–125, 2012.

248. Blank, M., Goodsaid, F., Harlow, P., Hausner, E., Jacobson-Kram, D., Taylor, W., Thompson, A., Throckmorton, D., and Xiao, S., Review of qualification data for biomarkers of nephrotoxicity submitted by the Predictive Safety Testing Consortium, Center for Drug Evaluation and Research U.S. Food and Drug Administration United States, 2009.

249. Consortium, S.-T., Summary Data Package—Novel clinical biomarkers of drug-induced kidney injury? The Drug induced kidney injury work package of Innovative Medicines Initiative SAFE-T Consortium, 2017.

250. Amin, R.P., Vickers, A.E., Sistare, F., Thompson, K.L., Roman, R.J., Lawton, M., Kramer, J., Hamadeh, H.K., Collins, J., Grissom, S., Bennett, L., Tucker, C.J., Wild, S., Kind, C., Oreffo, V., Davis, J.W. II, Curtiss, S., Naciff, J.M., Cunningham, M., Tennant, R., Stevens, J., Car, B., Bertram, T.A., and Afshari, C.A., Identification of putative gene based markers of renal toxicity, *Environ. Health Perspect.*, 112(4), 465–479, 2004.

251. Fay, M.J., Alt, L.A.C., Ryba, D., Salamah, R., Peach, R., Papaeliou, A., Zawadzka, S., Weiss, A., Patel, N., Rahman, A., Stubbs-Russell, Z., Lamar, P.C., Edwards, J.R., and Prozialeck, W.C., Cadmium nephrotoxicity is associated with altered microRNA expression in the rat renal cortex, *Toxics*, 6(1), 16, doi:10.3390/toxics6010016, 2018.

252. Slocum, J.L., Heung, M., and Pennathur, S., Marking renal injury: Can we move beyond serum creatinine?, *Transl. Res.*, 159(4), 277–289, 2012.

253. Zhang, A., Sun, H., Wang, P., Han, Y., and Wang, X., Recent and potential developments of biofluid analyses in metabolomics, *J. Proteomics*, 75(4), 1079–1088, 2012.

254. Carr, D.F., Ayehunie, S., Davies, A., Duckworth, C.A., French, S., Hall, N., Hussain, S., Mellor, H.R., Norris, A., Park, B.K., Penrose, A., Pritchard, D.M., Probert, C.S., Ramaiah, S., Sadler, C., Schmitt, M., Shaw, A., Sidaway, J.E., Vries, R.G., Wagoner, M., and Pirmohamed, M., Towards better models and mechanistic biomarkers for drug-induced gastrointestinal injury, *Pharmacol. Ther.*, 172, 181–194, 2017.

255. Miyazaki, H., Kitayama, T., Sekiya, K., Haruna, M., Mino, T., Suganami, H., Watanabe, H., and Yamamoto, K., Individual QT-RR correction and sensitivity to detect drug-induced changes in QT interval for canine telemetry assay, *J. Pharmacol. Toxicol. Methods*, 49, 224, 2004.

256. Bailey, J., Thew, M., and Balls, M., An analysis of the use of dogs in predicting human toxicology and drug safety, *Altern. Lab. Anim.*, 41, 335–350, 2013.

257. Trepakova, E.S., Koerner, J., Pettit, S.D., and Valentin, J.P., HESI Pro-Arrhythmia Committee, A HESI consortium approach to assess the human predictive value of non-clinical repolarization assays, *J. Pharmacol. Toxicol. Methods*, 60, 45–50, 2009.

258. Valentin, J.P., Bialecki, R., Ewart, L., Hammond, T., Leishmann, D., Lindgren, S., Martinez, V., Pollard, C., Redfern, W., and Wallis, R., A framework to assess the translation of safety pharmacology data to humans, *J. Pharmacol. Toxicol. Methods*, 60(2), 152–158, 2009.

259. Redfern, W.S., Waldron, G., Winter, M.J., Butler, P., Holbrook, M., Wallis, R., and Valentin, J.P., Zebrafish assays as early safety pharmacology screens: Paradigm shift or red herring?, *J. Pharmacol. Toxicol. Methods*, 58, 110–117, 2008.

260. Park, E., Gintant, G.A., Bi, D., Kozeli, D., Pettit, S.D., Pierson, J.B., Skinner, M., Willard, J., Wisialowski, T., Koerner, J., and Valentin, J.P., Can non-clinical repolarization assays predict the results of clinical thorough QT studies? Results from a research consortium, *Br. J. Pharmacol.*, 175, 606–617, 2018.

261. Pollard, C.E., Skinner, M., Lazic, S.E., Prior, H.M., Conlon, K.M., Valentin, J.-P., and Dota, C., An analysis of the relationship between preclinical and clinical QT interval-related data, *Toxicol. Sci.*, 159(1), 94–101, 2017.

262. Valentin, J.P. and Hammond, T., Safety and secondary pharmacology: Successes, threats, challenges and opportunities, *J. Pharmacol. Toxicol. Methods*, 58, 77–87, 2008.

263. Ando, K., Ikeda, H., Yamamoto, K., and Sagami, F., Dose (concentration)-response analysis of drug-induced QT interval prolongation in conscious monkeys (JPMA QT product), *J. Pharmacol. Toxicol. Methods*, 49, 220, 2004.

264. Ando, K., Sugiyama, A., Satoh, Y., Nakamura, Y., and Hashimoto, K., Predicting drug-induced QT prolongation using a new in vivo animal model: Comparison of risperidone and olanzapine, *J. Pharmacol. Toxicol. Methods*, 49, 221, 2004.

265. Miyazaki, H., Kitayama, T., Tashibu, H., Ando, K., Yamamoto, K., and Sagami, F., Comparison of the sensitivity between canine telemetry assay and isofluorane-anaesthetised model (JPMA QT Product), *J. Pharmacol. Toxicol. Methods*, 49, 224, 2004.

266. De Bruin, M.L., Pettersson, M., and Meyboom, R.H.B., Hoes, A.W., and Leujkens, H.G.M., AntihERG activity and the risk of drug-induced arrhythmias and sudden death, *Eur. Heart J.*, 26, 590–597, 2005.

267. Harmer, A., Valentin, J.-P., and Pollard, C., On the relationship between block of the cardiac Na⁺ channel and drug-induced prolongation of the QRS complex, *Br. J. Pharmacol.*, 164(2), 260–273, 2011.

268. Sewell, F., Edwards, J., Prior, H., and Robinson, S., Opportunities to apply the 3Rs in safety assessment programs, *ILAR J.*, 57(2), 234–245, 2016.

269. Ten Tusscher, K.H., Noble, D., Noble, P.J., and Panfilov, A.V., A model for human ventricular tissue, *Am. J. Physiol. Heart Circ. Physiol.*, 286, H1573–H1589, 2004.

270. Passini, E., Britton, O.J., Lu, H.R., Rohrbacher, J., Hermans, A.N., Gallacher, D.J., Greig, R.J.H., Bueno-Orovio, A., and Rodriguez, B., Human in silico drug trials demonstrate higher accuracy than animal models in predicting clinical pro-arrhythmic cardiotoxicity, *Front. Physiol.*, 8, 668, doi:10.3389/fphys.2017.00668, 2017.

271. Norinder, U., Sjoberg, P., and Osterberg, T., Theoretical calculation and prediction of brain-blood partitioning of organic solutes using molsurf parametrization and PLS statistics, *J. Pharm. Sci.*, 87, 952–959, 1998.

272. Garcia-Serna, R., Vidal, D., Renez, N., and Mestres, J., Large-scale predictive drug safety: From structural alerts to biochemical mechanisms, *Chem. Res. Toxicol.*, 28, 1875–1887, 2015.

273. Davila, J.C., Cezar, G.G., Thiede, M., Strom, S., Miki, T., and Trosko, J., Use and application of stem cells in toxicology, *Toxicol. Sci.*, 79, 214–223, 2004.

274. Odawara, A., Katoh, H., Matsuda, N., and Sozuki, I., Physiological maturation and drug responses of human induced pluripotent stem cell-derived cortical neuronal networks in long-term culture, *Nat. Sci. Rep.*, 6, 26181, doi:10.1038/srep26181, 2016.

275. Tukker, A.M., Wijnolts, F.M.J., de Groot, A., and Westerink, R.H.S., Human iPSC-derived neuronal models for in vitro neurotoxicity assessment, *Neurotoxicology*, 67, 215–225, 2018.

276. Junaid, A., Mashaghi, A., Hankemeier, T., and Vulto, P., An end-user perspective on Organ-on-Chip: Assays and usability aspects, *Curr. Opin. Biomed. Eng.*, 1, 15–22, 2017.

277. Rubinstein, A.L., Zebrafish: From disease modeling to drug discovery, *Curr. Opin. Drug Discov. Dev.*, 6, 218–223, 2003.

278. Zon, L.I. and Peterson, R.T., In vivo drug discovery in the zebrafish, *Nat. Rev. Drug Discov.*, 4, 35–44, 2005.

279. Parng, C., In vivo zebrafish assays for toxicity testing, *Curr. Opin. Drug Discov. Dev.*, 8, 100–106, 2005.

280. Spitsbergen, J.M. and Kent, M.L., The state of the art of the zebrafish model for toxicology and toxicologic pathology research—Advantages and current limitations, *Toxicol. Pathol.*, 31, 62–87, 2003.

281. Hunt, P.R., The *C. elegans* model in toxicity testing, *J. Appl. Toxicol.*, 37(1), 50–59, 2017.

282. Sarapa, N., Early human microdosing to reduce attrition in clinical drug development, *Am. Pharm. Outsourcing*, 4, 42–46, 2003.

283. Dixit, R., What non-clinical toxicology and safety pharmacology data are needed to accelerate phase I–II clinical trials?, *Am. Pharm. Outsourcing*, 5, 30–37, 2004.

284. Anon, Innovation or stagnation, challenge and opportunity on the critical path to new medicinal products, 2004.

285. Calamari, D., Strategic survey of therapeutic drugs in the rivers Po and Lambo in Northern Italy, *Environ. Sci. Technol.*, 37, 1241–1248, 2003.

286. Huggett, D.B., Khan, I.A., Foran, C.M., and Schlenk, D., Determination of beta-adrenergic receptor blocking pharmaceuticals in United States wastewater effluent, *Environ. Pollut.*, 121, 199–205, 2003.

287. Kopin, D.W., Furlong, E.T., Meyer, M.T., Thurman, E.M., Zaugg, S.D., Barber, L.B., and Buxton, H.T., Pharmaceuticals, hormones, and other organic wastewater contaminants in U.S. streams 1999–2000: A national reconnaissance, *Environ. Sci. Technol.*, 36, 1202–1211, 2002.

288. Alabaster, V., In vivo pharmacology training group, the fall and rise of in vivo pharmacology, *Trends Pharmacol. Sci.*, 23, 13–18, 2002.

289. Jobe, P.C., Adams-Curtis, L.E., Burks, T.F., Fuller, R.W., Peck, C.C., Ruffolo, R.R., Snead III, O.C., and Woosley, R.L., The essential role of the integrative biomedical sciences in protecting and contributing to the health and well-being of our nation, *Pharmacologist*, 40, 32–37, 1998.

290. Valentin, J.-P. and Price, S., Overview of the modular training programme in applied toxicology: Module on safety pharmacology in pre-clinical research and development, *Br. Toxicol. Newslett.*, 30, 63–65, 2007.

291. Pugsley, M.K., Authier, S., Koerner, J.E., Redfern, W.S., Markgraf, C.G., Brabham, T., Correll, K., Soloviev, M.V., Botchway, A., Engwall, M., Traebert, M., Valentin, J.P., Mow, T.J., Greiter-Wilke, A., Leishman, D.J., and Vargas, H.M., An overview of the safety pharmacology society strategic plan, *J. Pharmacol. Toxicol. Methods*, S1056-8719(17)30576-2, doi:10.1016/j.vascn.2018.01.001, 2018.

292. Authier, S., Curtis, M.J., Soloviev, M., Redfern, W.S., Kallman, M.J., Hamlin, R.L., Leishman, D.J., Valentin, J.-P., Koerner, J.E., Vargas, H.M., Botchway, A., Correll, K., and Pugsley, M.K., The Diplomate in Safety Pharmacology (DSP) certification scheme, *J. Pharmacol. Toxicol. Methods*, 75, 1–4, 2014.

293. Cavero, I. and Crumb, W., Safety Pharmacology Society: 5th annual meeting, *Expert Opin. Drug Saf.*, 5(1), 181–185, 2006.

294. Cavero, I., 10th annual meeting of the Safety Pharmacology Society: An overview, *Expert Opin. Drug Saf.*, 10(2), 319–333, 2011.

295. Cavero, I. and Holzgrefe, H., 15th annual meeting of the Safety Pharmacology Society: Focus on traditional sensory systems, *J. Pharmacol. Toxicol. Methods*, 83, 55–71, 2017.

296. Redfern, W.S. and Valentin, J.-P., Trends in safety pharmacology: Posters presented at the annual meetings of the Safety Pharmacology Society 2001–2010, *J. Pharmacol. Toxicol. Methods*, 64, 102–110, 2011.

17 Ethical Concerns in Clinical Research

Jonathan C. Young
West Virginia University

Lori Nesbitt
Compass Point Research

CONTENTS

17.1 INTRODUCTION

Drug development in the United States is both risky and expensive. The cost of developing a drug has increased from approximately $500 million in 1990[1] to more than $2.6 billion in 2015.[2,3] At the same time, the research and development (R&D) costs of the pharmaceutical industry have doubled from approximately 12% of sales to 21%.[1,4] As of 2017, only about 12% of drugs with clinical trials will be successful, and the drug development process takes at least 10 years.[4]

Given the growing number of clinical trials required for Food and Drug Administration (FDA) approval, opportunities are numerous in the provision of clinical research services. In addition, the FDA is becoming more vigilant in enforcing the ethical conduct of clinical research and the protection of research participants. Lastly, in an effort to avoid conflicts of interest or perceived improprieties, pharmaceutical and device manufacturers frequently outsource all or part of the clinical trial process to niche service providers. For these reasons, the clinical trial industry has become segmented. Each segment or service provider performs a necessary step in the clinical trial value chain. In addition to niche service providers, the growing clinical trial industry has created a need for service organizations, publications, and websites devoted to the specialized field.

To have a thorough understanding of the industry as a whole requires a working knowledge of the freedoms, constraints, and political environments in which each service provider must operate. Perhaps a better way to stress the complexities of the industry is to realize that drug development is characterized by a high rate of failure. It is a disorderly process where very few research efforts ever bear fruit. In fact, the typical pharmaceutical company spends about 40% of its R&D budget on compounds that do not make it to market. As one expert has stated, "Drug innovation is something that is sought but not known in advance … Only by aiming high can genuine innovation be coaxed into existence … Innovation must be able to pay the price of failure."[1] Unfortunately, even if the new compound under study is safe and efficacious, failure can occur because of undercapitalization of the pharmaceutical company, poor data quality on the part of the sponsor or investigative site, or disapproval by the FDA.

17.2 CLINICAL TRIAL SERVICE PROVIDERS

Although it is estimated that $90 billion was spent in human drug development (Phases I–III) in 2016, the industry is quite small and very specialized. In fact, most Americans have a limited knowledge of how new medications actually end up in their pharmacies.[1,5] The service providers challenged with bringing new treatments and cures to the masses include (a) FDA, (b) the clinical trial sponsor, (c) contract research organizations (CROs), (d) study monitors, (e) the clinical trial site, (f) site management organizations (SMOs), (g) the Institutional Review Board (IRB), and (h) study participants.

17.2.1 FOOD AND DRUG ADMINISTRATION

The FDA's very noble mission is "to promote and protect public health by helping safe and effective products reach the market in a timely way, and monitoring products for continued safety after they are in use."[2] The scope of this mission has grown substantially during the past century. In addition to its consumer protection role with regard to all prescription and over-the-counter drugs, the FDA establishes and enforces standards for all food (except meat and poultry), all blood products, vaccines, and tissues for transplantation, all medical equipment, all devices that emit radiation, all animal drugs and foods, and all cosmetics.

Given the complex and diverse tasks performed by the FDA, it is not surprising that there are critics on both sides of the drug approval fence. Many argue that the FDA has taken away the human right to make an informed choice regarding whether to take an investigational drug. Others argue that FDA acts too hastily in approving drugs that are later shown to be unsafe. Thus, there is a constant tension between those who want greater consumer protection and those who want greater freedom of choice.

In 1992, Congress passed the *Prescription Drug User Fee Act* (PDUFA).[3] The purpose was to establish a mechanism for financing the resources that would be needed to speed up the process of reviewing NDAs. PDUFA allows the FDA to collect user fees from pharmaceutical companies to support the review of applications. A recent report by the US General Accounting Office (GAO)[4] concludes that "PDUFA has been successful in providing FDA with the funding necessary to hire additional drug reviewers, thereby making new drugs available in the United States more quickly" (p. 6).

Supporting this conclusion is the fact that approval times have decreased from 27 to 10 months.[6] However, the GAO also reported a small increase in the drug-withdrawal rate since the implementation of PDUFA, although these results seem to be mitigated in more recent years. Initially, after PDUFA went into effect, a higher percent of approved drugs had been withdrawn from the market because of safety issues. FDA officials argued that the increase is insignificant – from 3.10% in the 8-year period before PDUFA to 3.47% in the 8-year period after. A more recent study showed that from 2001 to 2010, only 1.35% of 222 approved drugs were withdrawn from the market.[7]

This protection–freedom dynamic is even more intensive when it comes to drugs that are being tested for use in patients with terminal illnesses that have no other viable treatment options. In these situations, the FDA receives tremendous pressure to approve these drugs rapidly. The rationale is that the most serious risk of death from an experimental drug is no risk at all compared with the certainty of death in patients with a lethal disease.

17.2.2 CLINICAL TRIAL SPONSOR

The clinical trial sponsor is an individual, company, institution, or organization that assumes responsibility for the initiation, management, or financing of the clinical trial. The sponsor is required by the FDA to conduct clinical trials to determine the safety and efficacy of the investigational agent. Safety data are usually derived through documented occurrence of adverse pharmacokinetic or pharmacodynamic effects. Alternatively, efficacy data can be evaluated by the prevention of a medical condition or through improvement of specific symptoms of a disease process.[5]

In the conduct of the clinical trial, according to the Code of Federal Regulations (CFR), the study sponsor is responsible for all aspects of the study including, but not limited to, maintaining quality assurance and quality control, medical expertise, trial design, trial management, data handling, record keeping, investigator selection, allocation of duties and functions, determining compensation to subjects and investigators, financing, notification/submission to regulatory authorities, product information, preparing and supplying study medications, and monitoring and assuring that all clinical trial sites comply with federal regulatory requirements.[8]

The sponsor bears ultimate responsibility for the success, failure, and safety of the treatment under study, even after FDA approval. In addition, the sponsor is the true innovator in the clinical trial process. Innovation is expensive, causing newly available treatments to be costly to the end user. Thus, because of the escalating price of medications, the innovators are under increased scrutiny by consumers and policymakers. Paradoxically, as the population ages, the consumers are driving the demand for new cures and better treatments.

Given the high rate of "failure" in the drug industry, it is reasonable to hypothesize that drug development would take place in economies characterized by relatively free markets and prices.

The ideal environment would provide adequate incentives for investing in high-risk ventures. Such an environment exists in the United States. Although the United States is home to only about 5% of the world's population, roughly 36% of the worldwide pharmaceutical R&D is conducted in the United States on a yearly basis.[1]

The United States is the world's quantitative and qualitative leader in drug development,[1] but what is the price of that leadership? Many taxpayers and consumers are outraged at the high cost of prescription drugs compared with those available in nations with price controls, such as Canada, Mexico, and the United Kingdom. *USA Today* featured a front-page story that compared the price of ten innovator drugs that were still under patent.[9] According to the article, the sample of drugs was 100%–400% higher in the United States vs. Canada, Mexico, and a few European nations, where direct price controls exist. Having established this comparison, it was an easy step for the writer to conclude that what was a good deal for these other countries would be a good deal for the United States. The point that escapes the writer is that if such controls were in effect here, many of the sampled drugs would never have been developed and made available to price-controlled countries.

Citizens and policymakers misunderstand that although drug development is expensive, production costs of the pills are comparatively low. It is the formula, not the ingredients, that is costly. In addition, drug expenditure is just part of the overall expense of health care and must not be looked at in a vacuum. For example, it is estimated that, on average, US citizens spend about 29% more per capita on pharmaceutical goods – or about $228 per person per year more – than the most expensive price-controlled nation.[10]

17.2.3 CONTRACT RESEARCH ORGANIZATIONS

The daily attendance of the clinical trial process can require time, labor, and training that many sponsors feel do not match their current capabilities. Therefore, sponsors may elect to outsource any or all of their trial-related duties to a CRO. Full-service CROs provide data monitoring, data management, protocol development, medical writing, statistical analysis, contract management, site selection, and shipping and handling of investigational supplies. Niche CROs may elect to provide only a few of these services, such as data monitoring or medical writing. The CRO should maintain its own system of quality assurance and quality control. However, regardless of the duties assumed by the CRO, the final responsibility for the quality and integrity of the data always resides with the sponsor, and any duties not specifically transferred to a CRO remain the responsibility of the sponsor.

CROs are growing and mutating making it difficult to identify specific trends. However, one thing is certain, pharmaceutical, biotechnology, and medical devices sponsors expect to increase outsourcing to the several hundred CROs. In fact, pharmaceutical companies have already increased their use of CROs: from 28% of clinical studies in 1993 to 61% in 1999.[11] A more recent study suggests that, as of 2015, 100% of sponsors performed at least some amount of outsourcing to CROs.[12] This reflects increased spending on the CRO services of Phases I–IV study monitoring, data management, pharmacoeconomic analysis, and medical writing. Although these specialties remain the most frequently used, CROs are also offering new services to satisfy sponsors demands for faster trials and globalization.

To meet this challenge, CROs seem to be taking one of the two tracks: they are strategically planning to become either mega-CROs or niche providers. Industry observers believe the midsize CROs will disappear, mostly through merger and acquisition activity by larger CROs and by non-CROs with a strategic interest in entering the business. Analysts had forecasted that the midsize CRO will be gone, but niche players with special capabilities (statistical consulting, data management, monitoring) may survive.[11] Indeed, in 2016, merger and acquisition spending in the CRO industry had increased to $24 billion, and only nine CROs possessed 60% of the market share.[13] As the CRO industry consolidates, some large publicly traded CROs are making acquisitions that diversify the breadth of service beyond conducting studies. This move enables sponsors to do one-stop shopping instead of contracting with multiple companies throughout the discovery–development process. In addition, CROs are positioning themselves to gain access to populations in emerging markets such as Israel, Russia, Latin America, China, and India.

17.2.4 STUDY MONITORS

Study monitors or clinical research associates (CRAs) can be employed by the study sponsor, CRO, or independently contracted for a specific study, and according to the International Committee on Harmonisation (ICH) formalized by FDA in the *Guidance for Industry: Good Clinical Practice* (GCP), the purpose of a CRA is to[14]

1. Verify that the rights and well-being of human subjects are protected.
2. Verify that the reported trial data are accurate, complete, and verifiable from source documentation.
3. Verify that the conduct of the trial is in compliance with the currently approved protocol/amendment(s), with GCPs, and with applicable regulatory requirement(s).

CRAs achieve these tasks through frequent visits to the clinical trial site. During these visits, the monitor will verify source data, audit regulatory documents for accuracy and completion, perform drug accountability assessments, and communicate any concerns, problems, or new information with the study staff.

17.3 CLINICAL TRIAL SITE

The front line of clinical trials is the site. It is at the site level that participants are given informed consent, study-related procedures are conducted according to the clinical trial protocol, and data are collected and reported. It is these data, aggregated from all sites, that ultimately determine the fate of the investigational drug or device. With the rigor in which clinical trials must be conducted today, site research personnel usually include the principal investigator (PI), subinvestigators, study coordinators, and regulatory managers. However, depending on the amount of research being conducted at a given location, the study coordinators are often also responsible for the regulatory compliance.

The PI is the individual who is ultimately responsible for the clinical trial at the trial site and the one who verifies that the data reported to the study sponsor are accurate. Although not required by the FDA, the PI is usually a physician. In the event that the PI is not a physician, adequate physician oversight of the trial must be readily evident. As addressed by ICH and GCP guidance, the PI should be qualified through education, training, and experience to assume responsibility for the proper conduct of the trial and should meet all the qualifications specified by the applicable regulatory requirements (Section 4.1.1, p. 18).[14] Not all physicians are well suited for clinical research. A successful PI has distinct characteristics as listed in Table 17.1.

As defined by ICH and GCP guidelines, a subinvestigator is any individual member of the clinical trial team designated and directly supervised by the PI to perform trial-related procedures or make trial-related decisions (Section 1.56, p. 13).[14] Examples of subinvestigators include other physicians, pharmacists, pharmacologists, nurses, physician assistants, and study coordinators.

Clinical research coordinators (CRCs) are the research personnel who assist with patient visits, and perform study-related procedures that do not require a physician (phlebotomy, vital signs, adverse event, and concomitant medication discussions, etc.). CRCs provide the PI or physician with data required for interpretation, medical decisions (inclusion/exclusion, dosage adjustment, patient withdrawal, adverse event causality, etc.), and trial oversight. In addition, CRCs are usually responsible for transcribing source documentation (medical records, clinic notes, laboratory reports, etc.) into case report forms (CRF) supplied by the study sponsor.

TABLE 17.1

Qualities Needed by a Successful PI

Not Everyone Makes a Successful PI; a Good PI

- Has an intrinsic interest in science
- Is knowledgeable about the protocol
- Always places patient care above all other priorities
- Is willing to carve out time for the study
- Is very involved in medical oversight
- Knows his or her limitations and when to ask for help
- Is tolerant of the increased need for regulatory scrutiny
- Understands that being a respected clinician does not mean being a good researcher and is open to learning about the conduct of clinical research
- Is prompt in the turnaround on documentation
- Meets participant recruitment and enrollment goals established with the sponsor

Another important function of the CRC is to interact with the sponsor or CRO-appointed CRA. Because the CRA is an agent for the sponsor, the CRC–CRA relationship is one that can make or break a study. If a CRC is doing an excellent job, and the documents are available and accurate, the CRA's interactions with the site should be positive and productive. Unfortunately, this does not always happen. There are dynamics on both sides of the CRF. Some common complaints are

- The CRA assigned to a given study changes frequently. Each CRA communicates different directives to the site, causing the site to redo work.
- The CRA has a condescending attitude toward the site and the investigator.
- The CRA is not well trained.
- The CRC is inexperienced.
- The CRA cannot obtain rapid answers to questions, often creating patient-care issues.
- The CRC cannot obtain rapid answers to queries, often extending timelines for study closure.
- The CRC makes numerous errors in the CRF.
- The CRC does not seem dedicated to the study.

In an industry where there is virtually full employment, it is difficult to find trained CRCs and CRAs. So, conflicts can arise from interaction between untrained or inexperienced personnel. Sometimes, however, personality conflicts are the main culprit. Although technology can eliminate some of the need for CRA–CRC interaction, all parties need to understand the roles and pressures on the other person.[15] For example, many CRAs travel 4 days a week and see various levels of work quality at different sites. On the other hand, CRCs are often responsible for more than one study and have requests from multiple CRAs on any given day. In addition, the CRA must respond to the needs of the research participant first, which can cause time delays in completing data queries.

Regulatory managers are usually charged with submitting regulatory documents to the IRB and study sponsor and with maintaining a regulatory binder. A regulatory binder should contain a protocol, protocol amendments, IRB approvals and correspondence, all versions of the IRB-approved patient informed consent, investigators' brochure, sponsor correspondence, curriculum vitas and licensures of the PI and subinvestigators, and any safety reports.

17.3.1 Site Management Organizations

As the number of clinical research sites has grown, a new entity, the SMOs, has arisen. SMOs, in the traditional sense, were established to offer the sponsor consolidated services at the site level. SMOs took the CRO business model and brought it to the front lines. For example, CROs offer a variety of services for the sponsor, such as site monitoring, contract administration, shipping and receiving of study supplies, and data management. SMOs offer PI recruitment, patient recruitment, and regulatory and contract management for multiple sites. As sponsors often must recruit 50–200 clinical trial sites, SMOs offer a one-stop shop. SMOs can provide the sponsors with multiple PIs and centralized contract and regulatory services, expediting study initiation.

SMO models vary widely in the industry. Some SMOs hire physician investigators as employees of the company, and others subcontract for investigator services. However, few offer turnkey solutions for investigators who wish to be involved in clinical research, but lack the specialized training or necessary personnel. Full-service SMOs act as a liaison between the pharmaceutical, device, or biotechnology company (or CRO), and the research patient. Services often include patient and investigator recruitment, regulatory document preparation and compliance, and study coordination.

SMOs provide an interesting entry for investigators into the clinical trial business. Specifically, some SMOs can present new investigators with clinical trial opportunities, essential training, and qualified research personnel. In turn, the investigator assumes ultimate responsibility for the ethical conduct of the study. By alleviating the physician, hospital, and health care staff from time-consuming, nonclinical tasks, SMOs can make research not only feasible but also lucrative for investigators and hospitals. This risk-sharing model can be beneficial to all parties.

17.3.2 INSTITUTIONAL REVIEW BOARDS

Since the *Kefauver–Harris Amendment*[16] was adopted in 1962, pharmaceutical manufacturers have been held responsible by the FDA for providing new medications that are both safe and effective. In addition to the *Kefauver–Harris Amendment*, the *Belmont Report* (written in 1979)[17] established the ethical principles and guidelines for conducting research. The FDA requires that clinical trials be conducted in compliance with a protocol that has been approved by an IRB or Independent Ethics Committee (IEC). The terms "IRB" and "IEC" refer to any board, committee, or other group formally designated by an institution to review, approve initiation, and conduct periodic review of research involving human participants.[18] The FDA expects the IRB to review all research-related documents and activities that pertain directly to the rights and welfare of the participants of proposed research. The IRB has the authority to approve, require modification to, or disapprove all research activities as specified in the federal regulations. The primary purpose of the IRB and this formal review process is to protect the rights and welfare of human participants involved in these clinical trials. It is the federally mandated charge of the IRB to ensure the safety of the research participant.

However, the FDA is not the only agency governing the function of an IRB. The FDA only oversees clinical trials when they involve an FDA-regulated product (i.e., drug, device, or biologic). The Department of Health and Human Services (HHS) and Office for Human Research Protection (OHRP) oversee federally funded research. The *Federal Policy for the Protection of Human Subjects*,[19] known as the "Common Rule," is the basic HHS policy for protection of human subjects, now codified in the CFR.[20] Even though the FDA regulations are not part of the "Common Rule," the basic requirements for IRB's and informed consent are congruent. Please note that – as of the time of this writing – the HHS intends to revise the Common Rule, effective January 21, 2019. Even though the FDA has not yet harmonized with the HHS 2019 Common Rule revisions, the FDA has released guidance that states its intent to harmonize in the future.[21]

The differences between the "Common Rule" and the FDA regulations are differences in applicability. HHS regulations are based on federal funding, and FDA regulations are based on the use of FDA-regulated products. Examples of some of the differences in the FDA regulations and the "Common Rule" include the following:

- Differing definitions of a "human subject."
- HHS discusses "research," whereas FDA discusses "clinical investigations."

- FDA makes no provisions for waiving informed consent, whereas HHS regulations provide for certain conditions in which an IRB can waive or alter elements of informed consent.
- FDA regulations state that subjects must be informed about FDA inspections.
- FDA requires that subjects sign and date the informed consent.
- FDA allows the sponsor to request a waiver of IRB review; under HHS regulations, certain categories of research are exempt, and department heads can waive regulations.
- HHS and FDA both have adopted additional protections for children, but HHS also has additional protections for fetuses, pregnant women, and prisoners.
- FDA has responsibility and authority over all parties participating in FDA-regulated research and has regulations unique to product review responsibilities.
- The FDA requires annual review for all FDA-regulated research, even if the research is deemed to be approvable under one or more of the expedited IRB review categories, whereas, under the HHS 2019 Common Rule, studies approvable under the expedited review categories may not require annual review.

IRBs should be knowledgeable in both FDA and HHS regulations and should know which regulations to apply when conducting its review. FDA regulations apply when products are regulated by the FDA. The "Common Rule" applies when the research is federally supported or conducted or when it is being conducted in an institution that has agreed to review all research under the "Common Rule." Both rules apply if the research is federally funded and involves FDA-regulated products or if the FDA-regulated research is being conducted in an institution that has agreed to review all research under the "Common Rule." As a rule of thumb, no matter which regulations apply, IRB members should use the *Belmont Report*[17] and its guidelines in all of their daily decisions.

17.4 THE EVOLUTION OF ETHICAL PRINCIPLES IN CLINICAL RESEARCH

As the demand to develop new and improved medical treatments has grown so has the need for human participants to test the potential new treatments. The federal government has taken an increasingly prominent role in the oversight of human subject protection. The *Belmont Report*[17] established the ethical principles and guidelines for conducting research. It describes the three, basic, ethical principles (respect for persons, beneficence, and justice) that are relevant to clinical research involving humans. These principles dictate to the clinical research industry that it does not have the right to use people for scientific benefit without their permission – no matter how noble the cause may seem. The research must do something good, and it has to be fair – people cannot be exploited for research purposes. Applying these ethical principles brings the research industry closer to providing sound, ethical research. The regulations that govern the conduct and oversight of clinical research involving human participants were written to implement these ethical principles.

The history of the development of formal ethical principles of research with humans can be traced to World War II. Physicians in Nazi Germany performed "medical experiments" on thousands of concentration camp prisoners. These experiments included determining how long humans would survive in freezing water and high altitudes, injecting people with viruses, and forcing people to ingest poison. Whereas the German physicians argued that the experiments were medically justified, the Nuremberg Military Tribunal declared them to be "crimes against humanity." Seven of the physicians were sentenced to death. The judges who wrote the verdict in 1947 included a section on medical experiments, which became known as the *Nuremberg Code*.[22] This code was the first formal set of ethical principles for researchers. The first item in the *Nuremberg Code* emphasizes obtaining informed consent as a primary responsibility of the researcher:

The voluntary consent of the human subject is essential. This means that the person involved should have legal capacity to give consent; should be so situated as to be able to exercise free power of choice, without the intervention of any element of force, fraud, deceit, duress, over-reaching, or other ulterior form of constraint or coercion; and should have sufficient knowledge and comprehension of the elements of the subject matter involved as to enable him to make an understanding and enlightened decision. This latter element requires that, before the acceptance of an affirmative decision by the experimental subject, there should be made known to him the nature, duration, and purpose of the experiment; the method and means by which it is to be conducted; all inconveniences and hazards reasonably to be expected; and the effects upon his health or person, which may possibly come from his participation in the experiment.

The duty and responsibility for ascertaining the quality of the consent rests upon each individual who initiates, directs, or engages in the experiment. It is a personal duty and responsibility, which may not be delegated to another with impunity.[22]

In 1964, the World Medical Association adopted the "Declaration of Helsinki"[23] to guide researchers in the ethical conduct of medical research. This document has been revised several times since originally adopted. These guidelines reinforced the importance of informed consent, although it added a provision to allow for that consent process to be performed with a legal guardian for people unable to provide consent for themselves. The Declaration of Helsinki also delineated the difference between clinical research, defined as "medical research in which the aim is essentially diagnostic or therapeutic for a patient" and nonclinical biomedical research, defined as "medical research, the essential object of which is purely scientific and without implying direct diagnostic or therapeutic value to the person subjected to the research."

At the time of unfolding this history of medical research ethics internationally, a deplorable medical experiment was being conducted in the United States in Tuskegee by the US Public Health Service. The purpose of this research, which began in the 1930s, was to study the natural history of untreated syphilis. More than 400 African-Americans with syphilis were recruited into the study without their permission. They were also deceived; they were told that some study tests were "special free treatment." During the course of this study, the researchers learned that the mortality rate was twice as high in subjects with syphilis when compared with controls without syphilis. In the 1940s, penicillin was proven to be an effective treatment for syphilis. Despite this finding, the research continued. The subjects were not told of the available treatment, and it was not given to them.

Reports of the Tuskegee study began appearing in the media in 1972. Overwhelming and justified public shock and anger resulted in federal action to compensate for the egregious ethical breaches and to prevent them from recurring. The National Commission for the Protection of Human Subjects of Biomedical and Behavioral Research was established in 1974 to identify the basic ethical principles to guide all research with humans. The Commission's work was documented in the *Belmont Report — Ethical Principles and Guidelines for the Protection of Human Subjects*.[17] That report contained the three ethical principles mentioned above that guide the work of modern researchers and that became the foundation for federal regulations governing research with humans: respect for persons, beneficence, and justice. The principles are described in greater detail below.

17.4.1 RESPECT FOR PERSONS

The first of the ethical principles reflects the emphasis in the Nuremberg Code on informed consent. That is, respect for research participants means treating them as autonomous agents. The *Belmont Report* defines "autonomous agent" as "an individual capable of deliberation about personal goals and of acting under such deliberation…." This means that the informed consent process is needed so that the potential participants are given all the information they need to determine whether participating in the study is in their best interest. The principle of respect also means that researchers conducting trials with "vulnerable" populations (e.g., children or the cognitively impaired) should ensure that extra provisions are made to protect those participants who have diminished autonomy. Of course, the principle of respect includes the notion that the consent process must be completed with no coercion or pressure by the researchers to participate.

17.4.2   BENEFICENCE

The *Belmont Report*'s principle of beneficence refers to the need to ensure that all aspects of a study are designed to obtain the desired knowledge in a way that maximizes benefits and minimizes risks to the participants. The principle of beneficence also means that a risk–benefit analysis must be performed on every proposed study. In determining if this ratio is ethical, consideration must be given to the impact on both the participants and society.

17.4.3   JUSTICE

The principle of justice refers to the fairness or equity involved in the selection of research participants. The point is to ensure that risks or benefits incurred by research participants are not unfairly concentrated in one segment of the population.

17.4.4   HOW THE INVESTIGATOR APPLIES ETHICAL PRINCIPLES

An investigator applies these ethical principles daily by ensuring that the informed consent process is performed perfectly with every research participant. This means that all staff involved in the consent process must be well trained, not only in the research protocol in question but also in the proper elements of informed consent. A good rule of thumb for the people conducting consent discussions is for them to think about what they would want their mother, father, spouse, or friend to know about the study to enable them to make an informed decision. The investigator should never allow this process to be rushed. In addition, to reduce the possibility that the potential participant feels pressured by the power differential that often characterizes the doctor–patient relationship, some ethicists argue that the physician should not be the primary person conducting the consent discussion. This means that, ideally, other individuals should be on the study staff and they should be charged with carrying out the consent process with prospective subjects.

The researcher complies with the concept of beneficence in two ways. First, the investigator carefully reviews the study protocol to determine whether the design provides a generally favorable ratio between the potential risks to the subjects and the potential benefits to the subject and society. If the researcher decides to conduct the study, the second strategy is to review carefully each potential candidate to ensure that this ratio is favorable for each potential participant. Of course, the investigator also encourages the potential participant to weigh the risks and benefits for themselves through the consent process. Indeed, a central principle of the 1964 Declaration of Helsinki was that "Concern for the interests of the subject must always prevail over the interests of science and society." This means that societal benefits should always carry less weight than individual, subject-level benefits.

Finally, the investigator fulfills the principle of justice by ensuring that his or her subject-recruitment strategies do not systematically discriminate against any specific group. In the Tuskegee study, all of the risks were concentrated among African-American males. In other cases, the benefits might be concentrated in a particular group. One of the current ethical dilemmas facing research professionals is the conduct of research involving children. Because of the focus on protecting vulnerable populations, drug research in children has been very limited. Thus, the benefits of research were concentrated on adults. The result was that physicians had little scientific basis for prescribing medications in children. To improve the practice of pediatric medicine and to expand the benefits of research beyond adults, the federal government established incentives to pharmaceutical companies to sponsor research with children. This, in turn, led to additional regulations designed to protect the rights of minors who participate in research. This is just one example of the delicate balance of justice in selecting research populations.

17.5   REGULATIONS GOVERNING INSTITUTIONAL REVIEW BOARDS

Investigators embarking on research path might be tempted to focus their energies exclusively on those aspects of the study that involve direct patient care and medical decisions. They might be

DEPARTMENT OF HEALTH AND HUMAN SERVICES FOOD AND DRUG ADMINISTRATION **STATEMENT OF INVESTIGATOR** *(TITLE 21, CODE OF FEDERAL REGULATIONS (CFR) PART 312)* (See instructions on reverse side.)	Form Approved: OMB No. 0910-0014 Expiration Date: March 31, 2022 *See OMB Statement on Reverse.* **NOTE:** No investigator may participate in an investigation until he/she provides the sponsor with a completed, signed Statement of Investigator, Form FDA 1572 (21 CFR 312.53(c)).

1. NAME AND ADDRESS OF INVESTIGATOR

Name of Clinical Investigator

Address 1	Address 2		
City	State/Province/Region	Country	ZIP or Postal Code

2. EDUCATION, TRAINING, AND EXPERIENCE THAT QUALIFY THE INVESTIGATOR AS AN EXPERT IN THE CLINICAL INVESTIGATION OF THE DRUG FOR THE USE UNDER INVESTIGATION. ONE OF THE FOLLOWING IS PROVIDED *(Select one of the following.)*

☐ Curriculum Vitae   ☐ Other Statement of Qualifications

3. NAME AND ADDRESS OF ANY MEDICAL SCHOOL, HOSPITAL, OR OTHER RESEARCH FACILITY WHERE THE CLINICAL INVESTIGATION(S) WILL BE CONDUCTED   **CONTINUATION PAGE for Item 3**

Name of Medical School, Hospital, or Other Research Facility

Address 1	Address 2		
City	State/Province/Region	Country	ZIP or Postal Code

4. NAME AND ADDRESS OF ANY CLINICAL LABORATORY FACILITIES TO BE USED IN THE STUDY   **CONTINUATION PAGE for Item 4**

Name of Clinical Laboratory Facility

Address 1	Address 2		
City	State/Province/Region	Country	ZIP or Postal Code

5. NAME AND ADDRESS OF THE INSTITUTIONAL REVIEW BOARD (IRB) THAT IS RESPONSIBLE FOR REVIEW AND APPROVAL OF THE STUDY(IES)   **CONTINUATION PAGE for Item 5**

Name of IRB

Address 1	Address 2		
City	State/Province/Region	Country	ZIP or Postal Code

6. NAMES OF SUBINVESTIGATORS *(If not applicable, enter "None")*

CONTINUATION PAGE – for Item 6

7. NAME AND CODE NUMBER, IF ANY, OF THE PROTOCOL(S) IN THE IND FOR THE STUDY(IES) TO BE CONDUCTED BY THE INVESTIGATOR

FIGURE 17.1 Statement of Investigator: Form FDA 1572(2/16).

tempted to skip the remaining portions of this chapter that focus on the regulations governing IRBs. Why, after all, does an investigator need to know about IRBs?

The answer is that the FDA holds investigators responsible for ensuring that the studies they conduct are reviewed by an IRB that meets FDA regulations governing the function of IRBs. More specifically, when an investigator signs FDA Form 1572 (Figure 17.1), the investigator is making the following commitment (among others): "I will ensure that an IRB that complies with the requirements of 21 CFR Part 56 will be responsible for the initial and continuing review and approval of the clinical investigation. I also agree to promptly report to the IRB all changes in the research activity and all unanticipated problems involving risks to human subjects or others. Additionally, I will not make any changes in the research without IRB approval, except where necessary to eliminate apparent immediate hazards to human subjects." Therefore, it is prudent that the investigator contacts their IRB to make sure it complies with all requirements under 21 CFR Part 56 and that the PI maintains contact with the IRB staff to ensure adequate reporting and his or her own compliance with regulations.

17.5.1   What Institutional Review Board Members Should Know about Clinical Research

Although IRB members can and should come from a variety of backgrounds, all board members should know the basic elements of clinical research and should stay focused on their role in the ethical review process. Any board member who does not have a research background should take the time to learn the basics of the industry. Books, such as this one, can introduce the language, regulations, and ethical issues in the conduct of research with humans.

Board members should look carefully at an investigator's qualifications for conducting the research. If a protocol proposes to study the effects of a new treatment for hypertension, the investigator should not be a podiatrist. The investigator is responsible for overseeing the medical care and conduct of the trial and, therefore, must have the proper qualifications to accomplish that task.

Board members should understand the need to avoid conflicts of interest. Any board member with a conflicting interest in a study should not participate in the deliberations or voting on that study. The IRB may ask that member to leave the room during deliberations and voting to avoid a political environment that is not conducive to objective ethical review. The recusal and attendance requirements for IRB members with a conflict of interest mentioned above should be stated in the institution's IRB policies.

Finally, board members should understand that clinical research involves tremendous coordination of systems, staff, and supplies; however, it is *not* the responsibility of the IRB to plan, facilitate, or verify that the investigator has resolved the logistics of a given study. The logistics are something the investigator works out with the sponsor. The IRB members should stay focused on evaluating the risks and benefits of conducting the study, but sometimes, IRB members may find that the study is planned so poorly that it may affect scientific validity. In the case where scientific validity is affected, IRB members can (and should) make the argument that subjecting research participants to the study may not provide any benefit at all and thus should not be conducted. A study in which the PI is subjecting participants to risks while no benefit may be provided is not approvable by the IRB.

17.5.2   What Institutional Review Board Members Should Know about Their Responsibilities

IRB members should be aware that their duties include (but are not be limited to) evaluating proposed investigations and approving or disapproving the investigation after considering the medical soundness in light of the rights and safety of the human participants involved; determining compliance with acceptable standards of professional conduct and practice; and assessing community, ethical, and moral values. The members also evaluate the qualifications of the PI with emphasis placed on that individual's professional development as it relates to the degree of protocol complexity and the risk to human research participants. Because the primary purpose of an IRB is to ensure – in advance and by periodic review – that appropriate measures are taken to safeguard the rights, safety, and well-being of human participants involved in a clinical trial, the board members must determine that the following requirements are satisfied:

- Human participants are protected from ill-advised research or research protocols in light of both ethical and scientific concerns.
- Risks to participants are minimized.
- Risks to participants are reasonable in relation to anticipated benefits and the importance of the knowledge that may be expected to result.
- Selection of participants is equitable.
- Informed consent will be obtained from each prospective participant or the participant's legally authorized representative and will be documented in accordance with IRB, FDA, and ICH informed consent regulations and guidelines.
- Adequate provisions are made for monitoring the data collected to ensure the safety of participants.
- Adequate provisions are made to protect the privacy of participants and maintain confidentiality of data.
- Appropriate additional safeguards have been included to protect the rights and welfare of participants who are members of a particularly vulnerable group,[18] such as persons with acute or severe physical or mental illness, persons with limited capacity to consent, or persons who are economically or educationally disadvantaged.
- Participant selection and exclusion criteria – including justification of the use of special participant populations, such as children, pregnant women, human fetuses, and neonates or the mentally handicapped – are appropriately established.
- The study design includes a discussion of the appropriateness of the research methods.
- In addition to the above, the IRB has additional and special review requirements to protect the well-being of children when children are to be involved in the research (Subpart D, Sections 401–409).[20]

17.6 WHAT THE SITE SHOULD KNOW ABOUT INSTITUTIONAL REVIEW BOARDS

When choosing an IRB, the investigator may have the option of using a "local" IRB or a "central" IRB, depending upon the investigator's institutional policies. A local IRB is one that is housed within an institution and has been developed for overseeing research conducted within the institution or by the staff of the institution. For example, many community hospitals and academic medical centers have their own internal IRB, composed largely of clinicians who conduct research at that institution.

A central IRB is an "outside" or "independent" IRB. The central IRB serves to review research for non-IRB institutions, private practices, outpatient clinical trials, NIH studies where a central IRB is mandated, or institutions that allow the investigator to cede IRB review authority to the central IRB. Investigators conducting research in a non-institutional setting often choose to use an established central IRB rather than forming their own. In addition, some smaller hospitals and other institutions are choosing to eliminate their internal IRBs and outsource this function. The benefits of the decision to outsource include

- Reduced likelihood of conflict of interest or the appearance of bias because the board members reviewing the proposed protocols are unlikely to be friends, colleagues, or acquaintances of the researchers whose work they are reviewing.
- Reduced liability by selecting to outsource to an organization whose sole focus and expertise are in ensuring compliance with the many regulations governing the work of IRBs.
- Potential cost savings by eliminating staff currently dedicated to performing IRB duties by outsourcing to an IRB that achieves economies of scale through higher volumes of reviews.

However, some potential hazards of relying upon a central IRB include[24] the following:

- The central IRB may be less able to provide a complete and thorough review of the documents included in the IRB submission since local IRBs tend to have a better understanding of local study population norms, values, cultural context, and economic factors. Thus, when local IRB review is used, the consent and recruitment process may be enhanced.
- A local IRB may be better able to promote the Belmont principle of justice in that they can more readily seek input from the local community so that risks and benefits are distributed appropriately. Local IRBs tend to be more accountable and accessible.
- Local IRBs are more willing and better able to communicate with investigators at the site level, and also better able to provide information and training that may be needed.

An investigator must be qualified by education, training, and experience to assume responsibility for the proper conduct of a clinical trial and should meet all qualifications specified by the applicable regulatory requirements. Investigators must agree to abide by the decisions of their selected IRB and should comply with all governing rules and regulations.

17.6.1 What Research Requires Institutional Review Board Approval and Oversight

According to the *Belmont Report*,[17] items that must go before an IRB for review, for the protection of human participants, include any projects that include any element of research in an activity. The definition of research is described as "a systemic investigation, including research development, testing, and evaluation, designed to develop or contribute to generalizable knowledge."[20] Medical research involving human participants also includes research on identifiable human material and identifiable data.[23] The obvious material requiring submission to, and review by, an IRB would be industry-funded research, which would include biomedical, behavioral, medical devices, and humanitarian-use devices. If there is confusion about whether or not a project is research, several questions can be asked to help make this determination:

- Is there a hypothesis?
- Does it include research development?
- Will the knowledge be used outside the institution?
- Is there a question to be answered for reasons other than clinical care or routine evaluation?
- Is there an intent to generalize the information?
- Is there a specific intent?
- Is the purpose for the scientific community?
- If it were not going to be published, would you do it anyway?

Investigators should be aware that some research projects may qualify for expedited review. Research that would qualify for this type of review includes certain categories of research that involve no more-than-minimal risk. A list of research categories that qualify was published in the *Federal Register* in 1998 (Table 17.2).

17.6.2 Requirements of an Investigator for Institutional Review Board Research Approval

For an investigator to obtain IRB approval for a research proposal, required information should be provided to the prospective IRB for review and consideration in a timely manner. PIs should know what their IRBs expect and should ask questions early in the process to avoid any delays. Investigators should remember that an IRB review is based on human-participant concerns, and

TABLE 17.2
Categories of Clinical Investigations That Can Go through an Expedited Review[25]

1. Clinical studies of drugs and medical devices only when condition (a) or (b) is met.
   (a) Research on drugs for which an investigational new drug application (21 CFR Part 312) is not required. (Note: Research on marketed drugs that significantly increases the risks or decreases the acceptability of the risks associated with the use of the product is not eligible for expedited review.)
   (b) Research on medical devices for which (a) an investigational device exemption application (21 CFR Part 812) is not required; or (b) the medical device is cleared/approved for marketing, and the medical device is being used in accordance with its cleared/approved labeling.
2. Collection of blood samples by finger stick, heel stick, ear stick, or venipuncture as follows:
   (a) From healthy, nonpregnant adults who weigh at least 110 lb. For these subjects, the amounts drawn may not exceed 550 mL in an 8-week period and collection may not occur more frequently than two times per week;
   (b) From other adults and children, considering the age, weight, and health of the subjects, the collection procedure, the amount of blood to be collected, and the frequency with which it will be collected. For these subjects, the amount drawn may not exceed the lesser of 50 or 3 mL/kg in an 8-week period, and collection may not occur more frequently than two times per week.
3. Prospective collection of biological specimens for research purposes by noninvasive means. Examples: (a) hair and nail clippings in a non-disfiguring manner; (b) deciduous teeth at time of exfoliation or if routine patient care indicates a need for extraction; (c) permanent teeth if routine patient care indicates a need for extraction; (d) excreta and external secretions (including sweat); (e) un-cannulated saliva collected in either an unstimulated fashion or stimulated by chewing gum base or wax or by applying a dilute citric solution to the tongue; (f) placenta removed at delivery; (g) amniotic fluid obtained at the time of rupture of the membrane prior to or during labor; (h) supra- and subgingival dental plaque and calculus, provided the collection procedure is not more invasive than routine prophylactic scaling of the teeth and the process is accomplished in accordance with accepted prophylactic techniques; (i) mucosal and skin cells collected by buccal scraping or swab, skin swab, or mouth washings; (j) sputum collected after saline mist nebulization.
4. Collection of data through noninvasive procedures (not involving general anesthesia or sedation) routinely employed in clinical practice, excluding procedures involving X-rays or microwaves. Where medical devices are employed, they must be cleared/approved for marketing. (Studies intended to evaluate the safety and effectiveness of the medical device are not generally eligible for expedited review, including studies of cleared medical devices for new indications.) Examples: (a) physical sensors that are applied either to the surface of the body or at a distance and do not involve input of significant amounts of energy into the subject or an invasion of the subject's privacy; (b) weighing or testing sensory acuity; (c) magnetic resonance imaging; (d) electrocardiography, electroencephalography, thermography, detection of naturally occurring radioactivity, electroretinography, ultrasound, diagnostic infrared imaging, Doppler blood flow, and echocardiography; (e) moderate exercise, muscular strength testing, body composition assessment, and flexibility testing where appropriate given the age, weight, and health of the individual.
5. Research involving materials (data, documents, records, or specimens) that have been collected or will be collected solely for nonresearch purposes (such as medical treatment or diagnosis). (Note: Some research in this category may be exempt from the HHS regulations for the protection of human subjects. 45 CFR 46.101(b)(4). This listing refers only to research that is not exempt.)
6. Collection of data from voice, video, digital, or image recordings made for research purposes.
7. Research on individual or group characteristics or behavior (including, but not limited to, research on perception, cognition, motivation, identity, language, communication, cultural beliefs or practices, and social behavior) or research employing survey, interview, oral history, focus group, program evaluation, human factors evaluation, or quality assurance methodologies. (Note: Some research in this category may be exempt from the HHS regulations for the protection of human subjects. 45 CFR 46.101(b)(2) and (b)(3). This listing refers only to research that is not exempt.)
8. Continuing review of research previously approved by the convened IRB as follows:
   (a) Where (a) the research is permanently closed to the enrollment of new subjects; (b) All subjects have completed all research-related interventions; and (c) the research remains active only for long-term follow-up of subjects;
   (b) Where no subjects have been enrolled, and no additional risks have been identified;
   (c) where the remaining research activities are limited to data analysis.
9. Continuing review of research, not conducted under an investigational new drug application or investigational device exemption where categories two (2) through eight (8) do not apply, but the IRB has determined and documented at a convened meeting that the research involves no greater-than-minimal risk and no additional risks have been identified.

should expect questions. It is very helpful for the investigator to contact the IRB before making a submission to inquire about its document-submission requirements. Incomplete submissions to the IRB will definitely cause the review of the research project to be delayed. When IRBs have to request a number of modifications or seek additional information, IRB approval must be deferred, pending subsequent review by the full board upon receipt of the requested information. Providing the IRB with all required documents with the initial submission, however, will help eliminate undue delays and make the IRB review process more expedient. The pertinent information that the investigator will need to provide may vary from IRB to IRB; however, in general, the standard, required information would include the material listed in Table 17.3.

TABLE 17.3
Information that IRBs May Require from a PI

Standard Information Required	Written Information that Will Be Provided to the Participant
Brief letter, memorandum, or note requesting approval of the research	Provisions for managing adverse reactions
Completed, signed original of the IRB application or Information and Site Survey (if required by the IRB)	Justification for use of special/vulnerable participant populations, as well as additional safeguards that will be used to protect these participants (i.e., the mentally retarded, children, prisoners, pregnant women, etc.)
Investigational Device Exemption (IDE), Humanitarian Device Exemption (HDE), or Premarket Approval (PMA) numbers for device trials	A disclosure of any compensation provided to the participant for participating in the study
Copy of the protocol/amendments; (title of study; purpose, including expected benefits obtained by doing the study; participant selection criteria; participant exclusion criteria; study design, including as needed, a discussion of the appropriateness of research methods; description of methods performed)	A disclosure of extra costs to the participant because of participation in the study
Written informed consent document and consent form updates (which include all of the required elements of an informed consent – see Section 17.6)	Protection of the participant's privacy
Completed FDA 1572 form, if applicable (not required for device trials)	*Additional Items the PI Should Provide*
Investigator's brochure (if applicable)	Any changes in the study after initiation
Curriculum vitae and licensure for the PI and each subinvestigator	A report of any unexpected serious adverse reactions or information regarding similar reports received from the sponsor as soon as possible and, in no event later than 15 calendar days after the investigator discovers the information
Disclosure of any payments or compensation to a participant for participating in the study	Progress reports as requested by the IRB, but in any event, no less than on an annual basis, including the number of participants withdrawing from the study and the reasons for each withdrawal
Name of the sponsor	Any significant protocol deviation that considerably affects the safety of the participant, or the scientific quality of the study
Results of previous related research (i.e., investigator's brochure)	A final report
Participant recruitment procedures, materials, or advertisements	

Additional requirements of a PI include the following:

- The investigator should be committed to a trial before the IRB issues its written approval.
- The investigator should not deviate from, or initiate, any changes to the protocol, without prior written approval from the IRB for an appropriate amendment, except when necessary to eliminate immediate hazards to the participants or when changes involve only logistical or administrative aspects of the trial (e.g., a change of the monitor telephone number).

The investigator should promptly report the following to the IRB:

1. Deviations from, or changes in, the protocol to eliminate immediate hazards to the trial participants.
2. Changes that increase the risk to participants or affect significantly the conduct of the trial.
3. All adverse drug reactions that are both serious and unexpected.
4. New information that may adversely affect the safety of the participants or the conduct of the trial.

The investigator should be aware that the IRB has the authority to suspend or terminate approval of research that is not being conducted in accordance with IRB requirements or FDA regulations or that has been associated with unexpected serious harm to participants.

In cases in which an investigator engages in serious or continuing noncompliance, it is the responsibility of the IRB to report that activity to the sponsor, to the federal OHRP, and to the FDA. Noncompliance issues may include, but are not limited to, unreported changes in the protocol, misuse or nonuse of the informed consent document, failure to submit protocols to the IRB in a timely manner, and avoiding or ignoring the IRB.

17.7 RISKS vs. BENEFITS ANALYSIS: THE HUMAN ADVOCATE

To ensure that the rights and welfare of human research participants are protected, the IRB takes on the task of performing a risk vs. benefit analysis. Before a clinical trial can be initiated, foreseeable risks and inconveniences should be weighed against the anticipated benefit for the trial participant and for society (Section 17.2).[14] Risks associated with participation in research should be justified by the anticipated benefits; only then, can the trial be initiated. Because this requirement is clearly stated in the federal regulations, it is, therefore, one of the major responsibilities of the IRB to assess the risks and benefits associated with proposed research. Definitions of the terms that the IRB uses to assess risk include the following:

- *Benefit* – a valued or desired outcome, an advantage.
- *Minimal risk* – the probability and magnitude of harm or discomfort anticipated in the proposed research are not greater, in and of themselves, than those ordinarily encountered in daily life or during the performance of routine physical or psychological examinations or tests (45 CFR 46.102(j)).[20]
- *Risk* – the probability of harm or injury (physical, psychological, social, or economic) occurring because of participation in a research study. Both the probability and magnitude of possible harm can vary from minimal to significant.

The IRB must determine whether the anticipated benefit – either of new knowledge or of improved health for a research participant – justifies asking individuals to expose themselves to the potential risks. The IRB's assessment of risks and anticipated benefits involves the following steps:

- Identifying the risks associated with the research in comparison with the risks of treatments the participants would receive if not participating in research.
- Determining that the risks will be minimized to the extent possible.
- Identifying the probable benefits to be derived from the research.
- Determining that the risks are reasonable in relation to the benefits to the participants, if any, and the importance of the knowledge to be gained.
- Assuring that potential participants will be provided with an accurate and fair description of the risks or discomforts and the anticipated benefits.
- Determining intervals of periodic review and, where appropriate, determining that adequate provisions are in place for monitoring the data collected.

In addition, the IRB should determine that the provisions to protect the privacy of participants and to maintain the confidentiality of the data are adequate. When participants are members of a vulnerable population, the IRB should ensure that appropriate additional safeguards are being used to protect the rights and welfare of these participants.[26]

17.7.1 Identifying and Assessing the Risks

In the process of identifying what constitutes a risk, only risks that can result directly from the research should be considered, but before an activity involving participants is eliminated from the risk–benefit analysis as therapy, the IRB should be very sure that the activity is not actually research. IRB member should be aware that the potential risks faced by research participants could be associated with the design mechanisms used to ensure valid results and by the interventions that may be performed during the course of the research. Designs involving randomization to treatment groups have the risk that the participant may not receive a treatment that could be effective. Participants involved in a double-blinded study are at risk of the necessary information for individual treatment not being available to the proper individuals when needed. The added risk of invasion of privacy and violations of confidentiality are also possible within the methods used for gathering information in behavioral, social, and biomedical research.

Risks to which research participants may be exposed have been grouped as physical, psychological, legal, social, and economic harms.[27] The description of each classification is as follows:

- *Physical harms*: Medical research often involves exposure to minor pain, discomfort, injury from invasive medical procedures, or harm from possible side effects of drugs. All of these should be considered "risks" for purposes of the IRB review.
- *Psychological harms*: Participation in research may result in undesired changes in the thought processes and emotions. These changes may be temporary, recurrent, or permanent. IRB members should be aware that some research has the potential to cause serious psychological harm. Stress and feelings of guilt or embarrassment may arise from thinking or talking about one's own behavior or attitudes on sensitive topics. These feelings may occur when a participant is being interviewed or filling out a questionnaire. IRBs will confront the possibility of psychological harm when reviewing behavioral research that involves a component of deception, particularly if the deception includes false feedback to the participants about their own performance.
- *Legal: Invasion of privacy*: Access to a person's body or behavior without consent constitutes an invasion of privacy. The IRB must determine if the invasion of privacy is acceptable in light of the participant's reasonable expectations of privacy in the situation under which the study is being performed and if the research questions are of sufficient importance to justify this intrusion. The IRB should determine if the research could be modified so that the study could be conducted without the invasion of the participants' privacy.

- *Legal: Breach of confidentiality*: Confidentiality of data requires safeguarding information that has been given voluntarily by one person to another. Some research requires access to the participants' hospital, school, or employment records. Such access is generally acceptable as long as the researcher protects the confidentiality of that information. The IRB should be aware that a breach of confidentiality could result in psychological or social harm.
- *Social and economic harms*: Some invasions of privacy or breaches of confidentiality could result in embarrassment with one's business or social group, loss of employment, or criminal prosecution. Confidential safeguards must be strong in these instances. Examples of these particular sensitivities include information about alcohol or drug abuse, mental illness, illegal activities, and sexual behavior. Participation in research may also result in additional costs to the participant.

17.7.2 Minimal Risk vs. Greater-than-Minimal Risk

Once the risks have been determined, the IRB must then assess whether the research involves greater-than-minimal risk. Regulations governing the functions of an IRB allow approval through the expedited review process for research projects that contain no more-than-minimal risk and that involve participants only in one or more approved categories.

In research involving more-than-minimal risk, potential participants must be informed of the availability of medical treatment and compensation for a research-related injury, including who will pay for the treatment and the availability of other financial compensation.[28] Institutions are not required to provide care or payment for research injuries; however, some institutions provide hospitalization and necessary medical treatment in an emergency situation.

17.7.3 Vulnerable Populations and Minimal Risk

When research involves especially vulnerable populations (e.g., fetuses and pregnant women, prisoners, and children), regulations strictly limit research involving more-than-minimal risk. Special limitations are recommended when the research involves individuals who are institutionalized or mentally disabled. For these situations, it is recommended that minimal risk be defined in terms of the risks normally encountered in the daily lives or the routine medical and psychological examination of healthy participants. In these cases, the IRB should determine whether the proposed participant populations would be more sensitive or vulnerable to the risks involved by the research because of their general condition or disabilities. These concerns are also equally applicable to other participants (e.g., taking blood samples from a hemophiliac, outdoor exercises with asthmatics if the air is polluted, changes in diet for a diabetic, and giving over-the-counter drugs for minor ailments to pregnant women).

17.7.4 Determining Whether Risks Are Minimized

The IRB is responsible for assuring that risks to participants are minimized. In assuring that the risks are minimized, the IRB should obtain and review the protocol, including the investigational design, scientific rationale, and the statistical reason for the structure of the proposed research. Results from previous studies (e.g., the investigator's brochure) should also be reviewed during this process. The expected beneficial and harmful effects within the research, as well as the effects of any treatments that may be ordinarily administered, and those associated with receiving no treatment should also be analyzed. Whether potential harmful effects can be detected, prevented, or treated should be considered. Risks and complications of any underlying disease that may be present should be assessed as well.

The IRB should determine whether or not the investigators are competent in the area of the proposed research and whether they are serving in more than one role, which may complicate their interactions with participants. Potential investigator conflict-of-interest issues should be identified and resolved before IRB approval.

Deciding whether the research design will produce useful data will assist the IRB in determining whether risks are minimized. Participants may be exposed to risk without sufficient justification when the research design does not contain a sample size large enough to produce valid data or conclusions. Faulty or poor research design means that the risks are not likely to be reasonable in relation to the benefits. Sometimes, procedures that are included for purposes of good research design, but which add disproportionate risks to participants, may be unacceptable. Assuring that adequate safeguards such as data and safety monitoring are incorporated into the research design is a useful method of minimizing the risks.

17.7.5   Assessing Anticipated Benefits

Benefits of research are considered benefits to participants and to society. Research often involves the evaluation of procedures that may benefit the participants by improving their conditions or providing a better understanding of their diseases or disorders. In this type of research, participants undergo treatment for a particular illness or abnormal condition. Patients and healthy volunteers may choose to take part in research that is neither related to any illness or condition they might have, nor structured to provide any diagnostic or therapeutic benefit. This type of research is designed to gain knowledge about human behavior and physiology. Research that contains no immediate therapeutic intent may benefit society overall by providing increased knowledge, improved safety, advances in technology, and overall better health. Anticipated benefits to the participant and the expected knowledge to be gained should be clearly stated within the protocol.

17.7.6   Determining Whether Risks Are Reasonable

The IRB must consider a number of factors when determining whether the risks are reasonable in relation to the anticipated benefits. The evaluation of the risk–benefit ratio is an ethical judgment that an IRB must make, and each case must be reviewed separately. This judgment often depends on subjective determinations and community standards. When making its decision, the IRB relies on currently available information regarding the risks and benefits of the interventions from previous bench, animal, and human studies (e.g., from the investigator's brochure), and the extent of confidence in the knowledge. However, human responses may differ from those of animals; therefore, although that information may suggest possible risks and benefits to humans, it is not conclusive. Hence, first in human research is often considered riskier and the IRB should more carefully weigh the risks and benefits of such a study. Within its assessment, the IRB should also consider the proposed participants and be sensitive to the different feelings and views individuals may have about risks and benefits.

The risk–benefit assessments depend on whether the research involves the use of interventions that have the intent and reasonable possibility of providing a benefit to the participant, or whether it only involves procedures for research purposes. In research containing interventions expected to provide direct benefit to the participants, a certain amount of risk is justifiable. In research where no direct benefits are anticipated, the IRB should evaluate whether the risks presented by procedures only to obtain generalized knowledge are ethically acceptable.[26]

17.7.7   Continuing Review and Monitoring of Data

Regulations governing the functions of an IRB require that an IRB reevaluates research projects at intervals appropriate to the degree of risk but not less than annually. Note that, under the revised Common Rule effective January 21, 2019, only full board studies require renewal. However, the

IRB may determine upon review that expedited studies require renewal if there is an adequate and well-documented reason.[20] The reevaluation is performed to review the entire research project again and to reassess the risk–benefit ratio, which may have changed since the last review. During the course of a study, new information regarding the risks and benefits, unexpected side effects, unanticipated findings involving risks to participants, or knowledge resulting from another research project may become apparent. The IRB should determine whether these situations have occurred and whether there is any additional information regarding risks or benefits that should be revealed to the participants.

The investigator should be aware that the interval for IRB review can be more frequent than once per year. The approval period for a study can be shorter (i.e., less than a year) if the IRB feels that the study is of considerable risk and should be monitored by the IRB more frequently. If an investigator allows an IRB approval date to lapse, the research no longer has IRB approval. No approval means that all aspects of the research must stop. Extensions of approvals do not exist. Investigators should provide the IRB with an adequate detailed report in a timely manner to avoid expiration (and thus stoppage) of their study. Any monitoring reports (e.g., from a formally constituted data safety monitoring board) should be submitted with or before the annual renewal.

17.8 INFORMED CONSENT: NOT JUST A DOCUMENT

One of the most important pieces of a research trial is the informed consent of the research participant. However, a signature on an informed consent document does not constitute the end of the informed consent. It is very important for the research community (the sponsor, the investigator, and the IRB) to remember that informed consent is not just a document, but is, instead, a continuing process that carries through to the end of the individual's participation in the trial.

Informed consent involves the information presented to the participant, comprehension of that information by the participant, and the participant's voluntary and fully, continuously informed agreement to participate in the study. The informed consent process begins with the recruitment of potential participants, whether by radio advertisement, flyer, or initial contact with the PI or the research staff. The initial consent involves explaining to a potential participant the risks, benefits, alternatives, procedures, and purpose of the study; allowing that individual to ask questions and providing satisfactory answers to those questions; sharing new information with the participants as it becomes available during the course of the study because it may change their willingness to continue participation; and for long-term studies, revisiting the consent because of possible capacity changes.

The written informed consent form should be presented to potential participants in a language that they understand and written in terms that they can comprehend. This is one of the important functions of an IRB. The IRB members review every informed consent form to determine if it contains all of the required elements and any additional required elements of an informed consent form as set forth in the governing regulations. The informed consent form is also reviewed to determine that complete, accurate, and pertinent study-related information is being provided to the potential participants and that medical terms are clearly defined, in simple language that the study population can understand. A recommended approach is to – after presenting the consent form and study information – ask the participant to recall key facets of the study to assess study procedure comprehension.

Please remember that the consent process is not simply completed at the initial recruitment visit. The consent process continues throughout the participant's study enrollment. Thus, for example, if during the study, there is a newly discovered risk associated with the study drug, all participants who may have received the study drug should be notified of the risk in an appropriate manner. The consent process requires continuous and fully informed consent throughout the study.

Human-participant protection is a shared responsibility between the sponsor, the investigator, and the IRB. It is their responsibility as a team to ensure that the participants remain well informed and that their rights and welfare are protected. It is important that all members of the research team understand the informed consent form regulations,[29] their part of the informed consent process, and apply this knowledge to each informed consent form that is reviewed, presented, and distributed to a potential participant.

17.8.1 Elements of an Informed Consent

The investigator is responsible for providing every potential research participant with complete, accurate, and pertinent study-related information while adhering to the applicable governing rules and regulations. The process of providing this information to potential participants is informed consent. Except as provided in the regulations,[30] no investigator may involve a human being as subject in research unless the investigator has obtained the legally effective informed consent for the participant or the participant's legally authorized representative.[31] The exceptions include the participant being in a life-threatening situation that makes it necessary to use the test article; being unable to communicate with the participant to obtain a legally effective consent; insufficient time to obtain the consent from the participant's legally authorized representative; and no available alternative method of approved or generally recognized therapy that provides an equal or greater chance of saving the participants life.[32]

The governing rules and regulations require certain elements (information) be included within every informed consent form that is provided to a potential research participant. These required elements are outlined in Table 17.4.

17.8.2 Additional Required Elements

When appropriate, one or more of the following elements of information should also be provided to each participant. Note that additional required elements are not optional but are required when applicable:

- A statement that the particular treatment or procedure may involve risk to the participant (or to the embryo or fetus if the participant is, or may become, pregnant and to nursing infants) that are currently unforeseeable.
- Anticipated circumstances under which the participant's participation may be terminated by the investigator without regard to the participant's or the legally authorized representative's consent.
- Any additional costs to the participant that may result from participation in the research.
- The consequences of a participant's decision to withdraw from the research and procedures for orderly termination of participation by the participant.
- A statement that significant new findings developed during the course of the research, that may relate to the participant's willingness to continue participation, will be provided to the participant.
- The approximate number of subjects involved in the study.
- A statement that the subject's biospecimens (even if identifiers are removed) may be used for commercial profit and whether the subject will or will not share in this commercial profit.
- A statement regarding whether clinically relevant research results, including individual research results, will be disclosed to subjects, and if so, under what conditions; and
- For research involving biospecimens, whether the research will (if known) or might include whole-genome sequencing (i.e., sequencing of a human germline or somatic specimen with the intent to generate the genome or exome sequence of that specimen).[20]

In addition to the required elements of an informed consent form, an IRB can and may require other standard information or signatures be added to all informed consent forms being reviewed by the board.

TABLE 17.4
Elements Required in an Informed Consent

- A statement that the study involves research
- An explanation of the purpose of the research and the expected duration of the participant's participation
- A description of procedures to be followed and identification of any procedures that are experimental, including all invasive procedures
- The approximate number of participants involved in the trial and the participant's responsibilities
- A description of any reasonably foreseeable risks or discomforts to the participant
- A description of any benefits to the participant or to others that may reasonably be expected from the research; when there is no intended clinical benefit to the participant, the participant should be made aware of this fact
- A disclosure of appropriate alternative procedures or courses of treatment that might be advantageous to the participant and their important potential benefits and risks
- A statement describing the extent of confidentiality of records, which notes the possibility that the monitors, auditors, IRB, and the regulatory authorities may inspect the records for verification of clinical trial procedures and data without violating the confidentiality of the participant to the extent permitted by the applicable laws and regulations and that, by signing a written informed consent form, the participant or the participant's legally acceptable representative is authorizing such access; that records identifying the participant will be kept confidential and, to the extent permitted by the applicable laws or regulations, will be made publicly available; and that if the results of the trial are published, the participant's identity will remain confidential
- An explanation of whether compensation or medical treatment is available if injury occurs and, if so, what they consist of, or where further information may be obtained
- The anticipated prorated payment, if any, to the participant for participating in the trial
- A statement of dosage/frequency and the probability for random assignment to each treatment
- An explanation of whom to contact for answers to pertinent questions about the research and research participants' rights and whom to contact in the event of a research-related injury to the participant
- A statement that participation is voluntary, that refusal to participate will involve no penalty or loss of benefits to which the participant is otherwise entitled, and that the participant may discontinue participation at any time without penalty or loss of benefits to which the participant is otherwise entitled
- One of the following statements about any research that involves the collection of identifiable private information or identifiable biospecimens:
  - i. A statement that identifiers might be removed from the identifiable private information or identifiable biospecimens and that, after such removal, the information or biospecimens could be used for future research studies or distributed to another investigator for future research studies without additional informed consent from the subject or the legally authorized representative, if this might be a possibility;
  - ii. A statement that the subject's information or biospecimens collected as part of the research, even if identifiers are removed, will not be used or distributed for future research studies.
- A place for signatures by the participant, physician, person obtaining consent, and witness (less signatures may be required); it depends on the policies and procedures of the particular IRB

17.9 RESEARCH PARTICIPANTS

Research participants are what drive the entire clinical trial process. Without a sufficient number of volunteers, statistically significant conclusions about new drugs and devices would not be possible. Given that research volunteers have a wide variety of medical knowledge, there are mechanisms in place to ensure subjects are able to make educated decisions regarding study participation. The informed consent form alleviates the need for specialized knowledge on the part of the volunteer. In lay terms, the informed consent form describes, in detail, potential risks and benefits.

Regulations also exist to protect the confidentiality of the research participant. All information collected throughout the clinical trial remains with the study staff. For the purposes of data capture, each subject is identified by initials or study number only. In addition, the informed consent discusses who will have access to the trial documents.

Research participants are the true pioneers of medicine. Through their participation, novel therapeutic cures and treatments have been made possible. Furthermore, their participation also protects the public from approval of drugs that have a poor benefit-to-risk relationship. Thus, data obtained via research volunteers may be used to provide medical advances or to protect against insidious drugs entering the marketplace.

17.10   INDUSTRY TRADE ORGANIZATIONS AND SUPPORT SERVICES

The pharmaceutical industry must concede to a high failure rate. To minimize failures owing to poor performance on the part of any niche service providers, the need for structured education, training, and communication is clear. The largest trade organizations devoted to the clinical trial industry include the Associates of Clinical Research Professionals (ACRP, www.arcpnet.org), the Drug Information Associates (DIA, www.diahome.org), and the Pharmaceutical Research and Manufacturers Association (PhRMA, www.phrma.org). In addition, the FDA offers many training sessions and hosts a website for consumers as well as researchers. *CenterWatch* is the leading publication (and a leading website, www.centerwatch.com) for clinical trial information and industry news. Publications, such as the *Good Clinical Practice Handbook*, the *Code of Federal Regulations*, and the *ICH Guidelines* outline industry-specific standards and regulations. These publications are also available at no cost online.

17.11   SUMMARY: ETHICAL DILEMMAS IN CLINICAL RESEARCH

The rapid growth in the clinical research industry is met by a similar increase in knowledge, medical innovations, new technologies, and scientific breakthroughs. Yet despite these goods, many ethical dilemmas have also been raised. Such dilemmas as ethical violations, conflicts of interest, coercion, and misrepresentation date back to the 1930s. Ethical breaches include the study of untreated syphilis (commonly referred to as the Tuskegee Syphilis Study) conducted by the United States Public Health Services from 1932 to 1972, in which patients were not told they had syphilis, were not offered effective treatment, and were not allowed to be drafted because then they would receive treatment in the military. Clearly, these human subjects were deceived, coerced, and treated unjustly. In Nuremberg, Germany, 23 doctors were charged with crimes against humanity for performing incredibly egregious medical experiments on concentration camp inmates and others without their consent. In the 1950s' Willowbrook incident, mentally retarded children were deliberately infected with the hepatitis virus. The parents were coerced when they were told that if their child enrolled in the study, the child could occupy one of the very few beds in the hospital. In 1951, a physician obtained a sample of Henrietta Lacks' cervical cancer without her knowledge or consent and developed the HeLa cell line, creating decades of controversy. In the 1960s, live cancer cells were injected into 22 senile patients at the Jewish Chronic Disease Hospital. These patients were not told and were incapable of understanding the experiment. Between 1986 and 1990, there were as many as 3,000 women with high-risk pregnancies involved in experiments by the University of South Florida and Tampa General Hospital without their consent, a case that was settled for $3.8 million.[33] Other cases in which ethical or regulatory lapses have been sited include the 1999 death of an 18-year-old man participating in a gene therapy trial at the University of Pennsylvania and the death of a healthy 24-year-old woman participating in a clinical trial at Johns Hopkins using hexamethonium, a drug not approved by the FDA, to induce asthmalike symptoms in healthy volunteers.

Within these examples are unethical acts, coercion, conflicts of interest, and misrepresentation. Conflicts of interest are not limited to the rewards and stock options of investigators; they involve the entire research endeavor. Recently, financial conflicts of interest in medical research along with widely publicized episodes of scientific misconduct have been brought to the public's attention. In some episodes, researchers have been accused of falsifying or fabricating research data on therapeutic products in which they had substantial financial interests. There have now been many steps taken to keep such actions from occurring.

In April 2000, the American Society of Gene Therapy issued new guidelines controlling conflicts of interest in research. Among the reasons for the guidelines was the discovery that a researcher of a gene therapy trial, which involved the death of an 18-year-old subject, was heavily invested in the company that was funding the research. The American Society of Gene Therapy issued a statement making it clear that financial conflicts are unacceptable. It stated, "All investigators and team members directly involved with patient selection, the informed consent process, or clinical management in a trial must not have equity, stock options, or comparable arrangements in the companies sponsoring the trials."[34]

Because of several highly publicized deaths of patients involved in experimental studies at universities, the impetus for Congress and the federal bureaucracy to proceed with human research rules has become overwhelming. In 2011, HHS issued an advanced notice of proposed rulemaking (ANPRM) affecting subpart A of 45 CFR 46, known as the Common Rule. Commentary on the ANPRM was completed in 2015, and a noted of proposed rulemaking (NPRM) was issued on September 8, 2015. The NPRM to the Common Rule was revised and is now scheduled for implementation on January 21, 2019. The driving purpose of the Common Rule revisions is to enhance protection of human subjects while also decreasing the administrative burden on investigators and IRB staff. The major burden-reducing provisions include[35]

- The revised definition of "research," which explicitly deems particular activities not to be research.
- The allowance for no annual continuing review for certain categories of research.
- The elimination of the requirement that IRBs review grant applications or other funding proposals related to the research.

CROs are increasingly conducting drug company research privately. The number of private practice-based investigators has grown almost fourfold within a 5-year span. This change is a response to the pressures of being the first to get a new drug to market. So, managing conflicts of interest outside of academic settings increasingly take place on a national level, as opposed to following guidelines set by an academic center. On the national level, efforts are being made to control this issue. Sponsors now have their investigators as well as any other individuals involved in their trials complete financial disclosures. They are also requiring statements about which IRB members abstained from voting (for a conflicting interest) on the approval letter from the IRB.

Publicity surrounding these activities is often inflamed rather than informative, showing clinical research in a poor light instead of offering the public the big picture of clinical research, the realities of the present, and the possibilities of the future. Thousands of investigators and their staffs who participate in clinical trials are dedicated to advancing science and developing new therapeutic treatments, and they use the highest professional standards. Reporting the positive side of clinical research along with the negative publicity about events such as those described, the health care consumer would have more complete information to make an educated choice about participating in clinical trials. The clinical research community and federal government have worked hard to create and enforce guidelines and standards of practice for the protection of human subjects. ACRP offers training programs and certifications for CRCs and CRAs and is initiating certification programs for investigators.

The research community is charged with effectively protecting human subjects and ensuring that research is conducted ethically. Ethical violations in research are often caused by lack of awareness rather than malice. However, without clinical research, medical innovations and scientific breakthroughs would not be possible. So, practicing ethical conduct, seeking continued training, and complying with the governing regulations will promote good, sound, ethical research, which will, in turn, benefit society.

We have the many years of research and the hard work of the researchers to thank for providing the treatments that we now have available for the many diseases that plague our society. Keeping the trust, interest, and confidence of the public will allow for continued, successful clinical research with new scientific developments and breakthroughs.

REFERENCES

1. Orzechowski, W. and Walker, R., Dose of reality: How drug price controls would hurt Americans, Policy Paper No. 25, February 7, 2000.
2. DiMasi, J.A., Cost of developing a new drug. Briefing: Tufts Center for the Study of Drug Development, Boston, MA, November 18, 2014. Available at: http://csdd.tufts.edu/news/complete_story/cost_study_press_event_webcast. Accessed October 17, 2018.
3. Pammoli, F., Magazzini, L., and Riccaboni, M., The productivity crisis in pharmaceutical R&D. *Nature Reviews Drug Discovery*, 2011, 10(6) 429–438.
4. McClung, T., PhRMA member companies R&D investments hit record high in 2017- $71.4 billion; August 9, 2018. Available at: https://catalyst.phrma.org/phrma-member-companies-rd-investments-hit-record-high-in-2017-71.4-billion-0. Accessed October 17, 2018.
5. Research!America, U.S. Investments in Medical and Health Research and Development, 2013–2016, Arlington, VA, Fall 2017. www.researchamerica.org/sites/default/files/RA-2017_InvestmentReport.pdf.
6. Medical Research Lawsuit Settled, *Tampa Tribune*, March 11, 2000.
7. ASGT, Policy of the American Society of Gene Therapy Financial Conflict of Interest in Clinical Research, American Society of Gene Therapy, Milwaukee, WI, Adopted April 5, 2000.
8. Office of Public Affairs, Protecting consumers, promoting public health, in An FDA Overview: Protecting Consumers, Protecting Public Health, FDA, Rockville, MD, 2004.
9. Prescription Drug User Fee Act of 1992: PL102-571, 106 Stat 4991, primarily codified in 21 USC §379 (h), October 29, 1992.
10. Human Research Subject Protections Act of 2002, 107th Cong., 2nd session, H.R. 4697.
11. Tislow, J., Nesbitt, L., and Belcher, A., *Drug Injury. Liability, Analysis and Prevention*, Lawyers and Judges Publishing Company, Tucson, AZ, 2001, p. 70.
12. FDA, FY 2017 Performance Report to Congress for the Prescription Drug User Fee Act. September 20, 2017. Accessible at www.fda.gov/downloads/AboutFDA/ReportsManualsForms/Reports/UserFeeReports/PerformanceReports/UCM606719.pdf.
13. Downing, N.S., Shah, N.D., Aminawung, J.A., et al., Postmarket safety events among novel therapeutics approved by the US Food and Drug Administration between 2001 and 2010. *Journal of the American Medical Association*, May 9, 2017, 317(18) 1854–1863. Accessible at https://jamanetwork.com/journals/jama/fullarticle/2625319.
14. FDA, Food and drugs: additional safeguards for children, Code of Federal Regulations, Title 21, Part 50, Subpart D, Rev. April 2003, p. 293.
15. Cauchom, D., Americans pay more, here's why, USA Today, November 10, 1999, A1.
16. ACRP, Where we are and where we are going, 2000 White Paper, Association of Clinical Research Professionals, Alexandria, VA, 2000, www.acrpnet.org/whitepaper2/html/ii_contract_research_organizations.html.
17. *International Conference on Harmonisation of Technical Requirements for Registration of Pharmaceuticals for Human Use, Guidance for Industry: E6 – Good Clinical Practice; Consolidated Guidance*, Geneva, March 2018, Available at www.fda.gov/downloads/Drugs/Guidances/UCM464506.pdf.
18. ACRP, The CRC–CRA Relationship, 2000 White Paper, Association of Clinical Research Professionals, Alexandria, VA, 2000, www.acrpnet.org/whitepaper2/html/viii._crc_cra_relatioship.html.
19. Kefauver–Harris Amendment of 1962, PL 87-78, 76 Stat 780, codified in section 21 U.S.C. §502(n).
20. The National Commission for the Protection of Human Subjects of Biomedical and Behavioral Research, The Belmont Report: Ethical Principles and Guidelines for the Protection of Human Subjects of Research, Department of Health, Education, and Welfare, Washington, DC, April 18, 1979.
21. The Commonwealth Fund, National Trends in Per Capita Pharmaceutical Spending, 1980–2015. October 2017. Accessible at www.commonwealthfund.org/sites/default/files/documents/___media_files_publications_issue_brief_2017_oct_pdf_sarnak_paying_for_rx_exhibits.pdf.

22. FDA, Food and drugs: Institutional Review Boards, Code of Federal Regulations, Title 21, Part 56, Food and Drug Administration, Rockville, MD, Rev. April 2003.
23. HHS, Federal Policy for the Protection of Human Subjects, Federal Register, 56(117), Department of Health and Human Services, Washington, DC, 18, 1991.
24. ISR Reports, 2017 Edition of the CRO Market Size Projections: 2016–2021. Accessible at www.isrreports.com/reports/2017-edition-of-the-cro-market-size-projections-2016-2021/.
25. FDA, Protection of human subjects: General requirements for informed consent, Code of Federal Regulations, Title 21, Part 50, Subpart b 50.20, Rev. April 2004, p. 310.
26. Trials of War Criminals before the Nuremberg Military Tribunals under Control Council Law No. 10, Vol. 2, U.S. Government Printing Office, Washington, DC, 1949, pp. 181–182.
27. World Medical Organization, Declaration of Helsinki: Ethical Principles for Medical Research Involving Human Subjects, Adopted by the 18th WMA General Assembly, Helsinki, June 1964.
28. Hennig, M., Hundt, F. Busta, S., et al., Current practice and perspectives in CRO oversight based on a survey performed among members of the German Association of Research-Based Pharmaceutical Companies (vfa), German Medical Science, January 26, 2017. Accessible at www.ncbi.nlm.nih.gov/pmc/articles/PMC5278541/.
29. Amdur, R. and Bankert, E.A., *Institutional Review Board: Member Handbook*, 3rd ed., Jones and Bartlett Publishers, Boston, MA, 2011.
30. Levine, R., *Ethics and Regulation of Clinical Research*, 2nd ed., Urban and Schwarzenberg, Baltimore, MD, 1988, p. 42.
31. FDA, Protection of human subjects: Elements of informed consent, Code of Federal Regulations, Title 21, Part 50, Subpart b 50.25 (a) (6), Rev. April 2004, p. 314.
32. FDA, Institutional Review Boards: IRB functions and operations, Code of Federal Regulations, Title 21, Part 56, Section 108, Food and Drug Administration, Rockville, MD, Rev. April 2004, pp. 325–326.
33. FDA, Protection of human subjects: Exception from informed consent requirements for emergency research, Code of Federal Regulations, Title 21, Part 50, Subpart b 50.24, Rev. April 2004, pp. 313–314.
34. Kornetsky, S.A., IRB: 101, Physician Responsibility in Medicine and Research (PRIM&R), July 2002.
35. HHS and FDA, Impact of certain provisions of the revised common rule on FDA-regulated clinical investigations: Guidance for sponsors, Investigators, and Institutional Review Boards. October 2018. Accessible at www.fda.gov/downloads/RegulatoryInformation/Guidances/UCM623211.pdf?utm_campaign=Impact%20Guidance&utm_medium=email&utm_source=Eloqua&elqTrackId=B10F41E07EB29CFEE1F2ED0B62E45601&elq=ea9d7d8ca2c646ecb92fcb56108f3048&elqaid=5431&elqat=1&elqCampaignId=4386.

18 Clinical Trials Methodology

John Somberg
Rush University Medical Center

CONTENTS

18.1 OVERVIEW

The high risk of product attrition, costly projects, complex regulatory procedures, and lengthy timelines give clinical drug development its unique profile among other industrial and scientific development programs.[1,2] The grim statistics, that only one in five drugs entering clinical development will eventually yield a drug product, safe and effective enough to earn marketing approval, underscores a call for more understanding and improvement of clinical drug development,[3] especially given the disproportionate escalation of cost and time involved in this critical process.[4]

The ultimate target of clinical drug development is the delivery of safe and effective medicines for marketing approval by regulatory authorities. Clinical development is by definition conducted in humans, whether patients or healthy volunteers. The guiding ethics are based on the potential benefit that must outweigh the risks involved. These ethics underpin the Good Clinical Practice guidelines that mandate research subject and data protection.

For drugs to enter the first phase of clinical development, they have to be accepted by the Food and Drug Administration (FDA) under an IND (investigational new drug) application. The non-clinical data submitted with that application assure the FDA, and the clinical scientists involved in the subsequent phases of development, that the product in question is safe and effective enough to be tested in humans. One must remember that *in vitro* and animal model-derived information is limited in its potential extrapolation to humans. The purpose of clinical development is to explore and confirm the non-clinically acquired knowledge and expand upon it to the human condition.

Clinical trials can be divided into four stages. Phase I trials are often termed first-in-man. They aim to determine the pharmacokinetics of the drug. Phase I trials may also aim to determine activity of the drug: does it possess a pharmacologic property that has been observed in preclinical studies. Phase I studies are usually conducted in normal volunteers. However, with very toxic drugs, studies in patients, at times terminal patients, may be undertaken for safety reasons. Phase I studies often involve <50 subjects (Table 18.1).

Phase II studies constitute the next stage in drug development. These studies are controlled randomized trials with the aim of determining efficacy, usually employing biomarkers. These studies are also undertaken to determine the effective dose for larger definitive studies. Toxicity that occurs with some frequency may also be seen in these Phase II studies, but the sample size is often too small to determine drug toxicities. Phase II studies usually involve several hundred patients, and the studies are usually performed in stable patients.

Phase III studies aim at establishing efficacy, often long-term efficacy, as well as being of a large enough sample to determine safety, at least a preliminary gauge of safety, before the broad exposure of a drug during the initial marketing period. These studies usually involve several thousand patients. Administration of the study drug is often for 6–12 months, in order to expose an adequate number of patients to the drug for a longer duration than in Phase II studies, and long enough to identify low-frequency toxicity. The Phase III studies are usually parallel in design and constitute the trials that are deemed pivotal in a new drug application (NDA) to FDA. Often, natural history endpoints are employed.

Phase IV trials are initiated after an NDA has been submitted to the US FDA. Phase IV trials may be a comparison with another agent that is frequently employed to treat the disease being studied. Phase IV trials may be for marketing purposes, to provide experience to physicians deemed "thought leaders"; or these studies may be designated to explore higher or lower doses than studied previously. Additional studies may be undertaken to broaden "labeling" that was requested in the NDA or to study possible drug interactions, suggested from the results of Phase III trials.

Phase IV studies may be open-label or randomized controlled clinical trials, involving hundreds to thousands of patients. Special populations are often evaluated in the Phase IV trial process, including pediatric patients and geriatric patients. At times, sicker patients with hepatic and renal dysfunction may be evaluated as to efficacy, as well as toxicity of the agent, or at times to determine appropriate dosing. This is not to confuse pharmacokinetic evaluation in special populations of renally or hepatically impaired patients, which is often done under the Phase I designation, but late in the sequence of pre-NDA studies.

TABLE 18.1

The Clinical Trial Process

*Clinical trials for the development of drugs can be divided into different phases.

Phase I – Focuses on first-in-man trials, trials emphasizing kinetics and initial safety.

Phase II – Initial efficacy determinations while observing safety signals.

Phase III – Definitive evaluation of efficacy and a greater and longer exposure for safety.

Phase IV – Post-marketing trials to increase exposure, provide physician experience with use of a new agent, as well as study of special populations, and obtain comparative effectiveness information.

*The phases of clinical trials while often sequential may not always be rigid. Kinetic studies in patients with hepatic and renal impairment, for example, may be undertaken after Phase II trials are completed or after Phase III trials are initiated.

18.2   PHASE I CLINICAL TRIALS

18.2.1   Overview

Though Phase I studies are usually carried out initially in clinical development as *first-in-man protocols (FIM)*, the term describes the study methodology, i.e., clinical pharmacology, more than it is a chronology of where the study is performed. In fact "late" Phase I studies can be conducted concomitantly with late Phase II and at times Phase III studies.

For new chemical entities (NCEs) to be considered for human development, it is necessary to characterize them non-clinically as much as possible in terms of pharmacokinetic (PK) properties, pharmacodynamic activity, and toxicology profiles. Only compounds which have favorable pharmacodynamic (PD) toxicity ratios, with measurable, reversible, and non-serious toxicities are chosen for administration to man. It is required to obtain FDA and Institutional Review Boards (IRBs) approval for the conduct of first-in-man protocols (unless the studies are part of a bioavailability trial for generic drug development, i.e., a drug already FDA approved).

18.2.2   First-in-Man Trials

The objectives of these early studies are the identification of side effects, including their intensity, duration, and degree of reversibility, as a function of dose and plasma concentration relationships following the administration of single and multiple doses of an NCE, without compromising the participants' safety, whether patients or volunteers.[5]

While the identification of a safe dose or dose range for further studies is the primary purpose, these trials also provide PK data that would help in the design of clinical trials as well. It is also possible to have an initial assessment of the test compound's action and efficacy. Due to the limitations of preclinical *in vitro* and animal studies, these early trials can reveal very unexpected and serious adverse reactions. For this reason, it is especially important for investigators to be cautious, taking adequate safety precautions during early Phase I studies, especially with new novel compounds and biologics. Drug administration needs to be adequately spaced, allowing for toxicity to be evaluated before the subsequent patient, or volunteer, is dosed.

18.2.3   Starting Dose Estimation

The estimation of the starting dose for FIM studies has been largely an empirical exercise that depends on the preference of the clinicians involved and their practice environment. However, over the past decade, standardization of initial dose estimation appears to be gaining momentum, especially influencing antineoplastic therapeutic development. As the scaling from animal models to human poses difficulties for safe extrapolation of dose ranges from animal species to humans, it is important to find a starting dose that is low enough to be safe in humans, but not so conservative that excessive, costly, and time-consuming dose escalations are needed.

The estimation of the starting dose varies between cytotoxic compounds intended for cancer treatment, and drugs for other therapeutic indications, reflecting acceptance of higher incidence of adverse effects of drugs being developed for life-threatening illnesses.[6]

In terms of cytotoxics, two concepts from animal toxicology are particularly relevant. The first is *Toxic Dose Low (TDL),* which is defined as the lowest dose that produces drug-induced pathological alterations in hematological, chemical, or morphological parameters and which, when doubled, produces no lethality. The second concept is *Lethal Dose 10 (LD 10),* which is defined as the dose that is lethal to 10% of non–tumor-bearing mice. The initial starting dose for these compounds is calculated as the lower of one-third of TDL or one-tenth of LD 10.

The methods for estimating the starting doses for non-cytotoxic drugs are more varied, but generally based on the *No Observed Adverse Effect Level (NOAEL),* estimated from preclinical studies. The NOAEL is the highest dose at which no statistically significant and/or biologically relevant adverse effect is observed in the most sensitive animal species. The NOAEL can be extrapolated for humans using the following different approaches. In the dose by factor approach, NOAEL is empirically multiplied by one or more safety factors to estimate a safe human starting dose. This approach is limited by its simplicity, and it tends to ignore the PK/PD differences between animals and humans. It is also more conservative and thus may require larger studies with additional dose escalations.

The same drug approach uses another similar drug with a known safety profile as a reference for the drug in development. This similar drug is usually of the same chemical class, with similar or related chemical structure. The method assumes that the ratio of the optimal starting dose of the similar drug to its NOAEL would be equal to the ratio of the same parameters of the drug under development. The dose estimate obtained that way is then multiplied by an arbitrary safety factor. The assumption underlying this method assumes PK/PD differences between animal and man for both compounds are the same. This assumption should be validated by calculating the abovementioned ratio for another similar compound of the same chemical class to verify that the ratio remains reasonably constant. If there is a significant discrepancy, this method for dose selection should be abandoned.

The pharmacologically guided approach uses systemic exposure represented by the area under plasma concentration/time curve (AUC), instead of the dose, for extrapolation from animal to man, and estimates the starting dose by multiplying the AUC at NOAEL by clearance of the drug in humans as predicted by allometric scaling from animals. This approach has been gaining ground recently, but despite its utility, it is important to understand its shortcomings. This approach assumes that the concentration–effect relationship of the drug is the same in animals and man. If there is a known difference in this relationship, it should be accounted for in the calculation of the starting dose, perhaps by multiplying by a safety factor to reflect interspecies pharmacodynamic difference(s). Closely related is the inherent limitation of allometric scaling in predicting relevant human parameters. This approach can also be in error by ignoring the protein binding characteristics of the test compound. It is recommended to use the unbound fraction for the plasma drug concentration calculation to better reflect its PD effect.

The presence of non-linear PK and/or active metabolites can further undermine the utility of this method. The comparative dose estimation approach uses at least two methods to calculate the starting dose: comparing the results and interpreting the differences to come up with a safe starting dose. Generally, the initial dose for FIM is usually intended to be pharmacologically inactive, with the second dose having some activity. At times, to avoid starting with an active dose, the estimated initial dose may need to be lowered further.

18.2.4 STUDY PARTICIPANTS

Traditionally, early Phase I clinical trials are conducted in *healthy male volunteers*. The rationale behind using healthy subjects is their perceived ability to tolerate adverse reactions better than *patients*, owing to their better health. The ease of their recruitment and absence of confounding concomitant illnesses and drugs make healthy volunteers particularly appropriate. While the use of healthy subjects is the usual approach, their use has been questioned.[7] Using patients, in whom drug response can be demonstrated, could prevent the discontinuation of drug development due to toxicities observed. Additionally, the use of patients would shield healthy volunteers from risks for which they receive no benefit, except financial compensation. The use of patients in FIM trials is especially desirable if the pharmacodynamics and toxicity profiles pose significant clinical risk. If the pharmacodynamic parameters are not measurable in healthy volunteers, no toxicity parameters are available to follow in normal subjects, predicted toxicity can cause irreversible harm, or historical data may imply a different tolerance between patients and normal subjects, then patients should probably be used in early clinical trials.[8] Based on one or more of these factors, testing for drugs intended for use in HIV or cancer treatment is usually conducted in patients.

18.2.5 STUDY DESIGN

Generally, FIM studies are dose escalation protocols that aim to define *the maximum tolerated dose (MTD)* in the study participants. This MTD can be defined as the dose that provides the greatest potential for beneficial effects to the patient population in the absence of intolerable adverse effects. With dose escalation, the increasing pharmacologic effect of the drug is associated with increased incidence of adverse events until the *minimum dose (MID)* is reached. The MID is the dose at which greater than 50% of participants suffer limiting adverse events, or a medically unacceptable adverse event occurs in any one participant. The dose below this is defined as the MTD and can be thought of as the maximum dose having an adverse event profile in the population that is acceptable, based on indication-specific prospective criteria.[9] Clearly, this definition allows for dose levels to be accepted with higher probability of toxicities in Phase I cancer trials, conducted in cancer patients who failed all available treatments, compared to other therapeutic areas with available safe therapies.

The dose escalation scheme can be *arithmetic* by adding an equal amount of drug for each escalation (X, 2X, 3X, 4X, 5X), *geometric* by multiplying each dose by the same factor (X, 2X, 4X, 8X, 16X), or according to specific formula such as the *modified Fibonacci* scheme (X, 2X, 3.5X, 5X, 7X). This last scheme especially allows for fast escalation initially followed by more conservative increments as subjects get closer to MTD. This characteristic has identified the modified Fibonacci scheme as especially useful for cancer trials. However, in this particular therapeutic area two other escalation schemes have been gaining favor. *Pharmacologically guided dose escalation (PGDE)* is an adaptive dose escalation design that has been useful in oncology trials owing to its potential of cutting the timeline of the relatively long initial clinical development phases of antineoplastics.[10] This approach is based on the PK/PD hypothesis: when comparing animal and human doses, expect equal toxicity for equal drug exposure. Drug exposure is estimated by plasma concentration/time AUC, aiming for AUC in humans that mirror AUC for LD 10 in mice. One starts initially with one-tenth of LD 10 as mentioned previously, followed by continual evaluation of plasma concentration as the trial is going on, allowing for more rapid escalation till 40% of the AUC corresponding to LD 10 is reached, followed by slower escalation. The *continual reassessment method* by O'Quigley is another adaptive design that aims to identify probabilities of toxicities associated with various doses to ultimately select one with an acceptable probability.[11] These methods are gradually replacing the "3+3" design, which have been long associated with early cancer trials.

Phase I clinical trials' participants are usually organized into cohorts, each with a predefined number of subjects. The study designs follow two paradigms, parallel, and crossover.[12] In *parallel designs*, each study participant receives the same dose level, and this design can be a single-dose design in which each cohort of participants receives single dose of the same strength, or multiple-dose design in which each participant in one cohort receives multiple doses of the same strength. Multiple-dose escalation usually starts after safety has been demonstrated with single-dose designs, and a combined design allows for the administration in the same trial of multiple doses after single doses have been shown to be safe and tolerated. In *crossover designs*, participants receive multiple doses of different strengths, in either a grouped form, in which all participants in the same cohort get the same dose escalation scheme and a placebo is randomized into the dose process. In the alternate crossover design, dose escalation is alternated between cohorts and a placebo is randomized.

The number of doses evaluated in majority of FIM studies is usually <7, with a median of 5.[12] The highest dose in a single escalation study is determined by a number of factors including preclinical toxicology and pharmacology safety data, the number of capsules or tablets required to administer the dose, and the cost of goods. The rationale in deliberately, but safely, exceeding the anticipated human therapeutic dose is based on the desire to define dose-limiting side effects and margin of safety in a carefully controlled clinical environment. In addition, it allows for higher doses to be safely studied in Phase II efficacy trials should the projection of anticipated therapeutic dose in man based on preclinical information be underestimated.[5] Multiple-dose trials are usually based on the initial data obtained from single-dose designs, with the decision to proceed is based upon PK data, toxicity observed, and the possibility of a safe dose that can deliver the expected therapeutic effect. Dosing in multiple-dose studies should be continued till steady-state plasma concentrations of the drug have been achieved.[8] The treatment duration is usually 7–30 days. During these studies, at least one dose expected to produce mild, tolerable symptoms is studied to define dose-limiting side effects and margin of safety.

As adverse events are common, Phase I studies are usually, and probably should be, conducted using a placebo double-blind design, to statistically account for spontaneous changes in participants that are unrelated to drug effects.[13,14] The cohort sizes in Phase I clinical trials vary considerably, from 2 to 12 subjects, with no accepted standard number. The most frequent size is eight subjects, with six subjects allocated to active treatment and two to placebo. Small cohorts allow for reduced costs. Increasing the cohort size, while attractive for increased ability to elicit toxicities, will result in an increased risk for spontaneous non–drug-related adverse events to occur in one of the active subjects. An analysis of a large number of Phase I participants, in relation to laboratory liver abnormalities, has shown that to strike a balance between these competing effects, the active cohort size should be between 6 and 10 subjects.[15]

18.2.6 Routes of Drug Administration

The intravenous route offers significant advantages over other routes in early Phase I trials. It allows for the immediate discontinuation of the drug, in case of serious adverse reactions. It gives a complete and clear PK profile given complete bioavailability, and makes blinding easier. However, the information and advantages afforded by intravenous drug administration for FIM trials are offset by the cost and resources spent on toxicology and pharmacology studies required for IV drug administration. This is especially true considering that the majority of drugs do not make it to Phase I. For this reason, drugs in FIM phase are often given through the route that will be intended for patient care, and only dosed intravenously to establish the drug's PK profile when initial testing confirms the success potential of the drug candidate.[16] IV drug administration does require additional preclinical trials, since the Cmax obtained (highest serum concentration) is greater following rapid IV administration. Slow infusions following the timing of the oral dosing interval can be employed.

18.2.7 Late Phase I Trials

Many think that the different phases of clinical trials (I–IV) correspond to a rigid chronological sequence. However, the designation often pertains more to the methodology of the trial than its actual position in the drug development timeline. Any clinical pharmacology design is labeled a Phase I study, regardless of its time relationship to other clinical trial phases. "Late" Phase I studies are the clinical pharmacology studies that are conducted relatively late in clinical development, usually concomitantly with Phase II or III clinical trials. Their cost and the requirement to have a large sample, as in the case of population PK studies, are justified and feasible for drug candidates that have shown promise in the initial phase of clinical development. Several designs fall under this category and serve a number of important purposes.

18.2.7.1 Population Pharmacokinetic Studies

Population pharmacokinetics (PPK) is the quantitative estimation of PK parameters among the individuals who are the target patient population receiving clinically relevant doses of a drug of interest, analyzing and accounting for inter-patient sources of variability, and the residual intra-patient variability. This analysis is becoming more prevalent in clinical development programs, and is useful for designing dosing guidelines for drug labeling and understanding the effects of competing dosing regimens on outcomes of clinical trials.[17] Examples of potential sources on inter- and intra-subject variabilities include, but not limited to, demographic characteristics, genetic polymorphisms, and environmental factors. As this variability increases, it decreases the safety and efficacy of the drug, and may require dosing optimization by therapeutic drug monitoring. Therefore, PPK studies are most useful when the population for which the drug is intended is heterogeneous, and/or when the tested drug has a narrow therapeutic window. There are two common methods for PPK analysis: the two-stage approach and the non-linear modeling approach.[18] In the *two-stage approach*, data obtained from different individuals are used for the estimation of PK parameters. Subsequently, these parameters are used to estimate the corresponding sample PK values, with statistical analysis of the sources of variance. This two-stage approach can be used during Phase I and II studies, when it is possible to obtain many samples from the participants. The *non-linear modeling approach*, on the other hand, uses much less data obtained through "sparse" sampling of the population using the drug, generally under unbalanced and intermittent exposure. Non-linear regression is then used to estimate the PPK parameters and analyzed for variance. While this approach can be employed during early studies, it is very well suited for late clinical trials with large numbers of patients and when fewer samples are obtained for each patient. The sampling designs used for this method vary from single sampling designs in which one sample is obtained per patient to full population sampling designs with potentially more samples obtained per participant. PPK protocols of either approach can be a separate stand-alone protocol or an add-on protocol to another clinical study.

18.2.8 Bioavailability/Bioequivalence (BABE) Studies

Bioequivalence is established by showing the absence of a significant difference in the rate and extent, i.e., bioavailability, to which the active ingredient or active moiety in pharmaceutical equivalents or alternatives becomes available at the site of drug action when administered at the same molar dose under similar conditions in an appropriately designed study.[19] Drug products are considered pharmaceutical equivalents if they contain the same active ingredient(s), are of the same dosage form and route of administration, and are identical in strength or concentration (e.g., chlordiazepoxide hydrochloride, 5 mg capsules). Drug products are considered pharmaceutical alternatives if they contain the same therapeutic moiety, but are different salts, esters, or complexes of that moiety, or are different dosage forms or strengths (e.g., tetracycline hydrochloride, 250 mg capsules versus tetracycline phosphate complex, 250 mg capsules; quinidine sulfate, 200 mg tablets versus quinidine sulfate, 200 mg capsules). BABE trials are conducted to establish therapeutic equivalence between two drug formulations, usually a generic and an innovator product, and they are typically PK studies conducted as a two-treatment crossover design and can be conducted on a pilot scale (12 healthy volunteers) or on a pivotal scale (24–48 healthy volunteers) in fasting or fed conditions.[20,21] Bioequivalence is established in cases where the innovator and generic test product differ in their rate and extent of absorption by no more than 20% the upper boundary and 25% the lower boundary. Key parameters evaluated are the AUC concentration and maximum concentration (Cmax). The study design is usually a two-treatment crossover study.

18.2.9 Mass Balance Studies[22]

The PK (i.e. absorption, distribution, metabolism, and excretion) of a drug and its metabolites are usually investigated using a formulation labeled with a radioactive tracer. In these protocols, the radioactive tracer allows the drug and its metabolites to be detected in blood, urine, feces, and occasionally expired air. The concentration can be adequately quantified in relation to time, and collection is continued till near-total (90%) recovery of drug "mass" is achieved.[23] The ^{14}C isotope is usually used, due to its long half-life, and the relatively smaller difference in mass between naturally existing ^{12}C and ^{14}C which has a smaller impact on the metabolic stability of the radioactively labeled drug. The recent utilization of accelerated mass spectrometry (AMS) allowed the use of relatively minute amounts of the tracer-labeled drug to be used in these studies. Nonetheless, all personnel in the study should strictly follow the standard precautions involved in handling radioactive materials. The number of subjects needed for mass balance studies varies, but it is generally recommended to have between six and eight subjects. These subjects are usually patients when testing drugs for cancer, or as anti-arrhythmics. Mass balance studies yield PK (for unchanged drug) and pseudo-PK (for total radioactivity: drug plus metabolites) parameters, which should be used to add to the information being acquired as to the drugs pharmacologic and metabolic profile.

18.2.10 Thorough QT/QTc Studies

Because the most common reason for restricting or withdrawing a marketed drug is the prolongation of QT interval and associated polymorphic ventricular tachycardia,[24,25] it is most important and desirable to investigate a drug's potential for cardiac toxicity early and efficiently. Prolongation of QT interval, corrected for heart rate (QTc), as related to drug exposure is an important, and accepted, biomarker of prolonged ventricular repolarization and a clinically proven ECG surrogate associated with life-threatening ventricular tachyarrhythmias, often the Torsades de Pointes (TdP) morphology.[25,26] According to FDA guidance on "thorough QT/QTc Studies," these studies should be performed early in clinical development to provide guidance for later clinical trials.[27]

These studies should be conducted, as much as possible, in healthy volunteers, excluding individuals with baseline QT prolongation or a history of TdP. Parallel or crossover double-blind designs are used, with placebo and active controls. Data are obtained using standard 12-lead surface ECGs and are subsequently analyzed by expert readers, with or without computer assistance.

Recently, digital Holter continuous monitoring with manual over-read has been showing promising results[28] and is providing the added advantage of detecting infrequent ECG changes and arrhythmias. QT changes can be correlated with peak drug concentrations. If not employing continuous recording techniques, surface ECGs obtained should be timed to coincide with pharmacokinetically meaningful drug exposure and subsequently analyzed to determine peak drug effect on the QT interval.

Study results that statistically confirm QT/QTc prolongation by more than 5 msec are considered positive, and while this does not automatically confirm the drug's arrhythmogenic potential, it will almost always call for an expanded ECG safety evaluation during later stages of drug testing. Another important aspect of a thorough QT/QTc study is the use of a comparator group that receives miflofloxicin therapy, a known prolonger of the QT interval, to demonstrate that QT prolongation can be detectable in the cohort studied. This is an important test of assay sensitivity and underscores the importance of the concept of assay sensitivity in assessment of a drug testing protocol.

18.2.11 Bridging Studies

The MTD, as defined by studies conducted in healthy volunteers, can sometimes underestimate the MTD in the relevant patient population, and "bridging studies" have been suggested to circumvent this problem.[29] These studies explore safety and tolerance of study drugs in small patient populations and allow better understanding of the expected dose range in Phase II clinical trials. The bridging study can be performed according to two designs, the fixed-dose bridging study, and the titration bridging study.[9] In the *fixed-dose bridging study*, sequential dose panels are used, in which each panel of distinct participants receives a fixed dose of the drug if the previous dose was tolerated. This approach identifies the highest dose that can be safely employed in a fixed-dose efficacy Phase II clinical trial or a starting dose for dose titration. The *titration bridging study* has a starting dose of either the MTD or 25% below the MTD. This dose is escalated every 3 days until the MID is reached.

18.2.12 Drug–Drug Interaction Trials

A drug interaction is a pharmacologic or clinical response to the administration of a drug in combination with another agent different from that anticipated from the known effects of the two agents when each is given alone.[30] Because of their incidence and potential adverse effect in clinical practice, it is necessary to characterize drug–drug interactions in clinical trials. Additionally, this characterization uses the resulting information to appropriately label drugs, and to provide sound advice to medical practitioners.[31]

Clinically significant drug–drug interactions can be classified into the following:

1.Metabolism-based, when the metabolic routes of elimination of certain drugs are affected by the concomitant administration of other drugs, e.g., the induction or inhibition of liver cytochrome P450 (CYP) enzyme system would affect the drugs metabolized by this pathway.

2.Transporter-based drug–drug interaction, e.g., P-glycoprotein transporter proteins when drug induced or inhibited can affect the other drugs handled by these proteins.

The specific objective of drug–drug interaction clinical studies is to determine whether the interaction is sufficiently large to necessitate a dosage adjustment of the studied drug itself, or the drugs with which it might be used, or whether the interaction would require additional therapeutic monitoring.[32] Usually, *in vitro* and animal *in vivo* studies are the basis for further evaluation of a drug for potential drug–drug interactions. When these studies confirm that a particular drug is handled by a metabolic pathway and/or transporter susceptible to interference from other drugs, early human trials should be designed to explore potential interactions. If Phase I trials confirm and/ or identify drug–drug interactions, then these findings should be confirmed in larger clinical trials, which would also test dose modification and/or therapeutic monitoring strategies to avoid undesired clinical consequences.

Drug–drug interaction studies are usually conducted in healthy volunteers, using drugs that affect the mechanisms identified by non-clinical trials to be involved in handling the test drug. Crossover or parallel, and single- or multiple-dose designs, can be chosen according to safety and PK/PD considerations. Generally, the routes of administration in these studies are the routes clinically intended or used for the tested drug, and the doses should be selected to maximize the possibility of finding an interaction. The study should describe the potential interactions using PK endpoints, such as AUC, volume of distribution, clearance, and half-life. Relevant PD endpoints can provide useful information as to the clinical significance of the interaction. Examples of PD endpoints are QT interval changes and international normalized ration (INR) effects (an anticoagulation effectiveness measurement).

18.2.13   FOOD–DRUG INTERACTION CLINICAL TRIALS

Food is known to exert various effects on the absorption and bioavailability of orally administered drugs.[33] For this reason, it is often necessary to provide information on how to optimize absorption of a drug clinically using food or the lack of thereof.[34] Specially designed Phase I clinical trials are undertaken to investigate the interaction of food with different drugs and help make recommendations about the best way to utilize this interaction clinically. For drugs with predicted low therapeutic ratio, it is especially critical to determine the effect of food on PK parameters, to avoid loss of efficacy or exaggeration of toxic effects. The therapeutic ratio is the ratio of the dose at which 50% of subjects experience the toxic effect to the dose and 50% of patients experience the therapeutic effect.

It is also important to investigate the food-interaction effect on the bioavailability of compounds with special formulations, such as extended release compounds and enteric-coated drugs. The food–drug interaction trials can be parallel or crossover designs.[23] The *food spacing trial* utilizes inpatients, and controls the time between meals and drug administration. *The single-dose crossover food trials in fasted versus fed patients* compare various doses and formulations in patients that are fed or fasted in a crossover design. The *chronic dose food trials* can span weeks and evaluate numerous factors, including different formulations, meal types, and fast/fed conditions.

18.2.14   Recent Developments and Future Directions

18.2.14.1   Safety of FIM Trials in Healthy Volunteers

In March 2006, six healthy volunteers developed serious adverse reactions in an FIM clinical trial conducted in England, after being dosed with a newly developed monoclonal antibody. The adverse results of this study highlighted FIM clinical trials to the public and placed them in the spotlight. Although the study was conducted according to regulatory and professional standards, this study has caused protocol modifications for FIM studies.[35,36]

FIM trials need to be conducted in Phase I facilities in, or with, ready access to hospital facilities, and the protocols for new molecular entities (NME) will mandate sequential dosing to allow for safety assessment between doses. Another report provided insight about the potential adversity for healthy participants to participate in Phase I clinical trials in rapid sequence and in violation of trial protocol requirements.[37] More thought and caution is needed for trials that introduce new compounds for the first time in man.

18.2.14.2   Therapeutic Intent in Phase I Clinical Trials

Phase I clinical trials have been conducted primarily for safety, without any expectation of therapeutic benefit for participating patients. However, in the area of oncology trials, this seems to be changing. Multiple recent reports point out the possibility of clinical benefit from participating in early clinical trials for patient who have failed all available treatments.[38–41] It is likely that consent forms will reflect this in future Phase I clinical trials in cancer patients.

18.2.14.3   Exploratory Phase I Clinical Trials

As we have described, the initiation of Phase I clinical trials requires the commitment of time-consuming, scarce, and expensive preclinical resources to the development of drug candidates for further human development. The high failure rate of drugs entering Phase I clinical trials, due to unfavorable PK profile, makes it attractive to explore the PK of candidate drugs early on. It may be advisable to halt the development of drugs with poor absorption, distribution, metabolism, and excretion characteristics that predict problems in future clinical use.

The development of AMS technology allowed for the detection of minute amounts of radioactive tracer-labeled drugs in biological fluids, and the "micro-doses" that are used for the introduction of these minute amounts into humans are devoid of any significant pharmacologic activity.[42] The ability to use these pharmacologically inert micro-doses in "Phase 0" clinical trials may provide for exploratory studies with a reduced potential risks than do traditional Phase I clinical trials.[43] The micro-dose should be <1/100th of the dose of the test substance calculated, based on animal data, to yield a pharmacological effect, with a maximum dose of less than or equal 100 μm or 30 nmol for synthetic drugs. These studies can be designed as single- or multiple-dose studies, and provide information about the mechanism of action and interaction with therapeutic targets in humans, in addition to clarifying the PK characteristics of the drug.

18.3   PHASE II CLINICAL TRIALS

18.3.1   Overview

In Phase II clinical trials, the primary objective is to explore therapeutic effectiveness. These trials represent a shift of focus, from safety as the main concern, to the determination of efficacy in patients. Early on, these clinical trials utilize a variety of designs to determine the dose–response relationship of the test compound, frequently using *surrogate endpoints and biomarkers* to assess pharmacologic effects. Later in development, Phase II studies are typically controlled clinical studies conducted to evaluate the effectiveness of the drug, in different dose ranges, for a particular indication or indications in patients with the disease or conditions under study. These studies also help to determine the common or short-term side effects and risks associated with the drug.

Successful Phase II clinical trials prove the drug has the pharmacologic effect intended, provides therapeutic benefit, and provides working knowledge of the endpoints, usually surrogates, to the ultimate intended drug effect. These studies also provide a dose range and dose regimens for further testing in Phase III studies. Phase II studies aid in defining target populations (e.g., mild versus severe illness) for future clinical trials.

The initiation of this phase of drug testing rests on the data and results from critical studies both from non-clinical and clinical sources.43 Non-clinical studies should provide evidence of the therapeutic indication, with respective dose–response evaluations, as well as evidence that reasonably predicts the PK/PD behavior of the drug in man. Preclinical toxicology should support the phase of drug evaluation with the demonstration of a suitable therapeutic window. It is also necessary to demonstrate, in Phase I studies, the initial safety of the drug, and to have described the bioavailability and PK profile in healthy volunteers, and possibly patients, if indicated. Finally, pharmaceutical formulations that achieve efficacious concentrations in the plasma and at the target organ with reasonable variability must be bioavailable, and need to permit manufacture that is cost-effective and compliant with manufacturing procedures and standards.

18.3.2 Surrogate Endpoints

Clinical studies should ideally test drugs, or therapeutic interventions in protocols, that utilize the natural history outcome endpoints. For example, the evaluation of a drug that treats patients with coronary artery disease should show whether or not that drug would affect important clinical endpoints such as death or occurrence of a myocardial infarction. While this approach is ideal, it is often difficult to achieve.

Endpoints that are infrequent or take a long time to occur make these types of clinical studies difficult or impossible to reach a conclusion in a reasonable time period at a reasonable cost. The successful evaluation of new therapeutic interventions would lead to prohibitive costs and timelines, given the requisite number of patients needed to enroll in such trials. The introduction of "intermediate" or surrogate endpoints, which might or might not be clinically relevant, is a strategy aimed at the substitution of relatively rare clinical outcomes with ones that represent them. Surrogate endpoints are relatively easier to evaluate in a reasonable number of patients in a timely manner at affordable costs. Phase II studies routinely use surrogate endpoints.

18.3.2.1 Definition

"A surrogate endpoint of a clinical trial is a laboratory measurement or a physical sign used as a substitute for a clinically meaningful endpoint that measures directly how a patient feels, functions, or survives. Changes induced by a therapy on a surrogate endpoint are expected to reflect changes in a clinically meaningful endpoint."44

While surrogacy can usually be achieved by the measurement of a substitute naturally occurring physiologic or pathologic endpoint, a *biomarker*, surrogate models can be developed to evaluate the impact of a therapeutic intervention(s). A *model*, in this context, is an experimental system or paradigm, used in clinical trials to simulate some aspects of the disease of interest in which the effects of the drug are examined.45 Using asthma as an example, the desired clinical endpoint that reflects how the patient "feels, functions, or survives" would be the frequency of asthma attacks, or demonstration of an improvement of outcome when taking the experimental drug. These outcomes would argue for the drug's therapeutic effectiveness.

An example of surrogate endpoint in asthma would be the urinary excretion of leukotriene E_4, an asthma modulator that reflects asthma activity, and a surrogate model would be an allergen or cold air challenge inducing airway reactivity similar to an asthma exacerbation.

The demonstration of drug-induced reduction of the excretion of an asthma modulator, leukot-riene E_4, or the responsiveness to an allergen challenge, is then taken as proof of drug efficacy.

18.3.2.2 Correlation with Clinical Endpoints

Before biomarkers, and models, can be used to substitute for the sought-after clinical endpoints, they must be shown, with confidence, to reflect the treatment effect, or, in other words, are validated. Part of the controversy of using surrogate endpoints is that there is no agreed-upon definition of a validated surrogate endpoint.[46] The use of inadequately validated surrogate endpoints has led occasionally to the adoption of treatments that were later shown to be harmful or ineffective. The CAST (Cardiac Arrhythmia Suppression Trial) represents the evaluation of one of the most well-known examples of a surrogate endpoint failing to reflect a true effect of drug on a desired clinical endpoint (survival in the CAST trial).

Ventricular premature beats (ventricular premature complexes/VPCs) are associated with increased mortality after myocardial infarctions. Thus, it made sense to assume that suppression of VPCs would prolong survival. Suppression of ventricular premature beats after myocardial infarction was considered a surrogate endpoint for patients' survival. Anti-arrhythmic drugs were found to suppress VPCs and were thus used to treat patient after myocardial infarctions to prevent sudden death. CAST was a clinical trial that evaluated the effect of these anti-arrhythmic drugs on the natural history endpoint of patient survival, and showed that these drugs increased mortality.[47]

There are several examples of falsely assumed surrogates that have caused the medical community to be very skeptical of clinical trials based on surrogate endpoints. Clinical practitioners and regulators have called for closer scrutiny of surrogate endpoints in clinical trials and the need for their validation.

For a biomarker to have the greatest potential to be a validated surrogate endpoint, it must be involved in the causal pathway of the disease process. The intervention's entire effect on the true clinical outcome should be mediated through its effect on the suggested biomarker.[48] This is often not the case, and a surrogate endpoint thus only partially correlates with the natural history outcome. Thus, there are a number of reasons for the failure of a biomarker to function as a surrogate endpoint:

1. The biomarker might not involve the same pathophysiologic pathway that results in the disease outcome, although it might be a correlate of disease progression.
2. The effect of the intervention on the biomarker may be irrelevant to the clinical outcome, severely limiting the utility of the surrogate.
3. There might be several causal pathways of the disease, and the intervention affects only the one mediated through the biomarker, causing a false correlation at times and at other times yielding negative or harmful results.
4. The biomarker might not be in a causal pathway, which is affected by the intervention, or it might not be sensitive to that intervention, causing the apparent failure of the intervention.
5. Finally, the intervention might have different actions, which can affect the outcome of the disease, irrespective of the causal pathways, causing the net effect of the intervention to be positive or negative, but not reflected by the biomarker.

An example is the use of quinidine for the maintenance of sinus rhythm after the cardioversion of patients with atrial fibrillation. The maintenance of sinus rhythm can be beneficial, in terms of increasing the cardiac output and decreasing cerebral embolization. A meta-analysis evaluating the effect of quinidine in these patients indeed showed that the drug-maintained sinus rhythm longer than was the case for untreated patients, but the treated patients had a higher mortality than untreated patients.[49] The drug succeeded in favorably affecting sinus rhythm as a biomarker, but its proarrhythmic effect caused it to have a net negative effect on survival.

18.3.2.3 Validation of Biomarkers as Surrogate Endpoints

The use of statistical methods, based on the Prentice criteria, allows for the validation of a biomarker as a surrogate endpoint. These criteria should "yield unambiguous information about the differential treatment effects on the true endpoint."[50] This condition can be satisfied by meeting the following conditions[46]:

1. Treatment affects the surrogate endpoint.
2. Treatment affects the true endpoint.
3. The association of the surrogate endpoint and the true endpoint is the same for both treatment arms.
4. Using a null hypothesis based on the surrogate endpoint, of no difference between treatment and control groups, would serve as a valid test for the null hypothesis based on the true clinical endpoint.

As a result of satisfying all of these conditions, the surrogate endpoint will fully capture the net effect of treatment on the clinical endpoint. The ideal validation would thus be a very difficult and time-consuming process. Lesko and Atkinson summarized a more feasible approach to validating biomarkers as surrogates endpoints, and they can be summarized as the following criteria[51]:

1. Surrogate endpoints should be biologically correlated to the true endpoint on a mechanistic basis.
2. Epidemiologic studies provide evidence of correlation between the surrogate endpoint and true clinical endpoint under basal conditions.
3. Adequate, well-controlled clinical studies should provide an estimate of the clinical benefit in terms of clinical endpoints that can be derived mathematically or mechanistically from an estimate of the change in the potential surrogate endpoint.
4. Analysis should include a consideration of potential adverse reactions unrelated to the clinical endpoints as predicted by the surrogate endpoint.
5. An exposure-response model can be developed that mathematically describes, as well as predicts the relationships between, drug exposure, and surrogate endpoints with clinical outcome(s).
6. The development and validation of biomarkers and surrogate endpoints should be built into drug development phases, beginning with the preclinical phase.
7. Meta-analyses of multiple clinical trials can be helpful to look across and within studies to determine the consistency of effects following interventions with various drug classes and within different stages of disease.

It is also worth mentioning that other dimensions, besides validation, can also contribute to the success of biomarker/surrogate endpoint use in clinical trials. These include[45,51]

1. Practicality, simplicity, and ease of use.
2. Specificity and sensitivity, defined as the ability to detect the intended measurement or change in the target patient population via a given mechanism, without interference from other pharmacologic or clinical effects of the drug, unrelated to the drug's mechanism of therapeutic action.
3. Reliability, defined as the ability to measure analytically, the biomarker or change in the biomarker, with acceptable accuracy, precision, robustness, and reproducibility.

18.3.2.4 Development and Utility of Biomarkers and/or Surrogate Endpoints

The discovery and development of biomarkers for the subsequent use as surrogate endpoints is understandably difficult for diseases with poorly understood pathogenesis. Paradoxically, these are often the illnesses for which medicines are needed the most. Early on, the development of surrogate endpoints will require innovation with relatively rigorous validation. Innovative biomarkers in early clinical trials may be very useful for learning about different treatment interventions. These studies can guide further drug evaluation and make room for the confirmation of early findings, as well as providing data for the validation of these biomarkers, so they become surrogate endpoints.[52]

While surrogate endpoints are usually thought of as surrogates for efficacy, it is important to point out that surrogates can be used for safety endpoints as well. For example, the QT studies described earlier utilize electrocardiographic changes (QT prolongation) as a surrogate for risk of sudden death. In BABE (bioavailability and bioequivalence) Phase I clinical trials, drug absorption and exposure, as expressed by plasma concentration and AUC, are used as a surrogate for drug equivalency.

18.3.3 Design and Ethical Issues

The purpose of a well-designed clinical study is to avoid the introduction of *bias*, which can be defined as any systemic error in the design, conduct, analysis, and interpretation of the results that can cause study outcomes to deviate from their true values. To minimize such bias, clinical trials employ procedures to minimize bias. These approaches include control groups, placebo controls, randomization, and blinding.

18.3.3.1 Control Groups[53]

In a clinical trial, the control groups have one major function: to allow the discrimination of patient outcomes (e.g., changes of symptoms, signs, or other morbidity) caused by the test treatment, from outcomes caused by other factors, such as the natural progression of the disease, observer, or patient expectations, or other treatment(s).

A *concurrent control group* is one drawn from the same population as the test group and treated in a defined way as part of the same trial that studies the test treatment, over the same period of time. This is a superior approach than non-concurrent or *historical control groups,* in avoiding bias. This concurrent group can be administered no treatment, a placebo, or an active comparator agent.

It is also possible to use *multiple control groups* in the study; for example, both an active control and a placebo have been employed. The placebo group can be the usual placebo or an active placebo, in other words an inactive agent with side effects; for example, a bitter taste is often used to suggest an active drug. Similarly, trials can use several doses of the test drug, and several doses of an active control, with or without placebo. This design may be useful for active drug comparisons where the relative potency of the two drugs is not well established or where the purpose of the trial is to establish relative potency.

18.3.3.2 Blinding

Blinding is intended to minimize the potential biases resulting from differences in management, treatment, or assessment of patients, or interpretation of results that could arise as a result of subject or investigator knowledge of the assigned treatment or intervention. Clinical trials vary in the degree to which the trial participants and practitioners are blinded.

In *open-label trials*, the participants and the investigators know what intervention is being administered to each subject. While easier and more like actual clinical practice, open trials are more subject to bias than their blinded counterparts, as both the patients' and the investigators' behaviors are known to be changed by their knowledge of the nature of intervention.

Single-blind studies, in which the participants are unaware of the test article they are given, are better than open trials as they minimize the influence of the perception of the nature of treatment on the participants' responses. However, similar to open trials, they are subject to healthcare practitioners' biases in the evaluation of treatment responses, and what is known as compensatory treatment bias. When investigators are made aware that some patients are not receiving any treatment, or receiving a placebo, they may compensate for that by treating them with other available interventions, or more intensive standard therapy, possibly skewing the outcomes. For these reasons, it is better if the investigators are also unaware of the nature of the intervention, i.e., conducting a *double-blind study*.

A *triple-blind study* is an extension of this blinding to a committee (Data Safety Monitoring Board – DSMB) monitoring the response variables for patient safety. While this approach minimizes most of the potential for bias, it can be at times procedurally difficult, since the monitoring committee will find it more difficult to evaluate a toxicity that is reported. However, blinding the DSMB is recommended, since their bias may lead to premature termination of a trial for safety or efficacy reasons. DSMBs are subject to bias as are investigators.

18.3.3.3 Randomization

Randomization is a process that allocates, by chance, eligible trial participants to any single treatment or intervention group, including placebo. This process tends to produce study groups comparable with respect to known and unknown risk factors, removes investigator bias in the allocation of participants, and guarantees that the premises of statistical tests utilized in the study have been met.[54]

There are a number of types of randomizations that can be employed. There is fixed randomization with a pre-specified probability, usually equal. Randomization can be equal or unequal. Equal randomization to the different study groups ensures equipoise but has the disadvantage of exposing a relatively greater number of patients to the control intervention, while greater exposure to the active intervention would yield more information. The greater exposure, however, occurs at the expense of the impression that the intervention is favored with the resultant loss of equipoise. Still, increasing exposure to the intervention is a considerable advantage.

The randomization process can be simple, like a coin toss or using a random number table. However, these techniques have a disadvantage when used in small sized trials, where the distribution of subjects may be statistically unequal. Thus, block randomization is often used.

Using block randomization, participants are assigned in blocks with equal probability, preventing imbalance developing in small clinical trials. Block randomization can be subject to decoding by inquisitive study site investigators and staff. To prevent unblinding, variable block randomization is employed where one varies the blocking factor to prevent unblinding. Stratified randomization is undertaken to have equal balance in prognostic risk factors, factors that can influence outcome.

Large studies (thousands of patients) can adequately distribute risk factors equally, and smaller studies may not do this effectively. Stratification can be used to ensure equal distribution, according to pre-specified patient characteristics. One can also accomplish this stratification by using the technique of stratified analysis. However, stratified randomization can ensure balance, which may be an advantage, with less data manipulation. Bayesian studies may use adaptive randomization to compensate for imbalance as, or if, these problems develop.

Some techniques that use a Bayesian approach are the "Urn design" that enriches the pool from which randomization occurs. The procedure works by placing numbers, for instance, 1 or 2, let's say "1" for drug or "2" for placebo. Either a 1 or 2 is picked from the urn. If a number 2 ball is picked, then that ball is replaced in the urn along with a number 1 ball that increases the amount of number 1 balls in the urn. Using this technique, number 1 balls are more likely to be picked on the next draw.

Another technique is called "play the winner" in which randomization is repeated to the same study arm based on the prior success of that arm, and if a failure develops that causes crossing over to the alternative intervention. This technique requires that the outcome be known in a short period of time. Additionally, one can randomize in an adaptive way by using the "two-armed bandit" techniques where the randomization probability is changed (by a computer system) after the previous pick is made.

18.3.4 PLACEBO VERSUS ACTIVE CONTROLS IN CLINICAL TRIALS

The debate regarding the use of either placebo or active controls or standard therapy is a discussion involving both ethics and clinical trial designs that aim to effect differences between intervention and control arms of a study. Generally, the placebo orthodoxy advocates placebo-controlled clinical trials even in the case of medical conditions for which there are interventions known to be effective. The proponents of active-control trials maintain that it is appropriate to use active controls in superiority trials whenever such alternative treatments exist.

Both arguments are rooted in the ethics of conducting clinical research, and are based on the principle that ethical clinical research should be scientifically valid and should provide a favorable risk–benefit ratio for those involved to minimize potential patient harm.[55] It is clear that there will be circumstances that would mandate the use of available treatments as active controls, and comparing them against the new therapy in a superiority trial design is preferable. No clinical investigator would defend, for example, randomizing cancer patients to a placebo and denying them the benefit of any available treatment known to prolong life. In this scenario, the patients that are not receiving active treatments are certain to end up with serious and permanent harm, which would be clearly unethical.

At the other end of the spectrum, mild ailments left untreated can only cause minor and reversible discomfort, and are appropriate areas for the use and benefits of placebo-controlled clinical trials. Uncomplicated seasonal nasal allergies would be an example. The controversy involving the use of placebo versus active controls is centered mainly on disease conditions which lie in the middle of the spectrum. The "placebo orthodoxy" favors the use of a placebo comparator, with proponents of active controls advocating comparison to available standard treatments.

Placebo-controlled clinical trials are historically the gold standard for trial design, and their methodology has clear advantages.[56,57] The demonstration of benefit of a drug versus placebo is probably the best way to show that an intervention really works, as in some trials, an active treatment, which has been proven to be effective elsewhere, can be ineffective under the current trial's conditions. Without a placebo control group, a treatment that has been shown to be comparable to active treatment comparator can be in fact no better than a placebo. In addition, placebo-controlled trials require a smaller sample size and are thus more effective. However, despite the obvious scientific strength of placebo trials, placebo-controlled clinical trials suffer from the difficulty of defining the acceptable degree of harm that justifies placing patients on placebo.

A study performed by the Division of Cardiovascular and Renal Products of FDA dealing with this issue in the context of hypertension Phase II trials found that over several months, patients in trials for mild or moderate hypertension actually had less side effects and adverse outcomes on placebo than drug. This applies though to trials of relatively short duration. Thus, it is false to assume that drug therapy is always beneficial; in fact, a therapeutic interaction may actually be harmful.

Actively controlled clinical trials compare the new therapy versus established treatment, whenever available, in superiority or non-inferiority studies. Proponents argue that by using placebo instead on active controls, placebo-controlled clinical trials place scientific rigor and validity ahead of individual patients' interests, and that ethical clinical trials should almost always establish benefit against an alternative available intervention.[58–60] This position has limitations that interfere with its universal application. Trials using active controls can, paradoxically, expose more patients to harm than their placebo-controlled counterparts. In order to prove equivalence to an active comparator, statistical tests require samples that are much larger than those needed for similar trials involving a placebo arm. Such large sample sizes will expose more patients to the potentially harmful effects of a new drug or an older, less effective, agent. Also, non-inferiority trials can conclude that there is no difference (non-inferiority) where indeed the two groups are different, but the sample was too small, or the variability within the groups too great, to show a difference.

A middle ground has been suggested to resolve the controversy between these differing points of view. It focuses the attention on whether it is ethical to use placebo controls when there is a treatment known to be effective, and there is some potential of harm to participants receiving placebo. This view stipulates that compelling methodological reasons should exist before conducting placebo-controlled clinical trials in the presence of effective treatments. These reasons include

1. There is a high placebo-response rate.
2. The condition is typically characterized by a waxing-and-waning course, frequent therapeutic remissions, or both.
3. Existing therapies are only partly effective or have very serious side effects.
4. The low frequency of the condition means that an equivalence trial would have to be so large that it would reasonably prevent adequate enrollment and completion of the study.

If these reasons justify conducting a placebo-controlled clinical trial, then a placebo group can be used; if it is evident, the research participants in a placebo group would not be more likely than those enrolled in an active treatment arm to die, to have irreversible morbidity or disability, to suffer reversible but serious harm, or to experience severe discomfort.[61] Other criteria have also been proposed that incorporates these method-oriented approaches.[57] Placebo-controlled clinical trials can also be used in a sequential way. Only after the intervention is proven to be superior to placebo should active-controlled clinical trials be undertaken, usually with a much longer time frame for clinical exposure.

18.3.5 STUDY DESIGNS

Phase IIa clinical trials are often called *proof-of-concept* or *proof-of-principle* clinical trials. These trials are pilot designs aimed to detect a signal that the drug is active on a pathophysiologically relevant mechanism. These studies provide preliminary evidence of efficacy as reflected by the outcome of a relevant endpoint.[43] A *mechanistic trial* would use a biomarker or surrogate endpoint or model for efficacy analysis, while the typical clinical trial would use a clinical endpoint. Such studies may include proof of bioavailability in humans as well as an indication of tolerability and safety. These small, brief, pilot studies provide a link between Phase I and dose-ranging Phase II clinical studies. As the focus is primarily exploratory, designs may not include randomization and concurrent controls; instead, they can utilize baseline status as a comparator. Some designs even use a subjective approach in an *open-label uncontrolled intuitive study* design that depends on the impression of the investigators regarding the utility of the drug.[62] These types of studies are often used in the sickest patients with very severe disease, being treated with relatively toxic experimental agents.

Phase IIb **clinical trials** evaluate a drug's efficacy and safety in a larger patient population for which the drug is intended. The designs were historically largely affected by therapeutic area, with oncology trials usually uncontrolled compared to other therapeutic areas.[63] However, there has been a recent trend favoring controlled and randomized Phase IIb trials for cancer therapies. This is because of the perceived inefficiency of nonrandomized clinical trials to predict efficacy in subsequent confirmatory Phase III trials.[64–67]

Single-arm uncontrolled studies are the traditional design for cancer therapy studies and can also be adapted to other serious illnesses. In these protocols, all the patients receive the experimental treatment, and it is deemed effective when the proportion of responding patients exceeds a certain pre-specified value determined by response to standard treatment or placebo. Several methods can be employed to carry out this design, with various stopping rules that halt the trial because of treatment failure, or because of preliminary efficacy that can justify advancing to more definitive studies.

These methods take into account the exploratory nature of this phase of clinical trials. To maintain a small sample size, they adjust type 1 and type 2 error rates, allowing for more error compared to formal confirmatory Phase III trials. Also, carrying out these protocols through sequential stage 2 and stage 3 designs with an interim analysis allows for even smaller sample sizes.

In *comparative controlled designs*, patients are randomized between a single experimental treatment and a control, which may be either a standard treatment used as an active control or a placebo. As such, these designs are similar to Phase III clinical trials; however, the emphasis on providing preliminary exploratory information again allows for smaller sample size by tolerating a higher error rate. *Selection studies* compare different interventions, or formulations, of the same compound, to each other, with or without a control arm. These trials "screen" different interventions aimed to select one or two drugs for further evaluation by dropping less effective or less safe compounds.[68] Given the large number of drugs made available by high-throughput pharmaceutical research, it is clear that selection protocols need to be very cost-effective in evaluating multiple drugs in a limited patient pool.

18.3.5.1 Dose-Ranging Studies

18.3.5.1.1 Overview

Dose ranging, by definition, applies to any protocol calling for the evaluation of multiple dose strengths. In Phase I studies, multiple strengths are evaluated for tolerance and safety in volunteers, and occasionally in patients. Phase II trials, subsequently, evaluate a range of tolerated doses for efficacy. In this case, dose-ranging studies are usually carried out as part of proof-of-concept designs, or shortly afterwards, when proof-of-concept studies show promise. The aim of dose ranging in early Phase II is to describe a dose–response curve for one or more surrogate or clinical endpoints, which can be also called dose–response studies. The results would allow us to determine the no-effect dose range, the minimum effective dose, the mean effective dose, the maximum effective dose, and the optimal dose range.[69]

The maximum effective dose produces marked effects in the majority of patients, while the *minimum effective dose* produces a borderline effect in a small number of patients. As the dose increases in strength, the unintended side effects increase as well, thereby decreasing tolerability. *The therapeutic dose range* lies in between maximum and minimum effective doses, and in the midst of this range is the *optimal dose range*, at which the therapeutic efficacy is associated with the best tolerability. These concepts guide the analysis of exploratory Phase II dose-ranging studies, which yields information about a starting dose, rational response-guided titration, and a dose beyond which titration should not be attempted (because of lack of further benefit or unacceptable increase in undesirable effects). These studies yield information that then needs validation in confirmatory dose-ranging studies in late Phase II, and Phase III clinical trials.[70,71]

18.3.5.1.2 Subjects

Healthy volunteers can be used for some early Phase II clinical trials; in this case, the proof-of-concept or dose-ranging design will be evaluated against a surrogate, or endpoint that can respond to the drug's pharmacologic action. Volunteers are relatively easier to enroll. *Patients*, however, have to be enrolled in dose-ranging trials when the evaluated endpoint cannot be monitored in healthy participants. Also, patients often tolerate doses higher than those tolerated by normal volunteers.

A potential problem, especially in the case of innovative and new drugs, is demonstrating efficacy and dose response using "end-stage" patients. Very ill patients can potentially lead to selecting too high of a dose for subsequent studies or even the initial marketing of drugs at excessively high doses.[72] Therefore, the involvement of patients with different but well-defined stages of the disease is essential in dose-ranging trials, and interpretation of the results must take into account the various degrees of severity of the disease state being studied.[69]

18.3.5.2 Study Design: Crossover Studies

There are a number of designs for Phase II trials. Some Phase II trials are often short in duration. The favored design is a crossover study. In this design, subjects are randomized, usually in a double-blind fashion to receive either the active therapy or placebo. Alternatively, randomization could be to one of several doses or to active treatments. Often differences in therapy make blinding more difficult, but employing a "double-dummy" technique can be utilized to ensure blinding.

This technique uses a "dummy" placebo to blind both the patient and investigator. One can compare an intravenous placebo drug and an orally active drug. One group could receive the IV active therapy and the placebo orally. The other group would receive an IV, placebo therapy, and the PO active drug. Thus, both receive drugs orally (active or placebo), and both groups receive IV drugs (active and placebo). This procedure ensures blinding. Once a patient is randomized, they remain on the therapy for a designated period of time and are then "crossed over" to the alternative therapy.

The crossover design requires about a third less patients for comparable statistical power and can thus be accomplished at less cost and in less time. One major disadvantage is that there can be a "carryover" effect of the first therapy to the second segment of the study. This is a severe limitation that limits the design to shorter-acting therapies. Additionally, the disease process must be stable over time; otherwise, the second study arm may yield different results.

Clearly, a study in the treatment of pneumococcal phenomenon would not permit this design, since the effect of the therapy would modify the effect of subsequent exposure (therapy–disease interaction). Therapy could cure the disease initially, or ineffective therapy could make treatment in the crossover arm less likely to work due to the initial delay in therapy. For many of these reasons, regulatory authorities usually do not favor crossover designs but instead favor parallel studies.

18.3.5.3 Parallel Studies

Parallel studies compare groups that receive the same therapies, test, etc., except for the intervention. One group can be compared in a parallel fashion over time to a second group or to multiple "parallel groups" by dose or compared by alternative therapies or interventions. The parallel study uses larger samples because the variability is increased, as compared to a crossover study where individuals are compared to each other and thus, variability is decreased, requiring a smaller sample size for statistical significance determination.

There is no carryover effect in parallel design studies. However, an individual does not get the benefit of exposure to both interventions. Some patients in a parallel study never receive the active intervention(s), only the placebo. Thus, placebo parallel studies are limited in duration due to placebo exposure, since prolonged placebo can offer risk and be unethical. This problem is often overcome by overlaying the placebo on top of standard therapy and comparing the placebo plus active conventional therapy, to the intervention therapy. Thus all patients receive established therapy.

In the drug development process, Phase II studies are often crossover studies, while parallel studies are often employed in Phase III. Pivotal studies, studies that are the primary basis of drug approval, are usually Phase III, blinded, randomized parallel design studies. The FDA, as mandated by law, usually requires two studies (pivotal studies) that are positive and significant at the $p<0.05$ to approve a drug for marketing. There are exceptions to the "two-study rule" at times for rare diseases, "breakthrough treatments," and 505(b)2 development programs.

18.3.5.4 Factorial Design

Another study design is called a factorial design. This design is often used for combination drug development. If, for example, one was developing drug A and drug B, one could compare drug A to drug B. One could also compare the combination of drugs A and B to drug A alone and to drug B alone. Thus, a number of "factors" could be compared. One can expand the matrix to compare a number of individual comparators (factors) say A, B, C, and D to each other and then to a number of possible combinations of these factors A, B, C, and D. These factors, or cells, could then serve in multiple parallel studies for ongoing comparisons. The statistics are such that each comparison (cell) can be relatively small, but the whole comparison can still be adequate to provide the necessary statistical power for assessment. Not all factors need be compared, especially if combining some components would be ill-advised, due to excessive effect or adversity.

18.3.5.5   Withdrawal Studies

Different questions asked often require different study designs. If one wants to assess continued efficacy of a drug, the randomized placebo withdrawal study may be a very useful design. In this trial, patients have been on therapy for a long period of time. Patients are asked to sign a consent form and then are re-randomized to either remain on therapy or have the therapy withdrawn. Patients who have therapy withdrawn are contrasted in terms of efficacy variables to those who maintain therapy. If there is no difference between the groups, perhaps the drug was no longer effective and tolerance had developed. If the drug was thought to no longer be effective due to deterioration of those on therapy and the group with drug withdrawal does appreciably worse, then it is likely that clinical deterioration occurred for all patients, although the drug was still providing benefit. A blinded randomized withdrawal study may be very helpful in determining which scenario is correct.

18.3.5.6   Miscellaneous Considerations

A number of considerations in study design are requisite. One is study simplicity. The simpler the study, the more likely to obtain good data, less confounded by intervening variables, and the more likely you are to complete the study, facilitating patient entry, compliance, adherence to protocol, and study completion. Asking too many questions, too many sub-studies, is a formula for disaster, even for the most experienced investigator. Choosing the correct study endpoints is also critical. Ambiguity, complexity, and difficulty in assessments all can contribute to study failure. A primary endpoint is needed. Multiple endpoints or composite endpoints, while reducing sample size, at times can lead to confusion and a less significant endpoint, and can drive the study results. Carefully crafted study questions are perhaps the most important aspect of study design.

Other issues are the concepts of "population enrichment," "sub-population analysis," or "responder analysis." For a host of genetic reasons, drugs work in some patients and not in others. Should drugs be tested in a population which has the genetic composition that predisposes patients to respond? A study could have a "run-in" period where all patients are exposed to the drug and only patients who show response characteristics are then randomized to receive the drug or alternative therapy (placebo or drug). The population is thus "enriched" to include more or possibly only responders. One can administer a genetic test, for example, to determine if a patient with breast cancer has the HER2 tumor marker. Only those patients with the marker respond to Herceptin. Herceptin studied in an "all-comers" clinical trial would not show efficacy, since only 20% of breast cancer tumors show the HER2 marker and those tumors with the marker are the only ones expected to respond.

If we are to see the age of personalized medicine, we are going to have to modify the paradigm of the randomized trial evaluating all types of patients. Sub-population analysis that is both pre-specified and non-specified would find groups of responders. Then, studies that enrich the study population for responders will be needed to evaluate the proposed therapy. It is disheartening to think of how many therapies that may have been effective in a select population have not been identified in clinical trials and thus lost to therapeutics. Additionally, this approach of enrichment could be employed to identify patients likely to respond to therapy, but develop limiting adverse drug events. By identifying markers for adverse drug effects, a study could be developed to define a responder population without drug toxicity. None of this discussion is to suggest that the efficacy in the general population should not be defined in a clinical trial or the toxicity profile not identified. It may turn out that even a toxic agent or poorly effective agent for the general population may, in the patient who responds and does not show toxicity, be a valuable addition to our therapeutic drug armamentarium. What is critical is to be able to identify the patient likely to respond, so we can appropriately target therapy.

18.3.5.7 Dose–Response Studies

Dose–response studies evaluate different doses for efficacy, and it is an implied assumption that higher doses will be more effective in a proportional manner. However, it is important to determine when this assumption is incorrect. When there is no dose proportionally, the use of PK exposure parameters in relation to efficacy endpoints is needed. The major reasons for loss of correlation between dose strength and exposure are increased PK variability among the target population. After accounting for the sources of PK variability in *concentration-response studies*, one is then able to establish a dose–response relationship for further evaluation. This approach is especially important for drugs with a reduced *therapeutic index*.

The therapeutic index is defined as the ratio between the minimum toxic dose and the maximum effective dose, and it reflects the degree of overlap between the dose–response curves for efficacy and toxicity. Therapeutic drug monitoring may be recommended to ensure the safe and effective use of the drug in situations where there is no dose proportionality. It is important to consider that several factors can lead to the disassociation between exposure, as measured by drug concentration, and response.[73]

These factors are

1. The presence of other active intermediaries or compounds that partially or totally impact efficacy or toxicity, and are not detected by a drug concentration assay.
2. Tolerance, which means that the effectiveness of the drug decreases with prolonged exposure.
3. Time lag, when the measured response is delayed relative to plasma concentration.
4. A dose effect that involves a homeostatic system that is not directly responsive to dose.

The nature of surrogate endpoints used for dose-ranging studies depends on the phase of drug evaluation. Early on, during exploratory Phase II studies, more mechanistic endpoints can be used to accelerate and simplify the trial design, but later on, during confirmatory Phase II or III clinical trials, dose–response studies should incorporate clinically meaningful endpoints.

Different designs are used for dose-ranging efficacy studies.[70,74] In *parallel fixed-dose, dose–response studies,* subjects are randomized to several fixed-dose groups, and placebo. The fixed dose is the final or maintenance dose. Patients may be placed immediately on that dose or undergo forced titration, if it is safe to undertake. These studies are useful to assess long-term chronic responses, or responses that are not easily reversible. Parallel fixed-dose, dose–response studies are very familiar, have been historically successful, and have the added advantage of providing good information on safety. One of the disadvantages associated with this design is that it requires a relatively large sample size, given that patients receive one dose per patient. These studies provide information about the population mean for the dose response, without any data about individual dose–response relationships. There is the potential of ending up with doses that are too high or too low on the dose–response curve. This flaw can be discovered and corrected with an appropriate interim analysis.

Crossover designs can be used when the assessed responses are acute, immediate, and reversible, allowing for the drug effect to develop rapidly, and return to baseline status rapidly after cessation of therapy. As each patient receives several doses, crossover design provides information about individual dose–response curves in addition to mean population values. Another advantage would be a smaller sample size compared to parallel dose–response designs. The disadvantages include the potential confounding of the response by carryover effects (when a prior dose can affect the response to the new dose or placebo) and changes in baseline comparability between treatment periods. Also, crossover designs provide safety information that can be obscured by many factors such as time effects and tolerance.

Forced-titration studies, where all patients move through a series of rising doses, are similar to randomized multiple crossover dose–response trials, except that assignment to dose levels is mandated and thus not random. They have comparable advantages and disadvantages as described above. *Optional titration studies* are modified titration protocols to allow for patient titration until a well-characterized favorable or unfavorable response is obtained that is pre-specified.

18.4 PHASE III CLINICAL TRIALS

18.4.1 Overview

Phase III clinical trials are very resource intensive. Phase III trials are initiated when drugs and interventions show definite promise when tested in earlier phases. The time, money, and effort required for Phase III trials must be based on demonstrated initial safety and efficacy in Phase II studies. Phase III studies represent a shift in the testing model from a very structured protocol, to the evaluation of the drug in settings closer to actual clinical real-world exposure including a larger and broader patient population. Under these conditions, Phase III studies aim to confirm initial clinical safety and efficacy. Phase III studies usually accomplish this by comparing the intervention in question with other standard therapy(s).

Prolonged placebo trials are often not appropriate, both on ethical and practical grounds. Phase III trials are far longer in duration than the earlier Phase I and II trials – 6 months to a year in duration (if not longer). Ongoing preclinical animal studies must continue to provide evidence of safety, especially in terms of the long-term exposure for toxicologic, carcinogenic, and teratogenic potential of the drug to permit the prolonged drug administration expected in Phase III trials.

While the previously described paradigm fits new and innovative therapies,

Phase III clinical trials are often used as well to evaluate established treatments for new indications, new dosages, and additional safety information. Phase III clinical trials usually study large numbers of patients, many hundreds to thousands of patients, and they are usually performed at multiple investigative sites, at times in different countries simultaneously. Due to the sheer number of the involved participants, both patients and healthcare practitioners, and the voluminous data these studies generate, optimal design, coordination, and management are critically important.

18.4.2 The Concept of Clinical Equipoise

Phase III clinical trials are almost always randomized controlled clinical trials that compare a new intervention to a placebo or to standard therapy in the form of an active comparator. By definition, randomization allocates patients *randomly* to the different treatment and placebo groups. It is clear that in order to ethically use such a design, there must be reasonable uncertainty as to whether the intervention is effective. Otherwise, if the investigators are certain that the treatment is beneficial or harmful, it would be ethically unacceptable to deny or expose patients to the treatment. Clinical investigators might have a different judgment whether a treatment is effective or not, and this may make some investigators uncomfortable and hesitant in assigning patients randomly to different treatment arms.[75]

If an investigator has a clear conviction regarding the treatment benefit or harm, he or she should not take part in a clinical trial evaluating that treatment. In most circumstances, the investigator has some opinion regarding the value of the tested intervention. An opinion is fine as long as doubt exists as to the intervention(s) benefit.[54] The concept of *clinical equipoise* is predicated on the presence of uncertainty as to the benefits or harm from an intervention among the expert medical community. If equipoise exists, the trial can go forward, but an individual investigator, who has made up his or her mind, should not participate.[76]

The *uncertainty principle*[77,78] is a concept that has evolved to address the same issue at the individual investigator level. This principle takes into consideration that most clinicians and patients do have "hunches" about a treatment's effectiveness, but that the boundaries or "confidence intervals" around their hunches may run all the way from extremely effective (a wonder drug), to zero (ineffective drug) or into the realm of frank harm.[79,80] Those supporting the "uncertainty principle" argue that it takes into account individual investigator biases and this makes it ethically acceptable to conduct clinical trials given the uncertainty. The uncertainty principle is a practical approach to manage "equipoise" at the level of the individual physician–patient relationship, affording flexibility, within an ethical approach to clinical investigation.

18.4.3   Phase III Study Designs

The design of Phase III clinical trials can be described in terms of the different approaches to patient allocation and trial analysis.

18.4.3.1   Efficacy Analysis Designs

Typically, clinical trials scientifically demonstrate efficacy through a *superiority design*, where the intervention can be shown to be better than a placebo or active control. Another alternative approach is the *non-inferiority or equivalence design*, in which case the primary objective is to show that the response to the two treatments differs by an amount which is clinically unimportant.[81] *Dose–response trials* can also be designed to show efficacy.[74]

18.4.3.2   Superiority Clinical Trials

In this design, the aim is to show whether a particular intervention is superior to other comparator(s). Comparators might include placebo, utilizing a null hypothesis that states there is no difference between the intervention and the comparators, and the alternative hypothesis states that both interventions are not equal, may be worse, or may be better. Statistical tests are then applied to refute the null hypothesis. If tests refute, we accept the alternative hypothesis and conclude that there is a difference between the interventions. However, if the tests fail to reject the null hypothesis, we simply conclude that there was no evidence to show that one of the interventions is better than the other. This does not allow us to accept the null hypothesis or conclude that the interventions were the same. In this case, we cannot use the absence of evidence of difference to prove no difference between interactions.[82] If one intervention was tested previously against placebo and the current trial shows no difference between interactions, then the "putative placebo" concept would let one infer that the interaction is not superior to the therapy tested, but would be different from placebo.

18.4.3.3   Equivalence Clinical Trials

This approach attempts to prove that an intervention is as efficacious as its comparators, rather than proving that an intervention is superior. It utilizes the null and alternative hypotheses that are the flipside of the ones used in superiority trials. The null hypothesis states that the intervention is worse than its comparators, so rather than assuming that there is no difference, the null hypothesis assumes that there is a difference and that the intervention is actually worse.

The alternative hypothesis in this scenario states that the intervention is *at least as good as* the comparator. If the statistical tests reject the null hypothesis, we can then conclude that the intervention is at least as good as the other comparators in the trial.[83] Usually, this is accepted within a pre-specified range, called the equivalence margin. This approach is very useful when it is used to demonstrate that a drug is as effective as standard treatment(s). The drug may be better than its comparator in terms of other issues such as less side effects or costs. This approach can be useful when it would be considered unethical to use a placebo in a particular clinical trial.

Equivalence trials are also appropriate when investigators seek to establish that bio-chemically related compounds achieve clinical results similar to those of the standard therapy, this is the essence of bioequivalence aiming to provide evidence for equal efficacy of generic products to gain marketing approval. Generics are not identical but similar by a pre-specified margin.

There are several caveats associated with equivalence designs.[84] One is the pre-specified equivalence margin, which establishes that a "cushion" is selected to be the maximal difference clinically accepted when comparing the event rates for patients taking the experimental intervention and the comparator(s). Both treatments are considered equivalent if the event rate difference is within this margin. Thus, it is not that there is no difference but an acceptable difference given a certain variance. Another potential problem is that a comparator used for equivalence testing in a particular trial might be ineffective given that trial's circumstances, while it has been proven efficacious in previous trials. In this case, accepting that the intervention is equivalent to this comparator does not mean that it is indeed effective.

18.4.3.4 Non-Inferiority Clinical Trials

The non-inferiority design is a modification of equivalence protocols. In equivalence trial designs, just described, statistical tests evaluate the experimental intervention against the comparators. The trial tests the comparison in two directions, i.e., a two-tailed test. This means that the tests assume the intervention can be better or worse than the comparator by a margin of equivalence. By accepting the alternative hypothesis, one believes that the new intervention is *at least as good as* the comparator; that is, it may be equal to it or even better. In non-inferiority trials, on the other hand, statistical tests are applied only in one direction, a "one-tailed test," to demonstrate whether or not the intervention is not worse than the active comparator by the margin of equivalence.[85] The alternative hypothesis would state that the treatment is *not inferior to* the comparator. It is not established as to whether the intervention is better or just equal to the comparator.

The question of non-inferiority trial is different than one posed in a superiority trial, affecting the design and conduct of these types of trials. In a superiority trial, poor adherence will decrease the power to detect a difference that has been defined as meaningful. In a non-inferiority trial, poor adherence will diminish important differences and thus favor the results, revealing a positive, non-inferiority claim. Also, a small sample size will bias the outcome towards non-inferiority.

In interpreting a non-inferiority trial, a number of conditions are requisite. The active control must be established as an accepted standard of therapy. The studies showing this effect of the standard therapy must be current, and there must not be a significant change in medical care from those earlier studies to the time the non-inferiority study is undertaken. Additionally, one must be able to effectively estimate the control group event rate. The response variable(s) must be sensitive to control and intervention therapy (assay sensitivity), as well as being similar to the number employed in the prior study establishing the standard of care therapy.

The use of equivalency and non-inferiority studies remains controversial in drug development, but is increasingly employed. A key concept is assay sensitivity. The outcome(s) chosen must be a true measure of the effect and can be influenced to the same extent by the control and active interventions. If the variables are insensitive, no difference could be monitored, and non-inferiority concluded. Even though the variable did not measure the effect of the drug, the result would be a false-positive outcome.

Non-inferiority analysis, when undertaken with an "intention-to-trust" approach, dilutes whatever difference may exist between the interventions and thus can bias a non-inferiority study, suggesting non-inferiority when it does not exist. "On-treatment analysis" compares only those who adhere to the protocol-directed treatment and thus provides a more accurate assessment of the true drug effect. On-treatment analysis is the preferred approach for the analysis of non-inferiority trials, since patient adherence is critical to the validity of non-inferiority studies.

18.4.3.4.1 The Equivalence or Non-Inferiority Margin and
the Concept of Putative Placebo

The basic assumption underlying an equivalence or non-inferiority trial is that the experimental, or new intervention, by being equal to the standard treatment, previously proven to be better than placebo, can be estimated to be better than placebo, the so-called *putative placebo*.[86]

The estimation of the equivalence or non-inferiority margin incorporates statistical methods and clinical judgment to allow for an analysis based on the putative placebo concept.[87] This margin is estimated using the magnitude of efficacy offered by the standard active comparator over placebo, using historical trials and meta-analyses. This is usually framed in the form of an interval, with the upper limit of benefit derived from the standard treatment used to calculate the margin. This margin is then arbitrarily reduced further by a factor to reflect the maximum amount of potential loss of efficacy we are willing to accept. It is understandable that this factor would be higher in disease conditions with serious and irreversible complications.

The FDA, for example, stipulates a factor of 0.5 in clinical trials of thrombolytic and oncologic agents including mortality endpoints. The adjusted non-inferiority margin is used to calculate the sample size of the trial and, when the trial is completed, is used to evaluate the estimated interval of difference between the tested new intervention and the standard treatment. If the difference between both interventions is less than the non-inferiority or equivalence margin, the interactions can be judged as equivalent. The portion of treatment effect preserved by the new experimental treatment can be estimated, as well as its efficacy relative to a putative placebo.[88]

18.4.3.5 Patient Allocation Designs[23,54,81]

18.4.3.5.1 Single Group of Patients

In this design, a single group of patients is exposed to the intervention, which can be later modified according to adverse events or therapeutic response. It is an open-label and nonrandomized design that is subject to the bias inherent in these methods.

18.4.3.5.2 Parallel Group Designs

Parallel designs are the most common clinical trials designs for confirmatory Phase II and III clinical trials, in which subjects are randomized to one of two or more arms, each arm being allocated a different treatment. These treatments will include the investigational product at one or more doses, and one or more comparator treatments, such as placebo and/or an active comparator. The assumptions underlying this design are straightforward and less complex than those underlying other designs. Parallel designs can have different variations. For example, one variation of parallel design is for each group to receive alternating, and occasionally escalating, doses of the same drug.

A parallel trial can compare a single group of patients exposed to one intervention to another historical group. The historical group should resemble the prospective group as much as possible, especially in terms of factors considered to influence outcome. While *historical control designs* can be practical and easy to undertake, they pose several problems because of their lack of randomization and thus the possibility of bias. Also, using historical data can at times make for an uneven comparison if the disease pattern, or diagnostic criteria, may have changed over time.

Another variation of parallel designs is *matched design*. In this scenario, a pair of patients is matched on all the characteristics known or suspected to influence the outcome of the disease. Then, one of the pair is assigned to an intervention, and the other patient to the other treatment, or placebo. This approach matches controls for possible confounding variables and uses a relatively smaller sample size. The advantage of a parallel design is that while the disease state may change over time, this will occur in both groups, allowing for valid comparisons. Also, a parallel design has no carryover effect of a drug from one group to another.

18.4.3.5.3 Crossover Designs

In crossover designs, each subject is randomized to a sequence of two or more treatments and hence acts as his or her own control for treatment comparisons. The paradigm provides that after taking one of the medicines and a pre-specified period elapses, the participant is switched over to the other intervention. This approach, because it uses the same subject as its own control, reduces variability between patients and allows for a smaller sample size. However, the use of this design assumes that there will be no carryover *effect* from one intervention to the other, meaning that being exposed to one intervention will not affect the outcome after the participants' exposure to the other intervention. It is possible to decrease the likelihood for a carryover effect by including a *washout period* in the design interposed between interventions, to allow for the effects of the previous intervention to disappear. Also, carryover design requires a stable disease that does not change over a period of time, so as to allow for accurate assessment of the treatment effect, without the confounding effect of the disease's natural history.

A *withdrawal trial design* mimics the crossover design and calls for description of patients' responses after taking them off the intervention, again allowing them to serve as their own controls.

Another variation is the *multiple crossovers design* in which each group of patients receives each treatment twice or more during the trial. *Latin square design* is one multiple crossover group design. In this design, there are multiple groups of patients and multiple therapies, and each group of patients receives one sequence of all the available interventions, crossing over from one intervention to the other, the other groups receive different sequences of the available therapies. In a complete Latin square, all patient groups receive all the available treatments, but it is also possible to use an incomplete version, in which each group of patients is exposed only a sequence including some of the available interventions.

18.4.3.5.4 Factorial Designs

In a factorial design, two or more treatments are evaluated simultaneously through the use of varying combinations of treatments. The simplest example is the 2 by 2 factorial design in which subjects are randomly allocated to one of the four possible combinations of two treatments. One downside of the factorial design is the potential for *interaction*, meaning that one intervention can affect the response to another intervention. If one cannot reasonably rule out an interaction, the required statistical tests call for increased power, and sample size, to sort out this problem. On the other hand, this type of clinical trial can be very useful when it is desirable to explore different interactions, such as dose–response characteristics of the simultaneous use of two treatments. It is important to bear in mind, however, that complex diseases, and many concomitant drugs and interventions, would make this approach exceedingly difficult to analyze.

18.4.3.6 Large Simple Clinical Trials and Multi-Center Trials

Recently, trialists have advocated the concept of large, simple clinical trials to uncover modest but significant benefits in large cohorts of patients. When the intervention is fairly simple, and the outcome is easy to ascertain, it is possible to complete large clinical trials in a relatively short period of time. The ability to enroll large numbers of patients increases significantly the information yield of the trial. Multi-center trials are designed to achieve the objectives of the large simple clinical trial, and because they involve multiple investigators and clinical sites and recruit patients from a wider population and different practice settings, these trials make the generalization of the results scientifically more valid. Additionally, large simple trials can answer an important question at a much-reduced cost. Those trials have addressed important clinical issues such as the benefit of aspirin in primary prevention when a multi-center trial would never have been undertaken due to cost limitations.

18.4.3.7 Adaptive Clinical Trials Designs[89–91]

There is an emerging trend towards using a Bayesian statistical approach to design adaptive clinical trials. These designs continually assess the accumulating results, with the possibility of modifying the design of the trial, for example, by slowing (or stopping) or expanding patient accrual, correcting an imbalance in randomization, or increasing randomization to favor better-performing therapies. These approaches can drop or add treatment arms, or change the trial population to focus on a patient subset(s) that are responding better to the experimental therapies. These approaches have the potential benefit of conducting faster and smaller clinical trials that yield comparative information that is useful in reaching a rapid assessment of the utility of a new therapy. These trials respond to the ethical concerns of indiscriminate assignment of patients to different treatment arms and placebo that may be of no benefit or harmful. However, these studies pose special problems, as they obviously require the outcome response to an intervention to be readily available and discernable, and that the natural history of the disease to be stable over the period of the trial. Therapies with a long latency of effect are difficult to evaluate with these studies.

18.4.3.8   Pivotal Clinical Trials

A pivotal clinical trial is an adequate and well-controlled study that will form the basis for a regulatory agency marketing approval for the investigational pharmaceutical agent. Usually, at least two adequate and well-controlled studies are necessary to obtain the FDA's approval for a new drug. The requirement for well-controlled clinical investigations has been interpreted to mean that the effectiveness of a drug should be supported by more than one well-controlled trial and carried out by more than one independent investigator studying an adequate number of subjects. It is best if the pivotal clinical studies are not identical, as results that are obtained from studies that are of different design and independent in terms of execution, perhaps evaluating different populations, endpoints, or dosage forms may provide support for a conclusion of effectiveness that is as convincing as, or more convincing than, a repeat of the same study.

Although less common, regulatory agencies have accepted single clinical trials as a pivotal study to achieve a marketing approval. These single studies are usually large, multi-center studies and often involve multiple sub-studies. Often these studies evaluate multiple endpoints covering different events and show consistency across study subsets. These large diversified studies yield findings that are more likely to be accepted by FDA. With a single study, FDA usually expects the "P value" to be <0.0025, essentially the equivalent of two studies significant at the $p < 0.05$ study level. Reliance on a single study generally is limited to situations in which a trial has demonstrated a clinically meaningful effect on mortality, irreversible morbidity, or prevention of a disease with a potentially serious outcome, in that the conformation of the result in a second trial would be ethically difficult or impossible.

Generally, for a study to be considered pivotal it must satisfy certain *criteria*. A pivotal study must be a controlled trial, must have a blinded design, when such design is practical and ethical, must be randomized, and must be of adequate size to be considered appropriately powered. *Other characteristics* of a pivotal study include

* A clear statement of the objectives of the study.
* A design that permits a valid comparison.
* A method of subject selection that provides adequate assurance that enrolled patients have the condition being studied or they show evidence of susceptibility and exposure to the condition against which therapy is directed.
* Adequate measures to minimize bias through patients' selection and evaluation.
* Adequate analysis of the study results to assess the effects of the intervention.

18.4.4   Intention-to-Treat Analysis

The concept of intention to treat is that all entrants in a study are reported in the study analysis. Patients who are randomized are counted. This is often viewed as a careful, conservative approach to the analysis of a clinical trial. By counting "all entrants," one avoids many biases of the investigator or the study design. If patients are not counted in the study analysis, the omission can affect outcome and thus bias the study. Additionally, if some patients can tolerate the study therapy, that is believed an important factor and should be represented in the study outcome. For this reason, the intention-to-treat analysis is considered a bedrock of good clinical trial practices.

However, lately, the intention-to-treat analysis has been questioned as to its validity and viewed as a conservative approach to clinical trial design. One concern is that efficacy is being conflated with toxicity. If a therapy is very effective in a minority of patients, it still could be very useful. The signal of efficacy could be diluted because of adhering to intention-to-treat analysis. An example would be a trial of patients taking amiodarone to prevent sudden death. In the population on amiodarone for a year, the reduction in sudden death is noteworthy. But since approximately a third of patients tolerate amiodarone prolonged therapy, the drug is reported ineffective or marginally effective. The true fact is that if a patient can be treated with a certain dose, the drug prevents sudden death. The results should not be conflicted with the second important observation that amiodarone is a poorly tolerated treatment.

Another example is the drug Herceptin in breast cancer. It is only effective in the minority of breast cancer patients with a tumor marker (HER2), which is seen in about 20% of breast cancer. If one has a test for the maker (which is now available), the population that responds to Herceptin can be readily identified. The former example argues for reporting of both on-treatment analysis and intention-to-treat analyses, an analysis with a discussion of why the results differ depending on analysis method. The later example with Herceptin suggests that study results need to be carefully reviewed to see if a subgroup responds differently and if there is a plausible explanation. If there is a plausible explanation, then a second confirmatory study testing the hypothesis is needed, a time-consuming, but essential part of the trial process.

Shrier and associates came at the intention-to-treat analysis evaluation from an even more varied and highly valuable approach. The analysis began by pointing out that "the randomized trial design can be thought of as a means to answer two types of questions: (a) What is the effect of assigning a treatment? or (b) What is the effect of receiving treatment?" Intention-to-treat analysis answers the former question. The "real-world" question as to if a treatment is given, will it work in patients is not answered by the intention-to-treat analysis. This approach to analysis biases the study to the null hypothesis by making it harder to show a treatment effect if only a fraction of those assigned to treatment are receiving therapy.

At times, the intention-to-treat analysis may bias to a positive outcome and be a less conservative analysis. This is true if adherence rates differ between study groups when both treatments are equally effective. If adherence in a usual care group was inferior to the novel therapy, but both were equally effective, the novel therapy would appear far superior despite being of equal effectiveness. Non-inferiority trials showcase the problem with intention-to-treat analysis. A novel therapy with low dropout rates, but less activity would appear as effective as a less tolerated standard therapy, biasing the study away from the null hypothesis.

There is no easy solution to the analysis of clinical trials. Reporting the "on-treatment" or "as-treated" analysis can also be biased if there was a difference in response to treatment between those who did adhere to therapy and those who did not adhere to therapy. Often patients not conforming to therapy would have responded differently and thus an on-treatment analysis can be misleading. To be able to truly evaluate a study, Shrier et al. recommend providing the necessary information to calculate the average causal effect size, as well as the compliance and the average casual effect. The complier average causal effect is the difference in results between those who would and could receive the new treatment versus those who would and could receive the standard care. This excludes patients in the study who can't receive the normal therapy for reasons of toxicity and compliance, or for other reasons.

While the debate as to the most appropriate way to analyze a clinical trial continues, it becomes clear that the investigator can't be dogmatic and many approaches may be needed to reach an appropriate assessment and analysis of the study.

18.5   PHASE IV CLINICAL TRIALS

18.5.1   OVERVIEW

Phase IV trials are important for a number of reasons. They provide post-marketing safety profiling of pharmaceutical products. The importance of this safety information is periodically brought to the public's attention when a drug or product is associated with a newly discovered adverse side effect or, even worse, when a drug is withdrawn from the market due to adverse toxicity. Phase IV clinical trials also address other issues that arise during earlier phases of clinical development such as drug interactions, therapy-related quality of life, and as a way to broaden a drug's indications. It is important to note that a new indication for a product requires an IND and the initiation of studies for a supplemental NDA application.

Phase IV studies can also provide the clinical information to compare the marketed product with other interventions available for the same illness. Drugs are approved in the USA on the basis of safety and efficacy, while a drug's use as compared to similar agents or alternative therapies often remains untested. Phase IV studies provide the opportunity to compare agents to alternative therapy(s) and develop appropriate patient guidelines. While some call these studies marketing exercises, they are essential for the proper use of the drug. However, Phase IV studies can be designed for marketing purposes, giving physicians experience with the new therapy. At other times, these post-marketing studies are mandated by the regulatory authorities, to answer new questions that have developed since the Phase III program has been completed, or that were not well studied, but are not of a nature to prevent initial approval of the NDA.

18.5.2   POST-MARKETING DRUG SAFETY MONITORING

Regulatory authorities decide to approve a pharmaceutical product based on the safety profile obtained from administering the product to few thousands of patients. Even with large multinational Phase III clinical trials, it is very unlikely that more than 10,000 patients have been included in the safety analysis of the drug.[92] This means that a serious, or even fatal, side effect that occurs at a rate of <1 in 5,000 can be initially missed only to be apparent when tens of thousands to millions of patients receive the drug. To detect a case of an adverse toxicity whose incidence is one in a thousand would take 3,000 patients to be exposed, and to detect three cases would require 6,000 exposures.

Following the mass marketing of a drug, people usually receive the therapy for a longer period than the drug was tested, at times at doses higher than those studied, or in combinations with other medicines that have not been previously combined in clinical trials. These prolonged exposures and potential new drug interactions can lead to a safety profile quite different from the one initially observed during earlier phases of clinical research.[93] These problems highlight the most important reasons for completing Phase IV clinical trials, which is safety monitoring. Public debate is periodically stirred by drug withdrawals for safety reasons,[94] and voices call for more involvement of regulatory agencies in assuring that pharmaceutical sponsors execute their Phase IV commitments.

Some proposals call for the establishment of an independent drug safety monitoring center to monitor safety signals, as well as enforcing the completion of mandated Phase IV studies.[95–97] Others call for an approval system for new medicines conditional on the execution of these Phase IV clinical trials.[92] A provisional approval system that requires re-evaluation of a drug after 2 or 3 years post-marketing would be advisable since greater safety exposure would have been obtained. A drug's toxicity profile may change and thus may require labeling modifications to account for new knowledge about the drug,[98] or it is possible that with more extensive knowledge gained, the drug could be withdrawn from the market or its use severely restricted. Giving the FDA, the ability to re-adjust the approval decision would go a long way to accelerate the drug approval process, since the initial decision would not be an "all or nothing" decision with very little recourse left to FDA, except terminating marketing of an approved product.

18.5.3 SEEDING TRIALS

Phase IV clinical trials at times aim to accomplish the post-marketing objectives of increased drug exposure while having practitioners become familiar with the pharmaceutical product, so as to affect their prescribing habits.

18.5.4 DESIGNS OF PHASE IV CLINICAL TRIALS

Phase IV clinical studies can be descriptive studies or controlled clinical trials.

18.5.4.1 Descriptive and Analytic Studies[99]

These can be initially *passive epidemiologic surveillance studies* concerned with monitoring disease patterns and population demographics, as well as passively observing if signals related to drug safety and efficacy appear. This information can then help generate a hypothesis that can be tested by more formal study designs and statistical data analysis.

The design of these subsequent studies can be

- *Cross-sectional*, analyzing the relation of exposure and outcome at a particular point in time.
- *Retrospective case–control studies*: in which a sample is described in terms of outcomes, i.e., adverse events, followed by evaluation of drug exposure in the past, prospective case versus control studies.
- *Prospective cohort studies*: in which a patient sample is determined in terms of a certain exposure, then followed to determine outcome.

This can also be achieved in a *retrospective cohort study* in which a sample is described in terms of a past exposure and then evaluated for the outcomes in the present. Placebo-controlled clinical trials can be used as well to achieve the objectives of Phase IV studies, with the study drug contrasted to conventional therapy.

18.5.5 REGISTRIES

The use of registries, especially in the medical device area, has become a frequent vehicle for Phase IV post-marketing studies. Registries have the advantage of easy patient accession and are simpler to initiate and execute than controlled studies. However, registries have inherent limitations being nonrandomized, nonblinded study with all the ensuing biases. Registries often exaggerate benefit due to biased patient selection, but the biases implicit in patient selection can also involve the side effect profile, since only selected patients are entered who often are less likely to be subject to the drug's adverse profile. At times, the registry can work against the product because there is no control group to compare to in terms of adverse reports. Many different adversities can be reported, and all must be attributed to the drug. With at least a contemporaneous control group (nonrandomized), one may be able to discern a signal of toxicity from background chance side effects, although cause and effect can't be evaluated, in a nonrandomized, nonblinded registry.

18.5.6 SPECIAL POPULATIONS

Phase IV studies are often the time to develop exposure information in populations not previously studied. While exposure to drugs of women and children is mandated by FDA, Phase III trials due to disease prevalence, referral center patterns and reluctance to expose pediatric patients and women, find these groups inadequately represented. Phase IV studies specifically designed to increase exposure to women, women of childbearing potential, patients with severe liver and/

or renal impairments, patients on multiple other drugs, and patients that are diverse (biogenetic diversity) can be undertaken. Exposure to these populations can help in developing dosing adjustments, new labeling on side effect profiles, or the expanded use of a new drug in these underexposed groups. Pediatric dosing and indications often are added after a drug's approval for a condition that predominantly affects adults. The information can be obtained from special population studies initiated in the Phase IV context and update the limited pediatric labeling information included in the initial product label. However, FDA is mandating a pediatric study program before NDA submission when the therapy is expected to be used in children.

18.5.7 Pharmacoeconomic Analysis

The evaluation of the benefit of a new agent is not only limited to safety and efficacy but may also involve cost. In the US regulatory process, the pharmacoeconomic analysis is usually performed post-marketing and does not play a role in the NDA and approval process. Pharmacoeconomic studies compare the value of pharmaceutical products to existing products. These studies evaluate the cost, expressed in monetary terms and effects. Also, the studies can measure an effect as a monetary value to the drug effects in terms of efficacy or quality of life parameters. There are a number of methods of analysis that can entail measuring direct costs (price, staff, capital expense); indirect costs, such as loss of earnings, loss of productivity, cost of tests and/or treatment; and costs of hospitalizations.

Benefits can include years of life saved, quality of life, mobility, or psychological outcomes. Often the benefit of a therapy can be defined as a "quality" or the quality of adjusted life years and that can include both quality and quantity of life. A pharmacoeconomic analysis can include cost minimization, cost-effectiveness, cost–utility, cost–benefit, or cost-disability analyses.

Cost minimization analysis requires that the therapies being evaluated have similar health outcomes. The comparison is based on cost. Cost-effectiveness analysis can be undertaken for two or more therapies that have the same treatment objectives, but different degrees of efficacy. The question asked is how much additional benefit is achieved for the incremental cost. One calculates the increased cost-effectiveness of therapy A over therapy B expressed as the incremental cost-effectiveness ratio. The finding of higher effectiveness and lower cost is the highest level in pharmacoeconomic analysis. Higher cost, but with higher effectiveness, is a societal choice and may be an accepted finding. Lower cost and lower effectiveness are sub-optimal, and low or no effectiveness and higher cost are an unacceptable pharmacoeconomic outcome.

Cost–utility analysis provides a way to consider cost on quality of life and survival by converting to common units of measure. The outcome measure most frequently employed is quality-adjusted life years (QALY). The benchmarks are arbitrary and need to be accepted by society. Interventions costing between $50,000 and $90,000 are often considered cost-effective, while interventions below $50,000 are considered highly cost-effective. Those interventions above $90,000 are considered less cost-effective. The disability-adjusted life year (DALY) is another way of looking at costs in economic terms. This index is a measure of overall disease burden combining death and disability. Cost–benefit analysis (CBA) measures both cost and benefits measured in financial terms.

Some object to putting a monetary value on life, but precedence in the compensation area exists and has been used for centuries. The cost of a life is often set by insurance companies at 2 million dollars (upper lifetime insurance limit). This lifetime cap is a de facto cost assessment of a life in financial terms.

Pharmacoeconomic analysis needs to define costs involved in healthcare, combine these costs with acceptable societal evaluation of life, death, and morbidity, in dollars assigned, and factor these costs against the cost of a therapeutic intervention. As healthcare becomes ever more expensive, Phase IV studies in drug development will need to undertake pharmacoeconomic analysis of a drug's role before the acceptance of a new therapy or therapeutic paradigm is adopted. By not reimbursing for a therapy, insurers and the government can control the therapeutic decision process.

18.5.8 COMPARATIVE EFFECTIVENESS RESEARCH

On February 17, 2009, the American Recovery and Reinvestment Act was signed into law. The bill, among other considerations, allocates $1.1 billion to comparative effectiveness research. The allocation was $300 million for the agency for Healthcare Research and Quality, $400 million for the National Institutes of Health, and $400 million for the Office of the Secretary of Health and Human Services; all to promote and carry out comparative effectiveness research.

Comparative effectiveness research is designed to inform healthcare decision-makers by providing evidence on the effectiveness, benefits, and names of different treatment options. Evidence is generated from research studies that compare drugs, medical devices, tests, surgical procedures, or ways to deliver healthcare. The evidence is generated by researchers evaluating all available data about the benefits or harms of each claim for different groups of people with a given disease process from existing clinical trials, clinical observational studies, or other databases. This data collection is culled from research reviews or systemic clinical data reviews of existing evidence. Additionally, researchers can conduct prospective studies that generate **new** evidence of effectiveness or comparative effectiveness of a treatment, procedure, or healthcare service.

The Agency for Healthcare Research and Quality is charged with the mission of facilitating comparative effectiveness research by

1. Identifying new and emerging clinical interventions.
2. Reviewing and synthesizing current medical research.
3. Identifying gaps between existing medical research and the needs of clinical practice.
4. Promoting and generating new scientific evidence and analytic tools.
5. Training and developing clinical researchers in this area.
6. Translating and disseminating research findings to diverse stakeholders.
7. Reaching out to stakeholders through citizens forum(s).

Systematic reviews of existing research are perhaps the first line of comparative effectiveness research. If the data exist, the most cost-effective and timely way to obtain and analyze is by a systematic review. The reviewer must operate with pre-specified methods, obtain most, if not all, existing data, and provide a fair, standardized analysis of pre-specified response variables and reported toxicity.

Original research may be needed when existing data are not available, do not exist, or are incomplete. Undertaking with observational studies or randomized controlled clinical trials is far more expensive and time-consuming. However, these data are often more reliable and better controlled, possesses less bias, and compares comparable populations, treatments, and measures of outcomes. As with all forms of clinical research, there exists a hierarchy of validity, based on trial methodology. Which type of data is used depends on the availability of data, the time frame available, and the resources that can be applied to the question asked. Who undertakes comparative effectiveness research depends on the question asked and the resources available. While much of Phase IV research can be considered comparative effectiveness research and these trials have often been initiated by industry or academic physicians, in the future much of comparative effectiveness research will be more rigorously organized by Federal Agencies. Two government agencies involved are the DECIDE – Developing Evidence to Inform Decisions about Effectiveness Network and CERTS – Centers for Education and Research on Therapeutics. With the enormous sums the government is going to invest in this area, these groups will become even more important in guiding healthcare therapeutics and decision-making.

Key questions involved in comparative effectiveness research are

1. What is the societal impact of a therapy on health?
2. What is the economic impact?

3. Does the comparative effectiveness data currently exist?
4. Are the short- and long-term comparative effectiveness outcomes known? Other critical questions that need to be addressed are whether the existing data are generalizable to "real-world" patient settings and whether the information is generalizable to special populations: women of childbearing potential, ethnic minorities, children, and patients with mild disease or very severe disease. Often these patient groups are not included in the Phase I, II, and III clinical trials or even in Phase IV studies making assessment of generalizability difficult to impossible. Another critical area is the standardization of the process. This entails a systemic assessment of all studies and all existing data. Dose standardization in comparative effectiveness trials is a critical issue. Are we comparing similar doses, or a low dose of standard therapy that has low efficacy, to a higher dose of a new therapy with greater efficacy? The lack of dose standardization has been the bane of comparative effectiveness trials, making the conclusions vulnerable to bias. The designs of these studies by industry are often contrived, thus providing misleading information.

The future of comparative effectiveness research includes the rapidly developing area of data mining. With large databases, electronic research with sophisticated computer programs will allow for outcomes research to obtain efficacy and safety data from databases from hundreds of thousands to millions of patients. This will be "real-world" exposure in such large numbers as to provide very useful information on efficacy and toxicity. The size of the databases markedly diminishes the issues of bias and generalizability. Still, the quality of the database, and its reliability, is critical to the future utility of approaches.

Randomized controlled clinical trials and, to a lesser extent, observational studies still provide the majority of clinical effectiveness data. What questions to ask and how to use the newly created resources of the electronic medical record needs to be debated. Perhaps the "large simple trial," or a large pragmatic trial, will best lend themselves to aiding in the comparative research paradigm. These studies examine important issues in large, "real-world" populations in terms of cost-effectiveness and usually provide results in a short period of time. However, one must ensure adequate adverse event reporting, as well as rigorous follow-up of patients with near-total reporting of outcome and adverse events. Still not all outcomes can be easily measured using the "simple trial" paradigm, and more sophisticated, expensive, studies are needed to make a comparative effective assessment. One approach to increase the speed of assessment, diminish cost, but still undertake a randomized controlled trial with sophisticated endpoint assessment is to employ adaptive clinical trial designs. These trials can adjust for imbalances in randomization or outcome with previously planned changes in design, patient accession, and stratification as the trial enrolls patients. Unanticipated clinical outcomes are addressed in real time by protocol adjustments. Trial arms can be discontinued, doses adjusted, more patients entered, and all adjustments made on the information collected as to success, failure, or futility. These adjustments can facilitate an answer questions regarding comparative effectiveness. When all is said and done about comparative effectiveness research, it still comes down to a judgment. What margin defines a comparative effectiveness gain or loss, what defines inferiority and how does one judge efficacy against toxicity? A drug can be more effective, but more toxic: is it useful or not? These are clinical judgment issues, and comparative effectiveness research can only go so far in aiding clinical decision-making.

18.6 RECENT DEVELOPMENTS AND FUTURE DIRECTIONS

18.6.1 Pharmacogenetics

Pharmacogenetics involves the use of genetic analyses to predict drug response, efficacy, and safety.[100] The different human phenotypes make for wide individual differences in PK/PD of drug profiles, and these differences can impact an intervention's safety and efficacy. "Slow metabolizers," for example, have a cytochrome P450 enzyme system that clears certain drugs less efficiently, permitting an increase in the concentration of the metabolized drug. Also, the same phenotypes can sometimes allow the classification of the population into different groups, based on degree of response to the drug effect. Pharmacogenetics permits the identification of the various phenotype determinants of PK and PD, which may provide information to optimally, select the drug or dose for an individual patient. The development of "individualized medicine" can potentially improve the therapeutic intervention by administering drugs to patients most likely to benefit. This model, however, requires the concomitant development of practical assays and tests, to identify the different phenotypes, relevant to a particular intervention. Pharmacogenetic studies are increasingly incorporated in the design of clinical trials to increase their efficiency.

The outcome of exploratory Phase II clinical trials can yield

1. Clear efficacy, leading to further confirmatory testing.
2. Clear failure, in which case further testing stops.
3. Varying degrees of efficacy, a problematic scenario when some patients appear to show an effective response and others don't.

The application of *pharmacogenetics* can identify the genetic determinants of the responders, hypo-responders, as well as non-responders. By excluding early on, predicted non-responders, it is possible to design smaller more efficient clinical trials to demonstrate success of a medicine, which would have failed testing in the general population. An example is the failure of the beta blocker bucindolol in clinical trials. Subsequent analysis found that some patients responded, and indeed, a genetic polymorphism has been identified that is a marker of excellent drug response. Patients with this marker and who have heart failure (HF) respond best to bucindolol over other therapies. Early identification of the marker, in early Phase II studies, would have saved time and money in development and not required the treatment of a population of patients with HF that could not respond to the therapy.

A second example is in the treatment of breast cancer. Herceptin is effective for HER2-positive breast cancer. Given to all patients, the drug would not show efficacy, but administered to HER2-positive tumors, the drug is a most effective, life-saving, agent. An example of a toxicogenomic marker is patients with HLA B*57:01-phenotype that are more likely to develop a hypersensitivity to abacavir (fever, rash, GI symptoms, and even death). Checking for HLA B*57:01 is becoming routine before starting the anti-viral abacavir.

18.6.2 Data Safety Monitoring Board

While the principal investigator has primary responsibility for the study he or she supervises, multi-center studies and large studies that are blinded, need to be monitored by a group of trained and experienced individuals, independent of the study, the principal investigator, and study sponsor. A DSMD needs to have data access, tabulate the data, perform preliminary statistical analysis periodically, and evaluate the study progress in terms of both safety and efficacy. If an intervention becomes clearly harmful, a DSMB needs to stop the study. If an intervention is clearly beneficial, the study needs to be stopped and results need to be reported.

If it is futile to continue a study because a difference can't be shown, a study ethically needs to be stopped. If poor data are coming out of the study, they need to be identified and corrected. If toxicity is being seen and greater vigilance is needed by the study staff, this too must be communicated. These are important functions, but a DSMB poses problems.

Repeated outcome measures need to be taken into account and could compromise the statistical power of a study. Early stopping of a study can lead to reporting positive results that are not convincing to the medical community and thus fail to influence practice. Early termination due to an adverse event may be premature, a useful therapy for a serious disease may not be available, and a true risk–benefit evaluation may not be obtained. For these reasons, the deliberations of a DSMB may be some of the most critical aspects of a clinical trial with significant ramifications. Having the DSMB blinded may reduce premature stopping and termination for adverse reports, but can become difficult to manage, and some complain that patient safety may be compromised. However, a blinded DSMB is the optimal approach.

18.6.3   Master Protocols

To facilitate drug development and the introduction of new evidence-based therapies into clinical practice, the FDA in collaboration with the National Cancer Institute (NCI), academic investigators, and other "stake holders" has proposed the utilization of Master Protocols.[101] These protocols are extensions of conventional approaches to clinical research, but offer a method that facilitates complex studies. Master Protocols can fall under three general rubrics: umbrella trial protocols, basket trials, and platform trials.

Umbrella trials aim to study multiple therapies for a single disease, or closely related conditions, while basket trials study a single therapy for evaluation in multiple diseases, or disease subtypes. Platform trials can study multiple targeted therapies in the context of a single disease in an ongoing manner, with therapies allowed to enter the process or exit the evaluation process, on the basis of a carefully devised and thought-out decision algorithm.

An example of a basket trial is the B222 Master Protocol where a common biomarker treatment drug combination therapy is evaluated in multiple diseases. The National Cancer Institute Molecular Analysis for Therapy Choice (NCI-MATCH) Master Protocol is an umbrella trial evaluating multiple genetic markers and targeted therapies in cancer of differing histologic types with targeted mutations. Other examples include I-SPY2, a Master Protocol where various therapies are evaluated as neoadjuvant treatments for biomarker-identified subtypes in early-stage breast cancer. The study employs a response adaptive design randomization, assigning patients to the most promising therapies with Bayesian decision rules that determine whether therapies with low probabilities of success, or unacceptable side effects, are to be eliminated. Lung-MAP is an example of a Phase II, III Master Protocol evaluating a series of genetically targeted therapies against carefully characterized advanced squamous, non–small cell lung cancer.

These trial designs may be very complex and incorporate the latest in advanced trial design concepts, all aimed at facilitating testing of new treatments at the lowest cost, maximizing patient benefit. However, the potentially most innovative aspect is the establishment of a trial structure that can exist beyond a given research question. The trial network, its participants, the infrastructure, rules of governance, study sites, and study systems all can be established and exist long after a given therapy or disease is no longer part of the study. The advantages are that the tremendous time, effort, and expense going into creating these essential components of clinical research are an investment that can be used over and over, facilitating clinical research. The downside is that the process is initially more difficult, time-consuming, and costly. Like all investments, the initial gathering of finances and the requisite time and effort requires considerable work, but the long-term outcome may be more rewarding. An additional benefit of the Master Protocol is the use of a single control group, reducing the need for multiple controls and thus increasing drug exposure which can facilitate patient recruitment and reduce costs. Use of single control groups with multiple drug testing comparisons can be supported by the use of the same sites, and thus, other factors, confounders affecting outcome that one often tries to control for, will remain constant. Additionally, concomitant therapies and approaches are standardized, since the same physicians, centers, and approaches to therapeutics are repeatedly employed (standardized). Still there are factors such as elapsed time that can't be compensated for that can affect outcome. The requirement for multiple comparisons and repeated testing of significance raises the issue of multiplicity, which is a problem that adds to the statistical complexity of Master Protocols.

In summary, Master Protocols facilitate the development of evidence-based medicine introducing new effective therapies and their comparative effectiveness.

18.7 FINAL THOUGHTS

Clinical research is an exciting area of medicine with the potential to translate the breakthroughs of basic science into medical therapy. Phases I, II, III, and IV of clinical research clarify the steps in clinical trials aimed at the approval of new therapeutic entities into a sequential development paradigm that facilitates the generation of an extensive body of knowledge needed for the drug approval process. However, the clinical trials process does not stop with approval, but must continue the surveillance of new therapeutic entities to ensure efficacy, properly assess toxicity and perhaps, and, most importantly, integrate the new therapeutic modality into clinical practice. Most therapies have a place in clinical practice, the art is to find the proper role, and this art requires a rigorous scientific process of preclinical and clinical trial design and execution through the different phases of clinical therapeutic development and throughout the entire life cycle of the therapeutic agent.

REFERENCES

1. Barton JH, Emanuel EJ. The patents-based pharmaceutical development process: Rationale, problems, and potential reforms. *JAMA*. 2005;294(16):2075–2082.
2. Mervis J. Productivity counts—but the definition is key. *Science*. 2005;309(5735):726.
3. Dimasi JA. Risks in new drug development: Approval success rates for investigational drugs. *Clin Pharmacol Ther*. 2001;69(5):297–307.
4. DiMasi JA, Hansen RW, Grabowski HG. The price of innovation: New estimates of drug development costs. *J Health Econ*. 2003;22(2):151–185.
5. Posvar EL, Sedman AJ. New drugs: First time in man. *J Clin Pharmacol*. 1989;29(11):961–966.
6. Reigner BG, Blesch KS. Estimating the starting dose for entry into humans: Principles and practice. *Eur J Clin Pharmacol*. 2002;57(12):835–845.
7. Colburn WA. Controversy V: Phase I, first time in man studies. *J Clin Pharmacol*. 1990;30(3):210–222.
8. Reele SB. Decision points in human drug development. In: John O' Grady PHJ, ed. *Handbook of Phase 1/2 Clinical Drug Trials*. Boca Raton: CRC Press; 1996.
9. Cutler NR, Sramek JJ. Investigator perspective on MTD: Practical application of an MTD definition–has it accelerated development? *J Clin Pharmacol*. 2000;40(11):1184–1187; discussion 1202–1184.
10. Collins JM, Grieshaber CK, Chabner BA. Pharmacologically guided phase I clinical trials based upon preclinical drug development. *J Natl Cancer Inst*. 1990;82(16):1321–1326.
11. Zhou Y. Choice of designs and doses for early phase trials. *Fundam Clin Pharmacol*. 2004;18(3):373–378.
12. Buoen C, Bjerrum OJ, Thomsen MS. How first-time-in-human studies are being performed: A survey of phase I dose-escalation trials in healthy volunteers published between 1995 and 2004. *J Clin Pharmacol*. 2005;45(10):1123–1136.
13. Rosenzweig P, Brohier S, Zipfel A. The placebo effect in healthy volunteers: Influence of experimental conditions on physiological parameters during phase I studies. *Br J Clin Pharmacol*. 1995;39(6):657–664.
14. Sibille M, Deigat N, Janin A, Kirkesseli S, Durand DV. Adverse events in phase-I studies: A report in 1015 healthy volunteers. *Eur J Clin Pharmacol*. 1998;54(1):13–20.
15. Buoen C, Holm S, Thomsen MS. Evaluation of the cohort size in phase I dose escalation trials based on laboratory data. *J Clin Pharmacol*. 2003;43(5):470–476.
16. Thomas M. Study design and assessment of wanted and unwanted drug effects in phase 1/2 trials. In: John O' Grady PHJ, ed. *Handbook of Phase 1/2 Clinical Drug Trials*. Boca Raton: CRC Press; 1996.
17. Ette EI, Williams PJ. Population pharmacokinetics I: Background, concepts, and models. *Ann Pharmacother*. 2004;38(10):1702–1706.
18. Guidance for industry: Population pharmacokinetics. In: U.S. Department of Health and Human services FaDA, ed; 1999.
19. CDER. Guidance for industry: Bioavailability and bioequivalence studies for orally administered drug products-general considerations. In: FDA, ed; 2002.
20. Balthasar JP. Bioequivalence and bioequivalency testing. *Am J Pharm Educ*. 1999;63(Summer 1999):194–198.

21. Monica Takoo MM, Sahoo U. Bioavailability and bioequivalence trials: A promising business opportunity in India. *The Monitor.* 2006;20(3):43–46.

22. Beumer JH, Beijnen JH, Schellens JH. Mass balance studies, with a focus on anticancer drugs. *Clin Pharmacokinet.* 2006;45(1):33–58.

23. Spilker B. *Guide to Clinical Trials.* New York: Raven Press; 1991.

24. Lasser KE, Allen PD, Woolhandler SJ, Himmelstein DU, Wolfe SM, Bor DH. Timing of new black box warnings and withdrawals for prescription medications. *JAMA.* 2002;287(17):2215–2220.

25. Roden DM. Drug-induced prolongation of the QT interval. *N Engl J Med.* 2004;350(10):1013–1022.

26. Gussak I, Litwin J, Kleiman R, Grisanti S, Morganroth J. Drug-induced cardiac toxicity: Emphasizing the role of electrocardiography in clinical research and drug development. *J Electrocardiol.* 2004;37(1):19–24.

27. Guidance for industry: E14 clinical evaluation of QT/QTc interval prolongation and proarrhythmic potential for non-arrhythmic drugs. In: CDER F, ed; 2005.

28. Molnar J, Ranade V, Cvetanovic I, Molnar Z, Somberg JC. Evaluation of a 12-lead Digital Holter system for 24-hour QT interval assessment. Chicago: American Institute of Therapeutics, Rush University; 2006.

29. Cutler NR, Sramek JJ, Greenblatt DJ, et al. Defining the maximum tolerated dose: Investigator, academic, industry and regulatory perspectives. *J Clin Pharmacol.* 1997;37(9):767–783.

30. Tatro DS. *Drug Interaction Facts.* 2004 ed. St. Louis, MO: Facts and Comparisons; 2004.

31. Alfaro CL. Drug interactions. In: Atkinson AJ, ed. *Principles of Clinical Pharmacology.* San Diego, CA: Academic Press; 2001:xvi, 460 p.

32. Guidance for industry: Drug interaction studies-study design, data analysis, and implications for dosing and labeling. In: CDER F, ed; 2006.

33. Welling PG. Effects of food on drug absorption. *Pharmacol Ther.* 1989;43(3):425–441.

34. Santos CA, Boullata JI. An approach to evaluating drug-nutrient interactions. *Pharmacotherapy.* 2005;25(12):1789–1800.

35. Wadman M. London's disastrous drug trial has serious side effects for research. *Nature.* 2006;440(7083):388–389.

36. Wood AJ, Darbyshire J. Injury to research volunteers–the clinical-research nightmare. *N Engl J Med.* 2006;354(18):1869–1871.

37. Evans D, Smith M, Willen L. Big pharma's shameful secret. *Bloomberg Markets.* 2005;14:36–62.

38. Roberts Jr. TG, Goulart BH, Squitieri L, et al. Trends in the risks and benefits to patients with cancer participating in phase 1 clinical trials. *JAMA.* 2004;292(17):2130–2140.

39. Horstmann E, McCabe MS, Grochow L, et al. Risks and benefits of phase 1 oncology trials, 1991 through 2002. *N Engl J Med.* 2005;352(9):895–904.

40. Markman M. "Therapeutic intent" in phase 1 oncology trials: A justifiable objective. *Arch Intern Med.* 2006;166(14):1446–1448.

41. Khandekar J, Khandekar M. Phase 1 clinical trials: Not just for safety anymore? *Arch Intern Med.* 2006;166(14):1440–1441.

42. Garner RC, Lappin G. The phase 0 microdosing concept. *Br J Clin Pharmacol.* 2006;61(4):367–370.

43. Schmidt B. Proof of principle studies. *Epilepsy Res.* 2006;68(1):48–52.

44. Temple R. A regulatory authority's opinion about surrogate endpoints. In: Nimmo WS, Tucker GT, eds. *Clinical Measurement in Drug Evaluation.* Chichester; New York: Wiley; 1995: xi, 329 p.

45. Rolan P. The contribution of clinical pharmacology surrogates and models to drug development–a critical appraisal. *Br J Clin Pharmacol.* 1997;44(3):219–225.

46. Baker SG. Surrogate endpoints: Wishful thinking or reality? *J Natl Cancer Inst.* 2006;98(8):502–503.

47. Echt DS, Liebson PR, Mitchell LB, et al. Mortality and morbidity in patients receiving encainide, flecainide, or placebo. The cardiac arrhythmia suppression trial. *N Engl J Med.* 1991;324(12):781–788.

48. Fleming TR, DeMets DL. Surrogate end points in clinical trials: Are we being misled? *Ann Intern Med.* 1996;125(7):605–613.

49. Coplen SE, Antman EM, Berlin JA, Hewitt P, Chalmers TC. Efficacy and safety of quinidine therapy for maintenance of sinus rhythm after cardioversion. A meta-analysis of randomized control trials. *Circulation.* 1990;82(4):1106–1116.

50. Prentice RL. Surrogate endpoints in clinical trials: Definition and operational criteria. *Stat Med.* 1989;8(4):431–440.

51. Lesko LJ, Atkinson Jr. AJ, Use of biomarkers and surrogate endpoints in drug development and regulatory decision making: Criteria, validation, strategies. *Annu Rev Pharmacol Toxicol.* 2001;41:347–366.

52. Rolan P, Atkinson Jr. AJ, Lesko LJ. Use of biomarkers from drug discovery through clinical practice: Report of the Ninth European Federation of Pharmaceutical Sciences Conference on Optimizing Drug Development. *Clin Pharmacol Ther.* 2003;73(4):284–291.
53. ICH. E10: Choice of control group and related issues in clinical trials. 2001.
54. Friedman LM, Furberg C, DeMets DL. *Fundamentals of Clinical Trials.* 3rd ed. New York: Springer; 1998.
55. Emanuel EJ, Wendler D, Grady C. What makes clinical research ethical? *JAMA.* 2000;283(20):2701–2711.
56. Temple R, Ellenberg SS. Placebo-controlled trials and active-control trials in the evaluation of new treatments. Part 1: Ethical and scientific issues. *Ann Intern Med.* 2000;133(6):455–463.
57. Ellenberg SS, Temple R. Placebo-controlled trials and active-control trials in the evaluation of new treatments. Part 2: Practical issues and specific cases. *Ann Intern Med.* 2000;133(6):464–470.
58. Freedman B, Weijer C, Glass KC. Placebo orthodoxy in clinical research. I: Empirical and methodological myths. *J Law Med Ethics.* 1996;24(3):243–251.
59. Freedman B, Glass KC, Weijer C. Placebo orthodoxy in clinical research. II: Ethical, legal, and regulatory myths. *J Law Med Ethics.* 1996;24(3):252–259.
60. Rothman KJ. Declaration of Helsinki should be strengthened. *BMJ.* 2000;321(7258):442–445.
61. Emanuel EJ, Miller FG. The ethics of placebo-controlled trials–a middle ground. *N Engl J Med.* 2001;345(12):915–919.
62. Packer M. Current perspectives on the design of phase II trials of new drugs for the treatment of heart failure. *Am Heart J.* 2000;139(4):S202–206.
63. Stallard N, Whitehead J, Todd S, Whitehead A. Stopping rules for phase II studies. *Br J Clin Pharmacol.* 2001;51(6):523–529.
64. Turrisi III AT. Creeping phase II-ism and the medical pharmaceutical complex: Weapons of mass distraction in the war against lung cancer. *J Clin Oncol.* 2005;23(22):4827–4829.
65. Lee JJ, Feng L. Randomized phase II designs in cancer clinical trials: Current status and future directions. *J Clin Oncol.* 2005;23(19):4450–4457.
66. Estey EH, Thall PF. New designs for phase 2 clinical trials. *Blood.* 2003;102(2):442–448.
67. Van Glabbeke M, Steward W, Armand JP. Non-randomised phase II trials of drug combinations: Often meaningless, sometimes misleading. Are there alternative strategies? *Eur J Cancer.* 2002;38(5):635–638.
68. Rubinstein LV, Korn EL, Freidlin B, Hunsberger S, Ivy SP, Smith MA. Design issues of randomized phase II trials and a proposal for phase II screening trials. *J Clin Oncol.* 2005;23(28):7199–7206.
69. Schmidt R. Dose-finding studies in clinical drug development. *Eur J Clin Pharmacol.* 1988;34(1):15–19.
70. CDER. Guidance for industry: Exposure-response relationships–study design, data analysis, and regulatory applications. 2003.
71. Freston JW. Dose-ranging in clinical trials: Rationale and proposed use with placebo or positive controls. *Am J Gastroenterol.* 1986;81(5):307–311.
72. Johnston GD. Dose-response relationships with antihypertensive drugs. *Pharmacol Ther.* 1992;55(1):53–93.
73. Rowland M, Tozer TN. *Clinical Pharmacokinetics: Concepts and Applications.* 2nd ed. Philadelphia, PA: Lea & Febiger; 1989.
74. ICH. E4: Dose-response information to support drug registration. 1994.
75. Cassileth BR, Lusk EJ, Miller DS, Hurwitz S. Attitudes toward clinical trials among patients and the public. *JAMA.* 1982;248(8):968–970.
76. Freedman B. Equipoise and the ethics of clinical research. *N Engl J Med.* 1987;317(3):141–145.
77. Shapiro SH, Glass KC. Why Sackett's analysis of randomized controlled trials fails, but needn't. *CMAJ.* 2000;163(7):834–835.
78. Sackett DL. Equipoise, a term whose time (if it ever came) has surely gone. *CMAJ.* 2000;163(7):835–836.
79. Sackett DL. Why randomized controlled trials fail but needn't: 1. Failure to gain "coal-face" commitment and to use the uncertainty principle. *CMAJ.* 2000;162(9):1311–1314.
80. Peto R, Baigent C. Trials: The next 50 years. Large scale randomised evidence of moderate benefits. *BMJ.* 1998;317(7167):1170–1171.
81. ICH. E9: Statistical principles for clinical trials. 1998.
82. Sackett DL. Superiority trials, noninferiority trials, and prisoners of the 2-sided null hypothesis. *ACP J Club.* 2004;140(2):A11.
83. Ware JH, Antman EM. Equivalence trials. *N Engl J Med.* 1997;337(16):1159–1161.
84. Glasser SP, Howard G. Clinical trial design issues: At least 10 things you should look for in clinical trials. *J Clin Pharmacol.* 2006;46(10):1106–1115.

85. Piaggio G, Elbourne DR, Altman DG, Pocock SJ, Evans SJ. Reporting of noninferiority and equivalence randomized trials: An extension of the CONSORT statement. *JAMA*. 2006;295(10):1152–1160.

86. Gotzsche PC. Lessons from and cautions about noninferiority and equivalence randomized trials. *JAMA*. 2006;295(10):1172–1174.

87. EMEA. Guideline on the choice of the non-inferiority margin. 2006.

88. Kaul S, Diamond GA. Good enough: A primer on the analysis and interpretation of noninferiority trials. *Ann Intern Med*. 2006;145(1):62–69.

89. Berry DA. Bayesian clinical trials. *Nat Rev Drug Discov*. 2006;5(1):27–36.

90. Parmar MK, Spiegelhalter DJ, Freedman LS. The CHART trials: Bayesian design and monitoring in practice. CHART Steering Committee. *Stat Med*. 1994;13(13–14):1297–1312.

91. Spiegelhalter DJ, Myles JP, Jones DR, Abrams KR. Methods in health service research. An introduction to Bayesian methods in health technology assessment. *BMJ*. 1999;319(7208):508–512.

92. Strom BL. How the US drug safety system should be changed. *JAMA*. 2006;295(17):2072–2075.

93. Roden DM. An underrecognized challenge in evaluating postmarketing drug safety. *Circulation*. 2005;111(3):246–248.

94. Friedman MA, Woodcock J, Lumpkin MM, Shuren JE, Hass AE, Thompson LJ. The safety of newly approved medicines: Do recent market removals mean there is a problem? *JAMA*. 1999;281(18):1728–1734.

95. Woosley RL, Flockhart DA. Evaluating drugs after their approval for clinical use. *N Engl J Med*. 1994;330(19):1394–1395.

96. Ray WA, Griffin MR, Avorn J. Evaluating drugs after their approval for clinical use. *N Engl J Med*. 1993;329(27):2029–2032.

97. Wood AJ, Stein CM, Woosley R. Making medicines safer–the need for an independent drug safety board. *N Engl J Med*. 1998;339(25):1851–1854.

98. Somberg JC. The Provisional approval step. *Am J Ther*. 2005;12:1–2.

99. Hulley SB. *Designing Clinical Research: An Epidemiologic Approach*. 2nd ed. Philadelphia, PA: Lippincott Williams & Wilkins; 2001.

100. Roses AD. Pharmacogenetics and drug development: The path to safer and more effective drugs. *Nat Rev Genet*. 2004;5(9):645–656.

101. Woodcock J, LaVanger L. Master protocols to study multiple therapies, multiple diseases, or both. *N Engl J Med* 2017;377:62–70.

19 The Academic Research Enterprise

Crista Brawley, Mary Jane Welch, Jeff Oswald,
Erin Kampschmidt, Jennifer Garcia,
Allecia Harley, Shrijay Vijayan, John McClatchy,
Stephanie Guzik, and Stephanie Tedford
Rush University Medical Center

CONTENTS

19.1 INTRODUCTION

The complexities of taking a drug from conception to FDA approval is manifested in the complexities of managing a research site in general, and, an academic medical center (AMC) specifically. The multiple requirements of the pharmaceutical industry, the various agencies that regulate research, the continued fiscal viability of the AMC, and, most importantly, the medical needs of the patient, create a tension that results in partnerships that often have competing outcomes. To manage these many competing demands, AMCs have created research systems across their enterprise.

This chapter describes many of the activities of the Office of Research Affairs (ORA) at the Rush University Medical Center in Chicago. The research contributions of faculty generate meaningful publications, value-driven care, lifesaving therapies, and leading-edge innovations ranging from omics-based discoveries to the next breakthrough drug or device. Continuously, our research teams progress towards measured, peer-affirmed outcomes that improve quality of life and our understanding of the human condition. The ORA exists to partner with faculty and staff as they seek funding, propose clinical studies, establish collaborations, steward funds, submit grants, negotiate industry contracts, and secure patents and licensing agreements. At the hub of this complex enterprise, shouldering the ultimate responsibility is the principal investigator (PI).

The PI takes on this ultimate responsibility because he believes in the research, and he wants access to the developing research for his clinical practice. The PI shoulders the responsibility but also gets to engage in the development of new therapies for his patient population. PIs that contribute to new research ideas often emerge as leaders within their field. Without appropriate PI engagement in the research, conducting compliant clinical trials is often not achievable. This is why the PI commits to conducting the research in a specific manner.

The PI must sign an FDA Form 1572 (www.fda.gov/downloads/aboutfda/reportsmanualsforms/forms/ucm074728.pdf) in order to conduct the study. This form certifies that the PI will remain compliant with all federal, state, and organizational policies and procedures as they apply to the research. This agreement also certifies that the PI assumes ultimate responsibility for every aspect of the study being conducted under his/her name. This responsibility is regardless of the documented delegation of specific duties to other members of the study team. This involves assuring the team members are appropriately qualified, current with training, that all aspects of the protocol are followed, the data are accurate and documented, the study is being overseen by an institutional review board tasked with human subjects protection, and most importantly, that the subjects are diligently managed and cared for. The consequences of not conducting a quality, compliant research study have negative impact on the PI but also causes financial and reputational risk to the AMC. It is, thus, important for an organization to provide systematic support to the PI and the research team throughout the entirety of the research study.

There are multiple approaches to provide this support. Each organization must consider their specific circumstances and develop a research enterprise that fits their needs. Additionally, the enterprise system is dynamic and should be altered as the research that is conducted by the organization's PIs changes.

The overarching considerations, as the research systems are developed, are the amount of local (the PI and team) control or central control (the organization) of the multiple administrative responsibilities in conducting research at an AMC. Some organizations leave all of the responsibility with the PI and only become involved when there is a request for assistance, or there are identified problems, usually by an external source. This approach is most often taken when the organization has minimal research being conducted and does not require a robust centralized research administration. Likewise, some organizations are completely centralized, and all administrative activities are managed for the PI. This is most useful for a large research institution that has frequent rotation of PIs and staff.

A reasonable approach is when the PI and the organization partner to share the many regulatory and administrative responsibilities; a middle road. This allows the PI to remain autonomous, given the level of accountability assumed in that role, yet allows the organization to provide support and oversight in the management of the research. The many considerations to this approach are described below.

19.2   ACADEMIC RESEARCH: BUDGETING AND FINANCIAL OVERSIGHT

Research is typically a separate and distinct financial reporting entity at AMCs. Other primary entity categories would include clinical, education, and corporate. Since research occupies a large segment of the overall financial reporting landscape, institutional finance leaders, such as the Chief Financial Officer (CFO), naturally focus significant attention on closely monitoring research financial results.

There are two primary processes to the financial management of research operations. The first process is the annual budget in which the end product is the establishment of the research operational financial plan for the upcoming fiscal year period. The other process is financial oversight which measures the extent to which the budget financial operation plan is being attained or not attained.

Developing the annual budget for research operations is a collaborative process that's conducted in concert with the other institutional entities, as opposed to working in isolation. The process is initiated by the corporate finance department communicating net operating income financial target expectations to each institutional reporting entity including research. Once received, the research financial and administrative leaders meet to discuss and agree on an overall general plan for attaining the budget financial expectations. The general plan that is agreed to at this time is translated into individual budget target expectations for each research operating area in which the total for all operating areas equals the overall budget target expectation for the entire research entity. For some areas, the budget target expectation is based on an approved formula calculation.

The budget target expectations for each research operating area are loaded into budget workbooks. There are typically two types of budget workbooks (cost centers and funds). The budget workbooks are then released to the responsible person in each research area to complete with the understanding that the finalized budget workbook net income amount will reconcile to the budget target amount loaded into the budget workbook. In theory, if all research areas adhered to this approach, the budget process would be completed and the cumulative rolled up budget amount for the entire research entity would reconcile to the research entity net operating income financial target developed by corporate finance. In practice, however, it never works out this way when the initial rollup of institutional entity budget amounts takes place during the budget process. For explainable reasons, most of the areas in research are unable to meet the budget target that is set for them. The same processes are followed in the other institutional entities, and when the overall institutional initial budget rollup is developed, there is always a large budget gap amount that needs to be resolved.

When this point in the budget process is reached, corporate finance pivots into a reset of the budget net operating income financial target expectations for each reporting entity. This in turn kicks off an ongoing repeating process in which the entities receive updated budget target amounts. Each entity meets and develops an updated plan for achieving the new budget target amount. Areas impacted by the updated plan receive communications that include revisions to their individual budget target amount for their area. Corporate finance develops a rollup of all institutional entity budgets and measures the status of reaching the overall net income budget target. If the overall institutional net income budget target is reached, then the budget process is complete. If not, the above process is repeated until the overall institutional net income budget target is reached.

Once the budgets are finalized, the approved budget workbook amounts are loaded into the financial general ledger system. The annual budget amounts are divided into monthly increments covering the 12 months of the annual fiscal year period and in total represent the overall research entity financial plan for the fiscal year. This establishes the baseline for assessing how well the research entity is performing from a financial perspective and leads into the financial oversight process that is initiated when the new fiscal year begins.

The financial oversight process is conducted every month with greater emphasis focused on each of the four quarterly reporting periods. Financial results are categorized as budget, actual, and variance. Budget results are the budget amounts that were discussed previously. Actual results represent the amounts that actually occur for the month. Variance results represent an assessment of actual results compared to budget results. For example, if the budget expectation was to incur $100 in salary expense, but the actual result was $120 in salary expenses, the variance result would be an unfavorable amount of $20 since the actual salary expenses in this example exceeded the planned/budget amount.

The monthly financial results for research are developed at macro and micro levels which are analogous to the forest and the trees. At the macro level, research financial results reports are defined by broad groupings of administrative, operational, and funded areas in comparison to the micro level in which results are developed by individual department, cost center, or fund. Using these reports, financial and administrative leaders in the research entity meet to discuss the research financial results every month. During the conduct of these meetings, the leaders are able to gain an understanding of the various drivers that led to the financial results. Discussions of the monthly financial results are shared with senior leadership at the corporate level to ensure that everyone is in sync with the messaging being communicated to explain the financial results. There is typically a follow-up intervention with all areas that are identified as having material unfavorable variances. The intervention usually starts in the form of an email or memo requesting an explanation for how the unfavorable variance occurred. This can escalate to scheduling a face-to-face meeting in situations where there is a recurring unfavorable variance or if the unfavorable variance explanation response from the area is inadequate.

There are a variety of monthly financial reports supporting research operations that are available to departments to assist them in managing their research financial operations. Departments are empowered to utilize this information to make any necessary financial adjustments that will enable the department to remain as close as possible to being on track to meet the budget expectation for their area.

19.3 PRECLINICAL RESEARCH

If preclinical research is to be conducted utilizing animal models, the regulations around the use of animals in biomedical research must be considered. The most onerous of these is the Animal Welfare Act (AWA) which does not include the use of rats and mice bred for research but does cover the use of most other species. To comply with the AWA, institutional registration with the United States Department of Agriculture (USDA) is required, which conducts annual, unannounced inspections by a USDA Veterinary Medical Officer (VMO). If the proposed work is funded by any branch of the National Institutes of Health (NIH) or receives other government funding, the institution conducting such projects must have an approved Animal Welfare Assurance on file with the Office of Laboratory Animal Welfare (OLAW) and adhere to the Public Health Service's (PHS) Policy on the Humane Care and Use of Laboratory Animals. OLAW does not generally perform routine inspections but will make "for cause visits" when deemed necessary. Adhering to the PHS Policy also carries with it, following the National Research Council's *Guide for the Care and Use of Laboratory Animals*. Many institutions that follow the *Guide* are accredited by the Association for Assessment and Accreditation of Laboratory Animal Care (AAALAC) International, a third party, for-profit agency that conducts triennial site visits of institutions to maintain accredited status. The USDA, OLAW, and AAALAC International all require institutions to file annual reports with varying complexity.

Together, the AWA, the PHS Policy, and the *Guide,* create the regulatory foundation for the Institutional Animal Care and Use Committee (IACUC) to operate. According to the regulations, all preclinical research using animal models must be approved and monitored by the IACUC. The appointment, composition, and duties of the IACUC are mandated by the AWA and the PHS policy, and best practices of IACUCs can be found in several texts including the *Guide*. According to regulations, IACUCs may not use an expedited review process but may review preclinical research projects via either "full committee review" or "designated member review," the latter generally being a less lengthy process. Changes to approved IACUC protocols can be reviewed and approved by the IACUC via "full committee review," "designated member review," or if appropriate via the "veterinary verification and consultation" process. A best practice to ensure an accelerated review of IACUC protocols and changes to approved IACUC protocols is to have in place a veterinary pre-review process in the IACUC's workflow. Veterinary pre-review of IACUC documents is an effective way to address potential animal welfare concerns at the time of project development and to attend to other concerns including anesthesia and analgesia, perioperative care, and humane endpoints.

After IACUC approval and a preclinical research project is underway, some form of post-approval monitoring of the study by the IACUC or its agents (such as the attending veterinarian) should be conducted. The institutions attending veterinarian or their designee(s) are required to monitor the health and well-being of the animals assigned to each specific study daily to assure that the study is being conducted according to the IACUC-approved protocol and any associated IACUC-approved changes to such. Maintenance of animal welfare should be paramount. Any deviations to the study protocol should be discussed with the PI or the study director. Appropriate amendment requests for changes in the conduct of the project should be submitted to the IACUC for review and approval. Adverse events or unanticipated outcomes in the conduct of the study should be reported to the IACUC for consideration in a timely manner.

19.4 FINANCIAL CONSIDERATIONS

19.4.1 CONTRACT REVIEW AND NEGOTIATION

Since this book is about drug discovery and development, please assume that the following information regarding contracts is about clinical trial agreements between an industry sponsor and an AMC; however, much of this can also be applied to nonclinical trial research agreements between an industry sponsor and an academic institution.

The contract is the formal agreement between the parties outlining all of the expectations of the project, including the protocol, the budget, payment terms, confidentiality, publications, intellectual property ownership, compliance requirements, indemnification, and liability. All terms must be included in the written agreement signed by the parties. The contract should specifically reference the protocol to be performed, who is responsible for the performance, where it will be performed, and any reimbursement for the performance.

Each academic institution will have individualized requirements related to research contract review. However, generally, the contract must be reviewed by the institution's general counsel or the Office of Research Administration's contract review team. Academic institutions require legal review by knowledgeable counsel because unlike a private company, an academic institution has specific requirements to comply with in order to maintain its accreditations and status as an academic center. These requirements may include, but are not limited to, Association for the Accreditation of Human Research Protection Programs (AAHRPP) accreditation guidelines,[1] intellectual property assignment restrictions due to its 501(c)(3) tax-exempt status, licensing considerations to data or results to maintain academic integrity, opportunities for teaching, internal research integrity, and publication rights. The institution's legal reviewer should be knowledgeable about not only the requirements of an academic institution but also the regulations, guidelines, and industry standards related to research in addition to an understanding of the institution's level of risk-tolerance.

All terms must be in the written agreement signed by both parties; so, it is essential that the study team is involved in the contract review process to ensure that the contract fully encompasses the requirements and expectations of the research project. The clinical research team is a critical point of contact for the contract reviewer during the negotiation process. The contract reviewer will use the clinical researchers as a source for how the study will be performed at the institution, for what materials the sponsor is expected to provide to the institution to perform the study, for expectations related to data ownership and the likelihood of development of new intellectual property, and for anticipated timelines for study start-up. The contract reviewer is focused on ensuring not only that the institution is protected and in compliance with all relevant requirements but also that the business terms favor the institution and study teams' interests.

After the contract is initially reviewed by the institution's general counsel office or contract reviewer in the research office, the "redlines" will be sent to the sponsor for negotiation. Ideally, the sponsor will accept all of the institution's changes. However, it is likely that the agreement could go back and forth between the parties until it is finalized and acceptable to both parties. This process could take anywhere from weeks to months.

A recurrent frustration for study teams is the amount of time it takes to start-up a study at an academic institution. One of the main reasons for the delay is the contract review and negotiation process. There are a variety of methods to reduce the timeline of contractual review. An easy option to speed up the process is to use previously negotiated terms from a prior agreement with the same parties. Another option is the Accelerated Clinical Trial Agreement[2] (ACTA). The ACTA is a standardized agreement template developed by a working group of legal experts that may voluntarily be used to expedite the contracting process. Alternatively, a master agreement is an excellent option if the same sponsor and institution are continually working together on several studies. Each study to be performed in conformance with the terms of the master agreement will be added to the master agreement as an amendment. Instead of negotiating a contract for each study, only one contract would need to be negotiated. The study-specific amendments will identify the study to be performed, the PI, and the budget for the study.

Key Guidance:

- Seek out advice from legal counsel who have experience reviewing research agreements and who have knowledge of research regulations and industry standards.
- The study team including the PI should review the agreement to ensure the business terms are in line with the team's expectations and are feasible. The team should feel comfortable enough in their relationship with their counsel to request any revisions they require and outline timeline expectations.
- Understand that an academic institution must comply with policies, guidelines, and regulations (e.g., accreditations and/or tax-exempt status), which a private company does not have to consider.
- Look for opportunities to accelerate the process by using the ACTA, using pre-negotiated terms, or entering into a master agreement with a sponsor that the academic institution contracts with frequently.

19.4.2 Grants

Grants are different from contracts/agreements. A sponsor contract/agreement is a legally binding document where an entity provides a service in exchange for payment. A fee for a service typically. A grant is when one party grants funds to another party in hopes that the research spelled out can be accomplished and results can be achieved.

Preparing a grant application has many different aspects, from the science to the budget. Every grant is different; some will require more time to prepare than others. If we look at an application being submitted to an investigator-initiated R01, the NIH recommends that you allow at least 2 months to write the grant, 1 month for feedback from mentors and colleagues, and 2 weeks to perform the final edits.

Once the scientific portion of the grant is conceptualized, the investigator can begin to focus on the budget and the nonscientific aspects of the application. The first step in any grant application is identifying the program announcement or solicitation in which the investigator wants to apply. Once the opportunity is identified, the investigator should read the solicitation carefully. Most funders specify the format for the grant application, special requirements, such as page limits, formatting requirements, and budget limits. The investigator will want to be sure to pay close attention to the sponsor's instructions, as well as the specific program announcement in which he or she is applying when preparing the application and budget. Adherence to font size, type density, line spacing, and text color requirements is necessary to ensure readability and fairness. Although font requirements apply to all attachments, they are most important and most heavily scrutinized in attachments with page limits. The announcement will have guidelines on budgetary limits and restrictions, the investigator should look for limits set on certain categories, such as salaries or overall funding limits.

The budget is a key element of every proposal; it should be viable, complete, reasonable, and become a useful tool to manage the award once it is received. It is important for the budget to reflect the needs of the project and to keep in mind allowable and unallowable costs. The investigator should walk through the scope of work and aims of the project to help determine any potential expenses. What personnel and staff will be needed? What materials and supplies will be needed? Are there any special needs such as animals, data storage, or travel? Depending on the specific project and the investigator, it may be necessary to include Co-Investigators, collaborators, and expert consultants. For example, a new investigator without a history of independent research on an NIH grant may want to add a highly experienced investigator to the team to show the reviewers that there is a participant who will be able to step up and fill in gaps of expertise and/or training. There may also be a need to include a consortium site, if the expertise does not exist at the prime institution of the PI; perhaps a colleague at another institution can fill that need through a subcontract arrangement. Another thing to consider would be, institutional resources – equipment that may already exist that can be shared among projects or core facilities offering leading-edge tools and technologies that can be used for the project. If these things do not already exist, the investigator would want to build them into the budget and justify the need to the particular project in the budget justification.

It is important to remember that any item included in the proposal, be it personnel, equipment, supplies, or consortium, be appropriately justified based on allowability and need for the project. Although reviewers do not use the budget during the scoring of an application, they will know if the level of funding being requested is in line with the scope of the project being proposed. The NIH grant application scoring system uses a 9-point rating scale (1 = exceptional; 9 = poor) for Overall Impact and Criterion scores for all applications. The scoring system was developed to provide consistency across all applications, yet distinguish the cohort of applications being compared to each other. Therefore, it is important to present a budget that is carefully planned.

Many months after the submission of a grant application, it will be scored against other grant submissions to the same announcement. The NIH grant application scoring system uses a 9-point rating scale of overall impact and individual criterion where a score of 1 is exceptional and a score of 9 would be poor. The same scoring system is used to provide consistency across all applications, yet distinguish the cohort of applications being compared to each other.

Start early, read the program announcement carefully, understand the requirements of the sponsor, and ask for the money required to complete the aims of the project.

19.4.3 BILLING

The use of the term "billing" in clinical research can have a variety of meanings depending on your role in the research enterprise or patient revenue cycle. When billing is well managed, it is a feather in the cap of a strong academic research enterprise but must be actively attended to and never taken for granted.

What does "billing" mean? Which of the following scenarios are we talking about:

1. Generating invoices and or collecting payments from research sponsors?
2. Billing fees for routine care to third-party payers when the care is defined in a research protocol?
3. The decision-making and the intersection between (1) and (2) above?

The correct answer is all of the above. Let's start with number three.

19.4.4 COVERAGE/BILLING ANALYSIS

In order to determine to whom a bill should be sent for items and services in the context of clinical research, it is not uncommon for academic research institutions to conduct a coverage (billing) analysis. This approach is taken to mitigate the risks of billing both an insurance company and the study sponsor for the same services. This analysis commonly begins as an electronic version of the protocol's schedule of events, schema, or schedule of assessments. These schedules list what occurs and when those services are provided. Billing designations are assigned to each service and each time point to determine which items are sponsor paid, "standard of care," or not billable at all. For more information on coverage analysis, please use your favorite internet search engine to research "understanding coverage analysis" for articles, tools, and whitepapers on the topic.[3,4]

19.4.5 BILLING TO THIRD-PARTY PAYERS

Using the coverage analysis as a guide, institutions will follow the claims processing guidelines published by Medicare and insurance companies to bill the fees related to routine care provided in the context of a clinical trial when these services are not paid by the study sponsor.

19.4.6 INVOICING SPONSORS

In those instances where the study sponsor is covering fees for items and services provided, academic institutions either trust that the sponsor will make the payment on time, as promised, or for industry and/or nongovernmental payment sources, they can generate invoices and track the money owed (accounts receivable) using approved accounting procedures. The latter approach is helpful when the need arises to project revenue available to support staff salaries, expand into additional space, or to monitor operations.

19.5 INTELLECTUAL PROPERTY: DIVISION OF INNOVATION AND TECHNOLOGY TRANSFER

The role of the Technology Transfer Office (TTO) is to manage intellectual property assets generated by research and educational activities. The TTO plays a central role in bringing ideas from the lab to the marketplace by transferring discoveries developed through research to companies that can turn them into products and services that benefits society. The TTO seeks to guide technologies through the various stages of the commercialization process by providing services that include scouting, evaluation, protection, marketing, and licensing of intellectual property. The TTO protects faculty interests while advancing discoveries towards commercial development.

Upon disclosure of Inventions created by University employees, the TTO carries out an assessment of intellectual property as well as market potential by utilizing public and private databases and other resources. If the assessment indicates that the disclosed Invention has the potential for strong intellectual property as well as is marketable, the TTO, in collaboration with the Office of Legal Affairs, will work towards protection of intellectual property by filing a patent application or any other suitable form of intellectual property, such as copyright or trademark. Soon thereafter, a marketing strategy will be developed to help partner with companies that have interest in developing products or services that will be enabled by the University's Invention. If a company is interested in exploring a partnership, confidentiality agreements will be put in place to facilitate exchange of required information including those that are confidential to fully support the due diligence process. Once it is determined that the company and the University want to enter into a partnership, a license agreement will be structured and negotiated between the TTO of the University and the business development and licensing division of the company. A license agreement typically includes business terms, upfront license execution fees, royalty payments on net sales, as well as milestone payments by the company to the University in exchange for permission to exploit the University's intellectual property to develop a product or service.

After the license is executed, the TTO is responsible for management of the license portfolio and to ensure that all obligations are met. License revenues are distributed between the Inventors and the University according to the University's intellectual property policy. Sharing license revenues with the Inventors is designed to incentivize the Inventors as well as to spur innovation. License revenues are used to support new research and thereby encouraging new Invention disclosures and boosting the innovation cycle.

Products and services resulting from the TTO activities continue to have a huge impact on improving lives, creating jobs, and supporting new research discoveries.

19.6   RESEARCH COMPLIANCE

Research Compliance Offices are responsible for promoting a culture of compliance, research integrity, and high-quality research. Successful compliance programs have a shared model that functions for the organization but partners with University, corporate, and legal offices.

19.6.1   PROGRAMMATIC ELEMENTS

Oversight of the regulatory, ethical, and compliance aspects of all preclinical and clinical research conducted is a complex, multidimensional undertaking. Research Compliance Offices need deep expertise to advise and consult the research community on navigating regulatory complexities. Research Compliance Offices should ensure in their missions that supporting each operational area is its primary responsibility to ensure compliance. The following content areas should be considered in establishing a research compliance program:

- Human subject protections (biomedical and social/behavioral)
- Animal use protections
- Research privacy and security
- Regulatory compliance
- Assessments of scientific integrity
- Investigate noncompliance
- Conflicts of interest (individual and institutional)
- Research misconduct
- Questionable research practices
- Financial management associated with funded research.

An effective research compliance program needs to incorporate the eight elements prescribed by Health and Human Services Office of Inspector General (HHS OIG) Guidelines.[5] They are the following:

1. High-level company personnel who exercise effective oversight. Responsible parties must have the authority to report directly to the Board or appropriate subcommittee: (a) at least annually on the effectiveness of the program and (b) promptly if criminal conduct including fraud is discovered. This can be the Chief Compliance Officer or Research Compliance Officer or other responsible party.
2. Written policies and procedures including a Code of Ethical Conduct policy.
3. Mandatory compliance and ethics training and education.
4. Open lines of communication, including anonymous reporting lines and a policy of non-retaliation.
5. Standards enforced through well-publicized disciplinary guidelines.

6. Internal compliance monitoring/auditing.
7. Response to detected offenses and corrective action plans.
8. Periodic risk assessments.

The research compliance program at Rush University Medical Center includes a few key elements. The following describe in detail an example of how and what a research compliance program includes.

19.6.2 EDUCATION AND TRAINING

It's a fundamental belief that education, training, and knowledge transfer are key to promoting compliance. Facilitating education and training for the research community is a proactive approach to mitigate risk. Investigators and research personnel must complete mandatory training prior to beginning research. Training on research compliance elements for human subject and nonhuman research is provided to each researcher before access is given to local software programs that support research. In addition to local training on institutional policy and practices, the Collaborative Institutional Training Initiative (CITI) is used to provide instruction on the content area related to the category of research conducted and must be completed before the individual commences research (Figure 19.1).[6] Depending on the funding source and type of research, education requirements vary and the need for re-training varies but is typically required every 3 years.

19.6.3 RESEARCH QUALITY IMPROVEMENT AUDIT/REVIEW

Quality improvement (QI) programs are responsible to assess and monitor research compliance to promote high-quality research conducted within the research community. Culturally, some programs support this aim primarily through educating the research community about best practices, local policy, and regulations based upon detected deficits during assessments. Data from audits provides the organization with important information and a better understanding for developing education programs and materials to benefit the research community.

QI programs should be a systematic and objective review of research activities. The audits may include regulatory, subject, animal, financial, conflict of interest (COI), training records, data collection forms, data records, and source documentation, and any records associated with

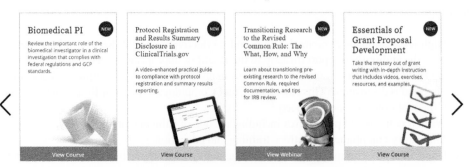

FIGURE 19.1 CITI program website.

the research project. Active protocols should be considered eligible for audit regardless of funding source. The attempt is to distribute the audits evenly among the various protocols.

Identification of noncompliance may trigger institutional policy on reporting and investigating research noncompliance. In addition, detected noncompliance may result in formal corrective and preventative action plans (CAPA).

19.6.4 CORRECTIVE AND PREVENTIVE ACTION PLAN (CAPA)

The purpose of a CAPA is to assist the research team when areas of compliance or regulatory concern are discovered. The CAPA allows the research team to develop processes, forms, standard operating procedures (SOPs), etc. to ensure that the deficiencies are not only rectified for the particular study under audit review but also for other current and future research studies. A CAPA is an educational tool to assist researchers in ensuring the quality of their research and data.

If a CAPA is issued, the research team should return the plan to the overseeing body (e.g., Research Compliance Office or IRB (Institutional Review Board)) outlining steps they will take to ensure that any identified issue or problem is rectified and that processes have been put in place to ensure that the issue or problem will not occur again. This may entail development of forms, additional training, or other mechanisms to assist the research team in accomplishing the requirements of the CAPA.

19.6.5 EXTERNAL RELATIONSHIPS AND CONFLICTS OF INTEREST IN RESEARCH

COIs in research may occur when interests compromise, or have the appearance of compromising, the professional judgment of a researcher. Maintaining objectivity in research and education is a fundamental academic value. Institutions and researches should strive to ensure a transparent research process by requiring a disclosure of external profession interests at least annually. Institutions that are recipients of federal funds must have policies that articulate their position on COIs. Through well-described policies, research institutions maintain the balance among competing interests that have the appearance or ability to bias the design, conduct, or reporting of the research. COI committees are established to review disclosed significant financial interests (SFIs) of individual researchers and determine whether outside professional relationships rise to the level of a COI. When COIs are detected, the Committee is tasked with managing, reducing, or eliminating COIs related to research. This includes sponsored (e.g., federal) and non-sponsored research, start-up ventures, or other activities that require objectivity under circumstances that could be influenced by personal financial gain.

19.6.6 DISCLOSURE OF EXTERNAL INTERESTS

A successful practice of disclosure allows an opportunity to disclose external interests at opportune times. Annual disclosure and transactional disclosure (protocol specific) archives the requirement and allows effective review and adjudication to assess COIs. Once a COI is determined, the institution is required to manage, reduce, or eliminate COIs. COI is a routine review consideration and sometimes the subject of intense interest by the IRB.

19.6.7 FEDERALLY MANDATED COI TRAINING (FCOI)

PHS-funded research requires training on Financial Conflict of Interest (FCOI). Mandatory training on FCOI is required for all PIs and key personnel *prior to* the expenditure of funds on any newly funded projects, including noncompeting continuation awards. This applies to all PHS-sponsored research projects as of August 24, 2012. The regulations require re-training on COI every 4 years.

19.6.8 Public Disclosure of Financial Conflicts of Interest

PHS regulations require that, prior to expenditure of any funds under a PHS-funded research project, Rush ensure the public can access certain information by submitting a written request for information concerning any SFI that is a COI or the information can be posted on the publicly accessible website of the institution that is the recipient of the federal funds.

19.6.9 Open Payments

Open Payments is a national disclosure program that promotes transparency by publishing the financial relationships between the medical industry and healthcare providers (physicians and hospitals) on a publicly accessible website developed by the Center for Medicare & Medicaid Services (CMS). Physicians should check data reported in the system on a routine basis to access accuracy of payments associated to them. The Open Payments schedule below articulates key dates (Figure 19.2). Open Payments link can be found here: https://openpaymentsdata.cms.gov

19.6.10 Research Misconduct

Research misconduct is defined[7] as fabrication, falsification, plagiarism, or other serious deviation from accepted practices in proposing, carrying out or reporting results from research.

- Fabrication is making up data or results and recording or reporting them.
- Falsification is manipulating research materials, equipment, or processes, or changing or omitting data or results such that the research is not accurately represented in the research record.
- Plagiarism is the appropriation of another person's ideas, processes, results, or words without giving appropriate credit.[8]

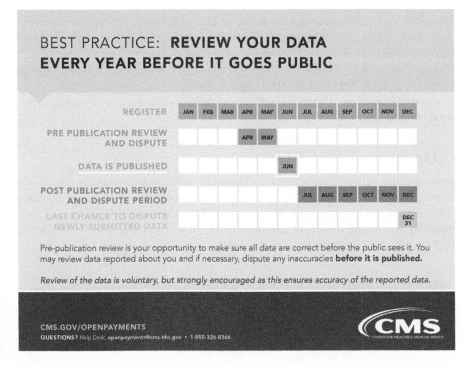

FIGURE 19.2 Open payments schedule.

Research misconduct does not include honest error or honest differences in interpretations or judgments in evaluating research methods and results. Research misconduct is a serious deviation from accepted practices which includes, but is not limited to, abusing confidentiality by use of ideas and preliminary data gained from:

- Accessing privileged information via editorial review
- Peer review of proposals
- Stealing, destroying, or damaging the research property to derail an investigation or tamper with evidence
- Directing, encouraging, or knowingly allowing others to engage in fabrication, falsification, or plagiarism.

Recipients of PHS funds are required to have policies and take the matter of research misconduct seriously. Regulations (42 CFR 93) require federally funded research centers to have a policy, procedure, and a Research Integrity Officer (RIO) to address allegations of research misconduct. Allegations should be made in good faith and be substantiated by information or some form of proof, such as documents or statements that would enable the RIO to begin the initial assessment. Institutions should have anti-retaliation policies and procedures for those who raise allegations in good faith.

19.6.11 RESPONSIBLE CONDUCT OF RESEARCH (RCR)

Responsible Conduct of Research (RCR) is defined by the NIH as "…the practice of scientific investigation with integrity. It involves the awareness and application of established professional norms and ethical principles in the performance of all activities related to scientific research."[9]

RCR provides the foundation for conducting sound, ethical research, and reinforcing public trust. The promotion of RCR within a research community through educational opportunities for initial and continuing education relevant to a broad spectrum of research matters increases research integrity.

RCR includes the following:

- Research misconduct
- Protection of human subjects
- Animal welfare and use
- Data management
- Conflict of interest
- Collaborative research
- Authorship and publication
- Peer review
- Mentor and trainee relationships.

Certain federal funding agencies require training on RCR before the issuance of funds. Specifically, the National Science Foundation and certain NIH training and other awards require RCR.

19.7 CONCLUSIONS

A strong academic research enterprise is the foundation for any recognized institution considered a top leader in strategic research initiatives and discovery. A research enterprise will be composed of a highly proficient and interdisciplinary workforce, will create access to quality and dependable resources, and will provide expert-level basic research applicable to all major areas of science. This support includes a wide range of qualified individuals, departments, and organizations that are strongly committed to academic research excellence at the institution. Leading scientific experts, compliance regulators, and funding specialists are among many individuals recruited to these enterprises who work together to maximize the benefit of the institution's research to the scholarly community. Scientific investigation is paramount to community growth and is critical for the translation of benchtop discoveries to the clinic. The Academic Research Enterprise activities of the Rush University Medical Center are described herein by key individuals involved in administering and shepherding research into and through the University. Sound research compliance programs contribute to this cause indirectly, and effective research compliance programs help protect research subjects and protect institutions and their researchers from fines and penalties, false claims, and reputational risk. Altogether, these activities are critical for academic existence and continued growth in all areas of research.

ACRONYMS

AMC: Academic medical center
ACTA: Accelerated Clinical Trial Agreement
AWA: Animal Welfare Act
AAHRPP: Association for the Accreditation of Human Research Protection Programs
AAALAC: Association for Assessment and Accreditation of Laboratory Animal Care
CAPA: Corrective and Preventative Action Plans
CFO: Chief Financial Officer
CITI: Collaborative Institutional Training Initiative
CMS: Center for Medicare & Medicaid Services
COI: Conflicts of interest
FCOI: Federally Mandated Conflict of Interest
HHS: OIG Health and Human Services Office of Inspector General
IACUC: Institutional Animal Care and Use Committee
IRB: Institutional Review Board
NIH: National Institutes of Health
OLAW: Office of Laboratory Animal Welfare
ORA: Office of Research Affairs
PI: Principal Investigator
PHS: Public Health Service
QI: Quality Improvement
RIO: Research Integrity Officer
RCR: Responsible Conduct of Research
SFI: Significant Financial Interests
TTO: Technology Transfer Office
USDA: United States Department of Agriculture
VMO: Veterinary Medical Officer

REFERENCES

1. https://admin.share.aahrpp.org/Website%20Documents/Tip%20Sheet%2025%20-%20Contract%20 Language%20(2016-06-28).pdf. Association for the Accreditation of Human Research Protection Programs, Inc. Accessed 6-28-19.
2. www.ara4us.org/. Accelerated Research Agreements. Published 2019. Accessed 6-28-19.
3. www.srainternational.org/learn/on-line-education/webinar/understanding-coverage-analysis-and-research-billing-rules.
4. https://pfsclinical.com/coverage-analysis/. PFS Clinical. Accessed 6-28-19.
5. http://oig.hhs.gov/compliance/compliance-guidance/index.asp. Office of Inspector General. Accessed 6-28-19.

6. https://about.citiprogram.org/en/homepage/. CITI Program. Accessed 6-28-19.
7. https://ori.hhs.gov/sites/default/files/42_cfr_parts_50_and_93_2005.pdf. Federal Register. Department of Health and Human Services. Published 5-17-05. Accessed 6-28-19.
8. www.plagiarism.org/article/what-is-plagiarism. Plagiarism.org. Published 5-18-17. Accessed 6-28-19.
9. https://grants.nih.gov/grants/guide/notice-files/not-od-10-019.html. Update for the Requirement for Instruction in the Responsible Conduct of Research. Published 4-19-11. Accessed 6-28-19.

ADDITIONAL RESOURCES

www.niaid.nih.gov/grants-contracts/overview-r01-process
https://grants.nih.gov/grants/how-to-apply-application-guide/format-and-write/format-attachments.htm
https://grants.nih.gov/grants/how-to-apply-application-guide/format-and-write/develop-your-budget.htm

20 Clinical Testing Challenges in HIV/AIDS Research

Vincent Idemyor
University of Port Harcourt

CONTENTS

20.1 INTRODUCTION

The disease known as Acquired Immune Deficiency Syndrome (AIDS) was first reported in 1981. The first cases of young, previously healthy homosexual men diagnosed with Pneumocystis pneumonia were reported in Los Angeles, followed by case reports of Kaposi sarcoma, an uncommon malignancy, occurring in gay men in New York City and San Francisco.

The Center for Disease Control and Prevention (CDC) established a case definition of Kaposi sarcoma/opportunistic infection. The virus that causes it, known as human immunodeficiency virus (HIV), was described in 1983. There are two known distinct virus types – namely HIV-1 and HIV-2. HIV-1 is further classified into one of three phylogenetic groups: M (main), O (outlier), and N (non-M/non-O). The M group, which accounts for most of HIV-1 infection, is further broken into ten subtypes called clades.

HIV continues to create formidable challenges to the biomedical research and public health communities around the world. The profile of the epidemic has shifted over the past 30 years. We have seen declining numbers of new infections and AIDS-related mortality throughout the 2000s. The expansion of antiretroviral therapy has improved survival in individuals infected with HIV.

HIV prevalence in many low-to middle-income countries is derived from statistical models based primarily on either sentinel surveys among pregnant women or household surveys. From 2000 to 2012, HIV incidence among adults in sub-Saharan Africa decreased by more than half, corresponding to an estimated 1 million fewer new HIV infections in 2012 compared to 2000.[1] AIDS-related deaths have decreased from ~1.4 million in 2000 to 1.2 million in 2012.[1]

The main mode of transmission contributing to the HIV epidemic in sub-Saharan Africa continues to be unprotected heterosexual intercourse. Risk is usually increased with multiple sex partners and when there is concurrent sexually transmitted infection. Among HIV-discordant couples in sub-Saharan Africa, the male has traditionally been viewed as the infected partner. Most education, prevention, and testing programs have focused on reducing risks for male-to-female transmission. However, a meta-analysis by Eyawo and colleagues[2] showed that in approximately 47% of heterosexual HIV-discordant couples in stable relationships, the infected partner was the female.

Men who are at risk for acquiring HIV infection through heterosexual intercourse can reduce their HIV risk by ~50%–60% through voluntary medical male circumcision.[3,4]

Another risk factor for HIV infection in sub-Saharan Africa is mother-to-child transmission of HIV. The global trends in mother-to-child transmission of HIV have led to a disproportionate number of children living with HIV in sub-Saharan Africa. Approximately 88% children in the world who are younger than 15 years old and infected with HIV live in sub-Saharan Africa.[1] In the 20 sub-Saharan African countries with the highest pediatric HIV burden, 56% of the HIV-positive children are not on antiretroviral drugs.[5] The *MMWR* highlights the continuing gaps in pediatric antiretroviral coverage in the United States President's Emergency Plan for AIDS Relief (PEPFAR) – supported sub-Saharan countries with high HIV burden, despite expanded World Health Organization (WHO) antiretroviral eligibility criteria.

Asia has the second largest number of HIV-infected individuals after sub-Saharan Africa. China and India have the highest burden in Asia. The modes of transmission are primarily injection drug use, heterosexual and homosexual transmission, with considerable variations within these countries.

Some areas of the world have increasing HIV prevalence in the context of decreasing new infections and AIDS-related mortality, primarily reflecting the scale-up of antiretroviral therapy programs in these regions. Continued efforts to expand access to antiretroviral therapy require identification of HIV-infected persons and linkage to HIV care. Increased access to effective use of antiretroviral medications may contribute to "treatment as prevention."

20.2   PATHOGENESIS OF HIV INFECTION AND VIRAL PERSISTENCE WITH RESERVOIRS

One of the most difficult hurdles to overcome in the global effort to treat HIV infection is the pathophysiological phenomena of viral reservoirs in the infected individual. The degree to which an antiretroviral medication is effective in reducing viral titer can vary greatly from one type of organ tissue to another within the same patient.

An improved understanding of the complexities of the interaction of the virus or virion and the cells being attacked, may help solve this stubborn research challenge.

HIV-1 binds to the cell membrane through interaction of the viral envelope glycoprotein-120 (gp-120) with CD4 molecules on the cell surface. HIV-1 then fuses with the cell membrane as a result of interaction of gp-120 with chemokine receptors (CCR5 or CXCR4). CCR5 and CXCR4 each bind to different viral isolate.

CCR5 binds the macrophage-tropic, *non*–syncytium-inducing viral isolates, while CXCR4 binds the *T cell-tropic,* syncytium-inducing isolates.

Classified as a retrovirus, HIV's baseline genetic material is RNA. After the initial attachment, the next step requires fusion of the viral and cell membranes, allowing the viral proteins to enter the cytoplasm. After cell entry as a retrovirus, the virus RNA template transcribes into a double-stranded viral DNA in the presence of the enzyme reverse transcriptase.

The virus is uncoated as it penetrates the cell. Viral RNA and reverse transcriptase are released into the cell cytoplasm, where HIV transcribes its RNA into viral DNA. The viral double-stranded DNA produced after reverse transcription is then transported into the cellular nucleus. This viral DNA is then integrated into the host cell's human DNA via an enzyme known as integrase. In the presence of the integrase enzyme, a multistep process allows the integration of viral DNA into the host genome and ultimately the formation of proviruses.

Once integrated, the viral DNA serves as a template for replication of viral RNA and synthesis of viral proteins and polyproteins. Viral components assemble on the inner wall of the cell membrane and bud off into new virus particles. The HIV Gag protein is critical to the budding process. The viral enzyme protease plays a key role in the formation of infectious virions. During this budding process, protease cleaves long polyproteins into their component enzymes and structural proteins to yield a mature virus particle.

Persistent, high-level viral replication is now established as the driving factor of HIV pathogenesis with about 10 billion virions (viral particles complete with RNA and envelope) in an infected person daily.[6,7]

There are some factors that promote viral replication and some factors that inhibit replication. There have been significant advances in the identification of cellular cofactors that promote viral replication after entry of the virus. While several cellular proteins are steadily being identified, we have yet to graze the surface of the identity of cellular ligands for HIV-1. There are also cellular factors that inhibit viral replication.

Human T cells possess a natural defense against retroviral invasion. This is a rapidly evolving area in molecular biology. In 2003, a unique antiviral pathway – cellular protein known as apolipoprotein B mRNA-editing enzyme, catalytic polypeptide-like 3G (APOBEC3G), also known as CEM15, was described.[8] APOBEC3G is a member of the APOBEC family of editing enzymes that is expressed in lymphocytes and macrophages.

APOBEC3 proteins are packaged within viral particles and in the target cell. APOBEC3 proteins inhibit reverse transcription by a mechanism yet to be elucidated and induce extensive G-to-A hypermutation in viral complementary DNA (cDNA).[9] APOBEC3 acts as a broadly active innate intracellular antiretroviral factor, except when counteracted by the viral infectivity factor (Vif) protein of lentiviruses.

In addition to APOBEC3 family of cytidine deaminase, other host restriction factors such as tetherin (bone marrow stromal antigen 2, coded by BST-2) and SAMHD1 (sterile alpha motif [SAM] domain and histidine–aspartate [HD] domain–containing protein 1) that antagonize HIV-1 in human have also been identified.

Tetherin is a transmembrane protein that tethers assembling virions to the surface of the virus producing cell. These tethered variants are inept at infecting other cells. However, when encountered by viral protein U (Vpu) of HIV-1, tetherin is neutralized. SAMHD1 is a deoxynucleotide triphosphate (dNTP) hydrolase that reduces levels of intracellular dNTP, thereby creating suboptimal conditions for reverse transcription of viral cDNA. These cellular factors continually regulate the interplay between virus and host.

As the research field embarks on finding a cure for HIV-1 infection, there has been substantial effort made in understanding how the virus is able to persist in the face of suppressive antiretroviral therapy. In early 1998, there were new discoveries about the pathogenesis of HIV-1 infection, and the existence of a latent reservoir for HIV-1 in resting CD4+ T cells was established. This reservoir is a significant cause of concern because it provides a potential mechanism for the virus to persist in an infected individual despite years of effective antiretroviral therapy. The debate that started years ago continues as to whether viral reservoirs that persist in the face of antiretroviral therapy are quiescent or whether a component of a viral reservoir can harbor the virus in an active state.[10] Another question that persists, that also needs to be delineated is: "Does cell-associated viral RNA act as a surrogate for viral reservoir activity during antiretroviral therapy?"[10] Chomont and colleagues[11] published that reservoir size and persistence are driven by T cell survival and homeostatic proliferation.

There are at least three potential reservoirs. The first phase reflects virus produced predominantly from activated CD4+ T cells. Extracellular virus particles can be trapped on specialized cells in the germinal centers of the peripheral lymphoid tissues. These cells, known as follicular dendritic cells (FDCs), can retain antigenic material on their surfaces for long periods of time. This reservoir declines rapidly with a half-life of about 2 weeks.[12]

Persistently infected macrophages represent a second potential reservoir for the virus. Because HIV-1 does not kill infected macrophages, these cells can continue to release virus for their normal life span. It is worth noting that macrophages are a heterogeneous population of cells residing in various locations, including the liver, lung, bone marrow, spleen, lymph nodes, and gastrointestinal tract. Therefore, infection status in one location may not reflect what is happening in other populations located in different organs. Perelson and colleagues stated that in treated individuals with viral titer decline, that this phase of decline in plasma virus is due to the turnover of infected macrophages, with an estimated half-life of 14 days.[13]

The third, and potentially most significant reservoir for HIV-1, consists of resting memory CD4+ T cells carrying an integrated copy of the viral genome. These cells must survive for long periods of time in order to provide the host with immunologic memory; in other words, the capacity to respond rapidly to previously encountered infections. Due to the potential of latently infected CD4+ cells to survive for months to years, the existence of reservoirs represents the major barrier to virus eradication. This possible mechanism is being actively explored by a number of laboratories. Finzi and colleagues[14] reported the mean half-life of latent reservoirs of HIV-infected cells in adults with undetectable plasma viral loads to be 43.9 months.

We have made progress understanding viral pathogenicity and the mechanism of viral latency. Numerous mechanisms contribute to HIV persistence and the viral reservoirs, many of which are currently being researched therapeutically. So far, the unifying theme in these technically demanding research efforts is trying to determine the theoretical explanation that could form the basis for a therapeutic approach that reduces the size of the HIV reservoir.

However, although this is a critical area of research, it represents only a small fraction of what remains unknown in the field of HIV research. And after solving the science questions in the laboratory, there still will remain the tremendous challenges of clinical testing.

20.3   ROLE OF HIGHLY ACTIVE ANTIRETROVIRAL THERAPY (HAART) AND ITS CHALLENGES IN LOW-TO MIDDLE-INCOME COUNTRIES

The discovery and elucidation of the replicative life cycle of HIV in human CD+ T cells led to the identification of potential antiviral drug targets to slow the replicative process. This also led to safer, more effective, and more convenient antiretroviral drug regimens during the past 30 years. There are currently six categories of antiretroviral agents available for the management of HIV disease.

The process of HIV entry into target cells is complex and can be divided into different phases: attachment, co-receptor binding, and membrane fusion. Medications are specific to a particular phase:

- Entry inhibitors: Agents that can interfere with the fusion or binding of HIV particles to host cells
- Reverse transcriptase inhibitors: Block the action of reverse transcriptase
- Integrase inhibitors: Designed to interfere with the integration of the viral genome into the cellular genetic material
- Protease inhibitors: Target the viral protease, the enzyme required for the cleavage of precursor protein (gag and gag-pol), thereby permitting the final assembly of the inner core of viral particles.

As of June 2018, more than 30 antiretroviral drugs have been approved by the United States Food and Drug Administration (FDA) for the treatment of HIV infection. The United States and Europe have reached a consensus regarding guidelines for the timing of the initiation of antiretroviral therapy for HIV-infected individuals. However, the same cannot be said regarding agreement among the low-to middle-income countries.

Initiation of antiretroviral therapy is currently recommended for essentially all infected adults regardless of CD4+ lymphocyte count. The current standard of care recommends the use of several different potent drug combinations. HIV is well known for its capacity to mutate rapidly. By attacking an HIV infection with several different potent drugs at one time, practitioners seek to avoid the emergence of resistant strains within the patient.

A systematic analysis by Lee and colleagues of 114 antiretroviral drug trials, with up to 144 weeks of follow-up through 2012, found that the number of individuals who achieved an undetectable viral load on their initial drug regimen increased from 43% in the mid-1990s to more than 80% by 2010.[15]

The unprecedented benefits resulting from highly active antiretroviral therapy (HAART) have been well described in the medical literature and there is agreement that symptomatic patient and those with AIDS require antiretroviral therapy. However, the timing of antiretroviral therapy in the asymptomatic patient remains a controversial issue.

The points of view regarding when to initiate therapy have evolved as a result of several factors, among them the current hypothesis that viral reservoirs do exist. The hypothesis regarding viral reservoirs has diminished the enthusiasm and rationale for early therapy. Also, evidence from multiple cohorts has demonstrated that CD4+ cell count is a reliable predictor of clinical progression, mortality, and benefit from antiretroviral therapy. Some research has shown that outcomes are comparable for a spectrum of HAART initiation times.

The goals of HAART are:

- Achievement and suppression of viral load to below the limits of detection of plasma HIV-1 RNA, <20–50 copies/mL, using an ultrasensitive viral load assay
- Reduction of HIV-associated morbidity
- Reduction of HIV transmission to others
- Improvement of immune function leading to improved overall quality of life.

However, it is not unusual for patients doing well on antiretroviral therapy to experience occasional "blips" of transient viremia with subsequent re-suppression, all while maintaining the same medication regimen. The success of HAART is highly dependent upon patient adherence. Adherence is a challenge for many HIV-infected individuals for various reasons, including pill burden, adverse effects, and lifestyle factors.

Funding from sources such as PEPFAR and the Global Fund paved the way for getting more than 20 million individuals on HAART treatment globally in 2017. The availability in 2018 of more than 30 antiretroviral drugs, for those individuals or country programs that can afford them, and their use in potent antiretroviral regimens have allowed for a level of control of HIV replication that was not possible before. The durability of that control, however, is threatened by the emergence of viral resistance to antiretroviral medications.

Antiretroviral drug resistance occurs on a molecular level. The nucleotide sequence of a gene determines the amino acid sequence of the protein encoded by the gene. Mutations arise during replication when changes occur to the nucleotide sequence at a given codon, resulting in the incorporation of a different amino acid that is normally found at that position in the protein.

HIV mutates rapidly, averaging one to two mutations per replication cycle. This means that in a relatively short period of time, all the potential mutations that can code for single-drug resistance can easily be generated. Multiple mutations can arise in response to selective drug pressure, in which only those viral strains that possess specific mutations are able to survive in the presence of the drug.

HIV exists *in vivo* as quasi-species in which many highly related, yet genetically distinct, viral variants coexist. The HIV resistance testing platforms now available evaluate the predominant viral quasi-species currently circulating. Using these platforms, mutant strains present in <20%–30% of the viral population are less likely to be detected by testing, causing concern and posing clinical testing challenges.

Genotypic resistance assays use DNA sequencing methods to examine the reverse transcriptase and protease regions of the HIV genome for all resistance-associated mutations. A major drawback of this testing method is that results are difficult to interpret, and expert consultation is necessary.

Phenotypic resistance assays directly measure the ability of HIV-1 to replicate in a cell culture in the presence of different antiretroviral drug concentrations. This process is similar to that used to determine antibiotic resistance and is therefore more familiar to most clinicians. In a phenotypic resistant assay, the recombinant virus, composed of a virus's reverse transcriptase and protease genes, is inserted into a standard reference strain of virus. The recombinant virus is then tested *in vitro* for the amount of drug needed to inhibit virus replication by 50%, relative to the amount of drug needed to inhibit a reference strain of virus. One major limitation of phenotypic resistance testing is that it is conducted *in vitro* and not *in vivo.*

Research has documented that the development of antiretroviral resistance is significantly less likely to occur when a patient's viral load is maintained at an undetectable level.[16] However, reduced susceptibility in viruses from untreated patients has also been documented. Up to 18.5% of recently infected individuals have been infected with a strain of virus bearing one or more well-characterized drug resistance mutations.[17] Although HIV resistance can develop rapidly both *in vitro* and *in vivo,* the incidence of primary resistance thus appears to vary by geographical region and with time. The development of resistance is dependent on the timing of initiation of therapy and on the stage of the disease, highlighting the focus on determining optimal starting times. It is worth noting that in the context of HAART, resistance is most often the consequence of initial treatment failure.[18]

As noted before, HIV isolates identified worldwide have been divided mainly into three groups, M, N, and O. Most isolates are in group M. Based on the sequences of the envelope and gag genes, ten genetically distinct subtypes have been identified within the M group. These genetically distinct subtypes pose therapeutic challenges in the design of antiretroviral therapy.

The management of HIV-infected individuals is now more complex than before; not only because of expanding choices of combination therapy but also because of the growing recognition of longer-term toxicities of HAART and the clinical implications of HIV resistance.

The International AIDS Society – USA Drug Resistance Mutations Group delivers accurate, unbiased, and evidence-based information on drug resistance–associated mutations for HIV clinical practitioners. The group reviews new data on HIV drug resistance to maintain a current list of mutations associated with clinical resistance to HIV-1. The lists that are periodically generated normally include mutations that may contribute to a reduced virologic response to the antiretroviral agent.

The two central treatment challenges that remain are ones we have struggled with for decades: adherence and adverse reactions. Although long-term toxicities are commonly cited as reasons for poor adherence; strategies need to be developed that will help HIV-infected individuals adhere to their medications, even during periods when experiencing severely unpleasant physical and emotional/psychological effects. It is also necessary to anticipate adverse effects of these medications with a patient before they occur and develop strategies for managing these adverse effects on a case-by-case basis so as to achieve better outcomes.

20.4   PREVENTION

The global HIV/AIDS pandemic has quickly evolved from a major health issue to a complex international emergency that undermines the social and economic fabric of nations, especially in sub-Saharan Africa, where decades of development have been reversed. In this region, nearly all the gains in life expectancy that have been made since 1950 have been wiped out. The epidemic in the other regions of the world does not have the levels of prevalence as in sub-Saharan Africa.

Unlike other microbial scourges such as malaria and tuberculosis, for which there is little that people can do to prevent infection, HIV infection in adults is almost entirely preventable by behavior modification. Very rare cases of contaminated blood transfusion products or contaminated hospital hypodermic needles are the exceptions.

There are several HIV hot spots of populations heavily impacted by high transmission rates. Primary prevention must be reinvigorated and targeted to vulnerable populations. Men who have sex with men (MSM) remain one of the populations most heavily impacted by HIV disease worldwide. Sex between men remains highly stigmatized in many parts of the world including sub-Saharan African region, where many MSM also maintain sexual relationships with women.[18]

In addition to MSM, there are other key populations at risk for HIV infection globally, such as female sex workers (FSWs), individuals who pay for sex, and injection drug users (IDUs). Women bear a disproportionate burden of HIV infection globally. Factors associated with future HIV acquisition include:

- Being younger than 25 years old
- Being unmarried
- Having a primary partner who does not provide material support or has other partners
- Having a curable sexually transmitted infection at screening such as chlamydia, gonorrhea, trichomoniasis, or syphilis
- Occasional to heavy alcohol consumption
- Having bacterial vaginosis with loss of healthy mucosal defenses and local inflammation in the vaginal mucosa.

Research indicates that the most important modes of HIV transmission in the North African and Middle East regions are IDUs and unprotected sexual intercourse, including MSM.[19]

Several approaches to prevention, if properly executed, can be effective. As with antiretroviral therapy, combinations of approaches are needed to achieve effective prevention results in a population.

The use of antiretroviral agents in pregnant women with HIV infection and their infants is a beneficial and successful prevention strategy. The rate of mother-to-child transmission of HIV in the United States has been cut to negligible levels among women and infants treated with an extended regimen of specific antiretroviral agents. In contrast, many low-to middle-income countries lack a strategic HIV/AIDS plan. In order to most effectively fight HIV/AIDS, all countries must have a strategic plan in place, and political leaders need to be active vocal proponents of science-based HIV prevention policies.

20.5   VACCINES AND MICROBICIDES INITIAL DEVELOPMENT CONCEPTS

On the global scale, HIV infections are recognized as several epidemics of genetically distinct types, each with characteristic geography and predominant viral strain.[20] Due to its dynamic nature of replication and high error rate during reverse transcription, HIV-1 has evolved into multiple subtypes, also called clades. *Intra*-subtype genetic diversity may be as high as 20%, while *inter*-subtype diversity might be as high as 35%.[21]

Subtypes A, C, and D are found all over sub-Saharan Africa, whereas subtype B is dominant in the United States and most of Europe. A heterogeneous virus population normally creates difficulties in vaccine development, as seen in the development of vaccine against influenza virus infection with as little as 2% viral diversity.[21] Difficulties are greatly compounded when variability rates reach 20%–35%.

It remains difficult for researchers to quantify exactly how much protection a specific vaccine provides. A major scientific obstacle to the development of a safe and effective vaccine is the difficulty in establishing the precise correlates of protective immunity against HIV infection. A number of factors such as lack of an adequate animal model, lack of incentives, genetic variability, coupled with the relatively poor understanding of the immunologic mechanisms that confer protection, have stymied the development of an effective HIV vaccine as of this time.

The key issues in the development of an HIV vaccine include identification of:

- Protective immune responses to infection and
- Viral gene products against which the host products are directed.

When HIV infects a cell, the host responds with both B-cell (humoral) and T-cell (cell-mediated) immune mechanisms. A central target of HIV infection is a group of T cells called CD4+T lymphocytes. These CD4+ T lymphocytes are central to a successful immune response. Loss of the CD4+ T-lymphocyte response is an important characteristic feature of HIV disease.

Additionally, decisions about when to start treatment with a variety of drugs are most often determined by CD4+ T levels. The CD4+ T cells normally recognize foreign antigens bound to host proteins and assist B cells through the production of various cytokines in the production of antibodies. The CD4+ T cells also assist in the recruitment of another subset of T cells known as CD8+ T cells. These CD8+ T cells are commonly known as cytotoxic T lymphocytes (CTLs).

Most of the initial work with HIV vaccines was directed at developing vaccines that elicited neutralizing antibodies. Unfortunately, these neutralizing antibodies have proven to be very narrow in focus as it relates to their action, i.e., specific almost entirely to the strain of the inoculating virus.[22] It has been apparent that developing a vaccine that elicits broad neutralizing activity is therefore an important quest for HIV vaccine developers.[23] However, it is very difficult to induce broadly reactive neutralizing antibodies to HIV by immunization because HIV envelope gp has loop domains with high variability. This allows the virus to evade antibody recognition. The gp-120 is also a flexible protein, assuming many novel configurations, which makes it a more difficult target for antibodies. Because HIV exhibits rapid escape from neutralizing antibodies, a number of strategies for improving neutralizing antibody responses are being researched.

Some research is now focused on the possibility of increasing the CTL responses that target virus-infected cells.[24] Even though antibodies can play an important role in preventing the infection of susceptible cells, CD8+ CTL remain a key element of the immune system for recognizing and lysing virus-infected cells.[25] CD8+ CTL lyse infected cells by recruiting natural killer (NK) cells in a process known as antibody-dependent cell-mediated cytotoxicity.

HIV-specific CD8+ responses have been detected in cervicovaginal fluid from both HIV-infected women and HIV-exposed uninfected women. Because HIV usually enters the body at mucosal surfaces, most often vaginally or rectally, and replicates in lymphoid tissue, induction of immune responses at the mucosal surfaces might be an effective preventive approach. However, one major challenge to this approach is that active memory cells must be present in sufficient amounts in addition to being able to induce immune responses at the mucosal surfaces involved in HIV transmission.

In the past, traditional approaches to viral vaccine development have focused on using live-attenuated virus or whole-killed virus. Because of safety concerns, these methods cannot be employed in HIV vaccine development.[26] Currently, one of the big obstacles facing candidate vaccines is the inability to induce antibodies that will neutralize a broad range of primary HIV strains. Ideally, the antibodies should have broad cross-reactivity with strains of HIV isolated from infected persons in the geographic region of interest. Controversy about both the design of assay systems to measure the neutralization of such isolates and the interpretation of the results did persist for more than a decade.[27]

There is greater need for research on assays used to measure immune responses. Earlier in the epidemic, enzyme-linked immunospot assays, in which T cells that make interferon gamma on peptide challenge, were the standard.[28] However, Hanke and colleagues[29] reported that interferon gamma may not be the best cytokine to measure because of its minimal or possible lack of anti-HIV effect.

Earlier vaccine development efforts predominantly used clade B isolates, which represent the subtype most common in North America and Western Europe. There was also increased interest in the development of clades A and C vaccines for the expanding pandemic in Asia and sub-Saharan Africa. A concern in the clade-specific vaccine strategy is the potential inability to produce sufficient amounts of vaccine specific for distinct clades. This leaves open the question of specific versus cross-clade effectiveness. Choices regarding immunogen(s), adjuvant, dose, and mode of administration are additional variables that must be addressed in candidate vaccine research.

The field of microbicide research initially started with several large trials among high-risk women in both the developed and the developing nations. Microbicide research and development is a complex and lengthy process. Conducting vaginal microbicide trials in the developing world sometimes poses complex ethical challenges such as difficulties in obtaining informed consent, maintaining the confidentiality of the study participant's HIV status, and also difficulty in providing care for those who may become infected with HIV during the trial period.[30]

In some countries, when women are infected with HIV, they often face physical violence, aggressive criticism, and severe emotional stress. They are often abandoned by their families and friends and ostracized by their respective communities. Fear of violence not only prevents women from accessing information about HIV/AIDS, it also prevents them from getting tested.

It is very important that HIV/AIDS prevention programs involve both men and women so as to effectively address gender inequality. Most of the ethical precepts for experimentation on humans are designed in the developed world where the legislative arms are in place, to some degree, to take care of the interests of the study participants. The same is not applicable in the developing world, thereby posing additional clinical challenges.

20.6 SUMMARY AND CONCLUSIONS

People of sub-Saharan Africa descent living in wealthy nations have intense concerns about the HIV epidemic. For example, in the United States, because of the disproportionate incidence of HIV/AIDS among people of Africa descent and the Latino population, special emphasis must be placed on reaching these groups with effective education, treatment, and risk reduction programs despite the challenges. The situation is still critical that not a few persons compare it to the 40-year Tuskegee medical study of untreated syphilis in African Americans, the longest nontherapeutic experiment on human beings in history.[31] At the beginning of the epidemic, many people of sub-Saharan Africa descent felt that the sluggish response to HIV/AIDS could even be a form of genocide.

The disease known as AIDS was first reported in 1981, and the HIV that causes it continues to create formidable challenges to the biomedical research and public health communities around the world. Most of the initial work with HIV vaccines was directed at developing vaccines that elicited neutralizing antibodies. These neutralizing antibodies have been narrowed in focus as it relates to their action and specific almost entirely to the strain of the inoculating virus (the specific clade of virus used to create that batch of vaccine). Additionally, there has been controversy reported about both the design of assay systems to measure the neutralization of such isolates and the interpretation of the results. Researchers are now looking for a "broad-spectrum" vaccine; however, the high variability of the HIV envelope glycoprotein and its rapid rate of mutation creates an elusive target. Safety concerns have reduced interest in live-attenuated virus or whole-killed virus vaccines.

Vaccine research considerations must include understanding the role of mucosal immunity, the importance of clades, and the continuing search for immune correlates of protection, so as to be able to quantify and communicate in a consistent way the degree of a vaccine's effectiveness. Also, microbicides research must move forward with trials in some of the same populations that HIV vaccine trials are being considered. Success of vaccines and microbicides will take vision, scientific breakthroughs, political will, and mobilization of far more resources than are now being made available.

Given the limitations of currently available antiretroviral drugs, strategies to tackle the long-term management of HIV disease should continue to be evaluated. Research evaluating the control of HIV replication through the use of pharmaceuticals that avoid long-term health complications of the available antiretroviral agents should be encouraged.

Early detection of HIV infection is vital in controlling its spread. However, some individuals in resource-limited settings still believe they have no reason to find out about their HIV status if there are no free treatment interventions. The availability of treatment will act as incentive for these individuals to seek testing and counseling. There clearly remains a strong need to create and sustain a more robust medical and public health infrastructure in the developing world.

Despite developments in the field of molecular biology, virology, immunology, and pharmacology, the control of HIV-1 still awaits effective vaccines and microbicides. While we await the significant technological advances that are still required to overcome the unique obstacles posed by HIV-1, we must, as stated, find ways to scale-up proven prevention strategies, while providing access to HIV treatment for the infected individuals in the world.

After 35 years of aggressive HIV research, challenges still confront researchers, sometimes at a basic level and sometimes at the frontiers of science. Research design has led to terminating trials earlier than was originally thought necessary. Challenges to methodology have made late-stage trials problematic. Estimating expected incidence in the trial population has sometimes proven difficult.[32]

Curiously, one challenge that has arisen as a result of the success achieved with antiretroviral therapy. If a patient is doing well and is stable and not infective, the motivation to participate in a trial with risks is lower than if that patient's progress had not been made.[33]

As an article on risk/benefit of HIV research notes, there are not just a few risks and uncertainties. Approaches with stem cells, new interventions with immunocompromised patients, approaches requiring interruption of antiretroviral therapy, and protocols that seek biopsies of reservoirs all come with risks, often in uncharted territory and certainly with no guarantees.[33]

In spite of the progress that has been made, there are still about 5,000 new HIV infections per day globally. Microbicides have resulted in a number of high-profile disappointments. Statistical modeling has shown that a microbicide that was 60% effective and used by only 20% of the women in touch with health care providers, in the world's 73 lowest-income countries, would prevent 800,000 cases per year.[34]

Following a health care pattern that is not uncommon, the antiretroviral medications being most widely used in the lower-income countries are those that were developed earlier. The earlier versions are not as advanced nor as cost effective as more recent versions. Income levels are such, that for essentially all patients who want the more effective antiretrovirals, the cost would exceed their total household income.

Anyone who has looked at a diagram of the life cycle of HIV replication has surely wondered if within this complex multi-stage process, there is some aspect that does not mutate in a prohibitive manner or is not hidden behind a to-date impenetrable physiological barrier and will prove vulnerable to attack from vaccine stimulated antibodies.

The elusive search for a vaccine continues. The United States National Institutes of Health announced on HIV Vaccine Awareness Day in 2017, that development of an effective vaccine remains the "highest priority for AIDS research" of the United States National Institute of Allergy and Infectious Diseases.[35]

The challenges to clinical testing for improved treatment and preventive approaches are multifaceted, ranging from limited funding to frustration with clinical trials at a level rarely seen in the history of scientific research funded at this level.

Just as we have recently passed through unprecedented advances in computer science, many now believe that medicine will build on these advances, using high-speed computing with the increasing knowledge of genetics, genomics, and immunology and the vast amount of knowledge gained in a myriad of failed attempts.

Governments must continue to do their part in funding efforts to find rational and effective solutions. It is not part of human nature to give up in the face of this type of medical challenge that affects every person on earth.

Annotation: *The development of drugs to treat and control AIDS has seen tremendous growth in 30 years, yet many challenges exist to provide protection against resistance emergence, vaccines for protection, and more widespread availability of agents for treatment, especially in low-to middle-income countries. The author is a Clinical pharmacologist/Pharmacist, Infectious Disease Pharmacotherapy Specialist, with expertise in HIV/AIDS and the Clinical Testing of Antiretroviral pharmacotherapy. He is currently a Visiting Professor at the University of Port Harcourt in Nigeria.*

REFERENCES

1. Global report. UNAIDS report on the global AIDS epidemic 2013, 2013.
2. Eyawo O, de Walque D, Ford N, et al. HIV status in discordant couples in sub-Saharan Africa: A systematic review and meta-analysis. *Lancet Infect Dis* 2010; 10(11): 770–7.
3. Bailey RC, Moses S, Parker CB, et al. Male circumcision for HIV prevention in young men in Kisumu, Kenya: A randomised controlled trial. *Lancet* 2007; 369(9562): 643–56.
4. Gray RH, Kigozi G, Serwadda D, et al. Male circumcision for HIV prevention in men in Rakai, Uganda: A randomised trial. *Lancet* 2007; 369(9562): 657–66.
5. Burrage A, Patel M, Mirkovic K, et al. Trends in antiretroviral therapy eligibility and coverage among children aged <15 years with HIV infection—20 PEPFAR-supported sub-saharan African countries, 2012–2016. *MMWR Morb Mort Wkly Rep* 2018; 67(19): 552–5.
6. Wei X, Ghosh SK, Taylor ME, et al. Viral dynamics in HIV-1 infection. *Nature* 1995; 373: 117–22.
7. Ho DD, Neumann AU, Perelson AS, et al. Rapid turnover of plasma virions and CD4+ lymphocytes in HIV-1 infection. *Nature* 1995; 373: 123–6.
8. Stopak K, de Noronha C, Yenemoto W, Greene WC. Vif blocks the antiviral of APOBEC3G by impairing both its translation and intracellular stability. *Mol Cell* 2003; 3: 591–601.
9. Stevenson M. CROI 2014: Basic science review. *Top Antivir Med* 2014; 22(2): 574–8.
10. Stevenson M. CROI 2016: Basic science review. *Top Antivir Med* 2016; 24(1): 4–9.
11. Chomont N, El-Far M, Ancuta P, et al. HIV reservoir size and persistence are driven by T cell survival and homeostatic proliferation. *Nat Med.* 2009; 15(8): 893–900.
12. Cavert W, Notermans DW, Staskus K, et al. Kinetics of response lymphoid tissue to antiretroviral therapy of HIV-1 infection. *Science* 1997; 276: 960–4.
13. Perelson AS, Essunger P, Cao Y, et al. Decay characteristics of HIV-1-infected compartments during combination therapy. *Nature* 1997; 387: 188–91.
14. Finzi D, Hermankova M, Pierson T, et al. Identification of a reservoir for HIV-1 in patients on highly active antiretroviral therapy. *Science* 1997; 278: 1295–300.
15. Lee FJ, Amin J, Carr A. Efficacy of initial antiretroviral therapy for HIV-1 infection in adults: A systematic review and meta-analysis of 114 studies with up to 144 weeks' follow-up. *PLOS One.* 2014; 9(5): e97482.
16. Wainberg MA, Friedland G. Public health implications of antiretroviral therapy and HIV drug resistance. *JAMA* 1998; 279: 1977–83.
17. Little SJ, Holte S, Routy JP. Antiretroviral resistance and response to initial therapy among recently HIV-infected subjects in North America. *Antivir Ther* 2001; 6 (suppl. 1): 21.
18. Clavel F, Hance AJ. HIV drug resistance. *N Engl J Med.* 2004; 350: 1023–35.
19. Smith AD, Tapsoba P, Peshu N, et al. Men who have sex with men and HIV/AIDS in sub-Saharan Africa. *Lancet* 2009; 374(9687): 416–22.
20. Global HIV/AIDS response: epidemic update and health sector progress towards universal access. Progress report, 2011.
21. Gaschen B, Taylor J, Yusim K, et al. Diversity considerations in HIV-1 vaccine selection. *Science* 2002; 296: 2354–60.
22. Dolin R. Human studies in the development of human immunodeficiency virus vaccines. *J Infect Dis.* 1995; 172(5): 1175–83.
23. Burton DR, Montefiori, DC. The antibody response in HIV-1 infection, *AIDS* 1997; 11(suppl. A): 87–98.
24. Letvin NL, Barouch DH, Montefiori, DC. Prospects for vaccine protection against HIV-1 infection and AIDS. *Annu Rev Immunol* 2002; 20: 73–99.
25. Byrne JA, Oldstone MB. Biology of cloned cytotoxic T lymphocytes specific for lymphocytic choriomeningitis virus: Clearance of virus in vivo. *J Virol* 1984; 51: 682–6.
26. Baba TW, Liska V, Khimani AH et al. Live attenuated, multiply deleted simian immunodeficiency viruses causes AIDS in infant and adult macaques. *Nat Med* 1999; 5:194–203.
27. Hanson CV. Measuring vaccine-induced HIV neutralization: Report of a workshop. *AIDS Res Hum Retroviruses* 1994; 10: 645–8.
28. Currier JR, Kuta EG, Turk E et al. A panel of MHC class 1 restricted viral peptides for use as a quality control for vaccine trial ELISPOT assays. *J Immunol Methods* 2002; 260: 157–72.
29. Hanke T, McMichael AJ. Design and construction of an experimental HIV-1 vaccine for a year 2000 clinical trial in Kenya. *Nat Med* 2000; 6: 951–5.
30. Ramjee G, Morar NS, Alary M et al. Challenges in the conduct of vaginal microbicide effectiveness trials in the developing world. *AIDS* 2000; 14: 2553–7.

31. Stebbing J, Bower M. Lessons for HIV from Tuskegee. *J HIV Ther* 2004; 9(3): 50–2.
32. Lagakos SW, Gable AR. Challenges to HIV prevention—seeking effective measures in the absence of a vaccine. *N Engl J Med* 2008; 358:1543–5.
33. Eyal N. The benefit/risk ratio challenge in clinical research, and the case of HIV cure: An introduction. *J Med Ethics* 2017; 43(2):65–6.
34. UNAIDS. Microbicides: challenges to development and distribution (Part 2), 21 Feb 2008. www.unaids.org/en/resources/presscentre/featurestories/2008/february/20080221microbicidespart2 (Accessed June 4, 2018).
35. NIH statement on HIV Vaccine Awareness Day, 2017. www.nih.gov/news-events/news-releases/nih-statement-hiv-vaccine-awareness-day-2017 (Accessed June 4, 2018).

21 The Evolving Role of the Pharmacist in Clinical, Academic, and Industry Sectors

Gourang Patel
Rush University Medical Center

Stephanie Tedford
LumiThera Inc

CONTENTS

21.1 INTRODUCTION

The evolution of the Pharmacist has transformed the position from a limited product and task-oriented role to expert leads in research development and integrated, essential members of the healthcare team, critical for ensuring best practice outcomes for patients. The dynamic training received by the Pharmacist allows for a varying range of applicability in many healthcare environments, whether in hospital, academic, or industry sectors. These evolutionary changes in the practical aspects of the Pharmacist's role to a research and/or patient-oriented position have raised the caliber of the profession and continue to expand on the potential impact the Pharmacist can have in the healthcare field.

21.2 FUNDAMENTALS AND EVOLUTION OF THE PHARMACY PROFESSION

Pharmacy has traditionally been a product and task-oriented occupation, with a primary focus on compounding and dispensing prescribed medications for patients. This "apothecary" type role has been maintained for decades. The standard practices that occur in retail pharmacies are most often associated with the pre-existing Pharmacist role; however, advancements in medication delivery systems, i.e., the introduction of drugs in tablet and capsule form, have reduced the pharmacy needs for medicine preparation. Outside of medication services, patient interaction has been largely limited to only educational needs for the prescribed compounds as requested. Evolution of the role dictates that it is not enough to only dispense the appropriate drug but to critically evaluate the entire medication therapy to identify any pharmacotherapeutic interventions [1]. Pharmacists have sought to further their contribution to healthcare largely through their own perception of societal needs that could be best addressed by the pharmacy profession. This may include areas of medicine that focus on furthering pharmaceutical research development and optimizing patient care plans involving drug treatment protocols. Their extensive training provides a unique opportunity to engage and provide significant impact in many areas of healthcare.

In the early 1980s, the National Pharmacy Association introduced the "Ask your Pharmacist" campaign as a tactic to increase recognition of the Pharmacist as a healthcare professional. In 1990, Hepler and Strand [1] introduced the term "pharmaceutical care," which exemplified a shift from the traditional role to a patient-centered, outcome-oriented pharmacy practice. This model was critical in promoting the Pharmacist as an important member of the patient healthcare provider team who held similar responsibility for patient outcomes. In 2000, the American College of Clinical Pharmacy extended this notion with an article titled "A vision of pharmacy's future roles, responsibilities, and manpower needs in the United States" [2], once again highlighting the move to a patient-centered role for the Pharmacist. Additional publications followed suit supporting the expansion of the Pharmacist's role and increased recognition of their expertise and the advancement of the profession. This shift by the Pharmacy community raised acknowledgment of the Pharmacist to a visible and vital member involved in patient care.

21.2.1 PHARMACY FOCUS ON PATIENT CARE

The role of the Pharmacist has evolved firstly to meet the growing needs and expectations of patients but secondly with the needs of other medical professionals in related fields. Many roles in the healthcare field have adapted to fit a more collaborative and interdisciplinary model. This is largely attributable to the healthcare demands of an aging population who typically exhibit many chronic conditions, leading to a greater societal burden across professional boundaries. Advancements in technology, overall healthcare system structure, sociopolitical changes, access to information, and a shift to multidisciplinary work also contribute to this evolution. Due to the ease of accessibility, Pharmacists are oftentimes the first and/or only point of contact in the healthcare system for many patients who do not seek out a physician's assessment. The information provided by the Pharmacist may very well be the only professional healthcare advice the patient receives. This direct line of contact has created an environment where Pharmacists are uniquely posed to have more contact with patients than other healthcare team members. These patients may include those who are hesitant to seek out medical advice and those with no health insurance or limited income who cannot afford to see a practicing physician or advanced practicing provider (e.g., Nurse practitioner or Physician's assistant). Pharmacists have long recognized this aspect of their practice and evolution of the role has included further training on patient interactions and education efforts. Furthermore, they are uniquely positioned to prescribe over-the-counter (OTC) medications for common ailments and to assess whether the desired drug therapy outcomes are being achieved while ensuring a balance of efficacy and safety. The shift in a patient-forward interaction by the Pharmacist has the potential to foster better healthcare services to a large population of patients.

Pharmacists are uniquely qualified to assist patients with any potential side effects or adverse drug reactions (ADRs). ADRs represent a significant clinical challenge that affects patients in the clinic and home setting potentially resulting in serious health complications and loss of life. A meta-analysis of prospective studies showed that 6.7% of hospitalized patients treated with pharmaceutical agents show an overall incidence of serious ADRs and 0.32% have fatal ADRs [1]. The incidence of serious and fatal ADRs demonstrates the clinical impact ADRs pose to patients. Drug-related morbidities and mortality are oftentimes preventable. Increased pharmacy services can aid in the reduction of ADRs, thereby improving patient outcomes, reducing the length of hospital stays, and reducing the overall healthcare costs. Enhanced patient-focused care can foster an environment where communication aids in identifying ADRs that may not be readily reported to primary care physicians and could potentially have life-threatening complications.

21.2.2 PHARMACISTS FOCUS ON RESEARCH

Interdisciplinary and collaborative efforts among scientific and medical communities have expanded the role of the traditional Pharmacist to allow for additional opportunity and oversight throughout the various steps of drug research. Pharmacy-led initiatives are becoming more commonly utilized in this process by enlisting "investigational" research-oriented Pharmacists who act to integrate pharmacy expertise and practice within the preclinical and clinical research settings to optimize drug therapies for patients. Investigational Pharmacists are part of an integrated drug research team that provides valuable contributions to each step of the investigational research process. These Pharmacists take on roles in both the industries, i.e., academic and hospital settings where their extensive training is influential in assisting research endeavors evaluating new drugs and their side effects and ultimately their effectiveness and safety in targeted patient populations.

The task of performing research requires many key elements such as the following: patient population with a specific question, organized/structured plan, and motivated research investigators. The core roles needed to conduct prospective research studies include the following:

- Research proposal/protocol review and submission to Institutional Review Board (IRB)
- Patient/population screening
- Obtaining informed consent
- Investigational drug procurement/dispensing/reconciliation
- Case Report Form (CRF) completion
- Lab/specimen chain of custody, when applicable
- Statistical analysis
- Manuscript preparation/submission
- Maintenance of IRB

In most academic environments, the initial study proposal review and submission is performed by the primary research investigator, research coordinator(s), and the Pharmacist. The primary research investigator can be from any healthcare discipline; however, the profession is often dictated by the sponsor (when applicable) and resources at the study facility. The advantage of a research coordinator is being an advanced practice provider that can assist with screening, informed consent, and lab collection when needed. The investigational drug procurement and dispensing is performed by the Pharmacist.

Traditional roles of the clinical Pharmacist for research studies included primarily data extraction/collection. However, as multidisciplinary teams evolved, clinicians quickly recognized the value and appreciation for that role to be expanded to include patient recruitment/screening and consent. Obtaining consent from a patient for a critical care study can be challenging, especially those prospective studies focused on life-saving disease states in which the patient may not be able to consent prior to enrollment (i.e., septic shock). Consent in prospective research studies can be provided by an available surrogate for the patient, as approved by the local IRB. The Food and Drug Administration (FDA) has approved criteria for waiver of informed consent in special circumstances. Implementing this waiver does not violate the ethical codes as the life-threatening disease state makes it difficult to determine the patient's preferences. IRBs may grant a waiver of informed consent for emergency research provided the study meets the following criteria: the participant has a life-threatening condition that necessitates the engagement of the experimental protocol, the participant has impaired decisional capacity, and there is time to obtain consent from a surrogate for the patient being treated.

In order to facilitate emergency protocols, it is recommended that the interpretation of a life-threatening condition include any physiologic or anatomic derangement that could result in mortality or morbidity. An independent data and safety monitoring board (DSMB) should be established in the study planning to provide oversight of the research study [3, 4]. However, retrospective research studies can often obtain a waiver of consent based on the nature of the review after medical care has already been provided. The expansion of the role for clinical Pharmacists has allowed healthcare providers to focus on other aspects of the study and simultaneously opened the door to pursue research grants as well. Another example of clinical Pharmacist role expansion has been the transition to provide statistical analysis support. Statistical analysis within the medical field is challenged with the influx of new medications/therapies, changing clinical guidelines, limited enrollment in prospective studies, and shorter time windows for study completion. Data collection and statistical analysis require a good foundation of scientific epidemiology, literature review/critique, and statistical analysis. Building a foundation for the clinician can be accomplished with on-the-job training/ mentoring.

Additional routes to facilitate foundational research training for clinical Pharmacists and members of the healthcare team are the following: pursuing a Masters in Clinical Research or development workshops via professional societies (i.e., American College of Clinical Pharmacy/ Research Institute) with training series focused on performing, executing, and publishing research.

21.3 PHARMACIST ROLES IN THE PHARMACEUTICAL INDUSTRY ENVIRONMENT

In the industry setting, Pharmacists are valued not only for their expertise in pharmacy but also for their understanding of patients and the real-world application and usage of therapeutics. Their extensive clinical training has expanded the Pharmacist's potential roles and responsibilities within many industry sectors. Conventionally, industry investigational Pharmacists become involved in the management and monitoring of drug manufacturing and supervision of related processes which include usage of the drug, quality control, and continued research endeavors. An investigational Pharmacist assists in identification of appropriate therapeutic targets and determinations of which drug therapies best treat various diseases to find viable alternatives to current drug treatments. Their expertise in the understanding of the medicine and its use in the clinic further support team-led initiatives targeted towards design and development of more effective drug delivery systems. These Pharmacists also review drug-usage patterns to assess which drugs are most effective and which drugs may be overprescribed or underutilized [4].

Investigational Pharmacists may become involved in all aspects of the investigational drug process through services in the Drug Regulatory Affairs sector where they participate in the logistics of drug trial conduction, drug registration, and regulatory body approval. In this role, an investigational Pharmacist would contribute to regulatory strategy; submissions to regulatory authorities (e.g., FDA) including Investigational New Drug Applications and New Drug Applications; label development and revision; and maintenance of approved drugs through annual reports, submissions, labeling, and line extensions.

Investigational Pharmacists also lend their expertise to research into new compounds and drug discoveries and are often employed in both early- and late-phase drug clinical development. In the early phases, an investigational Pharmacist works closely with regulatory affairs; data management; drug supply management; and preclinical safety scientists to design study protocols, determine dosing for medications, implement and manage clinical trials, evaluate serious adverse events, and to write up clinical study reports. Later phases of drug development employ investigational Pharmacists to evaluate the pharmaceutical product utilization, treatment modalities, pharmacokinetic/dynamic relationships, and potential drug–drug interactions.

Responsibilities in later phases of industry-led drug development may also include a role for the investigational Pharmacist in sales and marketing. Their expertise aids in development of the medical information, scientific sales materials, and product advertising to the public. Additional responsibilities include medical communication and information gathering where the investigational Pharmacist utilizes their clinical knowledge for content development for peer-reviewed publications, conference materials and digital media for a range of audiences including other medical professionals and consumers. An investigational Pharmacist may also contribute to the drug process by acting as a Medical Science Liaison who coordinates the communication of clinical information between the pharmaceutical company and medical experts in the drug's field of study.

21.4 PHARMACIST ROLES IN ACADEMIA/HOSPITAL ENVIRONMENTS

The implementation of expanded roles for clinical Pharmacists offers a viable strategy in the clinic that addresses the critical need for primary care assistance in the context of the medical professional workforce shortage. In the hospital setting, a clinical Pharmacist has traditionally resided in the Department of Pharmacy. The Department of Pharmacy functions to provide adequate medication needs throughout the hospital but may also be critically involved in human clinical research involving study drugs and the safety and care for patients in both the inpatient and outpatient care areas. The concept of an investigational Pharmacist is proposed as an approach to better oversee and regulate investigational drugs and patient outcomes [5, 6]. The investigational Pharmacist ensures that drug studies in the hospital are executed in a safe, effective, and efficient manner. These responsibilities include the distribution and control of study drugs, clinical services, research activities, and management of clinical studies as well as patient education. The precise responsibilities of an investigational Pharmacist will vary between hospitals reflecting institutional resources, research commitments, and the direct needs of the hospital. The investigational Pharmacist commonly works alongside clinical investigators, nursing staff, hospital risk management, sponsors of clinical research, and patients and their families [7].

During a clinical trial, investigational Pharmacists may be involved in the initial trial setup through collaborations with the sponsor and healthcare team to design study protocols that are in keeping with the pharmacy best practices. A primary role of the investigational Pharmacist is to ensure the appropriate management of the investigational medications as well as study drug accountability and prevention of study errors involving drug preparation, administration, and disposal of the investigational products [3, 8]. The investigational Pharmacist commonly prepares, compounds, manufactures, and dispenses investigational drugs and materials for ongoing research studies. They are responsible for preparing and collecting accurate study documentation including drug data sheets for study medications; drug usage and administrative records; and other required record-keeping, shipping, ordering, and inventory activities. Additional responsibilities include appropriately following the study design, protocol and data acquisition, subject randomization, and the integrity of study subject blinding (i.e., which subjects receive the active drug or placebo). The investigational Pharmacist may also be held accountable for ensuring compliance with institutional, federal, and state regulatory guidelines throughout the initiation, monitoring, and auditing of all investigational drug studies. They commonly serve as a liaison for study investigators, research sponsors, the IRB, and the pharmacy to ensure complete adherence to study protocols and/or policies [3]. The investigational Pharmacist may also work directly with study patients to provide educational tools and training related to investigational drug protocols, clinical checking, drug counseling, or drug interaction checking in support of study investigators.

Clinical Pharmacists have expanded to multiple sub-specialties of care such as the following: medical, cardiac, cardiovascular, oncology surgical, neuroscience, trauma/burn, emergency medicine, and perioperative care (i.e., anesthesia). In addition, the presence of clinical Pharmacists at the bedside has improved the scientific research contributions to improve patient outcomes and safety [9]. Furthermore, the addition of a clinical Pharmacist to the critical care research team has expanded from data extraction to more recently including patient recruitment/consent, data collection, statistical analysis, and manuscript preparation/submission.

One strategy being explored to support chronic disease management involves embedding clinical Pharmacists in primary care to facilitate patient education, supplemental patient interaction, and population management activities. Some of the expanded clinical responsibilities a clinical Pharmacist can assume around pain care include: ongoing reassessment, monitoring, and management of opioid therapy in conjunction with medication renewal, review of state prescription drug monitoring programs, and medication education. Some clinical Pharmacists are specializing in pain management to support teams with complex pain patients, opioid risk management, opioid education, opioid titration, opioid screening, and naloxone distribution [10]. When clinical Pharmacists are involved in an interdisciplinary team, clinics have reported a decrease of burden on Primary Care Providers (PCPs) and improved patient satisfaction [11, 12].

Table 21.1 contains the position description for the Investigational Drug Pharmacist at Rush. Note that 20% of the Pharmacists' salary is paid by IRB funds. In addition to the roles described in the text and in Table 21.1, Investigational Drug Pharmacists at Rush University play a key role in reviewing the informed consents for research studies.

TABLE 21.1
Rush University Medical Center Job Description

Title:	Investigational Pharmacist
Department:	Pharmacy
General Summary:	The scope of practice of the Investigational Pharmacist covers both inpatient and outpatient investigational studies, patient education, quality assurance, staff education, and consultation and is often in association with or in support of physicians and other healthcare practitioners. It is an integral component of a high-quality pharmaceutical care program and consists primarily of drug control, protocol management, clinical service, and education. In addition, the Investigational Pharmacist will exemplify the Rush mission, vision, and values and act in accordance with Rush policies and procedures.

Principal Duties and Responsibilities

Investigational Drug Control:	Communicates with investigators and sponsors to discuss the procurement, storage and anticipated use, and other issues regarding investigational substances
	Maintains adequate inventory levels of investigational drugs in the pharmacy
	Provides investigational supplies on a regular basis and maintains complete and accurate records
	Assures proper storage of investigational drugs
	Assures the proper disposition of used, unused, or completed study drug inventories
	Develops and implements procedures for preparation, labeling, and safe handling of study drugs
	Assures the proper procedures for preparation, labeling, and dispensing of investigational drugs are adhered to
	Works with information services to create epic orders for investigational drugs
	Maintains appropriate pricing of Investigational Drug Supplies (IDS) services and evaluates annually

(Continued)

TABLE 21.1 (*Continued*)
Rush University Medical Center Job Description

Protocol Management:	1. Assists investigators and their designees, when requested, with submission, approval, implementation, and coordination of study protocols and procedures
	2. Assures that the requirements of the manufacturer, FDA, and research protocol are met regarding completion of investigational drug records
	3. Reviews, as an agent of the pharmacy department, protocols and proposals and provides commentary with respect to investigational drug therapies; is an ex-officio member of an IRB committee
	4. Bills appropriate departments for time spent setting up, coordinating, and storing investigational studies
Clinical Service:	1. Services as a drug information resource for the medical center in matters concerning investigational drug studies
	2. Communicates regularly with nursing, Pharmacists, and medical personnel concerning questions and clarifications of investigational study protocols
	3. Communicates with investigators, their designees, study sponsor, and clinical monitor with regard to investigational drugs and clinical trials
	4. Inservices the pharmacy and nursing staff, when appropriate, in the procedures for dispensing and administering investigational drugs; at a minimum, this will be in the form of an orientation for all new Pharmacists
	5. Coordinates and/or assists in the education of the patient to drug therapy when requested
	6. Coordinates and/or assists in the monitoring of investigational drug therapy when requested or the need arises
	7. Prepares and distributes drug information specific to drugs used in investigational studies for use by the pharmacy department's technical and professional staff
	8. Attends and participates in local meetings regarding investigational drug studies; functions as a liaison of the IDS and Department of Pharmacy
	9. Works periodically in pharmacy operations on weekdays and weekends in order to keep abreast of distribution and operations
Education:	1. Provides educational and training programs to staff Pharmacists, pharmacy residents, and PharmD candidates from affiliated Colleges of Pharmacy (Introductory to Pharmacy Practice (IPPE) students, Advanced Pharmacy Practice Experience (APPE)) students as needed
	2. Provides education to residents about IRB submission for their research projects
Universal Pharmacist Responsibilities:	• Order approval
	• Aseptic technique
	• Document I-vents
	• Clean room procedures
	• Distribution systems (Pyxis, carousels, etc.)
	• Narcotics
Shared Support of IDS Positions:	Approximately 1 day per week paid for by Office of Research Affairs for participation as ex-officio members of IRB

21.5 PHARMACIST ROLES IN CLINICAL PHARMACY RESEARCH CONDUCTED IN THE CRITICALLY ILL

Emergency medicine and critical care centers utilize the expertise of many medical professions to provide multidisciplinary approaches and services to critically ill patients. A critically ill patient may present significant challenges for treatment due to the nature of the condition, the changing outlook of the condition, comorbidities, ongoing medication requirements, and potential interactions. Due to these moving factors, the integration of a Pharmacist onto the patient team ensures the appropriate treatment plan and best practice care in a situation that is often changing [13].

The critical care and perioperative environment are fortunate to have multidisciplinary health-care members collaborating on a therapeutic plan, balancing positive outcomes while ensuring patient safety. While each member of the multidisciplinary critical care team focuses on their specialty, a portion of that time also promotes and invites further critical care research. This area of medicine is in high demand, and emergency medical care is now being incorporated into many pharmacy school curriculums.

21.5.1 DOSE–RESPONSE RELATIONSHIPS IN THE INTENSIVE CARE UNIT

Dose–response relationships are the core of pharmacologic relationships. One barrier that is readily encountered within the critical care arena is having a controlled environment to obtain this critical information. The patient populations for which the FDA requires medication approval generally do not have data which is applicable for critically ill patients. Patient populations in the critical care setting have significant pharmacokinetic (pK) changes relating to absorption, distribution, metabolism, and elimination [14]. Two examples in which medications administered via the oral route can be impacted, resulting in delayed blood/plasma concentrations, are the following: vasopressor infusion support for hemodynamic instability or gastrointestinal edema secondary to volume resuscitation. Both of the abovementioned clinical scenario's result in the decreased drug absorption translating into lower bio-availability for oral medications. The subsequent impact on plasma concentrations can be observed and clinical outcomes can be impacted [15]. Another variable that is impacted in critically ill patients is distribution. Medication properties are primarily divided among hydrophilic and lipophilic aspects. During the course of volume resuscitation, the hydrophilic medications are distributed more widely as opposed to those that tend to possess more lipophilic properties. The nature of the distribution can be observed with antimicrobials administered to critically ill patients in the intensive care unit environment [10, 16]. The next variable that is observed in critically ill patients are various protein concentrations. Medications that are known to be more protein bound (i.e., albumin) tend to have elevated plasma concentrations and therefore need to be accounted for during patient assessments. Metabolism is another variable impacted in critically ill patients. Numerous drug interactions via the Cytochrome P450 system are observed and demonstrated to have a significant impact on patient plasma levels (i.e., International Normalized Ratio (INR) with warfarin in a patient presenting with atrial fibrillation requiring the rapid initiation of amiodarone), and adjustments are required to ensure safe levels of warfarin, if the therapy was to continue. Drug elimination is another aspect that is impacted in critically ill patients. Total elimination (renal + hepatic) can be impacted by changes in circulation/blood flow to the kidneys/liver and any intrinsic damage from an active disease state (i.e., septic shock) in which multiple organs can be affected either from an active infection and/or medications used to manage aspects of a patient's disease state (i.e., aminoglycosides, vasopressors, etc.) [11].

21.5.2 FOOD AND DRUG ADMINISTRATION (FDA)

The vast majority of medications utilized in the intensive care unit setting are for non-FDA-approved indications. The statement should not be a surprise, as clinicians practicing in the critical care environment are aware that the original submitted studies for FDA approval of a medication do not enroll or include the critically ill population. The reality presents a significant challenge to clinicians as the patient populations submitted for original FDA approval for a medication require more scientific discussion regarding the indication, dose, route, frequency, and duration of multiple therapeutic drug classes. Smithburger et al. presented data from three medical centers that evaluated 16,391 medications. Investigators reported that 43% of the medications were being utilized off-label with a total of 167 ADRs. A very important finding was that the ADR rate increased by 8% for every one additional off-label medication [17].

Additional research is needed and vital in the critically ill population, specifically focusing on approved indications. Prospective clinical studies in the critically ill should be encouraged, require more time to complete, coordinate additional resources, and often require funding (i.e., institutional funding, professional societies, and federal resources).

Another strategy would be for healthcare providers to ensure the consultation of a clinical Pharmacist on specific medication therapies in which the variables of indication, dose, route, frequency, and duration presented during the course of patient care discussions require more detailed pK analyses. The clinical Pharmacist can assist with clinical literature evaluation and safe extrapolation of data to ensure a balance of patient outcomes while maintaining the utmost patient safety.

21.5.3 CLINICAL PHARMACOLOGY DEVELOPMENT OPPORTUNITIES

21.5.3.1 Antibiotics

The increasing rise in antibacterial resistance has prompted the development of several new antibacterial therapies. However, the rate of resistance has overpowered many of the newer antimicrobial agents already available on the market. In addition, a review of older antimicrobial therapies (i.e., rifampin, sulbactam, etc.) has increasing interest to combat different types of multidrug-resistant (MDR) bacteria. The area of antimicrobial development yields increasing opportunity to research investigators with adequate laboratory equipment/staff, samples from the microbiology lab, and sufficient funding resources to continue research [14].

21.5.3.2 Analgesics

Analgesia still remains the cornerstone for pain management for critically ill patients. However, the advent of one new analgesia therapy (remifentanil) has paved the way to improved anesthesia recovery time after surgery, decreased adverse events, and quicker emergence time. In addition, the utilization of remifentanil allows otherwise higher risk surgical procedures to take place safely. Remifentanil possess unique pharmacokinetics of rapid onset and offset of action while providing a safe titration for patient comfort as its metabolism is via plasma esterases in the bloodstream [10, 15]. However, additional therapies are still needed for patient care in the operating room and intensive care unit areas so this area also lends itself for future research.

21.5.3.3 Sedatives

Sedative therapies have remained fairly underdeveloped over the last decade. Dexmedetomidine offers the ability to minimize analgesia/sedation both in the perioperative and intensive care unit patient care areas; however, there has not been additional developments [10]. One recent research development has been remimazolam. Remimazolam is an ultra-short-acting benzodiazepine that is currently being investigated for its potential role both in the perioperative and intensive care unit areas. The rapid metabolism of remimazolam transforms into an inactive metabolite. This pharmacokinetic property is attractive to intensive care unit patients who already possess significant organ dysfunction [16]. Remimazolam is currently awaiting further Phase III clinical trials.

21.5.3.4 Vasoactive Therapies

The mainstay of circulatory shock encompasses modalities such as epinephrine, norepinephrine, phenylephrine, and vasopressin. One recent therapeutic development includes the FDA approval of angiotensin II infusion for refractory septic shock. Angiotensin II provides an alternative mechanism compared to traditional catecholamine derivatives and vasopressin. In addition, its rapid response towards shock resolution, limited direct physiologic effects on the myocardium, and improved morbidity (i.e., continuous renal replacement therapy, CRRT days) all have been reported in clinical trials as well. One main limitation to the therapy is the acquisition cost of the medication [11, 17]. However, future clinical evidence can assist in providing a justification for therapy as the research studies are published. The area of research for vasoactive therapies still remains a necessity for intensive care unit and perioperative healthcare providers.

21.6　CONCLUSIONS

The role of the Pharmacist has undergone a significant evolution expanding the profession from traditional medication compounding and manufacturing to a central role in patient care and research development initiatives. Medication management and education have been critical areas in which Pharmacists have expanded their utility. Due to the increasing societal burden that a continuously expanding and aging population presents, this role has rightfully extended their expertise to enhance multifunctional and interdisciplinary approaches to better serve healthcare needs across the continuum of care. The Pharmacist is equally equipped to move into significant roles contributing to academic and industry-initiated research endeavors. Investigational and clinical Pharmacists can assist in providing a critical bridge to complete research studies with the appropriate organizational structure, coordination of resources, and mentoring. Medications utilized to treat life-threatening infections, to maintain safe and adequate analgesia/sedation, and support critically low blood pressure are potential drug development opportunities of high interest for research investigation. It is likely that the role of the Pharmacist will continue to evolve, expanding further into a well-outlined and substantial contributing member of any multidisciplinary patient care or research team.

REFERENCES

1. Hepler CD, Strand, LM. Opportunities and responsibilities in pharmaceutical care. *Am J Hosp Pharm.* 1990;47(3):533–43.
2. American College of Clinical Pharmacy. A vision of pharmacy's future roles, responsibilities, and manpower needs in the United States. *Pharmacotherapy.* 2000;20(8):991–1020.
3. U.S. Department of Health and Human Services. Office for Human Protections. Revised Common Rule.
4. U.S. Department of Health and Human Services. Guidance for Institutional Review Boards, Clinical Investigators, and Sponsors. Exception from Informed Consent Requirements for Emergency Research. 2011.
5. Maclaren R, et al. Critical care pharmacy services in United States hospitals. *Ann Pharmacother.* 2006;40(4):612–8.
6. Rudis MI, Brandl KM. Position paper on critical care pharmacy services. Society of critical care medicine and American college of clinical pharmacy task force on critical care pharmacy services. *Crit Care Med.* 2000;28(11):3746–50.
7. Dager W, Bolesta S, Brophy G, Dell K, Gerlach A, Kristeller J, et al. An opinion paper outlining recommendations for training, credentialing, and documenting and justifying critical care pharmacy services. *Pharmacotherapy,* 2011;31(8):135e–75e.
8. Preslaski CR, et al. Pharmacist contributions as members of the multidisciplinary ICU team. *Chest.* 2013;144(5):1687–95.
9. Tangden T, et al. The role of infection models and PK/PD modelling for optimising care of critically ill patients with severe infections. *Intensive Care Med.* 2017;43(7):1021–32.
10. Barr J, et al. Clinical practice guidelines for the management of pain, agitation, and delirium in adult patients in the intensive care unit. *Crit Care Med.* 2013;41(1):263–306.
11. Antonucci E, et al. Angiotensin II in refractory septic shock. *Shock.* 2017;47(5):560–6.
12. Giannitrapani KF, et al. Expanding the role of clinical Pharmacists on interdisciplinary primary care teams for chronic pain and opioid management. *BMC Fam Pract.* 2018;19(1):107.
13. Dasta JF, Evolving role of the Pharmacist in the critical care environment. *J Clin Anesth.* 1996;8(3 Suppl): 99s–102s.
14. van Geelen L, et al. (Some) current concepts in antibacterial drug discovery. *Appl Microbiol Biotechnol.* 2018;102(7):2949–63.
15. de Boer HD, Detriche O, Forget P. Opioid-related side effects: Postoperative ileus, urinary retention, nausea and vomiting, and shivering. A review of the literature. *Best Pract Res Clin Anaesthesiol.* 2017;31(4):499–504.
16. Wesolowski AM, et al. Remimazolam: Pharmacologic considerations and clinical role in anesthesiology. *Pharmacotherapy.* 2016;36(9):1021–7.
17. Jentzer JC, et al. Management of refractory vasodilatory shock. *Chest.* 2018;154(2):416–26.

22 Intellectual Property in the Drug Discovery Process

Martha M. Rumore
Maurice A. Deane School of Law at Hofstra University

William Schmidt
Sorell, Lenna & Schmidt, LLP

CONTENTS

22.1 INTRODUCTION

Intellectual property (IP) protections have affected drug development and are considered one of the most important assets of any corporate entity or research organization. The four main types of IP rights are patents, trademarks, copyrights, and trade secrets. IP affects everything from the choice of brand name to stock valuation. Patents are a *quid pro quo* "right to exclude" which encourage innovation by providing financial incentives for research and development (R&D) activities. The three requirements for patenting are usefulness, novelty, and nonobviousness. Patent protection allows authorities to take active measure to prevent counterfeit versions of medications. Additionally, few of the new medications for diseases such as AIDS that have caused deaths to plummet would have come into existence without patent incentives and the prospect of a return on investment provided by that incentive.

On the other hand, the fact that the spread of HIV is rampant among third world countries is a strong argument for omitting medications to treat it from patenting, ensuring cheaper access to the people for such drugs. Much of the controversy over pharmaceutical patents relate to the provision of pharmaceuticals in the developing world. Patents play an essential role in drug development because they allow pharmaceutical companies to earn profits and price pharmaceuticals to recover their R&D investments, which analysts now estimate is over 2.5 billion per new drug.[1] Brand companies market only about one out of a hundred of the products for which they have developed patents. FDA has estimated that the price of a brand pharmaceutical decreases by 55%

with two competitors and another 33% with five generic competitors once the product loses patent protection.[2,3] Drug manufacturers typically file patents for compounds while they are still being developed and evaluated. Therefore, the faster companies are able to develop a new drug and receive FDA approval, the longer they have to market the drug without facing competition.

Traditionally, the length of patent terms was 17 years. In 2010, with the enactment of the American Invents Act, patent lengths were changed to 20 years from filing.[4] The US patenting system changed from a "first to invent" standard to a "first to file" one. The amount of patent protection remaining after receiving FDA market approval is known as the effective life of a patent. Once a patent is granted, other drug companies are excluded from making, using, or selling the patented aspect of the drug during the term of the patent. Patents and market exclusivity periods are two ways brand pharmaceutical manufacturers may recoup their R&D investments by limiting competition for specified periods of time.[5] Since new drugs must be approved before marketing, the effective patent life can be shorter than 20 years. As a result of the lengthy time period between filing a patent and marketing a product, pharmaceutical manufacturers receive far shorter periods of patent protection than many other industries.

In some cases, patents can be extended to compensate for the time lost because of FDA review of a drug prior to approval. However, these extensions are limited to approved use or method of manufacturing for one patent, the term must not have been previously extended, the extension only applies to the first marketing of a new active ingredient, and the extension is for a maximum of 5 years or half the time period for regulatory approval, and for a maximum effective patent term of 14 years,[6,7] and therefore, do not equal the time lost to market. Further, the maximum period is minus any time during which the applicant did not act with due diligence. No generic manufacturer can even submit an Abbreviated New Drug Application (ANDA) until the end of the 5 years. Since it takes about 1 year for an ANDA to be approved, brand-name drugs have a minimum of 6 years or more of exclusivity. The exception is if the generic company is filing a Paragraph IV ANDA in which it can file its application at the 4-year mark. However, generic manufacturers may conduct testing while the patent is still in effect, thereby permitting a generic product to be market, literally the day the patent expires.

Market exclusivities, whereby FDA cannot approve a generic version of the drug for marketing, enhance market protection provided by patents but are independent of the rights granted under patents. Exclusivity can serve as the sole protection where patents are not available. IP protection via exclusivity is much broader than a patent since it provides protection from entirely different products, for example, different use or dosage form, from entering the market. In general, no ANDA filings and approvals are permitted until the end of the appropriate clinical exclusivity period (although FDA may accept and approve a second full NDA on the drug). Patent protection and market exclusivity are independent of one another and can run concurrently or not. These exclusivities can relate to chemical entities never approved before by FDA (5 years of exclusivity);[8] new biologics (12 years);[9] approval of a supplement for a new condition or use or other change to a previously approved chemically synthesized drug based on new clinical studies (3 years);[10] orphan drugs – drugs designated to treat rare diseases or conditions (7 years);[11] and pediatric indications (6 months). Pediatric exclusivity is an add-on to existing marketing exclusivity or patent protection in exchange for companies obtaining a pediatric indication by conducting clinical trials. Only one 6-month extension is awarded per drug. Hundreds of drugs have been awarded pediatric exclusivity since the program was started. When brand products' patents expire and exclusivity periods end, generics may enter the market.

The Drug Price Competition and Patent Term Restoration Act of 1984, known as the Hatch–Waxman Amendments, created a sensitive balance between brand and generic pharmaceutical companies. It provided patent extension and marketing exclusivity for brand companies, while allowing for market entry of generic drugs which must demonstrate equivalency to the brand drug product in active ingredient, dosage form, safety, strength, administration route, quality, performance characteristics, and intended use.[12] Hatch–Waxman established the ANDA process.

For biologics, the Biologics Price Competition and Innovation Act (BPCIA) of 2009 provided an abbreviated pathway for companies to obtain approval of "biosimilar" and "interchangeable" biological products. A biosimilar must demonstrate high similarity to an already approved biological product and have no clinically meaningful differences in terms of safety and effectiveness.

22.2 PATENT PROTECTION STRATEGIES

22.2.1 BEST PRACTICES FOR PATENT PROSECUTION AND PORTFOLIO DEVELOPMENT

A substantial portion of the value of a company is wrapped up in its patent portfolio. Pharmaceutical companies aim to obtain robust and defensible patents for the entire lifecycle of a product. The pharmaceutical sector is characterized by an extremely high propensity to patent, causing great expenditure.[13] Not all innovations are patented, and patents differ in their economic impact. In fact, most patents are not used in practice. Additionally, the original patent is rarely the last one that covers the drug to expire.

Typically, early in the R&D process, companies developing a new brand-name drug apply for a patent on the active ingredient and may additionally apply to the US Patent and Trademark Office for patents on other aspects of the drug, such as the method of use. Most patents are filed at the end of the basic research phase. This practice of early patenting reduces the effective life of the pharmaceutical, which now averages just 11.5 years. Companies attempt to obtain the broadest patent as early as possible during the R&D process and during clinical trials file for method of use and formulation patents. To receive a patent, a claimed invention must be both novel and nonobvious over previously disclosed information in the same field, commonly referred to as "prior art." Patents for pharmaceutical products are especially important compared to other industries because manufacturing processes are often easy to replicate with minimal capital investment.

Writing a patent is a highly specialized skill of patent attorneys. Parts of a patent include the specification and the claims. The claims define the "metes and bounds" of the invention and define the scope of the patent and must be supported by the written description in the specification. As 35 U.S.C. § 112 states, the specification shall conclude with one or more claims particularly pointing out and distinctly claiming the subject matter which the applicant regards as his invention. The broadest claims are called independent and must define the invention in such detail that the patent examiner and prospective infringers can understand what claimed subject matter is. As a matter of law, if an accused product does not infringe an independent claim, then it does not infringe any claim that depends from that claim. For there to be literal infringement, all the elements contained in the claim appear in the product under consideration. A product that does not literally infringe may infringe under something called the doctrine of equivalents. Under the doctrine of equivalents, we are looking at whether the differences between the accused product and the claimed invention are insubstantial. The patentee may be prevented from making certain assertions regarding the scope of the claims based upon amendments or arguments made during prosecution of the corresponding application. In some cases, to interpret the claims, expert witnesses will be called upon or dictionary meaning might be considered to aid in ascertaining the true meaning of the language employed in a patent.

There are several types of detailed opinions that patent attorneys prepare: invalidity, noninfringement, and freedom to operate. For noninfringement opinions, each claim is dissected and put into a claim table. Invalidity opinions are prepared to show the opponent's patent is invalid. A client who is concerned about infringing a particular patent may ask for an opinion on its validity. Like an infringement opinion, the first step is claim construction. Some grounds for invalidity include anticipation or obviousness. The grounds for each are found in Table 22.1.

Combinations of known pharmaceuticals are patentable if the substances have some working relationship when combined together. Additionally, a pharmaceutical may not be new, but it may be possible to patent it in combination with another known substance if it has been found to have a new property.[14]

TABLE 22.1

Anticipation and Obviousness

Section 102 Anticipation	Section 103 Obviousness
Each and every element in the claim is found, either expressly or inherently in a single prior art reference	Subject matter as a whole would have been obvious at the time the invention was made to a person of ordinary skill in the art

Pharmaceutical innovation has been directed to products meeting the needs of patients in developed countries, and strong patent protection with a market of free price controls is certainly a rationale for this. Looking at the more than 300 drugs on the WHO Essential Medicines List that is deemed essential to developing country public health systems less than a third are patented in any jurisdiction.[15] In fact, until the Trade Related Aspects of Intellectual Property Rights (TRIPS) Agreement in 1994, many countries provided no patent protection for pharmaceuticals. Even under the TRIPS Agreement, compulsory licensing is permitted for health emergencies. Additionally, it was not until 2016 that many less developed countries were required to commit to the Would Trade Organizations' strong patent principles. One obvious consequence of this has been cost shifting to more developed countries, such as the United States. Studies have shown that domestic innovation rises dramatically following the introduction of patent protection.[16]

Patent rights are territorial and exist only in the jurisdiction in which the patentee has applied for and received patents. Therefore, any infringement actions can be sought only in that country. Pursuing patents in many countries is both complicated, as each country has its own unique patent laws and practices, and expensive. Under the European Patent Convention, a single patent application may be filed for all European countries. In 1985, the Patent Cooperation Treaty (PCT) allowed the filing of a single international patent application. This reduced costs of filing as close to 90% of PCT patent applications are then filed in the United States, Europe, or Japan.[17] The time available for securing priority in other countries ranges from 20 to 31 months from the date of initial filing.

Compulsory licensing is not, however, a modern tale of Robin Hood. Abuses abound where some middle-income countries, with threats to use compulsory licensing, demand prices tantamount to those given to third world countries. For example, Brazil has repeatedly used this tactic.

22.2.1.1 Portfolio Development

While most in industry appreciate the value of patent information as a means of protecting IP, many do not realize how much more the data has to offer. Patent data is an underutilized source for details of new scientific research and evaluating the competitive landscape. Patent documents often contain information that has never been published and which is not fully available from any other source. It is important for pharmaceutical companies to obtain good portfolio management development and maintenance advice to provide competitor insight and fuel appropriate R&D. Investors conduct due diligence analysis of a company's patent portfolio to ensure that the pharmaceutical is properly protected in both scope and key markets. Management of a patent portfolio involves monitoring, strategic patent procurement, and commercialization. Monitoring includes both the prior art and the competition. Prior art can prevent patenting and commercialization of a product. Strategic patent procurement involves a determination of how and where the pharmaceutical will be patent protected. It is important to ensure good coverage in the markets where the product will be sold. However, this must be balanced with the cost of patent protection in all those markets. Commercialization involves coordinated marketing and/or seeking investors or collaborators.

The benefits of filing multiple related patent applications versus a single patent must be considered in terms of costs and scope. Some medications are protected by a maze of patents. It is not unusual for brand companies to erect a wall of patents, e.g. over 100 patents which must be taken down by competitors (see Table 22.2). The greatest number of patents filed today are for "method of use"

TABLE 22.2

Aspects of Pharmaceutical Patents

- Compounds and different forms (e.g. enantiomers, crystal polymorphs, solvates, salts, esters)
- Formulations (e.g. mixture of excipients, pharmacokinetic release profile)
- Dosage forms (e.g. controlled-release tablets, patches)
- Uses (e.g. indications)
- Processes and intermediates (e.g. synthesis pathways with increased efficacy)

or patents directed to indications, dosing regimens, or pharmacokinetic properties. The strongest patents are those protecting the active ingredient or molecule and are less likely to be invalidated. Method-of-use patents tend to be more vulnerable to challenge. In fact, generic manufacturers may often avoid infringement by omitting certain indications covered by patents. A patent that protects a controlled-release formulation may be easier to design around by either finding it invalid or finding a different way to create extended release of the drug.

These patent thickets are sometimes invalidated by the Patent Trial and Appeal Board (PTAB), the patent adjudicating agency created in 2012, and courts as not representing true innovation. In some cases, this is the consequence of initial development and in-licensing from academic institutions where the foundational patents are rudimentary and require optimization and reformulation before they can under clinical trials. In 1980 and 1986, Congress enacted the Bayh–Dole Act[18] and the Federal Technology Transfer Act,[19] both of which were aimed at facilitating technology transfer from universities and other research institutions such as NIH to industry.

Over the past few years, there have been large increases in the number of pharmaceutical licensing agreements and alliances. Licensing and assignment of patent rights are means for transfer of patent rights. Like any other property, IP is an asset which can be bought, sold, licensed, or gifted. A patent can be applied for by an inventor or any other person/company it is assigned to by the inventor. An assignment is an unconditional transfer of title which must be recorded in the US Patent and Trademark Office. A license, on the other hand, does not convey ownership, but the patentee (licensor) grants the licensee a right to use and is often used in joint ventures or partnerships. Licenses can be of different types such as cross licenses or material transfer agreement and they may divide rights, e.g. right to sell but not manufacture. Benefits to the licensee include savings on R&D and risk elimination, and quicker exploitation of markets. They may also arise as attempts to work around a patent and avoid a patent infringement action. Milestone payments or penalties are common in licensing agreements to compensate or punish the licensor for such activities as filing or failing to file a NDA.

22.2.2 ANTITRUST AND UNFAIR COMPETITION ISSUES

Profit mongering in the form of increasing the pricing of a readily available medication by 5,000% and generic drug blocking strategies such as reverse payment agreement and product hopping have earned the pharmaceutical industry a reputation that cares more about profits than patient healthcare.[20] Three practices that companies have engaged in with regard to IP have antitrust implications: reverse payments, product hopping, and citizen petitions.

In a strange twist, brand manufacturers are suing generic manufacturers for patent infringement and then agreeing to pay them millions to keep their generic equivalents off the market. Reverse payment agreements involve payments from a brand company to a generic company to abandon its patent challenge or delay marketing of its generic version. The Supreme Court had held that reverse payments may sometimes violate antitrust laws.[21] This is because it results in postponement of the start of their 180-day exclusivity period, blocking all other generic competitors. A "reasonable and justified" standard is applied to ensure the behavior is not anticompetitive.[22]

Patent hopping occurs when a company markets a new version of its product, while discontinuing the older, cheaper version. This has implications under state generic substitution laws inasmuch as the brand version no longer exists and the generic version is no longer "therapeutically equivalent" and cannot be substituted for brand-name prescriptions, thereby impeding generic competition.

Citizen petitions are another way to delay generic entry when filed near the date of patent expiration. FDA recently stated that brand manufacturers filed 92% of all citizen petitions. These ask FDA to delay action on a pending generic drug application and once filed succeed in doing just that for at least 150 days.

Some progress has been made in eliminating these tactics. Currently, the Federal Trade Commission reviews all agreements made between brand and generic companies to prevent anticompetitive practices.

Consumer advocates and some pharmaceutical industry analysts have expressed concerns that certain IP protections do not encourage innovation. Critics contend that companies can easily obtain new patents by making minor changes to existing products regardless of whether the drugs offer significant therapeutic advances. Additionally, pharmaceutical companies may develop new uses for previously approved drugs that have no patent protection and receive an additional 3 years of "market exclusivity." These activities which result in patents with different expiration dates are commonly referred to as "evergreening." Critics further contend some small improvements are too trivial to support even narrow patent protection, and these IP protections enable companies to earn significant profits, while reducing the incentive to develop more innovative drugs. However, some research indicates that secondary patents help brand companies sustain competition against other brand companies in the specific product segment.[23] The relatively high percentage of non-NMEs (new molecular entities), and standard NMEs, in particular, that have been approved over the past decade is evidentiary that development efforts have focused on making changes to existing drugs. Secondary or follow-on patents, typically made shortly prior to patent expiration, such as producing line extensions – deriving new products from existing compounds by making small changes to existing products, such as changing a drug's dosage, or changing a drug from a tablet to a capsule, redirect resources that otherwise could be applied to developing new and innovative drugs. An example of a secondary patent is the patenting of the S-isomer of omeprazole with no compelling clinical advantage, esomeprazole. The pharmaceutical industry and protagonists have provided more than non sequiturs as a defense and contend that due to the rising costs and complexity of developing new drugs, these IP protections are crucial to maintaining drug development efforts. Additionally, the revenues generated from incremental innovation are needed to fund the riskier ongoing research (estimates are that one out of six drugs that make it to clinical trials are eventually FDA approved and before that between 5,000 and 10,000 substances are assessed)[24,25] and development efforts, which can lead to innovation. There is no doubt that pharmaceutical companies differ from other industries in their cost structure.

They further contend that new drugs produced by modifying existing compounds are the result of incremental innovation, and such drugs can result in important therapies. Small changes to existing medications can offer clinically and economically meaningful benefits. For example, by reducing a medication's dosage schedule to once-daily or combining two or more drugs in an oral dosage form, patients are more likely to comply with their prescription's instructions. Changing a buffer so that a drug has an extended shelf life at room temperature, where perhaps, the original formulation was unstable and required refrigeration, is of real benefit to pharmacies stocking pharmaceuticals that too often expire before they are dispensed.

A number of IP reforms have been proposed over the years. One suggestion was the use of a royalty system when pharmaceuticals would be protected by copyright and licensed by pharmaceutical companies who want to market them. Another suggestion was amendments to the Orphan Drug Act to reduce patent incentives to discourage companies from drilling every dry hole needed to discover oil. The federal government could consider providing financial incentives or disincentives to affect the innovative potential of drugs produced by the industry. The government could achieve this by extending or reducing the period of patent protection associated with a drug based on its therapeutic value. A patent could be extended to 25 or 30 years for drugs considered innovative or offering high therapeutic potential; while patents for drugs offering less innovative benefits could be only 10 years.

Recently, to counter the high cost of some patented pharmaceuticals, it has been suggested that a type of "eminent domain" for pharmaceuticals be implemented under 28 U.S.C. § 1458. For some medications to treat hepatitis C, the manufacturers have priced the therapy at $1,000 per tablet and about $84,000 for a full course. A number of new chemotherapeutic agents cost patients over $100,000. These examples have drawn the ire of the US Senate and Medicaid programs. Under §1458, the federal government has the ability carve out exemptions from patents for public use and may use or manufacture any patented product and must provide "reasonable" compensation to the patent holder via a license. Alternatively, the government could elect to contract with another manufacturer to produce a cheaper generic version. This provision has been used in the past, and most recently, during the Anthrax scare in 2001, it served the purpose of having the pharmaceutical manufacturer cut the price of ciprofloxacin. It may also be used to provide an end run around Medicare's pharmaceutical pricing limitations, requiring manufacturers to provide discounted licenses to the government. Opponents of § 1458 argue that the provision stifles innovation. It remains to be seen if the current administration will at least try to use it as a hammer to obtain pharmaceutical price reductions.[26]

Another suggestion is to have pharmaceutical companies and/or the government cross subsidizing expensive drugs, or price medications so as to benefit the poor, especially those medications used to treat life-threatening conditions. This is already occurring in India.

22.2.3 PATENT LITIGATION ESSENTIALS

In recent years, the pharmaceutical industry has experienced a wave of broad patent challenges from both brand and generic companies. Strategies for patent litigation have evolved over time with weaker later expiring patent-attracting challenges. Increasingly, generic drug litigation outcomes have resulted in market entry prior to patent expiration.[27] Additionally, Paragraph IV patent challenges have been filed earlier than ever and even where success probability is unlikely, as generic manufacturers move to be the first ANDA filer. Patent challenges can be brought on the basis of, *inter alia*, noninfringement, invalidity in terms of and nonobviousness, and inequitable conduct. The requirements for patentability, itself, are continually refined in an effort to avoid allowance of weak patents which can subsequently be found invalid. For example, in April 2018, the US Patent and Trademark Office addressed new evidentiary requirements for nonobviousness and, specifically, for establishing that something is "well-understood, routine or conventional," resulting from the federal court decision in *Berkheimer*.[28]

Rather than pursuing costly and lengthy litigation, since the America Invents Act of 2011, companies have been seeking patent invalidation through *inter partes* review (IPR).[29] IPR is an administrative process used to challenge overly broad patents which replaces the previous systems of *ex parte* re-examination and *inter partes* re-examination, both of which were underused and deficient. Advantages of IPR review over litigation are found in Table 22.3. When IPR decisions are appealed, they are usually upheld; a recent American Bar Association study reporting this number as 82%.[30] Some believe the process invalidates patents overzealously. Others contend the change was needed in view of patent hopping and "evergreening" of patents. Unfortunately, this process has also been abused as a mechanism for negatively influencing pharmaceutical stocks when IPR challenges become public. Betting a pharmaceutical company's stock will fall as a result of an IPR challenge being made public or invalidating a patent, hedge fund companies use IPRs for financial gain via "short selling." The PTAB has stated that an economic motive for challenging a patent does not itself raise abuse of process issues.[31] Reforms of the *inter partes* process have been suggested.

TABLE 22.3

Advantage of *Inter Partes* Review over Litigation

Advantages of IPR over Litigation

- No need to supply any bioequivalence data to file
- Not appealable to PTAB versus ANDA litigation is appealable (under Fed Rules of Civil Procedure)
- Less expensive and time consuming
- Reviewed by a panel rather than a single judge (not technologically proficient patent judges)
- Damages not available
- Not estopped from bringing PTAB decisions into District Court
- A better option for late filing generic companies (no 180-day market exclusivity incentive)
- Does not trigger a 30-month stay
- Lower presumption of patent validity (broadest reasonable interpretation of patent claims)
- Preponderance of the evidence standard

Disadvantages of IPR over Litigation

- No 180-day exclusivity period for the first to file
- No Article III standing requirement (anyone can bring an IPR for any reason)
- Filing an IPR, without being the first to file a Paragraph IV Certification, opens the generic company to market exclusion by the first generic, even if successful
- Patents can only be challenged under obviousness or novelty (less avenues for attack)

Several court decisions concerning interpretation of the provisions of Hatch–Waxman, resulted in changes making it easier for companies to obtain the 180-day exclusivity. One change allowed for companies to receive the exclusivity on the basis of a negotiated settlement rather than just a court decision. Another allowed exclusivity period even where the brand manufacturers did not file suit against the ANDA filer.

The Medicare Prescription Drug Improvement and Modernization Act (MMA) of 2003[32] added some provisions to prevent actions by brand companies to delay generic market entry. The MMA closed loopholes such as late listing of additional patents in the Orange Book and forfeiture penalties for generic companies for failure to exercise the 180-day exclusivity period. The circumstances for forfeiture are (a) failure to market, (b) withdrawal of the application, (c) amendment of the certification, (d) agreement with another applicant or patent holder, or (e) expiration of all patents.[33] Several studies have demonstrated a correlation between patent challenges and shorter market exclusivity periods. However, MMA did not close all loopholes. Authorized generics have continued to be a strategic response to impending Paragraph IV entry. This occurs where the brand company would rather cannibalize their brand product by launching its own generic version. A brand company will only release an authorized generic if it has to in order to undercut the generic competition. Some believe this affects the incentive given to generic manufacturers to challenge drug patents; others believe it actually lowers the price of generic versions faster. The threat of an authorized generic does affect the negotiating power in settlement discussions where the generic manufacturer is faced with continuing to incur cost and risk of continuing patent challenge and facing at least 50% fewer returns during the 180-day period. FDA publishes a list of authorized generics on its website.

22.3 HATCH–WAXMAN AND BALANCING INNOVATION AND PRICE

Patent information is required to be submitted at the time of FDA submission of the NDA or supplemental NDA (sNDA). FDA conducts only a brief review of the patents submitted, listing in the Orange Book patents for active ingredients, formulations, metabolic intermediates, drug delivery systems including medical devices, and tablet designs but not process of making patents. Since FDA has no expertise with which to resolve complex patent issues, their role in listing patents is mostly ministerial. The Orange Book provides a convenient means to monitor drug patents. Patent information for listing in the Orange Book must be submitted within 30 days of NDA or sNDA approval. Patents issued after NDA or sNDA approval must be submitted within 30 days after approval of the patent. For patents that are not timely filed, previously submitted ANDA applicants need not amend their applications to file additional certifications. Under Hatch–Waxman, generic companies receive 180-day exclusivity period if they are the first to file an ANDA with a patent challenge. That is, FDA cannot approve any other generic application until 6 months after the first applicant has entered the market. ANDA applicants may assert one of four certifications: the drug is not patented (Paragraph I), the patent has expired (Paragraph II), the drug will not be marketed until the patent has expired (Paragraph III), or there is no patent infringement or the patent has expired (Paragraph IV).[34] Only Paragraph IV is considered a patent challenge. If the ANDA applicant waits until the listed patents expire, a brand company will not know that an ANDA application has been filed. Hatch–Waxman provided an incentive to generic companies to challenge brand pharmaceutical patents.

Once a Paragraph IV certification is filed, the ANDA applicant must send a notice letter to the patent owner. The patent owner can file a patent infringement lawsuit within 45 days. After 45 days, the FDA is free to approve the ANDA. Once a brand manufacturer files a patent infringement lawsuit, the Act does not allow FDA to approve any generic versions for 30 months or earlier upon a ruling in favor of the ANDA applicant by the US Federal District Court. The 30-month stay allows time for the litigation of the patent infringement without delaying approval of the ANDA.

In 2018, the Hatch–Waxman Improvement Act was introduced.[35] The amendment would attempt to restore the balance between generic and brand companies that some say has been disturbed by the IPR process.[36] Under the current regulatory scheme, generic companies may challenge patents under parallel proceedings. The amendment would require ANDA applicants challenging a patent to either choose the Hatch–Waxman process in Federal District Court or the IPR process. However, in choosing the IPR process, the generic applicant would not be able to use the safety and effectiveness data in the brand companies' NDA for approval of their generic version. This would effectively result in sole use of the Hatch–Waxman process as generic companies would not be willing or able to produce their own safety and efficacy data.

22.4 BPCIA AND THE PATENT DANCE

Pharmaceuticals cover more than half of all biotechnology patents. Without strong patent protection, it is difficult to attract the type of investment needed to develop a biologic or biosimilar.

In the United States, the BPCIA created an abbreviated pathway under the so-called §351(k) applications for submission and approval of biosimilars.[37] A manufacturer seeking to market a biosimilar must submit an abbreviated biologic license application (aBLA). FDA has issued many guidances on the requirements for aBLA, including for naming, interchangeability, quality considerations, questions and answers, and quality considerations.[38]

The BPCIA has not worked as intended, and like the early days of Hatch–Waxman, many of the cases revolve around first blush interpretation of the language. Some of the problematic provisions involve the requirement for the biosimilar applicant to wait until it is licensed by FDA before it can provide a 180-day notice to the reference drug manufacturer. Ongoing patent litigation and "the patent dance" have resulted in the Supreme Court weighing in.

As far as patents, the BPCIA is going to be quite interesting and will keep patent attorneys very busy. There are lots and lots of patents on each biologic agent. Humira (adalimumab) has more than 200 patents, and the manufacturers will likely defend them all.

BPCIA created a litigation process to navigate patent infringement concerns which is now being called the "patent dance." This dance kept the first approved biosimilar off the market as two companies "danced." Further to the revelation of proprietary data issue, the BPCIA has its own proscribed patent litigation process that is shown in Figure 22.1. The green box in the process is problematic. The patent dance provision states that "Applicant must provide information and manufacturing

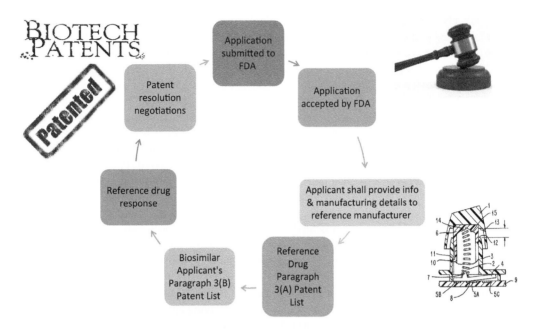

FIGURE 22.1   BPCIA patent litigation process.

details to the reference manufacturer within 20 days of submitting its application." This exchange is designed to enable the resolution of patent disputes before a biosimilar enters the market. When Sandoz refused to provide this info to Amgen, the court ruled "shall" does not necessarily mean "must." The box in Figure 22.1 – "Reference drug response" – involves both the reference drug and biosimilar being required to reveal trade secrets. This may also lead to antitrust issues. Companies may be more likely to go the second-generation or biobetter route to decrease demand for an inferior first-generation biosimilar. Some companies have stated they would rather go the full BLA route and obtain exclusivity than ask each other to dance. Some of the patent and BPCIA interpretation issues that have been litigated are found in Figure 22.2.

The FDA has published the "Purple Book" which lists all biologics, their date of licensure, and any biosimilar products that have been approved. Unlike the Orange Book for drugs, the

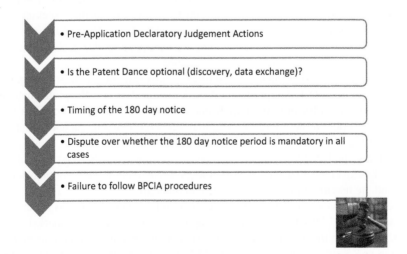

FIGURE 22.2   Biosimilar litigation BPCIA interpretation.

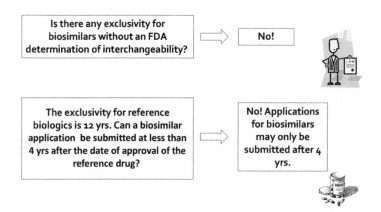

FIGURE 22.3 Exclusivity under BPCIA.

Purple Books does not include a listing of patents. So, brand (reference) biologics and biosimilar manufacturers must exchange proprietary information if they wish to identify which of the biologic manufacturer's patents may be infringed. It is this exchange that has been problematic.

Under BPCIA, manufacturers are aware of the patents at issue, so the argument can be made that the infringement was willful, resulting in increased patent damages.

The BPCIA exclusivity provisions are found in Figure 22.3. While close to ten biosimilars have been FDA approved, the patent litigation has resulted in only a few being marketed.

22.5 OTHER FORMS OF INTELLECTUAL PROPERTY

Other forms of IP for pharmaceuticals include trade secrets, copyrights, and trademarks.

22.5.1 TRADE SECRETS

Absent patent inventors would rely on trade secrets to protect their discoveries. This, however, is not feasible as it is frequently possible to reverse-engineer a pharmaceutical product. Trade secrets are becoming an increasingly useful tool for the pharmaceutical industry, especially in the United States where there are limitations of patentability of natural products. Trade secret law was recently strengthened through US Congressional passage of the Defend Trade Secrets Act (UTSA) of 2016.[39] Trade secret protection for misappropriation arises in both federal and state laws.[40] Trade secret information is generally not known to the public and is maintained as a secret. It is defined under the UTDA as "Information, including a formula, pattern, compilation, program, device, method, technique or process that: (a) derives independent economic value, actual or potential, from not being generally known to and not being readily ascertainable by proper means by other persons who can obtain economic value from its disclosure or use, and (b) it is the subject of efforts that are reasonable under the circumstances to maintain its secrecy."

In today's interconnected pharmaceutical environment with joint ventures and outsourcing agreements and mobility of employees, trade secrets have taken on increased importance. Within the last few years, American companies have been successful in pursuing trade secret theft claims in the US International Trade Commission rather than in US District Court or State Court. In Eli Lilly v. Emsphere,[41] when a collaborative research arrangement went sour, the later sued for trade secret misappropriate when Eli Lilly continued the research and filed patents on it. Eli Lilly settled the dispute for about 18 million dollars.

Unlike patents, trade secret protection lasts indefinitely. Trade secret protection is mutually exclusive from patent protection, and the two are often used together to protect different parts of pharmaceutical.

22.5.2 Trademark

A trademark provides rights to use symbols, logos, particular words, or other markings that indicate a source of a product.

22.5.3 Copyright

Copyright is a form of IP granted to works of authorship. Relevance in the new drug development process pertains to publications of research, e.g. clinical trials or advertising pieces such as charts and models.[42] Ownership of copyright usually rests with the author at the time the work is created. The exception is "work made for hire", i.e. a work made by an employee within the scope of the employment relationship or a work specially ordered or commissioned for use as a contribution to a collective work. Copyrights may last to 70 years beyond an author's lifetime and are easily and inexpensively obtained.

Copyright infringement requires a showing of copying, which can be proven circumstantially. "Fair use" is an exception and involves use for such things *inter alia* as teaching, scholarship, or research.

22.6 CONCLUSIONS

Pharmaceutical IP rights are essential for continued development of new medications. Further, pharmaceutical organizations must stay on top of the patent landscape to survive. Pharmaceuticals are the best example of where the patent system has achieved success. Even though some have criticized the patent regime, the system serves as an enabler of innovation. A large proportion of patents annually granted are to pharmaceutical products. The IP landscape has become more complex, with more patents per innovation and more patenting for upstream inventions. IP protection provides social and technological benefits in encouraging investors and creators, thereby encouraging new drug development and bringing social benefits.

NOTES

1. Tufts Center for the Study of Drug Development. 2017. The cost to develop a new drug now exceeds $2.5 billion. https://csdd.tufts.edu (accessed September 22, 2018).
2. Stolz, N. 2014. Reverse payment agreements: Why a "quick look" properly protects patents and patients. *St. Louis L. J.* 1189, 1190.
3. U.S. Food & Drug Administration. 2015. Generic Competition and Drug Prices. www.fda.gov/about fda/centersoffices/officeofmedicalproductsandtobacco/cder/ucm129385.htm (accessed September 22, 2018).
4. 35 U.S.C. § 154 (a)(2).
5. 35 U.S.C. § 111, 154.
6. 35 U.S.C. §156.
7. 35 U.S.C. § 156.
8. 21 U.S.C. § 355(c)(3)(E)(ii).
9. 42 U.S.C. § 262(k)(7)(A).
10. 21 U.S.C. § 355(c)(3)(E)(iv).
11. 21 U.S.C. § 360cc.
12. 21 U.S.C. § 355(j)).
13. Nagaoka, S., Motohashi, K., Goto, A. 2010. *Handbook of the Economics of Innovation.* 2:1083–1127.
14. Saha, C.N., Bhattacharya, S. Intellectual property tights: An overview and implications in pharmaceutical industry. 2011. *J Adv Pharm Tech Res* 2:88–93.
15. World Health Organization, Essential Medications List, 20th ed, www.who.int/medicines/publication/essentialmedicines/en/ (accessed September 16, 2018).
16. Gamba, S. 2017. The effect of intellectual property rights on domestic innovation in the pharmaceutical sector. *World Dev* 99:15–27.

17. Hall, B.H. 2012. Recent research on the economics of patents. *Ann Rev Econ* 36:585–592.
18. P.L. 96–517.
19. P.L. 99–502.
20. Thomas, J.R. 2015. Pharmaceutical Patent Antitrust: Reverse Payment Settlements and Product Hopping. Cong. Research Service, R44222: 1.
21. FTC v. Actavis, Inc., 133 S. Ct 2223, 2227 (2013).
22. Carrier, M.A. July 25, 2013. Actavis and 'large and unjustified' payments. SCOTUS Blog www. scotusblog.com/2013/07/actavis-and-large-and-unjustified-payments (accessed September 16, 2018).
23. Kiran, S., Kulkarni M. 2018. Secondary patents in the pharmaceutical industry: Missing the wood for the trees? *Expert Opin Ther Pat* 28:3.
24. Rinehart, W. 2014. Intellectual property underpinnings of pharmaceutical innovation: A primer. www. americanactionforum.org/experts/will-rinehart (accessed September 22, 2018).
25. Scherer, F.M. 2010. Pharmaceutical innovation. *Handbook of the Economics of Innovation.* 1:539–574.
26. Dodge, J. 2017. The government can legally commandeer drug patents. https://peoplespolicyproject. org/2017 (accessed September 19, 2018).
27. Hemphill, C.S., Sampat, B.N. 2011. When do generics challenge drug patents? *J Empirical Legal Stud* 8(4): 613–49.
28. Berkheimer v. HP (Case No. 2017-1437, Feb. 8, 2018).
29. P.L. 112–29.
30. Suarez, C. 2017. Navigating *inter partes* review appeals in the Federal Circuit: A statistical review. *Landslide Mag* 9:3.
31. Coalition for Affordable Drugs VI, LLC v. Celgene Corporation, No. 571.272.7822, Paper 19 at 3-4 (PTAB September 25, 2015).
32. P.L. 108–173.
33. 21 U.S.C. § 355(j)(5)(D)(I)-(VI).
34. 21 U.S.C. § 355(j)(2)(A)(vii)(I)-(IV).
35. S. 974. Senator Orin Hatch. Introduced July 2018.
36. Shepherd, J. 2016. Disrupting the balance: the conflict between Hatch Waxman and *Inter Partes Review.* N.Y.U. *J. Intell Prop.* 6:15.
37. P.L. 111–148.
38. FDA Guidances on Biosimilars. www.fda.gov/Drugs/GuidanceComplainceRegulatory Information/ Guidances/ucm290967.htm (accessed September 10, 2018).
39. P.L. 114–153.
40. Nealey, T., Daignault, R.M., Cai, Y. 2015. Trade secrets in life science and pharmaceutical companies. *Cold Spring Harbor Perspect Med* 5(4):a020982.
41. 408 Fed. Supp. 2d 668 (S.D. Ind. 2006).
42. 17 U.S.C.A. § 101, et seq.

23 Drug Repurposing
Academic Clinician Research Endeavors

Kathleen Heneghan
American College of Surgeons

Stephanie E. Tedford
LumiThera, Inc.

CONTENTS

23.1 INTRODUCTION

The process of developing drugs de novo is extensive, laborious, expensive, and time consuming. Estimates for the transition of a potential therapeutic molecule to an approved drug product average 10–15 years with a success rate of only ~2% for developing a viable product [1]. Associated costs can advance into the billions with no guarantee for approval pending outcomes in the later phases of clinical testing [2]. In an effort to reduce the cost, risk, and time constraint associated with bringing a new drug to market and ultimately integration into beneficial patient care, increased focus has been placed on developing new paths for existing pharmaceutical agents in industry and academic settings alike. Drug repurposing, also referred to as repositioning, is an innovative strategy that capitalizes on prior investments, established research and shows a more favorable risk-benefit ratio. These drugs may already be approved for other applications or are currently shelved having not succeeded in previous preclinical/clinical testing for other targeted indications. The considerable research these drugs have already undergone may provide important details on the drug's profile including pharmacology, safety, and efficacy information, thereby reducing the cost of necessary research activities. Thus, the recycling of these already established drugs makes repurposing an appealing avenue for drug discovery initiatives. While the end goal is to receive regulatory approval for repurposing efforts on a new indication, this does not necessarily need to be the hallmark for success. A repurposed drug with enough evidence to demonstrate efficacy of an approved drug for a new indication provides medical professionals with enough information to consider off-label use of the drug.

Drug repurposing significantly reduces the cost, risk, and time compared to traditional drug development strategies. The repurposing strategy can be summarized in four linear steps [3]:

1. Compound Identification: The investigator must first identify the candidate drug for selected targets and/or disease.
2. Compound Acquisition: The investigator must acquire the appropriate licensing for the candidate drug.
3. Compound Development: The investigator must provide sufficient evidence demonstrating the safety and efficacy of the candidate drug for the potential indication. This may include reanalyses of existing data and additional preclinical and clinical studies.
4. Food and Drug Administration (FDA) Post-Market Safety Monitoring: The safety of the drug will continue to be evaluated once approved for the new indication.

In recent years, drug repurposing accounts for up to 30% of newly approved drugs and vaccines by the US FDA [4]. The pharmaceutical industry has long been the driving force behind drug discovery initiatives, however; changes in legislation and research enterprise have created more opportunity for academia to play an increasingly significant role. The outsourcing of pharmaceutical jobs has led to the widespread exodus of expert scientists from industry to academia. This migration has brought skilled scientists with knowledge of drug discovery to the academic setting shifting drug discovery and drug repurposing activities from the industry to the academic sector. In addition, academia presents many interdisciplinary advantages which are cultivated through research collaboration among basic and clinical scientists. These interactions set the foundation for innovative discussions on translational research efforts with activities in drug repurposing such as research for new modes of action and clinical targets for approved drugs. The resources available in academic settings and the multidisciplinary research environment present an invaluable opportunity to explore drug discovery initiatives. For an investigator and health care provider not associated with a pharmaceutical company, knowing the process of the investigational new drug (IND) submission can help facilitate discovery. This chapter will outline the key considerations from discovery, to formulation, to FDA approval and will provide evidence from an academic investigator-led study evaluating the repurposing of the over-the-counter (OTC) drug loperamide, an anti-diarrheal agent with a strong affinity for Mu opioid receptors, into a topical form for pain reduction during repeat finger lancing.

23.2   IDENTIFYING A CANDIDATE DRUG

Advancements in technology present a vast amount of biomedical information available to interested investigators on an infinite number of topics. Scientific information is now readily accessible for the putative molecular underpinnings of various disease states and the pharmacology of developed pharmaceutical compounds [3]. This wealth of knowledge allows for a framework that can utilize literature mining to develop novel concepts in biology and medicine. Literature-based approaches can be implemented to drive hypothesis generation in tandem with drug discovery initiatives. Academic sectors are full of expert scientists and clinicians who are encouraged to engage in interdisciplinary activities. In addition to the immense knowledge available to scientific personal, these relationships aid in fostering an atmosphere where discussions are likely to facilitate novel and innovative approaches to drug discovery and patient care. The identification of a candidate drug for a new indication likely involves developing the scientific rationale for its efficacy and clinical weigh in. Thus, academic sectors are posed to be a hub for this interplay driving many drug repurposing activities.

23.3 THE INTELLECTUAL PROPERTY OFFICE AND INVESTIGATIONAL DATA SHEETS

When a potential new discovery of a drug or other intellectual property is considered, an Invention Disclosure Report to the Academic University Intellectual Property Office is submitted. Each medical center and University office has terms regarding patents, copyrights, and licenses resulting from discoveries, inventions, writings, and other work products. The application requests a concise title, your name, and any co-investigator names; a description with clear detail on the nature, purpose, and operation of the invention; and the chemical, biological, medical, physical, and electrical characteristics of the invention. Objectives, commercial application of any new features, and then a timeline of exactly when and with whom the inventions were discussed are disclosed. The Intellectual Property Office will then determine if they will support a patent and describe the terms of the intellectual property agreement.

23.4 INVESTIGATIONAL NEW DRUG APPLICATION

An IND application is a request for authorization from the FDA to administer an investigational drug or biological product to humans. Over the past 3 years, there has been an average of five submissions annually to the FDA [5]. The data gathered during the animal studies and human clinical trials of an IND become part of the New Drug Application (NDA). The NDA is the vehicle through which drug sponsors formally propose that the FDA approve a new pharmaceutical for sales and marketing in the United States. This is different from an Abbreviated New Drug Application (ANDA) which is submitted to FDA for the review and potential approval of a generic drug product.

23.4.1 INVESTIGATIONAL NEW DRUG PROCESS

The FDA provides very clear guidelines for the IND submission [5]. The protocols give detailed guidance on the required topics, including pharmacology and toxicology data, chemistry, manufacturing, and control data (shelf life, lab site safety, etc.); study protocol; and background of the researcher with human subjects. Other forms that need to accompany the IND application include research statements, certification, financial disclosures, mandatory reporting to med watch, and Investigational Review Board (IRB) approval forms. A medical doctor needs to be listed on the approval forms, as they are the oversight and have prescription authority for the treatment. The medical doctor has to be available while the studies are being conducted, and the studies need to be conducted preferably in a medical center facility.

Following submission of all materials to the FDA, an acknowledgment of receipt will be sent out and within several days an email from an appointed FDA safety advisor. Questions and additional requirements/information needed to support the IND are sent and the response time is quick. If there are delays, the application may fall into another review cycle requiring resubmission delaying the process another year. An area for consideration is the compounded materials. If applicable, the materials needed from the formulation lab require information on protection, compatibility, and stability of the compound and the compound container closure system. If the studies on shelf life and safe container storage have not yet been completed for the compound, recommend to the compounding agency that they use plastics or container tubes that have been previously identified as safe. Also recognize that the FDA may only allow a 30-day shelf life, so if that time passes, a new drug batch may need to be made. Coordination of human subjects to complete all experiments within the short time frame is another consideration. Planning and scheduling for two batches of the compound, one for a trial on a few case subjects followed by a second batch for the full trial should be included in the overall plan.

An area that can cause delay is the simultaneous request by the FDA to have IRB approval and the IRB requesting IND approval. Submitting the protocol to the IRB and including the IND submission will result in the IRB response of pending following IND approval. The IND accepts the IRB with the submission of status pending knowing that all other elements of the protocol, patient enrollment/recruitment forms, and safety standards have been reviewed by the IRB. Once the IND is approved, follow-up with each agency accordingly.

23.4.2   IND EXEMPTIONS

There are exemptions to the IND [6]. The exemptions are based on how you plan to use a marketed drug, the intent of the investigation, and the degree of risk associated with the use of the drug. To meet exemption criteria, (a) the drug has to be lawfully marketed in the United States; (b) the investigation should not be intended to be reported to the FDA as a well-controlled study, and support a new indication of the drug or support new advertising of use; and (c) the use of drug should not involve a route that significantly increases the risk. There are also exemption forms for individual physician use of a drug for an off-label treatment. These are usually single-use requests.

In the case of repurposing OTC loperamide from an oral substance to a topical, in correspondence with the FDA, there was consideration given for exemption status due to the OTC safety status and the small subject sample. While obtaining exemption status can decrease the time needed to prep for the trial by 6–12 months, if the new formulation is successful, exemption status prohibits you from reporting and marketing those results.

23.5   FDA INVOLVEMENT: POSITIVE INTERACTIONS THAT PROMOTE THE IND RESEARCH INITIATIVE

The FDA is an agency that is dedicated to public service and will provide guidance and assistance throughout the process of drug discovery. It is beneficial to enlist these resources as it promotes the likelihood that the product will be accepted. Including the FDA in discussions on study design, outcome measures and clinical trial (if applicable) setup is critical to ensure that the appropriate standards have been met which the FDA would approve of. This also significantly cuts back on the product development timeline. The mean product development time (2010–2012) without an FDA IND meeting resulted in a 5.7-year increase in time spent towards approval. For example, if a trial is designed to study the efficacy of a drug on visual function, it is beneficial to discuss what visual outcome measures and the magnitude of change the FDA would approve or consider sufficient to demonstrate a positive, significant effect. In this case, the standard outcome measure of visual acuity is deemed necessary by the FDA to show proof of effect and subsequent discussions with the FDA can be made to determine what the minimum letter score improvement is considered beneficial to a particular patient population. These discussions demonstrate your interest in working with the FDA and can be crucial to ensure that the most appropriate study design is created that will be approved by the FDA. It is important to include the FDA through meeting requests early in the process. A meeting request will require the following information [7]:

1. Product name and application number (if applicable)
2. Chemical name and structure

3. Proposed indication
4. Dosage form, route of administration, and dosing regimen (frequency and duration)
5. An updated list of sponsor or applicant attendees, affiliations, and titles
6. A background section that includes:
    a. A brief history of the development program
    b. The events leading up to the meeting
7. The status of product development
8. A brief statement summarizing the purpose of the meeting
9. A proposed agenda
10. A list of the final questions for discussion grouped by discipline and with a brief summary for each question to explain the need or context for the question
11. Data to support discussion organized by discipline and question.

A Regulatory Project Manager will be the primary contact at the FDA throughout this process. Discussions and concerns regarding a meeting with the FDA should be directed towards the Regulatory Project Manager.

23.6 ACADEMIC DRUG DISCOVERY – INVESTIGATOR-LED REPURPOSING OF LOPERAMIDE

This section will focus on the clinical development of loperamide, an anti-diarrheal agent with a strong affinity for Mu opioid receptors, into a topical form for pain reduction during repeat finger lancing.

23.6.1 ABSTRACT

Numerous studies have identified acute pain and inflammation associated with the common procedure of finger/heel lancing and emphasize the critical need to find an easy-to-use clinical method that reduces all aspects of the pain response. Direct activation of peripheral opioid receptors with loperamide opens the possibility of targeting opioid receptors at the source of the pain stimulus, alleviating the immediate pain, hyperalgesia, and exaggerated pain state from increased afferent activity. Since topical opioids medicate pain directly at the source, the drawbacks of systemic opiates are avoided. As topical opiates affect both nociceptors and mechanoreceptors and reduce the release of pro-inflammatory neuropeptides, they are ideal for consideration in models of tissue lancing. This IND trial evaluated the analgesic efficacy of topical loperamide 5% gel in reducing pain during repeat finger lancing in healthy adults. Two aims were pursued.

Aim 1 was to determine if topical application of 5% loperamide gel, a peripheral selective Mu opioid agonist, will produce analgesia in a repeat finger lance model in adults. To accomplish Aim 1, 34 adult volunteers received two lancings from a surgilance applied to the fifth digit. Ten minutes following lance one, loperamide gel or gel alone was rubbed into the fingertip for 1 min and left in place for 30 min followed by a second lance. Heart rate and pain scores were compared with each lance. Pain and sensitivity scores were also compared at 24 h. The results were that loperamide gel vs. placebo produced a significant decrease in visual analog scale (VAS) pain score 2.6 vs. 4.0 ($p = 0.005$) and comparison pain score (2.4 vs. 3.64 $p < 0.0001$). At 24 h, the proportion of participants who stated no pain (L 82%, P 29%), no sensitivity to light touch (L 82%, P 35%), and no sensitivity to pressure (L 88%; P Z9%) was significantly greater for the loperamide group. There was no difference in heart rate at any point during the lance procedure or for comparative groups.

Aim 2 determined if the topical application of 5% loperamide gel produces any local or systemic effects. Adults were exposed to an application of loperamide gel or the gel itself to the finger and forearm. A standard erythema scale was used to check for any reaction at three time points:

immediate, 30 min and 24 h. Systemic absorption was determined by the commonly reported constipation effect of Mu opioid agonists. Participants reported any changes in stool patterns or abdominal cramping at 24 h. The results were no erythema at any time following application of the loperamide gel or gel alone and no abdominal cramping for either group with no reported differences in stool patterns.

In conclusion, loperamide 5% topical gel safely and effectively decreases the pain and 24-h hypersensitivity in a repeat finger lancing model in healthy adults. Building on rodent and primate studies, the loperamide 5% gel penetrated the skin with no evidence of systemic absorption. This is the first study to utilize the Mu opioid loperamide as a topical agent in adult subjects. Considering that nearly 10% of the adult population (28.9 million) are diabetic and possibly doing finger lancing for blood glucose monitoring, along with one in ten newborns who require prolonged hospitalization and frequent heel lancing as part of their treatment plan, topical loperamide may be the ideal drug for prevention of pain with lancing.

23.6.2 BACKGROUND

With the recent discovery of opioid receptors and opioid peptides on sensory peripheral neurons, opioid agonists (morphine, fentanyl, loperamide) have demonstrated a significant analgesia when locally injected or topically applied to a site of inflammation/injury. See the review by Stein and Zöllner [8].

Peripheral opioids reduce nociceptor excitability and the propagation of action potentials through their effects on calcium, potassium, and sodium channels as well as through glutamate receptors and G protein-coupled receptors (GPCRs). Opioid receptors extend along the sensory neurons on C- and Ao-fibers and vanilloid receptor-1-positive visceral fibers, and on neurons expressing isolectin B4, substance P, and calcitonin gene–related peptide. Single Mu (MOR), Delta (DOR), and Kappa (KOR) opioid receptors are present with MOR being most abundant. Similar to the central nervous system (CNS), activation of peripheral opioids produces a dose-dependent analgesia. As summarized in Table 23.1, calcium channels, sodium channels, glutamate receptors, K+ channels, and other GPCRs are involved in the peripheral mechanism of action. The inhibition of calcium channels through Gi protein coupling seems to be the primary mechanism of action of opioid peripheral receptors [9, 10].

TABLE 23.1
Response to Peripheral Opioid Receptor Activation

Channel/Receptor	Response to Peripheral Opioids
Calcium channels	Opioids inhibit Gi protein coupling, resulting in decreased propagation of action potentials.
Tetrodotoxin-resistant voltage-gated sodium channels	Opioids inhibit adenylate cyclase resulting in decreased spontaneous ectopic impulse generation at the injury site. Found only in chronic pain states.
Glutamate receptors	Opioids suppress glutamate-evoked nociception on A and C fibers. Inhibit N-methyl-d-aspartate (NMDA), alpha-amino-3-hydroxy-5-methyl-4-isoxazolepropionic acid (AMPA), and kainite (KA) localized on unmyelinated and myelinated sensory axons in the skin.
Vanilloid VR1 receptors	Opioids inhibit the triggering of vanilloid VR1 receptors, which are upregulated with inflammation.
GIRK channels	Opioids stimulate GPCRs interacting with GIRK channels to open and become more permeable to inward K+, resulting in hyperpolarization.
Pro-inflammatory peptides	Opioids reduce substance P, calcitonin gene–related peptide, and cytokine release from peripheral sensory nerve endings resulting in reduced edema, vasodilation, and extravasation.

Opioid receptor numbers are markedly upregulated in the periphery only in inflammatory states, where tissue damage results in increased permeability of the perineural barrier and receptor availability within 10–30 min of tissue damage. This increased expression of opioid receptors as a result of inflammation suggests that peripherally administered opioids may be effective analgesics for certain types of injuries.

Locally administered opioids also have an anti-inflammatory effect. The local application reduces the release of pro-inflammatory neuropeptides (substance P, calcitonin gene–related peptide) and cytokines from peripheral sensory nerve endings, thus reducing the edema, vasodilation, and plasma extravasation [11]. Additional anti-inflammatory action is due to the effect on immune cells where the synthesis and release of pro-inflammatory mediators are inhibited.

23.6.3 CLINICAL PROBLEM

Lancing the finger and heel is a common procedure for blood specimen collection in the adult diabetic and the infant population [12, 13]. The lancet, composed of a blade and spring load devise, directly splices into the dermal layer to reach the capillaries while preventing penetration to the bone [13]. The fingertips in adults and the heel in infants are the most commonly lanced sites due to rich perfusion. However, there is also a high concentration of nerve ending surrounding the capillaries; thus, lancing results in a sharp, pain response, along with localized hypersensitivity from the release of chemical inflammatory mediators [13–21].

There are 28.9 million adults who are diabetic and possibly doing finger lancing for blood glucose monitoring, with one-third reporting anxiety around lancing, which can be a major barrier to improved glucose control [22]. In an effort to decrease the pain and increase compliance in the adult diabetic population, newer lancets which penetrate only to 0.3 mm, as well as quick skin vacuum devises which do not splice the skin at all, continue to be developed. These new devises work well when only a single drop of blood is needed. For a larger blood volume, the blade must penetrate deeper into the tissue.

In the infant, the heel vs. the finger is lanced. Heel lancing represents the most common procedure in the high-risk newborn with an average of 39 lancings per admission (range 1–171) in a very limited area of the heel [16]. A single lance results in bruising (84%–100%), visible inflammation (53%–79%), ankle/leg bruising (53%–92%) along with increased oxidative stress, reduced oxygen saturation, tachycardia, blood pressure fluctuations, increased cerebral pressure, and a long-term exaggerated pain response and change in cortisol expression vert [15, 17, 19, 23,24]. Topical analgesics, such as amethocaine gel, lignocaine ointment, or topical lidocaine-prilocaine (EMLA), are not typically used in the adult and ineffective to alleviate the pain and physiologic distress during infant heel lancing [24–30]. Given that nearly 10% of the US population is diabetic (28.9 million adults) and 15% of newborns required prolonged hospitalization and repeat lancing, there is a critical need to find a safe localized drug that decreases the pain experienced with lancing.

23.6.4 SPECIFIC AIMS

AIM 1. Determine if topical application of 5% loperamide gel, a peripheral selective Mu opioid agonist, will produce analgesia in a repeat finger lance model in adults.

Hypothesis 1A: Participants receiving topical loperamide gel prior to a second finger lance will have significantly decreased pain intensity when compared to gel alone and pre-application lance one.

Hypothesis 1B: Participants receiving topical loperamide gel prior to a second finger lance will have a significantly reduced allodynia (sensitivity to touch and pressure) at 24 h when compared to gel alone.

Rationale and Approach: Lancing the finger and heel to obtain capillary blood for specimen collection directly splices the capillaries along with free nerve endings resulting in a sharp localized pain response. There is currently no topical agent available to decrease the pain from repeat lancing in the infant population. Peripherally applied loperamide has demonstrated significant antihyperalgesic activity in inflammatory animal models and a small sample of human case studies with no systemic absorption or behavioral side effects. This study will develop loperamide hydrochloride as a topical gel and complete the first IND trial on healthy adults. We will (a) develop loperamide as a topical gel meeting all FDA requirements; (b) utilize a double-blind, repeated measures design to examine the effect of loperamide gel vs. gel alone on the immediate pain response following a finger lance utilizing the visual analogue scale, the pain comparison scale, and heart rate fluctuations; and (c) examine the antihyperalgesic effect of loperamide gel vs. gel alone 24 h post lancing to touch and pressure sensitivity using a numeric rating scale and adapted questions from the Pain Quality Assessment Scale.

AIM 2. Determine if topical 5% loperamide gel produces any adverse local or systemic effects.

Hypothesis 2A: Adult exposure to an application of 5% topical loperamide gel and the gel itself to the finger and forearm will produce no immediate or post 24-h skin reaction.

Hypothesis 2B: Topical exposure to 0.2 g of 5% loperamide gel for 30 min will not result in any changes in stool patterns or abdominal cramping when compared to gel alone.

Rational and Approach: The 5% cream concentration was chosen due to its antihyperalgesic effectiveness and the ability for skin penetration through the stratum corneum. The compound is formulated to penetrate the skin with no systemic absorption and utilized standard formulas which are nonirritating and/or non-sensitizing and possess a neutral pH. All of the solvents used for the topical cream loperamide formulation are approved by the FDA as inactive ingredients [6]. The topical formulation that will be used for this trial is 5% loperamide HCI, 30% propylene glycol, 34% ethanol (190 proof USP), 30% ethyl acetate, and Klucel HF 1%. The safety of loperamide gel was assessed by local skin reactions (forearm) and stool patterns. We will determine: if loperamide gel or the gel itself applied in a thin layer to the forearm skin using the mold and left in place for 30 min results in any immediate or 24-h erythema using a standard erythema scale; if after 24 h, there are any reported changes in stool patterns or abdominal cramps to reflect potential systemic affects from Mu opioid anti-diarrheal action on the gut.

23.6.5 Loperamide

Loperamide, 4-(p-chlorophenyl)-4-hydroxy-N, N-dimethyl-a, a-diphenyl-1-piperidinebutyramide monohydrochloride, is a piperdine derivative with a structure similar to the synthetic opioid meperidine. It was approved by the FDA in 1969 as an anti-diarrheal agent with a strong affinity for Mu opioid receptors to mimic the constipating effects of opioids and has been available as an OTC agent with a strong safety record for use in adults, pregnant women, and children [31,32]. Loperamide has markedly reduced CNS effects due to its affinity for p-glycoprotein, preventing crossing the blood–brain barrier under normal circumstances. Loperamide is 15-fold selective for the MOR vs. the DOR and 350-fold more potent for the MOR vs. the KOR [33]. The action of loperamide is competitively inhibited with naloxone.

Morphine and loperamide have been the most frequently tested agents in peripheral analgesia studies. Animal studies of inflamed paws in thermal, mechanical, and chemical models revealed that application of the opioids morphine and fentanyl reduced pain through a local action [34–36]. However, fentanyl and morphine also reduced pain in the contralateral paw suggesting systemic distribution through a topical application [34–36].

Loperamide 5% cream lightly rubbed on the rat paw for 1 min and then left in place for 10 min produced significant analgesia as well when applied both before and after injury in injury models using tape-stripping, capsaicin, formalin, Freund's complete adjuvant, and thermal injury [37]. The 5% cream lightly rubbed on the paw for 1 min and left in place for 10 min significantly increased paw withdrawal latency (PWL) compared with vehicle and 0.5% at 30 min which is the peak injury time. The dose-dependent, naloxone reversed analgesia occurred when loperamide was applied post-treatment and 1 and 2h. No rats displayed any behavior changes.

To date, all clinical studies on human subjects have utilized morphine liquid or gel directly applied to the site of injury/inflammation. Human case reports with application of morphine placed locally during surgical procedures, applied to wound ulcers and burns, all resulted in improved pain relief with no signs of tolerance with repeat use or systemic absorption [8, 38–43].

Support for loperamide includes (a) a greater efficacy in comparison to morphine for peripheral analgesia; (b) a 40-year safety record in adults, children, and pregnant women when taken orally; (c) radioisotope labeling of paste formulation confirming skin penetration and a localized response; (d) the limited ability of loperamide to cross the blood–brain barrier which is important in a repeat use model; and (e) antihyperanalgesic action when applied as a paste on inflamed and non-inflamed skin.

23.6.6 LOPERAMIDE PHARMACOKINETICS AND SAFETY

The IND requires extensive safety information on the current and new formulation. Loperamide has been available for nearly 50 years as an FDA-regulated OTC oral, anti-diarrheal agent safe for use in adults, children, and during pregnancy heel [31,32]. The standard oral dose for children is 0.1 mg/kg, up to 3 mg daily for toddlers and up to 4 mg daily for children. The adult dose is 4 mg with an additional 2 mg for each unformed stool, up to recommended maximum of 16 mg daily. Distribution remains local following oral and topical application. After standard oral dosing, the majority of drug is localized to the stomach and intestines with only 0.22% of the administered oral dose detected in the plasma and <0.022 µg/g found in the brain at peak concentration in rats [37]. In humans, the plasma levels of unchanged drug remain low at 1.18 ± 0.37 ng/mL after the intake of a 2 mg loperamide hydrochloride capsule.

The LD 50 for the oral dose is 185 mg/kg in rats and 105 mg/kg in mice [44]. In rats and mice given more than 30 times the human dose, there were no birth defects, mutations, or harm to offspring [44]. In humans, there is no association with an increased risk of major malformations when given to over 350 pregnant women, even with exposure during the first trimester [45]. In adults given intentional ingestion of up to 60 mg of loperamide to check for side effects, there were no significant adverse events. Infants given eight times the recommended oral dose (0.8 mg/kg/day vs. the recommended 0.1 mg/kg/day) did demonstrate a 3% increased risk of vomiting, a 3% increased risk of ileus, and 13% (4 of 30 infants) developed drowsiness which resolved rapidly upon drug discontinuation [46]. In 1990, loperamide 2 mg drops were removed as an OTC option in third world countries due to 19 infants (ages of 1–6 months) developing ileus with six deaths following administration of ten times the recommended dose [47]. A meta-analysis in 2007 on the use of oral loperamide (Imodium®, McNeil PPC.) reported increased adverse events in children who are younger than 3 years, malnourished, moderately or severely dehydrated, systemically ill, or have bloody diarrhea and that the adverse events (sedation and ileus) outweigh benefits even at oral doses 0.25 mg/kg/day [48].

Elimination of oral and intravenous loperamide occurs by oxidative N-demethylation metabolizing loperamide to N-demethylated loperamide. Loperamide metabolism involves multiple P450 enzymes including isoenzymes CYP2C8 and CYP3A4 and a minor role by CYP2B6 and CYP2D6 [49]. Excretion of approximately 1% of the dose in the urine is unchanged loperamide observed 1 day after intravenous or oral administration [32]. The mean biological half-life is 10.2 ± 0.6 h for syrup formulation and 11.2 ± 0.8 h for the capsules.

Topical loperamide C14 cream applied to the surface of intact skin following mild thermal injury resulted in radioactivity in the surface, epidermis, and dermal layers, with none found in serum samples or tissue from the contralateral paw. Loperamide cream applied to 6 mm diameter of paw for 10 min and then rubbed off results in an increase in radioactivity in 4 mm skin biopsy taken of the skin and the subcutaneous tissue. After application, there was peak concentration at 30 min, reflecting 0.4% of the applied radioactivity. Topical C14 30 mg loperamide cream demonstrated significant uptake in the epidermis and dermis with a peak concentration at 30 min. The half-life is biphasic with a rapid half-life of 2.3 h followed by a slow component with a half-life of 27 h. The duration of action is between 2 and 4 h which mirrors the first phase of the elimination half-life. When applied to a single paw, there was no detectable radioactivity in serum samples or in the contralateral paw. The local concentration of loperamide cream necessary to induce an antihyperalgesic action is in the range of 200–370 ng/mg/tissue [37]. Subcutaneous loperamide prevents allodynia and produced significant antinociception at a dose range of 0.1–1 mg/kg in rodents and rhesus monkeys [33, 50–52].

23.6.7   DEVELOPING THE FORMULATION – KEY CONSIDERATIONS

Topical application must be able to do the following: (a) cross the stratum corneum of the epidermis, (b) deliver sufficient concentration of loperamide to reach and activate the peripheral opioid receptors at the nerve tissue located at the dermal/epidermal junction, and (c) safely maintain the minimum effective concentration for a minimum effective period of time to sustain the desired response.

Based on the 5% concentration which is 5 gm per 100 ml applied to an 8 mm diameter mold, an application of 0.2 cc or 0.2 gm is equivalent to 10 mg or 10,000 μg. The 5% concentration could result in absorption increase from 0.3 to 1.5 μg/cm²/h or 0.015 mg. Considering the biphasic half-life with a rapid phase 1 of 2.3 h and a slow phase 2 of 23 h [53], even if the loperamide continued to be absorbed across the skin over a 24-h period, the maximum dose that could be absorbed during that time is 0.36 mg (1.5 μg × 24 h), which is far below the adult maximum oral dose of 16 mg in 24 h. Thus, topically applied, loperamide gel used in this study would be expected to produce minimal systemic effects even if it was found to distribute body wide. Moreover, as a p-glycoprotein substrate, its entry into the brain would be virtually nil [54].

Another safety consideration in this model is the inflammatory source. A lance is being introduced and loperamide is being applied to a site where the stratum corneum has been damaged with the potential of increase in absorption of the drug. In models specifically designed to break the stratum corneum with an array of micro-needles for the purpose of enhancing drug delivery, there is a 3.8-fold increase in drug delivery [55]. Assuming the enhanced absorption could apply to this model, there could be an increased penetration from 0.015 to 0.057 mg of loperamide, which is still significantly below the 2 mg starting oral dose.

23.7   PATENT REVIEW

The development of loperamide from an oral agent to a topical included a patent search for a review on any prior formulations. In the case of loperamide gel, all previous patent owners provided feedback on their experiments, suggested ways to improve on their work, and identified why the patent or drug development had not proceeded further. The lack of progress with new drug development is often due to the original patent being sold to another pharmaceutical company who then has other priorities and thus, focus on a different discovery.

23.8   FORMULATION

The formulation for this study was based upon US patent 6,355,657 B1.56 [58]. The solvent mixture was tested using skin permeation techniques with human abdominal cadaver skin. The "657" patent defined the scope of a formulation with loperamide at least 1% concentration having an in vitro rate of penetration through the stratum corneum which deposits the drug into the stratum corneum and allows the drug to diffuse at a flux rate of at least about 0.3 μg/cm²/h. The mixed solvent provided twice the skin permeability delivery than the ADL-2–1294-B cream originally developed by Yaksh and Maycock [56].

Using solvents in the percent already approved by the FDA is an important step in the IND approval process, as they already have extensive safety data on the solvent. In the case of the

TABLE 23.2
FDA-Approved Inactive Ingredients

Inactive Ingredient	Route: Dosage Form	CAS Number	UNII	Maximum Potency
Propylene glycol	Topical; emulsion, cream	57556	6DC9Q16V3	71.08%
Ethyl acetate	Topical solution	141786	76845O8NMZ	31%
Ethanol	Topical gel/lotion	64175	3K9958V90M	84.95% gel
				80.5% lotion

development of loperamide, since the intent was to eventually use the product with children and infants, checking for solvents approved for special populations adds in the safety factor for future use. It is also a cost-saving measure, since phase 1 and 2 studies require stability tests documenting shelf life if the original formulation is use; using a new formulation can save 2 years of.

All of the solvents used for the topical gel loperamide formulation or gel alone are approved by the FDA as inactive ingredients [6].

Propylene glycol is an approved inactive ingredient on the FDA-approved drug list. The approved amount for topical cream is 30% and topical gel is 98%. Ethyl acetate is an approved inactive ingredient on the FDA-approved drug list. The approved maximum potency is 31%. It is also a permitted direct food additive CFR 182.60. Ethanol (ethyl alcohol) is an approved inactive ingredient on the FDA-approved drug list. The approved maximum potency is 84.95% for gels and 80.5% for lotions. Klucel™ hydroxypropylcellulose (Ashland) is nonionic water-soluble cellulose ether with a thickening and stabilizing property. Food grade, designated with an F, complies with the requirements of the US FDA for direct addition to food for human consumption (title 21, section 172.870). Food-grade Klucel conforms to the specifications for hydroxypropylcellulose set forth in the Food Chemicals Code. Toxicity testing indicates that Klucel is physiologically inert, and repeat insult patch tests on humans disclose no evidence of either a primary skin irritant or skin sensitizing agent (Table 23.2).

23.9 TOPICAL LOPERAMIDE EFFECTIVENESS

Loperamide hydrochloride powder (Sigma Aldrich), developed as a topical solution at a concentration of 5% (5 gm/100 mL), was applied to the fingertip and forearm of a convenient sample of 34 healthy adult volunteers. Each adult volunteer received two lancings from a surgilance in the fifth digit of the nondominant hand. Ten minutes after the first lance, loperamide gel or the gel alone was rubbed onto the lanced fingertip and left in place for 30 min followed by a second lance. A VAS (0–10) pain scale was compared after lance one, two, and at 24 h. Pain ratings were obtained and recorded using a 0–10 (100 mm). VAS and the pain comparison scale were used following lance two. A 24-h post-lancing follow-up text/call was sent to determine pain intensity (0–10 numeric scale), sensitivity, and pressure ratings using questions adapted from the pain comparison scale (much less, a little less, about the same, a little more, or much more) following lance two.

Loperamide (L) 5% gel vs. placebo (P) produced a significant decrease in numeric rating scale (2.6 vs. 4.0 $p = 0.005$) and comparison pain scale (2.4 vs. 3.64 $p < 0.0001$). At 24 h, the proportion of participants who stated no pain (L 82% P 29%); no sensitivity to light touch (L 82%, P 35%), and no sensitivity to pressure (L 88%, P 29%) was significantly greater for the loperamide group.

Descriptive cross-tabulations identified that there were no participants in the placebo group rating their pain following the second lance as a little less (0%) or much less (0%) in comparison to the loperamide group where there were 3/17 (17%) rating the pain as much less and 5/17 (and 30%) rating the pain as a little less. The proportion of participants who identified having less pain with the second lance was greater for the loperamide group (8/17, 47%) in comparison to the placebo

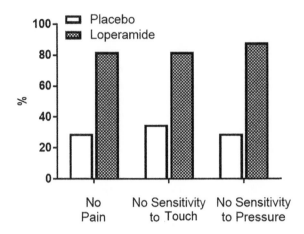

FIGURE 23.1 Effects of topical loperamide application following two finger lancings on the proportion of participants who felt no pain (numeric scale), sensitivity to touch, and sensitivity to pressure at 24h. Non-parametric chi-square population percentiles are represented, $p = <0.005$ for all comparisons.

group (0%), and the proportion of patients who had a comparison scale of more pain or much more pain was greater in the placebo group (8/17, 47%) in comparison to the loperamide group (2/17, 11%) (Figure 23.1).

The safety of loperamide solution was determined by assessment of local skin reactions and stool pattern. Loperamide gel or the gel itself was applied in a thin layer to the forearm skin using the mold and left in place for 30min. The solution was removed with gauze and the skin was immediately assessed for erythema using an erythema scale. The skin was similarly assessed after 24h along with any reported changes in stool patterns or abdominal cramps to reflect potential systemic effects from Mu opioid anti-diarrheal action on the gut. Loperamide or placebo gel did not produce any forearm redness at any point (immediate, 30min, or at 24h). There were no cramps reported in the loperamide or placebo group, and there was no difference between groups for any change in bowel patterns at 24h post gel application.

Loperamide, acting as a peripherally selective Mu opioid, agonist can act as an antinociceptive agent and prevents the allodynia that occurs with repeat finger lancing. The study also supports the safety of the gel formulation with no participants reporting any skin redness or itching and no reports of any signs of systemic absorption. While these results are promising, and they reach clinically significant results for the pain comparison scale and 24-h pain score, they do not reach clinical significance for adult repeat measure VAS scores. The clinically significant difference in VAS scores is reported as 0.5 for single subject pain scores 5min or less apart and 1.3cm for pain scores 20–30min apart [57,58]. Results of studies among children with acute pain suggest the minimal clinically significant difference in pain scores to be 1 on a VAS of 1–10 [57]. The results from this study were a mean score decrease for loperamide of 0.8 (first lance 2.7 and second lance 1.9) and an increase in VAS scores for placebo of 0.7 (first lance 3.5 and second lance 4.2). Additional studies which allow longer penetration time or using higher doses may be needed.

23.9.1 Key Take Home Points

- Utilize your University environment by creating interdisciplinary groups that consist of basic scientists and clinicians who specialize in overlapping fields.
- Work with your University resources (e.g., Intellectual Property Offices) to cultivate relationships that will enable you to further your research opportunities and to explore promising avenues.

- Review all existing and expired patents and reach out to the patent holder for insight on their progress and methods. Often the patent holds effective formulations but does not progress because they are sold as part of a group of patents and then not prioritized within the company.
- Contact the FDA scientific officers to discuss potential research endeavors. Pre-submission meetings with the FDA can ensure that your study is compliant in design and will improve your likelihood for approval. The FDA offers helpful guidance with the submission and directs you to required resources and forms.

23.10   CONCLUSIONS

Advancements in technology are continuing to further our understanding of disease pathology, thereby increasing our capacity to optimize targeted therapeutic agents. These advancements will create further opportunities for the repurposing of current drugs for alternative indications as new molecular pathways and targets are identified. As evidenced by the academic-led repurposing of loperamide as an anti-diarrheal agent to the use of pain management in repeat finger lancing, academic-driven drug discovery can significantly contribute to drug discovery initiatives.

The emerging sector of academic drug discovery has increased research activities into drug repurposing. There are significant advantages of conducting research in academia which includes the multidisciplinary team, integrated research environment, and unique access to hospital and healthcare practitioners. Against this backdrop, it is anticipated that much research focused on repurposing drugs for alternate uses will be accomplished in the academic setting proving an immediate and direct effect on clinically meaningful therapeutics practice.

ACKNOWLEDGMENTS

With thanks to Dr. David Osborne who was instrumental in sharing his formulation from the patent; this work would not be possible without his guidance. We would also like to thank Dr. Paul Carvey who was instrumental in helping formulate the concepts on the original studies on peripheral opioids.

REFERENCES

1. H. Xue, J. Li, H. Xie, Y. Wang, Review of drug repositioning approaches and resources, *International Journal of Biological Sciences*, 14 (2018) 1232–1244.
2. Roundtable on Translating Genomic-Based Research for Health; Board on Health Sciences Policy; Institute of Medicine. Drug Repurposing and Repositioning: Workshop Summary, National Academies Press (US), Washington, DC (2014).
3. C. Andronis, A. Sharma, V. Virvilis, S. Deftereos, A. Persidis, Literature mining, ontologies and information visualization for drug repurposing, *Briefings in Bioinformatics*, 12 (2011) 357–368.
4. G. Jin, S.T. Wong, Toward better drug repositioning: Prioritizing and integrating existing methods into efficient pipelines, *Drug Discovery Today*, 19 (2014) 637–644.
5. U.S. Food and Drug Administration, Investigational New Drug (IND) Application.
6. FDA, Guidance for clinical investigators, sponsors and IRB's. Investigational New Drug Status Applications, Determining whether human research studies can be conducted without an IND. US Department of Health and Human Services FDA (2013).
7. U.S. Food and Drug Administration, Formal Meetings with FDA, Engaging with the FDA During New Drug Development.
8. C. Stein, C. Zollner, Opioids and sensory nerves, *Handbook of Experimental Pharmacology*, 194 (2009) 495–518.
9. K.M. Standifer, G.W. Pasternak, G Proteins and opiod receptor-mediated signalling, *Cell Signal*, 9 (1997) 237–48.
10. J.E. Schroeder, P.S. Fischbach, D. Zheng, E.W. McCleskey, Activation of mu opioid receptors inhibits transient high- and low-threshold Ca^{2+} currents, but spares a sustained current, *Neuron*, 6 (1991) 13–20.

11. Z. Khalil, K. Sanderson, M. Modig, F. Nyberg, Modulation of peripheral inflammation by locally administered endomorphin-1, *Inflammation Research: Official Journal of the European Histamine Research Society... [et al.]*, 48 (1999) 550–556.

12. C.F.D.C.A. Prevention, National diabetes statistics report: estimates of diabetes and its burden in the United States, US Department of Health and Human Services Atlanta, GA (2014).

13. J. Arena, J.I. Emparanza, A. Nogues, A. Burls, Skin to calcaneus distance in the neonate, *Archives of Disease in Childhood. Fetal and Neonatal Edition*, 90 (2005) F328–F331.

14. H. Fruhstorfer, G. Schmelzeisen-Redeker, T. Weiss, Capillary blood volume and pain intensity depend on lancet penetration, *Diabetes Care*, 23 (2000) 562–563.

15. M.E. Owens, E.H. Todt, Pain in infancy: Neonatal reaction to a heel lance, *Pain*, 20 (1984) 77–86.

16. D.P. Barker, N. Rutter, Exposure to invasive procedures in neonatal intensive care unit admissions, *Archives of Disease in Childhood. Fetal and Neonatal Edition*, 72 (1995) F47–F48.

17. H. Vertanen, V. Fellman, M. Brommels, L. Viinikka, An automatic incision device for obtaining blood samples from the heels of preterm infants causes less damage than a conventional manual lancet, *Archives of Disease in Childhood. Fetal and Neonatal Edition*, 84 (2001) F53–F55.

18. L. Slater, Y. Asmerom, D.S. Boskovic, K. Bahjri, M.S. Plank, K.R. Angeles, R. Phillips, D. Deming, S. Ashwal, K. Hougland, E. Fayard, D.M. Angeles, Procedural pain and oxidative stress in premature neonates, *The Journal of Pain: Official Journal of the American Pain Society*, 13 (2012) 590–597.

19. A. Taddio, V. Shah, C. Gilbert-MacLeod, J. Katz, Conditioning and hyperalgesia in newborns exposed to repeated heel lances, *JAMA*, 288 (2002) 857–861.

20. R.E. Grunau, Neonatal pain in very preterm infants: Long-term effects on brain, neurodevelopment and pain reactivity, *Rambam Maimonides Medical Journal*, 4 (2013) e0025.

21. K.J. Anand, F.M. Scalzo, Can adverse neonatal experiences alter brain development and subsequent behavior? *Biology of the Neonate*, 77 (2000) 69–82.

22. M.D.F. Amy Shlomowitz, Anxiety associated with self monitoring of capillary blood glucose, *The British Journal of Diabetes*, 14 (2014) 60–63.

23. C.V. Bellieni, A. Burroni, S. Perrone, D.M. Cordelli, A. Nenci, A. Lunghi, G. Buonocore, Intracranial pressure during procedural pain, *Biology of the Neonate*, 84 (2003) 202–205.

24. J.C. Evans, E.M. McCartney, G. Lawhon, J. Galloway, Longitudinal comparison of preterm pain responses to repeated heelsticks, *Pediatric Nursing*, 31 (2005) 216–221.

25. B. Lemyre, R. Sherlock, D. Hogan, I. Gaboury, C. Blanchard, D. Moher, How effective is tetracaine 4% gel, before a peripherally inserted central catheter, in reducing procedural pain in infants: A randomized double-blind placebo controlled trial [ISRCTN75884221], *BMC Medicine*, 4 (2006) 11.

26. B.A. Larsson, L. Jylli, H. Lagercrantz, G.L. Olsson, Does a local anaesthetic cream (EMLA) alleviate pain from heel-lancing in neonates? *Acta anaesthesiologica Scandinavica*, 39 (1995) 1028–1031.

27. B. Stevens, C. Johnston, A. Taddio, A. Jack, J. Narciso, R. Stremler, G. Koren, J. Aranda, Management of pain from heel lance with lidocaine-prilocaine (EMLA) cream: Is it safe and efficacious in preterm infants? *Journal of Developmental and Behavioral Pediatrics: JDBP*, 20 (1999) 216–221.

28. L. O'Brien, A. Taddio, D.A. Lyszkiewicz, G. Koren, A critical review of the topical local anesthetic amethocaine (Ametop) for pediatric pain, *Paediatric Drugs*, 7 (2005) 41–54.

29. A. Taddio, H.K. Soin, S. Schuh, G. Koren, D. Scolnik, Liposomal lidocaine to improve procedural success rates and reduce procedural pain among children: A randomized controlled trial, *CMAJ: Canadian Medical Association Journal (Journal de l'Association Medicale Canadienne)*, 172 (2005) 1691–1695.

30. A. Taddio, C. Lee, A. Yip, B. Parvez, P.J. McNamara, V. Shah, Intravenous morphine and topical tetracaine for treatment of pain in [corrected] neonates undergoing central line placement, *JAMA*, 295 (2006) 793–800.

31. R.C. Heel, R.N. Brogden, T.M. Speight, G.S. Avery, Buprenorphine: A review of its pharmacological properties and therapeutic efficacy, *Drugs*, 17 (1979) 81–110.

32. J.M. Killinger, H.S. Weintraub, B.L. Fuller, Human pharmacokinetics and comparative bioavailability of loperamide hydrochloride, *Journal of Clinical Pharmacology*, 19 (1979) 211–218.

33. D.L. DeHaven-Hudkins, L.C. Burgos, J.A. Cassel, J.D. Daubert, R.N. DeHaven, E. Mansson, H. Nagasaka, G. Yu, T. Yaksh, Loperamide (ADL 2–1294), an opioid antihyperalgesic agent with peripheral selectivity, *The Journal of Pharmacology and Experimental Therapeutics*, 289 (1999) 494–502.

34. V. Kayser, D. Gobeaux, M.C. Lombard, G. Guilbaud, J.M. Besson, Potent and long lasting antinociceptive effects after injection of low doses of a mu-opioid receptor agonist, fentanyl, into the brachial plexus sheath of the rat, *Pain*, 42 (1990) 215–225.

35. S. Perrot, G. Guilbaud, V. Kayser, Differential behavioral effects of peripheral and systemic morphine and naloxone in a rat model of repeated acute inflammation, *Anesthesiology*, 94 (2001) 870–875.

36. S. Perrot, G. Guilbaud, V. Kayser, Effects of intraplantar morphine on paw edema and pain-related behaviour in a rat model of repeated acute inflammation, *Pain*, 83 (1999) 249–257.

37. N. Nozaki-Taguchi, T.L. Yaksh, Characterization of the antihyperalgesic action of a novel peripheral mu-opioid receptor agonist–loperamide, *Anesthesiology*, 90 (1999) 225–234.

38. C. Stein, M. Pfluger, A. Yassouridis, J. Hoelzl, K. Lehrberger, C. Welte, A.H. Hassan, No tolerance to peripheral morphine analgesia in presence of opioid expression in inflamed synovia, *The Journal of Clinical Investigation*, 98 (1996) 793–799.

39. H.L. Rittner, D. Hackel, R.S. Yamdeu, S.A. Mousa, C. Stein, M. Schafer, A. Brack, Antinociception by neutrophil-derived opioid peptides in noninflamed tissue–role of hypertonicity and the perineurium, *Brain, Behavior, and Immunity*, 23 (2009) 548–557.

40. S.A. Mousa, Morphological correlates of immune-mediated peripheral opioid analgesia, *Advances in Experimental Medicine and Biology*, 521 (2003) 77–87.

41. R.K. Twillman, T.D. Long, T.A. Cathers, D.W. Mueller, Treatment of painful skin ulcers with topical opioids, *Journal of Pain and Symptom Management*, 17 (1999) 288–292.

42. M.D. Ribeiro, S.P. Joel, G. Zeppetella, The bioavailability of morphine applied topically to cutaneous ulcers, *Journal of Pain and Symptom Management*, 27 (2004) 434–439.

43. L.C. Cerchietti, A.H. Navigante, M.W. Korte, A.M. Cohen, P.N. Quiroga, E.C. Villaamil, M.R. Bonomi, B.M. Roth, Potential utility of the peripheral analgesic properties of morphine in stomatitis-related pain: A pilot study, *Pain*, 105 (2003) 265–273.

44. U. Pharmacopeia, USP NF 2007 with Supplement (2007).

45. A. Einarson, P. Mastroiacovo, J. Arnon, A. Ornoy, A. Addis, H. Malm, G. Koren, Prospective, controlled, multicentre study of loperamide in pregnancy, *Canadian Journal of Gastroenterology*, 14 (2000) 185–187.

46. C. Motala, I.D. Hill, M.D. Mann, M.D. Bowie, Effect of loperamide on stool output and duration of acute infectious diarrhea in infants, *The Journal of Pediatrics*, 117 (1990) 467–471.

47. W. Rane, Dangerous antidiarrhoeals, *Economic and Political Weekly*, 25 (1990) 2649.

48. S.T. Li, D.C. Grossman, P. Cummings, Loperamide therapy for acute diarrhea in children: Systematic review and meta-analysis, *PLoS Medicine*, 4 (2007) e98.

49. K.A. Kim, J. Chung, D.H. Jung, J.Y. Park, Identification of cytochrome P450 isoforms involved in the metabolism of loperamide in human liver microsomes, *European Journal of Clinical Pharmacology*, 60 (2004) 575–581.

50. E.R. Butelman, T.J. Harris, M.J. Kreek, Antiallodynic effects of loperamide and fentanyl against topical capsaicin-induced allodynia in unanesthetized primates, *The Journal of Pharmacology and Experimental Therapeutics*, 311 (2004) 155–163.

51. D.L. DeHaven-Hudkins, A. Cowan, L. Cortes Burgos, J.D. Daubert, J.A. Cassel, R.N. DeHaven, G.B. Kehner, V. Kumar, Antipruritic and antihyperalgesic actions of loperamide and analogs, *Life Sciences*, 71 (2002) 2787–2796.

52. L. Menendez, A. Lastra, A. Meana, A. Hidalgo, A. Baamonde, Analgesic effects of loperamide in bone cancer pain in mice, *Pharmacology, Biochemistry, and Behavior*, 81 (2005) 114–121.

53. D.L. DeHaven-Hudkins, R.E. Dolle, Peripherally restricted opioid agonists as novel analgesic agents, *Current Pharmaceutical Design*, 10 (2004) 743–757.

54. M.I. Jumbelic, Deaths with transdermal fentanyl patches, *The American Journal of Forensic Medicine and Pathology*, 31 (2010) 18–21.

55. J.S. Kochhar, W.J. Goh, S.Y. Chan, L. Kang, A simple method of microneedle array fabrication for transdermal drug delivery, *Drug Development and Industrial Pharmacy*, 39 (2013) 299–309.

56. T. Yaksh, Maycock A., US Patent, Patent Storm (1999).

57. C.V. Powell, A.M. Kelly, A. Williams, Determining the minimum clinically significant difference in visual analog pain score for children, *Annals of Emergency Medicine*, 37 (2001) 28–31.

58. D.W. Osborne. Patent No. US 6355657 B1. US. (2002). www.google.com/patents/US6355657. Accessed June 1, 2014.

Section IV

Regulations

24 The Role of the Regulatory Affairs Professional in Guiding New Drug Research, Development, and Approval

S. Albert Edwards
eSubmissions University

CONTENTS

24.1 A BRIEF OVERVIEW OF FDA AND OTHER MAJOR DRUG REGULATION CONTRIBUTORS

Before embarking on the description and discussion of the roles of Regulatory Affairs Professionals (RAPs) in drug development, it seems appropriate to briefly review the United States Food and Drug Administration (FDA) and other major contributors that often have major roles in influencing the regulatory path and framework of drug development.

While the legislative and legal mandates, given to the FDA, require it to ensure that new drugs are safe and effective, it does not have the responsibility to develop new drugs itself. FDA's purpose, regarding new drugs, is to determine if the drug is safe enough to be tested in humans during the early phases of drug development. During the later stages, FDA decides on whether or not there is *substantial evidence from two adequate and well-controlled studies* [1,2] to grant marketing approval, such that the drug can be sold to the public and what the label should say about directions for use, side effects, warnings, etc.

Most recently, in December, 2016, FDA collaborated with Congress on the 21st Century Cures Act [3], regarding Drug Development Tools (DDTs) [4] for:

- Biomarkers – three guidances,
- Clinical Outcome Assessment – four subcategories, and
- Animal Models – in progress.

These DDTs represent FDA's efforts to establish and implement programs to aid drug developers and the drug development paradigm. In addition, a general, interactive, and thorough program, "The FDA's Drug Review Process" can be viewed on FDA's website [5]. Finally, FDA is recognizing that advances in biotechnology can be the source of new drugs products and is ready to evaluate these advances in the context of its own extensive, global, safety expertise while relaxing unnecessary regulatory burdens and barriers to new drug development [6].

FDA carries out other legislative and legal mandates beyond the regulation of drug products, in the areas of:

- Cosmetics
- Devices
- Foods
- Tobacco
- Vaccines–blood–biologic products
- Veterinary (drug) products.

FDA is also a participating member of the International Council for Harmonization (ICH) of Technical Requirements for Pharmaceuticals for Human Use, which began in 1990 and brings together the regulatory authorities and pharmaceutical industry to discuss scientific and technical aspects of global drug development and registration [7]. Three regions, the United States, Europe, and Japan (along with Canada, as an observer), have developed guidelines and standards for drug development to respond to the increasingly global need for new drugs in the most resource-efficient manner. At this time, little is known about what effect Brexit may have on the ICH initiatives.

FDA routinely publishes ICH guidance documents, for example, the structure of clinical studies reports, ethnic factors in foreign clinical data acceptance, and good clinical practices, as a result of its collaboration with the ICH.

Thus, the FDA has a broad span of regulatory duties, well beyond drugs, via it legislative and legal mandates and also plays a critical role in the ICH activities related to harmonization of requirements for global drug development.

24.2 THE LABYRINTH OF FDA LAWS, REGULATIONS, GUIDANCES, ETC.

With the deaths related to diethylene glycol, a toxic ingredient in the formulation of the Elixir of Sulfanilamide in 1937 and the European birth defects, phocomelia, related to thalidomide in 1962, the Food, Drug, and Cosmetic Act and the Kefauver–Harris Amendments became law in 1938 and 1962, respectively. This latter legislative effort marked the beginning of modern scientific design principles being applied to drug trials; drugs needed to be both safe and effective. FDA laws have been, subsequently, amended, repealed, and enacted, as new legislation, by Congress over the years. "Milestones in U.S. Food and Drug Law History" [8] on FDA's website can take the reader on a somewhat brief, yet interesting, chronological trip from the 1800s to 2013 regarding FDA law.

FDA responds to its legal and legislative mandates by promulgating regulations and guidances on a broad variety of drug development topics. For regulations, either new or amended, there is normally a comment period devoted to seeking the drug development industry's feedback. For guidance documents, FDA routinely uses a disclaimer that the document reflects FDA's current thinking on a topic, but alternate approaches can be used. For example, the 23-page, 1998, FDA guidance on Clinical Evidence of Effectiveness [9] uses the following disclaimer as footnote 1 on page 4:

"This guidance document represents the agency's current thinking on providing clinical evidence of effectiveness for human drug and biological products. It does not create or confer any rights for or on any person and does not operate to bind FDA or the public. An alternative approach may be used if such approach satisfies the requirements of the applicable statute, regulations, or both."

This approach fosters the notion that newer science, since the drafting of the guidance, may change the research and the design of human studies, needed to produce the substantial evidence required for marketing a new drug product. In contrast, many guidances are so thoughtfully written that they routinely stand the test of time and provide inherently good advice to drug developers even though they may, at their first read, appear to be dated. The 1998 FDA Guidance on Clinical Evidence of Effectiveness is one example.

Prior decision-making by the FDA in the application of these regulations and guidances may also be an important factor for drug developers to consider. This is especially true if the drug is not the first of its kind in a pharmacologic or therapeutic class.

This complex labyrinth of information about how drug development is regulated brings us to the doorstep and domain of the RAP.

24.3 DUAL, GLOBAL RESPONSIBILITIES OF THE REGULATORY AFFAIRS PROFESSIONAL

The RAP has dual, global roles. One role is to be knowledgeable about the aforementioned labyrinth of information regarding drug development and provide a meaningful interface with the FDA and other regulatory authorities. The second global role is to interface with the drug development team to both inform the team of the regulatory pathways and requirements necessary to successfully develop a new drug and to obtain, review, shape, and submit the necessary scientific and regulatory documentation that FDA requires.

Often, this second role initially entails the development of a Target Product Profile (TPP) for the drug. The submission of a TPP, in accord with FDA's Guidance on TPPs [10], is voluntary and not required of the drug developer. However, case studies in antibacterial and osteoporosis drug development [11] have shown the value, *both to the developer and FDA*, of the TPP at early and late stages of drug development. Depending on the drug's progress through the various phases of clinical studies, the RAP's role will grow substantially in Phases 2, 3, and 4 of the drug's lifecycle. *As the drug makes progress through this lifecycle, more than one Regulatory Affairs Professional is often required to support the development process. For example, a RAP may support Chemistry, Manufacturing, and Controls (CMC) with the remainder of the development activities supported by another RAP.*

The TPP is tailored to the requirements of FDA's labeling regulations [12, 13] for a drug product. The drug developer, after a complete read of these requirements, may become overwhelmed; the RAP's role is to chart a regulatory pathway that will incrementally, over time, meet these requirements. This pathway, over the course of the drug's development lifecycle, may be modified by:

- The results of the ongoing research on the drug,
- New or amended regulations and guidances,
- The economic and scientific resources available to the drug developer, and
- FDA decision-making on potential competitor products.

Finally, use of the word "global" in describing the RAP's role is appropriate as many drug development companies do business worldwide with many diversely skilled contract research organizations (CROs), and many drug development executives want their drug marketed in countries outside the United States. Hence, the strategic positioning of the drug's submission content for efficient re-use and re-purposing for markets outside the United States is often needed. If the drug is intended for markets beyond the United States, it is often best to allow the RAP the flexibility to plan a global strategy, beginning as early as Phase 1. This may initially appear to be a daunting and undoable task; however, the ICH continues to produce harmonized guidances and standards for the major markets, that is, the United States, Europe, and Japan, with the tenet of efficient use of drug development resources. This can create efficiencies for the drug developer by minimizing the number of duplicate or repeat studies and different report formats for different regions.

24.4 CHRONOLOGICAL PHASES OF DRUG DEVELOPMENT AND THE RAP'S ROLES

Both drug development and the RAP's various roles share two common traits, that is, diversity and complexity. The drug development process brings together a huge variety of scientific disciplines that contribute to the body of knowledge on a drug product. The RAP's role takes many forms and requires knowledge of many regulatory information sources so that the data and information generated by the various scientific disciplines meet FDA's regulatory requirements. Thus, the RAP's role is one that is derived from knowing enough about the sciences and, at the same time, having command of many FDA regulations and guidances to *give the drug product its best chance for approval.*

24.4.1 Phase 1

During Phase 1, one of the roles of the RAP is to submit the required data and information to the FDA in an Investigational New Drug (IND) Application. This data package or submission sent to the FDA, that is, the CMC, animal pharmacology and toxicology (Pharm-Tox), and the proposed clinical trial protocol (aka, the study or the protocol) must provide the evidence that FDA reviewers need to positively answer the key question in initial Phase 1 – *Is it safe?* Assuming the contents of the drug developer's submission yield a positive answer, the FDA 30-day review clock will expire, with no adverse comments coming from the FDA review division. The clinical study can then start with usually 10–20 normal human subjects, receiving single doses of the drug and 24-h medical monitoring. If the drug is found to be safe and well tolerated as a single dose, multiple escalating doses will be administered to normal subjects to better assess the safety and tolerability with a more prolonged exposure. The data obtained during this Phase 1 trial may include information on the drugs:

- Metabolism,
- Pharmacology,
- Safe dosing range,
- Side effects, and
- Possibly some evidence of effectiveness.

In some cases, the RAP may need to play another role in organizing the Pre-IND meeting with the FDA review division. During this meeting, key questions about the initial IND filing and the drug's regulatory pathway can be posed by the drug developer and potentially answered by the FDA review staff. These meetings may be conducted by phone, in-person at FDA headquarters, or by electronic media. Sometimes a skillfully drafted set of questions shaped by the RAP can elicit all the answers from FDA review staff and there is, then, no need for this meeting.

Depending on the drug's preclinical attributes and the TPP, that is, the intended use or labeling claim desired by the drug developer, the RAP may need to advocate for special regulatory pathways such as:

- Orphan Drug Status [14] for diseases or rare conditions affecting fewer than 200,000 persons in the United States [15],
- Fast Track Designation [16] for life-threatening conditions where the drug can provide an unmet medical need, and
- Qualified Infectious Disease Product (QIDP) [17] an antifungal or antibiotic drug product intended to treat serious or life-threatening infections, for example, methicillin-resistant *Staphylococcus aureus*, MRSA.

The above are example pathways and not an exhaustive listing. The legislative agenda for reauthorizing funding for the FDA may produce additional unique pathways depending on a variety of factors, for example, consumer/patient advocacy for a particular disease state. The RAP's role is to stay up-to-date on the changes to existing and new pathways and merge that information to meet the interests of the drug developer, where there is a significant regulatory or business advantage. For example, the Orphan designation carries with it a tax credit of 25% for qualified clinical research expenses; any excess credit, not able to be used in a given tax year, may be used on a carryback or carryforward basis, for other tax years. A QIDP designation gives the drug developer an additional 5 years of marketing exclusivity, past patent expiry, before generic competition can begin. Again, these are examples of various incentives weaved into various legislative actions that the RAP can discuss in strategy sessions with the drug developer. The award of a particular designation is totally in FDA's hands and may or may not be achievable for a particular drug product. Some designations are better left to the latter stages of drug development where FDA has had time to study the submissions on the drug's full capabilities.

24.4.2 Phase 2

About 70% of the drugs tested in Phase 1 make it to Phase 2 [18]. In Phase 2, the RAP's main role will likely center around two clusters of studies:

- Phase 2a: where dosing requirements need to be met for the minimum through maximum doses, over longer time periods, based on the drug's side effect profile, and
- Phase 2b: where efficacy is first assessed, with an eye to using the dose(s) that demonstrate low, medium, and high efficacy with the fewest possible side effects.

The doses selected here will likely proceed into the Phase 3 efficacy studies. This is usually the first period of intensive submission work for the RAP. The results of these studies need to be sent to the FDA for review. At about this time, the long-term (2 year) animal carcinogenicity studies (two animal species, usually rats and mice) and human pharmacokinetic (body's effect on the drug) and pharmacodynamics (drug's effect on the body) studies are initiated; these reports will need to be available to support the drug's development path, in Phase 3. All along this path, the RAP has likely been making submissions on the drug's stability and Pharm-Tox, since the duration of the drug's stability and the animal safety studies must always meet or exceed the duration of dosing in humans. Human studies are now weeks if not months in duration. Both the animal safety and drug stability studies are important scientific pillars that support the human testing done in Phase 2. The RAP's role is not to write these documents but to thoughtfully review them before they reach the FDA reviewer. One example, from the author's experience, in this sort of review, showed that the dates of start and stop for the drug's stability charts had been incorrectly entered. If left uncorrected, the stability study start date would have been 2 months *after* the end date. These mistakes, left in documents, submitted to the FDA, usually lead to unnecessary questions and irritation from the FDA review staff and can lead to unnecessary delays in the clinical protocols.

24.4.3   ClinicalTrials.gov

In either Phase 2 or 3, depending on the scope of the clinical trials and the indication or intended use of the drug, reporting of clinical trial information will also need to be made to the National Library of Medicine's, (NLM's), ClinicalTrials.gov database [19]. This database's intent is to improve the public's access to information about certain clinical trials of FDA-regulated drugs, biological products, and devices. While the RAP may or may not be the responsible party in reporting these data to NLM, there is an additional certification step/reporting form that FDA requires [20] when the clinical trial information is sent to NLM.

24.4.4   Phase 3

About 25%–30% of the Phase 1 drugs make it to Phase 3 [21] and the filing of a New Drug Application or NDA. Before the initiation of Phase 3, the RAP has a role in conducting an end-of-phase 2 (EOP2) [22] meeting with the FDA review staff. All available safety and efficacy data, from Phases 1 and 2, for the drug will be collected and summarized to allow the FDA staff to have a thorough understanding of the drug. In conjunction with the drug developer, the RAP will be proposing critical, final developmental questions to the FDA that will set the stage for the final Phase 3 studies that will hopefully confirm the drug's long-term safety and efficacy. The RAP will likely work closely with a variety of other scientists, clinicians, and statisticians to finely tune these Phase 3 study proposals and questions for the EOP2 meeting.

Why is this EOP2 effort so important? It is from this group of studies, assuming positive safety and efficacy results, that the drug developer or sponsor, the FDA term, will propose which studies are *pivotal studies*. These are the studies that provide the "substantial evidence" of safety and effectiveness for the drug [23]. FDA, after review of the final Phase 3 study reports, may choose to agree or disagree with the sponsor's designated pivotal studies. This is another submission intensive period for the RAP. These studies may last another 1–4 years depending on the disease state and outcomes measured. *More often than not, multiple Regulatory Affairs Professionals are involved in the diverse and intensive activities involved in Phase 3.*

These pivotal studies form the basis for a finding of substantial evidence, which is a critical step in FDA's approval of the drug. With these multi-year, Phase 3, study reports having been completely reviewed by FDA and a finding of substantial evidence, the FDA will move on to creating an *action package* for the drug's formal approval. The RAP's role is to ensure that the action package is carried out in an efficient manner consistent with FDA's procedures [24].

A role in labeling negotiation also begins for the RAP, trying to tie the commercial desires of the drug developer and the evidence-based findings of the FDA into a single, seamless, series of phrases or sentences that let the caregivers and health professionals know what the drug can be used for and under what conditions. The period of time this takes can be brief (all parties agree) or it can be protracted (parties have major disagreements on labeling and negotiations are required).

24.5 FDA PRE-APPROVAL INSPECTION – THE PAI

Later in the NDA review period, depending on the available resources of the drug developer, the factory making the drug must be readied for FDA inspection; some drug developers postpone this inspection until after FDA makes its findings for approval. This later delay means a further delay to market entry for the drug. The factory inspection spawns another role for the RAP. Sometimes, this role is so specialized that it is off-loaded or designated to an expert in FDA compliance or CMC specialist, a specialized RAP; these are individuals who have spent many years just working on CMC and/or compliance issues related to FDA regulations in the manufacture of drug products. The factory must be in full-scale operation and have produced multiple lots of drug with subsequent stability testing to qualify for an FDA Pre-approval Inspection [25]. This is likely the first time, in the development lifecycle, that the RAP has had to interface with FDA local district offices. These offices require their own submissions of the CMC information contained in the original NDA. Usually, the FDA district office nearest to the factory conducts the inspection. The results of the inspection are fed into the main FDA review division at headquarters.

24.6 CLINICAL STUDY SITE AUDITS

It is also more likely than not that the FDA's Bioresearch Monitoring Program (BIMO) will be conducting audits of some of the clinical sites [26] that produced the clinical study data that FDA has claimed as pivotal. The FDA review division will want assurances that the clinical study data in the NDA, declared as pivotal, are authentic.

Thus, in late Phase 3, the RAP's role is divided into four distinct venues that include:

- Labeling negotiations, if disagreements arise,
- Expediting, if possible, FDA's action package for the drug's approval,
- Overseeing the remediation of any negative results of the factory inspection, and
- Any negative audit results from clinical sites.

Should any one of these be delayed, the drug's approval will be held until further results or satisfactory remediation can be obtained.

Depending on the diversity and complexity of various subtopics, triggered by these four distinct venues of responsibility, there may be, again, more than one RAP providing support in bringing these tasks to a positive conclusion.

24.7 ADVISORY COMMITTEE MEETINGS

Depending on a variety of factors, there can be a fifth dimension to the RAP's late Phase 3 tasks, that is, an Advisory Committee meeting. Beginning with the FDA Amendments Act of 2007, the FDA must hold an Advisory Committee meeting on all new drug products using the NDA path for approval or provide an explanation about why it will not hold such a meeting [27]. Unfortunately, for the drug developer, the notice of an Advisory Committee meeting is usually given within the last few months of the drug's final NDA review cycle. This is a short period of time to prepare, based on the now vast volume of data submitted on the drug and the fact that the meeting is open to attendance by the:

- Public (minimum 1 h),
- Competitor companies,
- Media,
- Investment professionals and companies, and
- Stakeholder/patient advocacy groups.

This is further complicated, for the drug developer and the RAP, by two factors; they are not in control of the agenda and FDA will likely demand a *briefing package*, usually consisting of multiple copies of several volumes of data and information, previously submitted to the NDA. In turn, FDA sends these packages to the Advisory Committee members four or more weeks in advance of the meeting date. This means that the RAP and drug developer have about 1–2 months to prepare for this meeting and prepare the briefing package. Here is a brief outline of the run up to an Advisory Committee meeting:

- FDA gives notice of Advisory Meeting to sponsor between Day 90 and Day 120,
- The RAP and/or sponsor work to prepare their own presentation and the briefing package Day 90/120 to 60,
- Briefing package sent to FDA at Day 60,
- Briefing package sent to Advisory Committee members at Day 30 or sooner,
- Advisory Committee meeting held on Day 0.

While the RAP and drug developer attend this meeting and multiple presentations may be made by them, Advisory Committee Members plus the FDA review staff will also make their own presentations. At the conclusion of the Advisory Committee Meeting, the FDA has 90 days to finally accept or reject the Advisory Committee's positive or negative findings regarding the drug's approval [28]. Thus, this event can consume a lot of time and effort and publicly expose the drug to regulatory, legal, and financial risks that are not under the control of the RAP or the sponsor, immediately before the potential NDA approval.

24.8   FDA MEETINGS

In addition to the Advisory Committee and EOP2 meetings, the FDA also offers three other types of meetings, designated as Type A, B, and C meetings, which support communications between FDA and the drug developer for an IND or NDA [29]. The role of and place in the drug development path for each of these meetings are as follows:

- Type A meetings are usually held within 30 days of FDA's receipt of a written request and briefing package, prepared by the sponsor. These meetings are for stalled development programs due to, for example, a Clinical Hold [30] on a study protocol or a Post-Action meeting, that is, a meeting held after an NDA has not been approved.
- Type B meetings are normally scheduled within 60 days of FDA's receipt of a meeting request. These meetings often precede important regulatory turning points in a drug's lifecycle, for example, filing an IND or NDA or starting Phase 3. The FDA will likely request a briefing package at least 30 days in advance of the meeting.
- Type C meetings include any drug development meeting with FDA that does not fit into a Type A or B meeting. These meetings are scheduled within 75 days of FDA's receipt of a meeting request and again may require a briefing package.

The RAP, in collaboration with the drug developer, is usually the key person that authors the meeting request and oversees the submission of any briefing package that is required. There are content requirements for both a meeting request and briefing packages that the RAP must follow in accord with FDA's guidance on meetings [31].

During the meeting, notes should be taken by the RAP or a designee; the FDA will also take notes and create an official record of the meeting. While not mandated by any guidance, it is often good practice for the RAP to interface with the FDA to hopefully harmonize the content of the FDA's and the company's meeting minutes.

24.9   HEALTH OUTCOMES

There is an additional, non-FDA or non-regulatory burden that may be added to the drug development task list; that is, the need for stand-alone health outcome studies or incorporation of health outcome measurements into Phase 3 studies. These studies satisfy insurers, not FDA, that the drug product can contribute positively to the health outcomes of patients such that insurers will see the need to reimburse or cover the drug under their insurance plans. In some cases, Phase 3 health outcomes data may be incorporated into the final product label of the approved drug. These studies may also be delayed until Phase 4; however, there is then a risk of problems with reimbursement, unless the drug's efficacy is particularly strong in its effect on the disease state.

24.10   DRUG APPROVAL AND PHASE 4

At the close of this strenuous path of label negotiations, factory inspections, clinical site audits, and an Advisory Committee meeting, the drug developer is hoping that the drug receives FDA approval. About 20% of the drugs entering Phase 1 make it successfully to final FDA approval [32].

Assuming a positive finding by FDA on the drug's safety and efficacy, the drug enters the marketing phase of its lifecycle. During this time, FDA may have imposed some Phase 4 requirements. This adds to the RAP's role in submitting the reports on Phase 4 progress. These Phase 4 requests often involve large population studies (10,000+) centered on the drug's safety profile. Phase 3 efficacy studies, while often employing large numbers (5,000+) of patients, may not totally be representative of the drug's routine use in real patient treatment situations involving multiple millions of prescriptions annually and may not reflect the drug's complete safety profile.

24.11   DRUG SAFETY REPORTS

Overseeing the submission of pharmacovigilance reports, that is, drug safety reports or Individual Case Safety Reports (ICSRs) is another role that the RAP may need to either complete directly or provide regulatory oversight for. Drug developers and their companies may elect to have these reports handled by in-house staff that specialize in this function or use contractors with special skills. The choice of resources to complete these reports is often dependent on the volume of these reports which is likely driven by the number of drugs marketed.

Post-marketing safety reports that are serious and unexpected must be reported to FDA within 15 calendar days [33]. Other, more routine reports can be submitted quarterly during the first 3 years after the drugs approval; after that time, these reports are submitted annually. Overall, drug safety reports or ICSRs and Periodic Safety Update Reports or PSURs must be completed for the drug over its marketing lifetime. These reports are part of a harmonized process via the ICH, on a worldwide basis. Depending on the drug's status outside the United States, these reports may be required to be sent to other global regulatory authorities.

24.12 SUPPLEMENTS

Research on the drug is often never totally complete at the time of the NDA's approval. New uses or new indications and various other changes come into the lifecycle of a drug product. These submissions are called Supplements, that is, supplements to the original NDA. The RAP's role here is similar to the initial role outlined previously, that is, to obtain, review, shape, and submit the necessary scientific and regulatory documentation that FDA requires for the supplement's approval and potential inclusion in the labeling. One example of a required supplement is driven by the Pediatric Research Equity Act [34, 35]. Unless there is a clear, compelling reason that the drug should not be used in children, infants, and neonates, the drug developer has an obligation to conduct studies on these populations and submit the results to FDA for extending the drug's indications or uses into the pediatric population.

24.13 MARKETING AND POST-MARKETING ACTIVITIES

During the marketing and post-marketing periods, the RAP will likely have additional roles in:

- Transitioning the drug to OTC, over-the-counter, status,
- Interfacing with the FDA regarding the drug's market withdrawal, and/or
- Delisting the drug from FDA's databases.

With an approaching patent expiry, drug developers may elect to extend the market potential of their drug by applying for over-the-counter or OTC status. The RAP's role is to research this option to see how the drug may fit into one of the OTC monographs [36]. There are also special label comprehension studies, assessing if a consumer, unassisted by the health care professional, can self-diagnose, treat, and manage their disease state, which must be satisfactorily completed. The drug developer, assisted by the research of the RAP, must select whether they will continue to use the NDA route or OTC monograph route to achieve this status. The NDA route is a confidential filing; the OTC route is a public process. The drug developer also has the option to sell the drug to any of a series of specialized companies with deep expertise in the OTC application process. This latter action would involve the RAP in transferring all IND, NDA, and any other intellectual property about the drug to the successor company and filing a change in ownership with the FDA [37].

Phase 4 studies and/or pharmacovigilance reports may reflect new safety information not found in the original NDA. If this tilts the safety/risk profile of the drug in the wrong direction, then the FDA may initiate withdrawal actions [38]. The RAP would be involved in responding to FDA's intended action, on behalf of the drug developer. The drug developer may also elect to have legal representation, depending on the market size and the remaining patent life. If the market share is small and there is little or no remaining patent life, the drug developer may elect to have the RAP voluntarily withdraw the drug's application [39].

At the close of the drug's lifecycle, whether that be due to declining market demand or withdrawal action due to safety, there comes a time in almost every drug's life when it should be delisted. Delisting the drug removes it from certain FDA databases, for example, FDA's Orange Book–Approved Drug Products with Therapeutic Equivalence Evaluations [40]. This delisting relieves the sponsor of the burden to pay User Fees for the drug product.

24.14 PRESCRIPTION DRUG USER FEES

Beginning in 1992, Congress passed the Prescription Drug User Fee Act (PDUFA) which has been reauthorized in 5-year intervals since 1992. This law allows the FDA to collect fees from drug manufacturers to fund the new drug approval process; taxpayer funding via the federal budget is not adequate to support all FDA needs. Thus, the drug developer must pay these fees with the NDA filing and throughout the drug's marketed lifecycle. The RAP may oversee the transfer of funds, as it is one more critical step in meeting the filing requirements for an initial NDA. In brief, User Fees are required for:

- NDAs,
- NDA supplements containing clinical (human) data, and
- Facilities (factories) that produce the NDA-approved drug.

The 2018 User Fee for an NDA or NDA supplement containing clinical data is $2,421,495 [41].

From the author's experience, lost or missing User Fee checks or unacknowledged funding for a User Fee can be a challenging problem for the RAP to solve; see the section on FDA and Drug Development Stories.

24.15   CHANGES TO THE SUBMISSION PROCESS; THE ECTD AND ESG

The FDA, in accord with an ICH agreed timeline, has moved from the paper submission format to requiring the electronic submission format [42] for the following applications on/after the following dates:

- May 5, 2017, NDAs,
- May 5, 2018, commercial INDs and Drug Master Files, including
- All amendments, supplements, and reports pertaining to these applications.

Further, the data standard used for all electronic submissions is the eCTD or electronic Common Technical Document. What is the eCTD? It is the recognized international standard format for transmitting regulatory information to the FDA and other ICH regions. The eCTD consists of five modules with the following examples of actual contents:

- Module 1: administrative (FDA forms), labeling, and patent information,
- Module 2: summaries for each of Modules 3, 4, and 5,
- Module 3: product quality information, formerly referred to as Chemistry, Manufacturing, and Controls or CMC,
- Module 4: nonclinical (animal) toxicology and pharmacology information,
- Module 5: clinical information, for example, protocols and study reports.

These modules replace the 20 item listing/description on FDA's Form 356h, used in paper NDAs and modify FDA's existing requirements on NDA format but not content [43].

The specifications surrounding the generation of the eCTD's electronic file structure involve both complicated computer software programming and computer keyboard skills for insertion of bookmarks and hyperlinks. The expected outcome is easing the access to submission information and data for FDA reviewers. Moving to the eCTD format dictates both a change in the submission compilation process with new computer specifications and a fundamental change to the structure of original documents, word.doc, and PDF. This provides a new set of challenges and circumstances for the RAP. The RAP must work back through the various authoring groups, that is, departments in the company or CROs, to change the way people work with their original, scientific documents; this is change management. Documents, at their origin, must comply with certain electronic submission requirements. Failure to comply leads to errors in the downline electronic publishing systems and the need to rework or redraft the document. This often doubles or triples the work effort as the anomaly must first be identified before it can be corrected. Since the submission is usually the end point of a particular research activity or FDA response, delays can be problematic. The RAP may often need the backing and support of the drug developer or company senior management for this change to take root within the workplace. Core work habits for preparing scientific documents are often difficult to change.

The advantage to the drug developer is that this approach allows for rapid reuse and repurposing of data for submission to other ICH region markets, mainly, in Japan and Europe. The RAP will likely have developed the knowledge and experience to correctly position or reposition IND and NDA content into the eCTD electronic structure; he or she may or may not have the computer skills to publish the regulatory documentation into the eCTD format.

Specialized software vendors with specialized systems are now available to assist with publishing regulatory information in the eCTD format. The following is a partial listing of these specialized vendors and their websites:

- Acuta, www.acutallc.com
- Amplexor, www.amplexor.com,www.aspireectd.com
- Arivis, www.arivis.com/en/clinical-regulatory-us/clinical-regulatory
- Aspire eCTD, www.aspireectd.com
- DXC, www.dxc.technology/life_sciences/offerings/84081/84085-ectdxpress
- Cunesoft, www.cunesoft.com
- Dossplorer, www.dossplorer.com (a Qdoor and Generis partnership)
- eCTD Office, www.ectdoffice.com
- eCTD 247, http://ectd247.com/
- eSubmissions Solutions, www.e-submissionssolutions.com
- Extedo, www.extedo.com
- Freyr Global Regulatory Solutions, www.freyrsolutions.com
- Lorenz Life Sciences, www.lorenz.cc
- Navitas Life Sciences, www.navitaslifesciences.com/submissions-and-report-publishing
- R&D Advisors, LLC, www.rndadvisors.com/
- Regexia Expert Regulatory Solutions, www.regxia.com/
- Sage Submissions, www.sagesubmissions.com
- Sarjen Systems (KnowledgeNet), www.sarjen.com
- Synchrogenix (Certara), www.certara.com
- Veeva, www.veeva.com

Some of these vendors also have technology solutions for the final step in the eCTD submission process that transmits the eCTD data using FDA's Electronic Submission Gateway (ESG). This Gateway, built and supported by PDUFA funds, can receive and direct submissions to the appropriate FDA Center or Office for review in very short periods of time. This has eliminated the need to move boxes of paper documents throughout FDA buildings. The ESG also sends a receipt for the submissions to the sender. The ESG is *the only way* that NDA and commercial IND regulatory information can be submitted to the FDA. Canada also uses the ESG, and Canadian drug submissions can be made through it as well. The RAP may not know every last detail of this electronic submission process but must know enough to successfully oversee and troubleshoot this complex process.

24.16 COMBINATION PRODUCTS

Sometimes new health care products do not exactly *fit* into the regulatory classification of a drug, device, or biologic. In these circumstances, the RAP draws on knowledge and experiences outside the realm of just drug regulations and guidances. Since 1990, FDA has had an Office of Combination Products (OCP) that has the responsibility to assign the FDA Center (Drug, Biologics, or Devices) as the *lead center* with primary jurisdiction over the combination product. Here are three common examples of products covered by FDA's combination product regulation [44]:

- Drug eluting stent,
- Drug packaged with a syringe or a prefilled (drug-filled) syringe, and
- Photo-sensitizing drug activated by a laser or light source.

Many combination products have two or more regulated components. This can give rise to ambiguity or a dispute over which FDA Center (Drugs, Biologics, or Devices) will take the lead in reviewing the product. It is the RAP's role to write or oversee the writing of an RFD, Request for Designation [45]. This document cannot exceed 15 pages and is intended to advocate for a decision by the OCP for a particular center to have jurisdiction. The following factors usually inform the OCP's decisions:

- Statutory definition of a drug and a device,
- If the health product is similar or related to a prior product, and
- The product's primary intended purpose [46].

Thus, the RAP's role is to articulate and advocate for a particular center's jurisdiction, taking the drug developer's wishes into account; however, the FDA, via the OCP, is the final decision maker.

24.17 OTHER DRUG DEVELOPMENT PATHWAYS: 505(B)(2)S AND ANDAS

The NDA is not the only pathway available to the drug developer. Two other pathways, the 505(b)(2) and the Abbreviated New Drug Application (ANDA), are also available. Each has its own unique regulatory path, although there is also some significant overlap with the IND–NDA pathway. Here are some examples for each submission type:

24.17.1 THE 505(B)(2)

- A new chemical entity (NCE) that has published literature available on its safety and efficacy or an application seeking to modify a previously approved drug,
- Modifications, for example, can made to:
  - The dosage form, changing from a tablet to a transdermal patch,
  - Increasing or decreasing the strength or dose within the existing dosage form,
  - Changing the route of administration of the drug, and
  - Substitution of another active ingredient in a combination product [47].

The 505(b)(2) pathway is likely to demand all the previously identified skills and abilities of the RAP.

24.17.2 THE ANDA

- A duplicate drug product, that is, a drug product that has the same active ingredient(s), dosage form, strength, route of administration, and conditions of use as a listed drug and
- Is only available to the drug developer after the patents on the listed drug have expired.
- The listed drug is still a significant source of the safety and efficacy for the duplicate drug.
- The duplicate drug is often referred to as the generic drug; the listed drug is referred to as the brand name drug.

The ANDA pathway draws upon the RAPs knowledge of CMC and bioavailability (BA) and bioequivalence (BE) studies and certain submission formats, consistent with the eCTD, prescribed by FDA's Office of Generic Drugs (OGD) [48].

Determining the proper path can, sometimes, be a challenge, depending on any prior FDA approvals for the drug, its intended use, and the modifications proposed. FDA has recently published a guidance [49] to assist the RAP and drug developer in their selection of a path within the 505(b) (2) and ANDA options.

24.18   SIGNATORY RESPONSIBILITY FOR FDA SUBMISSIONS

Who signs FDA forms that accompany FDA submissions? There are a range of answers for this question. The author is familiar with organizations that:

- Limit signatures to only senior leadership in Regulatory Affairs (RA) Departments,
- Require the company president or designee to sign, or
- Delegate the signature to the RAP most closely associated with the work product.

In each case above, there are usually standard operating procedures or special documents that denote the delegation or nondelegation of signature authority. The value of having senior leadership sign-off is their opportunity to review each submission and its contents before FDA reviews it. The value of delegation of the signature to the RAP closest to the final work product is driving authority downward to be consistent with the level of effort for a work product. This latter approach relies upon the integrity of the RAP to sign off on only work that is fit for FDA review. FDA will, in the experience of this author, accept any of these approaches to signature. FDA's regulations are silent on the issue of who must sign.

24.19   COMPANY RESPONSIBILITY FOR FDA COMPLIANCE AND CONDUCT

On the issue of who or what entity has responsibility for overall FDA compliance and conduct of the company regarding compliance with FDA laws, regulations, and guidances, there are two legal cases that every drug developer or sponsor or applicant or company president should become familiar with; they are commonly referred to as Dotterweich and Park.

In 1943, the Supreme Court ruled, in Dotterweich, that the responsible official(s) of a corporation, as well as the corporation itself, may be prosecuted and found guilty for FDA violations. It need not be proven that the officials intended, or even knew of, the violations [50]. In this case, Dotterweich was the president of Buffalo Pharmacal; this company repackaged and relabeled drugs for sale to physicians. The drugs were mislabeled violating FDA law.

In the Park case, FDA's legal authority to bring a criminal case against an individual stems from prior notice of unsanitary conditions in a food warehouse and distribution of adulterated food in interstate commerce. In 1975, the Supreme Court held the government can seek to obtain misdemeanor convictions of a company and/or responsible officials for alleged violations of FDA law, even if the official claimed to be unaware of the violation. No knowledge or intent is required in this type of prosecution; instead, the FDA law "imposes the highest standard of care and permits conviction of responsible corporate officials who, in light of this standard of care, have the power to prevent or correct violations of its provisions." [51]

Even though this latter case involves foods versus drugs, it follows the contours of logic that company leadership, that is, the drug developer, *can be held responsible for FDA compliance with or without direct knowledge of the violation.* The RAP's role is to impart to the drug developer the needed appreciation for this overall responsibility as well as timely updates when laws and regulations change. From the author's experience, periodic meetings between the RAP and the drug developer, especially as the drug moves into the late stages of it lifecycle, can be an important way to stay in touch with this responsibility.

24.20   ORGANIZATIONS FOR REGULATORY AFFAIRS PROFESSIONALS

The Regulatory Affairs Professionals Society (RAPS) is the largest global organization of and for those involved with the regulation of health care and related products, including medical devices, pharmaceuticals, biologics, and nutritional products. There are more than 16,000 RAPS members from over 80 countries worldwide.

Founded in 1976, RAPS helped establish the regulatory profession and continues to actively support the RAP and lead the profession as a neutral, non-lobbying nonprofit organization. RAPS offers education and training, professional standards, publications, research, knowledge sharing, networking, career development opportunities, and other valuable resources, including Regulatory Affairs Certification (RAC), the only post-academic professional credential to recognize regulatory excellence. RAPS is headquartered in suburban Washington, DC, with chapters and affiliates worldwide [52].

RA is a derived profession, that is, individuals practicing in RA come from diverse backgrounds and often possess significant prior education and training, for example, in medicine, pharmacy, nursing, engineering, and other science-based disciplines as well as law and business, before committing to the RA profession. Some RAPs are former FDA employees.

The Food and Drug Administration Alumni Association (FDAAA) is another nonprofit, non-lobbying organization dedicated to serving current and retired employees of the FDA, including the agency's many past and present RAPs. The FDAAA, founded in 2001, has both a national and international focus. The organization serves to connect those seeking regulatory expertise, including foreign governments, with knowledgeable former FDA employees on selected regulatory topics [53]. The FDA Alumni Association is considered the official global network of former and current FDA employees, connecting and supporting alumni, the FDA, and the public health community.

24.21   CAREER PROGRESSION: THE LEVELS OF PRACTICE

The RA profession today encompasses multiple levels, from the professional at entry level through the highly experienced professional with extensive technical knowledge and management responsibilities. The RAPS undertakes biennial studies on the general scope of practice of RAPs in the United States, Canada, and the European Union. This has been an ongoing effort by RAPS since its initial efforts in 2003 to study the profession.

Four career stages for the RAP have emerged:

- Level I: are new or relatively new to the profession with limited or no RA knowledge. Many have education and/or experience in science, clinical studies, or engineering.
- Level II: possess knowledge and skills in regulatory pathways and regulatory documentation in areas such as: risk-benefit analysis, submissions, registrations, product approvals, post-marketing safety, and compliance.
- Level III: have strong technical and management skills and are actively engaged in regulatory strategy concerning the areas listed in Level II.
- Level IV: have an extensive understanding of the role of the profession in the product lifecycle and the dynamics of regulatory processes; these professionals also influence regulatory policy through external groups and are leaders and mentors within their organization and for the profession; and, these professionals are able to work effectively in multinational/multicultural environments [54].

The drug developer may start with only needing a Level I or II practitioner for IND submissions but ultimately may need a Level III or IV practitioner for the NDA and the complexities of the drug's post-approval lifecycle.

24.22   ROLE SPECIALIZATION AND CREDENTIALING – THE RAC AND FRAPS

Specialization has become part of the RA profession. Specialized roles for the RAP have been and can be divided into the following prominent examples:

- Drug Promotion and Advertising: a role limited to the study and application of all the Guidances, Code of Federal Regulations, Warning Letters, and Untitled Letters pertaining to the advertising and promotion of marketed drugs in all forms of public media.
- Regulatory Intelligence: a role limited to the study and application of all FDA and any other regulatory documents pertaining to future research or business pursuits regarding investigational and marketed prescription drug products. RAPS in this role may be closely aligned with the company's legal counsel and may forecast or summarize pending regulatory changes for other RAPS and drug developers or company presidents/other officers.
- Drug Class–Specific Specialization: a role limited to a specific drug class area, for example, antibiotics.

There are many additional role specializations that touch on components or small portions of the drug development lifecycle, for example, cGMP, current Good Manufacturing Practice; GLP, Good Laboratory Practice; GCP, Good Clinical Practice as well as General Compliance through written policies and procedures. Most of these role specializations closely follow their own subgroup's publications and are linked to specialized membership groups.

In 1990, the RAPS developed the *RAC* or Regulatory Affairs Certification. The RAC is the only credential for RAPs, worldwide [55]. The RAC examination is a multiple-hour process that is based on subjects identified by job analysis and survey of the profession coupled with rigorous review of the exam content by RA experts. In addition, the experts take the exam before it is administered to candidates to further validate the questions. The author has participated in this process from 1993 to 1998 for the United States and a as member of the EU examination committee, 1999–2001. The exam is geared to working regulatory professionals, with at least 3–5 years of regulatory experience. There are four different RAC exams. The US, EU, and Canada exams test regional regulations and involvement with regulatory bodies. The global exam focuses on international standards and guidelines. All four exams test for regulatory knowledge, critical thinking, and analysis throughout the product's lifecycle.

The RAPS Fellows program, or *FRAPS*, Fellow, Regulatory Affairs Professionals Society, recognizes and honors regulatory professionals who have made significant contributions to the regulatory profession. Candidates for the RAPS Fellow designation are assessed against established criteria by taking into consideration their work and career achievements, and more importantly, their volunteer contributions above and beyond their employment commitments such as educating, training, writing, and mentoring to the benefit of other regulatory professionals [56]. There is also an expectation that Fellows will continue to make a commitment to the advancement of the profession into the future by continuing to share their knowledge, experience, and expertise. The author was among the first 27 candidates awarded FRAPS status in 2008 when the program began.

24.23   FDA STATISTICS – FDA BY THE NUMBERS FY 2017

FDA's Center for Drug Evaluation and Research (CDER) should be staffed with approximately 5,500 employees. However, there are currently over 700 vacancies [57], that is, one out of every eight jobs in CDER is vacant. In spite of this vacancy rate, CDER, in FY 2017, has managed to review 137 NDAs and approve 101 of them [58]. CDER also reviews ANDAs and OTC monograph products.

24.24   FDA AND DRUG DEVELOPMENT STORIES

Active Treatment Control Failure: During the author's time as an FDA reviewer, one of the reviewed NDAs was found to be "not approvable." This was not necessarily due to a lack of efficacy of the new drug, a promising nonsteroidal, rather it was due to a failure of the standard treatment control, aspirin; it was found that the aspirin was packed in too small a capsule; it failed dissolution testing; and it lacked its usual efficacy. The small capsule was used to "blind" the studies by making the aspirin capsule look like the nonsteroidal, new drug. Multiple efficacy studies were run using this "small capsule" paradigm to the detriment of the new drug's approval. Aspirin as an active treatment control failed to show its usual efficacy in this series of studies to the detriment of the new drugs approval. Drug developers need to be aware that if their new drug requires an active treatment control [59], then correct formulation with attendant laboratory studies, for example, dissolution testing, are needed to support the scientific design of studies conducted on the new drug.

FDA: We invite you to send in an NDA: FDA rarely makes unsolicited, extemporaneous calls to drug manufacturers. However, this was a rare case of the agency, FDA, tracking, and reviewing 13 separate investigator-sponsored INDs for the same marketed drug product, a nonapeptide with impressive safety margins. *The FDA asked the company to submit the NDA.* The author was tapped to assemble and submit the NDA. It consisted of over 50 pediatric cases with a cross-reference to the marketed drugs CMC and Pharm-Tox from previously approved NDAs. This approval was done in the mid-1990s, well before PREA, the Pediatric Research and Education Act of 2007. Due, in part to FDA's call for an NDA submission and the drug's impressive radiographic effects on this rare childhood disease of precocious puberty, the drug was approved for this use with:

- One clinical study and not the usually required two pivotal studies,
- No pre-approval inspection at the factory, and
- No Advisory Committee!

Upon approval, the drug manufacturer received an *Orphan* designation for the drug, entitling it to an additional 7 years of marketing exclusivity and tax incentives.

Lost User Fee Check at Christmas time: FDA personnel share some common traits with all other workers; they often get tied into their jobs and forget about their vacation time. This results in many FDA staffers taking extra days off during the Christmas holiday season lest they loose and not use their allotted vacation time. The author has always cautioned project planners and drug developers to not submit NDAs or major supplements at this time for the aforementioned reasons, knowing that FDA staffing is likely at a very low point. Nonetheless, a date was etched into a corporate goal for an NDA submission on December 28. The NDA was submitted on that date, and the required User Fee check was also sent to the FDA-required bank. The bank staff, also on Christmas holiday, failed to process and deposit the check in a timely fashion. This led to FDA rejecting the NDA on December 31. The author was called in on an urgent basis to help. Unfortunately, key FDA contacts were on vacation and not available during this time, and no fruitful appeal to hold the NDA until the check was "found" could be made. A second User Fee check was issued, which confused the banking personnel, and the corporation's local DC office agreed to house the rejected NDA temporarily. As FDA staff returned after the New Year and the banking personnel were instructed to cancel one of the two User Fee checks, the NDA was correctly filed sometime during the first 10 working days of the next year. Drug developers, sponsors, and applicants should take heed of this protracted experience and only send major filings in very early December or wait until January. FDA staff share a common need with many other workers; that is, a need to use their vacation days at or near the end of a year before it is lost.

24.25   IN SUMMARY

In this chapter, the diverse roles, career progression, and credentialing of the RAP have been described. This role's definition was coupled with a chronological presentation of important aspects of the drug development lifecycle using selected FDA laws, regulations, and guidances to further illustrate the RAP's role. Due to the diversity, complexity, and global nature of drug development and the RAP's roles, the descriptions should not be interpreted as *complete,* at each stage of drug development process. Rather the RAP's role, like the drug development process, is open to change and enhancement. Reducing the RAP's role to a single phrase would be – *giving the drug product its best chance for approval* – through adept use of diverse and complex FDA information sources.

ACKNOWLEDGMENTS

The author extends his appreciation and thanks to:

- Carol M. Cooper, MS, RAC, IM(ASCP), RM(AAM)-subject matter coach and manuscript review,
- Donald C. Palmer, MS, RAC-electronic submission vendor/product identification,
- David G. Wettlaufer, PhD-manuscript review, and
- The Regulatory Affairs Professionals Society-liberal access to the Regulatory Affairs Professional Development Framework [54].

REFERENCES

1. Code of Federal Regulations, Substantial Evidence, title 21, sec. 314.125 (b)(5).
2. Code of Federal Regulations, Adequate and Well-Controlled Studies, title 21, sec. 314.126.
3. U.S. Food and Drug Administration, 21st Century Cures Act, Accessed April 4, 2018. www.fda.gov/RegulatoryInformation/LawsEnforcedbyFDA/SignificantAmendmentstotheFDCAct/21stCenturyCuresAct/default.htm
4. U.S. Food and Drug Administration, 21st Century Cures Act: Qualification of Drug Development Tools, Accessed April 4, 2018. www.fda.gov/Drugs/DevelopmentApprovalProcess/DrugDevelopmentToolsQualificationProgram/ucm561587.htm
5. U.S. Food and Drug Administration, The FDA's Drug Review Process: Ensuring Drugs Are Safe and Effective, Accessed April 4, 2018. www.fda.gov/Drugs/ResourcesForYou/Consumers/ucm143534.htm
6. U.S. Food and Drug Administration, FDA's New Efforts to Advance Biotechnology Innovation (blog), FDA Voice, June 8, 2018, https://blogs.fda.gov/fdavoice/
7. International Council for Harmonization of Technical Requirements for Pharmaceuticals, Mission Statement, Accessed March 25, 2018. www.ich.org/about/mission.html
8. U.S. Food and Drug Administration, Milestones in U.S. Food and Drug Law History, Accessed March 25, 2018. www.fda.gov/aboutfda/whatwedo/history/forgshistory/ evolvingpowers/ucm2007256.htm
9. FDA Guidance for Industry: Providing Clinical Evidence of Effectiveness for Human Drug and Biological Products, May 1998.
10. FDA Draft Guidance for Industry and Review Staff: Target Product Profile — A Strategic Development Process Tool, March 2007.
11. Ibid, pages 11–12.
12. Code of Federal Regulations, Labeling Requirements for Prescription Drugs and/or Insulin, title 21, sec. 201.56.
13. Code of Federal Regulations, Requirements on Content and Format of Labeling for Human Prescription Drug and Biological Products, title 21, sec. 201.57.
14. Code of Federal Regulations, Orphan Drugs, title 21, sec. 316.
15. FDA Draft Guidance: Rare Diseases: Common Issues in Drug Development, page 6, December 2018, revision 1.
16. FDA Guidance for Industry: Expedited Programs for Serious Conditions – Drugs and Biologics, Fast Track Designation, pages 9–10, May 2014.

17. FDA Draft Guidance: Qualified Infectious Disease Product Designation Questions and Answers, January, 2018.
18. From Test Tube to Patient: New Drug Development in the United States. Rockville, MD: Dept. of Health & Human Services, Public Health Service, Food and Drug Administration, 1988, page 14.
19. Code of Federal Regulations, Clinical Trials Registration and Results, title 42, sec. 11.
20. FDA Guidance for Sponsors, Industry, Researchers, Investigators, and Food and Drug Administration Staff, Form 3674- Certifications To Accompany Drug Submissions, June 2017.
21. From Test Tube to Patient: op. cit., page 14.
22. Code of Federal Regulations, Meetings, title 21, sec. 312.47(b).
23. Katz, R. 2004. FDA: Evidentiary standards for drug development and approval. *NeuroRx* 1: 307–316.
24. FDA Manual of Policies and Procedures, NDAs/BLAs/Efficacy Supplements: Action Packages and Taking Regulatory Actions, MAPP 6020.8 Rev. 1, June 2016.
25. FDA Compliance Program Guidance Manual, Chapter 46- New Drug Evaluation, Pre-Approval Inspections, May 2012.
26. FDA Compliance Program Guidance Manual, Chapter 48- Bioresearch Monitoring Clinical Investigators and Sponsor-Investigators, December 2008.
27. Putnam, W.T. Prescription drug product submissions. In *Fundamentals of US Regulatory Affairs*, edited by Pamela Jones, 163. Rockville, MD: Regulatory Affairs Professionals Society, 2015.
28. FDA Guidance for Industry Advisory Committees: Implementing Section 120 of the Food and Drug Administration Modernization Act of 1997, October 1998.
29. FDA Draft Guidance Formal Meetings Between the FDA and Sponsors or Applicants of PDUFA Products, December 2017.
30. Code of Federal Regulations, Clinical Holds and Requests for Modification, title 21, sec. 312.42.
31. FDA Draft Guidance Formal Meetings Between the FDA and Sponsors or Applicants of PDUFA Products op. cit., pages 10–12.
32. From Test Tube to Patient: op. cit., page 14.
33. Code of Federal Regulations, IND Safety Reporting, title 21, sec. 312.32.
34. Code of Federal Regulations, Pediatric Use Information, title 21, sec. 314.55(a).
35. Pediatric Research Equity Act, 21 U.S.C. § 355c (2003).
36. Code of Federal Regulations, Over-The-Counter (OTC) Human Drugs, title 21, sec. 330.
37. Code of Federal Regulations, Change in Ownership of an Application, title 21, sec. 314.72.
38. Code of Federal Regulations, Withdrawal of Approval of an Application or Abbreviated Application, title 21, sec. 314.150.
39. Code of Federal Regulations, Determination of Reasons for Voluntary Withdrawal of a Listed Drug, title 21, sec. 314.161.
40. U.S. Food and Drug Administration, Approved Drug Products with Therapeutic Equivalence Evaluations, Accessed April 9, 2018. www.fda.gov/drugs/informationondrugs/ucm129662.htm
41. U.S. Food and Drug Administration, Prescription Drug User Fee Act (PDUFA), Accessed March 26, 2018. www.fda.gov/ForIndustry/UserFees/PrescriptionDrugUserFee/
42. U.S. Food and Drug Administration, Providing Regulatory Submissions in Electronic Format Using the eCTD Specifications, Accessed April 4, 2018. www.fda.gov/downloads/drugs/guidancecomplian-ceregulatoryinformation/guidances/ucm333969.pdf
43. Code of Federal Regulations, Content and Format of an NDA, title 21, sec. 314.50.
44. D'Amico, M. Combination Products. In *Fundamentals of US Regulatory Affairs*, edited by Pamela Jones, 342–343. Rockville, MD: Regulatory Affairs Professionals Society, 2015.
45. Code of Federal Regulations, Product Jurisdiction, title 21, sec. 3.7
46. U.S. Food and Drug Administration, Classification of Products as Drugs and Devices & Additional Product Classification Issues: Guidance for Industry and FDA Staff, Accessed April 11, 2018. www.fda.gov/RegulatoryInformation/Guidances/ucm258946.htm
47. U.S. Food and Drug Administration, Guidance for Industry Applications Covered by Section 505(b)(2), Accessed April 9, 2018. www.fda.gov/downloads/drugs/guidancecomplianceregulatoryinformation/guidances/ucm079345.pdf
48. U.S. Food and Drug Administration, Guidance for Industry ANDA Submissions—Content and Format of Abbreviated New Drug Applications, Accessed April 11, 2018. www.fda.gov/downloads/drugs/guidances/ucm400630.pdf
49. U.S. Food and Drug Administration, Guidance for Industry Determining Whether to Submit an ANDA or a 505(b)(2) Application, Accessed April 11, 2018. www.fda.gov/downloads/drugs/guidances/ucm400630.pdf

50. United States v. Dotterweich, 320 U.S. 277 (1943)
51. United States v. Park, 421 U.S. 658 (1975)
52. Regulatory Affairs Professionals Society, Accessed March 26, 2018. www.raps.org/who-we-are
53. Food and Drug Administration Alumni Association, Accessed March 26, 2018. www.fdaaa.org/about. php
54. Regulatory Competency Framework and Guide, Regulatory Affairs Professionals Society, 2016.
55. Regulatory Affairs Professionals Society, RAC Credential, Accessed March 26, 2018. www.raps.org/ rac-credential
56. Regulatory Affairs Professionals Society, Fellows Program, Accessed March 26, 2018. www.raps.org/ membership-community/raps-fellows/meet-raps-fellows
57. The Washington Post, November 1, 2016, Accessed April 11, 2018. www.washingtonpost.com/national/ health-science/despite-ramped-up-hiring-fda-continues-to-grapple-with-hundreds-of-vacancies/2016/1 1/01/9b6dc9b0-a067-11e6-8832-23a007c77bb4_story.html?utm_term=.5cff25d0e7fb
58. U.S. Food and Drug Administration, Freedom of Information Act Request 2018-2046, received March 30, 2018.
59. Code of Federal Regulations, Active Treatment Concurrent Control, title 21, sec. 314.126 (b)(2)(iv).

25 Orphan Drug Development and Regulations

A.M. Lynch
ToxPlus Consulting

CONTENTS

25.1 INTRODUCTION

A rare disease is characterized as (a) a serious or life-threatening disease and (b) has a significant impact on the lifespan as well as quality of life of the individual, and (c) may require extensive lifelong medical treatment to address symptoms of the disease (NORD 2017). Approximately 7,000 rare diseases have been identified and are believed to affect around 30 million people in the United States. This equates to one out of every ten Americans being affected by a rare disease (NIH 2018). Any one rare disease may affect as few as a dozen individuals to an established upper threshold of 200,000 people in the United States alone. Rare diseases include cancers, inherited metabolic diseases, degenerative disease, and many others. The causes of rare diseases may include exposure to infections or toxins, autoimmune responses, or adverse reactions to therapeutic interventions to name a few. It is thought that 80% of the rare diseases are caused by genetic mutations. Tragically, half (50%) of those affected by rare diseases are children (NORD 2017).

It can be quite challenging for physicians to diagnose rare diseases. On average, patients visit 7.3 physicians and can have symptoms for 4.8 years before receiving a diagnosis due to the lack of knowledge about the disease and available diagnostic methods (Engel et al. 2013). Currently, the number of rare diseases which have no treatment option available is estimated to be between 4,000 and 5,000 worldwide (Sharma et al. 2010).

The number of identified rare diseases continues to increase each year. In 2009, there were reported 5,857 identified rare diseases, and by 2016, there were 6,084, an approximate increase of 30 new rare diseases each year (Orphanet 2016, 2009). This can be attributed to scientific advancement and improved understanding of various aspects of disease and characterization of genetic mutations as well as the use of biomarkers to help segment disease pathways. From 2009 to 2015, 16% of orphan designated drugs were based on predictive biomarkers that stratified a disease into subsets of patients with disease (Kesselheim et al. 2017). For example, 250,000 new lung cancers are reported per year in the US of which non–small cell lung cancer (NSCLC) represents 85% of the total new lung cancers. A subset of patients with NSCLC have epidermal growth factor receptor (EGFR) mutations, and this represents 40%–80% of the total NSCLC patients or 75,000–150,000 cases per year (Molina et al. 2008). Having the EGFR mutation as a biomarker allows companies to target more specific treatments for this subset patients, and the establishment of the Orphan Drug Act has provided companies with the incentives to develop treatment options. This represents a win-win for both the patient and the companies developing treatment options.

25.2   REGULATORY HISTORY OF THE ORPHAN DRUG ACT

In 1978, the Department of Health Education and Welfare formed a committee, chaired by Marion Finkel, M.D., of Food and Drug Administration (FDA), to study the increasing disparity between available treatment options for common diseases and treatment options for patients with less common disease or "rare" disease. The committee was formed in response to patients not seeing research going into treatment options for less common diseases. The multidisciplinary committee issued a report in 1979 entitled, "Significant Drugs of Limited Commercial Value." The report concluded that the cost of research and development prohibited return on investment for drugs which could be used for less common diseases with limited populations. Therefore, there was no financial incentive for drug companies to develop such drugs. The issue of a lack of development for drugs that were considered less profitable was finally given attention by government and industry (Finkel 1980; Haffner 2016).

At about the same time, the National Organization for Rare Disorders (NORD) was formed by leaders of several rare disease patient organizations to advocate on behalf of patients and their families dealing with a rare disease (NORD, 2018). In 1982, thanks to the efforts of patient advocates, NORD, rare disease medical experts, NIH and FDA staff, and members of Congress, particularly Senator Orrin Hatch and Representative Henry A. Waxman, the US Congress passed the first-ever Orphan Drug Act (ODA). On January 4, 1983, it was signed into law. According to the law, a rare disease was considered one that occurs so "infrequently in the United States that there is no reasonable expectation that the cost of developing and making available in the United States a drug for such disease or condition will be recovered from the sales in the United States of such drug" (97th Congress 1983) The law in 1983 contained economic incentives for drug developers to make a return on their drug research and development investment and included a federal grants program for orphan drug research, a 50% tax credit for expenses related to research and development centered around clinical trials, and allowed for 7 years of marketing exclusivity for approved products. The 7 years of marketing exclusivity is an additional 2 years from the 5 years FDA grants for approved new chemical entities.

The ODA law was amended in 1984 due to the difficulty of being able to objectively identify drugs to manufacture with "no reasonable expectation" that sales in the US could support development. As a result, the 1984 amendment defined rare disease as a condition affecting fewer than 200,000 Americans. The number of 200,000 patients was based on a number of diseases, including narcolepsy and multiple sclerosis, in which drugs seemed to be promising, but companies were not interested in developing due to a lack of a return on investment (DHHS 2001). Going forward, companies applying for orphan drug designation to provide proof of disease prevalence fewer than 200,000 patients with the rare disease of interest within the United States or that the disease affected more than 200,000 persons in the US but that there was no reasonable expectation that the sales of the drug will be sufficient to offset the costs of developing the drug for the US market as well as the costs of making the drug available in the United States in order to qualify for orphan drug designation.

A second amendment to the ODA in 1985 led to an additional incentive, the waiver of the Prescription Drug User Fee (PDUFA) for orphan products. This amounts to an approximate $2.5 million savings to companies who receive orphan drug designation. Also, the amendments

TABLE 25.1

Basic Principles of United States FDA ODA

Principle	Description
Definition of a "Rare Disease"	• <200,000 patients in the United States • > 200,000 patients but with proof that no reasonable expectation that the cost of development will be recovered
Market Exclusivity	• 7-year market exclusivity for approved orphan drugs or products
Clinical Research Subsidies	• Orphan Product Grant Program • Program provides funding for clinical testing
Tax Incentives	• Orphan Drug Tax Credit – 25% as of 2017 • Companies with an orphan designation can collect tax credits for expenses related to US clinical trials conducted on the orphan indication prior to designation
Fees	• Exemption from the Prescription Drug User Fee Act (PDUFA), approximately $2.5 million

included protocol assistance in drug development and the provision that an orphan drug could be either patentable or not patentable (Haffner 2016). See Table 25.1 for a summary of the key elements of the ODA.

FDA received more application than expected over time, and on June 29, 2017, the US FDA released a plan to completely eliminate the agency's existing orphan designation backlog of unreviewed requests with the goals that (a) within 90 days, FDA would complete reviews of all orphan drug designation requests older than 120 days and (b) that after 90 days, 100% of all new orphan drug designation requests will receive a response by the agency within 90 days of receipt (US FDA 2017b). FDA published an update in September 2017 stating that the first goal had been met successfully ahead of schedule (US FDA 2017c). In addition, a new draft guidance for industry was also issued in December 2017 to clarify pediatric sub-populations, "Clarification of Orphan Designation of Drugs and Biologics for Pediatric Subpopulations of Common Diseases" (DHHS 2017).

Another change to the Orphan Drug Act occurred during the 2017 tax legislation reform. In the tax reform bill, US Congress reduced the orphan drug tax credit from 50% of research and development costs related to clinical trials to 25%, a move that Congress claims will save the government $32.5 billion from 2018 to 2027 (Tribble, 2017).

25.3 RESULTS OF THE ODA ENACTMENT

As of December 2018, FDA had approved over 600 orphan drugs and biologics since the ODA was signed into law (US FDA 2018a). Further, approximately 4,000 drugs were designated with orphan drug designations for rare diseases (as of 2017) (US FDA 2017c; NORD 2017). Eighteen more orphan drugs were approved in 2017 (US FDA 2018a), 34 orphan drugs approved in 2018 (US FDA 2019a), and eight orphan drugs approved between January and March of 2019 (US FDA 2019b). From 1983 to 2016, there have been a total of 5,792 designation requests. As of 2016, 449 designations were approved for orphan therapies for 549 orphan indications and 36% of novel drug approvals in 2016 were orphan drugs (QuintilesIMS Institute 2017; US FDA 2017a). Just in the last 10 years, there has been a steady increase in both US FDA designations and approvals of orphan drugs, and one would expect this trend to continue into 2020 and beyond. See Figure 25.1.

Several areas of growth in medicine have occurred as a result of the ODA. One area of sufficient growth has been cancer. There were 1,391 orphan designations for drugs to treat rare cancers

US FDA ORPHAN DRUG
DESIGNATIONS AND APPROVALS
2007-2017

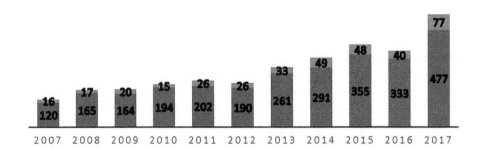

FIGURE 25.1   **US FDA Orphan Drug Designations and Approvals 2007–2017.** (Source: US FDA. Search Orphan Drug Designations and Approvals. 2018 March. www.accessdata.fda.gov/scripts/opdlisting/oopd/.)

from 1983 to 2015, and of those, 177 (36%) gained marketing approval (Stockklausner et al. 2016). According to a 10-year analysis of the ODA from 2000 to 2009, pediatric products increased from 17.5% to 30.8% of total orphan approvals with a mean prevalence of 8,972 patients in the US (Thorat et al. 2012). The growth continues in the area of pediatric product development.

Large pharmaceutical companies, however, were initially hesitant to move towards with development of drugs for rare diseases. This lead the way for new drug companies such as Genentech, Genzyme, and BioMarin to grow as a result of the incentives provided by the ODA. Also, when the ODA became law, biotechnology-derived products were just beginning to be developed but were difficult to patent with a product matter patent due to the molecular features being characterized and published in the public domain many years earlier. The inclusion of biologics as orphan drugs gave companies the incentive to develop biologic drugs for orphan diseases as well. For example, the orphan biological product recombinant erythropoietin (rEPO), made by Amgen, for example, was developed for chronic kidney failure which is considered a rare disease with a prevalence of 200,000 US patients. rEPO was approved by US FDA as an orphan drug in June 1989, giving Amgen the 7-year exclusivity to manufacture and sell tEPO to recoup development costs. (Haffner 2016; Coster 1992).

25.4   COSTS

According to the NORD-sponsored QuintilesIMS Institute report, the volume of prescriptions for orphan drugs is relatively low – in the US, 0.3% of total volume of pharmaceutical use in 2016 was for a rare disease. This number inched up to 0.4% in 2017 (IQVIA Institute 2018). Further, of total US drug sales in 2016, 60% of $450 million was for traditional drugs, while 7.9% was for drugs approved for orphan indications (QuintilesIMS Institute 2017). Because orphan drugs treat small populations, they can be expensive to patients and their families. In 2016, the median annual cost of an orphan drug was $32,000 per year (QuintilesIMS Institute 2017). The median annual cost of an orphan drug increased to over $46,800 in 2017; however, the annual cost for the ten rare disease therapies was lower than this, with a value slightly over $1000 (IQVIA Institute 2018). The ODA made possible treatments that resulted in cost savings. In California, for example, the California Public Health officials developed a treatment for infant botulism by taking advantage of the ODA grants program. The treatment has benefited more than 1,500 patients and resulted 90 years of avoided hospital stays and savings of more than $130 million of hospital costs (NORD 2017).

25.5 THE US ORPHAN DRUG APPLICATION PROCESS

The orphan drug application may be filed at any point in development prior to New Drug Application (NDA). To obtain an orphan designation in the US, companies must submit an application to the US FDA Office of Orphan Products Development (OOPD). The application then goes through the review process. First, the request for orphan drug designation is assigned a designation request number and logged into the OOPD database, and an acknowledgment letter is sent to the sponsor (or sponsor's agent). Second, OOPD assigns a reviewer of the application. Third, the review is forwarded to the Director of the Orphan Drug Designation Program for a second-level review. Following the OOPD Office Director's review, a designation letter, a deficiency letter requesting additional information, or a denial letter is issued to the sponsor. Once a drug has received orphan designation, the US FDA Center for Drug Evaluation and Research (CDER) or the US FDA Center for Biologics Evaluation and Research (CBER) will review applications for marketing approval.

The orphan drug application requires the company submit the following information (US FDA, 2018b):

- **Statement of Orphan Drug Designation:** The company should provide a statement requesting orphan drug designation for a rare disease or condition which will be identified with specificity further in the application.
- **General Information:** Contact information for the sponsor, information about the drug such as the generic name of the drug, tradename, and drug manufacturer should be included in this section of the application.
- **Information about the Rare disease:** A description of the rare disease or condition for which the drug is being or will be investigated is needed in this section of the application. The population size and characteristics of the population to be treated should be clearly stated. The proposed indication or use of the drug and the reasons why such therapy will benefit the patient population identified should also be included in this section of the application.
- **Information about the Drug:** A full description of the drug and its known risks as well as scientific rationale for medical plausibility for the use of the drug for the rare disease or condition indicated is needed in this section. All data the company has generated or that is publicly available in the literature should be included in this section of the application and the data reports as well as literature being cited should be provided to FDA.
- **Clinical Superiority Justification:** If the company believes their drug is clinically superior to one that is already approved for the same rare disease, then the company must provide a justification of why the proposed variation may be clinically superior to the first drug approved in this section. The justification may be based on issues such as less side effects experienced by the patient and provides a more convenient route of administration, etc.
- **Justification of a Subset:** If the company is seeking designation based on a subset of a common disease population, the sponsor must demonstrate that due to one or more properties of the drug, only the subset population would benefit from the drug and not the greater common disease population in this section of the application. This is common orphan drug designation pathway for sub-populations with a specific mutation.
- **Regulatory Status and Marketing History:** A summary of the regulatory status and marketing history of the drug in the US and foreign countries is needed in this section of the application. Indications for which the drugs are approved in other countries should be included as well as what, if any, adverse regulatory actions have been taken against the drug in any country.
- **Prevalence:** Documentation, with references, that the disease or condition for which the drug will treat affects fewer than 200,000 people in the US is needed in this section of the application. If there is no reasonable expectation that costs of research and development of

the drug for the indication can be recovered by sales, an estimate must be given of the cost of development and distribution of the drug, as well as an assessment of potential sales in the US.

The Office of Orphan Product Development will typically respond within 60 working days after submitting the orphan drug application; however, the process could take longer if additional information is needed from the company. After receiving the orphan designation and conducting more research, a company may seek marketing approval if the drug proves safe and effective in clinical trials (DHHS 2001).

25.6 REGULATION OF ORPHAN DRUGS OUTSIDE THE UNITED STATES

25.6.1 EUROPE

In 2000, the European Union (EU) established the EU Orphan Drug Program by passing Regulation (EC) No. 141/2000 (EU 1999). The EU Orphan Drug Program requires a ratio for prevalence (5/10,000) versus an exact number. Efforts have been jointly made at national and European levels by industry and health authorities such as the European Medicines Agency (EMA), in an effort to offer incentives required to stimulate the development of orphan drugs throughout Europe. The tax credit amount the company will receive is set by the individual member states within the EU. The EU allows a 6-year exclusivity period and if not challenged at the end of the 6th year allows additional 4 years (EU 1999; Haffner 2016).

From 2000 to 2017, more than 2,974 orphan drug designations requests have been submitted to the European Commission (EMA 2018). Based on a positive opinion from the Committee for Orphan Medicinal Products (COMP), 1,952 designations have been made as of 2017, and 238 orphan drugs have received marketing authorization in Europe (EMA 2018). The applicants are generally advised to have a pre-submission meeting with the EMA Secretariat. It typically takes 60–90 days to gain a designation in Europe. Because differences exist in legislation of each member state, the timeframe for review and designation is difference amongst member states.

25.6.2 JAPAN

In October 1993, the Japanese government, Ministry of Health, Labor and Welfare (MHLW) introduced special provisions related to research and development of orphan drugs into law that allowed a drug to be granted orphan drug status if it met the following two criteria (Japan MHLW 2018):

- No possible alternative treatment must exist for the disease for which the drug will be used or the efficacy and expected safety of the drug must be excellent in comparison with other available drugs.
- The number of patients affected by this disease in Japan must be less than 50,000 on the Japanese territory, which corresponds to a maximal incidence of four per ten thousand.

If orphan drug status is granted, there are the following incentives (Japan MHLW 2018):

- Reduced marketing authorization application for Japanese.
- An orphan products development grant worth up to 50% of the R & D cost per year for a maximum of 3 years after designation.
- Tax deductions.
- Fast-track review for approval.
- Consultation and assistance with development.
- Extension from a normal 5-year period to a 10-year period for reexamination period for orphan drugs and 7-year period for orphan devices.

The MHLW receives and determines orphan designation of a drug. The company must submit an application for a pre-designation hearing which typically lasts 30 min. If there are no concerns during the pre-designation hearing, then the company can submit a formal application for consideration as orphan designation. The Pharmaceuticals and Medical Devices Agency (PMDA) evaluates the application, and, if the product is determined appropriate for designation, the MHLW asks the opinion of the Pharmaceutical Affairs and Food Sanitation Council. When the Council determines that the criteria for designation are met, the MHLW designates the product as an orphan product and the applicant is notified accordingly (Japan MHLW 2018). As of December 2016, the MHLW had designated 336 products as orphan drugs and 276 have been approved for marketing (Pacific Bridge Medical 2017).

25.6.3 Australia

In 1998, Australian Government Department of Health Therapeutic Goods Administration (TGA) established the Australian Orphan Drugs Program. In June 2017, the TGA amended the policy to reflect an update that drugs are orphan drugs if they are used to treat diseases or conditions affecting fewer than 5 in 10,000 at any one time in Australia. Based on industry comments, other positive changes to the policy helped companies gain clarity on the orphan drug designation process in Australia (TGA 2017).

Incentives offered by Australia's orphan drug policy include the following (TGA 2017):

*   Fee waiver for the application and evaluation of the application
*   No annual registration fees
*   5-year exclusivity (under consideration by the Australian jurisdiction)
*   Once a drug is designated as orphan, fees for and evaluation are waived.

25.7   CHALLENGES FOR ORPHAN DRUGS GOING FORWARD

Knowledge of Rare Diseases. The information available for many rare diseases is non-existent. As a result, health professionals often lack appropriate training and awareness of signs and symptoms to be able to diagnose and adequately treat rare diseases. Additional research is needed to identify and development diagnostic criteria for rare diseases as we become aware of them through scientific advancements.

Diagnostic Methods for Rare Diseases: For many rare diseases, no diagnostic methods exist, and therefore, diagnosis is problematic. The process of developing diagnostic methods and validating them so that they are reproducible is time-consuming. Although the pace of gene discovery for rare genetic diseases has accelerated during the past decade, translation of these discoveries into the clinic is slow. A mechanism for faster translation to the clinic is necessary for progress.

Treatment Costs to Patient with a Rare Disease.: The cost of an orphan drug per treatment can be expensive for patients. The cost of treatment can range from $10,000 to $100,000s per year for patients with rare disease. The affordability of orphan drugs has become a major issue for payers and insurers. Some companies have responded by developing programs to help facilitate access to orphan drugs for patients with rare diseases, but more efforts are needed.

25.8   CONCLUDING REMARKS

The need for rare disease treatment options will remain great for future generations. Significant numbers of patients with rare diseases have been helped to date, and the economic incentives continue to encourage drug developers to work towards treatment options for patients with rare diseases. Patient biomarkers used to stratify individuals into subsets of populations illustrating differences in susceptibility to a disease and patient responses to treatment will lead towards the development of more treatment options for rare diseases in the not too distant future. Next-generation sequencing and prioritization of genome-wide studies will contribute significantly to the identification of additional rare diseases for which treatment options can be developed. In addition, access to global markets will also increase the opportunity for treatment options. As more countries develop and refine legislation for rare diseases and orphan drugs, the market potential for orphan drugs will increase worldwide.

REFERENCES

97th Congress. 1983. Public Law 97–414. https://www.fda.gov/downloads/ForIndustry/DevelopingProductsfor RareDiseasesConditions/HowtoapplyforOrphanProductDesignation/UCM517741.pdf. (Accessed on March 29, 2018).

Coster, J.M. 1992. Recombinant erythropoietin. Orphan product with a silver spoon. *Int J Technol Assess Health Care* 8(4):635–46.

Department of Health and Human Services (DHHS). 2017. Clarification of Orphan Designation of Drugs and Biologics for Pediatric Subpopulations of Common Diseases Draft Guidance for Industry. www.fda.gov/downloads/RegulatoryInformation/Guidances/UCM589710.pdf (Accessed on March 20, 2018).

Department of Health and Human Services (DHHS). 2001. Office of Inspector General. The Orphan Drug Act Implementation and Impact. https://oig.hhs.gov/oei/reports/oei-09-00-00380.pdf (Accessed on March 28, 2018).

Engel, P.A., Bagal, S., Broback, M., Coice, N. 2013. Physician and patient perceptions regarding physician training in rare diseases. The need for stronger education initiatives for physicians. *J Rare Dis.* 1(2): 1–15.

European Medicines Agency (EMA). 2018. Annual report on the use of the special contribution for orphan medicinal products: Year 2017. EMA/19529/2018. www.ema.europa.eu/docs/en_GB/document_library/Report/2017/02/WC500221159.pdf (Accessed March 27, 2018).

European Union (EU). Office Journal of European Communities. 1999. Regulation (EC) No 141/2000 of the European Parliament and of the Council. http://eur-lex.europa.eu/LexUriServ/LexUriServ.do?uri=OJ:L: 2000:018:0001:0005:en:PDF (Accessed on March 27, 2018).

Finkel, M.J. 1980. Drugs of limited commercial value. *N Engl J Med.* 302:643–644.

Haffner, M.E. 2016. History of orphan drug regulation – United States and beyond. *Am Soc Clin Pharmacol Ther.* 100(4):342–343.

IQVIA Institute; Orphan Drugs in the United States: Growth Trends in Rare Disease Treatments. October 17, 2018. www.iqvia.com/-/media/iqvia/pdfs/institute-reports/orphan-drugs-in-the-united-states-growth-trends-in-rare-disease-treatments.pdf?_=1559360172341 (Accessed May 31, 2019).

Japan, Ministry of Health, Labor and Welfare. 2018. Overview of Orphan Drug/Medical Device Designation System. www.mhlw.go.jp/english/policy/health-medical/pharmaceuticals/orphan_drug.html (Accessed March 27, 2018).

Kesselheim, A.S., Treasure, C.L., Joffe, S. 2017. Biomarker-defined subsets of common diseases: Policy and economic implications of orphan drug act coverage. *PLoS Med.* 14(1):e1002190.

Molina, J.R., Yang, P., Cassivi, S.D., Schild, S.E., Adjei, A.A. 2008. Non-small cell lung cancer: epidemiology, risk factors, treatment, and survivorship. *Mayo Clin Proc.* 83(5):584–94.

Orphanet. Orphanet Activity Report 2016. www.orpha.net/orphacom/cahiers/docs/GB/ActivityReport2016. pdf. (Accessed on March 21, 2018).

Orphanet. Orphanet Activity Report 2009. www.orpha.net/orphacom/cahiers/docs/GB/ActivityReport2009. pdf. (Accessed on March 21, 2018).

Pacific Bridge Medical. 2017. Orphan Drugs in Asia 2017: Guidelines and Regulatory Requirements To Help Orphan Drug Products Enter the Asian Market. www.pacificbridgemedical.com/wp-content/uploads/2014/03/Orphan-Drugs-in-Asia-2017.pdf. (Accessed March 27, 2018).

QuintilesIMS Institute. 2017. Orphan Drugs in the United States: Providing Context for Use and Cost. National Organization for Rare Disorders. https://rarediseases.org/nord-white-paper-quintilesims-report-download/ (Accessed March 20, 2018).

National Organization for Rare Disorders (NORD). 2018. History of Leadership. https://rarediseases.org/about/what-we-do/history-leadership/ (Accessed on March 31, 2018).

National Organization for Rare Disorders (NORD). 2017. Trends in Orphan Drug Costs and Expenditures Do Not Support Revisions in the Orphan Drug Act: Background and History. https://rarediseases.org/nord-white-paper-quintilesims-report-download/ (Accessed March 20, 2018).

NIH National Center for Advancing Translational Sciences. 2018. Genetic and Rare Disease Information Center. FAQs about Rare Diseases. https://rarediseases.info.nih.gov/diseases/pages/31/faqs-about-rare-diseases (Accessed on March 21, 2018).

Sharma, A., Jacob, A., Tandon, M., Kumar, D. 2010. Orphan drug: Development trends and strategies. *J Pharm Bio-allied Sci*. 2(4) 290–99.

Stockklausner, C., Lampert, A., Hoffman, G.F., Ries, M. 2016. Novel treatments for rare cancers: The U.S. orphan drug act is delivering a cross sectional analysis. *Oncologist*. 21(4):487–93.

Thorat, C., Xu, K., Freeman, S., Bonnel, R., Joseph, F., Phillips, M. 2012. What the orphan drug act has done lately for children with rare diseases: A 10-year analysis. *Pediatrics*. 129:3.

Tribble, Sarah Jane of Kaiser Health News. 2017. FDA chief open to rethinking orphan-drug incentives. December 28, 2017. Benefitspro.com.

United States Food and Drug Administration (US FDA). January 2017a. 2016 Novel Drug Summary. www.fda.gov/downloads/Drugs/DevelopmentApprovalProcess/DrugInnovation/UCM536693.pdf (Accessed March 20, 2018).

United States Federal Drug Administration (US FDA). June 29, 2017b. FDA unveils plan to eliminate orphan designation backlog. FDA News Release. www.fda.gov/NewsEvents/Newsroom/PressAnnouncements/ucm565148.htm (Accessed March 20, 2018).

United States Federal Drug Administration (US FDA). September 2017c. FDA is Advancing the Goals of the Orphan Drug Act. *FDA Voice*. https://blogs.fda.gov/fdavoice/index.php/2017/09/fda-is-advancing-the-goals-of-the-orphan-drug-act/ (Accessed on March 20, 2018).

United States Food and Drug Administration (US FDA). December 2018a. Developing Products for Rare Diseases & Conditions. https://www.fda.gov/industry/developing-products-rare-diseases-conditions (Accessed on May 30, 2019).

United States Food and Drug Administration (US FDA). January 2018b. 2017 New Drug Therapy Approvals. https://www.fda.gov/files/about%20fda/published/2017-New-Drug-Therapy-Approvals-Report.pdf (Accessed on May 30, 2019).

United States Food and Drug Administration (US FDA). January 2019a. 2018 New Drug Therapy Approvals. https://www.fda.gov/media/120357/download. (Accessed on May 30, 2019).

United States Food and Drug Administration (US FDA). 2019b. Total number of orphan drug approvals in the month requiring exclusivity determinations. https://www.accessdata.fda.gov/scripts/fdatrack/view/track.cfm?program=osmp&id=OSMP-OOPD-Number-orphan-drug-approvals-in-the-month (Accessed on May 30, 2019).

26 Development of Drug Products for Older Adults
Challenges, Solutions, and Regulatory Considerations

S.W. Johnny Lau, Darrell R. Abernethy,
and Chandrahas Sahajwalla
US Food and Drug Administration

CONTENTS

26.1 INTRODUCTION

According to the United States Census Bureau, persons aged 65 years and above will be the fastest-growing segment of the population in the United States for the next four decades due primarily to the migration of the Baby Boom generation into this age group with a steadily increasing life expectancy. In 2050, the projected number of persons in the United States aged 65 years and above will be 83.7 million, almost double its population estimate of 43.1 million in 2012.[1] Figure 26.1 shows the age distribution of the US population in the next four decades.

The oldest-old (age ≥85 years) segment of the United States is increasing even faster and will triple by 2060.[2]

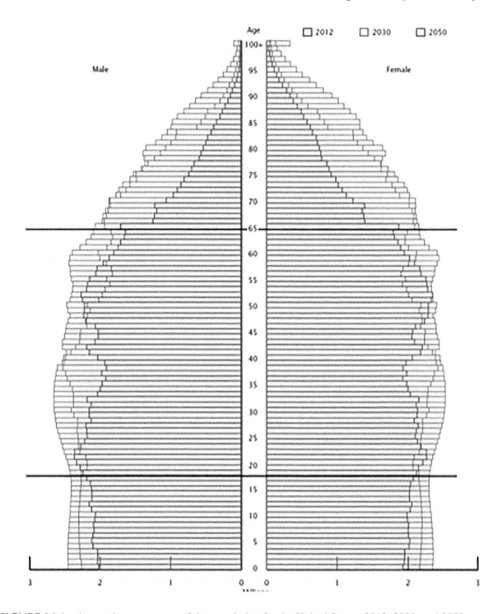

FIGURE 26.1 Age and sex structure of the population for the United States: 2012, 2030, and 2050.

The world is in the middle of a transition toward significantly older populations.[3] Figure 26.2 shows the pyramid for the less developed regions in the world in 2013 and a transformation from the wide base of a youthful population in 1970 to the more rectangular shape of an older population in 2050.

The age composition of the more developed regions in the world is also in a transitional phase from the already aged structure of 1970, which shows the demographic changes of the Second World War, to the even more aged structure expected for the year 2050. In the more developed regions, the 2013 pyramid shows a full mid-section, an indication that there is a predominance of young and middle-age adults, together with significant volume at the older ages, an indication of aging. But this structure is in a rather rapid transition to a more aged population in the more developed regions, with more than 30% of older persons by 2050.

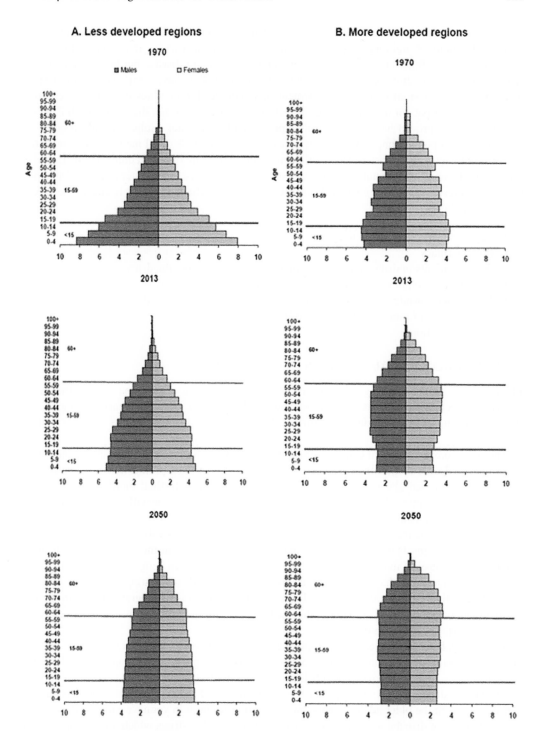

FIGURE 26.2 Population pyramids of the less and more developed regions in the world: 1970, 2013, and 2050.

Aging is a complex and multifactorial process that is an outcome of the accumulation of various functional deficits of multiorgan systems occurring over time at varying rates. As adults become older, their bodily functions change, and they tend to bear more disease burden. In general, this segment of the older adult population is frailer and is more likely to have significant sensory impairment (hearing and vision), cognitive impairment, and multiple chronic illnesses. Thus, it is common for an older adult patient (65 years of age and above) to have five to ten diagnoses, each of which has one or more proved beneficial drug therapies.[4,5] Although older adults currently account for 13.1% of the US population, they consume an estimated 30%–40% of all medications,[6] indicating that drug therapy is an important medical intervention for the care of older adult patients. However, these drug therapies may result in issues such as multiple drug use, drug interaction, and inappropriate drug prescription. This chapter discusses the following challenges as it relates to the care of older adult patients:

- Physiological changes due to aging
- Polypharmacy
- Age of patients in drug trials and age of actual patients taking the drugs
- Barriers and solutions to include older adult patients in drug trials
- Regulatory considerations on drug development for older adults in the United States
- Compression of morbidity to the end of life
- Future directions.

26.2   PHYSIOLOGICAL CHANGES DUE TO AGING

Physiological changes due to aging occur in all bodily organ systems. Due to the scope of the subject matter, this section only discusses the physiological changes that affect the pharmacokinetics and pharmacodynamics in older adults. For further discussions on this topic, readers can refer to three articles.[7–9]

The most common route of drug administration is oral. Aging results in various changes in the gastrointestinal tract such as increased gastric pH, delayed gastric emptying, decreased splanchnic blood flow, decreased absorption surface, and decreased gastrointestinal motility. Despite these changes, oral absorption does not appear to alter in advanced age especially for drugs that show passive diffusion mediated absorption.[7,8] However, older adults usually have comorbidities and require multiple drug therapies which are difficult for older adults to swallow multiple tablets daily. Difficulty swallowing is prevalent in older adults that can result from the medication itself, weak tongue, poor control of muscles in the mouth, and from the diseases being treated such as stroke, surgery after cancer, esophagus, and nervous disorders.[10] Orally disintegrating tablets may be a way to aid drug delivery for the older adults. These tablets dissolve or disintegrate rapidly in the oral cavity upon contact with saliva, without the need for chewing or additional water, thus facilitating administration of medication orally.[11]

With aging, body fat increases as a proportion of total body weight and lean body mass, but total body water decreases.[12] Accordingly, the volume of distribution per unit total body weight decreases for polar drugs such as digoxin, theophylline, and aminoglycosides, but increases for lipophilic drugs such as diazepam.[8] Assuming that the therapeutic goal is to achieve average plasma drug concentrations in the older adult patients that are similar to those in younger patients, the changes in drug volume of distribution will generally be relevant only for drugs which are administered as single doses or for the determination of loading doses.

Albumin and α1-acid glycoprotein are the major drug-binding proteins in plasma. In general, the blood albumin concentration is about 10% lower in older adults, but the α1-acid glycoprotein is higher in older adults.[13] These changes in plasma proteins are generally not due to aging itself but to the pathophysiological changes or disease states that may occur more frequently in older adult patients. Additionally, these changes in plasma proteins may not affect the clinical drug exposure of a patient. Thus, no adjustments in dosing regimens may be necessary in general except in rare case of a drug with a high extraction ratio and narrow therapeutic index that is parenterally administered such as intravenous dosing of lidocaine.[14]

The decreased splanchnic blood flow and reduced liver size in older adults may decrease the hepatic clearance of drugs with high extraction ratio. Additionally, the hepatic first-pass effect may be decreased, and oral bioavailability of these drugs may be increased. For drugs with low extraction ratio, the effects are less pronounced. Enzymatic activity of some drug-metabolizing cytochrome P450 (CYP) 1A2, 2D6, 2C9, 2C19, and 2E1 appears not changed in older adults, whereas CYP 3A4 seems to be decreased. In vivo hepatic drug clearance via CYP metabolism has been studied for many drugs in older adults and was found to be either unchanged or modestly decreased, with clearance reductions reported to be in the range of 10%–40% for young-old and old persons who were generally in good health.[15,16]

The sensitivity to the effect of benzodiazepines, warfarin, and hypotensives is increased, whereas the sensitivity to the effect of beta adrenergic function in cardiac tissues is decreased in older adults than those of younger adults. Additionally, the sensitivity to the effect of antipsychotics (anticholinergic and extrapyramidal effects), nonsteroidal anti-inflammatory drugs (gastric ulceration), and anticholinergics (central nervous system effect on early Alzheimer conditions) is increased, whereas the effect of calcium channel blockers on the PR interval is decreased in older adults than those of younger adults.[9,17]

For certain drugs, the pharmacodynamic changes associated with aging may be more clinically relevant than the pharmacokinetic changes associated with aging.[7]

26.3 POLYPHARMACY

Patients aged 65 and above are the most medicated group of patients. In general, about half of the population aged 65 and above in Sweden is exposed to five or more concurrent medications, and 11.7% of this population is exposed to ten or more concurrent medications.[18] Older adult patients usually have more disease burden and thus take multiple drug therapies that result in polypharmacy. Polypharmacy is most commonly defined as the concurrent use of five or more medications.[19] Because of the increased medication use, the term "excessive polypharmacy" (ten or more medications) has also emerged.

Polypharmacy should be separated from potentially inappropriate medication (PIM) use. PIMs describe medications that the risks generally outweigh the benefits, whereas polypharmacy usually defines the number of medications regardless of appropriateness.[19] However, polypharmacy is correlated with the prevalence of PIM use.[20]

Polypharmacy in older adults increases the risk of adverse drug reactions, inappropriate prescriptions, drug interactions, number of hospitalizations, costs, and even death.[21] Polypharmacy can also be a major contributor to the development of frailty in older adults.[22] Frailty in older adults is a condition characterized by the loss of biological reserves, failure of homeostatic mechanisms, and vulnerability to adverse outcomes.[23] Frailty eventually leads to disability and loss of independence of the individual.

Deprescribing is an emerging and promising approach to manage the medications of older adults. Deprescribing is the procedure of tapering or discontinuing medications to minimize polypharmacy and improve patient outcomes.[24] Deprescribing will be an important way forward in patient care for polypharmacy and research.[19]

Interindividual differences in drug response are large and difficult to predict in older adult patients. Thus, personalized medicine can be a way to minimize polypharmacy and guide prioritization for deprescribing medications.[19] However, a multiprofessional team approach is recommended to detect, manage, and prevent drug interactions as well as optimize drug therapy in older adult patients.[25]

26.4 AGE OF PATIENTS IN DRUG TRIALS AND THE AGE OF ACTUAL PATIENTS TAKING THE DRUGS

Underrepresentation of the older adult population in clinical trials has been reported to be very common across multiple therapeutic areas such as cancer, dementia, epilepsy, incontinence, transplantation, and cardiovascular disease.[26–31] For example, Figure 26.3 shows the clinical trial population of simvastatin and the actual target patient population of simvastatin that these two populations clearly differ from each other.[32]

The impact of the underrepresentation in clinical trials on risk–benefit relationships for older adult patients is unclear; however, it is known that the risk–benefit relationship for some important drug classes is shifted toward increased risk in older adult patients.[33–35]

Figure 26.4 shows signs that the underrepresentation of older adults in clinical trials is improving. A recent Food and Drug Administration (FDA)-conducted study shows that from 2013 to 2015, the percentage of Phase 3 trial participants aged 65 years and above increased from 19% to 40%.[36]

This increased reporting of participant age (<65 years and ≥65 years) in 2014 and 2015 compared with 2013 may be due to the 2012 update of the International Council for Harmonisation E-7 guidance, which emphasized the importance of including patients ≥65 years of age.[37] However, this increase is based on a limited sample of drugs surveyed in the 3-year study period, and no definite conclusions regarding percent participation for older patients may be drawn. Figure 26.5 shows that the therapeutic groupings with the greatest percentage of trial participants aged 65 years and above were Oncology/Nononcology Hematology, Medical Imaging, and Cardiovascular/Renal, which is likely due to cancer (oncology), cognitive decline (medical imaging), and heart conditions (cardiovascular) being more prevalent in older adult patients.[36]

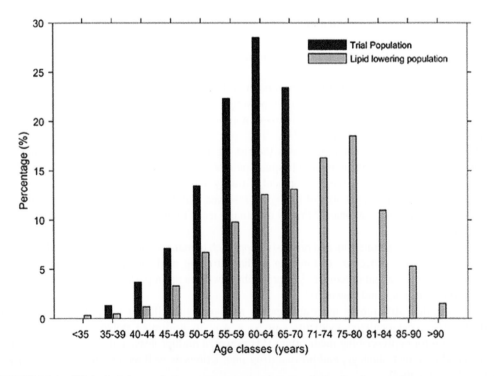

FIGURE 26.3 Clinical trials population versus actual patient population for simvastatin.

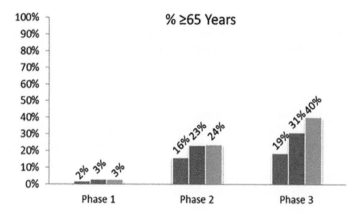

FIGURE 26.4   Age of clinical trial participants by phase of clinical trial and year: 2013–2015.

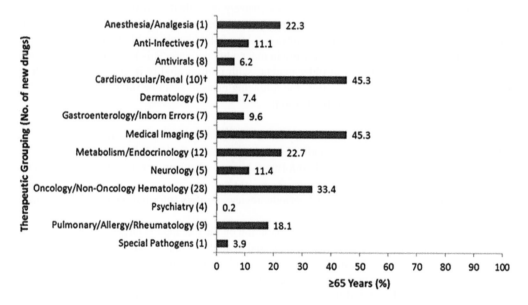

FIGURE 26.5   Percent older adult patients' participation in new drug trials by therapeutic grouping: 2013–2015.

26.5   BARRIERS AND SOLUTIONS TO INCLUDE OLDER ADULT PATIENTS WITH MULTIMORBIDITY IN DRUG TRIALS

More than one concurrent illness (multimorbidity) for which drug treatment is useful is very common in the older adult patient.[4] This leads to treatment of the multiple illnesses with multiple medications, and the term "polypharmacy" is often used for patients on five or more medications simultaneously which often leads to medication-related adverse events.[38] The evaluation of a drug for efficacy and safety is the core process for drug development, and its completion is necessary to make the drug available for patients. This requires the demonstration of a sufficiently positive effect on an agreed-upon clinical endpoint related to an illness for which the drug is being developed. This approach has served well from the time of a requirement in the United States (1962) that in addition to being safe, a drug be shown to be efficacious for the indication for the use that is being sought by the pharmaceutical sponsor of the drug candidate.

During the drug development process, patients are selected for inclusion into the late-stage (Phase 3) evaluation based on having the targeted disease and the likelihood of response to the drug under evaluation. Older adult patients with multimorbidity in addition to the disease for which the drug candidate is being developed may not be well represented in these evaluations for the following reasons[39]:

a. Multimorbidity may make drug response assessment difficult.
b. The patient receives multiple other medications for concurrent illnesses.
c. Logistical problems (such as transportation and need for supportive care) for patient participation.
d. Greater costs for medical management.
e. Perception by investigators that such patients are more difficult to study.

However, the older adult patient with concurrent illness is the frequent recipient of the newly developed medications in addition to the medications they are already receiving. Differences in pharmacokinetics in older adult patients are extensively described, and often the changes result in increased exposure to a given dose due to decreased clearance of drugs that undergo renal clearance and some metabolic clearance pathways.[40] Less is known about differences in drug effects (pharmacodynamics) at a given drug exposure in older adult patients. This leads to the need to overcome the challenges for inclusion of older patients with multimorbidity in clinical trials of drugs they are likely to receive. To achieve this, the following measures can be implemented, depending on the drug and the relevant clinical population:

a. Provision of transportation to the study site or in-home visits for trial participation
b. Between visit telephone access and immediate access for medical follow-up
c. Careful attention to obtaining truly informed consent, when appropriate witnessed consent or legal guardian consent
d. The Data Safety Monitoring Board periodic review of safety information for older patients

Other specific measures may be appropriate depending on the therapeutic target, potential toxicities from the drug exposure, and the multimorbid older adult patient population being evaluated. The value of participation by the multimorbid older adult patient who will receive a drug treatment following drug approval outweighs the additional effort and expense incurred by their inclusion into studies. There is recently a trend to use the virtual and decentralized clinical trial approaches to accelerate recruitment and improve patient retention for clinical trials.[41,42]

26.6 REGULATORY CONSIDERATIONS ON DRUG DEVELOPMENT FOR OLDER ADULTS IN THE UNITED STATES

26.6.1 REGULATORY DEVELOPMENT

The FDA has required geriatric data to be reported in New Drug Application (NDA) since 1985 when the content and format of NDA regulation was enacted.[43] The FDA published the guideline on format and content of clinical and statistical sections of the NDA in 1988 that requires reporting of age-related data.[44]

The inclusion of older adults in clinical trials of drugs under evaluation for registration in the United States was initially guided by the FDA's "Guideline for the Study of Drugs Likely to Be Used in the Elderly" published in November 1989.[45] The general theme of this guideline is "drugs should be studied in all age groups, including the older population, for which they will have significant age alone or the presence of any concomitant illness or medication, unless there is reason to believe that the concomitant illness or medication will endanger the patient or lead to confusion in interpreting the results of the study." This concept is further expanded with the statement that "attempts should therefore be made to include patients over 75 years of age and those with concomitant illness and treatments, if they are stable and willing to participate."

A common position among the regulatory bodies of the European Union, Japan, and the United State was later achieved and published in 1994 as the International Conference on Harmonisation of Technical Requirements for Registration of Pharmaceuticals for Human Use (ICH) E7 Guideline for studies in support of the older adult population.[46] This guideline noted the characteristics of the older adult patients that need specific attention such as the frequent occurrence of concomitant illness (and the comedications that entailed), the concomitant medications for these illnesses, and the importance of identifying pharmacokinetic differences that may derive from altered renal and hepatic function.[47] Of note, the ICH E7 Guideline recommends a minimum of 100 patients over the age of 65 for inclusion in a clinical drug development program.

In 1997, the FDA established the Geriatric Use subsection, as a part of the PRECAUTIONS section, in the labeling for human prescription drugs to include more comprehensive information about the use of a drug or biological product in persons aged 65 years and above.[48]

In 1998, the FDA issued a final rule (the "Demographic Rule") to clearly define in the NDA format and content regulations for the requirement to present effectiveness and safety data for important demographic subgroups in NDAs.[49] The Demographic Rule also requires sponsors to tabulate in the annual reports of their Investigational New Drug applications the numbers of subjects enrolled to date in clinical studies for drug and biological products according to age group, gender, and race.[49]

In 2001, the FDA published a guidance on the labeling of drug products for the older adults.[50]

Beginning in 2006, the FDA required the inclusion of specific sections in approved drug labeling for specific populations such as the geriatric ($\geq$65 years) population in compliance with the Physician Labeling Rule.[51]

Since the publication of ICH E7 Guideline, the rapidly changing worldwide demographics and patterns of drug use prompted regulators to examine whether the advice of a minimum of 100 patients over the age of 65 still provides appropriate explanation of the benefit to risk balance in this major group of drug users.[47] Particularly, people older than 75 years of age are the fastest-growing population in many countries, and they have the high likelihood of multimorbidities.

The ICH E7 guideline was updated in a question and answer format.[37] The main points of this ICH E7 update are[47]

1. A representative number of older patients should be included in the clinical trials. In the world of aging Baby Boomers, 100 patients are unlikely to be sufficient for most indications.
2. In the marketing application, data should be presented for four separate age groups: <65, 65–74, 75–84, and $\geq$85 to assess the consistency of the treatment effect and safety profile in these patients with the non-older adult patient population.
3. The guideline applies to both drugs intended for the older adult patient population and for diseases present in, but not unique to, the older adult patient population. For diseases specific to the older adults, older adult patients should constitute most of the enrollment.
4. The emphasis of studying patients aged 75 years and above.
5. Arbitrary upper age limits in the trial inclusion criteria should be avoided.
6. Inclusion of patients with concomitant illnesses is encouraged.
7. The preference is for inclusion of older adult patients in the pivotal Phase 3 trials, not in separate trials. This inclusion allows for the comparison of responses with younger patients in the same trial. However, this inclusion may not always be optimal.
8. The pharmacokinetics in older adult patients (over the entire spectrum of the older adult patient population) should be evaluated to identify age-related differences that are not explained by other factors, such as reduced renal function or weight differences. If a sufficient number of patients in different age ranges (including patients $\geq$65 and $\geq$75 years)

are included in the clinical trials, then population pharmacokinetic analysis could provide such data. Otherwise, a specific pharmacokinetic study comparing non-older adult and older adult participants in the same study (matched for relevant covariates such as weight and sex) could be performed.

In 2012, Section 907 of the Food and Drug Administration Safety and Innovative Act (FDASIA) directed the FDA to develop a report on the inclusion of demographic subgroups in clinical trials and data analysis in applications for drugs, biologics, and devices within 1 year.[52] In August 2013, the FDA released a report describing demographics and subset analyses included in 72 applications for drugs, biologics, and medical devices approved in 2011.[53] Section 907 of FDASIA also directed the FDA to publish an Action Plan to enhance the collection and availability of demographic subgroup data[54]:

- Priority 1: Improve the completeness and quality of demographic subgroup data (Quality).
- Priority 2: Identify barriers to subgroup enrollment in clinical trials and employ strategies to encourage greater participation (Participation).
- Priority 3: Making demographic subgroup data more available and transparent (Transparency).

To enhance transparency, the FDA implemented the Drug Trial Snapshots.[55] Drug Trial Snapshots present the participation of patients in trials that supported the approval of the drug by age, gender, and race, and highlight whether there was any difference in benefits or side effects among these subgroups. Readers can refer to two articles on the demographic information of clinical trials for drugs and biologics approved by the FDA from 2010 to 2015.[36,56]

26.6.2 FDA GUIDANCES FOR DRUG DEVELOPMENT

The FDA provides a multitude of guidances for the industry to develop drug products.[57] All regulatory guidances below are most relevant to guide the drug development for older adults. The following two regulatory documents provide guidance to show evidence of drug effectiveness and drug exposure–response relationships[58,59]:

- Providing Clinical Evidence of Effectiveness for Human Drugs and Biological Products, 1998
- Exposure–Response Relationships – Study Design, Data Analysis, and Regulatory Applications, 2005.

The absence of appropriate dosage form for older adults who are unable to swallow solids can pose a significant clinical challenge. In these situations, small amounts of liquids or soft foods that are shown to not alter the performance of the drug product can be used as described in the FDA-approved product labeling for immediate ingestion as the suitable vehicle(s) for oral administration of specific drug products. See the following regulatory document[60]:

- Use of Liquids and/or Soft Foods as Vehicles for Drug Administration: General Considerations for Selection and In Vitro Methods for Product Quality Assessments, 2018

The methodology to measure drug(s) or biologic product(s) in biological matrices for pharmacokinetic study and to measure biomarker(s) for pharmacodynamic study is essential to quantitate drug exposure and drug response. The following regulatory document provides guidance to validate bioanalytical methodologies[61]:

- Bioanalytical Method Validation, 2018

Many drug products are not small drug molecules but are biologics. The website of FDA's Center for Biologics Evaluation and Research contains useful information on developing biologic products.[62] The following three regulatory guidances provide some examples to guide development of biologic products[63–65]:

- Assay Development and Validation for Immunogenicity Testing of Therapeutic Protein Products, 2016
- Immunogenicity Assessment for Therapeutic Protein Products, 2014
- Clinical Pharmacology Data to Support a Demonstration of Biosimilarity to a Reference Product, 2016.

Organ impairments are common among older adult patients. The first two of the following regulatory documents provide guidance to conduct dedicated pharmacokinetic studies to assess the effect of either renal impairment or hepatic impairment on drug exposure. The third of the following regulatory document provides guidance to conduct population pharmacokinetic studies to assess the effect of organ impairment and other contributing factors (such as age, ethnicity, gender, and concomitant medications) on drug exposure through covariate analyses[66–68]:

- Pharmacokinetics in Patients with Impaired Renal Function: Study Design, Data Analysis, and Impact on Dosing and Labeling, 2010
- Pharmacokinetics in Patients with Impaired Hepatic Function: Study Design, Data Analysis, and Impact on Dosing and Labeling, 2003
- Population Pharmacokinetics, 1999.

There is a publication that complements the draft renal impairment study guidance on the regulatory aspect of designing pharmacokinetic studies for patients with chronic kidney disease.[69] Readers can refer to an article on the experience of applying the FDA draft renal impairment study guidance.[70] Readers can also refer to three articles on the population pharmacokinetic and pharmacodynamic modeling for drug development.[71–73]

Multiple drug use is common in older adult patients and prone to adverse drug interactions. The first two of the following regulatory documents provide guidance to conduct in vitro and clinical studies to assess the drug interaction potential of coadministered drugs. The third of the following regulatory document provides guidance on physiologically based pharmacokinetic (PBPK) analysis that can be applied to assess potential drug interactions[74–77]:

- In Vitro Metabolism- and Transporter- Mediated Drug-Drug Interaction Studies, 2017
- Clinical Drug Interaction Studies – Study Design, Data Analysis, and Clinical Implications, 2017
- Physiologically Based Pharmacokinetic (PBPK) Analyses – Format and Content, 2018.

Readers can refer to two articles on the PBPK modeling for drug development.[77,78] Scientists have applied the PBPK approach to predict the effect of renal impairment on the exposure of drugs to inform dosing recommendations in these patients.[79]

Drug product labeling is fundamental to the proper prescribing and use of medications. Besides the guidance for geriatric labeling mentioned above, the following four regulatory documents further provide guidance to effectively label the indications, adverse reactions, clinical pharmacology, and clinical studies sections of the drug product label[80–83]:

- Indications and Usage Section of Labeling for Human Prescription Drug and Biological Products – Content and Format, 2018
- Adverse Reactions Section of Labeling for Human Prescription Drug and Biological Products – Content and Format, 2006

- Clinical Pharmacology Section of Labeling for Human Prescription Drug and Biological Products, 2016
- Clinical Studies Section of Labeling for Human Prescription Drug and Biological Products – Content and Format, 2006.

A drug product may include a device for administration such as an injectable drug product. The device component for the drug product needs the FDA's Center for Devices and Radiological Health (CDRH) to evaluate the human factors or usability engineering of the drug product for marketing approval.[84] The CDRH section of the FDA website provides guidance on the design and documentation of the device as well as information for patients and care-givers.

26.7 COMPRESSION OF MORBIDITY TO THE END OF LIFE IS THE EVENTUAL GOAL OF HEALTHY AGING AND DRUG TREATMENTS

Healthier aging focuses on the compression of morbidity in older age.[85] The compression of morbidity hypothesis states that the age of onset of chronic illness may be postponed more than the age at death, squeezing most of the morbidity in life into a shorter period with less lifetime disability.[86,87] Thus, the ideal goal of healthier aging and drug treatment should help older adults maintain physical independence and biological reserve as well as enjoy psychosocial well-being well into older ages.

26.8 FUTURE DIRECTIONS

One of the workshops in the 2016 Annual Meeting of the American College of Clinical Pharmacology discussed the future directions of drug development for the older adults.[88] The following summarizes this workshop's discussions.

26.8.1 SHARING INFORMATION, DATA, AND MODELS

Sharing relevant information, data, and models to increase the knowledge base in size and quality requires initiative but offers the promise to greatly benefit the scientific community among academia, industry, and regulators for the patients. Besides the scientific community, patient initiatives may play an important role. Although these efforts are frequently structured around a disease or indication and relevant for pharmacodynamic considerations, age dependency of pharmacokinetics is primarily not indication dependent. As data under discussion may contain sensitive information, regulatory guidance on data protection, appropriate informed consent, and joined efforts together with patient initiatives toward legal hurdles are important although challenging.

26.8.2 PHARMACOMETRICS AND QUANTITATIVE SYSTEMS PHARMACOLOGY APPROACHES FROM BENCH TO BEDSIDE

Although pharmacometrics and systems pharmacology are increasingly used to analyze complex intertwined pharmacokinetic and pharmacodynamic models, information on covariate and combinatory age-dependent effects is limited. The application of these techniques to geriatric clinical pharmacology and drug development is still being explored and generally involves assumptions. However, these techniques can guide systematic evaluation and generate expectations via scenario or sensitivity analysis and should be used to make clinical trials more uniformly successful. Information technology and software solutions are rapidly evolving but need medical device standards to serve as decision support or to apply as primary tools for designing and conducting pharmacotherapeutic research and practice. Clear qualification and model acceptance criteria are necessary for an environment of accelerated technological development to the clinical setting. Readers can refer to four articles on the discussion of the role of modeling and simulation in drug development for older adults.[32,89–91]

26.8.3 PEDIATRIC DRUG DEVELOPMENT AS A BLUEPRINT TO SUPPORT OLDER ADULTS RESEARCH

Since the issuance of guidances on pediatric drug product development following the Best Pharmaceuticals for Children Act in 2002 which combines the concept of regulatory requirements with benefits in the form of a 6-month patent extension for conducting pediatric studies, there are many investigations for the young. However, this observation is not yet true for older adults. Consequently, the knowledge and investigation gap between these two age groups is widening. The recent activities to provide better care for the young may serve as a blueprint to improve care for older adults. However, pediatric and older adult patients have very different needs. Nevertheless, the solutions to meet these needs may use similar approaches. Understanding these different needs and developing suitable scientific methodology to qualify and quantify the impact of different drug development approaches are required to evaluate potential solutions. Readers can refer to an article on the discussion of the Geriatric Investigation Plan.[92]

26.8.4 EARLY CONSIDERATIONS IN DRUG DEVELOPMENT AND REGULATORY ALIGNMENT

The common concern that patient centric products and product design would increase the development times and manufacturing costs. In principle, drug products are developed for the concerned patient population, who is known at the very early stages of the development and can be addressed already before the drug product is being developed. As solutions to patient needs are similar between different patient populations, patient centric drug product design is not expected to increase the number of different product presentations. In contrast, it is expected to lead to a more "universal" design that is appropriate for patients with and without specific needs. Readers can refer to four articles on the development of patient centric drug products.[93–96] In addition, the early planning and communication between the sponsor and the regulators are essential for the success of the drug development program.

26.9 SUMMARY

Persons aged 65 years and above will be the fastest-growing segment of the US population in the next four decades. Aging affects both pharmacokinetics and pharmacodynamics of the medications that older adults consume. For certain drugs, the pharmacodynamic changes associated with aging may be more clinically relevant than the pharmacokinetic changes associated with aging. Because older adults have more comorbidities and they need concurrent multiple pharmacotherapies for their multimorbidities, which can lead to polypharmacy and/or potentially inappropriate medications. Deprescribing is an emerging approach to manage polypharmacy. However, a multiprofessional team approach is recommended to optimize drug therapy in older adult patients. Underrepresentation of older adults in clinical trials is common. Recent studies showed signs of improvement in older adults participating in clinical trials. Many measures can be implemented to encourage enrollment of older adults in clinical trials. The ICH E7 document and its Questions and Answers guide clinical studies for older adults. The FDA also publishes a multitude of guidances for the industry that are applicable for drug development in older adults. The ideal goal of healthier aging and drug treatment should help older adults maintain physical independence and biological reserve as well as enjoy psychosocial well-being into older ages. The future directions for drug development in older adults may include

- Sharing of data among academia, industry, regulators, and patients
- Modeling and simulation will play an important role
- Pediatric drug development may serve as an example
- Communicating with the regulators early in the drug development program.

DISCLAIMER

The views expressed in this chapter are the personal views of the authors and may not be understood or quoted as being made on behalf of or reflecting the position of the FDA.

REFERENCES

1. Ortman, JM, Velkoff VA, Hogan, H. An aging nation: The older population in the United States, Current Population Reports, P25-1140. U.S. Census Bureau, Washington, DC. 2014.
2. Mather M, Jacobsen LA, Pollard KM. 2015. Aging in the United States. *Popul Bull* 70. www.prb.org/pdf16/ging-us-population-bulletin.pdf.
3. United Nations. Department of Economic and Social Affairs, Population Division. 2013. World population ageing 2013. ST/ESA/SER.A/348.
4. Weiss CO, Boyd CM, Yu Q, Wolff JL, Leff B. 2007. Patterns of prevalent major chronic disease among older adults in the United States. *JAMA* 298:1160–2.
5. Boyd CM, Leff B, Wolff JL, Yu Q, Zhou J, Rand C, Weiss CO. 2011. Informing clinical practice guideline development and implementation: Prevalence of coexisting conditions among adults with coronary heart disease. *J Am Geriatr Soc* 59:797–805.
6. Qato DM, Alexander GC, Conti RM, Johnson M, Schumm P, Lindau ST. 2008. Use of prescription and over-the-counter medications and dietary supplements among older adults in the United States. *JAMA* 300:2867–78.
7. Reeve E, Wiese MD, Mangoni AA. 2015. Alterations in drug disposition in older adults. *Expert Opin Drug Metab Toxicol* 11:491–508.
8. Shi S, Klotz U. 2011. Age-related changes in pharmacokinetics. *Curr Drug Metab* 12:601–10.
9. Trifirò G, Spina E. 2011. Age-related changes in pharmacodynamics: Focus on drugs acting on central nervous and cardiovascular systems. *Curr Drug Metab* 12:611–20.
10. Stegemann S, Gosch M, Breitkreutz J. 2012. Swallowing dysfunction and dysphagia is an unrecognized challenge for oral drug therapy. *Int J Pharm* 430:197–206.
11. Quinn HL, Hughes CM, Donnelly RF. 2016. Novel methods of drug administration for the treatment and care of older patients. *Int J Pharm* 512:366–73.
12. Borkan GA, Hults DE, Gerzof SG, Robbins AH, Silbert CK. 1983. Age changes in body composition revealed by computed tomography. *J Gerontol* 38:673–7.
13. McLean AJ, Le Couteur DG. 2004. Aging biology and geriatric clinical pharmacology. *Pharmacol Rev* 56:163–84.
14. Benet LZ, Hoener BA. 2002. Changes in plasma protein binding have little clinical relevance. *Clin Pharmacol Ther* 71:115–21.
15. Le Couteur DG, McLean AJ. 1998. The aging liver. Drug clearance and an oxygen diffusion barrier hypothesis. *Clin Pharmacokinet* 34:359–73.
16. Schmucker DL. 2001. Liver function and phase I drug metabolism in the elderly: A paradox. *Drugs Aging* 18:837–51.
17. Bowie MW, Slattum PW. 2007. Pharmacodynamics in older adults: A review. *Am J Geriatr Pharmacother* 5:263–303.
18. Morin L, Johnell K, Laroche ML, Fastbom J, Wastesson JW. 2018. The epidemiology of polypharmacy in older adults: Register-based prospective cohort study. *Clin Epidemiol* 10:289–98.
19. Johnell K. 2018. The controversies surrounding polypharmacy in old age: Where are we? *Expert Rev Clin Pharmacol* 11:825–27.
20. Morin L, Laroche ML, Texier G, Johnell K. 2016. Prevalence of potentially inappropriate medication use in older adults living in nursing homes: A systematic review. *J Am Med Dir Assoc* 17:862.e1–9.
21. Maher RL, Hanlon J, Hajjar ER. 2014. Clinical consequences of polypharmacy in elderly. *Expert Opin Drug Saf* 13:57–65.

22. Gutiérrez-Valencia M, Izquierdo M, Cesari M, Casas-Herrero Á, Inzitari M, Martínez-Velilla N. 2018. The relationship between frailty and polypharmacy in older people: A systematic review. *Br J Clin Pharmacol* 84:1432–44.

23. Clegg A, Young J, Iliffe S, Rikkert MO, Rockwood K. 2013. Frailty in elderly people. *Lancet* 381:752–62.

24. Scott IA, Hilmer SN, Reeve E, Potter K, Le Couteur D, Rigby D, Gnjidic D, Del Mar CB, Roughead EE, Page A, Jansen J, Martin JH. 2015. Reducing inappropriate polypharmacy: The process of deprescribing. *JAMA Intern Med* 175:827–34.

25. Mallet L, Spinewine A, Huang A. 2007. The challenge of managing drug interactions in elderly people. *Lancet* 370:185–91.

26. Lewis JH, Kilgore ML, Goldman DP, Trimble EL, Kaplan R, Montello MJ, Housman MG, Escarce JJ. 2003. Participation of patients 65 years of age and older in cancer clinical trials. *J Clin Oncol* 21:1383–9.

27. Cohen-Mansfield J. 2002. Recruitment rates in gerontological research: The situation for drug trials in dementia may be worse than previously reported. *Alzheimer Dis Assoc Disord* 16:279–82.

28. Leppik IE. 2006. Epilepsy in the elderly. *Epilepsia* 47 (Suppl 1):65–70.

29. Morse AN, Labin LC, Young SB, Aronson MP, Gurwitz JH. 2004. Exclusion of elderly women from published randomized trials of stress incontinence surgery. *Obstet Gynecol* 104:498–503.

30. Blosser CD, Huverserian A, Bloom RD, Abt PD, Goral S, Thomasson A, Shults J, Reese PP. 2011. Age, exclusion criteria, and generalizability of randomized trials enrolling kidney transplant recipients. *Transplantation* 91:858–63.

31. Lee PY, Alexander KP, Hammill BG, Pasquali SK, Peterson ED. 2001. Representation of elderly persons in published randomized trials of acute coronary syndromes. *JAMA* 286:708–13.

32. Schlender JF, Vozmediano V, Golden AG, Rodriguez M, Samant TS, Lagishetty CV, Eissing T, Schmidt S. 2018. Current strategies to streamline pharmacotherapy for older adults. *Eur J Pharm Sci* 111:432–42.

33. Hernandez-Diaz S, Garcia-Rodriguez LA. 2001. Epidemiologic assessment of the safety of conventional nonsteroidal anti-inflammatory drugs. *Am J Med* 110 (Suppl 3A):20S–7S.

34. Hylek EM, Evans-Molina C, Shea C, Henault LE, Regan S. 2007. Major hemorrhage and tolerability of warfarin in the first year of therapy among elderly patients with atrial fibrillation. *Circulation* 115:2689–96.

35. Welch HG, Albertsen PC, Nease RF, Bubolz TA, Wasson JH. 1996. Estimating treatment benefits for the elderly: The effect of competing risks. *Ann Intern Med* 124:577–84.

36. Chen A, Wright H, Itana H, Elahi M, Igun A, Soon G, Pariser AR, Fadiran EO. 2018. Representation of women and minorities in clinical trials for new molecular entities and original therapeutic biologics approved by FDA CDER from 2013 to 2015. *J Womens Health (Larchmt)* 27:418–29.

37. U.S. Food and Drug Administration. Guidance for industry: E7 studies in support of special populations: Geriatrics, questions and answers. International Conference on Harmonisation of Technical Requirements for Registration of Pharmaceuticals for Human Use: ICH. 2012. www.fda.gov/downloads/Drugs/GuidanceComplianceRegulatoryInformation/Guidances/UCM189544.pdf.

38. Gnjidic D, Hilmer SN, Blyth FM, Naganathan V, Waite L, Seibel MJ, McLachlan AJ, Cumming RG, Handelsman DJ, Le Couteur DG. 2012. Polypharmacy cutoff and outcomes: Five or more medicines were used to identify community-dwelling older men at risk of different adverse outcomes. *J Clin Epidemiol* 65:989–95.

39. Shenoy P, Harugeri A. 2015. Elderly patients' participation in clinical trials. *Perspect Clin Res* 6:184–9.

40. Hilmer SN, McLachlan AJ, Le Couteur DG. 2007. Clinical pharmacology in the geriatric patient. *Fundam Clin Pharmacol* 21:217–30.

41. Smalley E. 2018. Clinical trials go virtual, big pharma dives in. *Nat Biotechnol* 36(7):561–2.

42. Sommer C, Zuccolin D, Arnera V, Schmitz N, Adolfsson P, Colombo N, Gilg R, McDowell B. 2018. Building clinical trials around patients: Evaluation and comparison of decentralized and conventional site models in patients with low back pain. *Contemp Clin Trials Commun* 11:120–6.

43. U.S. Food and Drug Administration. Code of Federal Regulation Title 21 Part 314 Section 50. 1986. www.accessdata.fda.gov/scripts/cdrh/cfdocs/cfcfr/CFRSearch.cfm?fr=314.50.

44. U.S. Food and Drug Administration. Guideline for the format and content of the clinical and statistical sections of an application. 1988. www.fda.gov/downloads/Drugs/GuidanceComplianceRegulatoryInformation/Guidances/UCM071665.pdf.

45. U.S. Food and Drug Administration. Guidance for industry: Guideline for the study of drugs likely to be used in the elderly. 1989. www.fda.gov/downloads/Drugs/GuidanceComplianceRegulatoryInformation/Guidances/ucm072048.pdf.

46. U.S. Food and Drug Administration. Guideline for industry: E7 studies in support the of special populations: Geriatrics. International Conference on Harmonisation of Technical Requirements for Registration of Pharmaceuticals for Human Use: ICH. 1994. www.fda.gov/downloads/Drugs/GuidanceComplianceRegulatoryInformation/Guidances/UCM073131.pdf.

47. Cerreta F, Temple R, Asahina Y, Connaire C. 2015. Regulatory activities to address the needs of older patients. *J Nutr Health Aging* 19:232–3.

48. U.S. Food and Drug Administration. 1997. Geriatric use subsection (62 FR 45313). *Fed Regist* 62:45313–45326.

49. U.S. Food and Drug Administration. 1998. Final rule. 63 FR 6854. 21 CFR 312 and 21 CFR 314: Investigational new drug applications and new drug applications (demographic rule). *Fed Regist* 63:6854–62.

50. U.S. Food and Drug Administration. Guidance for industry: Content and format for geriatric labeling. 2001. www.fda.gov/downloads/Drugs/GuidanceComplianceRegulatoryInformation/Guidances/ucm075062.pdf.

51. Code of Federal Regulations. Title 21, Section 201.56: Requirements on content and format of labeling for human prescription drug and biological products. 2006. www.gpo.gov/fdsys/pkg/CFR-2012-title21-vol4/pdf/CFR-2012-title21-vol4-sec201-56.pdf.

52. U.S. Food and Drug Administration Safety and Innovation Act. Public Law 112–144, July 9, 2012. www.gpo.gov/fdsys/pkg/PLAW-112publ144/pdf/PLAW-112publ144.pdf.

53. U.S. Food and Drug Administration. Report: Collection, analysis, and availability of demographic subgroup data for FDA-approved medical products. 2013. www.fda.gov/downloads/RegulatoryInformation/Legislation/FederalFoodDrugandCosmeticActFDCAct/SignificantAmendmentstotheFDCAct/FDASIA/UCM365544.pdf.

54. U.S. Food and Drug Administration. Report: FDA action plan to enhance the collection and availability of demographic subgroup data. 2014. www.fda.gov/downloads/RegulatoryInformation/Legislation/FederalFoodDrugandCosmeticActFDCAct/SignificantAmendmentstotheFDCAct/FDASIA/UCM410474.pdf.

55. Drug Trials Snapshots. www.fda.gov/Drugs/InformationOnDrugs/ucm412998.htm.

56. Eshera N, Itana H, Zhang L, Soon G, Fadiran EO. 2015. Demographics of clinical trials participants in pivotal clinical trials for new molecular entity drugs and biologics approved by FDA from 2010 to 2012. *Am J Ther* 22:435–55.

57. U.S. Food and Drug Administration. Webpage on guidances for drugs. www.fda.gov/Drugs/GuidanceComplianceRegulatoryInformation/Guidances/default.htm.

58. U.S. Food and Drug Administration. Guidance for industry: Providing clinical evidence of effectiveness for human drug and biological products. 1998. www.fda.gov/downloads/Drugs/GuidanceComplianceRegulatoryInformation/Guidances/UCM072008.pdf.

59. U.S. Food and Drug Administration. Guidance for industry: Exposure-response relationships: Study design, data analysis, and regulatory applications. 2005. www.fda.gov/downloads/Drugs/GuidanceComplianceRegulatoryInformation/Guidances/UCM072109.pdf.

60. U.S. Food and Drug Administration. Guidance for industry: Use of liquids and/or soft foods as vehicles for drug administration: General considerations for selection and in vitro methods for product quality assessments. Draft guidance. 2018. www.fda.gov/downloads/Drugs/GuidanceComplianceRegulatoryInformation/Guidances/UCM614401.pdf.

61. U.S. Food and Drug Administration. Guidance for industry: Bioanalytical method validation. 2018. www.fda.gov/downloads/Drugs/GuidanceComplianceRegulatoryInformation/Guidances/UCM070107.pdf.

62. U.S. Food and Drug Administration. Biologics development. www.fda.gov/BiologicsBloodVaccines/GuidanceComplianceRegulatoryInformation/Guidances/General/default.htm.

63. U.S. Food and Drug Administration. Guidance for industry: Assay development and validation for immunogenicity testing of therapeutic protein products. Draft guidance. 2016. www.fda.gov/downloads/Drugs/GuidanceComplianceRegulatoryInformation/Guidances/UCM192750.pdf.

64. U.S. Food and Drug Administration. Guidance for industry: Immunogenicity assessment for therapeutic protein products. 2014. www.fda.gov/downloads/drugs/guidances/ucm338856.pdf.

65. U.S. Food and Drug Administration. Guidance for industry: Clinical pharmacology data to support a demonstration of biosimilarity to a reference product. Draft guidance. 2016. www.fda.gov/downloads/Drugs/GuidanceComplianceRegulatoryInformation/Guidances/UCM397017.pdf.

66. U.S. Food and Drug Administration. Guidance for industry: Pharmacokinetics in patients with impaired renal function: Study design, data analysis, and impact on dosing and labeling. Draft Guidance. 2010. www.fda.gov/downloads/Drugs/GuidanceComplianceRegulatoryInformation/Guidances/UCM204959.pdf.

67. U.S. Food and Drug Administration. Guidance for industry: Pharmacokinetics in patients with impaired hepatic function: Study design, data analysis, and impact on dosing and labeling. 2003. www.fda.gov/downloads/Drugs/GuidanceComplianceRegulatoryInformation/Guidances/UCM072123.pdf.

68. U.S. Food and Drug Administration. Guidance for industry: Population pharmacokinetics. 1999. www.fda.gov/downloads/Drugs/GuidanceComplianceRegulatoryInformation/Guidances/UCM072137.pdf.

69. Zhang L, Xu N, Xiao S, Arya V, Zhao P, Lesko LJ, Huang SM. 2012. Regulatory perspectives on designing pharmacokinetic studies and optimizing labeling recommendations for patients with chronic kidney disease. *J Clin Pharmacol* 52(Suppl 1):79S–90S.

70. Paglialunga S, Offman E, Ichhpurani N, Marbury TC, Morimoto BH. 2017. Update and trends on pharmacokinetic studies in patients with impaired renal function: Practical insight into application of the FDA and EMA guidelines. *Expert Rev Clin Pharmacol* 10:273–83.

71. Mould DR, Upton RN. 2012. Basic concepts in population modeling, simulation, and model-based drug development. *CPT Pharmacometrics Syst Pharmacol* 1:e6.

72. Mould DR, Upton RN. 2013. Basic concepts in population modeling, simulation, and model-based drug development: Part 2-introduction to pharmacokinetic modeling methods. *CPT Pharmacometrics Syst Pharmacol* 2:e38.

73. Upton RN, Mould DR. 2014. Basic concepts in population modeling, simulation, and model-based drug development: Part 3-introduction to pharmacodynamic modeling methods. *CPT Pharmacometrics Syst Pharmacol* 3:e88.

74. U.S. Food and Drug Administration. Guidance for industry: In vitro metabolism- and transporter-mediated drug-drug interaction studies. Draft Guidance. 2017. www.fda.gov/downloads/Drugs/GuidanceComplianceRegulatoryInformation/Guidances/UCM581965.pdf.

75. U.S. Food and Drug Administration. Guidance for industry: Clinical drug interaction studies: Study design, data analysis, and clinical implications. Draft Guidance. 2017. www.fda.gov/downloads/Drugs/GuidanceComplianceRegulatoryInformation/Guidances/UCM292362.pdf.

76. U.S. Food and Drug Administration. Guidance for industry: Physiologically based pharmacokinetic analyses: Format and content. 2018. www.fda.gov/downloads/Drugs/GuidanceComplianceRegulatoryInformation/Guidances/UCM531207.pdf.

77. Jones H, Rowland-Yeo K. 2013. Basic concepts in physiologically based pharmacokinetic modeling in drug discovery and development. *CPT Pharmacometrics Syst Pharmacol* 2:e63.

78. Kuepfer L, Niederalt C, Wendl T, Schlender JF, Willmann S, Lippert J, Block M, Eissing T, Teutonico D. 2016. Applied concepts in PBPK modeling: How to build a PBPK/PD model. *CPT Pharmacometrics Syst Pharmacol* 5:516–31.

79. Yee KL, Li M, Cabalu T, Sahasrabudhe V, Lin J, Zhao P, Jadhav P. 2018. Evaluation of model-based prediction of pharmacokinetics in the renal impairment population. *J Clin Pharmacol* 58:364–76.

80. U.S. Food and Drug Administration. Guidance for industry: Indications and usage section of labeling for human prescription drug and biological products: Content and format. Draft Guidance. 2018. www.fda.gov/downloads/Drugs/GuidanceComplianceRegulatoryInformation/Guidances/UCM612697.pdf.

81. U.S. Food and Drug Administration. Guidance for industry: Adverse reactions section of labeling for human prescription drug and biological products: Content and format. 2006. www.fda.gov/downloads/Drugs/GuidanceComplianceRegulatoryInformation/Guidances/UCM075057.pdf.

82. U.S. Food and Drug Administration. Guidance for industry: Clinical pharmacology section of labeling for human prescription drug and biological products: Content and format. 2016. www.fda.gov/downloads/Drugs/GuidanceComplianceRegulatoryInformation/Guidances/UCM109739.pdf.

83. U.S. Food and Drug Administration. Guidance for industry: Clinical studies section of labeling for human prescription drug and biological products: Content and format. 2006. www.fda.gov/downloads/Drugs/GuidanceComplianceRegulatoryInformation/Guidances/UCM075059.pdf.

84. U.S. Food and Drug Administration. Human factors and medical devices. www.fda.gov/MedicalDevices/DeviceRegulationandGuidance/HumanFactors/default.htm.

85. Myint PK, Welch AA. 2012. Healthier ageing. *BMJ* 344:42–5.

86. Fries JF. 1980. Aging, natural death, and the compression of morbidity. *N Engl J Med* 303:130–5.

87. Fries JF, Bruce B, Chakravarty E. 2011.Compression of morbidity 1980-2011: A focused review of paradigms and progress. *J Aging Res* 2011:1–10.

88. Lau SWJ, Schlender JF, Abernethy DR, Burckart GJ, Golden A, Slattum PW, Stegemann S, Eissing T. 2018. Improving therapeutics to better care for older adults and the young: Report from the American College of Clinical Pharmacology workshop. *J Clin Pharmacol* 58:277–80.

89. Harnisch L, Shepard T, Pons G, Della Pasqua O. 2013. Modeling and simulation as a tool to bridge efficacy and safety data in special populations. *CPT Pharmacometrics Syst Pharmacol* 2:e28.

90. Saeed MA, Vlasakakis G, Della Pasqua O. 2015. Rational use of medicines in older adults: Can we do better during clinical development? *Clin Pharmacol Ther* 97:440–3.

91. Younis IR, Robert Powell J, Rostami-Hodjegan A, Corrigan B, Stockbridge N, Sinha V, Zhao P, Jadhav P, Flamion B, Cook J. 2017. Utility of model-based approaches for informing dosing recommendations in specific populations: Report from the public AAPS workshop. *J Clin Pharmacol* 57:105–9.

92. De Spiegeleer B, Wynendaele E, Bracke N, Veryser L, Taevernier L, Degroote A, Stalmans S. 2016. Regulatory development of geriatric medicines: To GIP or not to GIP? *Ageing Res Rev* 27:23–36.

93. Breitkreutz J, Boos J. 2007. Paediatric and geriatric drug delivery. *Expert Opin Drug Deliv* 4:37–45.

94. Hersberger KE, Boeni F, Arnet I. 2013. Dose-dispensing service as an intervention to improve adherence to polymedication. *Expert Rev Clin Pharmacol* 6:413–21.

95. Hanning SM, Lopez FL, Wong IC, Ernest TB, Tuleu C, Orlu Gul M. 2016. Patient centric formulations for paediatrics and geriatrics: Similarities and differences. *Int J Pharm* 512:355–9.

96. Stegemann S. 2018. Patient centric drug product design in modern drug delivery as an opportunity to increase safety and effectiveness. *Expert Opin Drug Deliv* 15:619–27.

27 Clinical Pharmacology and Regulatory Concerns for Developing Drug Products for Pediatric Patients

Janelle M. Burnham and Gilbert J. Burckart
US Food and Drug Administration

CONTENTS

27.1 INTRODUCTION

The understanding of the use of medicines for children over the past 50 years has dramatically increased. However, it has taken many decades to truly understand the science behind pediatric diseases, determine the best drug therapy to treat the disease, plan for effective dosing in children, and design a neonatal or pediatric clinical trial that would appropriately establish safety and efficacy of the drug. The regulatory process for including children in drug development studies has taken a parallel slow and plodding process. Together, the science and the regulatory process have made real progress in the understanding of the dosing, efficacy, and safety of drug use in pediatric patients since 1997. The complete timeline for pediatric drug development is provided in Table 27.1.

27.1.1 PEDIATRIC DRUG DEVELOPMENT AND THE SCIENCE OF CLINICAL PHARMACOLOGY

Pediatric drug development essentially did not exist in the 1950s to 1970s, and the inclusion of children in adult clinical trials or drug development rarely occurred. However, as new medicines were developed for adults, pediatric healthcare providers had to use drugs without FDA labeling (off label) which resulted in inaccurate dosing, ineffective therapy, and adverse effects.[1] A pharmacist and pediatrician, Dr. Harry Shirkey, in 1963 expressed that "By an odd twist of fate, infants

TABLE 27.1

A Timeline for Pediatric Drug Development

Date	Event
1963	Dr. Harry Shirkey introduced the term "Therapeutic Orphans"
1974	AAP report, commissioned by FDA, titled
	General Guidelines for the Evaluation of Drugs to be Approved for Use During Pregnancy and for Treatment of Infants and Children
1977	FDA Pediatric Guidance titled "General Considerations for the Clinical Evaluation of Drugs in Infants and Children"
1979	FDA adds the Pediatric Use Subsection to the Package Insert
1992	"Better Pharmaceuticals for Children Act" introduced in Congress
1994	Extrapolation of efficacy from adults to pediatrics is added to Pediatric Use
1997	FDA Modernization Act with initial pediatric incentive program (5 years sunset)
2002	BPCA renewed for 5 years
2002	1998 FDA Pediatric Rule was enjoined
2003	PREA passed to codify Pediatric Rule
2007	FDA Amendments Act, BPCA/PREA renewed and the FDA Pediatric Review Committee formed by the act
2012	FDA Safety and Innovation Act makes BPCA and PREA permanent
2017	FDA Reauthorization Act (FDARA) authorizes PREA to include pediatric Oncology agents based on molecular target, starting in 2020

and children are becoming therapeutic or pharmaceutical orphans."[2] He recognized the lack of inclusion of children in clinical studies and strived to have children included in future pediatric drug development programs. Dr. Shirkey contributed immensely to pediatric medicine and pharmacy by publishing his 1977 *Pediatric Dosage Handbook* and supporting the science of clinical pharmacology, and versions of this publication are still available. Dr. Shirkey became the Chair of the Committee on Drugs of the American Academy of Pediatrics (AAP) and continued to lobby for the inclusion of pediatric patients in drug development.

Dr. Sumner Yaffe is considered the "Father of Pediatric Clinical Pharmacology" and established the scientific underpinnings that form the basis for pediatric drug development today. From the 1960s through 1970s, Dr. Yaffe collaborated with scientists Dr.'s Gerhard Levy and William Jusko at the State University of New York at Buffalo and conducted studies at Buffalo Children's Hospital. Dr.'s Levy and Jusko established the science of pharmacokinetics which has become pharmacometrics, and Dr. Yaffe worked to apply these concepts to the science of pediatric clinical pharmacology. Additionally, Drs. Yaffe, Levy and Jusko' s research collaboration provided the basis for the modeling and simulation techniques that are frequently used to determine pediatric drug dosing and optimal trial designs.

From the 1970s to the 1990s, several regulatory activities established the ethical basis for the inclusion of children in drug development. The passage of the 1974 National Research Act and Title II of the Act, and the AAP report titled "General Guidelines for Evaluating Drugs Approved for Use During Pregnancy and the Treatment of Infants and Children" helped to protect human subjects enrolled in research but also to pave the regulatory path for inclusion of children in clinical studies and drug development.

27.1.2 FEDERAL REGULATIONS THAT HAVE CHANGED PEDIATRIC DRUG DEVELOPMENT

Table 27.1 provides the timeline for the regulatory actions that have established pediatric drug development. The initial incentive program for conducting pediatric trials requested by the FDA was included in the 1997 FDA Modernization Act, but had a 5-year sunset on the program. In 2002, the legislative passage of the Better Pharmaceuticals for Children Act (BPCA) renewed the incentive program, and in 2003, the Pediatric Research Equity Act (PREA) enacted the requirement to conduct pediatric studies when children had the same indication that was being developed by the sponsor for adults. BPCA offered a 6-month patent extension for conducting pediatric studies requested by the FDA. Additionally, the National Institutes of Health (NIH) conducts research programs and studies off patent drugs used in the pediatric population through the funding and authorization of BPCA.[3] However, PREA is an amendment to the Federal Food and Drug, and Cosmetic Act which requires pediatric studies to be conducted for a drug and biological product if there is a new active ingredient, new indication, new dosage form, or new route of administration. BPCA and PREA had to be renewed every 5 years and were renewed with the FDA Amendment Act of 2007. In 2012, BPCA and PREA became permanent with the passage of the FDA Safety and Innovation Act (FDASIA). In 2017, FDA Reauthorization Act (FDARA) was passed and allowed continuation of pediatric drug development, as well as requires adult cancer products that have a molecular target that is relevant to a pediatric cancer has to have a pediatric investigation.

All of this legislation was essential in requiring drug products to have pediatric studies and aided in increasing the amount of safe and effective medicines available to children. However, it is important to ensure that ethical principles are applied when investigating a pediatric product. As described in the code of federal regulations (CFR) to protect children participating in research, the drug product being studied in children should not present greater than minimal risk to the child.[4] Additionally, conducting a clinical investigation in the pediatric population may have ethical concerns, and if there is risk involved, that risk should be reasonable and anticipated as a direct benefit to the child.[5]

Through the parallel progression of science, regulatory changes, and the application of clinical pharmacology principles such as pharmacokinetics and pharmacodynamics, significant advancement in pediatric drug development has been made since 1997 and has provided positive outcomes for pediatric pharmacotherapy.

27.2 PEDIATRIC STUDY CHALLENGES AND REGULATORY CONCERNS

27.2.1 CLINICAL TRIAL DESIGN AND SUCCESS

During the first 10 years of BPCA studies, neonatal and pediatric drug development studies failed at the rate of 42%.[6] This failure rate is particularly notable since all of these products had demonstrated efficacy in adults. The concern about this failure rate came from considerations that active agents may have been missed due to poor study design, and recruiting pediatric patients into failing trials provided no benefit to the children.

There are several elements that could contribute to the pediatric trial failure rate which include dosing, placebo response, clinical trial design, and differences in the adult and pediatric disease pathophysiology.[7] In the review of trial failures, approximately 25% of the failures were associated with dosing issues. Determination of the appropriate pediatric dose can be challenging for drug developers. For some pediatric conditions, the adult studies for the approved drug products do not provide adequate data for the selection of a pediatric dose. If there is not adequate pediatric data available, then prior information from the adult population or an adult clinical trial can be leveraged to determine a pediatric dose.[8] For children aged 6 years or above, and possibly below 2 years, simple techniques such as allometric scaling are quite effective in determining a pediatric dose based upon the adult dose.[9] Dosing in infants and neonates may be much more complex, and may require tools such as physiologically based pharmacokinetics (PBPK), although experience with PBPK for this purpose is extremely limited at this time. Even the use of population pharmacokinetic (PK) methods to predict neonatal dosing requires studies in the infant population first to improve dose predictions.[10]

Depending on the rarity of the disease or other trial design limitations, the patient sample size can be very small. Therefore, utilizing data information from either the adult population or another pediatric age group has been useful in achieving a range of potential doses. For example, a dosage for canakinumab was able to be determined for periodic fever syndromes in infants down to birth even in the absence of detailed studies in this age group.[11]

27.2.2 EXTRAPOLATION

Extrapolation is a leveraging process whereby a new indication for use in the pediatric population can be supported by existing clinical efficacy data from another studied patient population.[12] Depending on the availability and validity of the efficacy data being leveraged, there are three approaches to the pediatric extrapolation of efficacy. The first approach is a pharmacokinetics (PK) only or "full extrapolation" where the assumption is that similarity in both the adult and pediatric populations exists for disease progression, disease treatment, and exposure-response or concentration-response relations, and that the drug concentration is predictive of a clinical response. Extrapolation of efficacy is possible for certain pediatric conditions where there is sufficient information that can be borrowed from the adult population. A common example is with anti-infective agents that are used in both the adult and pediatric populations. Another example is that clinical drug development programs were able to provide sufficient information to appropriately extrapolate effectiveness of drugs approved for the treatment of adult partial onset seizures for the approval of the drug in the pediatric population aged 4 years and above.[13] Extrapolating efficacy from adult data or other data including older pediatric populations can be used successfully, and decrease the number of pediatric patients required in clinical studies as well as increase the number of drug product approvals in children.

The second extrapolation approach is when there is uncertainty about the disease process being similar in adults and in pediatric patients. In this case, including pharmacokinetics and pharmacodynamics may be beneficial for "partial extrapolation" where the assumption is that the disease and treatment intervention may be similar in both populations, but the exposure-response relation may not be similar or is unknown.

To achieve partial or full extrapolation of efficacy, it is common for drug developers to match adult systemic exposures of drugs to the pediatric population to select an appropriate dose of the drug to use in the pediatric population. In a review of past pediatric clinical studies that used exposure matching for full extrapolation of efficacy in children, it was observed that when extrapolation of efficacy was used, it resulted in a higher percentage of products obtaining new FDA labeling for the pediatric indication compared to drug products that were not able to use extrapolation methods.[14] Additional findings were that a predetermined acceptance range for matching pediatric and adult exposures were not established in most programs and that exposures were poorly matched in many of the programs.

The third approach is involving PK and efficacy or "no extrapolation," where the pediatric population disease progression and/or response to the intervention are not similar to the adult population. Efficacy studies, in addition to the dosing and safety studies required in all programs, are required in a no extrapolation program. Each regulatory agency evaluates multiple types of quantitative methods and is attempting to use all of the existing data in the drug development program to minimize the number of children needed for these studies.

A recent evaluation of pediatric extrapolation has demonstrated that the percentages of pediatric programs under each of the three approaches are changing as our understanding of pediatric diseases is advancing.[15] The initial evaluation published in 2011 had about two-thirds of the programs in the partial extrapolation category, but the three approaches are now evenly divided. Therefore, the percentage of pediatric programs in the no extrapolation group, where the pediatric disease is different from the adult disease, and in the full extrapolation, where the diseases are similar, have both increased.

27.3 PATH TO OVERCOME PEDIATRIC DRUG DEVELOPMENT AND REGULATORY CHALLENGES

27.3.1 ENRICHMENT STRATEGIES IN CLINICAL TRIALS

Enrichment in clinical studies is focusing on a specific patient population that is more likely to respond or have an effect to the investigative drug product. There are several enrichment strategies that can be used to select the optimal sample of patients. Decreasing heterogeneity, prognostic enrichment, and predictive enrichment are three types of enrichment strategies used in pediatric drug development.[16] Clinical trials utilizing a strategy to decrease heterogeneity will select patients that will decrease the inter-patient variability and exclude patients that disease or symptoms improve randomly. When the program decreases heterogeneity in the clinical study, this strategy decreases placebo response. Conversely, prognostic enrichment strategies enroll patients that have a disease-related endpoint event or worsening of a disease which will likely increase the absolute effect difference. Lastly, predictive enrichment strategies are methods that select a patient population that is likely to respond to a drug treatment over another group of patients treated with the same disease. When predictive enrichment is incorporated into a development program, it can result in a larger effect size and allow for a reduced sample size being required.

Enrichment of pediatric drug development studies has been assessed by a group from the FDA.[17] All pediatric submissions were analyzed for the use of enrichment strategies in pediatric trials submitted to the US Food and Drug Administration (FDA) from 2012 to 2016. Seventy-six drug development programs were assessed for their overall success rates. Eighty-eight trials (76.8%) employed at least one enrichment strategy, and 66.3% of those employed multiple enrichment strategies. The highest trial success rates were achieved when all three enrichment strategies (practical, predictive, and prognostic) were used together within a single trial (87.5%), while the lowest success rate was observed when no enrichment strategy was used (65.4%). A drug development program can incorporate one or more enrichment strategies into the clinical trial and has been shown that when enrichment strategies are used in the study, there is a higher likelihood of trial success as compared to a program that did not use enrichment. Therefore, the use of enrichment strategies in pediatric trials was found to be associated with trial and program success.

27.3.2 INCLUSION OF ADOLESCENTS IN CLINICAL STUDIES

The inclusion of adolescents in adult studies was supported by an analysis of dosage similarities between these two patient populations in drug development.[18] This assessment examined 126 products with pediatric studies submitted to the FDA under FDAAA, of which 92 had at least 1 adolescent indication matching an adult indication. Of these 92 products, 87 (94.5%) have equivalent dosing for adults and adolescent patients. For 18 of those 92 products, a minimum weight or body surface area threshold was recommended for adolescents to receive adult dosing.

Since this study was conducted, a routine approach for initiating pediatric studies early in drug development has been to include adolescents in the adult trial whenever possible. Analysis of this approach is still ongoing, but concerns are not related to drug dosing. Instead, concerns have more to do with ethical concerns, since adolescents are still a protected pediatric population and safety concerns. A preliminary analysis found that most of these studies did not include a separate safety analysis for the adolescent population. This is a concern because of developmental concerns during the adolescent period, and analyses that have demonstrated that the incidence of adverse effects for some drug classes can be significantly different between adolescents and adults.[19]

27.3.3 ENDPOINTS AND SURROGATE ENDPOINTS

When clinical trials combine the adult and pediatric population into one clinical study or if the endpoint used in the study is the same for both populations, there is a higher likelihood of trial success.[20] When determining the endpoint for the individual or combined clinical study, consideration of the type of endpoint as either a clinical endpoint or a surrogate is crucial.[21] An endpoint is classified as a clinical outcome if it describes how a patient feels, functions, and survives. However, an endpoint is considered a surrogate if it is a substitute for how a patient feels, functions, and survives. Selecting a primary efficacy endpoint that is reliable, well defined, and interpretable is critical for trial success. Endpoints can also be classified as subjective (a scale or a symptom) or objective (laboratory measurement or clinical event). In drug development, the selection of an endpoint for the clinical study that will obtain the necessary information may determine whether a validated clinical outcome can be determined.

The endpoints used in adult drug development studies have frequently been used in multiple studies over years of studies, so are much more likely to be validated and be qualified for the purpose intended. However, the adult endpoint may not be appropriate for pediatric patients, such as in situations where the infant or child cannot communicate their level of discomfort. In these cases where a different endpoint is used in the pediatric trial than was used in the adult trial, these pediatric studies are much more likely to fail.[20] This same study found that there were no differences in pediatric trial failure when endpoints were subjective versus endpoints that were objective.

A few examples of surrogate endpoints accepted by regulatory agencies include blood pressure for antihypertensive medicines and low-density lipoprotein (LDL) cholesterol level for cholesterol-reducing medications. When pediatric endpoints in drug development trials are separated into clinical and surrogate endpoints, there is no difference in study success.[21]

Since multiple sponsors are involved in pediatric drug development trials, the choice of trial design endpoints may not always be based upon knowledge of past use and study success and failure. Therefore, a complete listing of study endpoints used in pediatric drug development trials has been included in the supplemental materials for the published pediatric endpoint analysis.[20]

27.4   REMAINING QUESTIONS AND THE FUTURE OF PEDIATRIC DRUG DEVELOPMENT

27.4.1   REMAINING QUESTIONS

There are a number of difficult questions that remain to be answered in pediatric drug development. One of those questions is about how to apply the concepts of pediatric ontogeny to pediatric drug development. The maturation of receptors, drug-metabolizing enzymes, and drug transporters all take place independently, so that our knowledge related to the disposition of an individual drug is severely lacking. This information usually comes from prior neonatal drug studies, and very few of these studies have been conducted.[22] While we are starting to be able to more accurately estimate renal drug elimination in premature infants and newborns,[23] drug metabolism is much more unpredictable as evidenced by the disparate metabolism of two CYP2C19 drugs in neonates and infants.[24]

Another question is how to apply drug development findings related to drug absorption, organ impairment, and drug interactions from the adult studies to pediatric patients. In each of these instances, we rarely conduct separate studies in pediatric patients for these specific aspects of drug development. For example, for applying drug interaction studies in adults to pediatric patients, a full discussion of the problem and the possibilities for the application of physiologically based pharmacokinetic modeling has been recently published.[25]

A question that is very pertinent to the pediatric population today relates to the epidemic of pediatric obesity, especially in adolescents. This societal problem is leading to an increase in type 2 diabetes mellitus and hypertension in this adolescent population. This need for drug therapy then demands the question of how to dose drugs in the obese adolescent population, and this problem has some unique aspects to it as compared to obesity in the adult population.[26]

Pediatric safety studies will continue to be an issue in pediatric drug development. Given the small pediatric patient population, these studies can be long and expensive. Therefore some pediatric regulatory agencies and people in industry think that adult safety studies are sufficient. However, the potential developmental toxicities of drugs in children cannot be ignored, and the incidence of adverse drug effects in pediatric patients has conclusively been demonstrated to differ considerably from that observed in adults.[19,27]

27.4.2 THE FUTURE OF PEDIATRIC DRUG DEVELOPMENT

Many challenges in pediatric drug development remain, but potential ways to overcome the difficulties include exploring a range of doses, integration of enrichment strategies, modeling and simulation, and determining optimal clinical trial design elements prior to conducting the trial. With experience and knowledge of trial design elements, drug developers, academicians, and regulators strive to determine the optimal pediatric clinical study that would be able to demonstrate the products safety and effectiveness. The failure rate for pediatric drug development trials has now fallen to about 20% from its high of 42% in a 10-year period of time. The routine inclusion of pediatric clinical trial simulation and an expansion of pediatric extrapolation can hopefully take the study failure rate down to 10%.

One of the opportunities for understanding pediatric drug development may come in surveying orphan drug studies in which pediatric patients were included. When a drug gets orphan designation (for a disease with <200,000 patients in the United States), it precludes PREA so that none of the prior assessments of pediatric studies mentioned in this chapter included orphan drugs. The exception may be studied under BPCA, since orphan drug products can still obtain a written request for studies under BPCA. Over 100 orphan drug development studies that include pediatric patients have been conducted since the year 2000.

Additionally, pediatric oncology drug development is an area that has expanded immensely over the past decade. With the passage of FDARA, the FDA has recently expanded the oncology drug development program by creating the US FDA Oncology Center of Excellence. This new regulation will allow the possibility to evaluate more products for the pediatric population and potentially increase the number of safe and effective medicines available for children with cancer.

One of the future challenges for pediatric drug development will come from sponsors and regulators who would like to label all drugs approved in adults for pediatric use, without adequate efficacy or safety studies. This is not surprising, since the economic reasons against conducting these pediatric studies prevented pediatric drug development from becoming a reality for more than 50 years. However, the science of pediatric drug development is now getting very efficient, and our society understands the need for providing safe and effective drug therapy for our children. Therefore, the future of pediatric drug development is very positive.[28]

REFERENCES

1. Wilson JT. An update on the therapeutic orphan. *Pediatrics.* 1999;104(3):585–590.
2. Shirkey H. Editorial comment: Therapeutic orphans. *Pediatrics.* 1999;104:583–584.
3. Ren Z, Zajicek A. Review of the best pharmaceuticals for children act and the pediatric research equity act: What can the obstetric community learn from the pediatric experience? *Semin Perinatol.* 2015;39(7):530–531.
4. Code of Federal Regulations Title 21 Section 50.51 Subpart D. 1978. www.ecfr.gov/cgi-bin/text-idx?SID=a7636d3aec39152edadf55182832fc4b&mc=true&node=se21.1.50_151&rgn=div8. Accessed September 2018.
5. Nelson R. Ethical considerations in the design and conduct of clinical lactation studies. www.fda.gov/downloads/Drugs/NewsEvents/UCM496973.pdf. Accessed September 2018.
6. Wharton GT, Murphy MD, Avant D, et al. Impact of pediatric exclusivity on drug labeling and demonstrations of efficacy. *Pediatrics.* 2014;134:e512–e518.
7. Momper JD, Mulugeta Y, Burckart GJ. Failed pediatric drug development trials. *Clin Pharmacol Ther.* 2015;98(3):245–251.
8. Abernethy DR, Burckart GJ. Pediatric dose selection. *Clin Pharmacol Ther.* 2010;87(3):270–271.

9. Edginton AN, Shah B, Sevestre M, Momper JD. The integration of allometry and virtual populations to predict clearance and clearance variability in pediatric populations over the age of 6 years. *Clin Pharmacokinet.* 2013;52:693–703.

10. Wang J, Edginton AN, Avant D, Burckart GJ. Predicting neonatal pharmacokinetics from prior data using population pharmacokinetic modeling. *J Clin Pharmacol.* 2015;55(10):1175–1183.

11. Zhuang L, Chen J, Yu J, et al. Dosage considerations for canakinumab in children with periodic fever syndromes. *Clin Pharmacol Ther.* 2018. doi: 10.1002/cpt.1302.

12. Dunne J, Rodriguez WJ, Murphy MD, et al. Extrapolation of adult data and other data in pediatric drug-development programs. *Pediatrics.* 2011;128(5):e1242–e1249.

13. US Food and Drug Administration. Guidance for industry: Drugs for treatment of partial onset seizures: Full extrapolation of efficacy from adults to pediatric patients 4 years of age and older guidance for industry. 2018. Access at: www.fda.gov/downloads/Drugs/GuidanceComplianceRegulatoryInformation/Guidances/UCM596731.pdf (accessed on December 27, 2018).

14. Mulugeta Y, Barrett JS, Nelson R, et al. Exposure matching for extrapolation of efficacy in pediatric drug development. *J Clin Pharmacol.* 2016;56(11):1326–1334.

15. Sun H, Temeck JW, Chambers W, Perkins G, Bonnel R, Murphy D. Extrapolation of efficacy in pediatric drug development and evidence-based medicine: Progress and lessons learned. *Ther Innov Regul Sci.* 2017;2017:1–7.

16. US Food and Drug Administration. Guidance for industry: Enrichment strategies for clinical support approval of human drugs and biological products. 2012. Access at: www.fda.gov/downloads/drugs/guidancecomplianceregulatoryinformation/guidances/ucm332181.pdf (Accessed on December 27, 2018).

17. Green DJ, Liu XI, Hua T, et al. Enrichment strategies in pediatric drug development: An analysis of trials submitted to the US Food and Drug Administration. *Clin Pharmacol Ther.* 2018;104(5):983–988.

18. Momper JD, Mulugeta Y, Green DJ, et al. Adolescent dosing and labeling since the Food and Drug Administration Amendments Act of 2007. *JAMA Pediatr.* 2013;167(10):926–932.

19. Liu XI, Schuette P, Burckart GJ, et al. A comparison of pediatric and adult safety studies for antipsychotic and antidepressant drugs submitted to the US FDA. *J Pediatr.* 2019;208:236–242.e3.

20. Green DJ, Burnham JM, Schuette P, et al. Primary endpoints in pediatric efficacy trials submitted to the US FDA. *J Clin Pharmacol.* 2018;58(7):885–890.

21. Green DJ, Sun H, Burnham J, et al. Surrogate endpoints in pediatric studies submitted to the US FDA. *Clin Pharmacol Ther.* 2018. doi: 10.1002/cpt.1117.

22. Wang J, Avant D, Green D, et al. A survey of neonatal pharmacokinetic and pharmacodynamic studies in pediatric drug development. *Clin Pharmacol Ther.* 2015;98(3):328–335.

23. Wang J, Kumar SS, Sherwin CM, et al. Renal clearance in newborns and infants: Predictive performance of population based modeling for drug development. *Clin Pharmacol Ther.* 2018. doi: 10.1002/cpt.1332.

24. Duan P, Wu F, Moore JN, et al. Assessing CYP2C19 ontogeny in neonates and infants using physiologically based pharmacokinetic models: Impact of enzyme maturation versus inhibition. *CPT Pharmacometrics Syst Pharmacol.* 2018. doi: 10.1002/psp4.12350.

25. Salerno SN, Burckart GJ, Huang SM, Gonzalez D. Pediatric drug-drug interaction studies: Barriers and opportunities. *Clin Pharmacol Ther.* 2018. doi: org/10.1002/cpt.1234.

26. Vaughns JD, Conklin LS, Long Y, et al. Obesity and pediatric drug development. *J Clin Pharmacol.* 2018;58(5):650–661.

27. Momper JD, Chang Y, Jackson M, et al. Adverse event detection and labeling in pediatric drug development: Antiretroviral drugs. *Ther Innovation Regul Sci.* 2015;49(2):302–309.

28. Green DJ, Zineh I, Burckart GJ. Pediatric drug development: Outlook for science-based innovation. *Clin Pharmacol Ther.* 2018;103(3):376–378.

28 Pharmacy Compounding Regulations

Loyd V. Allen, Jr.
International J Pharmaceutical Technology

Willis C. Triplett
W Triplett Consulting LLC

CONTENTS

28.1 INTRODUCTION

Pharmaceutics-based pharmaceutical compounding is an integral part of providing pharmaceuticals and is essential to the provision of contemporary health care.[1] Pharmaceutical compounding has an interesting relationship in the development of new drugs and is supported by the pharmaceutical sciences. Compounding is a professional prerogative that pharmacists have performed since the beginning of the profession. Compounding can be as simple as the addition of a liquid to a manufactured drug powder, or as complex as the preparation of a multicomponent parenteral nutrition solution or a multicomponent trans-dermal gel. In general, compounding differs from manufacturing in that compounding involves a specific practitioner–patient–pharmacist relationship, the preparation of a relatively small quantity of medication, and different conditions of sale (i.e., specific prescription orders). The pharmacist is responsible for compounding preparations of acceptable strength, quality, and purity with appropriate packaging and labeling in accordance with good pharmacy practices, official standards, and current scientific principles (pharmaceutics).

28.2 DEFINITIONS

Pharmaceutical compounding is defined as the act of preparing, mixing, assembling, packaging, and labeling of a drug or device as a result of a practitioner's prescription drug order, or the initiative based on the practitioner–patient–pharmacist relationship in the course of professional practice, or for the purpose of, or as an incident to, research, teaching, or chemical analysis and not for sale or dispensing. Compounding also includes the preparation of drugs or devices in anticipation of prescription drug orders based on routine, regularly observed prescribing patterns.[2–4]

Pharmaceutical manufacturing is defined as the production, preparation, propagation, conversion, and processing of a drug or device, either directly or indirectly, by extraction from substances of natural origin or independently by means of chemical or biological synthesis. This includes any packaging or repackaging of the substance(s), labeling or relabeling of its container, and the promotion and marketing of such drugs or devices. Manufacturing also includes the preparation and promotion of commercially available products from bulk compounds for resale by pharmacies, practitioners, or other people.[2,3]

28.3   RELATIONSHIP BETWEEN COMPOUNDING AND MANUFACTURING

The 5000-year history and heritage of pharmacy has centered around the provision of pharmaceutical products for patients. Pharmacists are the only healthcare professionals who possess the knowledge and skill required for compounding and preparing medications to meet the unique and individual needs of patients. The responsibility of extemporaneously compounding safe, effective prescription products for patients who require special care has always been fundamental to the pharmacy profession.

The 19th century did not see an end to compounding but was impacted by new technology. In the 20th century, a broad knowledge of compounding was still essential for about 80% of the prescriptions dispensed in the 1920s. Although pharmacists increasingly relied on commercially available products for dispensing, and chemicals were purchased from the manufacturers to make up prescriptions, there still remained much to be done *secundum artem.*[5]

The pharmaceutical industry began to assume the production of most medications, and in many ways, this has provided superior service and utilized new methods and a vast array of innovative products that could not have been provided on a one-on-one basis. Research and development have historically been the hallmarks of pharmaceutical manufacturers. However, the very nature of providing millions of doses of a product requires that the dosage forms (such as capsules, tablets, and suppositories) along with the dose amounts (individual strengths of each dose) be limited and result in a unilateral approach to therapy. Today, it is simply not economical for a pharmaceutical company to produce a product in numerous different doses or dosage forms to meet the needs of the entire range of individuals receiving therapy. Products are designed that will meet the majority of patient needs, but the very nature of the process cannot meet all patient needs.

Pharmaceutical compounding has recently been increasing for a number of reasons, including

- Availability of a limited number of dosage forms for most drugs
- A limited number of strengths of most drugs, home health care, and hospice
- The nonavailability of drug products or combinations, discontinued drugs, drug shortages, and orphan drugs
- New therapeutic approaches and special patient populations (such as pediatrics, geriatrics, bioidentical hormone replacement therapy for postmenopausal women, pain management, and dental patients)
- Environmentally and cosmetic-sensitive patients, sports injuries, and veterinary compounding, including small, large, herd, exotic, and companion animals.

Over the years, it has been interesting to note that many compounded products eventually become commercially available products. Recent examples might include fentanyl lozenges, minoxidil topical solution, nystatin lozenges, clindamycin topical solution, tetracaine adrenaline–cocaine (TAC) solution, dihydroergotamine mesylate nasal spray, buprenorphine nasal spray, buffered hypertonic saline solution, and erythromycin topical solution as well as numerous other dermatological and pediatric oral liquids and some premixed intravenous solutions. It is inevitable that a product will be manufactured when it becomes economically profitable for a pharmaceutical manufacturer to produce it. Likewise, when a commercially available product is no longer profitable for a manufacturer, they tend to drop it and pharmacists begin compounding it again.

28.4   ADVERSE COMPOUNDING EVENTS: THE NECC TRAGEDY

In 2012, an outbreak of fungal meningitis was traced to fungal contamination in three lots of compounded methylprednisolone suspension for epidural steroid injections. Doses from those three lots were administered to 14,000 patients. The New England Compounding Center (NECC) case is a tragedy for the 64 individuals that died, the hundreds that were sickened, and their families and loved ones.[6]

In December 2014, the federal government issued an indictment[7] against 14 NECC personnel, including the owners, supervising pharmacists, staff pharmacists, technicians, and the Director of Pharmacy. The seriousness of the charges in the indictment is commensurate with the deaths and injuries allegedly caused by the lapses in compliance with law, quality assurance, and quality control at NECC.

28.4.1 Charges

The charges included

- **Twenty-six charges of second-degree murder:** Second-degree murder is defined as an intentional murder with malice aforethought, but is not premeditated or planned in advance.[8]
- **Mail fraud:** Contaminated and mislabeled vials were sent by interstate courier.
- **Criminal contempt:** Defined as conduct that defies, disrespects, or insults the authority or dignity of a court. Often, contempt takes the form of actions that are seen as detrimental to the court's ability to administer justice.[9] These charges stem from the fact that the owners of NECC transferred millions of dollars in assets after the court issued a restraining order prohibiting transfer of assets.
- **Structuring:** The practice of executing financial transactions in a specific pattern calculated to avoid the creation of certain records and reports required by law. These charges relate to the withdrawal of cash from certain accounts in a manner designed to avoid legal reporting requirements.[10]
- **Racketeering:** The federal crime of conspiring to organize to commit crimes, especially on an ongoing basis as part of an organized crime operation.[11]

The second-degree murder and racketeering charges are based on the allegation that NECC personnel purposely did not comply with applicable laws and United States Pharmacopeia (*USP*) standards. Full compliance with *USP* standards, federal, and state laws is a significant undertaking, even for the majority of compounding pharmacies who are committed to compliance. Lapses in compliance with standards designed to assure the quality and safety of compounded medications, whether purposeful or unintentional, can result in patient harm or death.

The indictment also states

> All compounding personnel were responsible for understanding the fundamental practices and procedures outlined in *USP*-797 for developing and implementing appropriate procedures, and for continually evaluating the procedures and quality of sterile drugs.

This is a critically important point in this indictment. The owners, pharmacist-in-charge (PIC), supervising pharmacists, pharmacists, and technicians were all indicted and cited regarding lapses in compliance with applicable law and *USP* standards. This indictment makes it clear that in the interest of public safety, it is the responsibility of *all* personnel in the pharmacy to comply with quality and safety standards.

28.4.2 Regulatory Lapses

There are two major charges related to lapses in compliance with the Massachusetts Pharmacy Practice Act. These charges relate to (a) failure to obtain patient-specific prescriptions and (b) unlicensed personnel preparing sterile compounds.

28.4.3 Patient-Specific Prescriptions

In the State of Massachusetts, pharmacies may only perform compounding upon receipt of a patient-specific prescription. There are a number of charges related to prescriptions and/or lack thereof:

- Medications were dispensed in bulk for office use without a patient-specific prescription.
- The indictment also notes that the pharmacy falsely claimed to be dispensing all medications on the basis of a patient-specific prescription in order to avoid registering as a manufacturer with the US Food and Drug Administration (FDA).

Pharmacies must comply with the laws of each state in which they do business. Although NECC shipped medications into states that permitted non–patient-specific, office-use compounding, because it was based in Massachusetts, NECC was required to comply with Massachusetts law which prohibits non–patient-specific dispensing. Pharmacies do not get a choice of the laws with which they want to comply; they must comply with *all* applicable laws and regulations of each state in which they do business.

The Drug Quality and Security Act (DQSA), a direct result of the NECC tragedy, specifically prohibits non–patient-specific office use dispensing for traditional compounding pharmacies. Sterile compounding facilities that register as 503B Outsourcing Facilities may dispense non–patient-specific medications provided they meet several requirements, including compliance with Current Good Manufacturing Practices (cGMP).

Although NECC dispensed non–patient-specific medications, it appears the organization was aware of the need for patient-specific prescriptions because the NECC staff is accused of the following:

- Waiving the need for patient-specific prescriptions upon the receipt of an initial order from a particular prescriber.
- Dispensing medications in bulk, and then creating prescriptions after the fact when prescribers would submit lists of patients to whom the medications were administered. Patients' names on these lists were allegedly re-used to create prescriptions for subsequent orders.
- The indictment also alleges that prescriptions were falsified and/or prescribers falsified names. Celebrity, fictional, and office staff names were used to dispense prescriptions. Names such as Wonder Woman, Fat Albert, and Silver Surfer were among the names purportedly used.

Compounders must comply with all applicable laws and regulations. Compounding in anticipation is an appropriate, legal, and often necessary part of pharmacy practice. However, medications that are compounded in anticipation must be dispensed on the basis of a patient-specific prescription. Prescriptions cannot legally be created from patient lists submitted after the medications are dispensed. Of course, there is no ethical or legal rationale for creating or accepting prescriptions with names that are clearly false.

The indictment also charges that although the pharmacy received prescription orders from prescribers, the company's sales force was instructed to let customers know that medications would not actually be labeled with the patient's name, so that they could be used for any patient.

State pharmacy practice acts are all consistent in requiring prescription packages to be properly labeled in accordance with that state's particular requirements. In addition to a regulatory requirement, properly labeled medications are a fundamental pharmacy patient-safety requirement.

28.4.4 Personnel Licensure

The Massachusetts State Board of Pharmacy requires pharmacy technicians to register with the Board. The indictment accuses NECC's pharmacy leadership of permitting an individual who had previously surrendered his technician registration to prepare sterile medications. Furthermore, it also alleges that this individual was allowed to use another employee's name and password to access the computer system in order to hide the fact that he was compounding medications.

28.4.5   LAPSES IN COMPLIANCE WITH *US PHARMACOPEIA* STANDARDS

Of note to compounders is that many of the racketeering charges in the indictment relate to non-compliance with *USP* standards. In fact, the indictment explicitly recognizes that the *USP* sets standards for identity, quality, strength, and purity of medicines. It also notes that Massachusetts requires compliance with *USP* standards. The indictment also includes the two following notes:

All compounding personnel were responsible for understanding the fundamental practices and procedures outlined in *USP*-797 for developing and implementing appropriate procedures, and for continually evaluating the procedures and quality of sterile drugs.

USP-797's standards were meant to prevent harm, including death, to patients that could result from nonsterility of drugs.

The indictment contains a table, which is shown in Table 28.1 within this article, showing a consistent pattern of microbial and mold contamination of air, surfaces, and fingertips for 37 out of 38 weeks in 2012. In regard to microbiological contamination in the cleanroom, the indictment accuses pharmacy personnel of *not* performing the following procedures:

- Evaluating cleanroom procedures
- Conducting an evaluation into the source of the contamination
- Engaging a competent microbiologist, infection control professional, or industrial hygienist to evaluate the mold in cleanroom air, surfaces, and on fingertips

In addition to the charges above, the indictment also contends that the owners of the facility, PIC, supervising pharmacists, and staff pharmacists caused other personnel to

- Emphasize production over cleaning and disinfecting, resulting in these activities being poorly completed or not completed at all.
- Falsify cleaning and disinfecting logs to indicate these activities had been properly completed.

28.4.6   STERILITY TESTING

USP <71> Sterility Tests[6] establishes sterility testing methods, quantities, and volumes that must be tested. For example, for a batch of <100 containers, *USP* <71> requires the pharmacy to test 10% of the batch or four containers, whichever is greater. For a batch of 500 containers or greater, 2% or 20 must be tested, whichever is less. Batches must not be dispensed to patients until they pass the sterility test. The standard *USP* sterility test takes 14 days to return a result.

NECC is accused of routinely testing only one unit per batch regardless of the batch size, and, in some cases, performing no sterility testing at all prior to releasing a batch. In addition, proper sterility testing is performed on filled vials from the batch. NECC is accused of testing from batch containers rather than filled vials. Testing from batch containers is not appropriate because after the sampling is done, additional manipulation and transfers occur to fill the final containers. Contamination can occur during these steps, and while the sample from the batch containers may pass the sterility test, the actual finished vials may be contaminated during the transfer operation.

USP <797> allows the release of a compounding sterile products (CSP) prior to receiving sterility test results, provided that physicians and patients are notified of the potential risk if the testing indicates the medication may be contaminated. This provision is included in *USP* <797> to address emergent medical or medication needs – situations where the benefits outweigh the risks of using a medication that has not passed a sterility test. NECC is accused of failing to notify physicians and patients when medications failed sterility tests.

TABLE 28.1

From NECC Indictment; United States District Court, District of Massachusetts

Weeks	Surface Sampling Action/Alert	Microorganisms Listed	Air Sampling Action/Alert	Gloved-Fingertip Sampling Results
1/06/2012	Action	Mold	Not tested	One technician
1/12/2012	Action	Bacteria	Action	—
1/19/2012	Alert	—	Not tested	—
1/26/2012	Action	Bacteria/mold	Action	—
2/02/2012	Action	Bacteria/mold	Not tested	**(5) EVANOSKY**
2/08/2012	Action	—	Alert	—
2/16/2012	Action	Bacteria/mold	Not tested	—
2/23/2012	Action	Bacteria/mold	Action	**(2) CHIN** and three technicians
3/01/2012	Action	Bacteria/mold	Not tested	—
3/08/2012	Action	Bacteria/mold	Alert	One technician
3/15/2012	Action	Bacteria/mold	Not tested	One technician
3/22/2012	Action	—	Alert	**(3) SVIRSKIY**
3/29/2012	Action	Bacteria/mold	Not tested	**(3) SVIRSKIY**
4/5/2012	Action	Mold	Action	**(5) EVANOSKY**
4/12/2012	Action	Bacteria/mold	Not tested	**(4) LEARY** and one technician
4/20/2012	Action	—	—	One technician
4/26/2012	Action	—	Not tested	—
5/3/2012	Action	—	Action	—
5/10/2012	Action	Bacteria/mold	Not tested	—
5/17/2012	Action	—	Action	—
5/24/2012	Action	Bacteria/mold	Not tested	—
5/31/2012	Action	Bacteria/mold	Action	—
6/7/2012	Action	Bacteria	Not tested	One technician
6/13/2012	Action	Bacteria/mold	Alert	One technician
6/21/2012	Action	Mold	Not tested	—
6/28/2012	Action	Bacteria/mold	Action	Two technicians
7/5/2012	Action	Mold	Not tested	—
7/12/2012	Action	Mold	Action	**(3) SVIRSKIY** and two technicians
7/18/2012	Action	—	Not tested	Two technicians
7/26/2012	Action	Mold	Action	**(2) CHIN, (5) EVANOSKY**, and one technician
8/2/2012	Action	Bacteria/mold	Not tested	One technician
8/9/2012	Action	Bacteria	Action	Two technicians
8/16/2012	Action	—	Not tested	**(4) LEARY**
8/23/2012	Action	Mold	Action	One technician
8/30/2012	Action	—	Not tested	**(5) EVANOSKY**
9/6/2012	Action	—	Action	—
9/13/2012	Action	Bacteria/mold	Not tested	One technician
9/20/2012	Action	Bacteria/mold	Action	Two technicians

Source: United States District Court, District of Massachusetts; Criminal No. 14CR10363-RGS; Case 1:14-cr-10363-RGS *SEALED* Document 1 Filed 12/16/14 and Cabaleiro J. New England Compounding Center Indictment. IJPC, Vol. 19 No. 2/March/April 2015, pp. 94–103.

Compounding pharmacies must comply with *USP* <71> sterility testing requirements for high-risk sterile preparations when using BUDS (Beyond-Use Dates) that exceed *USP* <797> defaults. The number of units tested and the quantity tested from each unit must meet *USP* <71> requirements. If circumstances do not allow appropriate sterility testing, the *USP* defaults of 24 h at room temperature, 72 h refrigerated, or 45 days frozen may be used. Releasing batches prior to receiving sterility test results should be reserved for emergent situations where the *benefit* outweighs the *risk*.

The compounding record for each batch should indicate the number of units tested, the volume per unit, and the test results. Review of these results by a registered pharmacist should be documented on the compounding record.

28.4.7 EXPIRED INGREDIENTS

The indictment alleges that an ingredient known to be expired was used in preparations and that fictitious expiration dates were placed on documentation to hide this fact. These alleged acts resulted in several counts of mail fraud being added to the indictment.

Pharmacists have an ethical and legal duty to assure that all documentation is accurate and truthful.

28.4.8 NECC CONCLUSIONS

The tragic events associated with NECC are alleged to have occurred as a result of lapses in quality-assurance, quality-control, and legal requirements. The quality-assurance and quality-control requirements of *USP* <797>, *USP* <71>, and other applicable *USP* chapters are designed to establish systems to assure sterile medications are safe, and to verify their quality through sterility testing and other tests. These standards establish a redundant system of processes to assure CSP quality. However, when these processes are not performed properly or their results are ignored, as is alleged in the NECC case, serious injury or death to patients can occur.

28.5 LAWS, REGULATIONS, AND STANDARDS AFFECTING PHARMACEUTICAL COMPOUNDING

28.5.1 DRUG QUALITY AND SECURITY ACT: TITLE I: DRUG COMPOUNDING: COMPOUNDING QUALITY ACT (SEC. 102)

Amends the Federal Food, Drug, and Cosmetic Act (FFDCA) with respect to the regulation of compounding drugs. Exempts compounded drugs from new drug requirements, labeling requirements, and track and trace requirements if the drug is compounded by or under the direct supervision of a licensed pharmacist in a Registered Outsourcing Facility and meets applicable requirements.

- Establishes requirements for 503A Pharmacies.
- Establishes requirements for 503B Outsourcing Facilities.
- Establishes annual registration requirement for any Outsourcing Facility. Requires a facility to report biannually to the Secretary of Health and Human Services (HHS) on what drugs are compounded in the facility and to submit adverse event reports. Subjects such facilities to a risk-based inspection schedule.
- Requires the Secretary to (a) publish a list of drugs presenting demonstrable difficulties for compounding that are reasonably likely to lead to an adverse effect on the safety or effectiveness of the drug, taking into account the risk and benefits to patients, and (b) convene an Advisory Committee on Compounding before creating the list.
- Requires the Secretary to assess an annual establishment fee on each Outsourcing Facility and a reinspection fee, as necessary.
- (Sec. 103) Prohibits the resale of a compounded drug labeled "not for resale," or the intentional falsification of a prescription for a compounded drug. Deems a compounded drug to be misbranded if its advertising or promotion is false or misleading in any particular.
- (Sec. 105) Requires the Secretary to receive submissions from State Boards of Pharmacy: (a) describing any disciplinary actions taken against compounding pharmacies or any

recall of a compounded drug, and (b) expressing concerns that a compounding pharmacy may be violating the FFDCA.

- (Sec. 106) Revises compounding pharmacy requirements to repeal prohibitions on advertising and promotion of compounded drugs by compounding pharmacies and repeal the requirement that prescriptions filled by a compounding pharmacy be unsolicited.
- (Sec. 107) Requires the Comptroller General (GAO) to report on pharmacy compounding and the adequacy of state and federal efforts to assure the safety of compounded drugs.

28.5.1.1 503A Pharmacies

This section of the law probably affects more than an estimated 99% of the pharmacies in the USA, most of which do some traditional compounding.

1. FDA expects State Boards of Pharmacy to continue their oversight and regulation of the practice of pharmacy, including traditional pharmacy compounding.
2. FDA intends to continue to cooperate with State authorities to address pharmacy activities that may be violative of the FD&C Act, including Section 503A.
3. A drug must be compounded for an identified individual patient based on the receipt of a valid prescription order, or a notation, approved by the prescribing practitioner, on the prescription order that a compounded product is necessary for the identified patient.
4. The compounding of the drug product is performed:
   - By a licensed pharmacist in a State-licensed pharmacy or a Federal facility, or by a licensed physician on the prescription order for an individual patient; or
   - By a licensed pharmacist or licensed physician in limited quantities before the receipt of a valid prescription order for such individual patient when there is documentation of need.
5. The drug is compounded in compliance with the *USP* chapters on pharmacy compounding using bulk drug substances (Active Pharmaceutical Ingredients – APIs) that
   - Comply with the standards of an applicable *USP* or National Formulary (NF) monograph, if one exists.
   - If such a monograph does not exist, the drug substance(s) must be a component of an FDA-approved human drug product.
   - If a monograph does not exist and the drug substance is not a component of an FDA-approved human drug product, it must appear on a list of bulk drug substances for use in compounding developed by FDA through regulation.
6. The drug product is compounded using bulk drug substances that are
   - Manufactured by an establishment that is registered (including a foreign establishment).
   - Accompanied by valid certificates of analysis for each bulk drug substance.
7. The drug product is compounded using excipients that comply with the standards of an applicable *USP* or NF monograph, if one exists, and the *USP* chapters on pharmacy compounding.
8. The drug product does not appear on the list of drug products that have been withdrawn or removed from the market that have been found to be unsafe or not effective.
9. Drug products that are essentially copies of commercially available products are not compounded regularly or in inordinate amounts.
10. Drug products listed by the FDA that present demonstrable difficulties for compounding or that reasonably demonstrate an adverse effect on the safety or effectiveness of that drug product are not compounded.
11. The drug product is compounded in a State that has entered into a memorandum of understanding (MOU) with FDA that addresses the distribution of inordinate amounts of compounded drug products interstate and provides for appropriate investigation by a

State Agency of complaints relating to compounded drug products distributed outside such State; or in States that have not entered into such an MOU with FDA, the licensed pharmacist, licensed pharmacy, or licensed physician does not distribute, or cause to be distributed, compounded drug products out of the State in which they are compounded, more than 5% of the total prescription orders dispensed or distributed by such pharmacy or physician.

28.5.1.2 503B Outsourcing Facilities

Some points of 503B Outsourcing Facilities include the following.

In Section 503B, an Outsourcing Facility is defined as a facility at one geographic location or address that

a. Is engaged in the compounding of sterile drugs.
b. Has elected to register as an Outsourcing Facility.
c. Complies with all of the requirements of Section 503B.

Also, an Outsourcing Facility is not required to be a licensed pharmacy. An Outsourcing Facility may or may not obtain prescriptions for identified individual patients. A "sterile drug" is on that is intended for parenteral administration, an ophthalmic or oral inhalation drug in aqueous form, or a drug that is required to be sterile under Federal or State law.

A facility that chooses to register with the FDA as an outsourcing pharmacy involved with sterile compounding is required to use bulk drug substances:

1. Appearing on a list established by the FDA.
2. Appearing on the drug shortage list in effect at the time of compounding, distribution, and dispensing.
3. That comply with monographs in the *USP*, NF, or other compendium or pharmacopeia recognized by the FDA.
4. Manufactured by an establishment that is registered with the FDA.
5. Accompanied by valid certificates of analysis.

The facility must

6. Use other ingredients that comply with standards of the *USP*, NF, if such monograph exists, or of another compendium or pharmacopeia recognized by the FDA.
7. Not use drugs that have been withdrawn or removed by the FDA because they have been found to be unsafe or ineffective.
8. Not prepare drugs that are essentially copies of one or more approved drugs (except in the case of an approved drug that appears on the drug shortage list).
9. Not prepare drugs that present demonstrable difficulties for compounding.
10. Be in compliance with the FDA if preparing any drugs that are the subject of a risk evaluation and mitigation strategy.
11. Not sell or transfer drugs to a second entity by the Outsourcing Facility.
12. Pay all applicable fees.
13. Adhere to the requirements for the label, containers, and any other required information.
14. Register with the FDA between October 1 and December 31 of each year.
15. Provide reports during June and December of each year to the FDA of the drugs compounded during the previous 6 months.
16. Comply with Good Manufacturing Practices.
17. Be subject to FDA inspections.
18. Submit adverse event reports.

28.5.2 THE "TRIAD"

The Act states that a compounded product is exempt from meeting the "new drug" requirements if the drug product is compounded for an individual patient based on the unsolicited receipt of a valid prescription order, or a notation, approved by the prescribing practitioner, on the prescription order indicating that a compounded product is necessary for the identified patient – if the product meets certain requirements. A pharmacist may compound a drug when a prescription clearly requires compounding (because the drug is not commercially available in the form needed, or when a physician authorizes compounding). Also, a pharmacist may compound a drug if – with the physician's approval – the pharmacist determines that a compounded drug is necessary and notes that information on the prescription. This allows pharmacists to suggest therapeutic switches to a compounded drug, just as they do for other types of medications. The physician, the patient, and the pharmacist form the legal compounding "triad."

28.5.2.1 Licensed Pharmacist or Physician

The product must be compounded by a licensed pharmacist – in a state-licensed pharmacy or federal facility – or by a licensed physician or other licensed practitioner authorized by state law to prescribe drugs.

28.5.2.2 Anticipatory Compounding

Limited quantities of products can be compounded in advance – if there is a history of receiving valid prescription orders for the product, generated by an established relationship between the licensed pharmacist and individual patients for whom the prescriptions are provided. There must also be an established relationship with the physician or other licensed professional who wrote the prescription.

28.5.2.3 Substances That May Be Used in Compounding

Substances that may be used for compounding include the following:

- Bulk drug substances that contain monographs in the *USP*/NF, according to the *USP* chapter on Pharmacy Compounding.
- Drug substances that are components of FDA-approved drugs/drug products, including any ingredient that is contained in commercially available FDA-approved drug products.
- Bulk drug substances that appear on a list of approved bulk drug substances developed by the FDA.
- Substances used must be manufactured by establishments registered with the FDA – including foreign establishments – and comply with standards of any monograph in the *USP*/NF (if a monograph exists), as well as with the *USP* chapter on Pharmacy Compounding.

28.5.2.4 Substances That May Not Be Used in Compounding

Compounding cannot be done using substances or involving products that fall under the following categories:

- Products listed in Appendix 28.A should not be used in compounding, because they are on the list of drug products withdrawn or removed from the market (because they have been found to be unsafe or ineffective).
- Inordinate amounts of commercially available drug products (not including drug products in which there has been a change made for an individual patient, such as omitting a dye, flavor, sweetener, preservative, or the like) to which the patient may be sensitive. According to this law, pharmacists are allowed to compound copies of commercially available drug products within the definition the FDA provides. In other words, the quantity described

as "inordinate amounts" has yet to be defined. Much latitude is given to the prescribing practitioner in this area. It should be mentioned, however, that a small variation in strength, such as from 50 to 45 mg, would not be determined to be a significant difference.

28.5.3 Memorandum of Understanding (MOU)

Section 503A(b) (3) (B) established that to qualify for the exemptions in Section 503A, the drug product must be compounded in accordance with either of the following: (a) it was compounded in a state that had entered into an MOU with the FDA that addressed the interstate distribution of inordinate amounts of compounded drug products and provided for investigation by a State Agency of complaints related to compounded drug products distributed outside such state; or (b) it was compounded in a state that had not entered into such an MOU, but the licensed pharmacist, pharmacy, or physician distributes (or causes to be distributed) compounded drug products outside of the state in which they were compounded – in quantities that did not exceed 5% of the total prescription orders dispensed or distributed by such pharmacy or physician.

28.5.4 Definitions of Compounding

28.5.4.1 H.R. 3204

The term "compounding" includes the combining, admixing, mixing, diluting, pooling, reconstituting, or otherwise altering of a drug or bulk drug substance to create a drug. For the purposes of the Act, the term "compounding" does not include mixing, reconstituting, or other such acts that are performed in accordance with directions contained in approved labeling provided by the product's manufacturer or other manufacturer directions consistent with that labeling. From this, it is assumed, for example, that reconstituting with different volumes of water than are written on the label constitutes compounding. To clarify, it would not be compounding if an approved manufacturer's label states to add 88 mL of purified water to an antibiotic for suspension, and the pharmacist does just that. It could be considered compounding, however, if the pharmacist was directed by the physician to add 38 mL of purified water for an increased concentration per dose of the drug.

28.5.4.2 US Pharmacopeia

The preparation, mixing, assembling, altering, packaging, and labeling of a drug, drug-delivery device, or device in accordance with a licensed practitioner's prescription, medication order, or initiative based on the practitioner/patient/pharmacist/compounder relationship in the course of professional practice. Compounding includes the following:

- Preparation of drug dosage forms for both human and animal patients
- Preparation of drugs or devices in anticipation of prescription drug orders based on routine, regularly observed prescribing patterns
- Reconstitution or manipulation of commercial products that may require the addition of one or more ingredients
- Preparation of drugs or devices for the purposes of, or as an incident to, research (clinical or academic), teaching, or chemical analysis
- Preparation of drugs and devices for prescriber's office use where permitted by federal and state law.

28.5.5 Implementation of H.R. 3204

28.5.5.1 Draft Guidances

The US FDA has issued five Draft Guidance documents related to drug compounding and repackaging that will help entities comply with important public health provisions. The Draft Guidance documents are applicable to pharmacies, federal facilities, Outsourcing Facilities and physicians, as follows.

28.5.5.1.1 Draft Guidance for Industry: For Entities Considering whether to Register as Outsourcing Facilities under Section 503B of the Federal Food, Drug, and Cosmetic Act

The draft guidance provides an entity considering whether to register with the FDA as an Outsourcing Facility under the law with information about the regulatory impact of registering. For example, it explains that a facility engaged in only certain activities, including repackaging human drugs and compounding nonsterile drugs, should not register as an Outsourcing Facility because its drug products will not qualify for the exemptions provided in Section 503B, including the exemption from the new drug approval requirements.

28.5.5.1.2 Draft Guidance for Industry: Repackaging of Certain Human Drug Products by Pharmacies and Outsourcing Facilities

The draft guidance describes the conditions under which the FDA does not intend to take action for certain violations of the law when state-licensed pharmacies, federal facilities, or Outsourcing Facilities repackage certain drug products. Repackaging generally involves taking a finished drug product from the container in which it was distributed by the original manufacturer and placing it into a different container. Repackaged drug products are generally not exempt from any of the provisions of the FD&C Act related to the production of drugs, and the compounding provisions of the FD&C Act do not address repackaging. Therefore, the FDA is issuing guidance to describe how it intends to address repackaging when done in a state-licensed pharmacy, federal facility, or Outsourcing Facility.

28.5.5.1.3 Draft Guidance for Industry: Mixing, Diluting, or Repackaging Biological Products Outside the Scope of an Approved Biologics License Application (BLA)

The draft guidance describes the conditions under which the FDA does not intend to take action for violations of certain sections of the Public Health Service Act (PHS Act) and the FD&C Act when state-licensed pharmacies, federal facilities, or Outsourcing Facilities mix, dilute, or repackage specific biological products without an approved BLA, or when such facilities or physicians prepare prescription sets of allergenic extracts (used to treat allergies) without an approved BLA. The draft guidance notes that a biological product that is mixed, diluted, or repackaged outside the scope of an approved BLA is an unlicensed biological product under Section 351 of the PHS Act and may not be legally marketed without an approved BLA. Additionally, the compounding provisions of the FD&C Act do not address biological products subject to licensure under Section 351 of the PHS Act. Therefore, the FDA is issuing guidance to describe how it intends to address these practices.

28.5.5.1.4 Draft Guidance for Industry: Adverse Event Reporting for Outsourcing Facilities under Section 503B of the Federal Food, Drug, and Cosmetic Act

Entities registered as Outsourcing Facilities are required to report adverse events to the FDA. The draft guidance explains adverse event reporting for Outsourcing Facilities.

28.5.5.1.5 Draft Memorandum of Understanding between a State and the US Food and Drug Administration Addressing Certain Distributions of Compounded Human Drug Products

The draft MOU under Section 503A of the FD&C Act describes the responsibilities of a state that chooses to sign the MOU in investigating and responding to complaints related to compounded human drug products distributed outside the state, and in addressing the interstate distribution of "inordinate amounts" of compounded human drug products.

The above documents are the latest in a series of policy documents related to FDA oversight of drugs produced by state-licensed pharmacies, federal facilities, and Outsourcing Facilities.

28.5.5.2 Organization

The provisions of Section 503A require rulemaking or other action by FDA, including the formation of the Advisory Committee on Compounding:

The law (H.R. 3204) states

> Before issuing regulations to implement subsection (a)(6), the Secretary shall convene and consult an advisory committee on compounding. The advisory committee shall include representatives from the National Association of Boards of Pharmacy, the United States Pharmacopeia, pharmacists with current experience and expertise in compounding, physicians with background and knowledge in compounding, and patient and public health advocacy organizations.

The FDA, in consultation with the Advisory Committee on Compounding, will make decisions regarding the following lists to be developed.

28.5.5.3 Lists

In addition, the FDA, in consultation with the Advisory Committee on Compounding, will make decisions regarding the following.

1. **The "Negative List"**
   The first committee developed and approved a "negative" list consisting of drugs which have been removed for safety reasons. The new committee will review and recommend modifications to this list of drug products that may not be compounded because they have been withdrawn or removed from the market because they have been found to be unsafe or not effective. See Appendix.
2. **The "Positive List of APIs"**
   This activity will look at bulk drug substances that (a) do not contain a *USP* or NF monograph, (b) are not components of FDA-approved drug products, or (c) do not contain a monograph in another compendium or pharmacopeia recognized by the Secretary.
3. **Demonstrable Difficulties in Compounding**
   A list of drug products that present demonstrable difficulties for compounding is to be developed. This will be drug products that present demonstrable difficulties for compounding that reasonably demonstrate an adverse effect on the safety or effectiveness of the drug product.

28.5.5.3.1 State Boards of Pharmacy

The general mission of the individual State Boards of Pharmacy is *to protect the health, safety, and welfare of the citizens of the respective state* by regulating and enforcing the laws regarding the practice of pharmacy and the manufacturing, sale, distribution, and storage of drugs, medicines, chemicals, and poisons.

Boards of Pharmacy consist of several members appointed and/or elected and serve for a defined term.

Boards are responsible for promoting, preserving, and protecting public health, safety, and welfare by and through the effective control and regulation of the practice of pharmacy. Boards are also responsible for the licensing and regulation of pharmacists, pharmacy interns, and pharmacy technicians.

Boards are also responsible for the licensing and regulation of all pharmacies, medical gas suppliers, medical gas distributors, and prescription drug wholesalers, packagers, and manufacturers that do business in the state.

Boards have the power and duty to inspect all places handling drugs, medicines, chemicals, and poisons. Boards are charged with the enforcement of federal and state controlled dangerous substance and prescription drug laws.

Boards investigate complaints concerning pharmacists, technicians, and interns as well as pharmacies, medical gas suppliers and distributors, and prescription drug wholesalers, packagers, and manufacturers. Boards conduct hearings on all types of registrants and have the authority to reprimand, fine, suspend, or revoke licenses and permits. The Board also reviews and approves continuing education programs that pharmacists are required to complete in order to renew their license.

28.6 US PHARMACOPEIA: NATIONAL FORMULARY

28.6.1 *USP* CHAPTERS

In 1990, the US Pharmacopeial Convention approved the appointment of a Pharmacy Compounding Practices Expert Advisory Panel. The activities of the panel initially were to (a) prepare a chapter on compounding for the *USP*/NF and (b) to begin the process of preparing monographs of compounded products for inclusion in the NF.

The prepared Chapter <1161>, "Pharmacy Compounding Practices," was published and became official in 1996. With the mention of this chapter in FDAMA 1997, this chapter was renumbered as Chapter <795>; subsequently, its title was also changed to "Pharmaceutical Compounding—Nonsterile Preparations."[3] (It should be noted that *USP* chapters with numbers greater than <1000> are informational, while numbers less than <1000> can be enforceable.)

The first of the compounding monographs became official in November 1998, and these monographs are being published in the US Pharmacopeia section of the *USP*/NF. For each monograph published, a considerable amount of work is done, including a detailed, validated stability study. Some monographs that were previously published in the *USP*/NF are being reintroduced into the compendia.

A second chapter related to compounding in the *USP*/NF is Chapter <1206> "Sterile Drug Products for Home Use"; this chapter was renumbered as <797>, revised, and re-titled as "Pharmaceutical Compounding-Sterile Preparations."[4]

Two additional chapters, Chapter <1075>, "Good Compounding Practices" and Chapter <1160>, "Pharmaceutical Calculations in Prescription Compounding," have also been added. Recently, *USP* Chapters <795>, Pharmacy Compounding—Nonsterile Preparations, and <1075> Good Compounding Practices, were combined and expended into a new *USP* General Chapter <795>, Pharmacy Compounding-Nonsterile Preparations. *USP* Chapter <797>, Pharmacy Compounding—Sterile Preparations, has also been updated.

28.6.1.1 Chapter <795>: Pharmaceutical Compounding: Nonsterile Preparations

This is divided into discussions on the (1) Introduction, (2) Definitions, (3) Categories of Compounding, (4) Responsibilities of the Compounder, (5) Compounding Process, (6) Compounding Facilities, (7) Compounding Equipment, (8) Component Selection, Handling and Storage, (9) Stability Criteria and Beyond-Use Dating, (10) Packaging and Drug Preparation Containers, (11) Compounding Documentation, (12) Quality Control, (13) Patient Counseling, (14) Training, and (15) Compounding for Animal Patients.

The introduction to the chapter discusses the chapter's purpose, which includes information to enhance the compounder's ability to extemporaneously compound preparations that are of acceptable strength, quality, and purity. The next section provides a number of definitions, including the definition of compounding and manufacturing. Three categories of compounding are described in the third section: simple, moderate, and complex. This section details the differences between these three categories.

The fourth section describes the responsibilities of the compounder, which are detailed later in this chapter. The "Compounding Process" section describes 15 criteria to be followed when compounding each drug separation. The "Compounding Facilities" section discusses the design and maintenance of the facilities to be used and the equipment selected for compounding. It refers to other specific chapters in the *USP* and to other documents, including the OSHA Technical Manual-Section VI: Chapter 2, Controlling Occupational Exposure to Hazardous Drugs and the NIOSH Alert: Preventing Occupational Exposure to Antineoplastic and Other Hazardous Drugs in Health Care Settings. Section 7, Compounding Equipment, states that any equipment used for compounding must be of appropriate design and size for compounding and suitable for the intended use.

The section "Component Selection, Handling and Storage" describes, in detail, sources for drugs and excipients that are appropriate for compounding. Section "Stability Criteria and Beyond-Use Dating" discusses the packaging, sterility, stability criteria, and guidelines for assigning BUDS for compounded preparations. The chapter goes on to explain that these BUD limits can be exceeded if there is supporting valid scientific stability information that is directly applicable to the product being compounded. The product must of the same drug, in a similar concentration, similar pH, similar excipients, similar vehicle, similar water content, and so on.

The "Packaging and Drug Preparation Containers" relates several different *USP* general chapters that are to be considered in extemporaneous compounding. As described in Section 11, Compounding Documentation, record-keeping requirements of various states must be followed, and generally include a formulation record and a compounding record. The formulation record is a file of individually compounded preparations, including the name, strength, and dosage form of the preparation compounded, all ingredients and their quantities, equipment needed to prepare the preparation (when appropriate), and mixing instructions. The formulation record must also include an assigned BUD, the container used in dispensing, storage requirements, and any quality-control procedures. The compounding record contains documentation of the name and strength of the compounded preparation, the formulation record reference, and the sources and lot numbers of ingredients used in compounding. Among other items, it should also contain the results of quality-control procedures that were conducted on the lot of compounded product.

The chapter also places an emphasis on quality control and the responsibility of the pharmacist to review each procedure and observe the finished preparation. The chapter continues with the importance of patient counseling in the proper use, storage, and observation (for instability) of the dispensed product. Next, the training section states that all personnel involved in the compounding processes must be properly trained for the type of compounding conducted. Finally, the chapter ends with a short discussion on compounding for animal patients and the standards that should be considered. The overall emphasis of the chapter is to support the pharmacist in the compounding of products of acceptable strength, quality, and purity.

28.6.1.2   Chapter <797>: Pharmaceutical Compounding: Sterile Preparations

In general, this chapter separates sterile products compounding into different categories, depending upon the risk levels associated with different types of compounding. For example, reconstitution and combination of commercially available sterile products is considered a low-risk-level operation, whereas compounding sterile products (CSPs) from nonsterile ingredients is a high-risk-level operation, based upon a reference to microbiological quality and the potential for contamination. The chapter is divided into several sections: (1) Definitions, (2) Responsibility of Compounding Personnel, (3) CSP Microbial Contamination Risk Levels, (4) Personnel Training and Evaluation in Aseptic Manipulation Skills, (5) Immediate-Use CSPs, (6) Single-Dose and Multiple-Dose Containers, (7) Hazardous Drugs as CSPs, (8) Radiopharmaceuticals as CSPs, (9) Allergen Extracts as CSPs, (10) Verification of Compounding Accuracy and Sterility, (11) Environmental Quality and Control, (12) Suggested Standard Operating Procedures, (13) Elements of Quality Control, (14) Verification of Automated Compounding Devices for Parenteral Nutrition Compounding, (15) Finished Preparation Release Checks and Tests, (16) Storage and Beyond-Use Dating, (17) Maintaining Sterility, Purity, and Stability of Dispensed and Distributed CSPs, (18) Patient or Caregiver Training, (19) Patient Monitoring and Adverse Events Reporting, (20) Quality Assurance Program, (21) Abbreviations and Acronyms, and four appendices.

The overall objective of the chapter is to describe conditions and practices to prevent harm, including death, to patients that could possibly result from microbial contamination, excessive bacterial endotoxins, variability in the intended strength and composition, and unintended chemical and physical contaminants and ingredients of inappropriate quality in CSPs. All compounding personnel are responsible for understanding fundamental practices and precautions described within this chapter, for developing and implementing appropriate procedures, and for continually evaluating these procedures and the quality of final CSPs to prevent harm.

28.6.1.3 Chapter <1160>: Pharmaceutical Calculations in Prescription Compounding

The purpose of this chapter is to provide general information to guide and assist pharmacists in performing necessary calculations when preparing or compounding any pharmaceutical article (or when simply dispensing prescriptions).

Calculations are discussed as they relate to the amount of concentration of drug substances in each unit or dosage portion of a compounded preparation. Special emphasis is placed on calculations involving the purity and dosage portion of drugs, their salt forms, and equivalent potencies.

The chapter is generally divided into an introduction, basic mathematical concepts (significant figures, logarithms), and basic pharmaceutical calculations (calculations in compounding, buffer solutions, dosage calculations, percentage concentrations, specific gravity, dilution and concentration, use of potency units, base versus salt or ester forms of drugs, reconstitution of drugs using volumes other than those on the label, alligation alternate (an arithmetic method of solving problems relating mixtures of components of different strengths) and algebra, molar, molal and normal concentrations, isosmotic solutions, flow rates in intravenous sets, and temperature).

28.6.1.4 Chapter <1163>: Quality Assurance in Pharmaceutical Compounding

Quality assurance is paramount in any compounding pharmacy practice. A quality-assurance program for compounding should include at least the following eight separate, but integrated, components, as detailed in this chapter: (1) Training, (2) Standard Operating Procedures, (3) Documentation, (4) Verification, (5) Testing, (6) Cleaning and Disinfecting, (7) Containers, Packaging, Repackaging and Storage, and (8) Outsourcing. This chapter provides guidance for each of these topics as well as suggested compendial testing methods for bulk substances and various dosage forms. Additional information is provided on selected dosage forms. Physical, chemical, and microbiological testing is discussed.

28.6.1.5 Chapter <800>: Hazardous Drugs: Handling in Healthcare Settings (Proposed)

This proposed chapter describes practice and quality standards for handling hazardous drugs (HDs) to promote patient safety, worker safety, and environmental protection. Handling HDs includes, but is not limited to, the receipt, storage, compounding, dispensing, administration, and disposal of sterile and nonsterile products and preparations.

This chapter applies to all healthcare personnel who handle HD preparations and all entities which store, prepare, transport, or administer HDs (e.g., pharmacies, hospitals and other healthcare institutions, patient treatment clinics, physicians' practice facilities, or veterinarians' offices). Personnel who may potentially be exposed to HDs include, but are not limited to, pharmacists, pharmacy technicians, nurses, physicians, physician assistants, home healthcare workers, veterinarians, and veterinary technicians.

Entities that handle HDs must incorporate the standards in this chapter into their occupational safety plan.

28.6.2 Individual Compounding Monographs

USP Compounded Preparation Monographs (CPMs) contain formulations used in human and/or animal patients. They provide quality standards for specific preparations to assist practitioners in compounding formulations for which there is no suitable commercially available product.

USP CPMs include the following:

- Formulas (ingredients and quantities)
- Directions for compounding the preparation
- BUDS based on stability studies
- Packaging and storage information
- Acceptable pH ranges
- Stability-indicating assays.

28.6.3   Creating a *USP* Compounded Preparation Monograph

The process of creating CPMs begins with identifying a public health need involving the following considerations:

- Medications with the highest public health impact (i.e., affecting major population groups, disease states, and access needs)
- Medications essential to treat pediatric and geriatric patients where there are unmet needs
- Medications that need to be formulated to avoid allergic reactions and to be suitable for administration to patients with specific genetic anomalies
- Medications for currently unmet clinical and therapeutic needs.

There are currently two procedures for compounding monograph development:

- First is for *USP* to initiate the creation of the data and evidence necessary for the monograph addressing a priority public health need. Since 2012, *USP* has invested more than $1.8M for the development of monographs to meet public health needs.
- Second is via monograph donations. Here, information to support monograph development is provided by entities in academia, drug manufacturers, and other stakeholders who may have the data necessary for the creation of a monograph. The data is then provided to the Compounding Expert Committee for analysis, further development, and approval. Once approved, the CPM is published in the *USP–NF*.

28.6.4   Marketed but Not FDA-Approved Drugs

An area of confusion often arises over what is or is not "FDA-approved." There are actually many drugs on the market that Congress or the FDA allows that have never been "approved," as shown below. This relates to the substances that can and cannot be used in compounding.

1. Pre-1938 drugs that have never been FDA-approved; they were already on the market and "grandfathered" in at the time the FDA was created. Of the 59 items listed in the "Pre-1938" Products, eight (8) have no *USP* monographs. Source: "Section III: Listing of 'Pre-1938' Products". Approved Drug Products and Legal Requirements (USP DI 2004), 24th Ed. USP Drug Information, Volume III; III/1.
2. Orphan drugs: At the HRSA website, there are 262 pages of drugs the FDA has "designated" with orphan status. The vast majority of these do not have *USP* monographs. Compounding has been used to prepare these drugs for these patients for many years. (Source: FDA website).
3. Investigational/study drugs: New drugs (new chemical entities-NCEs) are not FDA-approved and will not have *USP* monographs. Some studies at academic/clinical centers, etc. may use APIs that are not commercially marketed or have *USP* monographs. At least 70

pharmaceutical companies use compounding in some part of their IND/NDA process, and it has been very beneficial for many years. (Source: Discussions at Innovation and Quality in Pharmaceutical Development-Consortium Workshop, February 2014, Washington DC).

4. Non-prescription or over-the-counter (OTC) drugs: According to the FDA website, OTC drugs are "not FDA-Approved." The OTC monograph system is a regulatory pathway for bringing an OTC drug product to market without FDA approval. There are over 300,000 marketed OTC drug products. There are over 80 therapeutic classes of drugs, and for each category or class, an OTC drug monograph is developed. The OTC drug monographs are "recipes" covering acceptable ingredients, doses, formulations, and labeling. Once a final monograph is implemented, companies can make and market an OTC product without the need for FDA pre-approval. In looking at the OTC Active Ingredients list, about 30% do not have *USP* monographs. (Source: FDA website).

5. Herbal/some dietary supplement/nutritional drugs: The *USP* has standards for a number of dietary supplements. This group is largely unapproved by FDA with some having *USP* monographs. (Source: *USP* 38/NF 33; FDA website).

6. Most commercial veterinary drug products

However, H.R. 3204 applies only to compounding for humans.

28.7 STATUS OF 503B OUTSOURCING FACILITIES

28.7.1 DQSA AND THE ADVENT OF "OUTSOURCING FACILITIES"

The DQSA of 2013 not only reinstated Section 503A of the FFDCA but also inserted a new Section 503B into the statute. Section 503B created a novel category of compounding establishment called a Registered Outsourcing Facility, the legislative purpose of which was to bring large-scale, patient non-specific sterile compounding under the scrutiny and oversight of the US FDA.

In the aftermath of the fungal Meningitis Outbreak, FDA commissioner Margaret Hamburg was called to testify before the House Energy and Commerce Committee and the Senate Health, Education, Labor, and Pensions ("HELP") Committee on consecutive days in November of 2012. Legislators from both houses attempted to place blame on FDA for the outbreak, but Dr. Hamburg steadfastly and correctly maintained that the Agency was limited in its regulatory powers by the Western States Medical Decision. She assured the legislators that FDA was eager to rigorous enforce federal regulations against pharmacy-based sterile compounders, but that in order to do so, Congress would be required to revise the statutes.

After NECC was identified as the source of the outbreak and state and federal investigators entered its Framingham facility, 17,000 vials were recalled from healthcare facilities in 20 different states. As the case count and the fatalities were tallied by the Center for Disease Control (CDC), there was not a single case reported in the Commonwealth of Massachusetts[13] and it became clear that the calamity had resulted from the company's cynical use of its state pharmacy license to masquerade as a sterile drug manufacturer. It also became clear that the Massachusetts Board of Pharmacy Regulation had been informed of NECC's improper practices but had failed to appropriately intervene.

Neither legislators nor regulators could accurately estimate how many other state-licensed pharmacies might be similarly skirting the system or whether another catastrophe was looming. The Congress was determined to close off the regulatory gap. If compounding organizations intend to produce large batches of sterile preparations without prescriptions for individually identified patients, the lawmakers wanted adherence to the quality-assurance level of cGMP and they wanted their compliance assessed and verified by FDA.

Registered Outsourcing Facilities were designed to operate at a much higher level of control than traditional sterile compounding pharmacies, which typically worked on small batches with a prescription in hand from a known prescriber, while being exempt from the costliest aspects of pharmaceutical manufacturers. Section 503B created exemptions from three critical sections of the FFDCA for organizations that are willing to register as Outsourcing Facilities with FDA and create their sterile compounds at the quality-assurance level of cGMP.

The sections from which Outsourcing Facilities are exempt are[14]

- The drug approval requirements in Section 505 of the FD&C Act (21 U.S.C. 355)
- The requirement to be labeled with adequate directions for use in Section 502(f)(1) of the FD&C Act (21 U.S.C. 352(f)(1))
- The track and trace requirements in Section 582 of the FD&C Act (21 U.S.C. 360eee-1).

In order to qualify for these three exemptions, the Outsourcing Facility must meet several detailed and specific conditions, including requirements to register and regularly report production, restrict compounding to certain ingredients, not to sell copies of approved drugs, not to compound dosage delivery forms beyond their level of sophistication, not subvert risk evaluation and mitigation strategies (REMS) requirements, pay the required fees, and adhere to the specific labeling requirements of 503B(a)(10).

28.7.2 LAUNCH AND TRAJECTORY OF THE OUTSOURCING FACILITY PROGRAM

Registration under 503B was deemed voluntary, but the Congress believed that the marketplace would drive enrollment for the program. The DQSA statute required FDA to publish a listing of Registered Outsourcing Facilities to its website,[15] which it has maintained in a dynamic form since 2014. At least 13 entities had registered as Outsourcing Facilities by December 2013 and that number had grown to 56 by the end of August 2014.

Newly registered Outsourcing Facilities appear frequently on the web page and defunct ones routinely vanish. Unfortunately, although most of the early registrants were busy, >>successful sterile compounding pharmacies owned by highly experienced pharmacists, these skillful healthcare professionals were unprepared for the rigors of cGMP and the dropout rate was very high. Twelve of the earliest registrants had already vanished from the web page by April 2015, and six more were removed by the end of that year.

As of the date of this writing, there have been 114 Outsourcing Facility registrants posted to the web page since September 22, 2014. Of these 114, 75 still appear on the latest web page update, while 39 have vanished (Table 28.2). The current status of four of these 75 listed has having ceased operations or being under a consent decree. The four with recent cessations and/or consent decrees were all veteran Outsourcing Facilities and had been operating since before August of 2015.

Six others of the 75 are operating under the status of "Regulatory Meeting," and the current status of seven more is listed as "Warning Letter." Both designations call the quality-assurance practices of the entities into question.

Twelve of the 75 currently registered sites are represented by a single parent organization with centralized control. PharMEDium has four Active sites, but one of those is Memphis, which is listed as having ceased sterile compounding operations. Central Admixture Pharmacy Services, Inc ("CAPS"), Avella, and QuVa Pharma each operate three sites and SCA Pharmaceuticals operates two.

Of the 75 current registrants, ten others are currently listed as "Not yet Inspected" (Table 28.3).

28.7.3 FDA INSPECTIONS OF OUTSOURCING FACILITIES

When FDA conducts an inspection, the investigator reviews the organization's standard operating procedures in fine detail and contrasts those written standard operating procedures (SOPs) against the behaviors observed as staff members conduct operational processes and against the written

TABLE 28.2

Registered Outsourcing Facilities that Appeared on the FDA List, but Subsequently Vanished

Outsourcing Facility	Vanished from List by:
ALK-Abelló, Inc., Port Washington, NY	April 11 2015
Allergy Laboratories, Inc., Oklahoma City, OK	April 11 2015
Greer Laboratories, Inc., Lenoir, NC	April 11 2015
Institutional Pharmacy Solutions, LLC, Irwindale, CA	April 11 2015
Institutional Pharmacy Solutions, LLC, Virginia Beach, VA	April 11 2015
IV Specialty Ltd, Austin, TX	April 11 2015
Pharmagen Laboratories Inc., Stamford, CT	April 11 2015
Pharmalogic CSP, Bridgeport, WV	April 11 2015
Professional Pharmacy & Compounding Services LLC, Miami, FL	April 11 2015
Synergy Pharmacy Services, Inc., Palm Harbor, FL	April 11 2015
Texas Health Infusion, The Woodlands, TX	April 11 2015
UCSF Home Therapy Services, San Francisco, CA	April 11 2015
Absolute Pharmacy, Lutz, FL	July 11 2015
ACS Dobfar Info S.A., Campascio, Switzerland	October 21 2015
Brown's Compounding Center, Inc., Englewood, CO	October 21 2015
Healix Infusion Therapy, Inc., Sugar Land, TX	October 21 2015
Region Care, Inc., Great Neck, NY	October 21 2015
Alexander Infusion LLC dba Avanti Health Care, New Hyde Park, NY	December 13 2015
PharMedium Services, LLC, Edison, NJ	December 13 2015
Pharmaceutic Labs, LLC, Albany, NY	April 22 2016
Downing Labs, LLC, Dallas, TX	August 8 2016
ILS Genomics, LLC Morrisville, NC + Laboratory Corporation of America, Morrisville, NC	August 8 2016
Medistat RX, LLC, Foley, AL	August 8 2016
Unique Pharmaceuticals, Ltd., Temple TX	December 19 2016
ImprimisRx TX, Inc., Allen, TX	July 25 2017
INCELL Corp., LLC, San Antonio, TX	July 25 2017
Jubilant HollisterStier LLC, Spokane, WA	July 25 2017
Meta Pharmacy Services, Las Vegas, NV	July 25 2017
Pharmakon Pharmaceuticals, Noblesville, IN	July 25 2017
Topicare Management LLC, Conroe, TX	July 25 2017
Kings Park Slope, Inc., Brooklyn, NY	August 9 2017
Bella Pharmaceuticals, Inc., Chicago, IL	December 6 2017
Banner Health, Chandler, AZ	May 10 2018
Bioserv Corporation, San Diego, CA	May 10 2018
California Pharmacy and Compounding Center, Newport Beach, CA	May 10 2018
Essential Pharmacy Compounding Division of Kohll's Pharmacy, Omaha, NE	May 10 2018
Isomeric Pharmacy Solutions, LLC, Salt Lake City, UT	May 10 2018
Resource Optimization and Innovation, Springfield, MO	May 10 2018
V Manufacturing & Logistics, Inc., Walnut, CA	May 30 2018

records of the activities. Additionally, the investigator gauges whether needed SOPs do not exist. The investigator makes extensive written notes of his/her perceptions during the several days on site. At the conclusion of the inspection, the investigator transmits this "Establishment Inspection Report" (EIR) to managers at FDA.

The managers convene one or more multidisciplinary meetings to review the results of the inspection, and if there are important variances from cGMP requirements, these "objectionable

TABLE 28.3

Registered but Listed by FDA as "Not Yet Inspected"

Outsourcing Facility

BMD SKINCARE INC, Canoga, CA

Central Admixture Pharmacy Services, Phoenix, AZ

F.H. Investments, Inc., DBA Asteria Health, Birmingham, AL

Fresenius Kabi Compounding, LLC, Canton, MA

INTACT PHARMACEUTICALS LLC, New Milford, CT

IntegraDose Compounding Services, LLC, Minneapolis, MN

Maitland Labs of Central Florida, Orlando, FL

Medcraft LLC, Mounds, OK

MedisourceRx, Los Alamitos, CA

Molecular PharmaGroup, New Providence, NJ

conditions" are communicated to the covered entity in the form of Observations recorded on an FDA Form 483. Every Form 483 carries a disclaimer that the Observations "do not represent a final Agency determination of your compliance."

FDA managers and inspectors are merely human and subject to making judgments. They have a strong internal culture, and they share stories with one another about what they've seen. In the early days after the Meningitis Outbreak, there was an adversarial air to FDA interactions with pharmacies. Their Agency had been heavily criticized, and they had read media stories of NECC executives sacrificing quality-assurance controls to enhance profits.

Conversely, there was a great deal of resentment from pharmacy toward the FDA as well because their appearance was often considered an unwelcome intrusion by pharmacy owners with strong professional personalities who took great pride in their judgment, track record, and their close relationship with their patients. Very few pharmacists were abusing their state licenses by the compounding they performed, and there was limited mutual respect or trust between FDA and pharmacy owners.

There is never a second chance to make a first impression, but the most central principle of the FDA is relentless improvement. No matter the starting point, if an organization's focus is on improving its operational processes every day, it will survive with FDA. The Agency expects diligent attention and measurable remediation of all aspects of all processes that are noted on an FDA Form 483 and repeat Observations have very poor connotations.

28.7.3.1 Initial Inspections

Initial FDA inspections have gone very poorly for the Outsourcing Facility sector. Out of 104 instances, only two Outsourcing Facilities have come through their initial inspection without generating an FDA Form 483. The two organizations with such good starts were Nephron Sterile Compounding Center, LLC (NSCC) of West Columbia, SC, in July 2016, and, very recently, Nubratori, Inc. of Torrance, CA.

Nephron had a tremendous advantage in that the parent company, although owned and managed by pharmacists was a full-fledged pharmaceutical manufacturer with long experience in cGMP and with FDA inspections. Nonetheless, they were issued 483's on their two subsequent inspections, albeit with few and easily managed Observations.

All the other registrants that underwent an initial inspection before dropping out were issued 483s, some of them with numerous Observations regarding very fundamental cGMP principles. For example, Observation #1 on a very recent initial inspections stated, "There is no Quality Control Unit," which is probably the most devastating Observation that can be made from FDA's perspective.

28.7.3.2 Recurring Inspections

FDA's inspectional approach for Outsourcing Facilities is "risk-based," which means that the riskier the compounding, the more frequently they inspect. If there is a quality event, they usually arrive soon after, but their routine rhythm appears to be approximately one inspection every 2 years.

Among the Outsourcing Facilities listed on the current webpage the breakdown of 483's is that one entity has received 6 483s, 4 entities have received 4, 16 have received 3, 23 have received 2, and 19 have received 1 and Nubratori has received none after its "clean" initial inspection.

Some Outsourcing Facilities have shown steady improvement with each subsequent inspection, but others have had repeated inspections with numerous observations, many of which are "Repeats."

Medi-Fare of Blacksburg, SC is an example of successful continuous improvement under intense FDA scrutiny. This firm had 12 Observations on its initial inspection (1/18/2013) and received a Warning Letter on 3/7/2013, which is a very short relative lag time. It was inspected a second time on 6/21/2013, also a very short lag time, and received only three Observations. It was inspected a third time on 9/12/2014, also a very short lag time, but with only a single Observation. It has been inspected twice more since without being issued a 483. This successful trajectory was probably the result of the engagement of an expert cGMP validation consulting firm.

Conversely, the firm that most recently suffered the imposition of a consent decree was Delta Pharma. This firm received a 483 on 10/2/2013, which triggered a Warning Letter on 12/9/2014. It received another 483 after an inspection ending 5/24/2016 and then a third 483 after an inspection on 2/23/2017. The 2/23/2017 inspection included 12 Observations, 9 of which were "Repeats." This failure to improve was apparently sufficient for FDA to convince the Department of Justice to sue for an injunction and obtain a consent decree.

It is important to understand that for FDA to take legal action against any firm, it must convince the Department of Justice that the observed violations of federal code represent a significant threat to the public health. Only the Department of Justice (DOJ) can represent the USA in federal court.

28.7.4 RESPONSE LETTERS BY OUTSOURCING FACILITIES TO FDA

In order for a firm's response to be published to the FDA web pages, appropriate waivers of privacy must be provided in writing to the Agency. The Outsourcing Facility business is highly proprietary and protective of trade secrets, and information release waivers are often not executed. As a result, there are only five response letters available for public viewing and evaluation on the FDA's web pages from current Outsourcing Facilities, and some of these were issued prior to the firm registering under 503B.

Often, response letters from Pharmacy-based Outsourcing firms fail to comprehend the cGMP approach and may appear argumentative as a result. Here is a sample of two exchanges from the most recently posted response letter from an Outsourcing Facility to a 483:

- FDA states in the 483: "A piece of apparent exposed particle board measuring approximately 3/4 in. thick, 44 in. wide, and at least 6 in. deep was observed through the front edge guard vent of the Baker EG-4320, SIN 55304 ISO 5 horizontal LFH in the ISO 7 Cleanroom. The apparent particle board was observed seated on two metal u-channels and adhered to the underside of the metal workbench of the ISO 5 horizontal LFH."
- The Outsourcing Facility respondent responds: "The apparent particle board is contained in the inner workings of the horizontal LFH and therefore should not be considered an exposed surface..."

If the respondent better understood the requirements of cGMP, he/she would know that wooden surfaces cannot be validated inside any qualified air environments, let alone inside ISO Class 5 primary engineering controls.

- FDA further states in the 483: "Two pieces of apparent plastic were observed missing from a frame installed between the ceiling and metal HEPA filter grate of the Baker EG-4320, SIN 55304 ISO 5 horizontal LFH which exposed a groove that appears to not be smooth or easily cleanable."
- The Outsourcing Facility respondent states: "Nothing is missing from the frame installed between the ceiling and metal HEPA filter grate. The apparent observation is erroneous and due to how the manufacturer painted the surface. A photograph has been taken with the protective grill removed for clarity. A photo of this area is attached."

This response does not address the finding that the investigator found a groove that appears not to be smooth or easily cleanable. FDA will evaluate all entity responses but will not address their adequacy until or unless a Warning Letter is issued.

28.7.5   FDA's Approach to Encourage Registration

It should be clear to all stakeholders that the launch of the Registered Outsourcing Facility approach has not yet been a success. Given the early optimism that led more than 60 entities to register by late 2016, the fact that we still have only 75 registered two-and-a-half years later does not represent wide acceptance. From a kinetic standpoint, rate out has approximately matched rate in.

Pharmacists whose size and scope might support registration as an Outsourcing Facility often note that at the rate such entities have been failed or even been sued into consent decrees, sitting on the sideline is the only rational course.

Skilled compounding pharmacists are accustomed to hard and fast rules and clear, universally accepted practices. Our state regulations usually provide concrete targets, as do *USP* standards.

cGMP represents virtually an opposite approach. We cannot refer to any FDA document and read "the rules" of cGMP because they are in constant evolution. The target of cGMP is our own operational practices, and we are required to use science and statistics to set our own rules.

Rather than hard and fast "Thou shalt" statements, cGMP requires us to statistically validate that our processes will dependably produce the desired result. In the case of sterile compounds, that desired result is an assurance that, within the BUD, our compounded sterile preparations will be sterile, pyrogen free, and stable within compendial excursions.

Every system that affects that final preparation must be validated. We cannot depend on compiled literature that cite studies in which an investigator has shown that, for example, cefazolin 2 g is stable in 100 mL normal saline for 7 days. Instead, under cGMP, we must demonstrate that in *our* facility, as compounded by *our* staff, who have been through *our* training and demonstrated their competency according to *our* SOPs, 2 g of *our* brand of cefazolin diluted in 100 mL of *our* brand of normal saline is dependably stable within compendial excursions, sterile, and pyrogen free under the conditions of *our* transportation and *our* storage.

Demonstrating these principles one time is not nearly enough to satisfy cGMP. After every system can be shown to have achieved a "state of control," we must monitor it with control charts to surveil for "special causes of variation," and if our results begin to trend "out of specification" or "out of trend," then we must assemble the appropriate team members to conduct a root cause analysis and then enter a cycle of "Plan-Do-Check-Act" until we have brought our system back into a state of control. We must document the fact of the statistical excursion and the entire process of how it was remedied in our Corrective and Preventive Action Log ("CAPA Log"), and our "Quality Unit" must review the results and document its contemplation of the implications.

These approaches are commonplace to FDA and to the large pharmaceutical manufacturers. This has been their evolving reality for several decades now, but it has akin to an unintelligible language to healthcare practitioners. Contributing to the very slow learning curve for pharmacists has been the fact that FDA cannot provide recommendations, consultations, or advice. FDA traditionally communicates the requirements of cGMP to regulated sectors via 483s, Warning Letters,

Guidance Documents, and Webinars, but they realize that for the Outsourcing Facility sector to succeed, they must modulate their approach.

In FDA's 2019 Budgetary Request, they have included $25 million to create a "Center of Excellence on Compounding for Outsourcing Facilities."[16] They concede that "After three years, this domestic sector is still relatively small (~70 entities), is experiencing growth challenges, and is not yet fulfilling its potential."

This proposed Center of Excellence will include two key initiatives:

1. $22.3 million and 17 FTE *to provide training on current good manufacturing practice (cGMP) – the quality standard applicable to Outsourcing Facilities. The cGMP training would include in-depth, hands-on instruction and demonstrations offered in small settings to members of the sector with minimal cost to participants. The Center of Excellence would also conduct market research to help inform regulatory decision-making by FDA and its external partners, including identification of key challenges and opportunities, as well as growth potential. FDA staff would work closely with a partner organization on engagement with Outsourcing Facilities, development of research initiatives and developing and executing cGMP training. Increased direct FDA engagement is also essential. Outsourcing Facilities consistently seek more in-depth information, prompt feedback, and timely inspections and site visits from the Agency. They frequently request FDA's views on, for example, facility design, production and testing methods, and new technologies. The requested resources will allow the Human Drugs Program to offer new programs for FDA review of method and process design and study protocols upon request, as well as conduct more meetings to provide prompt in-depth feedback. This approach has the potential to significantly reduce future compliance failures, thus improving confidence in the sector, and would also support technical advancements and encourage market entry and growth. As part of our increased engagement initiative, FDA will also expand efforts to work with states to harmonize and streamline their approach to the Outsourcing Facility sector and improve the quality of compounded drugs. State quality requirements for compounding pharmacies not registered with FDA as Outsourcing Facilities also vary, as do the frequency and duration of inspections, often due to budgetary constraints. FDA often observes insanitary conditions at state-licensed compounding pharmacies. Additional funding will support training and outreach initiatives to strengthen state oversight of compounding facilities, as well as a pilot program of contracted state inspections. This pilot program would fund eligible states to conduct inspections under federal standards, to help ensure that compounding pharmacies not registered as Outsourcing Facilities provide solely patient-specific compounded drugs prepared under appropriate quality conditions (not insanitary) and Outsourcing Facilities become the sole source of compounded drugs for office stock prepared under cGMP.*

2. $2.7 million and 10 FTE for ORA *to establish a specialized group of investigators who will spend a majority of their time on Outsourcing Facility inspectional activities. As discussed above, Outsourcing Facilities are in their early growth years and would benefit from more frequent FDA inspections and site visits, which Outsourcing Facilities in the past have requested. These visits would not only help the sector come into compliance, but also help address regulatory hurdles in states that refuse to license these facilities unless they receive annual inspections by FDA. Furthermore, Outsourcing Facilities are distinct from conventional manufacturers in numerous ways and require specialized knowledge to inspect. A specially trained group of investigators who spend a majority of their time on Outsourcing Facility oversight will develop a highly sophisticated expertise; will become intimately familiar with the facilities, systems, and technologies that they routinely inspect; and will provide timely, consistent, substantive feedback when compliance issues are identified. This initiative will also help FDA meet annual inspection targets and conduct additional facility visits when requested by the Outsourcing Facility.*

The ultimate success or failure of the Outsourcing Facility initiative depends entirely on a rapid transfer of cGMP understanding to the healthcare sector, and the Agency's main vehicle for the technology transfer will likely be this proposed "Center of Excellence."

28.8 SUMMARY

Pharmaceutics, pharmaceutical manufacturing, and pharmaceutical compounding share a symbiotic relationship to provide quality pharmaceuticals to patients. Manufacturers can most efficiently provide quality products on a large scale to the masses, and pharmaceutical compounding can provide specific preparations for individual patients. The new 503B Outsourcing Facilities add a new dimension to the provision of medications that are in short supply.

The use of pharmaceutics and most of the pharmaceutical or physicochemical factors involved in formulation of these products are identical. One exception, however, is the shorter BUDS assigned to compounded preparations as compared to expiration dates for manufactured products. A second notable exception is that commercially manufactured products are developed over several years, whereas an extemporaneously compounded preparation may be required for a specific patient in a matter of minutes or hours.

Pharmaceutical manufacturing will continue to grow as more new, innovative, and novel dosage forms and drug-delivery systems become routine. Pharmaceutical compounding will continue to grow and become an even more important part of pharmacy practice in the future as individualization of patient care becomes more important. The 503B Outsourcing Facilities, while still developing, may be able to be of great value in the future for drug shortages. Working together, the needs of most patients should be satisfactorily addressed.

APPENDIX 28.A   LIST OF DRUGS THAT WERE WITHDRAWN OR REMOVED
FROM THE MARKET FOR SAFETY REASONS

The Pharmacy Compounding Advisory Committee (PCAC) of the FDA voted to add 25 drugs to the list of drug products that may not be compounded under the exemptions provided by the Federal FDC Act because they have been withdrawn or removed from the market as a result of their components having been determined to be unsafe or ineffective. The list includes

- Adenosine phosphate: All drug products containing adenosine 5'-monophosphate (AMP), adenosine 5'-diphosphate (ADP), and adenosine 5'-triphosphate (ATP).
- Adrenal cortex: All drug products containing adrenal cortex.
- Alatrofloxacin mesylate (all drug products).
- Aminopyrine: All drug products containing aminopyrine.
- Astemizole: All drug products containing astemizole.
- Azaribine: All drug products containing azaribine.
- Benoxaprofen: All drug products containing benoxaprofen.
- Bithionol: All drug products containing bithionol.
- Bromfenac sodium: All drug products containing bromfenac sodium except ophthalmic solutions).
- Butamben: All parenteral drug products containing butamben.
- Camphorated oil: All drug products containing camphorated oil.
- Carbetapentane citrate: All oral gel drug products containing carbetapentane citrate.
- Casein, iodinated: All drug products containing iodinated casein.
- Cerivastatin sodium.
- Chloramphenicol (all oral drug products).

- Chlorhexidine gluconate: All tinctures of chlorhexidine gluconate formulated for use as a patient preoperative skin preparation.
- Chlormadinone acetate. All drug products containing chlormadinone acetate.
- Chloroform: All drug products containing chloroform.
- Cisapride: All drug products containing cisapride.
- Cobalt: All drug products containing cobalt salts (except radioactive forms cobalt and its salts and cobalamin and its derivatives).
- Dexfenfluramine hydrochloride: All drug products containing dexfenfluramine hydrochloride.
- Diamthazole dihydrochloride: All drug products containing diamthazole dihydrochloride.
- Dibromsalan: All drug products containing dibromsalan.
- Diethylstilbestrol: All oral and parenteral drug products containing 25 mg or more of diethylstilbestrol per unit dose.
- Dihydrostreptomycin sulfate: All drug products containing dihydrostreptomycin sulfate.
- Dipyrone: All drug products containing dipyrone.
- Encainide hydrochloride: All drug products containing encainide hydrochloride.
- Esmolol HCl (all parenteral drug products that supply 250 mg/mL of concentrated esmolol per 10-mL ampule).
- Etretinate (all drug products).
- Fenfluramine hydrochloride: All drug products containing fenfluramine hydrochloride.
- Flosequinan: All drug products containing flosequinan.
- Gatifloxacin (except ophthalmic solutions).
- Gelatin: All intravenous drug products containing gelatin.
- Glycerol, iodinated: All drug products containing iodinated glycerol.
- Gonadotropin, chorionic: All drug products containing chorionic gonadotropins of animal origin.
- Grepafloxacin: All drug products containing grepafloxacin.
- Mepazine: All drug products containing mepazine hydrochloride or mepazine acetate.
- Metabromsalan: All drug products containing metabromsalan.
- Methamphetamine hydrochloride: All parenteral drug products containing methamphetamine hydrochloride.
- Methapyrilene: All drug products containing methapyrilene.
- Methopholine: All drug products containing methopholine.
- Methoxyflurane.
- Mibefradil dihydrochloride: All drug products containing mibefradil dihydrochloride.
- Nitrofurazone: All drug products containing nitrofurazone (except topical drug products formulated for dermatologic application).
- Nomifensine maleate: All drug products containing nomifensine maleate.
- Novobiocin sodium.
- Oxycodone: All extended-release drug products containing oxycodone hydrochloride that have not been determined by the FDA to have abuse-deterrent properties.
- Oxyphenisatin: All drug products containing oxyphenisatin.
- Oxyphenisatin acetate: All drug products containing oxyphenisatin acetate.
- Pemoline.
- Pergolide mesylate.
- Phenacetin: All drug products containing phenacetin.
- Phenformin hydrochloride: All drug products containing phenformin hydrochloride.
- Phenylpropanolamine.
- Pipamazine: All drug products containing pipamazine.
- Potassium arsenite: All drug products containing potassium arsenite.

- Potassium chloride: All solid oral dosage form drug products containing potassium chloride that supply 100 mg or more of potassium per dosage unit (except for controlled-release dosage forms and those products formulated for preparation of solution prior to ingestion).
- Propoxyphene.
- Povidone: All intravenous drug products containing povidone.
- Rapacuronium bromide.
- Reserpine: All oral dosage form drug products containing more than 1 mg of reserpine.
- Rofecoxib.
- Sibutramine hydrochloride.
- Sparteine sulfate: All drug products containing sparteine sulfate.
- Sulfadimethoxine: All drug products containing sulfadimethoxine.
- Sulfathiazole: All drug products containing sulfathiazole (except those formulated for vaginal use).
- Suprofen: All drug products containing suprofen (except ophthalmic solutions).
- Sweet spirits of nitre: All drug products containing sweet spirits of nitre.
- Tegaserod maleate.
- Temafloxacin hydrochloride: All drug products containing temafloxacin.
- Terfenadine: All drug products containing terfenadine.
- 3,3′,4′,5-Tetrachlorosalicylanilide: All drug products containing 3,3′,4′,5-tetrachlorosalicylanilide.
- Tetracycline: All liquid oral drug products formulated for pediatric use containing tetracycline in a concentration greater than 25 mg/mL.
- Ticrynafen: All drug products containing ticrynafen.
- Tribromsalan: All drug products containing tribromsalan.
- Trichloroethane: All aerosol drug products intended for inhalation containing trichloroethane.
- Troglitazone: All drug products containing troglitazone.
- Trovafloxacin mesylate.
- Urethane: All drug products containing urethane.
- Valdecoxib.
- Vinyl chloride: All aerosol drug products containing vinyl chloride.
- Zirconium: All aerosol drug products containing zirconium.
- Zomepirac sodium: All drug products containing zomepirac sodium.
- All drug products containing the combination of polyethylene glycol 3350, sodium chloride, sodium bicarbonate, and potassium chloride for oral solution AND 10 mg or more of bisacodyl delayed-release tablet (actually combined in a "kit").

REFERENCES

1. U.S. Pharmacopeia 41-National Formulary 36. Rockville, MD: U.S. Pharmacopeial Convention, 2018: 6546.
2. National Association of Boards of Pharmacy. Good compounding practices applicable to state-licensed pharmacies, In: *Model State Pharmacy Act and Model Rules of the National Association of Boards of Pharmacy.* Park Ridge, IL: NABP, 1993.
3. The Food and Drug Administration Modernization Act of 1997, Public Law 105–115, 127, 1997.
4. National Association of Boards of Pharmacy. *Good Compounding Practices Applicable to State Licensed Pharmacies.* Park Ridge, IL: NABP, 1993.
5. Allen, L.V., Jr. A history of pharmaceutical compounding, *Secundum Artem*, 11, 3, 2003.
6. Cabaleiro J. New England compounding center indictment, *IJPC*, 19(2), 94–102, 2015.
7. The United States Department of Justice. Justice news: NECC indictment. [Department of Justice Website.] Available at: www.justice.gov/search/all/NECC%20Indictment?page=1. Accessed January 3, 2015.

8. FindLaw. FindLaw dictionary: Second-degree murder. [FindLaw Website.] Available at: public.findlaw. com/LCsearch.html?restrict=consumder&entry=second-degree+murder+definition. Accessed January 3, 2015.

9. FindLaw. FindLaw dictionary: Criminal contempt. [FindLaw Website.] Available at: public.findlaw. com/LCsearch.html?restrict=consumder&entry=criminal+contempt+definition. Accessed January 3, 2015.

10. Internal Revenue Service (2006-06-01). Part 4 examining process; Chapter 26 Bank secrecy act; Section 13 Structuring. Internal Revenue Manual. Washington, DC: US Treasury Department. OCLC 37305546. [Internal Revenue Service Website] Available at: www.irs.gov/irm/part4. Accessed January 3, 2015.

11. USLegal. Legal definitions and legal terms: Racketeering. [USLegal Website.] Available at: definitions. uslegal.com/search/?q=racketeering. Accessed January 3, 2015.

12. United States Pharmacopeial Convention, Inc. United States Pharmacopeia 37–National Formulary 32. General Information Chapter <797> Pharmaceutical Compounding: Sterile Preparations. Rockville, MD: US Pharmacopeial Convention, Inc., 2013: 71–77, 425–427, 430, 432.

13. Centers for Disease Control and Prevention. Multistate outbreak of fungal meningitis and other infections: Case count. Available at: www.cdc.gov/hai/outbreaks/meningitis-map-large.html.

14. U.S. Food and Drug Administration. Registration of human drug compounding outsourcing facilities under Section 503B of the FD&C Act: Guidance for industry. Available at: www.fda.gov/downloads/Drugs/GuidanceComplianceRegulatoryInformation/Guidances/UCM377051.pdf.

15. U.S. Food and Drug Administration. Registered outsourcing facilities: Facilities registered as human drug compounding outsourcing facilities under Section 503B of the Federal Food, Drug, and Cosmetic Act (FD&C Act) - Updated as of 6/22/2018. Available at: www.fda.gov/Drugs/GuidanceComplianceRegulatoryInformation/PharmacyCompounding/ucm378645.htm.

16. U.S. Food and Drug Administration. Justification of Estimates for Appropriations Committees, Food and Drug Administration – Fiscal Year 2019. pp. 79–81. Available at: www.fda.gov/downloads/AboutFDA/ReportsManualsForms/Reports/BudgetReports/UCM603315.pdf.

Index

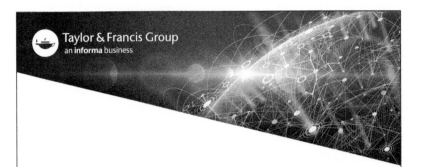

Printed and bound by CPI Group (UK) Ltd, Croydon, CR0 4YY

17/10/2024

01775698-0012